Wireless Sensor Networks and Energy Efficiency:

Protocols, Routing and Management

Noor Zaman
King Faisal University, Saudi Arabia

Khaled Ragab
King Faisal University, Saudi Arabia

Azween Bin Abdullah
University Technology PETRONAS, Malaysia

Managing Director: Lindsay Johnston
Senior Editorial Director: Heather Probst
Book Production Manager: Sean Woznicki
Development Manager: Joel Gamon
Development Editor: Myla Harty
Acquisitions Editor: Erika Gallagher
Typesetter: Adrienne Freeland
Cover Design: Nick Newcomer, Lisandro Gonzalez

Published in the United States of America by
Information Science Reference (an imprint of IGI Global)
701 E. Chocolate Avenue
Hershey PA 17033
Tel: 717-533-8845
Fax: 717-533-8661
E-mail: cust@igi-global.com
Web site: http://www.igi-global.com

Library of Congress Cataloging-in-Publication Data

Wireless sensor networks and energy efficiency : protocols, routing, and management / Noor Zaman, Khaled Ragab, and Azween Bin Abdullah, editors.
p. cm.
Includes bibliographical references and index.
Summary: "This book focuses on wireless sensor networks and their operation, covering topics including routing, energy efficiency and management"-- Provided by publisher.
ISBN 978-1-4666-0101-7 (hardcover) -- ISBN 978-1-4666-0102-4 (ebook) -- ISBN 978-1-4666-0103-1 (print & perpetual access) 1. Wireless sensor networks. I. Zaman, Noor, 1972- II. Ragab, Khaled. III. Abdullah, Azween Bin, 1961-
TK7872.D48W577 2012
681'.2--dc23

2011048212

British Cataloguing in Publication Data
A Cataloguing in Publication record for this book is available from the British Library.

Table of Contents

Detailed Table of Contents

This chapter provides a brief technical introduction to wireless sensor networks, with background and history and an understanding of sensor networks, design constraints, security issues, and a few applications in which wireless sensor networks are enabling. A brief discussion of the network topologies that apply to wireless sensor networks are also discussed.

Topology management is a key component of network management of wireless sensor networks. The primary goal of topology management is to conserve energy while maintaining network connectivity. Topology management consists of knowing the physical connections and logical relationships among the sensors and at the same time creating a subset of nodes actively participating in the network, thus creating less communication and conserving energy in nodes. Networks require constant monitoring in order to ensure consistent and efficient operations. The primary goal of topology management is to maintain network connectivity in an energy-efficient manner. Topology management is one of the key aspects of configuration management, which entails initial set-up of the network devices and continuous monitoring and controlling of these devices. The main objective of this chapter is to throw a light on the recent developments and future directions of research in these directions.

This chapter compiles some related information on the basis of studied literature regarding wireless sensors network management, including WSN background, issues and challenges, proposed solutions, and research trends.

The syntactic and semantic interoperability is a challenge of the Wireless Sensor Networks (WSN) with smart sensors in pervasive computing environments to increase their harmonization in a wide variety of applications. This chapter contains a detailed description of interoperability in heterogeneous WSN using the IEEE 1451 standard. This work focuses on personal area networks (PAN) with smart sensors and actuators. Also, technical, syntactic, and semantic levels of interoperability based on IEEE 1451 standardization are established with common control commands. In this architecture, each node includes a Transducer Electronic Datasheets (TEDS) and intelligent functions. The authors explore different options to apply the IEEE 1451 standard using SOAP or REST Web service style in order to test a common syntactical interoperability that could be predominant in future WSNs.

In this chapter, the authors present a literature review for MAC, routing, and cross layer design protocols proposed for WSN. This chapter consists of three sections. In the first section, the authors discuss in depth the most well-known MAC protocols for WSN. A comparison among theses protocols will be presented. Moreover, the major advantages and disadvantages of each protocol are discussed. The routing protocols for WSN are discussed in the second section. The discussed protocols are classified into data centric routing protocols, Hierarchical routing protocols, location based protocols, and QoS aware routing protocols. Moreover, A Classification of Routing Protocols based on the Application is presented in this section. In the third section, some cross layer design protocols are discussed. A comparison among the discussed protocols according to layers integrated, intended applications, cross-layer objectives, and the evaluation approach, is presented.

In this chapter, the authors present a review and comparison of different algorithms proposed recently for underwater sensor networks. Later on, all of these have been classified into different groups according to their characteristics and functionalities.

This research presents a survey of energy efficient routing protocols in sensor network by categorizing into a main classification as architecture based routing. Architecture based routing is further classified into two main areas: flat or location based routing protocol, and hierarchical based routing protocols. Flat based routing is more suitable when a huge number of sensor nodes are deployed, and location based routing is employed when nodes are aware of their location. Hierarchical routing look into alternative approach by placing intermediate nodes in terms of cluster heads, gateway nodes, or mobile entities for efficient handling of energy. The survey is presented in order to highlight the advantage of hierarchical based routing, mainly the deployment of mobility routing. As not many surveys have been conducted in mobility based routing, this chapter can be helpful for looking into a new perspective and paradigm of energy efficient routing protocols.

In this chapter, the authors investigate the use of various cooperative diversity techniques in wireless sensor networks to increase the transmission range, minimize power consumption, and maximize network lifetime.

This chapter analyzes the effect of density on inter cluster and intra cluster communication and evaluates a hybrid cross layer scheduling schemes to enhance the life time of the WSNs. In the conventional scheduling schemes at the application layer, all the nodes whose area are covered by their neighbors are put to sleep in order to prolong the life time of the WSNs. The hybrid cross layer scheme in this chapter suggests that instead of putting the redundant nodes to sleep if they are used for some other energy intensive tasks, for example the use of redundant nodes as relay stations in inter cluster communication, will be more energy efficient compare to the conventional application layer scheduling schemes in WSNs. Performance studies in the chapter indicate that the proposed communication strategy is more energy efficient than the conventional communication strategies that employ the sleep/wake up pattern at application layer.

In this chapter, the authors propose a multi-objective solution to the problem by using multi-objective particle swarm optimization (MOPSO) algorithm to optimize the number of clusters in a sensor network in order to provide an energy-efficient solution. The proposed algorithm considers the ideal degree of nodes and battery power consumption of the sensor nodes. The main advantage of the proposed method is that it provides a set of solutions at a time. The results of the proposed approach were compared with two other well-known clustering techniques: WCA and CLPSO-based clustering. Extensive simulations were performed to show that the proposed approach is an effective approach for clustering in WSN environments and performs better than the other two approaches.

This chapter presents the studies and analysis on the approaches, the concepts, and the ideas on data packet size optimization for data packets transmission in underwater wireless sensor network (UWSN) and terrestrial wireless sensor network (TWSN) communications. These studies are based on the related prior works accomplished by the UWSN and TWSN research communities. It should be mentioned here that the bulk of the studies and analysis would be on the data packet size optimization techniques or approaches rather than on the communication channel modeling, but the channel model is deemed essential to support the optimization approaches.

In this chapter, a novel approach is explored to employ high-altitude platforms (HAPs) to remove the relaying burden and/or de-centralize coordination from wireless sensor networks (WSNs). The approach can reduce the complexity and achieve energy efficiency in communications of WSNs, whereby applications require a large-scale deployment of low-power and low-cost sustainable sensors. The authors review and discuss the main constraints and problems of energy consumptions and coordination in WSNs. The use of HAPs in WSNs provides favorable communication links via predominantly line of sight propagation due to their unique position and achieves benefits of reduced complexity and high energy efficiency, which are crucial for WSN operations.

With this chapter, the authors focus on operational and architectural challenges of handling QoS, requirements of QoS in WSNs, and they discuss a selected survey of QoS aware routing techniques by comparing them in WSNs. Finally, the authors highlight a few open issues and future directions of research for providing QoS in WSNs.

This chapter presents low complexity processor designs for energy-efficient security and error correction for implementation on wireless sensor networks (WSN). WSN nodes have limited resources in terms of hardware, memory, and battery life span. Small area hardware designs for encryption and error-correction modules are the most preferred approach to meet the stringent design area requirement. This chapter describes Minimal Instruction Set Computer (MISC) processor designs with a compact architecture and simple hardware components. The MISC is able to make use of a small area of the FPGA and provides a processor platform for security and error correction operations. In this chapter, two example applications, which are the Advance Encryption Standard (AES) and Reed Solomon (RS) algorithms, were implemented onto MISC. The MISC hardware architecture for AES and RS were designed and verified using the Handel-C hardware description language and implemented on a Xilinx Spartan-3 FPGA.

This chapter provides an overall understanding of the design aspects of Medium Access Control (MAC) protocols for Wireless Sensor Networks (WSNs). A WSN MAC protocol shares the wireless broadcast medium among sensor nodes and creates a basic network infrastructure for them to communicate with each other. The MAC protocol also has a direct influence on the network lifetime of WSNs as it controls the activities of the radio, which is the most power-consuming component of resource-scarce sensor nodes. In this chapter, the authors first discuss the basics of MAC design for WSNs and present a set of important MAC attributes. Subsequently, authors discuss the main categories of MAC protocols proposed for WSNs and highlight their strong and weak points. After briefly outlining different MAC protocols falling in each category, the authors provide a substantial comparison of these protocols for several parameters. Lastly, the chapter discusses future research directions on open issues in this field that have mostly been overlooked.

The selection of candidate sensors is based on multi-criteria function, which is computed by using the predicted target position provided by the QVF algorithm. The proposed algorithm for the detection of malicious sensor nodes is based on Kullback-Leibler distance between the current target position distribution and the predicted sensor observation, while the best communication path is selected as well as the highest signal-to-noise ratio (SNR) at the CH. The efficiency of the proposed method is validated by extensive simulations in target tracking for wireless sensor networks.

This chapter covers applications and routing in WSNs, different methods for routing using ant colony optimization (ACO), a summary of routing algorithms based on ant systems, and the Improved Energy-Efficient Ant-Based Routing Algorithm approach. Simulations results were analyzed while also looking at open research problems and future work to be done. The chapter concludes with a comparative summary of results with IABR and EEABR.

Wireless sensor networks (WSANs) are increasingly being used and deployed to monitor the surrounding physical environments and detect events of interest. In wireless sensor networks, energy is one of the primary issues and requires the conservation of energy of the sensor nodes, so that network lifetime can be maximized. It is not recommended as a way to transmit or store all data of the sensor nodes for analysis to the end user. The purpose of this "Event Based Detection" Model is to simulate the results in terms of energy savings during field activities like a fire detection system in a remote area or habitat monitoring, and it is also used in security concerned issues. The model is designed to detect events (when occurring) of significant changes and save the data for further processing and transmission. In this way, the amount of transmitted data is reduced, and the network lifetime is increased. The main goal of this model is to meet the needs of critical condition monitoring applications and increase the network lifetime by saving more energy. This is useful where the size of the network increases. Matlab software is used for simulation.

Section 3
Application, Theoretical and General

In this chapter, the authors use the concepts of game theory to design an energy efficient MAC protocol for WSNs. This allows them to introduce persistent/non-persistent sift protocol for energy efficient MAC protocol and to counteract the selfish behavior of nodes in WSNs. Finally, the research results show that game theoretical approach with the persistent/non-persistent sift algorithm can improve the overall performance as well as achieve all the goals simultaneously for MAC protocol in WSNs.

This chapter surveys routing algorithms in Euclidean, virtual, and hyperbolic space for wireless sensor networks that use geometric structures for route decisions. Wireless sensor networks have a unique geographic nature as the sensor nodes are embedded and designed for employing in the geographic space. Thus, the various geometric abstractions of the network can be used for routing algorithm design, which can provide scalability and efficiency. This chapter starts with the importance and impulse of the geographical routing in wireless sensor networks that exploits location information of the nodes to determine the alternatives of the next hop node on the desired routing path. The scalability of geographical routing encourages more effort on the design of virtual coordinates system, with which geographical routing algorithms are built up and applied to route data packets in the network. The geometry of large sensor network motivates to calculate geometric abstractions in hyperbolic space. Thus the challenge is to embed the network virtually or hyperbolically, which affects the performance and efficiency in the geographical message delivery.

In this chapter, the author provides the basics of information security, with special emphasis on WSNs. The chapter also gives an overview of the information security requirements in these networks. Threats to the security of data in WSNs and some of their counter measures are also presented.

The future generation of vehicles on the road is going to be driven by wire. To aid in this 'electronic' revolution in the vehicle, the role of wireless sensors and their interaction amongst themselves and with the environment is gaining importance. It is an area where the majority of research resources is allocated and being spent. For successful interaction of information from environment / vehicle, there is a need for wireless networking of the information from different sources. To keep pace with the development of wireless networks for intelligent transport systems, newer network architectures, protocols, and algorithms are being developed. This chapter sheds light on all these issues.

The middleware is a very important component in the wireless sensor network system. It has major challenges that are generic for any distributed systems as well as specific ones that are inherent to

Preface

INTRODUCTION

Wireless Sensor Networks (WSN) are an emerging field, with special characteristics, includingthe capability of easy deployment with a number of different applications. Sensor networks normally consist of small nodes with limited capability for computation and communication. The major share of research in the last decade pertains to WSN. The main reason for the popularity of wireless sensor networks is because of ease in deployment and low cost. Its main function is to collect the data, resulting of any physical event occurrence in the network area and then send it to the sink. This book focuses on wireless sensor networks in general, and operations including routing, energy efficiency, and the management aspects.

This book is intended for anyone who wants to cover a comprehensive range of topics in the field of wireless sensor networks paradigms and developments. It is both for an academic audience (teachers, researchers, and students, mainly of post-graduate studies) and professional audience. Readers of this book are pre assumed familiar with the concepts and paradigms of wireless sensor network concepts and its related concepts. It provides guidance for technology solution developers from academia, research institutions, and industry, providing them with a broader perspective of wireless sensor networks.

This book contains 27 excellent chapters authored by a group of internationally experienced professionals and researchers in the field of computer science, communication, and networking. Contributors also include younger authors, creating a value-added constellation of dynamic authors. Concerning the environments from which the contributions are presented, the chapters came from academia, research institutions and industry.

Organization of the Book

This book is designed to cover a wide range of topics in the field of wireless sensor networks. It includes three sections that provide a comprehensive reference for wireless sensor network by covering all important topics, including an introduction, MAC, Routing Protocols, TCP, performance and traffic management, time synchronization energy efficiency, applications, and security. Each chapter is designed to be as a stand-alone as possible; the reader can focus on the interested topics only. The chapters are described briefly as follows.

Section 1: Introduction

Chapter 1 provides a brief technical introduction to WSN, with background and history and an understanding of sensor networks, design constraints, security issues, and a few applications sensor networks are enabling. A brief discussion of the network topologies that apply to wireless sensor networks are also discussed in this chapter.

Chapter 2 throws a light on the recent developments and future directions of research in WSN topology management that maintain wireless network connectivity in an energy-efficient manner.

Chapter 3 compiles some related information on the basis of studied literature, regarding WSN management, including WSN background, issues and challenges, proposed solutions, and research trends.

Chapter 4 includes a detailed description of interoperability in heterogeneous WSN using the IEEE 1451 standard. It focuses on personal area networks (PAN) with smart sensors and actuators. In this chapter, the authors explore different options to apply the IEEE 1451 standard using SOAP or REST Web service style in order to test a common syntactical interoperability that could be predominant in future WSNs.

Chapter 5 presents literature review for MAC, routing, and cross layer design protocols that proposed for WSN. This chapter discusses and compares the most well-known MAC protocols for WSN according to layers integrated, intended applications, and cross-layer objectives. The authors classified the routing protocols for WSN into data centric routing protocols, hierarchical routing protocols, location based protocols, QoS aware routing protocols, and application. Finally, this chapter discusses some cross layer design protocols.

Chapter 6 presents a review and comparison of different algorithms proposed recently for underwater sensor networks (UWSNs). The UWSNs are finding different applications for offshore exploration and ocean monitoring.

Chapter 7 presents a survey on the Wireless Sensor Network Testbed (WSN-Testbed). The WSN-Testbed is a platform for experimentation of development projects. It enables realistic and reliable experimentation in capturing the subtleties of the underlying hardware, software, and dynamics of the WSN. The authors adopt and describe a classification methodology for WSN-Testbeds. Consequently, they present a generic architecture for the different classes of WSN-Testbeds. Carefully, this chapter discusses and analyzes a variety of 30 WSN-Testbeds. Finally, the authors believe that this chapter is a contribution towards realizing the important role that a WSN-Testbed plays in hastening the industrial adoption for the promising WSN technology.

Chapter 8 provides an overview of techniques to mitigate Hot Spot impacts, such as the uneven distribution of sensors, routes that balance energy consumption, sink mobility, and the use of unequal clustering. Further, it depicts an approach for achieving mitigation of sink centered Hot Spots.

Chapter 9 provides an overview of different location aware algorithms to focus to save the energy resource of sensor network. The chapter describes the current available approaches, issue and challenges with current approaches and future directions for node localization, one by one. Node localization is highly important for large sensor networks where users desire to know about the exact location of the nodes in order to know the data location.

Section 2: Energy Efficiency of WSN

Chapter 10 presents a survey of energy efficient routing protocols in sensor networks by categorizing architecture based routing. Furthermore, it classifies the architecture based routing into two main areas: flat or location based routing protocols, and hierarchical based routing protocols. The authors present a survey of relevant literature in order to highlight the advantages of hierarchical based routing, particularly with respect to the deployment of mobility routing. Finally, this chapter is helpful in providing new perspectives and paradigms for the design and analysis of energy efficient routing protocols.

Chapter 11 investigates the use of various cooperative diversity techniques in wireless sensor networks to increase the transmission range, minimize power consumption, and maximize network lifetime.

Chapter 12 analyzes the effect of density on inter cluster and intra cluster communication and evaluates a hybrid cross layer scheduling schemes to enhance the lifetime of the WSNs. The authors suggests a hybrid cross layer scheduling scheme at the application layer that puts the redundant nodes to sleep if they are used for some other energy intensive tasks. Performance studies in the chapter indicate that the proposed communication strategy is more energy efficient than the conventional communication strategies, which employ the sleep/wake up pattern at application layer.

Chapter 13 proposes a multi-objective particle swarm optimization (MOPSO) algorithm to optimize the number of clusters in a sensor network in order to provide an energy-efficient solution. The proposed algorithm considers the ideal degree of nodes, and battery power consumption of the sensor nodes. The main advantage of the proposed method is that it provides a set of solutions at a time. This chapter performed extensive simulations to compare the proposed approach with two other well-known clustering techniques: WCA and CLPSO-based clustering.

Chapter 14 presents the studies and analysis on the approaches, the concepts, and the ideas on data packet size optimization for data packets transmission in underwater wireless sensor network (UWSN) and terrestrial wireless sensor network (TWSN) communications. This chapter starts off with the studies and analysis on prior arts found in UWSN and then moves on to the similar works found elsewhere in the TWSN communications counterparts. In addition, it summarizes comparison on some important issues related to data packet size optimization approaches used in UWSN and TWSN communications. The findings in this chapter may be helpful to readers who are interested in the R&D of data packet size optimization techniques with the intention to formulate new data packet size optimization framework or algorithms.

Chapter 15 proposes a novel approach to employ high-altitude platforms (HAPs) to remove the relaying burden and/or de-centralize coordination from wireless sensor networks (WSNs). The approach can reduce the complexity and achieve energy efficiency in communications of WSNs, whereby applications require a large-scale deployment of low-power and low-cost sustainable sensors. Moreover, the authors review and discuss the main constraints and problems of energy consumptions and coordination in WSNs.

Chapter 16 describes the importance of quality of service QoS and focuses on operational and architectural challenges of handling QoS, as well as requirements of QoS in WSNs. It discusses a selected survey of QoS aware routing techniques by comparing them in WSNs. Finally, the chapter highlights a few open issues and future directions of research for providing QoS in WSNs.

Chapter 17 presents low complexity processor designs for energy-efficient security and error correction for implementation on wireless sensor networks (WSN). This chapter describes Minimal Instruction Set Computer (MISC) processor designs with a compact architecture and simple hardware components. The MISC is able to make use of a small area of the FPGA and provides a processor platform for security and error correction operations. In this chapter, two example applications, which are the Advance Encryption Standard (AES) and Reed Solomon (RS) algorithms, were implemented onto MISC. The MISC hardware architecture for AES and RS were designed and verified using the Handel-C hardware description language and implemented on a Xilinx Spartan-3 FPGA.

Chapter 18 provides an overall understanding of the design aspects of Medium Access Control (MAC) protocols for Wireless Sensor Networks (WSNs). The authors first discuss the basics of MAC design for WSNs and present a set of important MAC attributes. Subsequently, they discuss the main categories of MAC protocols proposed for WSNs and highlight their strong and weak points. After briefly outlining different MAC protocols falling in each category, the authors provide a substantial comparison of these protocols for several parameters. Finally, this chapter discusses future research directions on open issues in this field that have mostly been overlooked.

Chapter 19 proposes an algorithm for the detection of malicious sensor nodes based on Kullback-Leibler distance between the current target position distribution and the predicted sensor observation, while the best communication path is selected, as well as the highest signal-to-noise ratio (SNR) at the cluster head (CH). The efficiency of the proposed method is validated by extensive simulations in target tracking for wireless sensor networks (WSN).

Chapter 20 presents an improved energy-efficient Ant-Based routing algorithm (IEEABR) in wireless sensor networks. The proposed IEEABR approach has advantages of reduced energy usage and achieves a dynamic and adaptive routing, which can effectively balance the WSN node power consumption and increase the network lifetime. This chapter covers applications and routing in WSNs, different methods for routing using ant colony optimization (ACO), a summary of routing algorithms based on ant systems, and the Improved Energy-Efficient Ant-Based Routing Algorithm approach.

Chapter 21 describes an "Event Based Detection" model to simulate the results in terms of energy savings during field activities like a fire detection system in a remote area or habitat monitoring, and it is also used in security concerned issues. The model is designed to detect events (when occurring) of significant changes and save the data for further processing and transmission. In this way, the amount of transmitted data is reduced, and the network lifetime is increased. The main goal of this model is to meet the needs of critical condition monitoring applications and increase the network lifetime by saving more energy.

Section 3: Application, Theoretical and General

Chapter 22 uses the concepts of game theory to design an energy efficient MAC protocol for wireless sensor networks (WSNs). This enables the authors to introduce persistent/non-persistent sift protocol for energy efficient MAC protocol and to counteract the selfish behavior of nodes in WSNs. Finally, the results show that game theoretical approach with the persistent/non-persistent sift algorithm can improve the overall performance as well as achieve all the goals simultaneously for MAC protocol in WSNs.

Chapter 23 surveys routing algorithms in Euclidean, virtual, and hyperbolic space for wireless sensor networks (WSNs) that use geometric structures for their route decisions. This chapter starts with the importance and impulse of the geographical routing in WSNs that exploit location information of the nodes to determine the alternatives of the next hop node on the desired routing path. The scalability of geographical routing encourages more effort on the design of virtual coordinates system, with which geographical routing algorithms are built up and applied to route data packets in the network.

Chapter 24 provides the basics of wireless sensor network (WSNs) security to help researchers and engineers in better understanding of this applications field. The authors provide the basics of information security, with special emphasis on WSNs. Moreover, this chapter gives an overview of the information security requirements in these networks. Threats to the security of data in WSNs and some of their counter measures are also presented.

Chapter 25 discusses the development of wireless sensor networks (WSNs) for intelligent transport systems. Due to the electronic revolution in the vehicle, the role of wireless sensors and their interaction amongst themselves and with the environment is gaining importance. For successful interaction of information from environment/vehicle, there is a need for wireless networking of the information from different sources. Mainly, this chapter sheds light on the development of WSNs for intelligent transport systems, newer network architectures, protocols, and algorithms that are being developed.

Chapter 26 The middleware is a very important component in the wireless sensor networks (WSNs) system. It has major challenges that are generic for any distributed systems as well as specific ones that are inherent to the physical nature of WSNs systems due to its resource constraint nature. The authors classify WSNs middleware systems as event based middleware, system abstraction based middleware, application and network aware system, and query based systems to gain a better understanding of such systems.

Chapter 27 presents a novel grid-based localization technique dedicated for forest fire surveillance systems. The proposed technique estimates the location of sensor node based on the past and current set of hop-count values, which are to be collected through the anchor nodes' broadcast. The proposed algorithm incorporates two salient features, grid-based output and event-triggering mechanism, in order to improve the accuracy while reducing the power consumption. The estimated computational complexity of the proposed algorithm is O(Na) where Na is the number of anchor nodes. Through computer simulation, results showed that the proposed algorithm shows that the probability to localize a sensor node within a small region is more than 60%.

The Editors,

Noor Zaman
King Faisal University, Saudi Arabia

Khaled Ragab
King Faisal University, Saudi Arabia

Azween Bin Abdullah
University Technology PETRONAS, Malaysia

Section 1
Introduction

Chapter 1
Introduction and Overview of Wireless Sensor Networks

Seema Ansari
Karachi Institute of Economics and Technology, Pakistan

Syeda Fariha Hasnain
Karachi Institute of Economics and Technology, Pakistan

Adeel Ansari
Karachi Institute of Economics and Technology, Pakistan

ABSTRACT

Wireless Sensor Networks are an exciting technology that can solve a variety of applications. Wireless sensor networks, coupled with the efficient delivery of sensed information, could provide great benefits to society. The surroundings can be the physical world, a biological system, or an Information Technology (IT) framework. The use of wireless sensors allows for fast setting up of sensing tools and access to locations that would not be realistic if cables were attached. This chapter provides a brief technical introduction to wireless sensor networks, with background and history and an understanding of sensor networks, design constraints, security issues, and a few applications in which wireless sensor networks are enabling. A brief discussion of the network topologies that apply to wireless sensor networks are also discussed.

INTRODUCTION

A sensor network is an infrastructure comprised of sensing (measuring), computing, and communication elements that gives an administrator the ability to instrument, observe, and react to events and phenomena in a particular environment. The supervisor typically is a civil, governmental, commercial, or industrial body. The environment can be the corporal world, a biological system, or an information technology (IT) structure.

Network(ed) sensor systems are seen by viewers as a vital technology that will practice major exploitation in the next few years for a surplus of applications, space study, vehicular movement and critical movement detection are few examples of its applications.

DOI: 10.4018/978-1-4666-0101-7.ch001

Today's wireless sensor networks are made up of a big number of inexpensive devices that are networked via low power wireless communications. It is the networking potential that basically distinguishes a sensor network from a mere collection of sensors, by enabling cooperation, coordination, and collaboration among sensor assets (Swami, Zhao, & Hong, 2007).

Whilst several sensors can be connected to controllers and processing stations directly (e.g., using local area networks), a large number of sensors send the gathered data wirelessly to a centralized processing station. This is needed since many network applications require hundreds or thousands of sensor nodes, often set up in remote and unreachable areas. Thus, a wireless sensor is equipped with not only a sensing element, but also on-board processing, communication, and storage capabilities. With these developments, a sensor node is usually responsible for data collection, for in-network analysis, correlation, and fusion of its own sensor data and data from other sensor nodes. Many sensors collectively monitoring large physical environments, form a wireless sensor network (WSN). Using their wireless radios, Sensor nodes communicate with each other and also with a base station (BS) allowing them to broadcast their sensor data to remote processing, visualization, analysis, and storage systems. For example, Figure 1 shows two sensor fields monitoring two different geographic regions and connecting to the Internet using their base stations (Dargie & Poellabauer, 2010).

The sensing and control technology comprises electric and magnetic field sensors; radio-wave frequency sensors; optical, electro-optic-, and infrared sensors; radars; lasers; location/navigation sensors; seismic and pressure-wave sensors; environmental parameter sensors (e.g., wind, humidity, heat); and biochemical national security–oriented sensors. Today's sensors can be described as "smart" inexpensive devices equipped with multiple onboard sensing elements; they are low-cost low-power multifunctional nodes that are logically homed to a central sink node (Sohraby, Minoli, & Znati, 2007).

BACKGROUND AND HISTORY OF WIRELESS SENSOR NETWORK

A Sensor is a device that uses a sensing technique to collect information about a physical entity or process, including the occurrence of actions such

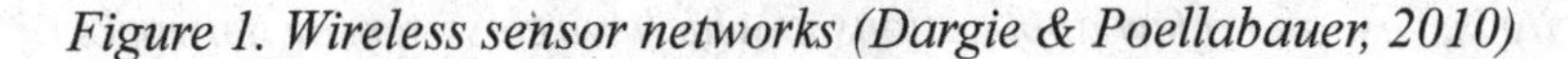

Figure 1. Wireless sensor networks (Dargie & Poellabauer, 2010)

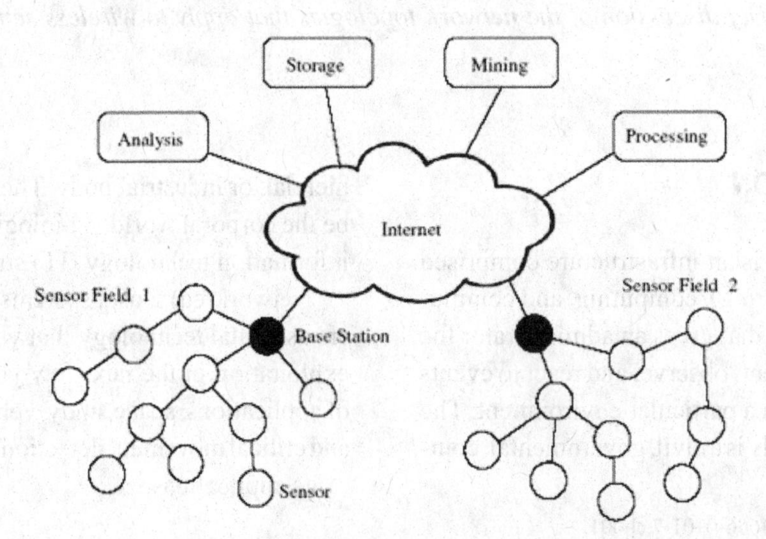

as, changes in state e.g. a drop in temperature or pressure). Other examples include natural sensors, for example the human body is equipped with sensors that are able to receive and capture optical information from the surroundings (eyes), audio information such as sounds (ears), and smells (nose). These are considered as remote sensors, that gather information of the monitored object without touching them.. However, from a technical viewpoint, a sensor is a device that translates occurrence of events and their parameters in the physical world into signals that can be precisely measured and examined.

Another frequently used word is a transducer, that converts energy from one form into another. A sensor, may thus be considered as a type of transducer that converts energy in the physical world into electrical energy that can be sent to a computing system or controller. An example of the steps performed in a sensing (or data acquisition) task is shown in Figure 2.

Phenomenon in the physical planet (often referred to as process, system, or plant) are observed by a sensor device. The electrical signals obtained are often not ready for immediate processing, thus they are passed through a signal conditioning stage. Here, a diversity of processes can be applied to the sensor signal to get it ready for further use. For example, signals may require amplification (or attenuation) to alter the signal magnitude to improve and match the range of the subsequent analog-to-digital conversion. Additionally, signal conditioning frequently applies filters to the signal to eliminate redundant and unwanted noise within certain frequency ranges (e.g., highpass filters can be used to remove 50 or 60 Hz noise picked up by surrounding power lines). After conditioning, the analog-to-digital converter (ADC) converts the analog signal into a digital signal for additional processing, storing, or analysis.

Nowadays *actuators* have been included in many wireless sensor networks which enable them to control the physical world directly. Examples of actuators could be a pump that controls the amount of fuel injected into an engine, a valve controlling the flow of hot water, or a motor that opens or closes a door or window. The processing device (controller) sends commands to a *wireless sensor and actuator network* (WSAN) which then converts them into input signals for the actuator, that acts with a physical process, thus forming a closed control loop (also shown in Figure 2).

Like several other technologies, the military has been a motivating force behind the development of wireless sensor networks. In 1978, the

Figure 2. Data acquisition and actuation (Dargie & Poellabauer, 2010)

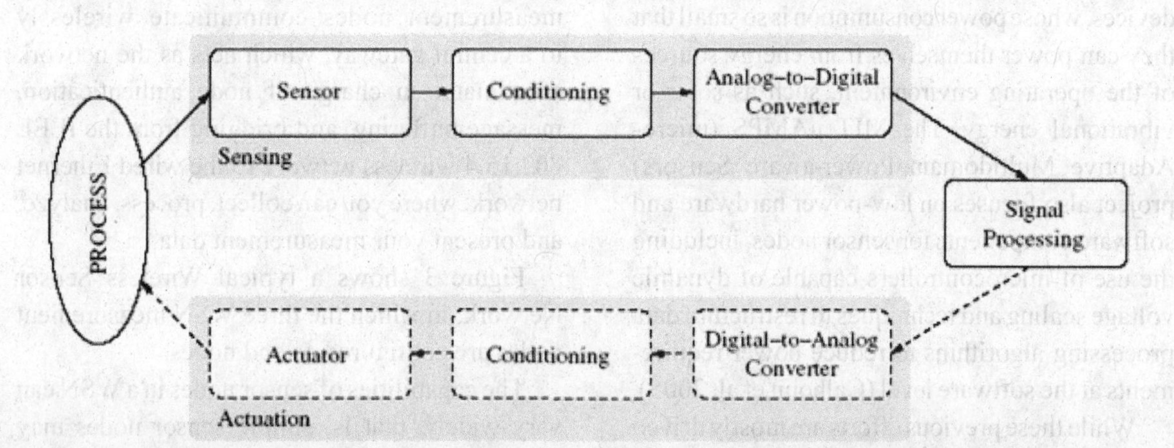

Defense Advanced Research Projects Agency (DARPA) arranged the Distributed Sensor Nets Workshop (DAR 1978), that focused on sensor network research challenges and constraints such as networking technologies, signal processing techniques, and distributed algorithms. DARPA also worked on the Distributed Sensor Networks (DSN) program in the early 1980s, which was later followed by the Sensor Information Technology (SensIT) program. The University of California at Los Angeles, in collaboration with the Rockwell Science Center proposed the concept of Wireless Integrated Network Sensors or WINS (Pottie 2001). One resulting product of the WINS project was the Low Power Wireless Integrated Microsensor (LWIM), produced in 1996 (Bult et al. 1996). This well turned-out sensing system was developed on a CMOS chip, that integrated several components like multiple sensors, interface circuits, digital signal processing circuits, wireless radio, and microcontroller onto a single chip. The Smart Dust project (Kahn et al. 1999) at the University of California at Berkeley focused on the design of extremely small sensor nodes called motes. The goal of this project was to demonstrate that a complete sensor system can be integrated into tiny devices, possibly the size of a grain of sand or even a dust particle. The PicoRadio project (Rabaey et al. 2000) by the Berkeley Wireless Research Center (BWRC) focuses on the development of low-power sensor devices, whose power consumption is so small that they can power themselves from energy sources of the operating environment, such as solar or vibrational energy. The MIT µAMPS (micro-Adaptive Multidomain Power-aware Sensors) project also focuses on low-power hardware and software components for sensor nodes, including the use of microcontrollers capable of dynamic voltage scaling and techniques to restructure data processing algorithms to reduce power requirements at the software level (Calhoun et al. 2005).

While these previous efforts are mostly driven by academic institutions, over the last decade a number of commercial efforts have also appeared (many based on some of the academic efforts described above), including companies such as Crossbow (www.xbow.com), Sensoria (www.sensoria.com), Worldsens (http://worldsens.citi.insa-lyon.fr), Dust Networks (http://www.dust-networks.com), and Ember Corporation (http://www.ember.com). These companies provide the opportunity to purchase sensor devices ready for deployment in a variety of application scenarios along with various management tools for programming, maintenance, and sensor data visualization (Dargie & Poellabauer, 2010).

WIRELESS SENSOR NETWORKS

A wireless sensor network (WSN) in general is made up of a base station (or "gateway") that can communicate through radio link with a number of wireless sensors. The wireless sensor node collects the data, compresses it, and transmits it to the gateway directly or, if needed, uses other wireless sensor nodes to forward data to the gateway. The transmitted data is then presented to the system via gateway connection. A WSN consist of spatially dispersed independent devices that use sensors to monitor physical or environmental conditions. These independent devices, known as routers and end nodes, together with a gateway combine to create a typical WSN system. The distributed measurement nodes communicate wirelessly to a central gateway, which acts as the network coordinator in charge of node authentication, message buffering, and bridging from the IEEE 802.15.4 wireless network to the wired Ethernet network. where you can collect, process, analyze, and present your measurement data.

Figure 3 shows a typical Wireless Sensor Network, in which the three WSN measurement nodes are configured as end nodes.

The capabilities of sensor nodes in a WSN can vary widely, that is, simple sensor nodes may monitor a single physical phenomenon, while

Figure 3. Basic WSN system with end nodes, ethernet gateway, and host PC (NI Developer Zone Tutorials, 2010)

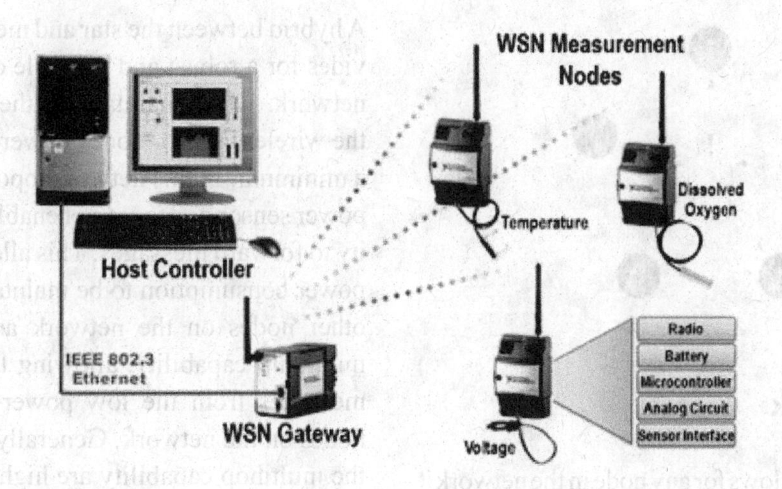

more complex devices may combine many different sensing techniques (e.g., acoustic, optical, magnetic). They can also differ in their communication capabilities, for example, using ultrasound, infrared, or radio frequency technologies with varying data rates and latencies. While simple sensors may only collect and communicate information about the observed environment, more powerful devices (i.e., devices with large processing, energy, and storage capacities) may also perform extensive processing and aggregation functions. Such devices often assume additional responsibilities in a WSN, for example, they may form communication backbones that can be used by other resource-constrained sensor devices to reach the base station. Finally, some devices may have access to additional supporting technologies, for example, Global Positioning System (GPS) receivers, allowing them to accurately determine their position. However, such systems often consume too much energy to be feasible for low-cost and low-power sensor nodes (Dargie & Poellabauer, 2010).

BASIC SENSOR NETWORK ARCHITECTURE

There are a number of different topologies for radio communications networks. A brief discussion of the network topologies that apply to wireless sensor networks are outlined below.

Star Network (Single Point-to-Multipoint)

A star network (Figure 4) is a communications topology where a single base station can send and/or receive a message to a number of remote nodes. The remote nodes can only send or receive a message from the single base station; they are not permitted to send messages to each other. The advantage of this type of network for wireless sensor networks is in its simplicity and the ability to keep the remote node's power consumption to a minimum. It also allows for low latency communications between the remote node and the base station. The disadvantage of such a network is that the base station must be within radio transmission range of all the individual nodes and is not as robust as other networks due to its dependency on a single node to manage the network.

Figure 4. Star network topology (Townsend & Arms, 2004)

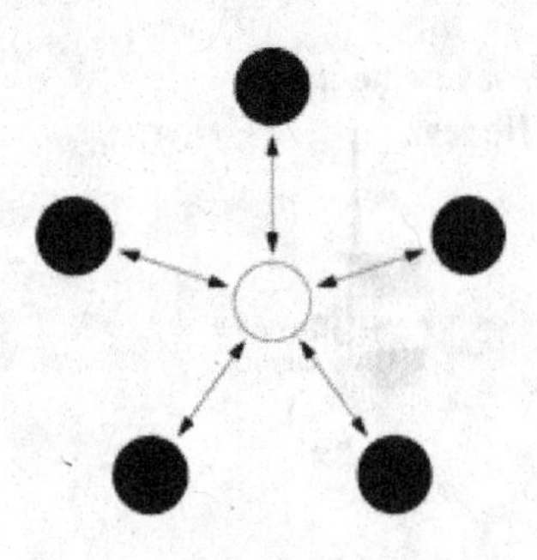

Mesh Network

A mesh network allows for any node in the network to transmit to any other node in the network that is within its radio transmission range. This allows for what is known as multihop communications; that is, if a node wants to send a message to another node that is out of radio communications range, it can use an intermediate node to forward the message to the desired node. This network topology has the advantage of redundancy and scalability. If an individual node fails, a remote node still can communicate to any other node in its range, which in turn, can forward the message to the desired location. In addition, the range of the network is not necessarily limited by the range in between single nodes, it can simply be extended by adding more nodes to the system. The disadvantage of this type of network is in power consumption for the nodes that implement the multihop communications are generally higher than for the nodes that don't have this capability, often limiting the battery life. Additionally, as the number of communication hops to a destination increases, the time to deliver the message also increases, especially if low power operation of the nodes is a requirement. Figure 5 shows an example of one possible mesh configuration.

Hybrid Star: Mesh Network

A hybrid between the star and mesh network provides for a robust and versatile communications network, while maintaining the ability to keep the wireless sensor nodes power consumption to a minimum. In this network topology, the lowest power sensor nodes are not enabled with the ability to forward messages. This allows for minimal power consumption to be maintained. However, other nodes on the network are enabled with multichip capability, allowing them to forward messages from the low power nodes to other nodes on the network. Generally, the nodes with the multihop capability are higher power, and if possible, are often plugged into the electrical mains line. This is the topology implemented by the up and coming mesh networking standard known as ZigBee (Figure 6).

CHALLENGES AND CONSTRAINTS

Although sensor networks have many similarities in common with other distributed systems, they face a variety of unique challenges and constraints. These constraints effect the design of a WSN, leading to protocols and algorithms that differ from their counterparts in other distributed systems. In this section the most important design constraints of a WSN are described.

Energy

The limitation most often related with sensor network design is that sensor nodes run with restricted energy budgets. Typically, they are driven by batteries, which must be either replaced or recharged (e.g., using solar power) when used up. For some nodes, neither option is suitable, that is, they will simply be discarded once their energy source is exhausted. Whether the battery can be recharged or not significantly affects the strategy applied to energy consumption. For non-

Figure 5. Mesh configuration (Townsend & Arms, 2004)

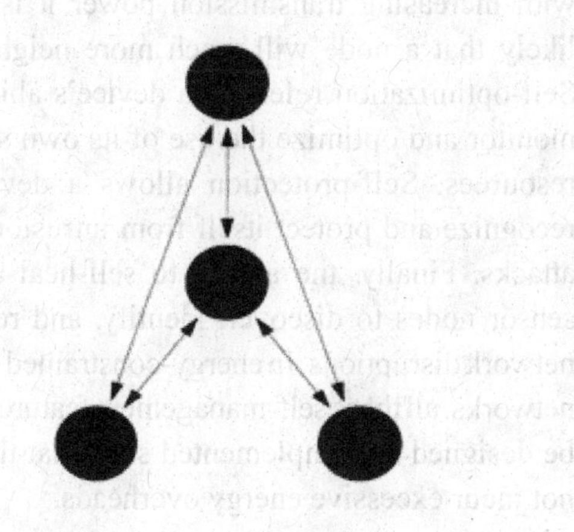

rechargeable batteries, a sensor node should be able to operate until either its *mission time* has passed or the battery can be replaced. The length of the mission time depends on the type of application, for example, scientists monitoring glacial movements may need sensors that can operate for several years while a sensor in a battlefield scenario may only be needed for a few hours or days. As a consequence, the first and often most important design challenge for a WSN is energy efficiency. This requirement permeates every aspect of sensor node and network design.

The medium access control (MAC) layer is responsible for providing sensor nodes with access to the wireless channel. Some MAC strategies for communication networks are contention-based, that is, nodes may attempt to access the medium at any time, potentially leading to collisions among multiple nodes, which must be addressed by the MAC layer to ensure that transmissions will eventually succeed. Downsides of these approaches include the energy overheads and delays incurred by the collisions and recovery mechanisms and that sensor nodes may have to listen to the medium at all times to ensure that no transmissions will be missed. Therefore, some MAC protocols for sensor networks are contention-free, that is, access to the medium is strictly regulated, eliminating collisions and allowing sensor nodes to shut down their radios when no communications are expected. The network layer is responsible for finding routes from a sensor node to the base station and route characteristics such as length (e.g., in terms of number of hops), required transmission power, and available energy on relay nodes determine the energy overheads of multi-hop communication.

Figure 6. Hybrid star – Mesh network (Townsend & Arms, 2004)

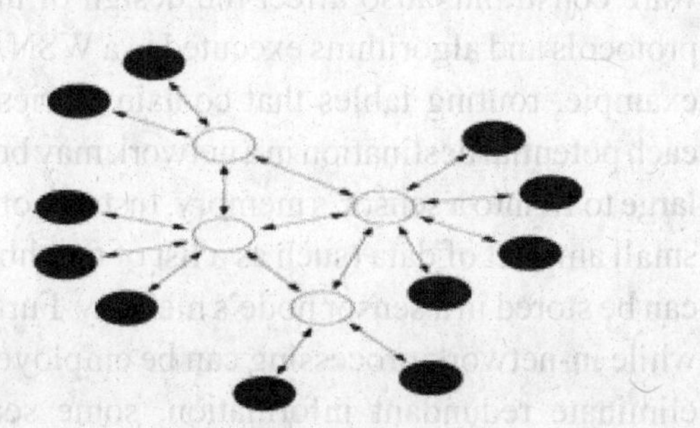

Besides network protocols, the goal of energy efficiency impacts the design of the operating system (e.g., small memory footprint, efficient switching between tasks), middleware, security mechanisms, and even the applications themselves. For example, in-network processing is frequently used to eliminate redundant sensor data or to aggregate multiple sensor readings. This leads to a tradeoff between computation (processing the sensor data) and communication (transmitting the original versus the processed data), which can often be exploited to obtain energy savings.

Self-Management

It is the nature of many sensor network applications that they must operate in remote areas and harsh environments, without infrastructure support or the possibility for maintenance and repair. Therefore, sensor nodes must be self-managing in that they configure themselves, operate and collaborate with other nodes, and adapt to failures, changes in the environment, and changes in the environmental stimuli without human intervention.

Ad Hoc Deployment

Many wireless sensor network applications do not require predetermined and engineered locations of individual sensor nodes. This is particularly important for networks being deployed in remote or inaccessible areas. For example, sensors serving the assessment of battlefield or disaster areas could be thrown from airplanes over the areas of interest, but many sensor nodes may not survive such a drop and may never be able to begin their sensing activities. However, the surviving nodes must autonomously perform a variety of setup and configuration steps, including the establishment of communications with neighboring sensor nodes, determining their positions, and the initiation of their sensing responsibilities.

The mode of operation of sensor nodes can differ based on such information, for example, a node's location and the number or identities of its neighbors may determine the amount and type of information it will generate and forward on behalf of other nodes.

Unattended Operation

Many sensor networks, once deployed, must operate without human intervention, that is, configuration, adaptation, maintenance, and repair must be performed in an autonomous fashion. For example, sensor nodes are exposed to both system dynamics and environmental dynamics, which pose a significant challenge for building reliable sensor networks (Cerpa and Estrin 2004). A *self-managing* device will monitor its surroundings, adapt to changes in the environment, and cooperate with neighboring devices to form topologies or agree on sensing, processing, and communication strategies (Mills 2007). Self-management can take place in a variety of forms. Self-organization is the term frequently used to describe a network's ability to adapt configuration parameters based on system and environmental state. For example, a sensor device can choose its transmission power to maintain a certain degree of connectivity (i.e., with increasing transmission power it is more likely that a node will reach more neighbors). Self-optimization refers to a device's ability to monitor and optimize the use of its own system resources. Self-protection allows a device to recognize and protect itself from intrusions and attacks. Finally, the ability to self-heal allows sensor nodes to discover, identify, and react to network disruptions. In energy-constrained sensor networks, all these self-management features must be designed and implemented such that they do not incur excessive energy overheads.

Design Constraints

While the capabilities of traditional computing systems continue to increase rapidly, the primary goal of wireless sensor design is to create smaller, cheaper, and more efficient devices. Driven by the need to execute dedicated applications with little energy consumption, typical sensor nodes have the processing speeds and storage capacities of computer systems from several decades ago. The need for small form factor and low energy consumption also prohibits the integration of many desirable components, such as GPS receivers. These constraints and requirements also impact the software design at various levels, for example, operating systems must have small memory footprints and must be efficient in their resource management tasks. However, the lack of advanced hardware features (e.g., support for parallel executions) facilitates the design of small and efficient operating systems. A sensor's hardware constraints also affect the design of many protocols and algorithms executed in a WSN. For example, routing tables that contain entries for each potential destination in a network may be too large to fit into a sensor's memory. Instead, only a small amount of data (such as a list of neighbors) can be stored in a sensor node's memory. Further, while in-network processing can be employed to eliminate redundant information, some sensor fusion and aggregation algorithms may require more computational power and storage capacities than can be provided by low-cost sensor nodes. Therefore, many software architectures and solutions (operating system, middleware, network protocols) must be designed to operate efficiently on very resource-constrained hardware (Dargie & Poellabauer, 2010).

Security Issues

Many wireless sensor networks collect sensitive information. The remote and unattended operation of sensor nodes increases their exposure to malicious intrusions and attacks. Further, wireless communications make it easy for an adversary to eavesdrop on sensor transmissions.

For example, one of the most challenging security threats is a *denial-of-service* attack, whose goal is to disrupt the correct operation of a sensor network. This can be achieved using a variety of attacks, including a *jamming attack*, where high-powered wireless signals are used to prevent successful sensor communications. The consequences can be severe and depend on the type of sensor network application. While there are numerous techniques and solutions for distributed systems that prevent attacks or contain the extent and damage of such attacks, many of these incur significant computational, communication, and storage requirements, which often cannot be satisfied by resource-constrained sensor nodes. As a consequence, sensor networks require new solutions for key establishment and distribution, node authentication, and secrecy (Dargie & Poellabauer, 2010).

From the discussion so far, it becomes clear that many design choices in a WSN differ from the design choices of other systems and networks. Figure 7 summarizes some of the key differences between traditional networks and wireless sensor networks. A variety of additional challenges can affect the design of sensor nodes and wireless sensor networks. For example, some sensors may be mounted onto moving objects, such as vehicles or robots, leading to continuously changing network topologies that require frequent adaptations at multiple layers of a system, including routing (e.g., changing neighbor lists), medium access control (e.g., changing density), and data aggregation (e.g., changing overlapping sensing regions). A heterogeneous sensor network consists of devices with varying hardware capabilities, for example, sensor nodes may have more hardware resources if their sensing tasks require more computation and storage or if they are responsible for collecting and processing data from other sensors within the network. Also, some sensor applications may have specific performance and quality requirements, for example, low latencies for critical sensor events or high throughput for data collected by video sensors. Both heterogeneity and performance requirements affect the design of wireless sensors and their protocols. Finally, while traditional computer networks are based on established standards, many protocols and mechanisms in wireless sensor networks are proprietary solutions, while standards-based solutions emerge only slowly. Standards are important for interoperability and facilitate the design and deployment of WSN applications; therefore, a key challenge in WSN design remains the standardization of promising solutions and the harmonization of competing standards.

APPLICATIONS OF WIRELESS SENSOR NETWORKS

Conventionally, WSNs have been used in the environment of high-end applications such as radiation and nuclear-threat detection systems; weapon sensors for ships; battlefield investigation and inspection; military command, control, communications, intelligence, and targeting systems; biomedical applications; habitat sensing; and seismic monitoring. Recently, interest has extended to networked biological and chemical sensors

Figure 7. Comparison of traditional networks and wireless sensor networks (Dargie & Poellabauer, 2010).

Traditional networks	Wireless sensor networks
General-purpose design; serving many applications	Single-purpose design; serving one specific application
Typical primary design concerns are network performance and latencies; energy is not a primary concern	Energy is the main constraint in the design of all node and network components
Networks are designed and engineered according to plans	Deployment, network structure, and resource use are often ad hoc (without planning)
Devices and networks operate in controlled and mild environments	Sensor networks often operate in environments with harsh conditions
Maintenance and repair are common and networks are typically easy to access	Physical access to sensor nodes is often difficult or even impossible
Component failure is addressed through maintenance and repair	Component failure is expected and addressed in the design of the network
Obtaining global network knowledge is typically feasible and centralized management is possible	Most decisions are made localized without the support of a central manager

for national security applications; furthermore, evolving interest extends to direct consumer applications. Applications with potential growth in the near future include military sensing, physical security, process control, air traffic control, traffic surveillance, video surveillance, industrial and manufacturing automation, distributed robotics, weather sensing, environment monitoring, and building and structure monitoring (Sohraby, Minoli, & Znati, 2007).

Structural Health Monitoring: Smart Structures

Sensors embedded into machines and structures enable condition-based maintenance of these assets.

Typically, structures or machines are inspected at regular time intervals, and components may be repaired or replaced based on their hours in service, rather than on their working conditions. This method is expensive if the components are in good working order, and in some cases, scheduled maintenance will not protect the asset if it was damaged in between the inspection intervals. Wireless sensing will allow assets to be inspected when the sensors indicate that there may be a problem, reducing the cost of maintenance and preventing catastrophic failure in the event that damage is detected. Additionally, the use of wireless reduces the initial deployment costs, as the cost of installing long cable runs is often prohibitive. In some cases, wireless sensing applications demand the elimination of not only lead wires, but the elimination of batteries as well, due to the inherent nature of the machine, structure, or materials under test. These applications include sensors mounted on continuously rotating parts, within concrete and composite materials, and within medical implants.

Industrial Automation

In addition to being expensive, leadwires can be constraining, especially when moving parts are involved. The use of wireless sensors allows for rapid installation of sensing equipment and allows access to locations that would not be practical if cables were attached. An example of such an application on a production line is shown in Figure 8. In this application, typically ten or more sensors are used to measure gaps where rubber seals are to be placed. Previously, the use of wired sensors was too cumbersome to be implemented in a production line environment. The use of wireless sensors in this application is enabling, allowing a measurement to be made that was not previously practical. Other applications include energy control systems, security, wind turbine health monitoring, environmental monitoring, location-based services for logistics, and health care.

Application Highlight: Civil Structure Monitoring

One of the most recent applications of today's smarter, energy-aware sensor networks is structural health monitoring of large civil structures, such as the Ben Franklin Bridge (Figure 8), which spans the Delaware River, linking Philadelphia and Camden, N.J. The bridge carries automobile, train and pedestrian traffic. Bridge officials wanted to monitor the strains on the structure as high-speed commuter trains crossed over the bridge (Townsend & Arms, 2004)/

Consumer applications include, but are not limited to, critical infrastructure protection and security, health care, the environment, energy, food safety, production processing, and quality-of-life support. WSNs are expected to afford consumers a new set of conveniences, including remote-controlled home heating and lighting, personal health diagnosis, and automated automobile maintenance telemetry, to list just a few. Near-term commercial applications include, but are not limited to, industrial and building monitoring, appliance control (lighting and HVAC), automotive sensors and actuators, home automation, automatic meter

Figure 8. Ben Franklin bridge (Townsend & Arms, 2004)

reading, electricity load management, consumer electronics and entertainment, and asset management. Specifically, these applications fall into the following categories:

- Commercial building control
- Environmental (land, air, sea) and agricultural wireless sensors
- Home automation, including alarms (e.g., an alarm sensor that triggers a call to a security firm)
- National security applications: chemical, biological, radiological, and nuclear wireless sensors (sensors for toxic chemicals, explosives, and biological agents)
- Industrial monitoring and control
- Metropolitan operations (traffic, automatic tolls, fire, etc.)
- Military sensors
- Process control
- Wireless automated meter reading and load management

Observers expect that in the medium term, one will be able to integrate sensors into commercial products and systems to improve the performance and lifetime of a variety of devices while decreasing product life cycle costs. The ultimate expectation is that eventually, WSNs will enable consumers to keep track of their belongings, pets, and young children (called quality-of-life support). Anywhere there is a need to connect large numbers of sensors, the approach of using WSNs with some well-established local and metropolitan area technology (e.g., IEEE 802.11/.15/.16) makes economic sense (Sohraby, Minoli, & Znati, 2007).

FUTURE DEVELOPMENTS

The most common and versatile operations of wireless sensing networks require the installation of batteries. Future work is being performed on systems that make use of piezoelectric materials to yield ambient strain energy for energy storage in capacitors and/or rechargeable batteries. By combining smart, energy saving electronics with advanced thin film battery chemistries that permit infinite recharge cycles, these systems could provide a long term, maintenance free, wireless monitoring solution (Townsend & Arms, 2004).

CONCLUSION

This chapter provides a preliminary survey of WSNs, including applications, communication, technology, network architecture, middleware, security, and management. It also includes some of the challenges to be faced and addressed by the growing practice. Wireless sensor networks have enabled applications that previously were impractical. While new standards-based networks are emerging and low power systems are constantly being developed, the widespread deployment of wireless sensor networks will soon be seen. The technology has reached to a point where one can begin exploring WSN applications with an eye to the fiscal return on savings that a company could anticipate with the deployment of sensor networks.

REFERENCES

Dargie, W., & Poellabauer, C. (2010). *Fundamentals of wireless sensor networks: Theory and practice*. West Sussex, UK: John Wiley & Sons Ltd.doi:10.1002/9780470666388

Developer Zone Tutorials, N. I. (2010). *Wireless sensor network topologies and mesh networking*. Retrieved December 28, 2010, from http://zone.ni.com/devzone/cda/tut/p/id/11211

Sohraby, K., Minoli, D., & Znati, T. (2007). *Wireless sensor networks, technology, protocols, and applications* (p. 3). Hoboken, NJ: John Wiley & Sons, Inc.doi:10.1002/047011276X

Swami, A., Zhao, Q., & Win Hong, Y. (2007). *Wireless sensor networks signal processing and communications perspectives*. West Sussex, UK: John Wiley & Sons Ltd.

Townsend, C., & Arms, S. (2004). *Wireless sensor networks: Principles and applications* (pp. 441–442). MicroStrain, Inc.

KEY TERMS AND DEFINITIONS

Actuator: A device that enables wireless sensor network to control the physical world directly. Examples of actuators could be a pump that controls the amount of fuel injected into an engine, a valve controlling the flow of hot water.

Conditioning: A diversity of processes that can be applied to the sensor signal to get it ready for further use. For example, signals may require amplification (or attenuation) to alter the signal magnitude to improve and match the range of the subsequent analog-to-digital conversion.

Self-Organization: The term frequently used to describe a network's ability to adapt configuration parameters based on system and environmental state.

Sensor: May be considered as a type of transducer that converts energy in the physical world into electrical energy that can be sent to a computing system or controller.

Signal Conditioning: Frequently applies filters to the signal to eliminate redundant and unwanted noise within certain frequency ranges.

The Medium Access Control (MAC) layer: Responsible for providing sensor nodes with access to the wireless channel.

Transducer: A device that converts energy from one form into another.

Wireless Sensor Network (WSN): Spatially dispersed independent devices (routers and end nodes) that use sensors to monitor physical or environmental conditions.

Chapter 2
Topology Management in Wireless Sensor Networks

Chiranjib Patra
Calcutta Institute of Engineering and Management, India

Arindam Mondal
Jadavpur University, India

Parama Bhaumik
Jadavpur University, India

Matangini Chattopadhyay
Jadavpur University, India

ABSTRACT

Topology management is a key component of network management of wireless sensor networks. The primary goal of topology management is to conserve energy while maintaining network connectivity. Topology management consists of knowing the physical connections and logical relationships among the sensors and at the same time creating a subset of nodes actively participating in the network, thus creating less communication and conserving energy in nodes. Networks require constant monitoring in order to ensure consistent and efficient operations. The primary goal of topology management is to maintain network connectivity in an energy-efficient manner. Topology management is one of the key aspects of configuration management, which entails initial set-up of the network devices and continuous monitoring and controlling of these devices. The main objective of this chapter is to throw a light on the recent developments and future directions of research in these directions.

BACKGROUND

One primary goal of network management in sensor networks is that it be autonomous. This is especially important in fault and configuration management. Configuration management includes the self-organization and self-configuration of the sensor nodes. Since WSN's involve very little human intervention after deployment, it is imperative that the areas of fault management be self diagnostic and self-healing. Another important issue to consider in fault management of WSN's

DOI: 10.4018/978-1-4666-0101-7.ch002

is that a single node failure should not impact the operation of the network, unlike a traditional network device failure causing impact to several users to potentially the entire network. There are several new functional areas of network management in sensor networks. Apart from topology management there are still new functional areas introduced for network management of WSN's are energy management and program management.

In (Akyildiz et al., 2002) energy management the most common way to conserve energy in WSN's is to power off a node when idle, but there have been many proposals in existing algorithms and protocols as well as establishing new protocols in order to be more energy efficient.

Program or code management (Wattenhoffer & Zollinger, 2004) is another aspect of network management in WSN's. The traditional method of updating a program in a sensor node is to attach the node to a programming interface of a laptop or PDA. This is not feasible in many WSN deployments. Transmitting an entire new program version to all sensors in a WSN is not practical as it consumes too much energy and will lead to a short network lifetime. There needs to be a way to transmit minimal packets to all nodes requiring the update while ensuring appropriate nodes receive the update reliably. There have been several proposals in the area of code update/management and it continues to be an active research area.

To begin with topology management (Zhang et al., 2009) there is six properties that should exist in the topology of WSN's: 1) symmetry, 2) connectivity, 3) spanner, 4) sparseness, 5) low degree, and 6) low interference. It is often observed the case that two properties, connectivity and sparseness conflict with each other. Despite all conflicts, the objective of topology management is to provide a backbone to various routing protocols so that there is an energy-efficient communication of the data. The algorithms responsible for building up the backbone may be categorized into three types (1) topology discovery, (2) sleep cycle management, and (3) clustering.

1. **Initialization Phase (Topology discovery Phase):** In this phase, nodes discover themselves and use their maximum transmission power to build the initial topology.
2. **Sleep cycle management:** To conserve energy, in a node is to only have it powered on when necessary; the node would be powered off or put to sleep all other times.
3. **Clustering:** Clustering algorithms are used to decrease the number of nodes that transmit data to the base station (BS). These algorithms arrange the nodes deployed in the WSN into groups or clusters. One node in each cluster is identified as the leader of the cluster or the cluster head (CH). The nodes that are in a cluster, but are not cluster head, become member nodes of that cluster. The member nodes will transmit their data to their cluster head, which is typically within only a short distance thus consuming less energy.

Over the entire life of the network topology control cycle will repeat many times until the energy of the network is depleted.

Topology discovery involves a network management station, or a base station, determining the organization or topology of the nodes in the sensor network. The physical connectivity and/or the logical relationship of nodes in the network are reported to the management station, which maintains a topology map of the WSN. The base station, or network management station, will send a topology discovery request to the network. Each node in the network will respond with its information. There are three basic approaches taken for topology discovery:

1. The first one is a direct approach. In this approach, a node will immediately sends a response back upon receiving a topology request. The node's response will contain information only about that particular node.
2. The second approach is an aggregated approach in which a node will forward the

request but will not send an immediate response. Instead, the node will wait until it gets responses to the request from its children. The node will then aggregate all the data received from its children, include its own information, and then send the response back to its parent or the initiating station.

3. The third approach is a clustered approach, which forms groups or clusters from the nodes. One node in each cluster is selected as the leader. Only the leader will reply to the topology request. The leader's reply will include the topology information about all the nodes in its cluster.

As all the approaches described above require the system energy to find the neighboring nodes and then aggregating or clustering which is one of the greatest drawback logically and physically for implementing static wireless sensor network. In order to improve upon, the authors in (Shahbazi, Araghizadeh, & Dalvi, 2008) proposed an intelligent method based on Self Organizing Map neural networks that optimize the routing in the terms of energy conservation and computation power of each node. This algorithm has been designed for a wireless sensor node called MODABER. The assumption is that every node has an importance due to its role in routing so that the nodes which are used more than other nodes in routing have more importance due to their positions. They defined a Network Life Time (NLT) parameter which is sum of the nodes importance in routing at time t and the amount of energy consumption of node for routing. They used a self-organizing (competitive) neural network to decide for every node containing the data packet and participate in routing or dropping the packet. The Self Organizing Map (SOM) learning algorithm is used for training of neural network. As soon as a packet arrives, its feature vector will be extracted and this vector is sent to self organizing NN of that node as input. The goal is to maximize NLT parameter. After winning of node in competition against other nodes, it is allowed to send the packet and participate in routing. Otherwise it should drop the packet. Since the learning algorithms of SOM's generally obey from linear computations, they believe that this method can be efficient to wireless nodes due to their limited computation and energy powers. While implementing SIR (Barbancho, Leon, Molina, & Barbancho, 2007), SOM neural network is introduced in every node to manage the routes that data have to follow. Here they have implemented SOM using QoS metrics viz. latency, throughput, error-rate and duty-cycle related to each node as input samples with an output layer neuron. The samples allocated in the SOM form groups, in such a way that all the samples in a group have similar characteristics (latency, throughput, error-rate, and duty-cycle). This way we obtain clusters with specific QoS values, lower value of QoS should be avoided as it depicts the worst case scenario.

All the above two schemes have shown their respective success but the parameters or metrics assumed are not difficult but time consuming to evaluate. This situation adds up to the cost of computation.

So in this regard we (Patra, Roy, Chattophaday, & Bhaumik, 2010) have designed an approach using Self Organizing Map (SOM) which makes the necessary high cost computation offline and only the implementation online. The scheme details can be found below.

This approach is selection by Self Organizing Map, in which a continuous input space of activation patterns (the spatial co-ordinates of sensors) is mapped onto a discrete output space of neurons (they would be selected spatial co-ordinate of sensors) by the process of competition among the neurons in the network. At the same time if the coordinate obtained do not map to the original ones, then by using k-nearest-neighbor algorithm the mapping is done as well we can iteratively remove the redundancy from the list of spatial coordinates.

Important advantages of this scheme is in addition to low power consumption include simplicity, inherent robustness to node or link failure and changing network geometry (in case of battery depletion), reduced redundant packet transmissions and implicit network reconfiguration. The only disadvantage is the need for sufficient density to maintain network operation.

The primary use of energy in WSN's is the transmission of data. Another way to conserve energy is to have fewer nodes transmit data to the base station, which is the device collecting the application data. Clustering algorithms are used to decrease the number of nodes that transmit data to the base station (BS). These algorithms arrange the nodes deployed in the WSN into groups or clusters. One node in each cluster is identified as the leader of the cluster or the cluster head (CH). The nodes which are in a cluster, but are not cluster head, become member node of that cluster. The member nodes will transmit their data to their cluster head, which is typically within only a short distance thus consuming less energy. The cluster head will then forward the data received from each of its member nodes to the base station. Only the cluster heads will transmit data to the base station. Many clustering algorithms will also aggregate or fuse the data received from the member nodes at the cluster head resulting in less data being transmitted from each cluster head to the base station. As less data is transmitted, less energy is used.

So in order to minimize the energy consumption authors (Patra, 2010) (Vesanto & Alhoniemi, 2000) have proved SOM to be an effective platform for visualization of high dimensional data and hence SOM as the first level of abstraction in clustering has some clear advantages. First the original data set is represented using smaller set of prototype vectors, which allows efficient use of clustering algorithms to divide the prototypes into groups. Secondly, reduction of the computational cost/transmission power is especially important for hierarchical algorithms allowing clusters of arbitrary size and shape.

Energy conservation is critical in all WSN's, but may be more critical in a long-term deployment. If the intended lifetime of the WSN is relatively short, then an algorithm that conserves less energy but sacrifices less latency may be appropriate.

Techniques of Topology Maintenance

There are three topology maintenance techniques

1. Static
2. Dynamic
3. Hybrid

- **Static Topology Maintenance:** In static topology maintenance technique all possible topologies are calculated and stored. The topology just switched from one to another when needed.
  Advantage: As all the topologies are pre-calculated, so the transition among the available topologies is fast. It also saves the overhead of topology construction every time the switching is happened.
  Disadvantage: But it has some disadvantages also, it cannot be known in advance that how the nodes will lose their energy. Due to extensive use of some nodes in one topology makes them unavailable for the next. It also takes some more time at the beginning to calculate all the possible topologies.
- **Dynamic Topology Maintenance:** Dynamic topology maintenance technique creates a new topology when necessary on the fly.
  Advantage: It has the current information about the network that helps it to make an appropriate reduced topology.
  Disadvantage: It takes more resource and time every time it runs.

- **Hybrid Topology Maintenance:** It uses both static and dynamic techniques. It calculates all the possible reduced topologies at the beginning that is during the first topology construction phase (static approach), but if it cannot implement it because of the node failure or connectivity failure, it creates a new topology on the fly (dynamic approach).
  It inherits the advantages and disadvantages of both the techniques.

Design Issues

The following are important considerations for effective topology maintenance mechanisms:

- **Distributed:** Being distributed rather than central, the algorithms can save more energy because in central approach some nodes may have to communicate long distances, whereas in distributed approach there are more sinks and base stations resulted in communications with only the closest ones. So there will be even distribution of energy among the sensor nodes.
- **Local information:** Nodes should be able to make topology control decisions locally. This reduces the energy costs and makes the mechanism scalable.
- **Need of location information:** The need of extra hardware or support mechanisms adds to the cost in terms of dollars and energy consumption. One example is the need of location information, which might be provided by GPS devices or localization protocols.
- **Robust to node failures and node mobility:** The algorithm will be more successful if it is robust to node failures. Sensor nodes are often prone to failure due to running out of energy, hardware failures or simply the node being destroyed due to harsh conditions. Sometimes these sensor nodes move by nature of application or by accidentally. The protocol or algorithm should be developed so it is robust to node mobility and node failures.
- **Low overhead:** Topology control mechanisms must work with very low message overhead, so they are energy-efficient and can be run many times as part of the topology maintenance cycle.
- **Low Complexity:** Topology control algorithms must have a low computational complexity, so they can be run in wireless sensor devices.
- **Low Convergence Time:** During the topology maintenance process a current topology will be replaced by a new one, therefore there will be a transition time during which the network might not be active. This time must be as small as possible. Static techniques offer a clear advantage in this aspect, as the new topology has already been calculated. In dynamic techniques, this time will be longer and depends on the convergence time of the topology construction mechanism.
- **Memory Consumption:** The memory of wireless sensors devices is limited. The topology maintenance static techniques need to have a considerable amount of memory to store all the pre-calculated topologies.

Triggering Criteria

The topology maintenance mechanism may be static, dynamic, or hybrid, global or local, there is one important question related to all: what is the criterion or criteria that will be used to trigger the process of changing the current topology? The triggering criteria, which may have important implications in terms of energy savings as well as coverage, reliability, and other important metrics, may be based on one of the following choices:

- **Time based:** In time-based topology maintenance, the current topology is changed every time a timer expires. The amount of time is usually fixed and pre-defined. This is a very critical variable. It can't be too short or too long. As being too short the switching of the topology maintenance algorithms will be very often, resulted in a waste of energy of the sensor nodes. On the other hand too long a time can make some important nodes unavailable.
- **Energy based:** Sometimes we can use the remaining energy of the nodes as the triggering criterion of the topology maintenance techniques. There should be a threshold value, on reaching it the topology will be triggered to change. Again this value is critical too for same reasons explained before.
- **Random based:** In random based topology maintenance a random variable is used to switch the current topology.
- **Failure based:** The failure based technique triggers the topology change only if one or some of the wireless sensor nodes failed. But failure detection and notification technique should be there.
- **Density based:** In density based triggering criterion the node degree of the nodes can be an important metric.
- **Combinations:** The criteria can be used in combination as well. Such as we can use energy and time or energy and failure to change the wireless network topology.

Existing topology maintenance algorithms are detailed below.

A. Topology Discovery

There are several solutions for topology discovery. The Topology Discovery Algorithm or TopDisc (Deb, Bhatnagar, & Nath, 2001) uses of a tree structure, with the root of the tree being the monitoring node, to find the network topology. There are three types of nodes: White (undiscovered), Black (cluster head) and Grey (neighbor of black node).

At first all the nodes remain white except the root. The root starts sending topology request packets to its neighbors. If a white node receives the packet from a black node, it turns grey. If a white node receives the packet from a grey node, then it will wait for a specific time. If it gets another request from a black node within that time, it will turn grey otherwise it will turn black. The black and grey nodes ignore any further request.

As the topology request is propagated, neighborhood sets will be generated. This is done by finding the set coverage with a greedy approximation algorithm.

Another topology discovery algorithm is Sensor Topology Retrieval at Multiple Resolutions or STREAM (Deb, Bhatnagar, & Nath, 2003). Using the Wireless Multicast Advantage, STREAM detects the presence of neighboring nodes by eavesdropping on the communication channel. This allows STREAM to create an approximate topology by getting neighborhood lists from a subset of nodes.

B. Sleep Cycle Management

There are many protocols that allow all nodes to sleep for a specific period taking a few at a time in order to save energy. After that they will wake up and join the network but some another set of nodes will go back to sleep. Thus these algorithms will take care of the sleep-wake cycle of the sensor nodes.

Sparse Topology and Energy Management (STEM) (Schurgers et al., 2002a, 2002b).Stem adds two radios with the sensor nodes. The first on is for forwarding and receiving messages. The second on is a low duty cycle radio and is used to send periodic "wake-up" messages to the neighboring nodes. A node will periodically wake up to see if another node is trying to communicate with it. If a node is trying to communicate with it, the node will wake up and receive the communication, otherwise the node will go back to sleep. Latency or delay is the main disadvantage of this protocol.

Geographic Adaptive Fidelity or GAF (Xu et al., 2003; Xu, Heidemann, & Estrin, 2001). GAF uses location information, typically from a GPS device, to organize redundant nodes into groups. GAF will divide the network area into virtual grids. A virtual grid is "defined such that, for two adjacent grids A and B, all nodes in A can communicate with all nodes in B and vice versa". (Xu, Heidemann, & Estrin, 2001) Since all nodes in adjacent grids can communicate with each other the nodes in these two grids are equivalent to the routing protocol.

The nodes can be in one of the three states, discovery, active or sleeping state. All nodes begin in the discovery state where it will send out discovery message and receive replies back in order to determine the nodes in its same grid. The node will then enter the active state. The node will stay in active state for a specified period of time and will then go back to the discovery state. If a node determines that it is a redundant node for the routing protocol, it will enter the sleeping state for a specified period of time.

Cluster-based Energy Conservation or CEC(Xu et al., 2003) is an algorithm based on GAF. CEC will directly measure the network connectivity, thus not requiring location information and finding redundancy in the network more accurately. Since CEC is more accurate at identifying redundant nodes it conserves more energy than GAF.

CEC will organize nodes into overlapping clusters. A cluster is created by grouping nodes that are at most two hops from each other. The node in the cluster that has the most residual energy will select itself to be the cluster head. Since the clusters overlap, some nodes will be members of multiple clusters. These nodes are gateway nodes and will connect the clusters preserving network connectivity.

C. Clustering

There are many different ways the algorithms form clusters, from using node id, node degree (number of neighbors), and location information.

Single-hop Clustering Algorithms: One of the most successful clustering algorithms is LEACH (Heinzelman, Chandrakasan, & Balakrishnan, 2000). In LEACH all the nodes die at almost same time. It selects the cluster heads based on the remaining energy in the nodes and also rotates the cluster heads periodically. Thus it guarantees a certain network lifetime while minimizing the energy consumption by the sensor nodes. Heinzelman, et. al. (Zhang et al., 2009) has shown that LEACH "successfully distributes the energy-usage among the nodes in the network such that the nodes die randomly and at essentially the same rate". Cluster members send the data to its cluster head. The cluster head will fuse all the data and then transmit one message to the base station, containing the data for its cluster.

The disadvantage of LEACH is the cluster heads forward the data to the base station which is in a single hop distance but may be long. Overhead regarding clusters creation is also a drawback.

LEACH can also be extended to be hierarchical, so that cluster heads communicate with a higher-level cluster head instead of directly with the base station. M-LEACH (Mhatre & Rosenberg, 2004) is an implementation of LEACH for multi-hop networks, where a node is multiple hops from its cluster head.

ABCP (Hou & Tsai, 2001) or Access-Based Clustering Protocol designed the clustering operation from a protocol point of view. It defines the message formats, describes how a node responds when a message arrives, and specifies how a node handles errors. This algorithm is a "simple broadcast request-response with first-come-first-serve selection". (Hou & Tsai, 2001) There are many advantages to using ABCP. It does not require any location information. Cluster heads will fuse the data of its member nodes before transmitting the data to the base station that shortens the message to be sent and it is stable even during topology changes, even during the cluster formation process.

Another request-response with first-come-first-serve selection for cluster formation is ABEE (Hong & Liang, 2004) or Access-Based Energy Efficient cluster algorithm. This algorithm is very similar to ABCP but is based primarily on location. ABEE will try to balance the residual energy in all the nodes by periodically rotating the role of the cluster head. The new cluster head is selecting by treating the "whole cluster as an entity and each node stands for particles with equal mass to form the entity" (Hong & Liang, 2004). ABEE improves the lifetime of the network when compared to ABCP. According to (Hong & Liang, 2004) there is "a 92.3% lifetime enhancement" over the ABEE protocol and "around a 50% gain in the lifetime of the network coverage".

*Multi-hop Clustering Algorithms: PEGASIS (Lindsey & Raghavend*ra, 2002) or Power-Efficient Gathering in Sensor Information Systems lowers the overhead of cluster formation in LEACH. The key idea of PEGASIS it to form a chain among the nodes and take turns transmitting the data to the base station. This allows each node to communication only with a closest neighbor, thus consuming less energy.

There are several assumptions in the PEGASIS algorithm. First, it assumes that all nodes have global knowledge of the network. This allows them to create the best chain using the greedy algorithm and each node will know its neighbor nodes. It also assumes that all nodes employ the greedy algorithm and that the radio channel is symmetric.

The advantage of PEGASIS is since nodes only receive and transmit to its neighbors, and they form a chain, each node will only transmit and receive one packet of data in each round. If a node fails the chain can be reconstructed with the remaining nodes. This makes PEGASIS robust to node failures.

Another clustering algorithm is the Energy Efficient Clustering Scheme or EECS (Ye et al., 2005). The goals were to create a fully-distributed, load-balancing clustering algorithm that had little overhead. It is very much like LEACH but can better balance the load among the clusters and cluster heads. In EECS, there is only one cluster head within a certain range with a high probability. According to (Ye et al., 2005), the control overhead across the network is O(n). This paper also indicated that EECS will prolong the network lifetime over 35% when compared to LEACH. The energy utilization rate is also better in EECS because "EECS always achieves the well distributed cluster heads while considering the residual factor; further, we consider to balance the load among the cluster heads with weighted function". (Ye et al., 2005)

Another energy-conserving technique for WSNs is data compression. Sending large amounts of data from sensor nodes to the base station is very energy-draining. In order to conserve some of the valuable energy, data compression can be used. Before sending data from a node, the data is first compressed, requiring less data to be transmitted.

FUTURE DIRECTIONS

Research is going on in all the aspects of Topology Management. Any clustering algorithm developed should be distributed and should rotate the cluster head so that nodes will die at approximately the same rate. During cluster head selection, residual energy should be one of the considerations. The cluster head should fuse the data received by the member nodes before sending it to the base station in order to help conserve energy. Maintaining a relatively short distance from each cluster head to the base station is also an important factor to consider. This will lessen the energy consumption required to transmit the data from the cluster head to the base station and prolong the life of each of those nodes.

Non-cluster head nodes in the network should select the best cluster to join based on a variety of information. One of these pieces of information should not be location information as that adds complexity and requires more energy consumption by the nodes. The nodes should join the cluster that allows them to consume the least amount of energy while achieving the goal of the application. The clusters should also be load balanced, again so that all nodes die at approximately the same time.

Almost all the protocols have some assumptions. We have to work on these assumptions so that the new topology can be the optimal one. Such as all transmission channels are assumed to be collision and error free, where in real that is not possible. The nodes can be destroyed due to some abnormal conditions. Nodes mobility should also be addressed. The protocols that will take care of these assumptions will surely have an edge over the existing ones.

CONCLUSION

The flexibility, fault tolerance, high sensing fidelity, low-cost and rapid deployment characteristics of sensor networks create many new and exciting application areas for remote sensing. In the future, this wide range of application areas will make sensor networks an integral part of our lives. However, realization of sensor networks needs to satisfy the constraints introduced by factors such as fault tolerance, scalability, cost, hardware, topology change, environment and power consumption.

After the creation of the reduced topology from a set of sensor nodes it has to be maintained properly. Here comes the importance of topology management techniques. In order to increase the total network lifetime, the set of active nodes, the ones in the reduced topology, cannot be active all the time. Rather, a topology maintenance mechanism should be in place to build a new reduced topology – with the collaboration of formerly inactive nodes – so that all nodes participate in the network, consume their energy in a fair manner, and increase the lifetime of the network accordingly.

Research is going on in all fields to find better algorithms for the welfare of mankind.

REFERENCES

Akyildiz, I. F. (2002). Wireless sensor networks: A survey. *Computer Networks*, *38*, 393–422. doi:10.1016/S1389-1286(01)00302-4

Barbancho, J., Leon, C., Molina, F. J., & Barbancho, A. (2007). Using artificial intelligence in routing scheme for wireless networks. *Computer Communications*, *30*, 2802–2811. doi:10.1016/j.comcom.2007.05.023

Deb, B., Bhatnagar, S., & Nath, B. (2001). *A topology discovery algorithm for sensor networks with applications to network management*. (Technical Report dcs-tr-441). Rutgers University, May 2001

Deb, B., Bhatnagar, S., & Nath, B. (2003). Multi-resolution state retrieval in sensor networks. *Proceedings of the First IEEE International Workshop on Sensor Network Protocols and Applications,* 11 May (pp. 9–29).

Heinzelman, W. R., Chandrakasan, A., & Balakrishnan, H. (2000). Energy efficient communication protocol for wireless microsensor networks. *Proceedings of the 33rd Annual Hawaii International Conference on System Sciences,* 4-7 January 2000.

Hong, X., & Liang, Q. An access-based energy efficient clustering protocol for ad hoc wireless sensor network. *15th IEEE International Symposium on Personal, Indoor and Mobile Radio Communications*, vol. 2, (pp. 1022-1026).

Hou, T.-C., & Tsai, T.-J. (2001). An access-based clustering protocol for multihop wireless ad hoc networks. *IEEE Journal on Selected Areas in Communications*, *19*(7), 1201–1210. doi:10.1109/49.932689

Lindsey, S., & Raghavendra, C. S. (2002). PEGASIS: Power-efficient gathering in sensor information systems. *IEEE Aerospace Conference Proceedings*, 2002.

Mhatre, V., & Rosenberg, C. (2004). Homogeneous vs heterogeneous clustered sensor networks: a comparative study. 2004 IEEE International Conference on Communications, vol. 6, pp. (646 - 3651).

Patra, C. (2010). Using Kohonen's self-organizing map for clustering in sensor networks. *International Journal of Computers and Applications*, *1*(24), 80–81.

Patra, C., Roy, A. G., Chattophaday, S., & Bhaumik, B. (2010). Designing energy-efficient topologies for wireless sensor network: Neural approach. *International Journal of Distributed Sensor Networks*, *2010*, 216716. doi:10.1155/2010/216716

Schurgers, C., Tsiatsis, V., Ganeriwal, S., & Srivastava, M. (2002a). Optimizing sensor networks in the energy-latency-density design space. *IEEE Transactions on Mobile Computing*, *1*(1). doi:10.1109/TMC.2002.1011060

Schurgers, C., Tsiatsis, V., Ganeriwal, S., & Srivastava, M. (2002b). *Topology management for sensor networks: Exploiting latency and density*. MOBIHOC'02, Lausanne, Switzerland, ACM, June 9-11 2002.

Shahbazi, H., Araghizadeh, M. A., & Dalvi, M. (2008). Minimum power intelligent routing in wireless sensors networks using self organizing neural networks. *IEEE International Symposium on Telecommunications*, (pp. 354-358).

Vesanto, J., & Alhoniemi, E. (2000). Clustering of self organizing map. *IEEE Transactions on Neural Networks*, *11*(3), 586–358. doi:10.1109/72.846731

Wattenhofer, R., & Zollinger, A. (2004). XTC: A practical topology control algorithm for ad-hoc networks. *Proceedings of the 18th International Parallel and Distributed Processing Symposium*, 2004, 26-30 April (p. 16).

Xu, Y., Bien, S., Mori, Y., Heidemann, J., & Estrin, D. (2003). *Topology control protocols to conserve energy in wireless ad hoc networks*. Technical Report 6, University of California, Los Angeles, Center for Embedded Networked Computing, January 2003.

Xu, Y., Heidemann, J., & Estrin, D. (2001). Geography-informed energy conservation for ad hoc routing. *Proceedings of the 7th annual international conference on Mobile computing and networking*. Rome, Italy (pp. 70–84).

Ye, M., Li, C., Chen, G., & Wu, J. (2005). EECS: An energy efficient clustering scheme in wireless sensor networks. 24th IEEE International Performance, Computing, and Communications Conference, (pp. 535- 540).

Zhang, Z. (2009). Resource prioritization of code optimization techniques for program synthesis of wireless sensor network applications. *Journal of Systems and Software*, *82*(9). doi:10.1016/j.jss.2009.05.018

KEY TERMS AND DEFINITIONS

Base Station: In the area of wireless computer networking, a base station is a radio receiver/transmitter that serves as the hub of the local wireless network, and may also be the gateway between a wired network and the wireless network. It typically consists of a low-power transmitter and wireless router.

Cluster: A cluster is a small group or bunch of something. Here it is a group of wireless sensor nodes.

Network Capacity: Network capacity is the amount of traffic that a network can handle at any given time.

Network Coverage: The geographical area covered by the network is called as the network coverage.

Network Topology: Network topology is the layout pattern of interconnections of the various elements (links, nodes, etc.) of a computer network.

QoS: Quality of service in network traffic engineering refers to resource reservation control mechanisms. Quality of service is the ability to provide different priority to different applications, users, or data flows, or to guarantee a certain level of performance to a data flow. For example, a required bit rate, delay, jitter, packet dropping probability and/or bit error rate may be guaranteed.

Self Organizing Map (SOM): A self-organizing map (SOM) or self-organizing feature map (SOFM) is a type of artificial neural network that is trained using unsupervised learning to produce a low-dimensional (typically two-dimensional), discretized representation of the input space of the training samples, called a map. Self-organizing maps are different from other artificial neural networks in the sense that they use a neighborhood function to preserve the topological properties of the input space.

Sensor Node: A sensor node, also known as a mote (chiefly in North America), is a node in a wireless sensor network that is capable of performing some processing, gathering sensory information and communicating with other connected nodes in the network.

Wireless Sensor Network: A Wireless Sensor Network (WSN) consists of spatially distributed autonomous sensors to monitor physical or environmental conditions, such as temperature, sound, vibration, pressure, motion or pollutants, and to cooperatively pass their data through the network to a main location.

Chapter 3
On Network Management of Wireless Sensor Networks:
Challenges, Solutions and Research Trends

Sana Khan
National University of Science and Technology (NUST), Islamabad, Pakistan

Sheikh Tahir Bakhsh
Universiti Teknologi PETRONAS, Malaysia

ABSTRACT

Wireless sensor networks have become a popular research area of distributed computing. It is the emerging area of pervasive computing, for supporting daily life applications of human usage in future and could provide large amount of benefits to the society, by efficient delivery of sensed information. Network management of WSNs plays an important role for the efficient working of whole network and application. It is the research area that is recently gaining the attraction from research community. This chapter compiles some related information on the basis of studied literature regarding wireless sensors network management, including WSN background, issues and challenges, proposed solutions, and research trends.

INTRODUCTION

A WSN is a network of devices, called *nodes (stationary or moving)*, having sensing capabilities and deliver the information collected from the monitored field with the help of wireless links Buratti et al. (2009). Wireless sensor networks have attained considerable attention recently. Many researchers have attempted to improve wireless sensor networks' management efficiency. Wireless sensors networks can sense and monitor information from the physical world, and are used in the scientific, medical, and commercial domains (Ma et al. 2010; Jiang et al. 2007).

Supporting convenient and effective network management is crucial in wireless sensor net-

DOI: 10.4018/978-1-4666-0101-7.ch003

works. Network management of WSNs play an important role for the efficient working of whole network and application (Zhang & Li 2009; Yu et al. 2006; Zhang, & Li 2008), because of having different architecture from normal data networks, it is more challenging task in WSNs (Buratti et al. 2009; Deb, Bhatnagar, & Nath, 2002; Zhang, Xu, & Sun, 2010). It is the research area that is recently gaining the attraction from research community (Buratti et al. 2009; Yu, Mokhtar, & Merabti, 2006; Li et al. 2005; Zhang, Xu, & Sun, 2010). Network management is the process that manages, monitor and control the network behavior. (Lee, Datta, & Oliver). Traditional network management techniques have become impractical for wireless sensor networks because of its unique network management requirements (Lee, Datta, & Oliver; Yu, Mokhtar, & Merabti, 2006). An efficient network management architecture is a challenge for proper working wireless sensor networks to support several sensor applications (Zhang, & Li, 2009; Yu, Mokhtar, & Merabti, 2006; Zhang, & Li, 2008; Frye, & Cheng, 2007; Townsend, & MicroStrain, 2004; Lee, Datta, & Oliver; Obaisat, & Braun, 2006; Ganesan et al. 2003; Zhang, Xu, & Sun, 2010).

This chapter aims at providing some relevant details regarding network management of wireless sensor networks. Objective is to provide the discussion regarding network management in perspective of wireless sensor networks' issues, challenges and proposed solutions.

BACKGROUND

Wireless sensor networks are increasingly gaining popularity today as a new category of networking technology because of their flexibility and adaptability in a variety of environments. Management of WSNs as a hot area of research these days has presented some significant management challenges to the research community, because of the unique characteristics of wireless sensor networks (Zhang, Xu, & Sun, 2010). Sensor network applications are diverse in their requirements, so research proved that network management architecture should be lightweight, robust, fault tolerant, adaptable, and scalable. Network management is the process that manages, monitor and control the network in terms of its behavior (Zhang, & Li, 2009; Lee, Datta, & Oliver). Function of the Network management system and protocol is to allow the sensor networks to self forming, self organization and self configuration in case of failures without having the prior knowledge of network topology (Frye, & Cheng, 2007: Zhang, & Li, 2008; Zhang, Xu & Gen 2010). In other words for wireless sensors networks, a network management system should provide a set of management functions for configuration, security, operation, administration and overall maintenance of sensor network elements and services (Lee, Datta, & Oliver). Several issues have been discussed that make the management of WSN a challenging task (Zhang, & Li, 2009; Yu, Mokhtar, & Merabti, 2006; Zhang, & Li, 2008; Li et al. 2005; Lee, Datta, & Oliver; Obaisat, & Braun, 2006; Ganesan et al. 2003) and authors also discussed the required design criteria for proposing an efficient management architecture/protocol for WSNs. (Zhang, & Li, 2009; Yu, Mokhtar, & Merabti, 2006; Zhang, & Li, 2008; Lee, et al. 2006; Obaisat, & Braun, 2006; Cerpa, et al. 2004)

Sensor network management is defined as a system or process for the management and coordination of sensor nodes in a dynamic and unsure environment to achieve specific objectives and perception performance with minimum energy (Li, et al. 2005). It is the research area that is recently gaining the attraction from research community (Li, et al. 2005; Yu, Mokhtar, & Merabti, 2006). Traditional network management techniques have become impractical for wireless sensor networks because of its unique network management requirements (Yu, Mokhtar, & Merabti, 2006; Lee, Datta, & Oliver). Efficient network management architecture is a challenge for proper working of wireless sensor networks to support several sensor applications (Townsend, & MicroStrain, 2004). Today wireless sensor networks are playing an indispensable role in the field of civil and military, having large number of nodes, requiring significant attention toward network management (Abbasi, & Younis, 2007). Various approaches have been practiced and deployed for efficient network management of sensor networks. The primary goal of wireless sensor network is to reduce the energy consumption through reducing traffic overload. Moreover wireless sensor networks should be smart, autonomous and self aware to support different applications for long period of time, so network management is very crucial for proper working of whole network (Zhang, & Li, 2008; Frye, & Cheng, 2007; Lee, Datta, & Oliver; Zhang, Xu, & Sun, 2010).

Key features of wireless sensor networks, which distinguished them from traditional wired networks, are as follows (Zhang, & Li, 2009; Zhang, & Li, 2008; Lee, Datta, & Oliver; Sorniotti et al. 2007).

Limited resources: Sensor nodes have resource limitations in terms of battery, memory, processing capability and achievable data rate. So efficient use of these limited resources, by reducing the redundant activities, saves power and maintains the longer lifetime of sensors, thus make the whole network more stable.

Dynamic topology: Usually wireless sensor network has dynamic topology, because of node's changes due to environmental obstacles and limited energy, so periodic reconfiguration of the network becomes crucial. Network topology is an important aspect of management structure for WSNs.

Application dependency: Resource limitations in wireless sensor networks don't allow accommodating a wide variety of applications in contrast to traditional networks. The design of applications and management architectures in wireless sensor networks are dependent upon application semantics, so designers have to develop special programs for node localization, data routing and data aggregation to meet the needs of a specific sensor application. Research is ongoing to solve this major issue. There is a need of solution that can accommodate a wide variety of sensor applications by integration of application semantics with management architectures in WSNs

Fault tolerance: Sensor node failure and communication failure are very common in WSNs because of energy shortages, connectivity interruptions and environmental obstacles. There is a need of management architecture for WSNs, to equip them with the ability to reconfigure and recover from faults itself, without too much human interference.

Different type of nodes: There are three types of nodes involved in WSNs. Common nodes, sink nodes and gateway nodes. Common nodes are mainly involved in sensor data collection or performing collaborated tasks with neighborhood nodes, usually don't have extra storage space to hold large amount of data. Sink nodes receive store and process data from common nodes. Gateway nodes connect sink nodes to different sensor applications.

Difficult to access: Because of large scale of network deployment, high mobility and prone to failure sensor nodes are difficult to access. So network reconfiguration, fault recovery, or handling other technical problems is a big challenge. Still there is a need of solution that must work without too much human interference.

Data in-network processing: In WSNs, information processing and delivery are not independent, and their interaction also impacts the required QOS. Application independent and self organizing architectures have become crucial to be developed to properly perform in network processing on raw data. In-network data processing is to perform processing on data by the network nodes in a distributed manner. The process is performed during data exchange, before the data is ready for use by higher layers besides data gathering functionality; In-network data processing allows WSNs to provide complex services to application layer. This approach allows several advantages like; less energy consumption is required for computation, and reduction in traffic overhead. However an efficient, cost effective and secure data aggregation scheme is still an open area of research.

ISO has identified five functional areas for traditional wired networks, which are also true for sensor network management, but in different terms (Frye, & Cheng, 2007). These are Fault management, security management, configuration management, performance management and accounting management (Frye, & Cheng, 2007; Alam, Rashid, & Hong, 2008; John Wiley, & Sons, 2005; Ma et al. 2010).

Fault management is the process of detecting faults or problem in the network. This process includes fault detection, its isolation and recovery if possible. There are different ways to report these faults like, a log file, email alert to network mgt, or alarm on the network mgt system Fault prioritization is necessary to deal with the network related problems in systematic way. It is important in sensor networks, that failure of single node should not affect the whole network.

Configuration management not only deals with establishing network device, but monitoring and controlling setting up an inventory to maintain the configuration of all network devices, plus report generation on periodic basis to collect information about current network stat is an important part of configuration mgt. An automated tool called auto discovery, detects the network devices currently installed and also collects the information about their current configuration, but it is not suitable for networks with band width constraints. Self configuration and self organization of sensor mode is very important, and it is the main table of configuration management.

Security Management

Security of information on the network is called security management but it is different from the security of an operating system, physical or application security. Security is a challenging task in WSNs because of limited processing power, energy constraints and transmission bandwidth. A single faulty node can easily upset the functioning of the entire network. Primary task of security management is to identify and keep secure the access points to critical data stored in the network devices. Sensitive data is to be protected by analyzing and minimizing the possible risks.

Performance management involves determining the vitalization, by monitoring network devices and lines. It is essential for capacity planning and future and current risks which can affect network performance. It involves determining the CPU load, network card utilization, packet forwarding rate, error occurrence rate as well as no of packets in a queue, to ensure the network capacity. Simply it determines the network coverage and its connectivity. The two main objectives of performance management are acquisition of quality information and distribution of services

Accounting Management tracks the utilization of network resources by each user. Its primary function is to charge the user for network and its resources, but it is not a common in practice. Several issues have to be considered for these functional areas. Self organization and self configuration of sensor nodes is to be made possible through configuration management. Fault management should be self diagnostic in sensor networks because a very little human interference is involved (Zhang, & Li, 2009; Frye, & Cheng, 2007; Zhang, Xu, & Sun, 2010). Several new functional areas identified for WSNs are topology management, energy management, and program management.

Energy management: Optimizing the energy consumption is the most significant performance objective in wireless sensor networks (Berman, et al. 2005; Jiang, et al. 2007) or in simple words it is the most crucial resource of WSNs (Chen, et al. 2006). Energy could be saved at different levels and in different ways, and so it is a separate area of network management. The most common way to conserve energy is to keep some nodes in sleep state and others in the active state for sensing and communication task (Wang, & Xiao, 2005) but some other schemes and protocols have also been proposed for energy conservation. Especially MAC and routing protocols are being introduced for energy efficient data transmission (Frye, & Cheng, 2007).

Program management (Frye, & Cheng, 2007) is a new functional area of wireless sensor networks' management and getting significant attention as a research area also. In the traditional program management, program is updated by attaching the node to programming interface of a laptop or PDA, which is not a reasonable mechanism in many WSN deployments. Moreover transmission of an updated program to all the sensors of WSNs is not practical because of much energy consumption, as it affects network's life time. Two solutions are recommended to handle with this situation.

- To send update to only those nodes, which require the update instead of sending it to entire network
- Another solution is to send only updated part of the code to the nodes

Topology management (Frye, & Cheng, 2007; Deb, Bhatnagar, & Nath, 2002) is to be discussed as the main management function for energy conservation in wireless sensor networks. The goal of topology management is to coordinate the sleep transitions of all the nodes, while ensuring adequate network connectivity, such that data can be forwarded efficiently to data sink (Schurgers, Tsiatsis, & Srivastava, 2002). It is the process of knowing physical connections and logical relationships among the sensors (Woungang, Subhas & Cheng, 2009).

Topology management involves topology discovery, sleep cycle management and clustering. In topology discovery a base station determines node organization in the sensor network. A topology discovery request is sent by the base station or management station to the network so that nodes in the network could respond with their information. In direct scheme, upon receiving topology discovery request, node will immediately send back response regarding its operations. If node waits for response from its children and send the aggregated response (children information + node information) to the parent or initiating station then this is called aggregated approach. Sleep cycle management causes some node to sleep for eliminating redundancy thus conserve energy, while maintain network connectivity. Different topology management algorithms are discussed in terms of topology discovery; sleep cycle management and clustering. Clustering algorithms are discussed and compared on the basis of different factors (Frye, & Cheng, 2007; Abbasi, & Younis, 2007). Clustering is regarded as very advantageous for network scalability and different clustering algorithms are discussed in detail followed by their comparison on the basis of convergence time, node mobility, cluster overlapping, location awareness, energy efficiency, failure recovery, balanced clustering and stability (Abbasi, & Younis, 2007).

Three types of Network management architecture are discussed in literature, namely Centralized, distributed and hierarchical (Frye, & Cheng, 2007; Lee, Datta, & Oliver; Alam, Rashid, & Hong, 2008; Zhang, Xu, & Sun, 2010).

In **Centralized architecture** base station serves as the manager station, collect information from all nodes and supervise the whole network. Its disadvantages are High bandwidth and energy, and limited scalability. In **Distributed architecture** multiple managers coordinated with each other to perform management functions. Distributed management architecture is beneficial in terms of lower communication costs and also provides better reliability and energy efficiency than centralized management. This type of architecture is too complex to be managed for sensor networks with constrained resources. In **hierarchical architecture** each intermediate manager is responsible for the management of its own sub network. There is no direct communication between these intermediate managers but supports two way communication between the sub network and higher level managers. This type of architecture is more suitable for WSNs and integrates the benefits from centralized and distributed architecture.

Several network management protocols have been emerged for wireless sensor networks. Like MANNA, RRP, SNMS, sNMP, WinMS, SNMP, NetConf. Details and analysis of these protocols on the basis of different design criteria is found in literature. Researchers (Zhang, & Li, 2008; Lee, Datta, & Oliver) have evaluated these protocols on the basis of energy efficiency, robustness, adaptability, memory efficiency, and scalability, features that are required by a management protocol and a system for proper working of wireless sensor network. Several factors have been discussed for communication architecture of wireless sensor networks like, reliability, density and scalability, network topology, energy consumption, hardware constraints, data aggregation, transmission media, security, self configuration, network dynamics, quality of service, coverage, and connectivity (Zhang, & Li, 2009; Yu, Mokhtar, & Merabti, 2006; Obaisat, & Braun, 2006).

ISSUES, CONTROVERSIES, PROBLEMS

There are several issues of wireless sensor networks that have been discussed in literature (Sorniotti, et al. 2007). The primary issue is energy conservation and secondarily the bandwidth. Network management is a crucial function for the maintenance and monitoring of all type of networks. Most important issue to be considered while designing a network management solution is energy conservation. It is one of the main obstacles for the successful exploitation of this technology, when long network lifetime and a high quality are the main requirement for certain applications. Efficient energy management is thus a key requirement for a credible design of a wireless sensor network (Alippi, et al. 2009).

Energy is a critical resource of a wireless sensor network, and all operations to be performed must be energy efficient (Li, et al. 2005).While covering the vast area, with the short lifetime of a battery operated sensors, and also having the possibility for damaged nodes during deployment, there is expected a large population of sensors in a particular WSNs application. So, scalable and flexible architectures and management schemes are indispensable for the designing and operation of such large scale networks having sensors with energy constraints and non rechargeable batteries. There is a need for designing energy efficient algorithms for extending sensor's lifetime (Abbasi, & Younis, 2007). Energy and bandwidth could be conserved by limiting the number of nodes management traffic has to travel. This can be done to retrieve network management information from one node instead of redundant sensor nodes in an area. Another method is to process the data and or perform data aggregation or compression before sending it to the management station. Nodes can collect data from neighboring nodes and convert the data into a single packet to be sent to the base station by performing data aggregation. By maintaining node process management data, nodes can determine the critical data that should be forwarded instead of a lot of irrelevant data. Data can also be compressed before transmission, thus transmission of less management data also conserve energy and bandwidth (Frye, & Cheng, 2007). In traditional wired networks the primary goal is to minimize response time along with providing comprehensive information but in sensor networks the primary goal is to minimize use of energy, primarily by reducing the communication between nodes (Lee, Datta, & Oliver). Several techniques have been proposed for energy conservation and bandwidth to increase the network lifetime for considerable period. Main focus of this chapter is to discuss the need for the efficient management architecture for wireless sensor networks in the light of issues and challenges currently addressed by the research community. According to studied literature (Alam, Rashid, & Hong, 2008) management protocol architecture should consider the, following design criteria to be scalable with the characteristics of sensor network: Low computation and message overhead, Minimal memory operations, Scalability, Application independence, Robustness and fault tolerance, and energy efficiency. According to (Obaisat, & Braun, 2006), energy depletion, robustness to dynamic environment and scalability to large networks have been mentioned as crucial design challenges for wireless sensor networks' protocols. Energy depletion challenge could be handled by reduction in active duty cycle for each sensor node, minimization of data transmission rate, and maximization of network life time. Clustering or organization of nodes in hierarchical way may enhance the scalability and robustness could be achieved by self organization, self configuration and self healing of sensor networks.

Sensor network applications are diverse in their requirements as discussed in the previous section, so following are the challenges, a network management architecture have to cope with, for efficient working of wireless sensor network (Yu, Mokhtar& Merabti; Zhang & Li, 2008; Zhang & Li, 2009; Lee, Datta & Oliver)

Lightweight management: Sensor Network management architecture should be lightweight in order to match the computation and communication requirements of sensor nodes. Light weight management architecture based on asynchronous model best suites the event driven asynchronous communication and "prone to failure" nature of wireless sensor networks.

Localized management and coordination: Well solution to save sensor nodes energy and reduce redundancy, because localized approach intelligently select and group sensor nodes subset and allow the coordination and communication between the neighboring nodes thus reducing the messages overhead.

Generic management functions: The design of traditional management architecture for WSN suffers from the problems of data routing, resource utilization and communication patterns. Each application has its own requirements for WSN management architecture. So it is not feasible to carry out existing management architecture from one application to another application, because of requirements diversity. Research (Yu, Mokhtar, Merabti; Romer et al. 2004; Vieira, 2005; Steffan, Cilia, & Buchmann, 2004) has got its focus toward the development of application independent WSN management architecture. There is a need to separate application semantics from basic hardware, operating system and network infrastructure.

Adaptive reconfiguration: Management architecture of WSN is required to reconfigure its operations and functions to comply with the changing environment and circumstances to help the sensor nodes to continuously work for long time to gather information for various applications. Management system should be able to response effectively to changing network topology, node energy level, as well as the convergence and exposure bounds of WSNs.

High level interfaces: Network management system should provide a high level interface for the configuration of sensor network behavior. Semantically rich querying algorithms are required to specify the quality of required data, its presentation before the user and application of in network data processing on the collected data by allowing the user to specify application specific algorithms.

Integration with IP architecture: For integration between WSNs and internet use of application level gateways and overlay IP networks is strongly recommended, as features of WSNs rule out the possibility of all–IP sensor networks.

Robustness and fault tolerance: Network management system and protocol should be compatible to changing network conditions like dropped packets, node dying, node mobility etc, to allow reconfiguration of the network as required, to smoothly supporting further developments of basic hardware and software.

Scalability and Heterogeneity: Network size should not affect the management system efficiency.

SOLUTIONS AND RECOMMENDATIONS

The energy cost of transmitting a single bit of information is approximately the same as required for processing a thousand operations in a typical sensor node (Anastasi et al. 2009). Regarding this, there are two main schemes to energy conservation: *in-network processing* and power saving through *duty cycling*. In-network processing involves reducing the number of information to be transmitted by means of compression or aggregation techniques. It typically makes use of the temporal or spatial correlation among data acquired by sensor nodes (Anastasi, et al. *Chapter 6 in Mobile Ad Hoc and Pervasive Communications, American Scientific Publishers*). Performing data operations inside the network, such as eliminating irrelevant records and aggregating raw data, can reduce energy consumption and improve sensor network lifetime significantly. In-network data processing is a useful technique to reduce the energy consumption significantly (Chen et al. 2006).

On the other hand, duty cycling schemes specify coordinated sleep/wakeup schedules among nodes in the network (Anastasi, et al *Chapter 6 in Mobile Ad Hoc and Pervasive Communications, American Scientific Publishers*). It involves waking up the sensorial system only for acquiring new sample set and immediately powered it off afterwards. This is an optimal energy management technique by considering the periodic sensing and assuming that process dynamics are stationary. It is an effective energy-conserving mechanism to put the radio transceiver in the sleep mode whenever communication is not required and should wake up as soon as a new data packet becomes ready. In this way nodes alternate between active and sleep periods depending on network activity. So duty cycle is defined as the fraction of time nodes are active during their lifetime. As sensor nodes perform cooperatively, they need to coordinate their sleep/wakeup times. A sleep/wakeup distributed scheduling algorithm thus accompanies any duty cycling scheme, allows sensor nodes to decide when to transition from active to sleep mode, and back. It allows neighboring nodes to be active at the same time, thus making packet exchange feasible even when nodes are mostly in sleep condition. Two different and complementary approaches are used to implement duty cycling with different granularity. To adaptively select only a minimum subset of nodes to remain active for maintaining connectivity and nodes that are not currently needed for ensuring connectivity can go to sleep mode thus saves energy referred to as topology control. the basic idea behind is to exploit the network redundancy to prolong the network lifetime by a factor of 2-3 with respect to a network with all nodes always on. On the other hand, active nodes selected by the topology control protocol, do not need to maintain their radio continuously on. They can switch off the radio when there is no network activity, thus alternating between sleep and wakeup periods referred as power management. (Anastasi, et al. 2009; Alippi, et al. 2009).

(Alippi et al. 2009), proposed energy management as an essential requirement for WSN credibility. Most of the proposed strategies targeted at minimizing radio activity are based on the assumption that data acquisition is less energy consuming process than data transmission, needs to be complemented with energy management techniques at sensor's level, as this assumption doesn't hold true for many practical application scenarios.

Adaptive sensing strategy dynamically adapts the sensor activity to the real dynamics of the process. It is being implemented by exploiting three different mechanisms. Hierarchical sensing, Model based active sampling, and Adaptive sampling.

Hierarchical sensing trades off accuracy for energy conservation by dynamically selecting, which of the available sensors needs the activation. Basic assumption is, that multiple sensors are installed on the sensor nodes observe the same phenomenon with different resolution and energy consumption. Hierarchical sensing involves triggered sensing and multi-scale sensing. *Triggered sensing* involves the activation of more accurate and power consuming sensors, after simple sensors, which are energy efficient but provide very low resolution. *Multi scale sensing* involves the identification of areas which require more accurate observation within the monitoring field, by depending on the coarse-grained description of the field with sensors having low accuracy and activation of other sensors having high resolution sensors, only in those areas, having the requirement for accurate acquisition.

Model based active sampling involves building a model on the top of initial set of sample data to predict the next data saving the required energy consumption for data sensing. Model can be updated for achieving the required accuracy to adhere to the new dynamics of physical phenomenon under observation. But selection of an appropriate model is the key design issue in the design of a model based active sensing strategy and mostly solutions based on this approach are expensive in terms of computation and needs to be implemented in centralized fashion.

Adaptive sampling technique take the advantage of temporal and spatial correlations among the sensed data referred as *activity driven adaptive sampling* and/or information related to the available energy whenever sensor node is able to harvest energy from the environment referred to as *harvesting aware adaptive sampling*, thus aimed at dynamically adapting the sampling rate on the basis of available energy.

Suggestions for Designing Network Architectures

Yu, Mokhtar, & Merabti (2006) presented some visions and suggestions for the design of efficient management architecture for WSNS.

Layered System Structure

Common approach adopted by existing management architectures is layered structure.

Key features of layered system structure.

- Designing individual functional components for different sensor applications and management functions is more feasible than designing a system with static integrated functions.
- Performing software changes locally regarding management functions is more easy,
- Layered approach offers light weight design, efficiently support management architectures so sensor- nodes only had selected function layers into their management architectures according to their roles or energy levels.

(Alam, Rashid, & Hong, 2008) proposed a low overhead three layered architecture, which allows efficient periodic reconfiguration of the sensor network on the basis of topology generation mechanism. Architecture is consisted of a manager at sink node, sub manager at cluster heads and agents which are normal sensor nodes. Manager has the global knowledge of network states, knowledge of underlying layers and sub-managers as well. Sub-managers independently distribute management functions on the basis of local network states and collect and collaborate management data and report the data to central manager on the behalf of entire cluster allowing minimum overhead for efficient functioning.

Management Functions Distributions

The traditional centralized architectures only support specific applications, and not applicable for networks with large size mainly because of traffic concentration problems because all the management decision have to come from or lade to centralized base station. This approach consumes more energy and expensive then localized approach in WSNs. So this requires distribution of operations and decisions for resource constrained WSNs, i.e. decentralized management approach. MANNA is the best example of decentralized management approach in which management functionalities are distributed among manager and agents in WSN by assigning individual roles with different functionalities for network maintenance.

Policy based management provides its services in designing self-adaptive decentralized management support in WSNs. MANNA defines the policies for real time operations by describing a set of desired behaviors for management components. These policies allow manager and agent to interact cooperatively to achieve overall management goal. TinyCubus and integration of reconfigurable group management service with Mire is another example of policy based management approach. (Li, Li, & Xingshe, 2009) proposed an integrated Policy-based sensor network management architecture keeping in view the characteristics of sensor networks. The PFMA management framework divides sensor network management into three levels: the task Layer, network layer and the node layer, allows the management and control of single sensor at node layer, management of wireless ad hoc network itself, including the network configuration of operating parameters, maintenance of network connectivity, topology discovery and control, mobile node management function at network layer and the distribution and coordination of perception tasks, scheduling of tasks and setting up clues between tasks etc at task layer, with management functions broken down into four functional areas: configuration management, performance management, fault management and state management improved the overall performance of sensor networks.

Information Model

Policies need to identify certain conditions that must be satisfied before the execution of management operations for achieving the desired goals, on this basis, applications select some suitable policy to tackle with a request or any change in the network. These conditions for executing certain management functions mostly depend on the information obtained from real time network state. MANNA describes two kinds of management information.

- Static information
- Dynamic information

Static information describes,

- Management service configuration
- Network information
- Network element information

Dynamic information covers,

- Sensing coverage area map
- Communication coverage area map
- Network topology

These aspects are described by WSN models containing all the information regarding the changes in the network state. But retrieving that information is an energy consumption process in WSN. Therefore suitable service oriented architectures are becoming popular in distributed computing.

Service oriented architecture allows the decomposition of large complex application and middle ware architecture into reusable services, or function units, which can be flexibly combined in a loosely coupled manner at either design level or runtime. This allows the application, by hiding the service operational description, by hiding the services implementation complexity from end uses.

SoA can deal with various challenges like WSN heterogeneity, mobility and adaptation etc, but still significant research is required for its appropriate implementation in WSNs because of its "resource-constrained" nature because SOA implementation and design is largely dependent on web services like WSDL, OGSA so direct application of these complex technologies is not possible for sensor modes with resources limitations.

Several network management protocols (Zhang, & Li, 2009; Zhang, & Li, 2008; Lee, et al. 2008; Ma, Chen, Huang, & Lee, 2010) have been proposed to meet the design challenges discussed in the previous section.

MANNA: It is a policy based management system. Policy based management has supported to design self adaptive decentralized management service in WSNs. MANNA is the best example of policy based management in which management functions like fault, security, performance and accounting are dependent upon configuration management. In MANNA, policies are defined for managers and agents for cooperative interaction, to achieve the desired goal of overall network management. It works by mapping dynamic management information into WSN models and based on these models management functions and services are executed. Management service consists of one or more management functions like coverage area supervision, topology map discovery, aggregation discovery etc. Network state information is defined in WSN models. MIB defines the relationship among different WSN models. MANNA handles changing WSN behaviors by analyzing and updating MIB, which is a centralized operation and not energy efficient. There is also a chance of inaccuracy regarding collected management information because of uncertainty and delay of WSNs. MIB update is not an easy process in terms of determining the right time for management information querying and right frequency to obtain the management information. MANNA network management protocol (MNMP) is a light weight protocol for the management of information exchange among management entities. It organizes the nodes into clusters and sends their states to the agent located in the cluster head. Cluster heads are responsible for the execution of local management functions; perform aggregation on received data from sensor nodes, forward management data directly to the base station and work cooperatively with each other to achieve the overall management goal. Manager as powerful entity, located outside the WSN is responsible for performing the complex management tasks on the basis of network global knowledge. This approach offers energy efficiency and management decisions accuracy. Automatic management services of MANNA could help in the fault management of event driven applications by providing coverage area maintenance service and failure detection service. The central manager is responsible to build a coverage area model for monitoring areas of sensing and communication coverage by using topology map model and energy model. Moreover central manager can commend the agent for executing failure detection management service.

RRP: It is a hybrid data dissemination framework, based on supply chain management. Its function is to manage data gathering applications common examples are habitat monitoring and battle field surveillance. Supply chain management in business terms can be defined as a coordinated system of entities and activities to deliver a product or service from supplier to customers, to satisfy the customer's needs through the efficient use of resources. Concept of supply chain is applied in RRP, by dividing a sensor network in to numerous functional regions, with different routing mechanisms to different regions, and provides cooperation among these regions to achieve the goal of better networks performance in terms of reliability and energy usage.

Advantages

Another concept in RRP is a zone flooding scheme to reduce the cost of topology maintenance and route discovery. It is a combination of flooding and geometric routing techniques. Zone flooding allows low messages overhead and size adjustment of flooding zone allows higher reliability.

Goal of desired energy consumption, end-to-end delay and routing overhead is achieved by allowing the end user to pre define the size of warehouse and flooding zones while maintaining reliability.

Limitations

RRP requires GPS-attached nodes for implementing zone flooding protocol. A human manager is required to support hierarchical network management.

SNMS is an interactive sensor networks management system to monitor sensor networks health. Its supports two primary management functions. These are query based network health data collection and event logging. Query system is used to collect and monitor physical parameters of the node environment for the user. Through event logging system, a user sends event parameters. It allows network nodes to report their data on reaching a particular event's thresholds.

Traffic patterns supported by SNMS are collection and Dissemination. Health data is acquired through collection pattern, and management messages, command, and queries are distributed through dissemination pattern. SNMS allows memory efficiency, because every node in the networks only maintains a neighborhood table, by keeping the single best parent based on the strongest received signal strength. Traffic is also minimized, because it allows tree construction only to respond the message sent from the base station. Tree construction protocol is also adaptive to dynamic network conditions, by allowing the node to only select a new parent, if the existing one dies.

Drip protocol is used in SNMS for the reliable dissemination of messages, commands and queries. When any user or a sensor node has to make a query, it selects a specific identifier to represent a reliable delivery channel. Then message or received replies are transported on that reliable channel to the component. Every sensor node is required to regularly check the channel, it subscribes to, cache and perform data extraction on the latest messages received on that channel and send a reply.

Advantages

- This protocol provides management functions, even application fails, because it is application independent.
- SNMS allows overhead only for human queries, with minimal impact on memory and network traffic.
- It is also energy efficient, because it makes the bundles of multiple queries results into a single message. There is no need to return each result individually.

Limitations

It is a centralized approach, so it requires managed nodes to pull network health data continuously to bas station, and limited to passive monitoring, because human intervention is required to submit queries and perform post mortem analysis of management data.

But sideways this approach requires a large key space and memory also, because a trade off is required between channel wage and node caching for a set of independent variables.

sNMP sensor network management protocol performs two functions.

- To define sensor models for the representation of current network state, as well as provides the definition of various network management functions.
- Allows network state retrieval through the execution of network management functions by providing algorithms and tools.

It has been suggested, to make use of sensor models for different management functions, to achieve the following benefits.

- Current state of network topology could be used for future node deployment by human manager.
- Network maintenance and monitoring could be performed periodically by the identification of low performance network parts accompanied with necessary corrective actions.
- Periodic network monitoring also allows human manager to analyze changing network conditions, for the prediction of network failures to take in time preventive actions.
- Network topology determines network state, which in turn useful in identifying the number of active nodes and node connectivity. Top disc and stream algorithm has been proposed for network topology retrieval.

The TopDisc algorithm of sNMP maintains networks connectivity by allowing minimum numbers of nodes to be active using clustering mechanism. Network topology is generated on the basis of nodes neighborhood information by selecting a set of distinguished nodes.

Only local neighborhood information is exchanged between adjacent clusters. Local neighborhood information is collected when a node listen to other nodes of their communication range. This algorithm provides three management functions.

- Network state retrieval
- Data dissemination and aggregation
- Duty cycle assignment

Cluster heads retrieves network states by determining network topology map, energy map, and usage pattern. Tree cluster is maintained to determine the number of sensor nodes, optimal in the number of hops from cluster head monitoring node. This approach offers several advantages.

- Allows deficiency in terms of data dissemination and data aggregation
- Fair duty cycling is performed by allowing a human manager to specify the rules for forwarding loads in distributed manner among nodes highly scalable, because network management functions are performed on the basis of local information.
- Highly scalable because network management functions are performed on the basis of local information

But also has a draw back in terms of latency and energy because of clustering overhead, due to cluster heads election and maintenance.

Another topology discovery algorithm is STREAM, stands for sensor topology retrieval at multiple resolutions.

Different application may have different topology resolution requirement so STREAM returns networks topology at different degrees. Stream creates an approximate topology, from nodes subsets by determining the nodes neighborhood lists. It returns the required network topology resolution at proportional cost.

WinMS stands for wireless sensor network, management system, is an adaptive policy based management system. Management parameter thresholds are predefined by an end user or sensor nodes to be used as event triggers, and management tasks are also specified, which are executed when an event occurs.

WinMs allows maintaining the network performance and effective networked node operations, by allowing the network to periodically configure its self on the basis of current events and prediction of future events as well.

WinMs provides two network management schemes

- Local
- Central

According to local scheme, individual sensor nodes can perform management functions self independently based on their neighbor hood's network state.

In a central network management scheme, central manager performs corrective and preventive management maintenance on the basis of global network knowledge.

An MIB is maintained by central manager WSN models are stored in it, to allow central manager to detect areas of weak network health, noisy areas, areas of where data changes occur rapidly etc.

Advantages

- Energy efficient management, data transport and local repair, because of light weight TDMA protocol
- Allow automatic self configuration and self stabilization locally and globally, by allowing the network to adapt to dynamic conditions without human involvement.

Limitations

- Initial setup cost is proportional to network density for building data gathering tree and node schedule.

WSNMP: A Network Management Protocol for Wireless Sensor Networks (Alam, Rashid, & Hong, 2008) is based on hierarchical network management scheme, uses three tier architecture consisted of a central manager, at the highest level and implemented at the sink node, intermediate managers at cluster heads and agents are placed at each sensor nodes. Intermediate managers collect and collaborates management information between central manager and agents and work independently, by executing the management functions on the basis of their local network states. Central manager has to keep the global check over the entire network states. Proposed management architecture performs the management functions with low overhead for WSNs and manages the network efficiently with periodic reconfiguration.

CONFIGURATION MANAGEMENT

Most of the existing techniques use flooding to create the clusters a major issue in hierarchical sensor networks, but flooding cause too much energy consumption for cluster reconfiguration and also causes data transfer to be interrupted. A re-clustering method is required to keep the energy usage of sensors in balanced state, so the author proposed configuration algorithm instead of using flooding, to model the network topology, and performs network reconfiguration with minimum energy usage after the topology generation. TopDisc algorithm is used for initial cluster formation that maintains the network connectivity by allowing minimum number of nodes to be active at a time. By selecting distinguished nodes, network topology is created, on the basis of nodes neighborhood information. In the proposed protocol, the topology generation algorithm works in two phases. In the first phase, a node first selects its cluster head, and sends the joining message along with its neighbors list. If a node hears transmission from any new node it is also included in that node neighbors list and send its information in the next packet. After a cluster head has received a joining message from the entire neighbors, it builds the topology. In the second phase, cluster manager receives the data from all intermediate managers (periodic update of topology) and generate the topology. Cluster reformation is required for balance energy consumption by changing the cluster heads periodically. For cluster reformation, CM is allowed to find the minimum number of cluster heads and gateway nodes on the basis of complete knowledge regarding network topology and energy level of all sensor nodes and at the same time if required, create sleep/awake period of the sensor nodes. Reconfigured topology information can be sent to existing cluster heads, and each cluster head make the announcement about new cluster head and its members and gateway nodes.

FAULT MANAGEMENT

Architecture also incorporates fault management mechanism, by fault detection and fault recovery. Fault detection involves identifying non response or near to be dead node and reconfiguration of routing path if required. Fault of the normal sensor nodes is detected by the intermediate managers. Proposed mechanism emphasize on the importance of cluster head fault detection. As the central manager is aware of the network topology, it maintains the timers T1 and T2 for cluster heads and gateway nodes. When a sink is receiving packet from a node timer restarts. After the time expires the node is considered as dead. A query packet is sent by cluster head to that node and waits for time T2; and node is decided as dead if it doesn't response. After the fault detection fault recovery for the cluster head and gateway node is very crucial for the efficient functionality and reliability of the sensor network. A new cluster head is to be selected and new path has to be created for the new cluster head, if a cluster head is affected by the fault and central manager has to select the new cluster head without disturbing the existing infrastructure so authors also provided the mechanism for fault recovery. In case of gateway node fault recovery, central manager has to perform the reconstruction of all upstream nodes.

Performance management of the proposed protocol performs network monitoring by minimum resource consumption for best service by keeping some nodes in the dead and others in the live state.

Advantages

- Low message overhead and energy consumption, Because of topology generation and cluster reformation
- Light weight Fault detection is performed both locally and centrally by intermediate manager and central manager respectively.
- Proposed event reliability mechanism periodically measures the event reliability and maintains the data generation rate.
- Alternative scheduling of cluster heads keeps the energy consumption at low level.

Limitations

- Energy dissipation increases with the increased number of resources, if the protocol is deployed for large networks
- All nodes have been assumed with same processing power, sensing, transmission range and data generation rate.

WSNManagement system: *WSNManagement system* (Ma, et al. 2010) is based on easy operation environment, to allow network manager to monitor and analyze network system. An event bus provides the commands and events, on the basis of which, network administrator can utilize the configuration, performance and fault function of the WSNManagement system user interface through communication technology TOSSIM or mote.

- **Configuration management** control and monitor sensor nodes status. Sink node has a link to each sensor node; determine the network status via communication function. Configuration management of the proposed system allows sensor nodes distribution uniformity by allowing network manager to deploy sensor nodes to areas that need additional sensors to meet system requirements.
- **Performance management** allows the administrator to analyze and predict the entire network development, to solve the problems and upgrade the system if it declines, on the basis of performance data, detected values, and recorded power consumption for secure data transmission. Two performance metrics are taken into consideration.
- Sensing coverage
- Prolonged network life time

For performance management the performance tab displays sensing coverage of the current designated sensor in the sensor node in the *WSNManagement* system. Analysis of sensing coverage and identification of areas in which sensor node coverage overlaps; system allows network managers to deactivate redundant sensor nodes temporarily until other nodes run out of power for limited power consumption.

Fault management enhances network reliability by monitoring network status. Fault administrator control process can be used to identify errors associated with sensor hardware faults, power shortages, lost connections, and environment dynamics. In the *WSNManagement* system, for fault management, a network manager can select the fault tab, to be aware of dead sensor nodes in grey, assess the effect of these faulty nodes on the entire system and make decision for the deployment of new nodes for system proper functionality.

OVERALL PERFORMANCE ANALYSIS

- Packet loss rate for data transmission is about 5% less with *WSNManagement* mechanism.
- Considerable enhancement in network data transmission services, and traffic stability

According to (Frye, & Cheng, 2007), not only the energy among the nodes should be minimized but also the energy required for running. This will definitely increase the network life time. Moreover a distributed solution is more beneficial as compared to centralized one, because in centralized solution a node has to communicate long distances because of multiple sinks or base stations and this communication head could be lesson by designing a distributed algorithm or protocol. Maintenance overhead of an algorithm is other important issue. It should be kept minimum for less energy consumption and processing power, which is also a limitation of sensor nodes. Designed algorithm should not require location information as well as time synchronization, to reduce communication overhead. Robustness and scalability to node mobility and node failure is another requirement because many sensor network applications suffer from node mobility, and wireless sensor networks are to prone to failure because of energy limitations, hardware failure or other environmental conditions.

Clustering (Frye, & Cheng, 2007; Abbasi, & Younis, 2007; Ganesan, Cerpa, Ye, Yu, Zhao, & Estrin, 2003) has been widely pursued by the researchers to achieve the scalability objective for wireless sensor networks. Clustering involves nodes arrangement into groups called clusters. Each cluster could have a leader called cluster head (CH). Cluster Head may be chosen by a sensor in a cluster or it may be pre assigned by a network designer. It is usually a node rich in resources. Cluster has numerous advantages besides network scalability.

- Clustering allows a subset of nodes to communicate with the sink nodes, allow energy conservation of those nodes that must no longer send data to the sinks. Thus abandon long paths required for communication for longer hops as nodes will communicate only with the neighbor nodes except for cluster head. This definitely helps in energy and bandwidth conservation
- It can reduce the size of routing table at each node by localization of route setup within the cluster.
- By avoiding redundant messages exchange among sensor nodes it helps in bandwidth conservation also by reducing the scope of inter cluster interactions to cluster heads
- Reduce topology maintenance overhead by stabilizing the network topology at the sensor's level
- Sensors would only have concern for their connection with CHs. Change at inter-CH tier would not affect them
- Clustering helps in enhancing the network operation and prolong battery lifetime of individual sensors as well as the whole network by allowing CHs to implement optimized management strategies.
- Helps in reducing energy consumption by activities scheduling within the cluster because some nodes can be switched to the low power sleep mode most of the time.
- Time for sensor transmission and reception can be determined to avoid sensor reties. This helps in limiting coverage redundancy and also prevents medium access collision.
- CH can perform data aggregation and thus reduce the no of relayed packets.

Clustering Algorithms (Frye, & Cheng, 2007; Abbasi, & Younis, 2007; Obaisat, & Braun, 2006). Low energy adaptive clustering hierarchy i.e. **LEACH**, is one of the most popular and studied clustering algorithm.

How LEACH Works

LEACH guarantee a specific network life time by minimizing energy consumption. First it forms cluster on the basis of received signal strength and cluster heads work as router to base station. Cluster performs data fusion and aggregation locally to reduce global communication, is an important feature of LEACH. Distributed algorithm is used to make clusters and nodes act automatically without any central control. LEACH minimize energy consumption by ensuring that all nodes run out of energy at about the same time for extending network life time, and very little energy left nodes when the network dies. LEACH rotate the role of cluster head and perform cluster head selection partly on the basis of remaining energy, to make dies the nodes at about the same time. LEACH performs the successful distribution of energy usage among the nodes, so that node dies randomly and at the same rate.

LEACH Operation

Operation of LEACH is divided into rounds. Each round consists to two phase.

- The setup phase
- The steady phase

In the setup phase, nodes are organized into clusters. A node decides to be as a cluster head for that round, independently of all other nodes. This is done, when a node selects a random number, and becomes a cluster head, if the value of that number is the less than the value of threshold. Threshold values is decided on the basis of

- Cluster heads percentage for that round
- Number of times, node has already been a cluster head
- Amount of residual energy in the node

Cluster head broadcast a message to be a cluster head. A non cluster head node will join the cluster, after sending join request to cluster head, from which it received the strongest advertised signal.

Normal data collection and routing phase is steady phase. Cluster nodes send data to cluster head. Cluster head will transmit to receive data of its cluster in one message to bas station.

Advantages

- ◦ Flexible and self adaptive
- ◦ Makes efficient transmission, uses TDMA at cluster head level
- ◦ Prevent interference between clusters by using CDMA for cluster communication
- ◦ Provide robustness failure, to maintain over all network operation.

Limitations

- ◦ Advertisement message is to be sent by all cluster heads to all nodes with in the communication range, makes LEACH inapplicable for time constrained applications
- ◦ Data transmission from all cluster heads to the base station, may be a long distance
- ◦ Large overhead necessary for cluster formation

HEED

In Hybrid Energy-Efficient Distributed clustering, CH nodes are selected from deployed sensors with taking both energy and communication cost into consideration. Sensor nodes with high residual energy are cluster head nodes and are well distributed in the network, unlike LEACH in which two nodes within each other's transmission range become CHs, is most probable. Energy consumption is also not assumed to be same for all nodes. Each node is mapped to exactly on cluster head for direct communication. HEED also ensure inter-CH connectivity, by adjusting the probability of CH selection with in a particular sensor's transmission range.

In **extended HEED** orphaned nodes (which don't hear from any CH) are allowed to be cluster heads per the finalization phase (CH selection). This modification reduces their data latency by significantly decreasing cluster head count, which in turn reduce routing tree size required during inter-CH communication.

(DWEHC) Distributed weight based energy efficient hierarchical clustering

- Allows generation of balanced cluster sizes
- Lower energy consumption in intra-cluster and intra-cluster communication

Each sensor node has to locate its neighbors, and calculate its weight, which is a function of sensor's energy reserve and proximity to the neighbors. Nodes with largest weight of neighborhood will be selected as cluster head, and other nodes are its members. These are first level members because of direct link with CH. This membership is successfully maintained by a node to reach its CH, by suing least amount of energy. To do this node uses the knowledge regarding the minimal cost required by its non CH neighbors to reach CH, knowing the distance to its neighbors, it can decide to keep itself as the first level member or second one. This process continues until a node settles itself on the most energy efficient intra cluster topology.

Similarities to HEED

- Every node participate in the clustering process
- No assumption is made regarding network site
- Energy conservation is considered for CH selection

EEHC Energy Efficient Hierarchical clustering is a distributed randomized clustering algorithm for maximizing WSNs life time.

This clustering algorithm involves two stages. In initial stage or single level clustering, first each announces itself as a CH with the probability *p* to their neighbor nodes of its communication range. These CHs are called volunteer CHs. All nodes within *k* hops range of a CH receive this message either by directly or by forwarding. Node that receives this announcement and not a CH becomes the member of a cluster, closest to that node. There are also forced CHs that are neither CHs and nor a member of any cluster. Node that doesn't receive announcement within a preset time interval t, the node will become a forced CH, with an assumption of not being within k hops of all volunteer CHs. t is calculated on the basis of duration for a packet to reach a node, k hops away. In multilevel clustering, clustering process is recursively repeated for h levels to ensure h-hop connectivity between CHs and base station. Each CH collects the data of its sensor nodes at its level and sends an aggregated report to the base station, like CHs at level-1 will send aggregated data to CHs at level-2 and so on. By using optimal parameter values, algorithm shows significant reduction in energy consumption.

PEGASIS: POWER EFFICIENT GATHERING IN SENSOR INFORMATION SYSTEM

PEGASIS has two basic ideas.

Chasing and Data Fusion

First a chain is formed among all the nodes in the network and each node can take turn to be a leader of the chain, transmitting data to the base station, allows less energy consumption as each node has to communicate only with close neighbor. Chain can be formed by using a greedy algorithm deployed by the sensor nodes.

PEGASIS assumptions

- Sensor nodes have global network knowledge
- Nodes are not moving nodes
- Nodes have location information of all other nodes
- Radio channel is symmetric

Each node will fuse its data with the data received, and forward it to the next one. Leader of that round will forward all the fused data to the base station. There is even distribution of energy load among all nodes, because nodes will take turn to be a leader.

Advantages

- Chaining mechanism will allow a node to receive and transmit one packet of data in each round and only one node transmit to base station in a particular round.
- Robust to node failure, when a node fails, chain could be reformed with remaining nodes
- Less overhead is required to form chain as compared to cluster in LEACH

Limitations

- Not applicable to large networks
- Do not consider residual energy while selecting a leader node.

FUTURE RESEARCH DIRECTIONS

Future work involves the designing of a total solution for the management of wireless sensor networks in terms of scalability, robustness, energy efficiency, resource limitation and other performance parameters. Development of an efficient generic management protocol is another important area requires significant research contribution. Areas of topology management also require development optimal techniques by eliminating the assumptions and advantages and combine the strengths, evidenced by the literature review.

CONCLUSION

In this chapter we have discussed about network management of wireless sensor networks in terms of some background information including wireless sensor network features, functional management areas, types of network architectures, as well as network management challenges, protocols and mechanisms proposed by research community for efficient management of these networks.

REFERENCES

Abbasi, A. A., & Younis, M. (2007). A survey on clustering algorithms for wireless sensor networks. *Computer Communications*, *30*, 2826–2841. doi:10.1016/j.comcom.2007.05.024

Alam, M. M., Rashid, M. O., & Hong, C. S. (2008). *WSNMP: A network management protocol for wireless sensor networks,* (pp. 17-20).

Alippi, C., Anastasi, G., Di Francesco, M., & Roveri, M. (2009). *Energy management in wireless sensor networks with energy-hungry sensors* (pp. 16–23). Instrumentation & Measurement Magazine.

Anastasi, G., Conti, M., & Francesco, M. D., & Passarella, A. (2006). How to prolong the lifetime of wireless sensor networks. In L. T. Deng & M. K. Denko (Eds.), *Mobile ad hoc and pervasive communications.* American Scientific Publishers.

Anastasi, G., Conti, M., Francesco, M. D., & Passarella, A. (2009). Energy conservation in wireless sensor networks: A survey. *Ad Hoc Networks*, 7(3), 537–568. doi:10.1016/j.adhoc.2008.06.003

Berman, P., Calinescu, G., Shah, C., & Zelikovsly, A. (2005). Efficient energy management in sensor networks. In Xiao, Y., & Pan, Y. (Eds.), *Ad hoc and sensor networks*. Nova Science Publishers.

Boulis, A., & Jha, S. (2005). Network management in new realms: wireless sensor networks. *International Journal of Network Management*, *15*(4). doi:10.1002/nem.569

Buratti, C., Conti, A., Dardari, D., & Verdone, R. (2009). *An overview on wireless sensor networks technology and evolution* (pp. 6869–6896). Open Access Sensors.

Cerpa, D. G. A., Ye, W., Yu, Y., Zhao, J., & Estrin, M. D. (2004). Networking issues in wireless sensor networks. *Journal of Parallel and Distributed Computing*, *64*(7), 799–814. doi:10.1016/j.jpdc.2004.03.016

Chen, Y., Leong, H. V., Xu, M., Cao, J., Chan, K. C., & Chan, A. T. S. (2006). *In-network data processing for wireless sensor networks*. 7th International Conference on Mobile Data Management (MDM'06), IEEE Computer Society.

Deb, B., Bhatnagar, S., & Nath, B. (2002). *A topology discovery algorithm for sensor networks with applications to network management*. DCS Technical Report DCS-TR-441, Rutgers University to appear in IEEE CAS Workshop.

Deb, B., Bhatnagar, S., & Nath, B. (2003). Multiresolution state retrieval in sensor networks. *IEEE International Workshop on Sensor Network Protocols and Applications* (pp. 19-29).

Frye, L., & Cheng, L. (2007). *Network management of a wireless sensor network*. Technical Report LU-CSE-07-003. Lehigh University.

Ganesan, D., Cerpa, A., Ye, W., Yu, Y., Zhao, J., & Estrin, D. (2003). Networking issues in wireless sensor networks. *Journal of Parallel and Distributed Computing*, *64*(7).

Jiang, X., Taneja, J., Ortiz, J., Tavakoli, A., Dutta, P., & Jeong, J. … Shenker, S. (2007). *An architecture for energy management in wireless sensor networks*. International Workshop on Wireless Sensor Network Architecture.

Lee, W. L., Datta, A., & Oliver, R. C. (2006). Network management in wireless sensor networks. In Deng, L. T., & Denko, M. K. (Eds.), *Mobile ad hoc and pervasive communications*.

Lewis, F. L., & Donnell, M.-O. (2006). *Wireless sensor networks: Issues, advances, and tools*. Automation & Robotics Research Institute (ARRI). The University of Texas at Arlington, Sponsored by IEEE Singapore Control Chapter.

Li, Z., Li, S., & Xingshe, Z. (2009). PFMA: Policy-based feedback management architecture for wireless sensor Networks. *5th International Conference on Wireless Communications, Networking and Mobile Computing,* (pp. 1–4).

Li, Z., Zhou, X., Li, S., Liu, G., & Du, K. (2005). Issues of wireless sensor network management. *ICESS, LNCS, 3605*, 355–361.

Ma, Y. W., Chen, J. L., Huang, Y. M., & Lee, M, Y. (2010). An efficient management system for wireless sensor networks. *Sensors (Basel, Switzerland)*, *10*, 11400–11413. doi:10.3390/s101211400

Obaisat, Y. A., & Braun, R. (2006). *On wireless sensor networks: Architectures, Protocols, applications, and management*. AusWireless Conference.

Romer, K., Frank, C., Marron, P. J., & Becker, J. (2004). Generic role assignment for wireless sensor networks. *Proceedings of the 11th Workshop on ACM SIGOPS.*

Schurgers, C., Tsiatsis, V., & Srivastava, M. B. (2002). STEM: Topology management for energy efficient sensor networks. *Aerospace Conference Proceedings,* vol. 3, (pp. 1099-1108).

Sorniotti, A., Gomez, L., Wrona, K., & Odorico, L. (2007). Secure and trusted in-network data processing in wireless sensor networks: A survey. *Journal of Information Assurance and Security*, *2*(3), 189–199.

Steffan, J., Cilia, M., & Buchmann, A. (2004). *Scoping in wireless sensor networks*. Workshop on Middleware for Pervasive and Ad-Hoc Computing. Toronto, Canada: ACM Press.

Townsend, C., & Arms, S. (2004). Wireless sensor networks: Principles and applications. In Wilson, J. S. (Ed.), *Sensor technology handbook*.

Vieira, M. S. (2005). *A reconfigurable group management middleware service for wireless sensor networks*. 3rd International Workshop on Middleware for Pervasive and Ad- Hoc Computing, Grenoble, France.

Wang, L., & Xiao, Y. (2005). Energy saving mechanisms in sensor networks. *2nd International Conference on Broadband Networks* (pp. 724–732).

Woungang, I., Subhas, C., & Cheng, L. (2009). Topology management for wireless sensor networks. In Misra, S., Woungang, I., & Misra, S. C. (Eds.), *Guide to wireless sensor networks*. London, UK: Springer.

Yu, M., Mokhtar, H., & Merabti, M. (2006). A survey of network management architecture in wireless sensor network. *Annual Postgraduate Symposium on the Convergence of Telecommunications, Networking & Broadcasting* (pp. 221-225).

Zhang, B., & Li, G. (2008). Analysis of network management protocols in wireless sensor network. *International Conference on Multimedia and Information Technology* (pp. 546–549).

Zhang, B., & Li, G. (2009). Survey of network management protocols in wireless sensor network. *International Conference on E-Business and Information System Security*, (pp. 1-5).

Zhang, W.-B., Xu, H.-F., & Sun, P. G. (2010). *A network management architecture in wireless sensor network. Communications and Mobile Computing* (pp. 401–404). CMC.

KEY TERMS AND DEFINITIONS

Cluster Head: Leader of a cluster, elected by sensor nodes or a network designer is called cluster head.

Clustering: Process of arranging the nodes into clusters.

Gateway Node: Node that connects sink node with external entities like sensor applications or traditional enterprise applications.

Network Topology: Arrangement of sensor nodes in the network

Sensor Network Management: Monitoring and maintenance of a sensor network and all associated components.

Sensor Nodes: Devices having sensing capabilities and deliver the information collected from the monitored field with the help of wireless links.

Sink Node: Node that is responsible for receiving storing and processing data from common nodes in the sensor network.

Topology Management: Mechanism to conserve energy while maintain network connectivity.

Wireless Sensor Networks: Networks that consists of many small sensors, limited in processing power and battery life.

Chapter 4
Interoperability in Wireless Sensor Networks Based on IEEE 1451 Standard

Jorge Higuera
Universitat Politècnica de Catalunya (UPC), Spain

Jose Polo
Universitat Politècnica de Catalunya (UPC), Spain

ABSTRACT

The syntactic and semantic interoperability is a challenge of the Wireless Sensor Networks (WSN) with smart sensors in pervasive computing environments to increase their harmonization in a wide variety of applications. This chapter contains a detailed description of interoperability in heterogeneous WSN using the IEEE 1451 standard. This work focuses on personal area networks (PAN) with smart sensors and actuators. Also, technical, syntactic, and semantic levels of interoperability based on IEEE 1451 standardization are established with common control commands. In this architecture, each node includes a Transducer Electronic Datasheets (TEDS) and intelligent functions. The authors explore different options to apply the IEEE 1451 standard using SOAP or REST Web service style in order to test a common syntactical interoperability that could be predominant in future WSNs.

INTRODUCTION

In our days, the interoperability of Wireless Sensor Networks (WSN) is an attractive goal to share metadata information across heterogeneous WSN deployments, using an effective common model to represent the information based on standardization rules. Nevertheless, in many cases different WSN based on smart sensors and actuators are inaccessible to extract the information in a common client application, because each network employs different data formats and their data cannot be accessed thought a standard wireless interface. The challenge for this situation is the inclusion of a

DOI: 10.4018/978-1-4666-0101-7.ch004

model based on levels of compliance with clear goals aimed to the harmonization of WSN, by introducing an open and global standard as IEEE 1451, which uses syntactic rules with metadata organized in a tuple structure. Also, it introduces different standard common commands to control efficiently each sensor node to share information with other heterogeneous devices. Motivated by this issue, our efforts are concentrated in standardized wireless physical transducer interfaces for smart sensors to encourage an efficient communication, allowing syntactic and semantic rules to exchange and to access metadata with accuracy and assurance.

The content of this chapter is organized into eight sections. In the related work section, a literature review on smart sensors and the IEEE 1451 standard is provided. Next, two sections provide an overview of smart sensor in WSN and the different interoperability levels, from technical level toward semantic level. Next, the family of IEEE 1451 standards in wired and wireless sensor networks is studied and a smart sensor model is defined based on the Network Capable Application Processor namely (NCAP) as coordinator node, and the sensor node as Wireless Transducer Interface Module (WTIM). Also, in this section is defined the Transducer Electronic Datasheet (TEDS) to store information related to each physical sensing channel, based on metadata information using a generalized tuple format. The subsequent section is dedicated to describe the main features of SOAP and REST Web Services for pervasive embedded devices that could be predominate in the future of WSNs. Finally, the chapter provides possible future research directions based on interoperable WSN.

RELATED WORK

The broad diversity of WSN hardware systems (Beutel, 2010) and the increase of different data formats raise the complexity in metadata representation for interoperable WSN. In effect, to share data information across heterogeneous network deployments requires an effective common model to represent the messages based on standardized rules. To address these problems, recent efforts are concentrated in standardized wireless physical interfaces (Gutierrez, Callaway & Barrett, 2003) to communicate and to process the information effectively allowing syntactic and semantic rules to exchange and to access metadata with accuracy and assurance.

Different initiatives toward interoperable sensor networks have been proposed in the past, for example, based on the Open Geospatial Consortium (OGC) standards to enable WSN for exchange and re-use the information in Service oriented architectures (SOA) as in (Klopfer, 2005), to provide an unified formal model and framework for sensor networks as in (Gracanin, Eltoweissy, Wadaa & DaSilva, 2005), or using metadata elements and context rules to improve the network interoperability (Ballari, Wachowicz & Callejo, 2009). These proposals take in account a standardized Physical (PHY) and Medium Access Layer (MAC) related with low rate Personal Area Networks (IEEE Std. 802.15.4, 2006). Likewise, in the higher layers, as in application layer, sensor network schemes focus on a conceptual structures to describe and exchange ordered data information, using reusable syntactic metadata by including standard languages and meta-languages as XML (Bray, Paoli, Sperberg-McQueen, Maler & Yergeau, 2008) by enabling the exchange of structured information between different platforms, allowing compatibility to share information in a safe, easy and reliable arrangement.

In this context, a WSN can increase the interoperability by introducing metadata structures and standard physical interfaces using a set of rules to encode the information that could be used over Service Oriented Architectures (SOA) (Sleman & Moeller, 2008), integrating heterogeneous services that execute discrete transaction without depending on the state of other processes or functions that resides on the Web or other network, as an intranet. SOA architecture is a way to share metadata with heterogeneous networks according to standards related to Web Services (Samaras, Gialelis & Hassapis, 2009) and facilitate the integration of different web technologies as SOAP or REST Web services.

The interoperability in WSN is a pragmatic assignment (Pokraev, Quartel, Steen & Reichert, 2006) and it requires appropriate treatment of the information, in a short time for an adequate response related with each request-response transaction using standardized rules to decode a message and also, standard web browsers for Web clients.

Historically, Standards that were developed for wireless communications have been used for WSN based on smart sensors as IEEE 802.11, IEEE 802.15.1 and IEEE 802.15.4 (Sliva, 2008). Others efforts in standardization for WSN have been produced by industrial forums and consortia, as ZigBee alliance, Wireless Hart or ISA 100 (Radmand, Talevski, Petersen & Carlsen, 2010), but in last years, with the development of standards for smart sensors in WSN, as the IEEE 1451 standard, a new generation of smart sensors was enabled to expand the use of technical interoperability over wired sensor networks using links point to point as in IEEE 1451.2, multipoint distributed network as in IEEE 1451.3, analog-mixed mode as in IEEE1451.4, or with wireless communication interfaces as in IEEE 1451.5 based on IEEE 802.15.4, Wi-Fi, Bluetooth, ZigBee, 6loWPAN and finally IEEE 1451.7 oriented for RFID tags (NIST, 2010). The IEEE1451 standard in a ZigBee and 6loWPAN sensor network using the RTOS TinyOS and Telosb platform has been successfully adapted to diverse scenarios, including environmental monitoring (Higuera & Polo, 2009) and (Higuera & Polo, 2010). In all cases, each device includes diverse Transducer Electronic Datasheets (TEDS) sub-modules that contain the node configuration. Also, the IEEE 1451 standard includes common commands to understand others devices that support a set of the same syntactical and semantic rules (IEEE Std. 1451.0, 2007).

SMART SENSORS AND INTELLIGENT SENSORS IN WSN

Technological advances in sensor manufacturing technologies and signal processing based on low-cost products open the door to design innovative smart sensors and intelligent sensors to apply in WSN deployments. In most of cases the term smart sensor is related with the technological construction aspects, whereas the intelligent sensor definition is focused on functional capabilities (Yurish, 2010). Figure 1 shows the difference between smart and intelligent sensor nodes. Smart sensors can be defined as a combination of primary sensor, analog front-end, signal conditioning, analog to digital conversion stage, digital processing, standardized digital communication interface, all in a unique single chip so called System on a Chip (SoC), (Frank, 2000). Otherwise, intelligent sensor is a sensor that includes some functional aspects, for example autonomic and cognitive capabilities in a high level.

In this case, the intelligent sensor can be a SoC or a discrete collection of parts in a complete system. The autonomic capabilities bring additional benefits such as self-configuration, self-management, self-healing, self-diagnosis, self-calibration (Ganek & Corbi, 2003) and others, for example, cognitive radio capabilities to detect the communications needs that depend on the context improvement, the channel allocation and the network radio resources. In some cases, smart sensors could not use intelligent functions, but in others, intelligent sensor cannot be a unique smart device or based in any sensors but not only smart sensors. These technological and functional aspects could be used for different scenarios in environmental WSN, but the innovations in this sense, also, requires interoperability aspects, for

Figure 1. Smart sensor vs an intelligent sensor

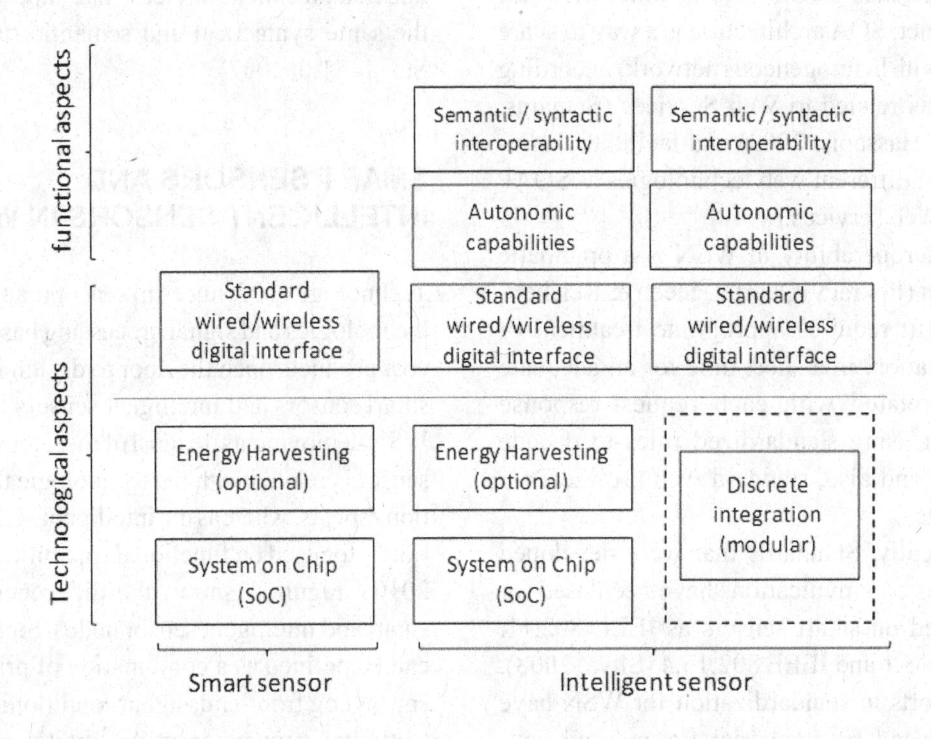

example, the inclusion of standards in a low and high level to connect different smart sensors and harmonize heterogeneous sensor networks using a same syntax and metadata to use a common semantic with open and global standards.

INTEROPERABILITY

One of the great challenges of wireless sensor networks in a high level is their interoperability. The term interoperability is defined as the ability of devices to communicate and exchange data effectively, among multiple heterogeneous networks (Pokraev, Reichert, Steen, & Wieringa, 2005). An interoperable smart sensor node in WSN context can be modeled by approval levels based on compliance standards or any suite of protocols and global standards, as the IEEE 1451 standard (NIST, 2010). The interoperability in WSN is guided to pragmatism using standardized interfaces, which is trained toward the solution of practical problems of communication by meeting measurable objectives and, thus, a WSN more interoperable depending on the technical, syntactic and semantic compliance levels for diverse sensor devices (Ballari, Wachowicz & Callejo, 2009). In many situations, to achieve an effective communication, it is suitable for all heterogeneous sensor nodes a common syntax and semantics to enable them exchange information accurately and securely. To address these changes, global standards are now beginning to be incorporated in physical layer and high layers of WSN.

Interoperability Levels for WSN

Consider a global infrastructure of WSN with different types of nodes and manufacturers with thousands of telemetric devices sending and receiving data over a smart city. Management of this volume of information on a global scale will be difficult for humans if does not introduced different levels of interoperability. In this context, the interoperability is divided in three levels to process metadata information on large WSN scale in order to share metadata: technical, syntactic and semantic level.

Technical Level

In technical level, interoperability focus on transparent network connectivity between sensor nodes, including the establishment of the communications in physical layer level (PHY) and Medium Access Control layer level (MAC). Technical level can be improved using global and open standardization. In this chapter, we are concentrated in WSN over IEEE 802.15.4 using different methods and frequency bands. Other standards that can be used to support technical interoperability with WSN include IEEE 802.15.1 (Bluetooth) and IEEE 802.11 (WiFi), IEEE 802.15.4-TG6 (BAN), (Moller, Newe, & Lochmann, 2009). In each case, sensor nodes employs a specific PHY and MAC layer to manage the radio transmissions, taking into account a robust and energy efficient operation, to maintain the performance in dense deployments.

In WSN deployments that are oriented to share data over Internet, the use of the Internet Protocol (IP) in network layer is an alternative for pervasive PAN networks. These WSN are oriented to global connection by employing the Internet Protocol version 6 (IPv6), named 6LoWPAN, and it is oriented to small embedded devices to operate in the internet of things (IoT).

Syntactic Level

Syntactic Interoperability involves standardized formats for messages and control commands by employing global standards in network and application layer. In this case, is mandatory a common structure to exchange messages between sensor nodes with the coordinator node in both directions. In this level, the interoperability can be enhanced using open standardization. Open standardization is related with freely and publicly available standards for all communities, with non discriminatory, no license fees, vendor neutral and agreed to by a formal consensus process. An example of syntactic interoperability is the IEEE1451 standard (Lee & Song, 2008), with smart sensors in wired or wireless sensor networks. IEEE 1451 standardization includes some advantages, for example, the inclusion of a common syntax to send and to receive messages using standard commands IEEE 1451.0 (IEEE Std. 1451.0, 2007). In IEEE1451 standard, each command is embedded within the message inside the payload and it standard is oriented for wired or wireless sensor networks with smart sensors and actuators. Some different physical wireless interfaces are compatibles with the family of IEEE1451 standards as IEEE802.15.4 (ZigBee and 6loWPAN), IEEE 802.11 (WiFi), IEEE 802.1 (Bluetooth), (IEEE 1451.5 std, 2007) or RFID tags (IEEE1451.7 std, 2010). Additional advantages of syntactic open interoperability are related with metadata integration by avoiding proprietary syntactic data formats in each message. In some cases, a common syntax in a message allows detecting errors in metadata translation with other sensor network systems, but in other cases, this ability not enough to translate the information accurately among heterogeneous systems that introduce different syntaxes and the interoperability could be limited in some cases. To address these limitations a semantic level based on Web services and a common ontology is other method to increase the WSN interoperability.

Semantic Level

The next stage of interoperability is semantic level that is responsible to give meaning to the information embedded in each syntactic structure. Semantic information allows a transparent communication between different entities, for example sensor nodes, coordinator node, and external clients to share the information without ambiguity over heterogeneous systems.

In this level, a sensor network pursues an effective communication scheme to exchange metadata using a clear syntax and controlled vocabularies (Neiswender, 2009). Therefore, there is compromise between a concise and a common syntax to perform metadata queries that return concise information over multiple crossing platforms of sensor networks.

Typical technologies that are used to achieve this objective are based on logic specifications by developing common ontologies (Yang & Miao, 2007) to describe the sensor network architecture and the environment. The inclusion of an ontology in semantic level allows to define and represent the sensor network in a specific context, for example, the definition of a common ontology related with a meteorological WSN.

Semantic interoperability aims to give meaning to the information and it includes logical sentences to link diverse objects. In a practical approach, the ontologies are not statics and usually are evolving from a common base to be used or expanded by user communities that pursue a common goal.

Figure 2 shows the expected time-line of WSN interoperability and the tendency to share metadata over Internet based on trends of smart sensors.

In a semantic WSN a common ontology based on a controlled vocabulary with syntactic rules as the IEEE1451 standard for smart sensor are an option to increase the interoperability in ZigBee and 6loWPAN networks. Finally, the cognitive and dynamics networks are related with autonomic sensor networks based on algorithms to increase the node intelligence for future sensor networks.

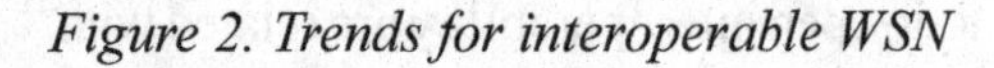

Figure 2. Trends for interoperable WSN

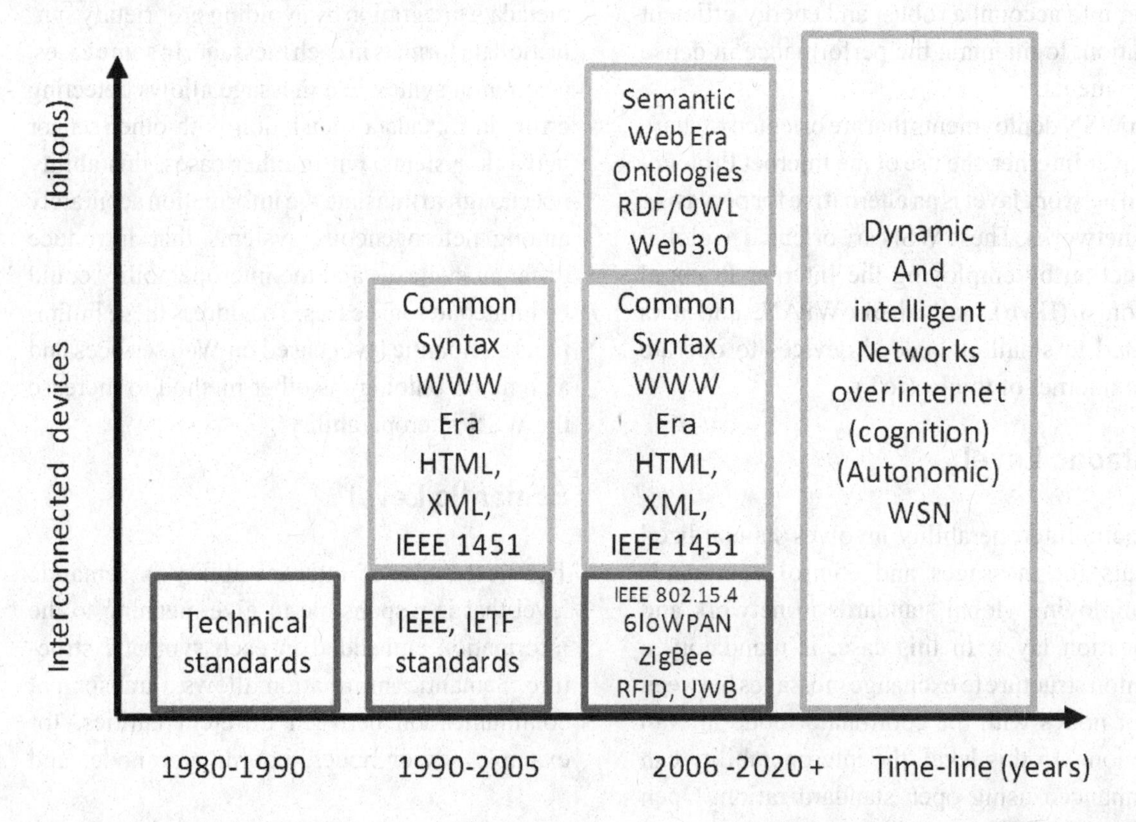

The tendency of interoperable WSN based on smart sensors includes an efficient operation over Internet with heterogeneous devices including small RFID tags, Reduced Function devices (RFD), Full function Devices (FFD), small autonomous devices, smart phones, laptops, netbooks, and other Internet of things IoT objects that are oriented to work together interconnected.

IEEE 1451 STANDARDIZATION

The IEEE 1451 is a family of high level standard organized around a set of common architectures and protocols to permit the interoperability of smart sensors with wired or wireless sensor networks and it belongs to the syntactic level of interoperability. In our context, a smart sensor based on IEEE 1451 is an intelligent sensor. These devices can be connected in a plug and play around any WSN to interpret correctly each operational parameter of their internal physical channels in order to automate in an standardized format the node messages. In a WSN, each IEEE 1451 smart sensor node employs a standard digital wireless communication interface and standard Transducer Electronic Data Sheets (TEDS) to store metadata for self-diagnosis, self-calibration, and node configuration. In these smart sensor nodes, each analog input signal from each physical channel is converted into a digital format using engineering units. An IEEE 1451.5 smart sensor node is named Wireless Transducer Interface Module (WTIM). WTIM nodes can be deployed in a dynamic manner inside of heterogeneous indoors and outdoors sceneries and these nodes may include intelligence capabilities in a high level. Furthermore, the IEEE 1451 standard could be included in small nodes as battery-free clever devices with power management options to increase lifetime in monitoring and control tasks on a large scale.

IEEE 1451 Smart Sensor Model

Figure 3 shows the IEEE 1451 model for smart sensor in wired and wireless sensor networks. In this chapter we employ the IEEE1451.5 standard (IEEE1451.5 std, 2007) for use in pervasive environments to deploy WSN applications. This model employs a Network Capable Application Processor (NCAP) that is the coordinator node, and it includes a Wireless Transducer Interface Independent (TII) to communicate with the other wireless devices named as Wireless Transducer Interfaces Modules (WTIM). WTIM devices include the End Devices and repeater nodes in a WSN. Furthermore, the NCAP is a bridge between the wired network and the WSN. It includes the IEEE 1451.0 standard commands to communicate with the other devices and clients in the sensor network or an external wired network. In WSN oriented toward Web technologies, the NCAP includes a standard Application Programming Interface API to increase the level of syntactic and semantic interoperability and to communicate with external devices as smart phones, Web servers and Internet of Thing (IoT) devices. A NCAP offers additional options as run a semantic Web service based on HTTP transactions.

The NCAP, as sink node, consists of the next parts: in a technical level, a communication submodule based on a wireless transceiver, MCU or Digital signal processing unit (DSP) to acquire and process the sensing information and finally, in a high level, the IEEE 1451.0 standard commands and IEEE 1451 API service to employ the HTTP protocol over IP sensor networks. In a practical implementation, the NCAP could be an edge router to connect with an external network, for example, a Local Area Network (LAN). In ZigBee deployments, the NCAP operates as a gateway between the WSN and wired networks and it performs the control of multiple WTIM nodes by employing common commands IEEE1451 with the networks topologies supported by the IEEE 80215.4 standard. In 6loWPAN

Figure 3. IEEE1451 smart sensor model

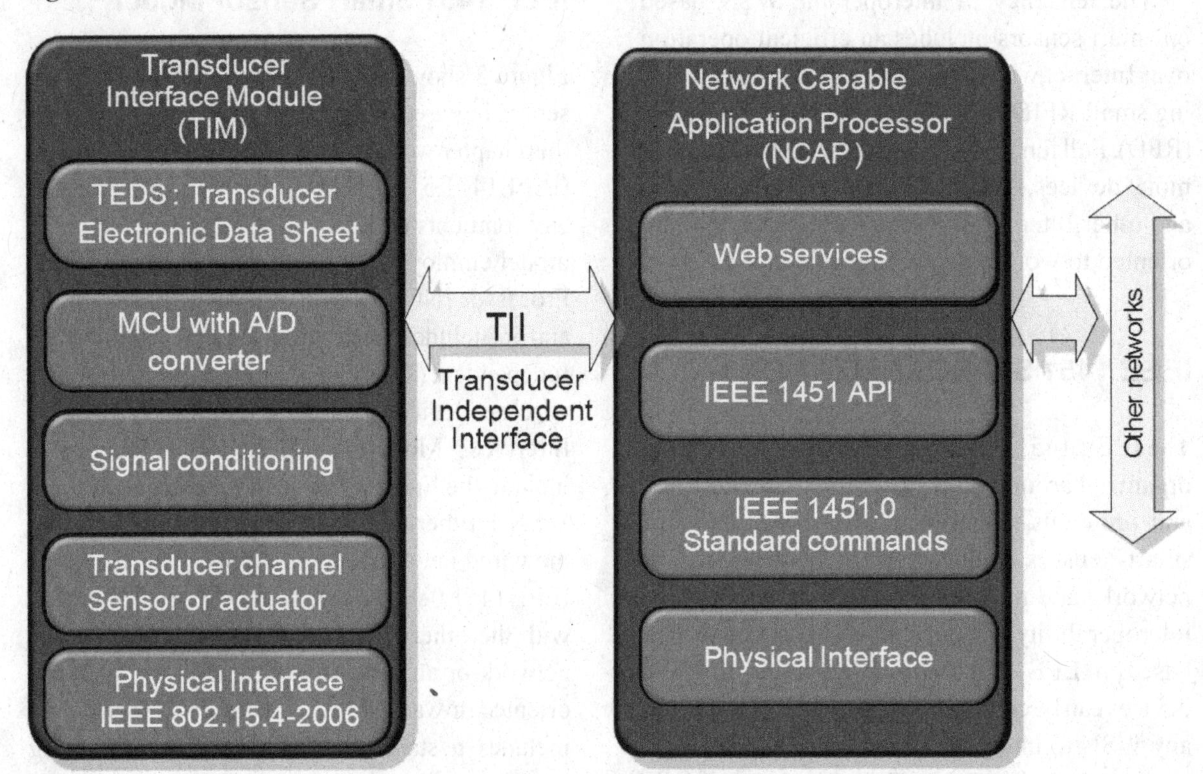

the NCAP operates as an edge router with the external IP networks.

In most cases, the NCAP power sub-module employs batteries or is plugged in another device that is connected to the electrical network, for example, a personal computer, server, laptop and others. The NCAP activities include the creation of the sensor network by discovering and associating each WTIM node around the sensor network coverage range.

A WTIM node includes in their hardware architecture channels to sense physical parameters. Also, it includes a signal conditioning submodule, Analog to Digital Converter (ADC). In many cases, the ADC converter is embedded in a Microcontroller Unit (MCU) and it includes some ADC channels for their operation. Other important element of a WTIM node is so-called, Transducer Electronic DataSheet (TEDS) that is embedded in a non volatile memory. IEEE 1451 standard has a great potential to interact in PAN networks with IEEE 802.15.4 PHY and MAC low layers, and IEEE 1451.5 standard to implement in the high layers of any WSN. Figure 3 shows the IEEE 1451 smart sensor model in a WSN.

In a typical WSN deployment based on ZigBee or 6loWPAN the two first tasks of the NCAP node includes the network creation, discover and associate each WTIM node in their network coverage range. Also, these features can be developed with different wireless physical interfaces that depend on the technical interoperability. By other hand, some tasks of any WTIM node includes the auto-description of their physical sensing channels based on TEDS metadata stored in a non volatile memory. Also, the WTIM employs a common structure for IEEE 1451 commands related with the TEDS access and configuration. Finally, for web clients exist an HTTP API is an option to communicate with the NCAP or any WTIM node in Service Oriented Architectures (SOA).

Figure 4. IEEE 1451.0 mandatory and optional TEDS

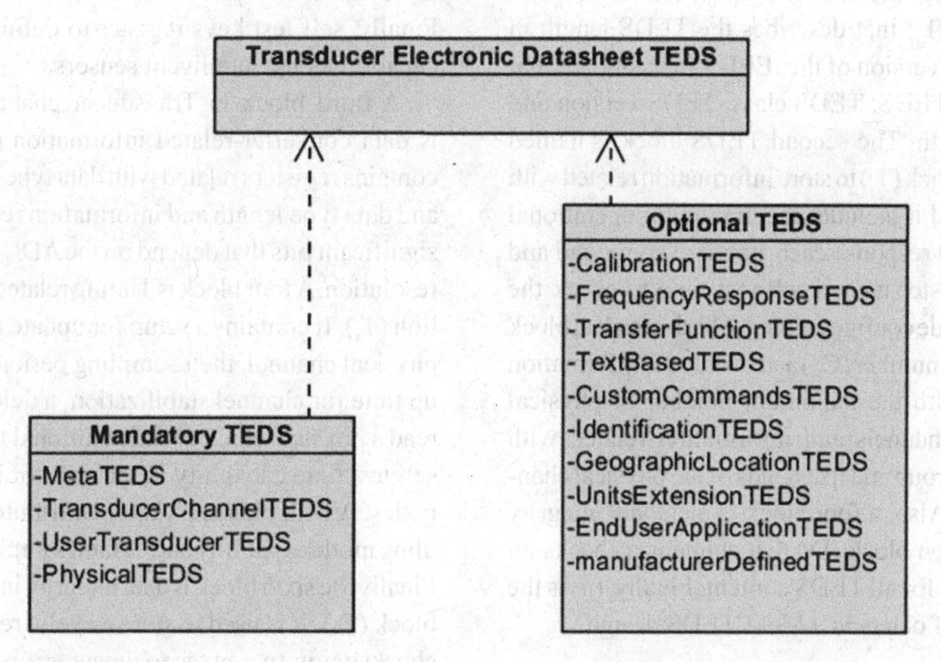

Transducer Electronic Datasheet (TEDS)

Smart sensors as NCAP and WTIM devices contain metadata stored in a group of Transducer Electronic Datasheets (TEDS). For each TEDS, an amount of EEPROM memory is written and reprogrammed depending on the node attributes, as physical node configuration related with their physical transducer channels. Each TEDS employs an ordered list of elements called tuple and their construction is organized in a specific metadata format. Unified TEDS format of this type allows increase a syntactic interoperability in deployments of heterogeneous smart sensors for WSN applications. The IEEE 1451.0 standard includes four mandatory TEDS and others TEDS optional as shown in Figure 4. Mandatory and optional TEDS could be modeled using two different approaches:

- IEEE 1451.0 TEDS based on Tuple format: TEDS tuple model defined in the IEEE 1451.0 standard and at present, this structure is employed in WTIM nodes that are oriented to Personal Area Networks (PAN). In this case, tuple structure is composed by a field type header, data length, and finally is introduced all data information content in the payload.
- IEEE 1451.0 primary TEDS: TEDS for RFID tags with a small non volatile internal memory. Their structure allows to store in a compressed format: TEDS type, sensor type, units, output data resolution, scale factors and security options.

Meta-TEDS

Meta-TEDS is a type of mandatory Transducer Electronic Data Sheet that contains four different metadata blocks. The following definition shows a compact format to model each Meta-TEDS.

$$MetaTEDS = \sum_{i=0}^{n} I_B + T_I + C_N + D_I$$

The first block is a metadata informational structure (I_B) that describes the TEDS length in bytes, the version of the IEEE 1451 standard that uses this TEDS, TEDS class, TEDS version and tuple length. The second TEDS block is named Timing block (T_I) to store information related with timing and it includes the maximum operational timeout to response each standard command and other register named self test time to check the WTIM node configuration. A Third metadata block is channel number (C_N) and it includes information related with the maximum number of physical sensing channels and information related with channel group and it depends of the physical channel type. Also, a four block is metadata integrity information block (D_I) that contains a checksum algorithm for all TEDS content. Finally, (n) is the number of octets in a Meta-TEDS design.

Transducer Channel TEDS

Transducer Channel TEDS submodule is a structure based on metadata that store the configuration related with the physical sensing channels coupled in a WTIM node. Their model and block structure includes the next parts:

$$Transd\_Ch\_TEDS = \sum_{i=0}^{n} I_B + T_{CH} + D_C + T_I + A + D_I$$

Where the structure is composed of the information block (I_B) related with total TEDS length, TEDS family, TEDS class, TEDS version and tuple TEDS identifier taking into account the IEEE 1451.0 standard. A second block is named Transducer channel related information (T_{CH}) and it includes a register for calibration capabilities, other register to define a transducer channel as operated by poll or any events. Additional registers are physical channel units to store the units to be measured. Also, registers to define the low and high values in a transducer channel, a register to describe the operational channel uncertainty. Finally, self test keys register to define self test capabilities for intelligent sensors.

A third block in Transducer channel TEDS is data converter-related information (DC) that contains registers related with data type definition and data type length and information related with significant bits that depend on the ADC converter resolution. A four block is Timing related information (T_I). It contains a setup for update and read a physical channel, their sampling period, a warm-up time for channel stabilization, a delay time to read a physical channel and additional features as self test time capability in seconds for intelligent nodes. A fifth blocks (A) store attributes as sampling mode capability and default sampling mode. Finally the sixth block is data integrity information block (D_I). It is used to store a cyclic redundancy checksum in two bytes to detect errors.

User's Transducer Name TEDS

In a WTIM with many different physical sensing channels, a User`s transducer name TEDS stores end user metadata information to identify each physical channel. Their model and block structure contains five parts.

$$User\_TEDS = \sum_{i=0}^{n} Length_{TEDS} + ID_{TEDS} + F_{TEDS} + TIM_I + CRC$$

TEDS length ($Length_{TEDS}$) is the first block and it is employed to store the overall TEDS length. A second block (ID_{TEDS}) contains a global TEDS identifier of user`s transducer TEDS.

A third block named Format TEDS (F_{TEDS}) stores the content type as user defined in a low level or text based content. A fourth block is named user TEDS information (TIM_I) that contains metadata for channel user identification. Finally a fifth block is a Cyclic Redundancy Checksum (CRC) to detect errors.

IEEE 1451.5 Physical TEDS (PHY-TEDS)

Mandatory Physical TEDS (PHY-TEDS) store the main parameters associated with the Transducer Interface Independent (TII). IEEE 1451.5 standard defines four TII to use with WTIM nodes (Wi-Fi, Bluetooth, ZigBee and 6loWPAN) to use in a PHY-TEDS configuration. PHY-TEDS Metadata increases the technical interoperability and it provides valuable information that can be updated in a dynamical manner. The following definition has been chosen to model each metadata PHY-TEDS block structure

$$PHY\_TEDS = \sum_{i=0}^{n} PHY_C + PHY_S$$

The first block (PHY_C) stores a first group of field type registers as independent of the physical layer to maintain the same configuration compliant with the IEEE 1451.5 standard. The (PHY_S) TEDS block store the radio options based on their specific Wireless transducer interface independent. For example, field type registers reserved to work with Wi-Fi, ZigBee, Bluetooth or 6loWPAN.

IEEE 1451.0 Primary TEDS

A model of TEDS primary is showed in Table 1.

In WTIM devices with small non volatile internal memory as passive, semi-passive and active RFID tags, the previous model for mandatory TEDS is merged in a unique TEDS format named primary. It is addressed to be included in RFID nodes and other WTIM nodes that are powered by energy harvesting techniques due the hardware limitations. A device of this type allows only one TEDS primary by transducer channel using a reduced format based on IEEE 1451.7 standard (IEEE1451.7 std, 2010) to store metadata information. Each Field type register is defined in a reduced format based on bits to increase the header compression. TEDS Primary model does not use a tuple format, but instead it uses a compact structure that includes few registers based on IEEE 1451.7 standard. In this approach, TEDS primary includes some physical data units that are shown in (IEEE1451.7 std, 2010) and others can be included as additional environmental variables, for example, rainfall (mm/H2O), total solar radiation and photosynthetic solar radiation (W/m2). Additional chemical substances like some pollutants as sulfur dioxide (SO2), hydrogen sulfide (H2S), ozone (O3) or others are includes in the units extension map for TEDS primary. In total, only 16 octets are necessary to store the node configuration based on primary TEDS model.

IEEE 1451 WTIM NODE IMPLEMENTATION

Key components of an autonomous and interoperable WTIM node include the next parts:

Power Management Submodule

A sub-module of this type permits capture and manages the node energy for their efficient operation. It includes a combination of energy harvesting methods to eliminate or reduce the need for batteries in some cases. In outdoor locations the most common source of energy available for energy harvesting is the sun light. In this case, a WTIM node allows a small solar panel to recharge the batteries. The output solar cell current usually produces a few milliamps that depend on the incident solar irradiance and location. In a practical circuit the incident sunlight charge the batteries and also, this module includes a small super-capacitor to maintain the voltage in the solar panel stable to charge the batteries. Furthermore, a comparator with hysteresis is used for controlling the charge of batteries. Also a DC/DC boost converter is employed to transfer efficiently the node energy.

Table 1. TEDS primary compliance IEEE 1451.7

Field	Name	Size	Example
1	TEDS type	3 bits	$[001]_2$ = RFID tag ; $[000]_2$ =WTIM harvesting node ; $[010]_2$ to $[111]_2$ = reserved
2	Sensor type	7 bits	$[11011]_2$ = wind speed m/s
3	Units extension	5 bits	Sub-type to use in chemical sensor
4	Supported measurement	16 bits	$[1111000000000000]_2$ Present, maxim, minim and average wind speed
5	Output Data resolution	5 bits	MCU with ADC 16 bits of resolution $[10000]_2$
6	Scale factor	11 bits	*R= Real number Nd= Value ADC* $R = Nd \times SF + SO$ SF = scale factor $SF = SFSd \times 10^{SFEd}$ SO = scale offset $SO = SOSd \times 10^{SOEd}$ Example wind speed sensor, range 0.0 m/s to 60m/s, with 16-bit ADC
7	Scale factor exponent	6 bits	Represents the decimal Scale Factor Exponent (SFEd)
8	Scale Offset Significant	11 bits	signed 11-bit binary number that represents the decimal Scale Offset Significant (SOSd)
9	Scale Offset Exponent	6 bits	signed 6-bit 1 binary number that represents the decimal Scale Offset Exponent (SOEd)
10	Data uncertainty	3 bits	value that represents the accuracy of the transmitted data
11	Sensor Reconfiguration Capability	1 bit	indicates whether or not the sensor supports the capability for the user to reconfigure the sensor
12	Memory Rollover Capability	1 bit	With memory rollover enabled, if the memory becomes full the earliest record is overwritten on a FIFO mode.
13	Air Interface Security Capability Code	3 bits	sensor supports air interface security and the allowed level of security options
14	Sensor Security Capability Code	3 bits	sensor supports its own internal security
15	Sensor Authentication Encryption Capability Map	7 bits	authentication encryption algorithms for example AES-128 or SHA1 and others
16	Sensor Data Encryption Capability Map	7 bits	Supported sensor data encryption algorithms. For example AES-128
17	Sensor Authentication Password/Key Size	3 bits	Authentication password key size
18	Sensor Data Encryption Key Size	3 bits	Data Encryption key size
19	Random Number Sizes Supported	3 bits	Random Number Sizes Supported
20	Continuing Authentication	2 bit	optional
21	Data Encryption Capability	2 bit	Optional feature
22	Clock Accuracy	3 bits	In percentage (%) or in parts per million (ppm)
23	RFU	17 bits	Reserved for future use
	TOTAL	128 bits	

Sensing Submodule

This module involve some physical channels to sense the environment, for example, channels to monitor temperature, relative humidity, solar radiation, wind speed, wind direction, rain, barometric pressure, acoustic noise, poison gases in an environmental WSN. Each physical sensing channel in WTIM node contains a series of signal conditioning circuits depending on the analog signal type and it introduces an anti-aliasing filter based on Nyquist-Shannon sampling requirements (Luke, 1999) to apply before the ADC converter in a classical approach or use any direct sensor to microcontroller interface as in (Reverter & Pallas-Areny, 2005).

A Computing Unit

A computing unit is based on low power microcontrollers (MCU) or, in other cases, it could be a Digital Signal Processor (DSP) to process the signal and design digital filters. In many cases, a low power MCU contains several internal analog to digital converter channels (ADC). Also, it contains some digital input/output depending on the application. Some MCU includes UART interfaces to communicate with different peripherals as RS232 serial ports and USB. Additional options in an MCU include several modes for sleep and hibernate to reduce the overall energy consumption. Furthermore, the use of internal or external 32 KHz clock system extends the WTIM lifetime.

The MCU contain serial interfaces as Serial Peripheral Interface bus I2C or any Serial Peripheral Interface bus SPI to read and write serial non volatile memories that contains each Transducer Electronic Data Sheet (TEDS).

Communication Submodule

WTIM nodes include a wireless transceiver that operates in a standard Industrial, Scientific and Medical ISM band with some type of antenna, for example, PCB, coaxial or integrated to use in indoor or outdoor places. In many cases, an external MCU transceiver is employed in a WTIM node. In this case the MCU uses a serial interface as I2C, SPI or UART, and several I/O pins for their configuration and operation. In other cases a transceiver and MCU are combined in the same package reducing their hardware architecture and turning it in a highly integrated System on a Chip (SoC). Furthermore, each WTIM node could be programmed using assembler language in a low level or C language in a high level. For WSN, a Real Time Operative System (RTOS) as TinyOS enables a schedule of modules and configurations to compile sensor networks applications (Levis & Gay, 2009). Other some popular RTOS for WTIM nodes are shown in (Kuorilehto, Kohvakka, Suhonen, Hämäläinen, Hännikäinen, & Hamalainen, 2007).

IEEE1451 STANDARD COMMANDS

In this section, we present the IEEE 1451.0 standard commands which are used in forward and incoming messages. An interoperable IEEE1451 WSN uses standard commands to increase the syntactic and semantic interoperability based on WTIM node capabilities. IEEE 1451 standard commands are composed of nine classes that include some functions with unique characteristics that are related with WTIM node states and transducer channel states. IEEE 1451 command classes related with WTIM includes: IEEE 1451 common commands, WTIM sleep state commands, WTIM active state commands and WTIM any state. IEEE 1451 command classes related with Transducer channel includes: common commands, transducer channel idle state, transducer channel operational state, transducer channel idle or operational state. Finally, the last two command classes are open to manufacturers or reserved to include new characteristics in the future. All IEEE 1451 command classes are shows in Figure 5.

In Figure 5 WTIM class commands are related with the global WTIM node while transducer channel commands are relate with a specific transducer channel of the WTIM node. The WTIM nodes in some cases are composed of some transducer channels.

IEEE 1451 Command Structure

IEEE 1451 Command structure are primarily intended to monitor and control smart sensors as WTIM nodes. A typical IEEE 1451 command structure is shown in Table 2.

In Table 2, the IEEE1451 command structure is compatible with small WTIM nodes that employ low duty cycles in their operation. For example less than < 1% in active state. The IEEE 1451 command structure includes two phases called request mode and response mode.

The request mode introduces four headers and one payload. Headers in request mode are related with any specific transducer channel, IEEE 1451 command class, command function and offset length. The payload in request mode stores the information to send in each message and it depends on the command type (write or read).

The response mode uses two headers and one payload. Headers in response mode are related with success or fail reception of the IEEE 1451 command by the WTIM node. The second header in this mode precise the length offset of the payload. The payload in response mode corresponds with the metadata information stored in any transducer channel or TEDS submodule, and the maximum length over an IEEE 802.15.4 transceiver is around 110 bytes. This effective payload can be reduced if the WTIM node operates in a ZigBee or 6loWPAN sensor network.

IEEE 1451 AND WEB SERVICES IN WSN

Web services are related with Web Application Programming Interfaces (APIs) that are hosted on a web server and these execute different functions that are capsulated through standard protocols in application layer. Web services can be used with Hypertext Transfer Protocol (HTTP) to transfer messages over Internet by using languages as HTML or XML.

In general, Web services involve two types of styles called Simple Object Access Protocol (SOAP) (Louridas, 2006), and Representational State Transfer (REST), (Haibo & Doshi, 2009). SOAP Web services adopt XML meta-language,

Figure 5. IEEE 1451 command classes

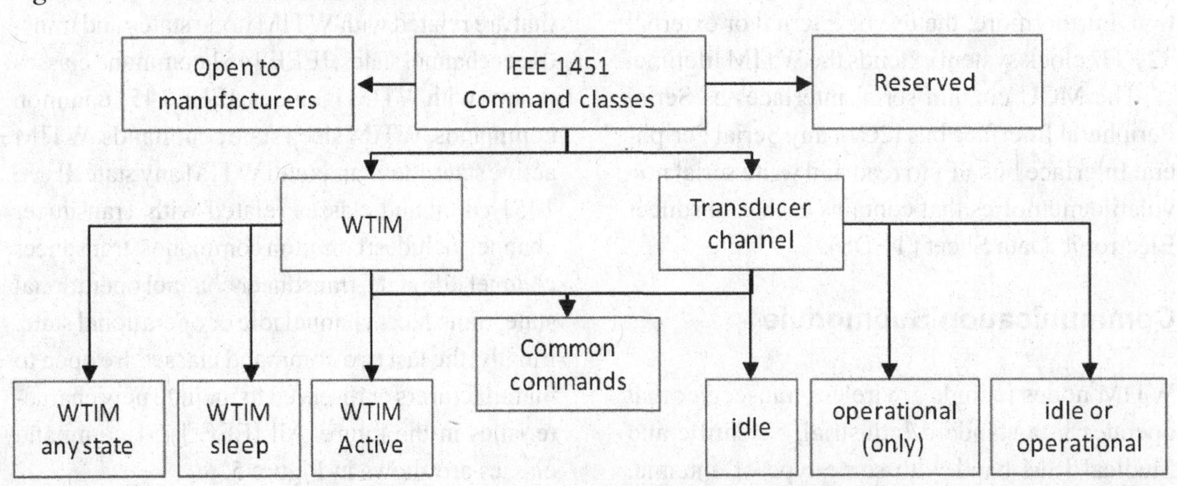

Table 2. IEEE standard request-response command structure

Request mode				
IEEE 1451 header				IEEE 1451 payload
Destination Transducer channel	**Command class**	**Command function**	**Offset length**	**Command dependent**
Number of Transducer physical channel [2 Bytes]	Code class [1 Byte]	Code function [1 Byte]	Offset [2 Bytes]	Variable read or write [~ 110 bytes] (IEEE 802.15.4)
Response mode				
IEEE 1451 header		**IEEE 1451 payload**		
Success flag	**Length offset**	**Reply dependent**		
Success /fail [1 Byte]	Offset 2 Bytes	Variable read or write [~ 110 bytes] (IEEE 802.15.4)		

and the Web Services Description Language (WSDL) that is relates with big Webs services, while REST style is based on the HTTP protocol that is oriented to small Web services that share a few amount of metadata with a simplicity structure.

Hence, an interesting question arises of whether Web service can be employed for WSN applications. An IEEE 1451 smart sensor could potentially adopt a message structure to travel in any Web service, especially between remote clients that employ only a web browser and the NCAP node to monitor the network operation. Motivated by this issue, this section will investigate the feasibility of the smart sensors based on IEEE 1451 that employ a syntactic structure to share messages using IEEE 1451 standard commands, which are wrapped in a Web service. Figure 6 shows an example of Web service for an IEEE 1451 sensor network.

Each external Web client has a web browser to access the WSN by addressing to the NCAP through a SOAP or REST Web service stored in a Web Server. In the sensor network, the NCAP requests each WTIM node and it returns the response with direction to the NCAP node. The NCAP is a bridge between the sensor network and the Web client. The NCAP is responsible to wrap each transaction with the Web clients using a SOAP or REST Web service style stored in the Web service provider, named Web server. In addition, different web clients will interact with the Web service to retrieve information as metadata related with the IEEE 1451 sensor network operation. These communication schemes between heterogeneous devices using Web services increase the flexibility to connect heterogeneous sensor network systems with efficiency and enabling different levels of interoperability between different sensor nodes and overcomes the limitations of earlier approaches as Common Object Request Broker Architecture (CORBA), (Polze, Richling, Schwarz & Malek,1999) or Distributed Component Object Model Architecture (DCOM), (Dezhgosha & Angara, 2005).

Standards for Web Services

The main features of static Web services are summarized in Table 3. In each case, Web services based on SOAP could be developed using frameworks as .NET environment (Skonnard, 2003) and Java web based applications (Oracle, 2011). In few years, REST Web service style is growing due to a dynamic development of open tools that are growing fast, although that REST Web services are not part of World Wide Web Consortium (W3C) organization, but all of them

Figure 6. IEEE 1451 smart sensors and Web clients interacting in a Web service

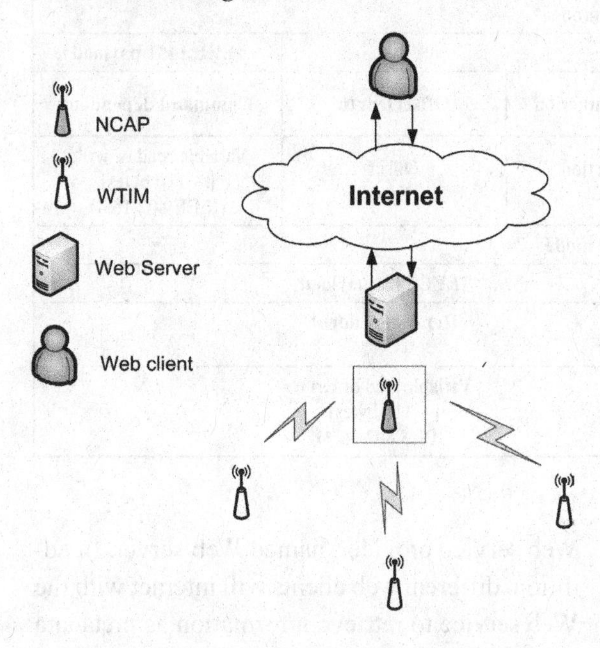

use standards of W3C. Also, the main novelty of SOAP and REST is their independence of the programming languages and Operative System. A Traditional SOAP Web service style represents all data information in XML format to exchange messages between the client and the Web service provider using HTTP, secure HTTP, File Transfer Protocol (FTP) or Simple Mail Transfer Protocol (SMTP). Indeed, SOAP will use the Web Services Description Language (WSDL) to describe the type of Web service stored in the Web service provider. In a similar way, WSDL employs the XML language to describe the Web service, their methods, bindings and data for input and output. Thus, SOAP Web services that use the WSDL language define a wrap that contains a header and the body. The header provides instructions on how the message must be processed and the SOAP body is related with the contents of the message.

The HTTP application protocol (RFC2616, 1999) defines the format and sequence of the messages that are passed between the NCAP and any Web client using a standard web browser (i.e., Internet Explorer, Mozilla Firefox). There are two types of HTTP messages, which are request and response type. The structure of an HTTP message includes a line for request that contains some method field as GET, POST, DELETE, PUT, the URL field and the HTTP version field, for example: (GET /principal/index.html HTTP 1.2). Also, the next lines are used as headers. An example of header line is (Host: http://www.upc.edu) that explains where was stored the object in the web address.

SOAP Web Services

As showed in Table 3, the Web service style uses XML as meta-language and the support of WSDL language to describe the service. The HTTP protocol is used to exchange messages between any client and the web server using one web browser in application layer. Figure 7 shows the layers of a SOAP message to deploy a Web service in a remote Web server.

Table 3. Web services based on SOAP and REST style

Web Services	Based on SOAP	Based on REST
Protocol/language	XML, HTTP, WSDL	HTTP, HTML
Framework	.NET, Java, Spring, Apache Axis2, Apache CXF	.NET, Java, PHP, Python, Ruby
Use Style	Remote procedure calls (RPC), Service-oriented architecture (SOA)	Restful paradigm
Devices	Laptops, netbooks, Servers, PCs, smart phones	tiny nodes
Current status	W3C recommendation	Still being developed

Figure 7. SOAP Web service layers

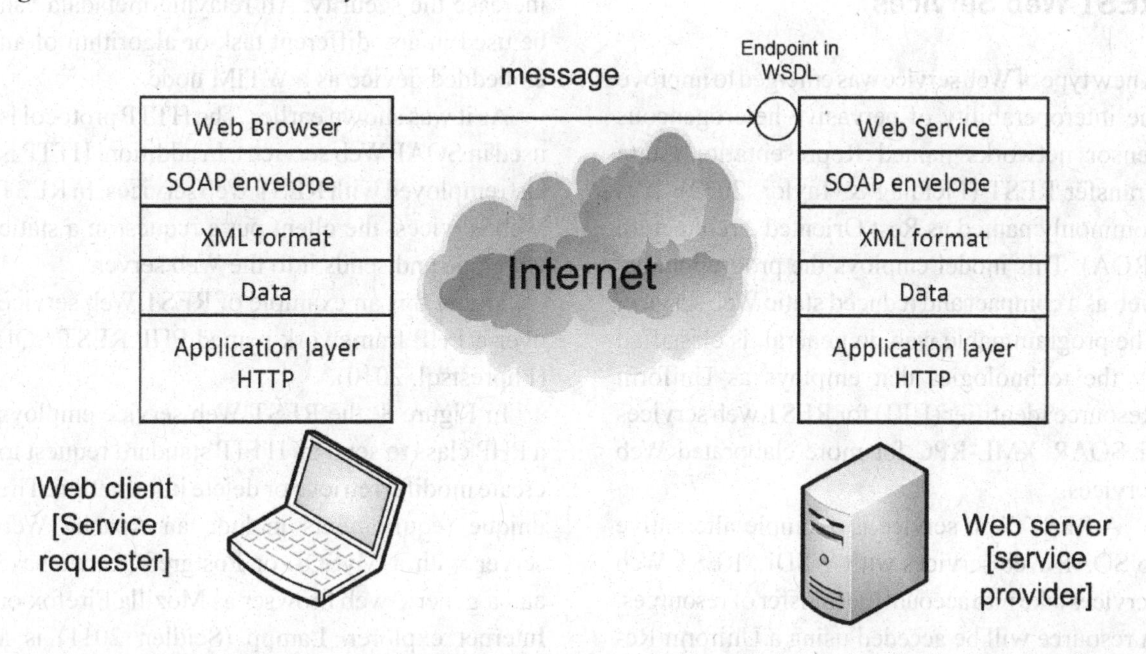

Any Web service provider evaluates the rules to determine if the message is valid or not. When the SOAP message is valid, the Web service provider evaluates the request and it executes a response using a SOAP message toward the service requester through a wrapped message in an HTTP transaction over the web browser. The SOAP envelope structure is composed of next parts: SOAP header that is optional, SOAP body part that is mandatory and it includes data payload. The last part is a fault element that is optional and it includes information about errors related with SOAP processing message. A more elaborated description of any SOAP Web service can be developed using the WSDL language that proposes an standardized manner to describe a Web service as a set of endpoints. Also, each endpoint contains a Web service description in a WSDL file. Any WSDL file contains the next elements that describe the SOAP message using XML metadata:

- **Types:** related with the type of messages that only will be used to send and receive SOAP messages.
- **Binding:** their elements specify how to access in the service using an application protocol.
- **Service:** this element describes where can be acceded the Web service by name or interface, and it includes the endpoint element.
- **Interface:** Includes a set of operations and a series of input and output fault messages of the Web service.

At present the different frameworks to implement SOAP Web services compete for mass deployment as .NET Framework or Java. Other programming frameworks, hides to the users, many of their SOAP elements and these have included different APIs to support the emergent programming languages. For example, the SOAP Web services based on Microsoft Visual Studio 2010 with a WSDL file that is created by the .NET Framework to be employed by a Web client using any standard web browser.

REST Web Services

A new type of Web service was emerged to improve the interoperability of pervasive heterogeneous sensor networks named Representation State Transfer REST (Fielding & Taylor, 2002). It is commonly named as Rest Oriented Architecture (ROA). This model employs the programmable web as a compact and reduced static Web service. The programmable web, in general, is classified by the technologies that employs as Uniform Resource Identifier (URI) for REST web services or SOAP, XML-RPC for more elaborated Web services.

A REST Web service is a simple alternative to SOAP Web services with WSDL. REST Web services take into account the transfer of resources. A resource will be acceded using a Uniform Resource Identifier (URI), for example, to access an object of a database hosted in a Web server. REST Web service style is built on top of the application layer and it employs the HTTP protocol without WSDL in a lightweight structure for using with embedded devices as WTIM nodes.

In many cases, the process to create an elegant and easy Web service is better with REST, since the client extracts the information from the URI resource using only the World Wide Web, with a browser using the HTTP or HTTPS protocol to increase the security. All relaying metadata can be used in any different task or algorithm of an embedded device as a WTIM node.

As it was shown earlier, The HTTP protocol is used in SOAP Web services. In addition, HTTP is also employed with REST Web services. In REST Web services, the client put a request in a static envelope and sends it to the Web server.

Figure 8 is an example of REST Web service over a PHP framework named PHP REST SQL (Phprestsql, 2010).

In Figure 8, the REST Web service employs a PHP class to send an HTTP standard request to create modify, retrieve or delete information. The unique requirements include an Apache Web server with a MySQL or PostgreSQL database and a generic web browser as Mozilla Firefox or Internet explorer. Lampp (Seidler, 2011) is a compact solution to install the Apache Web server, MySQL database and PHP in Linux as open source Operative System. An extension tool to test a REST Web service in Firefox is Poster (Milowski, 2010). The HTTP method POST in a REST Web service includes the URI path, request headers as host, user agent, an finally the accept headers. The entity body or representation is used in a POST request. Figure 9 shows a GET interaction with a REST Web service.

Figure 8. POST request – response using a PHP REST Web service

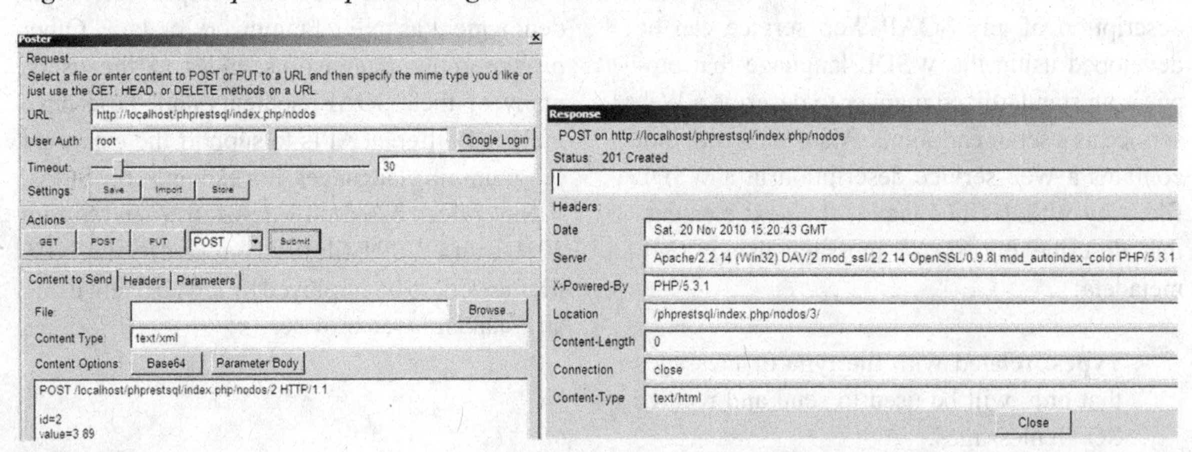

Figure 9. GET request – response using a PHP REST Web service

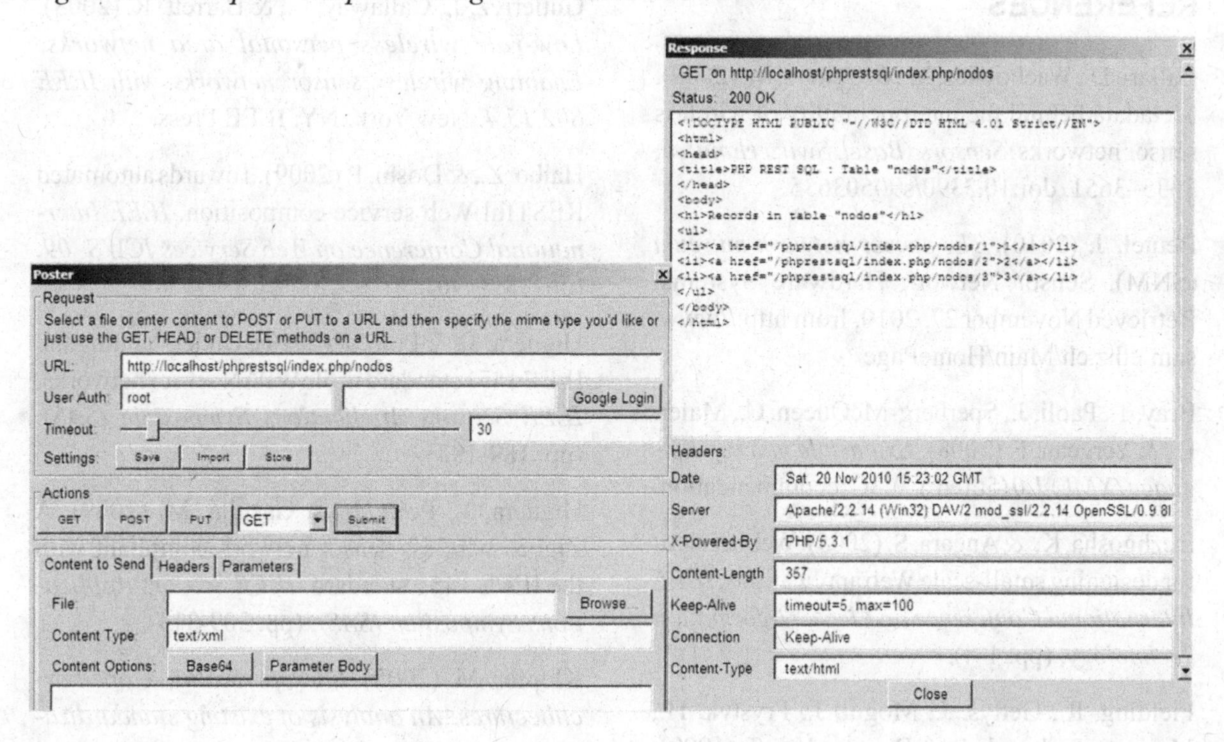

A Web service client requester sends an inquiry and then, a web service provider response sends an envelope with the response toward the web service client. In a GET request as shown in Figure 9, the envelope does not have entity body. The HTTP response is an envelope similar to the HTTP request. The GET response includes an HTTP response code (Status 200 OK), the HTTP response headers as date, server name and so on and finally the entity-body corresponds with the resource representation of the GET request.

FUTURE RESEARCH DIRECTIONS

There are several interesting open research lines in the field of interoperability of smart sensors based on IEEE 1451. Some tasks are focused on the development of a compact TEDS model to reduce the non volatile memory in a WTIM node and facilitate the assessment of the IEEE 1451 standard and their adoption in ubiquitous devices such as IoT devices, energy harvesting sensors and RFID nodes.

The rapid evolution of sensor networks is creating new opportunities to deploy sensor nodes in remote and heterogeneous environments and their interoperability based on SOAP Web services. Furthermore, others Web services based on REST style allocate a human-readable format using standard protocols as HTTP to increase the semantic interoperability. In the future, the convergence of heterogeneous devices tends toward Internet of things (IoT) but the main problems are related with the efficiency, security and transparent cooperation to achieve syntactic and semantic interoperability taking in account standardized metadata information generated by heterogeneous devices.

REFERENCES

Ballari, D., Wachowicz, M., & Callejo, M. (2009). Metadata behind the interoperability of wireless sensor networks. *Sensors (Basel, Switzerland)*, *5*, 3635–3651. doi:10.3390/s90503635

Beutel, J. (2010). *The sensor network museum* (SNM). Sensor Network Hardware Systems. Retrieved November 27, 2010, from http://www.snm.ethz.ch/Main/HomePage

Bray, T., Paoli, J., Sperberg-McQueen, C., Maler, E., & Yergeau, F. (2008). *Extensible markup language (XML) 1.0* (5th ed.). W3C recommendation.

Dezhgosha, K., & Angara, S. (2005). Web services for designing small-scale Web applications. *IEEE International Conference on Electro Information Technology*, (pp. 1-4).

Fielding, R., Gettys, J., Mogul, J., Frystyk, H., Masinter, L., Leach, P., & Berners-Lee, T. (1999). *Hypertext transfer protocol -- HTTP/1.1. RFC editor, USA,* (pp. 1-176).

Fielding, T., & Taylor, N. (2002). Principled design of the modern Web architecture. *ACM Transactions on Internet Technology*, 115–150. doi:10.1145/514183.514185

Ganek, A., & Corbi, T. (2003). The dawning of the autonomic computing era. *IBM Systems Journal*, *42*(1), 5–18. doi:10.1147/sj.421.0005

Gay, D., Levis, P., Behren, R., Welsh, M., Brewer, E., & Culler, D. (2003). The nesC language: A holistic approach to networked embedded systems. In *Proceedings of the ACM SIGPLAN Conference on Programming Language Design and Implementation (PLDI '03)*, (pp. 1-11). New York, NY: ACM.

Gracanin, D., Eltoweissy, M., Wadaa, A., & DaSilva, L. A. (2005). A service-centric model for wireless sensor networks. *IEEE Journal on Selected Areas in Communications*, *23*(6), 1159–1166. doi:10.1109/JSAC.2005.845625

Gutierrez, J., Callaway, E., & Barrett, R. (2003). *Low-rate wireless personal area networks: Enabling wireless sensor networks with IEEE 802.15.4*. New York, NY: IEEE Press.

Haibo, Z., & Doshi, P. (2009). Towards automated RESTful Web service composition. *IEEE International Conference on Web Services ICWS '09*, (pp. 189-196).

Higuera, J., & Polo, J. (2010). Understanding the IEEE 1451 standard in 6loWPAN sensor networks. *IEEE Sensors Applications Symposium (SAS)*, (pp. 189-193).

Higuera, J., Polo, J., & Gasulla, M. (2009). A Zigbee wireless sensor network compliant with the IEEE 1451 standard. *IEEE Sensors Applications Symposium (SAS)*, (pp. 309-313).

Klopfer, M. (2005). *Interoperability & open architectures: An analysis of existing standardization processes & procedures*. Open Geospatial Consortium. OGC document number 05-049r1.

Kuorilehto, M., Kohvakka, M., Suhonen, J., Hämäläinen, P., Hännikäinen, M., & Hamalainen, T. (2007). *Ultra-low energy wireless sensor networks in practice: Theory, realization and deployment*. Hoboken, NJ: Wiley. doi:10.1002/9780470516805

Lee, K., & Song, E. (2008). Understanding IEEE 1451-Networked smart transducer interface standard - What is a smart transducer? *Instrumentation & Measurement Magazine*, *11*(2), 11–17. doi:10.1109/MIM.2008.4483728

Levis, P., & Gay, D. (2009). *TinyOS programming*. Cambridge, UK: Cambridge University Press.

Louridas, P. (2006). SOAP and Web services. *IEEE Software*, *23*(6), 62–67. doi:10.1109/MS.2006.172

Luke, H. (1999). The origins of the sampling theorem. *IEEE Communications Magazine*, 106–108. doi:10.1109/35.755459

Milowski, A. (2010). *Poster Firefox extension*. Retrieved from https://addons.mozilla.org/en-US/firefox/addon/2691/

Moller, S., Newe, T., & Lochmann, S. (2009). Review of platforms and security protocols suitable for wireless sensor networks. *IEEE Sensors Conference* (pp. 1000-1003). Christchurch, New Zealand: IEEE.

Neiswender, C. (2009). What is a controlled vocabulary? In *The MMI guides: Navigating the world of marine metadata*. Retrieved November 27, 2010, from http://marinemetadata.org/guides/vocabs/vocdef

NIST. (2010). *IEEE 1451 standard*. Retrieved November 27, 2010, from http://ieee1451.nist.gov/

Oracle. (2011). *Metro Web services overview*. Retrieved January 17, 2011, from http://www.oracle.com/technetwork/java/index-jsp-137004.html

PHP REST SQL. (2010). *A HTTP REST interface to MySQL written in PHP*. Retrieved November 28, 2010, from http://phprestsql.sourceforge.net/download.html

Pokraev, S., Quartel, D., Steen, M. W. A., & Reichert, M. (2006). A method for formal verification of service interoperability. *ICWS '06 International Conference on Web Services*, (pp. 895-900).

Pokraev, S., Reichert, M., Steen, M., & Wieringa, R. (2005). Semantic and pragmatic interoperability: A model for understanding. *Open Interoperability Workshop on Enterprise Modelling and Ontologies for Interoperability*, (pp. 1-5).

Polze, A., Richling, J., Schwarz, J., & Malek, M. (1999). Towards predictable CORBA-based Web-services. *Proceedings 2nd IEEE International Symposium on Object-Oriented Real-Time Distributed Computing (ISORC '99)*, (pp. 182-191).

Radmand, P., Talevski, A., Petersen, S., & Carlsen, S. (2010). Comparison of industrial WSN standards. *4th IEEE International Conference on Digital Ecosystems and Technologies (DEST)*, (pp. 632-637).

Reverter, F., & Pallas-Areny, R. (2005). *Direct sensor-to-microcontroller interface circuits: Design and characterization*. Barcelona, Spain: Editorial Marcombo S.A.

Samaras, I., Gialelis, J., & Hassapis, G. (2009). Integrating wireless sensor networks into enterprise information systems by using Web services. *Third International Conference on Sensor Technologies and Applications SENSORCOMM*, (pp. 580-587).

Skonnard, J. (2003). *Understanding SOAP*. Retrieved January 17, 2011, from http://msdn.microsoft.com/en-us/library/ms995800.aspx

Sleman, A., & Moeller, R. (2008). Integration of wireless sensor network services into other home and industrial networks. *3rd International Conference on Information and Communication Technologies ICTTA*, (pp. 1–5).

Sliva, I. J. (2008). Technologies used in wireless sensor networks. *15th International Conference on Systems Signals and Image Processing IWSSIP*, (pp. 77-80).

IEEE Std. 802.15.4. (2006). *IEEE standard for Information Technology-Telecommunications and information exchange between systems-Local and metropolitan area networks-Specific wireless medium access control (MAC) and physical layer (PHY) specifications for low rate wireless personal area networks -Amendment of IEEE Std 802.15.4-2003.* IEEE Std 802.15.4-2006.

Std, I. E. E. E. 1451.0. (2007). Standard for a smart transducer interface for sensors and actuators - Common functions, communication protocols, and transducer electronic data sheet (TEDS) formats. *IEEE Std 1451.0-2007*, (pp. 1-335).

Std, I. E. E. E. 1451.5. (2007). Standard for a smart transducer interface for sensors and actuators wireless communication protocols and transducer electronic data sheet (TEDS) formats. *IEEE Std 1451.5-2007*, (pp. C1-236).

Std, I. E. E. E. 1451.7. (2010). Standard for smart transducer interface for sensors and actuators--Transducers to radio frequency identification (RFID) systems communication protocols and transducer electronic data sheet formats. *IEEE Std 1451.7-2010*, (pp. 1-99).

Yang, Q., & Miao, C. (2007). Semantic enhancement and ontology for interoperability of design information systems. *IEEE Conference on Emerging Technologies and Factory Automation*, (pp. 169-176).

Yurish, S. (2010). Sensors: Smart vs. intelligent. *Sensors & Transducers Journal*, *114*(3), I–VI.

ADDITIONAL READING

Berners-Lee, T., Handler, J., & Lassila, O. (2001). The semantic web, Scientific American, From: http://www.scientificamerican.com/article.cfm?id=the-semantic-web.

Brock, M., & Goscinski, A. (2008). Publishing dynamic state changes of resources through state aware WSDL. IEEE International Conference on Web Services, (pp.449-456).

Chen, C., & Helal, S. (2008). Sifting Through the Jungle of Sensor Standards. *IEEE Pervasive Computing / IEEE Computer Society [and] IEEE Communications Society*, *7*(4), 84–88. doi:10.1109/MPRV.2008.81

d'Aquin, M., Motta, E., Sabou, M., Angeletou, S., Gridinoc, L., Lopez, V., & Guidi, D. (2008). Toward a New Generation of Semantic Web Applications. *IEEE Intelligent Systems*, *23*(3), 20–28. doi:10.1109/MIS.2008.54

Frank, R. (2000). *Understanding Smart Sensors*. Artech House.

Haykin, S. (2005). Cognitive Radio: Brain-Empowered Wireless Communications. *IEEE Journal on Selected Areas in Communications*, *23*(2), 201–220. doi:10.1109/JSAC.2004.839380

Huebscher, M., & McCann, J. (2008). A survey of Autonomic Computing -Degrees, Models, and Applications. *ACM Computing Surveys*, *40*(3), 7–28. doi:10.1145/1380584.1380585

Hui, J., & Culler, D. (2008). 6LoWPAN: Extending IP to Low-Power, Wireless Personal Area Networks. *IEEE Internet Computing*, *12*, 37–45. doi:10.1109/MIC.2008.79

Internet Engineering Task Force (IETF). (2010). IPv6 over Low power WPAN (6lowpan). From: URL: http://www.ietf.org/dyn/wg/charter/6lowpan-charter.html

Manso, M., Bernabe, M., & Wachowicz, M. (2009). Toward an integrated model of interoperability for spatial data infrastructures. *Transactions in GIS*, *13*, 43–67. doi:10.1111/j.1467-9671.2009.01143.x

Mitola, J. III, & Maguire, G. (1999). Cognitive radio: making software radios more personal. *IEEE Personal Communications*, *6*(4), 13–18. doi:10.1109/98.788210

Moller, S., Newe, T., & Lochmann, S. (2009). Review of platforms and security protocols suitable for wireless sensor networks. *IEEE Sensors conference*, (pp.1000-1003).

Mozek, M., Vrtacnik, D., Resnik, D., Aljancic, U., Penic, S., & Amon, S. (2008). Digital self-learning calibration system for smart sensors. *Sensors and Actuators. A, Physical*, *141*(Issue 1), 101–108. doi:10.1016/j.sna.2007.07.006

Nakamura, E., Loureiro, A., & Frery, A. (2007). Information fusion for wireless sensor networks: Methods, models, and classifications. *ACM Computing Surveys*, *39*(issue 3), 9–55. doi:10.1145/1267070.1267073

Ohlsen, S., & Buschmann, C. Werner, C.(2008). Integrating a Decentralized Web Service Discovery System into the Internet Infrastructure. *IEEE European Conference on Web Services*, (pp.13-20).

Petrie, C. (2009). Practical Web Services. *IEEE Internet Computing*, *13*(6), 93–96. doi:10.1109/MIC.2009.135

Seidler, K. (2011). Apache Friends. XAMPP for Linux. Retrieved January 22, 2011, from http://www.apachefriends.org/en/xampp-linux.html

Song, E., & Lee, K. (2008). Sensor Network based on IEEE 1451.0 and IEEE p1451.2-RS232. *IEEE Instrumentation and Measurement Technology Conference Proceedings, IMTC '08*, (pp. 1728-1733).

Standard ISA-100. 11a. (2009). International Society of Automation ISA. Wireless systems for industrial automation: Process control and related applications, *Std ISA-100.11a* (pp. 1-816).

Sweetser, D., & Sweetser, V. Nemeth-Johannes, J. (2006). Modular approach to IEEE-1451.5 wireless sensor development. *IEEE Sensors Applications Symposium*, (pp. 82- 87).

Teilans, A., Kleins, A., Sukovskis, U., Merkuryev, Y., & Meirans, I. (2008). A Meta-Model Based Approach to UML Modeling. *Tenth International Conference on Computer Modeling and Simulation, UKSIM*, (pp. 667-672).

Weiser, M. (1998). The future of ubiquitous computing on campus. *Communications of the ACM*, *41*, 41–42. doi:10.1145/268092.268108

Wobschall, D.(2007). IEEE 1451—a universal transducer protocol standard. *IEEE Autotestcon*, '07, pp.(359-363).

Xinyang, F., Jianjing, S., & Ying, F. (2009). REST: An alternative to RPC for Web services architecture. *First International Conference on Future Information Networks*, (pp.7-10).

Yang, Q. Miao, C.(2007). Semantic enhancement and ontology for interoperability of design information systems. *IEEE Conference on Emerging Technologies and Factory Automation*. (pp.169-176).

Chapter 5
Literature Review of MAC, Routing and Cross Layer Design Protocols for WSN

Tayseer A. Al-Khdour
King Faisal University, Saudi Arabia

Uthman Baroudi
King Fahd University of Petroleum and Minerals, Saudi Arabia

ABSTRACT

In this chapter, the authors present a literature review for MAC, routing, and cross layer design protocols proposed for WSN. This chapter consists of three sections. In the first section, the authors discuss in depth the most well-known MAC protocols for WSN. A comparison among theses protocols will be presented. Moreover, the major advantages and disadvantages of each protocol are discussed. The routing protocols for WSN are discussed in the second section. The discussed protocols are classified into data centric routing protocols, Hierarchical routing protocols, location based protocols, and QoS aware routing protocols. Moreover, A Classification of Routing Protocols based on the Application is presented in this section. In the third section, some cross layer design protocols are discussed. A comparison among the discussed protocols according to layers integrated, intended applications, cross-layer objectives, and the evaluation approach, is presented.

INTRODUCTION

Although a WSN is a wireless multi-hop network, it has distinguished features over the traditional multi-hop wireless networks. These features are related to the deployment of sensor nodes, the resources constraint, and the QoS requirement of the WSN. These distinguished features must be considered when designing different protocols that control the operation of WSN such as MAC protocols and routing protocols. Sensor nodes are usually deployed randomly. There is not a pre-defined infrastructure of the established network.

DOI: 10.4018/978-1-4666-0101-7.ch005

Therefore, sensor nodes must organize themselves autonomously. Self organizing techniques must be integrated within MAC and routing protocols. On the other hand, Existing of many nodes close to each other generates redundancy in the data gathered from the environment. Therefore, it is not required to transmit all the sensed data, data aggregation can be performed at each intermediate sensor node. Data aggregation will reduce the data traffic in the network. The aggregation function depends on the application. Sensor node has limited resources, for example, all the sensor nodes have a limited power supply (batteries). In most WSN applications, all the sensor nodes are out of control, it is impossible to replace or recharge these batteries. All the control protocols of the WSN must be designed taking into account the energy constraint. These protocols must be energy efficient. Another example of limited resources in the sensor node is the radio transceiver. The transmission range for radio transceiver of the sensor node is very limited. Therefore, not all the sensor nodes in WSN can hear all other nodes. All sensor nodes must collaborate to transfer data from the source sensor node to the sink. Different WSN applications needs different QoS requirement. For example, Data latency in WSN is very critical in some applications and not so much in other applications. Taking into account these distinguished features, a lot of MAC, routing and cross layer design protocols for WSN are proposed.

A WSN is composed of a large number of sensor nodes that are communicated using a wireless medium (air) as shown in Figure **1**. The sensor nodes are deployed in the environment to be monitored in ad hoc structure. In WSN, there is sink node that collects data from all sensors, and usually not all nodes hear all other nodes. WSN is considered a multi-hop network.

Figure 1. A wireless sensor network

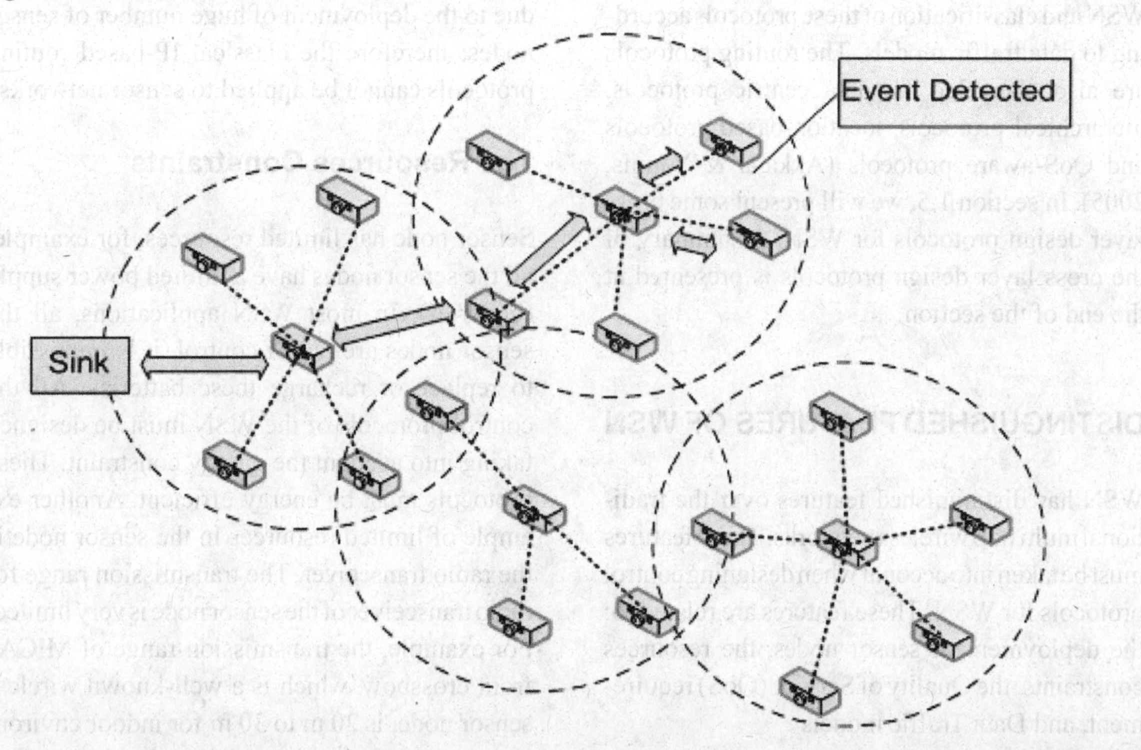

Although a WSN is a wireless multi-hop network, it has distinguished features over the traditional multi-hop wireless networks. These features are related to the ease of deployment of sensor nodes, the system lifetime, the data latency, and the quality of the network. These features must be taken into account when designing different protocols that control the operation of WSN such as MAC protocols and routing protocols. Therefore, Many MAC and Routing protocols are proposed for WSN. These protocols take into account the distinguished features of WSN. Moreover, Cross layer design protocols are proposed for WSN. In cross layer design protocols, different layers interact to optimize the performance of the WSN protocol.

In this chapter, we will present a survey of the most well known protocols for WSN. In section 1.2, we will present the distinguished features of WSN that motivate proposing control protocols dedicated to WSN. A survey of the most well-known MAC protocols is presented in 1.3. Section 1.4 presents discussion of routing protocols of WSN and classification of these protocols according to data traffic models. The routing protocols are also classified as: data centric protocols, hierarchical protocols, location-based protocols and QoS-aware protocols (Akkaya & Younis, 2005). In section 1.5, we will present some cross layer design protocols for WSN. A summary of the cross layer design protocols is presented at the end of the section.

DISTINGUISHED FEATURES OF WSN

WSN has distinguished features over the traditional multi hop wireless networks. These features must be taken into account when designing control protocols for WSN. These features are related to: the deployment of sensor nodes, the resources constraints, the Quality of Service (QoS) requirement, and Data Traffic models

The Deployment of Sensor Nodes

Sensor nodes are usually deployed randomly. There is not a predefined infrastructure of the established network. Some of the nodes are very close to each other, and other nodes may be farther than others. Existing of many nodes close to each other generates redundancy in the data gathered from the environment. Therefore, it is not required to transmit all the sensed data, data aggregation can be performed at each intermediate sensor node. Data aggregation will reduce the data traffic in the network. The aggregation function depends on the application. Examples of aggregation functions are SUM, AVG, MEAN and MAX. For example; if it is required to measure the maximum temperature in the monitored area, the MAX function will be used as an aggregation function. Unlike the traditional ad hoc network, the number of sensor nodes in the WSN is very large. Hundreds or even thousands of sensor nodes are usually deployed in the monitored environment. It is not possible to build a global addressing scheme due to the deployment of huge number of sensor nodes; therefore the classical IP-based routing protocols cannot be applied to sensor networks.

The Resources Constraints

Sensor node has limited resources, for example, all the sensor nodes have a limited power supply (batteries). In most WSN applications, all the sensor nodes are out of control, it is impossible to replace or recharge these batteries. All the control protocols of the WSN must be designed taking into account the energy constraint. These protocols must be energy efficient. Another example of limited resources in the sensor node is the radio transceiver. The transmission range for radio transceiver of the sensor node is very limited. For example, the transmission range of MICAz from crossbow, which is a well-known wireless sensor node, is 20 m to 30 m for indoor environment, and it is less than 100m for the outdoor environment (Sensor Modules, n.d.). Therefore, not all the sensor nodes in WSN can hear all other nodes. All the sensor nodes must collaborate to transfer data from the source sensor node to the sink. The limited transmission range of the sensor node must be considered when designing routing protocols for WSN.

The Quality of Service (QoS) Requirement

Different WSN applications needs different QoS requirement. For example, Data latency in WSN is very critical in some applications and not so much in other applications. For example, if the status of sensed object is changing very fast, then the data latency must be very low. Otherwise, the data collected from the environment will not be valid when it reaches the sink node. In other applications, the events that are monitored in the environment may continue for sometime and they are not so frequent; it takes a lot of time between events. In this case, the data latency can be high. On the other hand, In WSN, the end user does not need all the data in network. Data collected from neighboring node will be highly correlated. It is not required to send all these data to the end user. Usually, a user requires a high-level description of events being monitored in the environment.

Data Traffic Models

In WSN, data traffic can be classified into: periodic, event-driven, and query based. In periodic traffic, the sensor nodes send their measurements to the sink once every fixed time interval. In this model, all sensor nodes must be synchronized. In the event driven model, data traffic flows in the network when a special event is detected. The events must be reported immediately to the sink when they are detected. In the query-based model, data traffic flows from sensor to sink in response to the query generated from the sink. The sink generates a specific query then the relevant sensor node will respond to the query with the requested data. A route must be computed for the query and for the data transmission. In most applications of WSN, data traffic usually flows from multiple sources to one destination, which is not the case in the traditional ad hoc wireless network.

MAC PROTOCOLS FOR WSN

In designing a MAC protocol for a Wireless Sensor Network (WSN), some of the unique features of WSN must be taken into consideration. Low-power consumption must be the main goal of the protocol. The coordination and synchronization between nodes must be minimized in the protocol. The MAC protocol must be able to support a large number of nodes. It must have a high degree of scalability. The MAC protocol must take into account the limited bandwidth availability. Since sensor nodes of a WSN are deployed randomly without a predefined infrastructure, the first objective of the MAC protocol for a WSN is the creation of the network infrastructure. The second objective is to share the medium communication between the sensor nodes (Akylidis et al., 2002).

IEEE 802.11 is a well-known MAC protocol for Ad hoc network (IEEE, 1999). In IEEE 802.11 protocol, each node will be in one of the three states, sending, idle-listening, and receiving. In the idle-listening state, the node does not do anything except sensing the medium to check if any node sends RTS to it. Sensing the medium will consume a power. The energy constraints in the sensor nodes make it is unpractical to apply the IEEE 802.11 protocol directly in WSN. IEEE 802.11 has a power save mode. The power save mode in IEEE 802.11 is designed for a single hop network, where all nodes can hear each other. This is not the case in WSN.

A set of MAC protocols for the WSN were proposed. Most of the existing protocols aimed to save power consumption in the sensor nodes. Since most of WSN MAC protocols aimed to reduce the power consumption we will describe firstly a Power Aware Multi-Access protocol with signaling for Ad hoc Networks (PAMAS)(Singh & Raghavendra, 1998). We explain this protocol because it is used as a base for some WSN MAC protocols.

Power Aware Multi-Access Protocol with Signaling for Ad Hoc Networks (PAMAS) Protocol

Singh et al propose PAMAS(Singh & Raghavendra, 1998). PAMAS is a channel access protocol that reduces the power consumption at each of the nodes in a general Ad-Hoc network. In an Ad Hoc network, if a node transmits a packet all the close nodes will hear its transmissions even if the packet is intended to one of them only. Overhearing the unwanted packet by these nodes will consume a power without gaining any useful data. PAMAS aims to reduce the power consumed due to receiving packets that are intended to another node. PAMAS protocol is a combination of the MACA protocol and the idea of using a separate signaling channel. In PAMAS the RTS-CTS message exchange occurs over a signaling channel (control channel) separated from the packet transmission channel. The separate channel enables the nodes to determine when and for how long they can power themselves off. Each node in PAMAS protocol can be in one of six states:

- **Idle:** the node currently does not transmit or receive and it has not packet to send, but its RF receiver is on to be able to receive packets.
- **Await packet:** the node is waiting for data packets from a source node after it transmits Clear-To-Send (CTS) packet to that node. The node firstly receives a Request-To-Send (RTS) from the source node then it waits one time step before replying with CTS.
- **Await CTS:** the node sent RTS and it is currently waiting for the CTS.
- **Receive packet:** the node is currently receiving a packet.
- **BEB:** Binary Exponential Backoff time; the node is waiting for random backoff time.
- **Transmit packet:** the node is currently transmitting a packet.

Initially a node will be in the idle state. Upon receiving an RTS message, the node will wait for one time step, if no neighbor node is currently transmitting, i.e. it can receive data without causing collision, and then the node sends the CTS message. The node waits for a one time step to ensure that no neighbor node is waiting for the CTS form another node. After sending CTS, the node will go to the Await packet state. If it receives a new RTS packet, the node will resend the CTS packet. If the data packets start arriving the node will transmit a busy tone over the control channel and it will go to the Receive packet state. If the node receives an RTS directed to another node over the control channel while it is in the Receive packet state, the node will transmit a busy tone over the control channel. The node that sends the RTS will be blocked when it hears the busy tone. When all the packets are received, the node will go back to the idle state. If the node is in the Await packet state, and it does not receive data packets within the expected time (round trip time to the transmitter plus some processing delay at the receiver) or it receives a noise, it will go back to the idle state.

If the node is in the idle state, and it receives packets from upper layer, and its queue size becomes greater than 0, it will send RTS to destination node and goes to Await CTS state. If it receives the CTS, it goes to Transmit packet state and it starts transmitting its data. While transmitting data the node will ignore RTS/CTS packets. When the transmission ends, the node goes back to the idle state. If the node receives RTS while it is in Await CTS state, it will send CTS and it will go to the Await packet state. If the node in the Await CTS state and it does not receive CTS within the expected time or it receives a busy tone, it will go to BEB state. It selects a random backoff time for waiting. When the selected random time is expired, the node will send RTS and go back to the Await CTS state. If the node receives RTS while it is in the BEB state, it will send CTS and go to the Await packet state.

A node will turn itself OFF, it cannot receive or send data, if it has no packet to transmit and if one of its neighbors starts to transmit to another node, or if one of the neighboring nodes is transmitting and another one is receiving. The node knows that one of its neighbors is transmitting by hearing data in the transmission channel. It knows that one of its neighbors is receiving by hearing the busy tone in the control channel. To determine a node's OFF duration, there will be two cases, the first case; a neighbor of the node starts transmission while the node is ON, the node determines the duration of its OFF period from the RTS duration field. The second case; its neighbor starts transmission while the node is OFF, the node will uses a probe protocol to determine the OFF duration(Singh & Raghavendra, 1998). If a node wishes to transmit a packet to its neighbor while that node turns itself OFF, then the transmitter node must wait until its neighbor wakes up. This will not increase the delay since the destination node is OFF because one of its neighbors is currently transmitting so it cannot receive data correctly.

Singh et al. measure the performance of the PAMAS by simulation for different network topology: random, line, and fully connected with dense and spare networks. They note that there is a large energy saving in PAMAS. They also derive bounds and approximations on energy saving in the different network topology.

Although there is power saving in nodes due to power off mode on nodes, using a separate channel to transmit control messages will cost some power. This is considered as a disadvantage of the PAMAS. PAMAS is extended by Wei Ye and others to propose the Sensor MAC protocol S-MAC (Ye, Heidemann, & Estrin, 2004).

S-MAC Protocol

The main goal of S-MAC protocol is to reduce energy consumption while supporting good scalability and collision avoidance. Wei Ye et al extend PAMAS protocol by using a single channel for transmitting data packets and control packets. In designing S-MAC protocol they assume the following about sensor networks and applications:

- Sensor network composed of many small nodes deployed in an Ad Hoc fashion. They take the advantage of physical proximity to simplify signal processing.
- Most communication will be between nodes as peers rather than one base station.
- The sensor nodes must be self configured.
- The sensor network is dedicated to a single application or a few collaborative applications. The focus will be on maximizing system-wide application performance. The single packet delay will be a secondary goal.
- The sensor network has the ability of in-network processing. Data aggregation can reduce the traffic in the network.
- The intended application can tolerate some latency. The monitored object has long idle periods.

Ye et al identify four sources for energy wasting. The first source is collisions which will cause retransmission the packet. Transmission will consume power. The second source is overhearing; picking a packet intended to another node. The third source of energy consumption is transmission of control packets. The final source of energy consumption is idle listening. S-MAC reduces the energy waste due to these reasons. The basic idea of S-MAC is to let the node sleep and listen periodically. In sleeping mode, the node turns its radio off. The listening period is fixed according to physical layer and MAC layer parameters. The complete cycle of listening and sleeping periods is called a frame. The duty cycle is defined as the ratio of the listening interval to the frame length. Neighboring nodes can be scheduled to listen and sleep at the same time. Two neighboring nodes may have different schedules if they are synchronized by different two nodes. Nodes exchange their schedule by broadcasting a SYNC packet to their immediate neighbors. The period to send a SYNC packet is called the synchronization period. If a node wishes to transmit a packet to its neighbor it must wait until its neighbor becomes in its listening period. Figure 2 shows 4 neighboring nodes A, B, C, and D. Nodes A and C are synchronized together (they have the same schedule, they listen and they sleep at the same time) while nodes B and D are synchronized together.

S-MAC forms nodes into a flat, peer-to-peer topology. To choose a schedule the node firstly listens for a fixed amount of time (at least the synchronization period). If the node does not receive a schedule within the synchronization period, the node chooses its own schedule and starts to follow it, and then it announces its schedule to its neighbors by broadcasting the SYNC packet. If it hears a schedule from one of its neighbors before it chooses or announces its own schedule, it follows that schedule. If a node receives a different schedule after it announces its own schedule, then there will be two cases, in the first case, the node has not other neighbors, then it discard its own schedule and it will follow the new schedule. In the second case, the node already follows a schedule with one of its neighbors; therefore it will adopt both schedules by waking up at the listening intervals of the two schedules. To maintain the schedule, each node maintains a schedule table that stores the schedules of all its known neighbors. To prevent case two in which neighbors miss each other forever when they follow two different schedules, a periodic neighbor discovery is introduced. Each node periodically listens for the whole synchronization period. Figure 3 shows the timing relationship of three possible situations.

If multiple nodes wish to talk to the same node that is in listening period, then all of them must contend for the medium. IEEE 802.11 scheme with RTS and CTS is used to avoid collision, which will save energy consumption due to the packets collision and retransmissions.

To avoid overhearing which is one of the sources of energy consumptions, each interfering nodes must go to sleep after they hear RTS and CTS. All immediate neighbors of both sender and receiver should sleep after they hear RTS or CTS. To reduce the delay due to sleeping, a technique called adaptive listening is integrated in S-MAC. Each node will wake up for a short period at the end of the transmission. In this way, if the node is the next-hop node, its neighbor is able to pass the data immediately to it instead of waiting for its scheduled listening time.

To reduce energy due to control packet overhead a message passing technique is included in S-MAC. If a node wishes to transmit a long message, the long message is fragmented into fragments and the node will transmit them in burst; one RTS and one CTS are used for all the

Figure 2. S-MAC: Neighboring nodes A and B have different schedules. They synchronize with nodes C and D, respectively.

Figure 3. Timing relationship between a receiver and different senders, CS stands for carrier sense

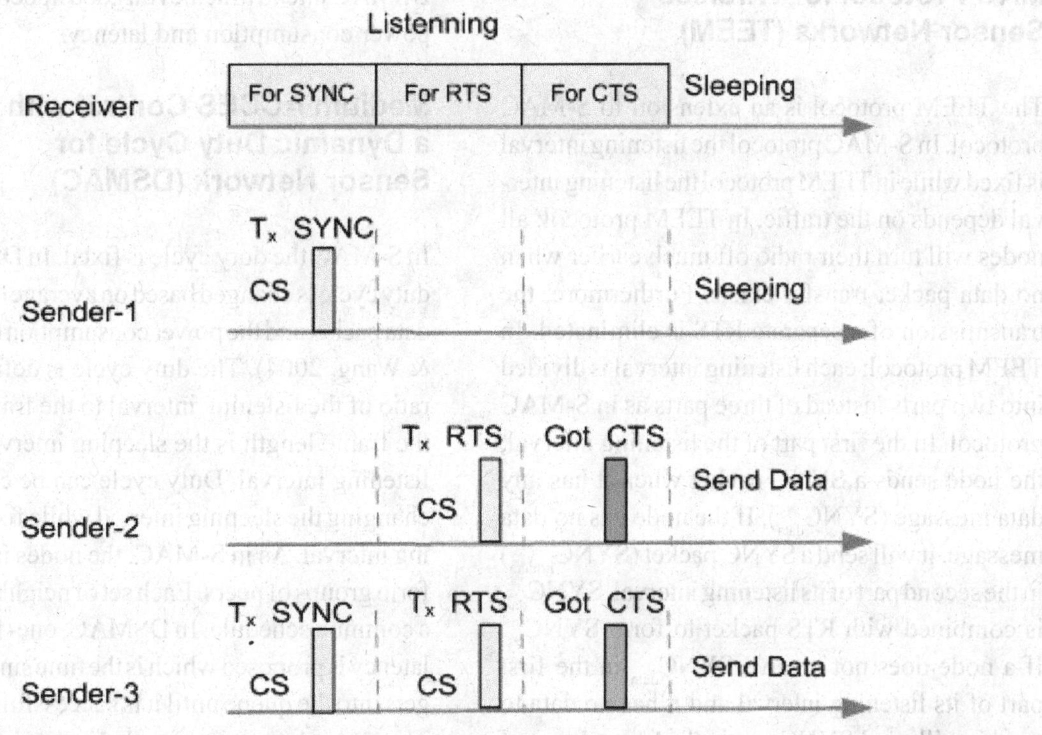

fragments. When a node sends data, it waits for ACK. The ACK is useful to solve the hidden terminal problem. Data fragment and ACK packets have a duration field. If a node wakes up or joins the network and it receives a data or ACK packet, it will go to sleep for the period in the duration field in data or ACK packet.

Synchronization among neighboring nodes is required to remedy their clock drift. Synchronization is achieved by making all nodes exchange a relative timestamps and letting the listening period is longer than clock drift.

The average latency of S-MAC without adaptive listening over N hops is

$$E[D(N)] = NT_f - T_f + t_{cs} + t_{tx} \quad (1)$$

While the average latency of S-MAC with adaptive listening over N hops is

$$E[D(N)] = \frac{NT_f}{2} - T_f + 2t_{cs} + 2t_{tx} \quad (2)$$

Where

N: Number of hops.

T_f: frame time (complete cycle of listening and sleep)

t_{cs}: The carries sense delay

t_{tx}: The transmission delay for a packet with fixed length.

A disadvantage of S-MAC is that the listening interval is fixed regardless whether the node has data to send or there are data intended to it. Suh et al proposed a Traffic Aware, Energy Efficient MAC protocol for Wireless Sensor Networks (TEEM) (Suh & Ko, 2005). They extend the S-MAC protocol by reducing the listening interval.

A Traffic Aware, Energy Efficient MAC Protocol for Wireless Sensor Networks (TEEM)

The TEEM protocol is an extension to S-MAC protocol. In S-MAC protocol the listening interval is fixed while in TEEM protocol the listening interval depends on the traffic. In TEEM protocol; all nodes will turn their radio off much earlier when no data packet transfer exists. Furthermore, the transmission of a separate RTS is eliminated. In TEEM protocol; each listening interval is divided into two parts instead of three parts as in S-MAC protocol. In the first part of the listening interval, the node sends a SYNC packet when it has any data message ($SYNC_{data}$). If the node has no data message, it will send a SYNC packet ($SYNC_{nodata}$) in the second part of its listening interval. $SYNC_{data}$ is combined with RTS packet to form $SYNC_{rts}$. If a node does not receive $SYNC_{data}$ in the first part of its listening interval and it has no data to send it will send $SYNC_{nodata}$ in the second part of its listening interval. If a node receives a $SYNC_{rts}$ that is intended to another node, it will turn its radio off and goes to sleep until its successive listening interval starts. The intended receiver will send CTS in the second part of its listening interval. The performance evaluation of TEEM protocol shows that the percentage of sleeping time in TEEM is greater than the percentage of sleeping time in S-MAC. The number of control packets in TEEM protocol is less than the number of control packets in S-MAC protocol. Energy consumption in TEEM protocol is the least compared with S-MAC and IEEE 802.11. Although the power consumption is reduced in the TEEM protocol by decreasing the listening interval, the latency will increase since decreasing the listening interval depends only on the local traffic, traffic in the node itself and in the neighboring node, and does not take into account the traffic in the whole network. To take into account the delay in the whole network, Lin et al propose a sensor medium access control protocol with a dynamic duty cycle, DSMAC(Lin, Qiao, & Wang, 2004). DSMAC intend to achieve a good tradeoff between power consumption and latency.

Medium ACCES Control with a Dynamic Duty Cycle for Sensor Network (DSMAC)

In S-MAC the duty cycle is fixed. In DSMAC the duty cycle is changed based on average delay of the data packet and the power consumption (Lin, Qiao, & Wang, 2004). The duty cycle is defined as the ratio of the listening interval to the frame length; the frame length is the sleeping interval plus the listening interval. Duty cycle can be changed by changing the sleeping interval while fixing listening interval. As in S-MAC, the nodes in DSMAC form groups of peers. Each set of neighbors follow a common schedule. In DSMAC, one- hop packet latency is proposed which is the time since a packet gets into the queue until it is successfully sent out. The packet latency is recorded in the packet header and sent to the receiver. The receiver calculates the average packet latency. The average packet latency is an estimation of the current traffic. If the average packet latency is larger than a threshold delay (D_{max}), and if the energy consumption level greater than a threshold energy (E_{max}), then the duty cycle will be doubled by decreasing the sleeping interval such that the new frame length is half of the original frame length. Otherwise the duty cycle will be halved by doubling the sleeping interval, doubling the sleeping interval will double frame length. The purpose of changing the duty cycle by two (or half) is to maintain the old schedule, which enables neighboring nodes to communicate using the old schedule. Figure 4 shows the schedule before and after doubling the duty cycle.

It is shown analytically in (Lin, Qiao, & Wang, 2004) that the average delay in the case of one hop in DSMAC is less than the average delay in S-MAC. It is shown also that the average delay

Figure 4. Neighboring nodes which adopt different duty cycles can still communicate with old schedule

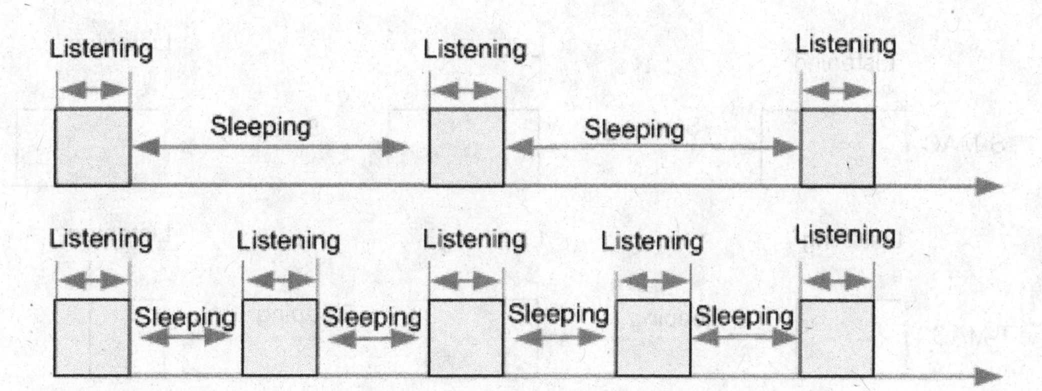

in the case of multiple hops with adaptive listening is:

$$E[D(N)] = \frac{NT_f}{8} - \frac{T_f}{8} + 2t_{cs} + 2t_{tx} \quad (3)$$

Comparing equation 3 with equation 2 we note that the delay in DSMAC is less than the delay in S-MAC.

Comparing DSMAC with T-MAC protocol, we note that the purpose of the DSMAC protocol is to decrease the delay by decreasing the sleeping interval between the listening intervals. This may increase the power consumption. On the other hand in TEEM protocol the main objective is decrease the power consumption by making the node sleep earlier and maintaining the frame length fixed. The delay will not be decreased in the TEEM protocol

In S-MAC protocol, the listening and sleeping intervals are fixed. Dam et al propose a Timeout-MAC (T-MAC) protocol (Dam & Langendoen, 2003). In T-MAC, the listening interval of the node may end earlier in some situations.

Timeout-MAC (T-MAC)

In T-MAC protocol, the node will keep listening and transmitting as long as it is in an active period. An active period ends when no activation event has occurred for a specific time TA. An activation event may be firing of a periodic frame timer, reception of any data on the radio, sensing of communication on the radio, end-of-transmission of a node's own data packet or acknowledgement, or the knowledge that a data exchange of a neighbor has ended. A comparison of sleeping and listening periods of the S-MAC and T–MAC is shown in Figure 5

Communications between nodes in T-MAC is performed using RTS/CTS mechanism. The node that wishes to transmit data must send an RTS and wait for the CTS. If it does not receive CTS within the TA period the node will go to sleep. The node does not receive CTS in three cases; the receiver has not received the RTS, the receiver receives RTS but it is prohibited from replying, or the receiver is sleeping. It is accepted and recommended for the node to go to sleeping in the third case. But it is not an optimal decision to go to sleeping in the first two cases. To take into account all the three cases; when the node does not receive CTS to the first RTS it will resend another RTS and if it does not receive a response to the second RTS then it will go to sleeping. Sending two RTS packets without getting a CTS indicates that the receiver cannot reply now so it is convenient for the sender to go to sleeping. TA

Figure 5. Listening sleeping periods in both S-MAC and T-MAC

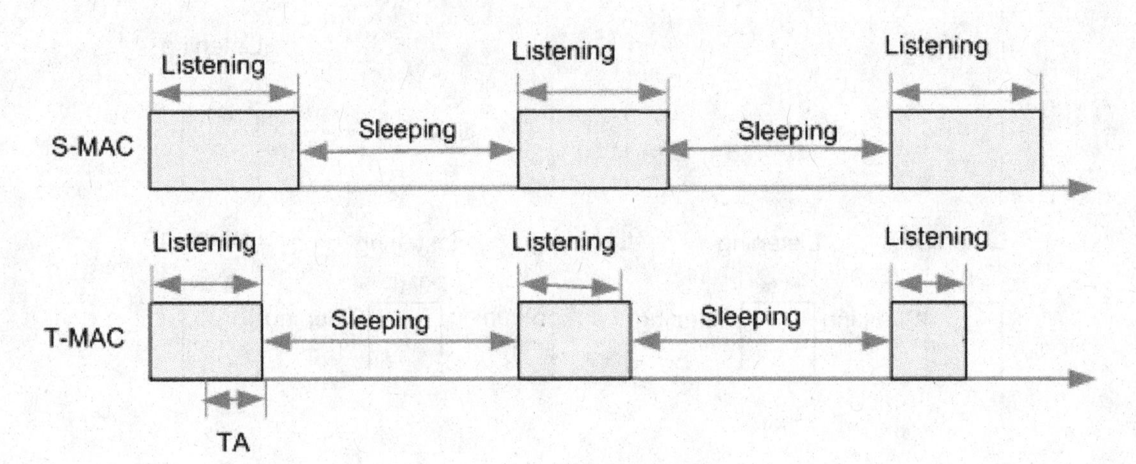

must be long enough to receive at least the start of the CTS packet. So TA must be

$$TA>C+R+T \quad (4)$$

Where C is the length of contention interval, R is the length of an RTS packet, and T is the turn-around time. Overhearing avoidance is achieved by the same technique used in S-MAC protocol.

One problem of the T-MAC is the early sleeping problem, which occurs in case of asymmetric communication where there are four consecutive nodes: A, B, C, and D as shown in Figure 6. node A sends data to B which its final destination is C, at the same time C wishes to send data to node D but it cannot transmit data since a collision will occur at node B with the transmission from A to B, so node C will go to sleeping. Moreover, node D will go to sleeping. Later when node B wishes to forward the data to node C, it will find that node C is sleeping which will make node B to go to sleeping and transmit its data later which will increase the delay and decrease the throughput. Two solutions are proposed: future request-to-send and taking priority on full buffers (Dam & Langendoen, 2003).

The main idea of the future request-to-send technique is to let another node know that we still have message for it but we are prohibited from using the medium (Dam & Langendoen, 2003). If a node overhears a CTS packet destined for another node it may send a future-request-to-send (FRTS) packet to that node, for example, node C sends FRTS to node D as shown in Figure 7. A node that receives an FRTS packet will know that it will be the future target for an RTS packet and must be awake up by that time.

The second technique for the early sleeping problem is called the full-buffer priority. In this technique, when a node's transmit/routing buffers are almost full, it will prefer sending to receiving. When this node receives an RTS packet destined for it, it sends its own RTS packet to another node instead of replying with a CTS packet. In the previous example, when node C receives an RTS packet from node B it will send an RTS to node D instead of replying with CTS to node B as shown in Figure 8.

There are some applications, in which most of the traffic in the nodes is a forwarding traffic. For these network models, Biaz et al propose a MAC protocol (GANGS) in which the nodes are organized into clusters (Biaz & Barowski, 2004). The communication within the cluster is contention based and the communication between cluster heads is TDMA based.

Figure 6. The early sleeping problem. Node D goes to sleep before C can send RTS to it

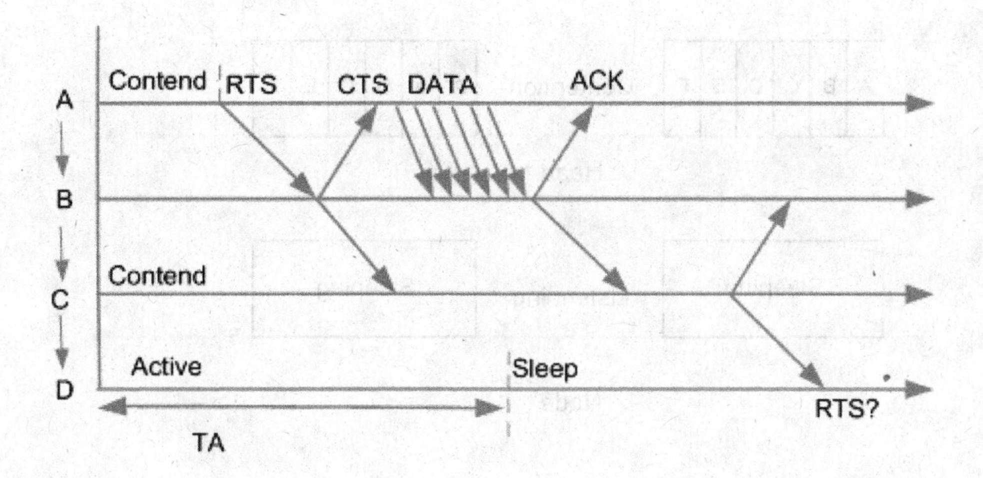

Figure 7. The future-request-to-send packet exchange keeps Node D awake

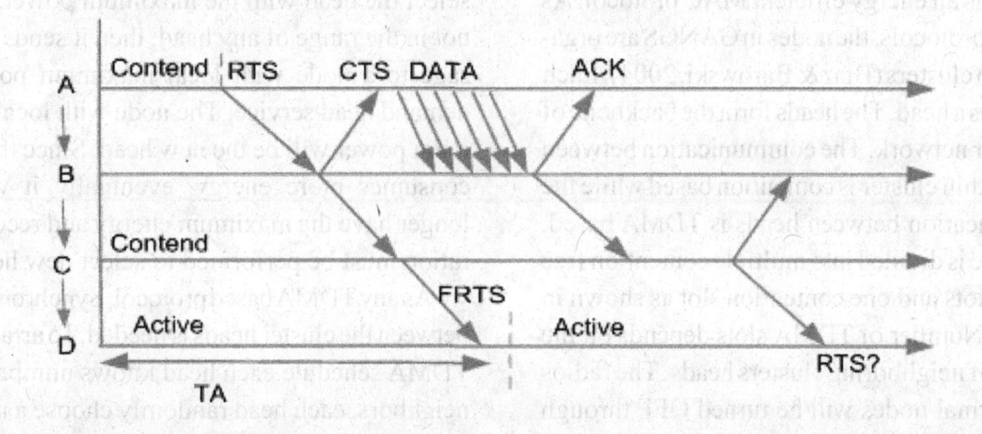

Figure 8. Taking priority upon receiving RTS

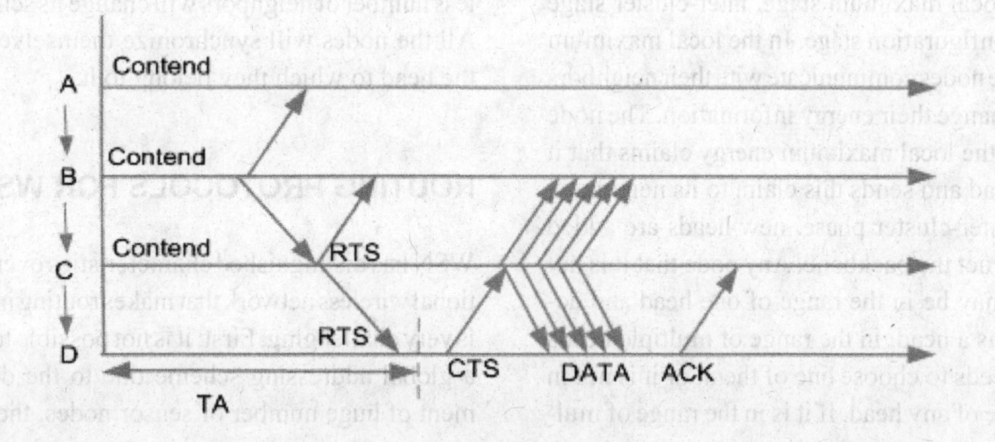

Figure 9. Time frame for cluster head/node

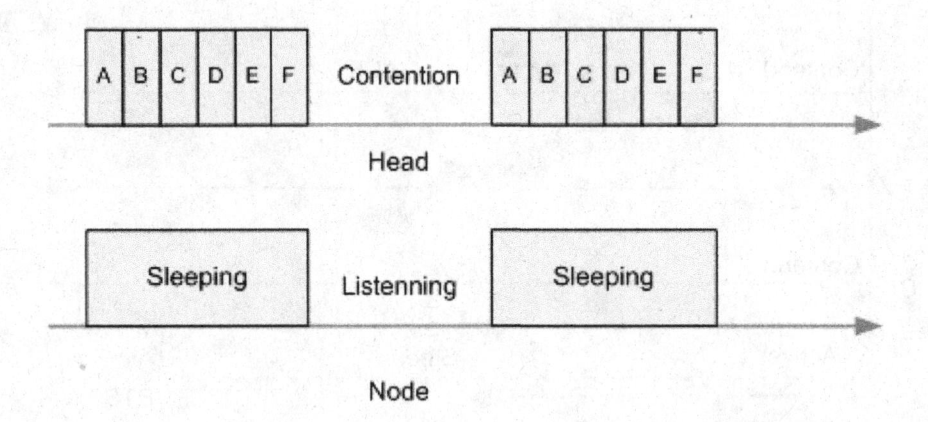

GANGS Protocol

GANGS is an energy efficient MAC protocol. As the other protocols, the nodes in GANGS are organized into clusters (Biaz & Barowski, 2004). Each cluster has a head. The heads form the backbone of the sensor network. The communication between nodes within cluster is contention based while the communication between heads is TDMA based. The frame is divided into multiple contention free TDMA slots and one contention slot as shown in Figure 9 Number of TDMA slots depends on the number of neighboring clusters heads. The radios of all normal nodes will be turned OFF through TDMA slots while the radios of all heads are turned ON through the entire frame.

Establishing the cluster consists of three stages: local maximum stage, inter-cluster stage and reconfiguration stage. In the local maximum stage, the nodes communicate with their neighbors and exchange their energy information. The node that has the local maximum energy claims that it is the head and sends this claim to its neighbors. In the Inter-cluster phase, new heads are added to construct the backbone. Any node that it is not a head may be in the range of one head and accepts it as a head, in the range of multiple heads and it needs to choose one of them, or it is not in the range of any head. If it is in the range of multiple heads and if it has a maximum energy, then it will be the new head, otherwise the node will select the head with the maximum power. If it is not in the range of any head, then it sends a message to a node with local maximum power to demand head service. The node with local maximum power will be the new head. Since the head consumes more energy, eventually it will no longer have the maximum energy and reconfiguration must be performed to select new heads.

As any TDMA based protocol, Synchronization between the cluster heads is needed. To arrange the TDMA schedule each head knows number of its neighbors, each head randomly choose a number in the range [1, number of neighbors+1]. Each head sends the chosen number to its neighbors. If the chosen number is the same, the head with less number of neighbors will change its schedule. All the nodes will synchronize themselves with the head to which they belong to it.

ROUTING PROTOCOLS FOR WSN

WSN has distinguished characteristics over traditional wireless network that makes routing in WSN is very challenging. First; it is not possible to build a global addressing scheme due to the deployment of huge number of sensor nodes, therefore the classical IP-based routing protocols cannot be applied to sensor networks. Second, Most applications of the sensor networks require the data flow from multiple sources to a particular sink. Third, the generated data has significant traffic redundancy in it. Furthermore, sensor nodes have limited power resource and processing capacity. Due to such differences many routing protocols for WSN are proposed. The routing protocols are classified as data centric, hierarchical, or location based (Akkaya & Younis, 2005). Data-centric protocols are query-based and depend on naming of desired data. Hierarchical protocols aim at clustering the nodes so that cluster heads can do some aggregation and reduction of data to reduce energy. Location based protocols utilize the position information to relay data to the desired region rather than the whole network.

Flooding is a classical mechanism to relay data in sensor network without using any routing protocol. In flooding, each sensor node receives a data packet; it will broadcast data to all its neighbors (Hedetniemi & Liestman, 1988). Eventually the data packet will reach its destination. To reduce the data traffic in the network, gossiping is implemented in which a receiving node send packet to a randomly selected neighbors. In flooding and gossiping, a lot of energy is wasted due to unnecessary transmissions. In addition to energy loss, flooding and gossiping have many drawbacks such as implosion where duplicated message sent to the same node, and overlap where many nodes sense the same region and send similar packets to the same neighbors (Heinzelman, Kulik, & Balakrishnan, 1999).

Data-Centric Protocols

In data-centric routing protocol, the sink sends queries to specific regions and the sensor nodes located in the selected region will send the corresponding data to the sink (Akkaya & Younis, 2005). To specify the properties of the requested data, attribute-based naming is usually used. Many data centric routing protocols are proposed such as: SPIN (Heinzelman, Kulik, & Balakrishnan, 1999), Directed Diffusion(Intanagonwiwat, Govindan, & Estrin, 2000), Energy-aware routing (Shah & Rabaey, 2002), Rumor routing (Braginsky & Estrin, 2002), CADER(Chu, Haussecker, & Zhao, 2002), COUGAR (Yao & Gehrke, 2002), ACQUIRE (Sadagopan, Krishnamachari, & Helmy, 2003), and Gradient based routing (Schurgers & Srivastava, 2001).

Sensor Protocols for Information via Negotiation (SPIN)

In SPIN protocol, the data is named using high-level descriptor or meta-data(Heinzelman, Kulik, & Balakrishnan, 1999). When a node receives data, it will advertise the meta-data to all its neighbors by broadcasting an ADV message. The neighbors who do not have the advertised data, and interested in the data can reply with a REQ message to request it. In SPIN protocol, since a sensor node will request the desired data only, the problems of flooding such as data redundancy and resource blindness will be solved. In addition to reducing the data redundancy in SPIN, the energy consumption is reduced. However, SPIN protocol is not a reliable routing protocol. For example if the nodes that are interested in the data are far away from the source node and the intermediate nodes are not interested in that data, then the intermediate nodes will not request the data advertised by the source node. Therefore, the data will not arrive the interested node. The SPIN protocol is not suitable for applications that require reliable delivery of data packets such as intrusion detection.

Directed Diffusion

In Directed Diffusion, a naming scheme for the data is used; attribute-value pairs for the data are used (Intanagonwiwat, Govindan, & Estrin, 2000). The sensor nodes are queried on demand using attribute-value pairs. To create a query, an interest is defined using a list of attribute-value pairs such as name of objects, interval, duration and geographical area. The interest is broadcasted by the sink. Each node receives the interest will cache it along with the reply link to a neighbor from which the interest is received. The reply link which is called a gradient is characterized by data rate, duration and expiration time. To establish the path between the sink and source, each node will compare the attribute of received data with the values in the cached interest. Using the gradients, the receiving node will specify the outgoing link. Path repairs are possible in Directed Diffusion, when a path between a source and sink fails, a new path should be identified. Multiple paths are identified in advances so that when a path fails one of the alternative paths is chosen without any cost of searching for another path. Directed Diffusion has many advantages; since all communication is neighbor-to-neighbor there is no need for addressing mechanism. Using caching will reduce processing delay. Moreover, Direct Diffusion is energy efficient since the transmission is on demand and there is no need for maintaining global network topology. On the other hand, directed diffusion cannot be applied to all sensor network-application since it is based on query-driven data delivery model. It cannot be used for applications that require continues data delivery such as environmental monitoring. In addition, the data naming scheme used in Directed Diffusion is application dependent, it must be defined in advance.

Energy Aware Routing

Shah et al. proposed to use a set of sub-optimal paths occasionally to increase the lifetime of network (Shah & Rabaey, 2002). The paths are chosen by a means of probability function. The probability function depends on the energy consumption. Instead of using the minimum path energy all the time one of the multiple paths is used with a certain probability. In the proposed protocol, it is assumed that each node is addressable through a class-based addressing that includes the location and type of the node. The proposed protocol consists of three phases: Setup phase, Data Communication phase, and route maintenance phase. In setup phase, routes are found and forwarding tables are created. The total energy cost is calculated in each node. The destination node initiates the connection by flooding the network in the direction of the source node. It sets the cost field to 0 before sending the request. Each intermediate node forwards the request only to the neighbors that are closer to the source node than itself and farther away from the destination node. Upon receiving the request, the energy metric for the neighbor that sent the request is computed and is added to the total cost of the path. Paths that have very high cost are discarded and not added to the forwarding table. Each node assigns a probability to each of its neighbors in the forwarding table. The assigned probability is inversely proportional to the cost of the path. Each node will have a number of neighbors through which it can route packets to the destination. Each node will then calculate the average cost of reaching the destination using its neighbors. The average cost is set to the cost field in the request packet and sent to the source node.

In the Data Communication phase, each node sends the data packet to one of its neighbors, which is selected randomly. The probability to select a neighbor equals to the probability of the neighbor in the forwarding table that is assigned at the setup phase.

In the route maintenance phase, localized flooding is performed infrequently to keep all the paths alive.

Rumor Routing

Rumor Routing is another variation of the Directed Diffusion (Braginsky & Estrin, 2002). It is based on a query-driven data delivery model. In Rumor Routing, the queries are routed only to the nodes that have observed a particular Instead of querying the entire network as in Directed Diffusion. In Rumor Routing protocol, each node maintains a list of neighbors and events table with forwarding information to all the events it knows. When a node senses an event, it adds it to its event table with a distance of zero to the event, and it generates an agent. An agent is a long-lived packet that travels the network in order to propagate information about local events to all the nodes. The agent contains an events table similar to the table in the nodes. Any node may generate a query for an event; if the node has a route to the event, it will transmit the query. If it does not, it will forward the query in a random direction. This continues until the query TTL expires, or until the query reaches a node that has observed the target event. If the node that originated the query determines that the query did not reach a destination it can retransmit or flood the query.

Constrained Anisotropic Diffusion Routing (CADR)

CADR(Chu, Haussecker, & Zhao, 2002) is proposed as a general form of Directed Diffusion. The idea is to query sensors and route data in a network in order to maximize the information gain while minimizing the latency and bandwidth. Only the nodes that are close to a particular event will be active and routes will be adjusted dynamically. Two techniques are proposed: Information-driven sensor querying (IDSQ) and Constrained Anisotropic Routing (CADR). In CADR, each node evaluates an information/cost objective and routes data based on the local information/cost gradient and end-user requirement. In IDSQ, the querying node can determine which node can provide the most useful information while balancing the energy cost. IDSQ provides a way of selecting the optimal order of sensors for maximum incremental information gain.

COUGAR

In COUGAR protocol, the network is viewed as a huge distributed database system (Yao & Gehrke, 2002). Declarative queries are used to abstract query processing from the network layer functions. A new query layer between the network and application layers is proposed to support this abstraction. Architecture for the database systems is proposed where nodes select a leader node to perform aggregation and transmit the data to the sink. The sink generates a query plan, which specifies the necessary information about the data flow, and in-network computation for the incoming query and send it to the relevant nodes.

Active QUery Forwarding In SensoR nEtworks (ACQIRE)

In ACQIRE protocol, the sensor network is viewed as a distributed database that well suited for complex queries that consist of several sub queries(Sadagopan, Krishnamachari, & Helmy, 2003). The sink broadcast the query to the network. Each node receiving the query tries to respond partially by using its pre-cached information and forward it to another sensor. If the information stored in the cache is not up-to-date, the node gathers information from its neighbors within a look-ahead of d hops. Once the query is resolved completely, it is sent back either through the reverse path or through shortest path to the sink.

Gradient-Based Routing (GBR)

GBR (Schurgers & Srivastava, 2001) is a slightly changed version of the Directed Diffusion, when the interest is diffused through the network, the number of hops to the sink is kept at each intermediate node. Each node can discover the minimum number of hops to the sink, which is called height of the node. The difference between a node's height and that of its neighbor is considered as the gradient on that link. At each intermediate node, a packet is forwarded on a link with the largest gradient. Data aggregation and traffic spreading are used with GBR to balance the traffic over the network. Three different data spreading techniques are presented; Stochastic scheme, Energy-based scheme and Stream-based scheme. Stochastic scheme is used when more than one next hop has the same gradient; the node chooses one of them randomly. In energy-based scheme, when a node's energy drops below a certain threshold, it increases its height so other nodes will not select it to send data. In Stream-Based scheme, new data streams will be diverted away from the nodes that are currently part of the path of other streams.

A New Gradient Based Routing Protocol

Li Xia proposes a new gradient-based routing protocol (Xia & Chen, 2005). The proposed protocol takes into account the minimum hop count and remaining energy of each node while relaying data from source node to the sink. The optimal routes can be established autonomously with the proposed protocol. A simple acknowledgement scheme, which is implemented without extra overheads, is proposed. Data aggregation is performed to save transmission energy. To handle the frequent change of the topology of the network, a scheme for frequent change of the topology of the network is provided.

O(1)-Reception Routing Protocol

Bachir et al. proposes a technique that enables the best route selection based on exactly one message reception. It is called O(1)-reception (Bachir et al., 2007). In O(1)-reception, each node delays forwarding of routing messages (RREQs) for an interval inversely proportional to its residual energy. This energy-delay mapping technique makes it possible to enhance an existing min-delay routing protocol into an energy-aware routing that maximizes the lifetime of sensor networks. They also identify comparative elements that help to perform a thorough posteriori comparison of the mapping functions in terms of the route selection precision. The O(1)-reception routing enhances the basic diffusion routing scheme by delaying the interests forwarding for an interval inversely proportional to the residual energy: nodes compute a forwarding delay based on their residual energy and defer the forwarding of interest messages for this period of time. As maximum lifetime routing should combine the min and the max–min metrics, in the energy-delay mapping function, nodes with high residual-energy forward interests without delay to make diffusion equivalent to the min energy routing, and nodes with low residual-energy delay forwarding of interests for a time interval to make diffusion equivalent to the max–min residual energy routing.

Energy-Balancing Multipath Routing (EMPR)

The basic idea of EMBR is that the base station finds multipath to the source of the data and selects one of them for data transmission (Chen & Nasser, 2006). The base station dynamically updates the available energy of each node along the path based on the amount of packets being sent and received. The base station then uses the updated energy condition to periodically select a new path from multiple paths. The base station takes the role of the server and all sensor nodes work as clients. Base station does everything from querying specific sensing data, broadcasting control packets, routing path selection and maintenance to work as the interface to the outside networks. Sensor nodes are only responsible for sensing data and forwarding packets to the base station. Topology construction is initiated by the base station at any time. The base station broadcasts Neighbor Discovery (ND) packet to the whole network. Upon receiving this packet, every node records the address of the last hop from which it receives and stores it in the neighbors list in ascending order of receiving time. The node changes the source address of the packet to itself. Then it broadcasts the packet. If the new packet is already received the node drops the ND packet and does not rebroadcast. After the completion of Neighbors discovery, the base station broadcasts another packet, Neighbors collection (NC) to collect the neighbor information of each node. Upon receiving the NC packet, the node replies a NCR (Neighbors Collection Reply) packet by flooding. The base station now has a vision of the topology of the networks through the neighbor's information of all nodes. After the topology construction, the base station constructs a weighted directed graph. The weight of each edge is the available energy of the head node. In the data transmission phase, the base station broadcasts enquiry (DE) for sensing data with specific features. Then the sensor nodes satisfying an enquiry will reply with Data Enquiry Reply (DER) packet. On the other hand, the sensor node does not satisfy the enquiry will rebroadcast DE. The base station calculates the shortest path to the desired node in the weighted node.

Hierarchical Protocols

In hierarchical routing protocols, clusters are formed. For each cluster, a head node is assigned dynamically, a set of nodes will attach the head node, and the head nodes can communicate with the sink either directly or through upper level of heads. Data aggregation is usually performed at each head. Many hierarchical routing protocols are proposed such as LEACH (Heinzelman, Chandrakasan, & Balakrishnan, 2002), EAD (Boukerche, Cheng, & Linus, 2005), TinyDB(Madden et al., 2005), Hierarchical-PEGASIS (Lindsey, Raghavendra, & Sivalingam, 2001), TEEN (Manjeshwar & Agrawal, 2001), and APTEEN (Manjeshwar & Agrawal, 2002).

Low-Energy Adaptive Clustering Hierarchy LEACH

Heinzelman et al propose a Low-Energy Adaptive Clustering Hierarchy protocol (LEACH). LEACH is application-specific protocol architecture for wireless micro sensor network(Heinzelman, Chandrakasan, & Balakrishnan, 2002). In LEACH protocol the nodes organize themselves into clusters. In designing the LEACH protocol, it is assumed that all the nodes in the network can transmit with enough power to reach the base station (BS) of the network and each node has sufficient computational power to support different MAC protocols and perform signal processing functions. Regarding the network model it is assumed that the network consists of nodes that always have data to send to the end user and the nodes which are located close to each other have correlated data.

In LEACH protocol, the nodes organize themselves into local clusters. One of the nodes is identified as a cluster head and all other nodes in the cluster send their data to the cluster head. The cluster head is responsible for processing the data received from the nodes and transmit the resulted data to the base station. Since the cluster head performs data processing and transmission, it will consume more power than normal nodes. The cluster head must be changed through the system life time. Each node must take its turn to act as a cluster head. Operation of LEACH protocol is divided into rounds. Each round begins with a set-up phase followed by a steady-state phase as shown in Figure 10. In set-up phase the clusters are formed and the cluster head is assigned. In the steady state phase the nodes will transmit their data.

The algorithm to select a cluster head is a distributed algorithm. Each node makes autonomous decision to be a cluster head. During each round, there are k clusters so there must be k heads. At round r+1 which starts at time t, each node selects itself to be a cluster head with probability $P_i(t)$. $P_i(t)$ is chosen such that the expected

Figure 10. Time line showing LEACH operation

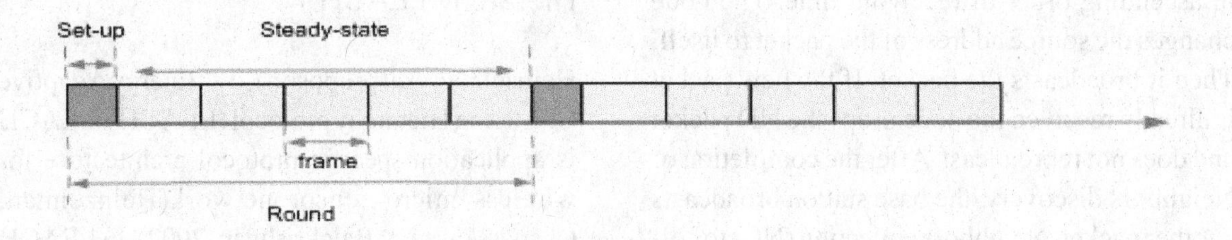

value of the cluster head must be k. To ensure that all nodes will act as cluster head equal number of times, each node must be a cluster head once in N/k rounds. If $C_i(t)$ is an indicator function determines whether a node i has been a cluster head in the most recent ($r \mod \frac{N}{k}$) rounds, then the probability that the node is a cluster head will be:

$$p_i(t) = \begin{cases} \dfrac{k}{N - k(r \mod \frac{N}{k})} & C_i(t) = 1 \\ 0 & C_i(t) = 0 \end{cases} \quad (5)$$

$C_i(t)=1$ indicates that a node i has not been a cluster head in the most recent ($r \mod \frac{N}{k}$) rounds, and $C_i(t)=0$ indicates that a node i has been a cluster head in the most recent ($r \mod \frac{N}{k}$) rounds. The probability given in equation 5 is a good estimation for the power. All nodes that have not assigned as a cluster head in the last ($r \mod \frac{N}{k}$) rounds ($C_i(t)=1$) will have more energy than the other nodes and so it is more likely to be selected as cluster heads. In (Heinzelman, Chandrakasan, & Balakrishnan, 2002) a new probability is proposed to take into account the energy in each node

$$p_i(t) = \min\left\{\frac{E_i(t)}{E_{total}(t)}, 1\right\} \quad (6)$$

Where $E_i(t)$ is the current energy of node *i* and $E_{total}(t)$ is the summation of the current energy at each node. To calculate the probability using equation 6 each node must know the power of all other nodes. All nodes must broadcast its energy level to all other nodes this can be performed directly for the neighboring nodes and using routing protocol for the non reachable nodes. Broadcasting the energy information will consume additional energy.

N and *k* are parameters that are programmed into the nodes. However, *k* is a function of the number of nodes *N* distributed throughout an a region of dimension *M* by *M*. *k* can be calculated by a distributed algorithm, each node sends a hello message to all neighbors within a predetermined number of hops. Each node counts the number of hello messages received (*N*) then *k* calculated based on *N*. this algorithm will cost additional energy but it is useful for networks with changing topology.

After identifying the clusters heads, each node must determine the cluster to which it belongs. Each cluster head broadcasts advertisement message containing the head's id using non-persistent CSMA scheme. Each node determines its cluster by selecting the head whose advertise signal is the strongest signal. This head is the closest head to the node. The node will transmit a joint request message to the chosen cluster head using CSMA. Upon receiving all the joint request messages the cluster head sets up the TDMA schedule and transmit this schedule to the nodes in the cluster. Each node will turn OFF its radio all the time slots except their assigned slots. This will end up the set-up phase and start the steady state phase.

The steady state phase is divided into frames; each node sends its data to the cluster head once per frame during its assigned slot. All nodes must be synchronized and start their set-up phase at the same time. This can be done by transmitting a synchronization pulse by the base station to all nodes. To reduce energy dissipation each non head node use power control to set the least amount of energy in the transmitted signal to the base station based on the received strength of the cluster head advertisement. When a cluster head receives the data from all nodes, it performs data aggregation and the resultant data will be sent to the base station. Processing the data locally within the cluster reduces the data to be sent to the base station; therefore the consumed energy will reduced. This is an advantage of the LEACH protocol. To reduce inter-cluster interference, each cluster communicates using direct sequence spread spectrum DSSS. Each cluster uses a unique spreading code.

The distributed cluster formulation algorithm does not offer guarantee about placement and number of cluster head nodes. An alternative algorithm is a central cluster formation; base station (BS) cluster formation. The central cluster formation produce better clusters by dispersing the cluster head nodes throughout the network. In the central algorithm, each node sends information about its current location and its energy level to the BS. The BS computes the average energy level. Any node has energy level less than the average cannot be a cluster head, other nodes can be clusters heads. The BS use simulated annealing to find the cluster heads. The solution must minimize the amount of energy for non-cluster head and find k the optimal number of clusters k_{opt}. When the cluster heads and associated clusters are found the BS broadcasts a message that contains the cluster head ID for each node.

Heinzelman et al derive a formula to find the optimum number of clusters that minimize the total consumed energy (Heinzelman, Chandrakasan, & Balakrishnan, 2002).

$$k_{opt} = \frac{\sqrt{N}}{\sqrt{2\pi}} \sqrt{\frac{\varepsilon_{fs}}{\varepsilon_{mp}}} \frac{M}{d^2_{toBS}} \qquad (7)$$

The frame size in LEACH is fixed regardless of the active nodes in the cluster since it is assumed that all nodes have data to send. This is not the real case all the time, sometimes some of the nodes are active and other nodes are not active.

An Adaptive Low Power Reservation Based MAC Protocol (ALPR MAC)

Although ALRP protocol is named as MAC protocol, I consider it as a routing protocol since the routes for the data to reach the sink is identified in it. ALRP MAC likes the LEACH algorithm in which both are based on cluster-hierarchical network organization and the communication in each cluster is based on TDMA-like frame structure. The difference between two protocols is that ALPR adapts the TDMA frame size based on active nodes to maintain high channel utilization (Mishar & Nasipuri, 2004).

- **Cluster formation and cluster head identification:** to form the cluster and to identify the cluster head, each node upon power ON waits for a random period before broadcasting a claim to become a cluster-head. The first node capture the medium will be the cluster head in the neighbor. All nodes that hear the broadcast before they transmit their claim will accept the first node as a cluster head. If a node receives more than one claim, it will select the node whose signal is the strongest.

Figure 11. Superframe structure in ALPR

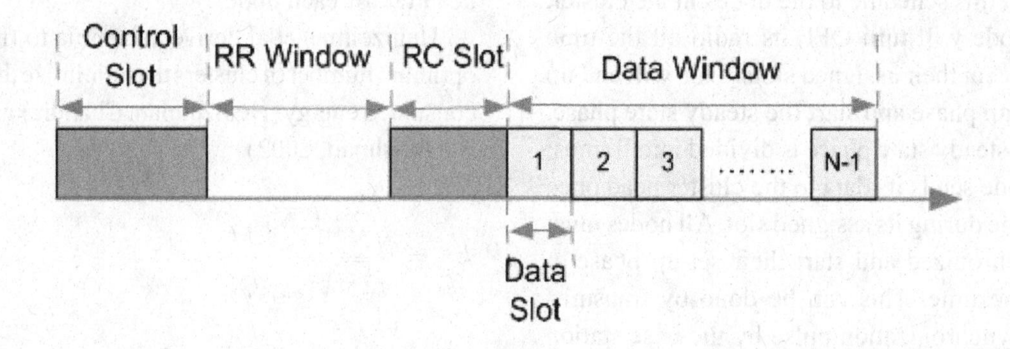

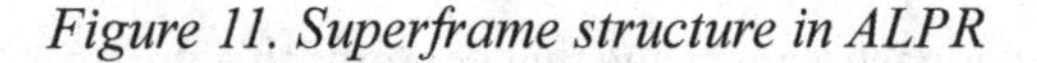

The job of cluster head in ALRP is similar to the job of cluster head in the LEACH protocol; maintaining the TDMA schedule and maintaining the synchronizations among all nodes. We note that the probability to be a cluster head depends only on the probability to capture the medium, which is unfair. A node may be a cluster head twice while another node is not identified as cluster head. This can be considered as a disadvantage of this protocol.

- **Channel structure:** the shared channel is divided into superframes. The boundaries between superframes are maintained by beacon signals that are transmitted by the cluster head. Each superframe is divided into four parts as shown in Figure 11.
- **A short control slot:** it is used by the cluster head to broadcast the control information such as the length of next superframe and the request for a new cluster-head.
- **Reservation Request (RR) window:** an unslotted contention based window. In this window, all the nodes that have data to transmit will send reservation request (RR) packet to the cluster head. All the nodes will contend the medium to send their RR packets. The RR packet contains the identities of the source and the intended receiver. To avoid collision, a non-persistent CSMA scheme is used. Due to the constraint in the RR window size the backoff time must be in the interval [0,RRwindowtime]. If a node fails to send RR packet in the current RR window, it will transmit it in the next window. If the cluster head successfully receives the RR packet, it will reserve a data slot for the source to transmit a data packet in data window. During RR window the cluster head and the nodes which want to transmit RR packet must be awake up
- **Short Reservation Confirmation (RC) window:** the cluster head will send the Reservation Confirmation (RC) packet that contains data transmission schedule of all nodes whose RR packets were successfully received during RR window. All nodes will be awake up during this window.
- **A slotted data window:** all nodes will be awake up in their assigned slots. The receiving nodes will be awake up also during their assigned slot. Other nodes will not be awake up; they will be in sleeping mode.

The superframe size is adapted based on the traffic intensity. The traffic intensity increases due to the addition of new nodes to the network or due to the increasing in the activity of a specific node. Increasing the traffic intensity will increase the number of RR packets, which increase the number of failure transmissions of RR packets. The number of failure transmissions of RR packets can be used as approximation for the traffic intensity. If the number of failure transmissions of RR packets increases to be larger than a threshold value, the superframe size must be increased. Otherwise, the superframe size must be decreased. To calculate the number of failure transmissions of RR packets each node counts the number of failure in transmissions of its RR packet. This count is sent with the RR packet that is successfully transmitted at the end. The cluster head calculates the average number of failure transmission of RR packet and decide whether to increase or to decrease the superframe size. However, this approach is not accurate. It is possible that a node fails to transmit RR packets many times and cannot successfully transmit any RR packet. These numbers of failure transmissions will not be considered in calculating the average number of failure transmissions.

Other disadvantages may rise. Firstly, the superframe size is too long. If a node fails to transmit an RR packet in the current RR window, it has to wait for the next RR window, which increases the delay. Secondly, although the number of slots in TDMA schedule will be equal to the number of active nodes only, TDMA schedule will be shorter but there will be additional slots in RR window. It is possible for a node to try to send RR packet many times which will consume more energy.

A Bit-Map-Assisted (BMA) MAC Protocol

Another extension of the LEACH protocol is proposed by Li and Lazarou (Li & Lazarou, 2000). They proposed a bit-map-assisted (BMA) MAC protocol for large-scale cluster based WSNs. BMA is intended for event-driven applications where the sensor nodes transmit data to the cluster only if significant events are observed. The main idea is to reduce energy wastes due to idle listening and collision while keeping good low latency. The operation of BMA is divided into rounds as in LEACH. Each round consists of cluster set-up phase and steady-state phase as shown in Figure 12. The set-up phase is identical to the setup phase in the LEACH protocol. Clusters heads are identified and clusters are formed. The steady state phase is divided into sessions with fixed durations. Each session consists of contention period, data transmission period and idle period.

For N nodes in the cluster, the contention period consists of N slots, and the transmission period is variable and less than N. the data transmission period plus the idle period is fixed. In the contention period all nodes keeps their radio ON. Each node is assigned a specific time slot and transmits 1-bit control message if it has data to send (source node), otherwise the scheduled slot remains empty. The cluster head knows the nodes that want to transmit. It will construct the transmission schedule and broadcasts it to the source nodes. In data transmission period, each source node turns its radio ON and sends its data to the cluster head during its allocated time slot. The node keeps its radio OFF during all over the remaining time. All non-source nodes turn their radios OFF all over the time. Li and Lazarou introduced an analytical model for the average system energy consumed during each round and an analytical model for the average time required for a packet to be transmitted by a source node and received by the cluster-head for TDMA, E-TDMA, and BMA. The results show that BMA is superior for the cases of low and medium traffic load, relatively few sensor nodes per cluster, and relatively large data packet sizes.

Energy-Aware Data-Centric Routing Algorithm (EAD Protocol)

Boukerche *et al* proposes an Energy-Aware Data-Centric Routing Algorithm (EAD) (Boukerche, Cheng, & Linus, 2005). EAD protocol is designed for event driven application. In EAD protocol, a tree rooted at the base station is constructed. The tree consists of leaf and non-leaf nodes. A non-

Figure 12. A single round of the BMA protocol

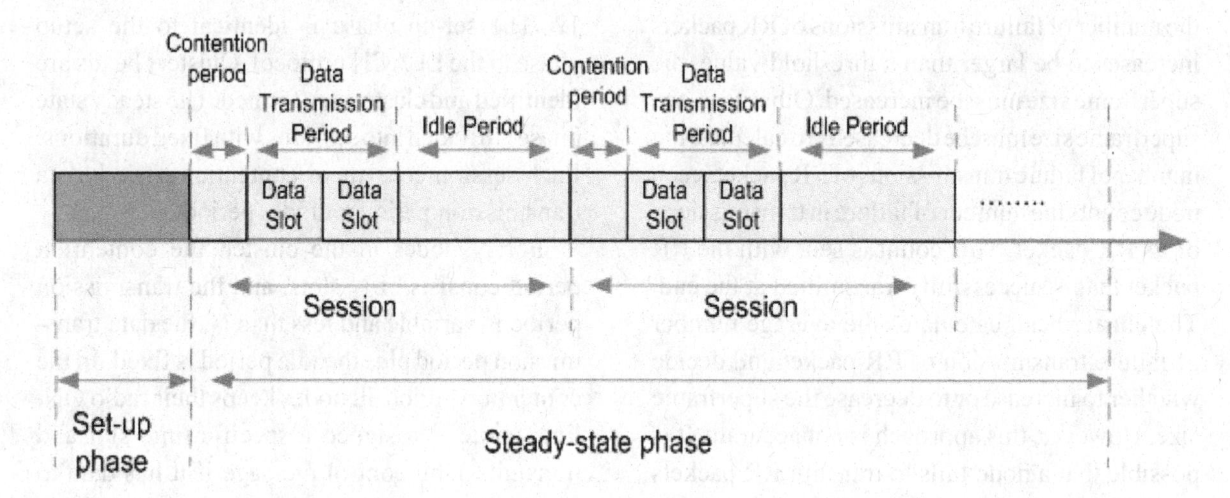

leaf node is a node that has at least one child. On the other hand, a leaf node is a node that has no child. All the leaf nodes of the tree will turn their radio OFF most of the time. On the other hand, all the non-leaf nodes will turn their radio ON all the time. When an event occurs, the leaf nodes will collect the related data and turn its radio ON to transmit the data to its parent. When a non-leaf node receives data from all its children, it will aggregate the data and send it to its parent. All the nodes use CSMA/CA for transmitting the data. Since the radio of the non-leaf sensor nodes will always be ON, they will lose much power than the leaf nodes. The tree will be reconstructed from time to time. Boukerche *et al* proposes an energy aware algorithm to build the tree. One of the disadvantages of EAD protocol is that the non-leaf nodes will be awake all the time even though there are not events to detect. This makes EAD unsuitable for applications with periodic data traffic.

To build a tree rooted at the sink, the sink initiates the process of building the tree. Building the tree is performed by broadcasting control messages. Each control message consists of four fields: *type, level, parent, power*. For the sender node *v*, $type_v$ represents its status; 0: undefined; 1: leaf node; 2: non-leaf node. $level_v$ refers to the number of hops from v to the sink. $parent_v$ is the next hop of *v* in the path to the sink; $power_v$ is the residual power E_v. Initially each node has status 0. The sink broadcasts $msg(2, 0, NULL, \infty)$. When a node *v* receives $msg(2, level_u, parent_u, E_u)$ from node *u*, it becomes a leaf node, sense the channel until it is idle, then waits for T_2^v time, if the channel is still idle, *v* broadcasts $msg(1, level_u+1, u, E_v)$. If *v* receives $msg(1, level_u, parent_u, E_u)$ from *u*, it senses the channel until it is idle, waits for T_1^v if the channel is still idle, *v* broadcasts $msg(2, level_u+1, u, E_v)$. And it becomes non-leaf node. If node *v* receives more than one message from different nodes before it broadcasts its message, it will select the node with larger energy as its parent. If both nodes have the same energy, it will select one of them randomly. The waiting node will go back to sensing state, if another node occupies the common channel before it times out. If a node *v* with status 1 receives $msg(2, level_w, v, E_w)$ from node *w* indicating that *v* is its parent, *v* broadcasts $msg(2, level_v, parent_v, E_v)$ immediately after the channel is idle. The process will continue until each node becomes leaf or non-leaf node. A sensor with status 2 becomes a leaf node if it detects that it has no children. Both T_1^v and T_2^v are chosen such that no two neighboring broadcasts are scheduled at the same time. On the other hand, to force the neighboring sensors with higher energy to broadcast earlier than those nodes with a lower residual power, both T_1^v and T_2^v must be monotonically decreasing functions of E_v (Boukerche, Cheng, & Linus, 2005) chooses $T_1^v = 2 * t_0 + \frac{c}{E_v}$ and $T_2^v = t_0 + \frac{c}{E_v}$ where t_0 is the upper bound of the propagation time between any pair of neighboring sensors and c>0 is an adjusting constant.

A Generalized Energy-Aware Data Centric Routing For Wireless Sensor Network ($EAD_{General}$)

EAD is generalized such that any node can act as a gateway (Al-Khdour & Baroudi, 2007). To generalize EAD, they assume that each node has the ability to transmit its data for long distance, i.e. its transmission can reach the sink. Each node has power control capability such that the transmission energy depends on the distance to the destination node. When a node sends data to its nearest neighbor, the transmission energy will be small compared with the transmission energy required to transmit data to the sink. In $EAD_{General}$, a new phase; Selecting Gateways (SG), is added. In this phase, gateway nodes are selected. It is assumed that the network is virtually divided into tiers. Each tier includes all nodes that can hear a signal transmitted with specific energy from the sink. For example, $tier_0$ includes all nodes that can hear the signal transmitted from sink with transmission energy equals to E_0. $Tier_1$ includes all nodes that can hear the signal transmitted from sink with transmission energy equals to E_1, where $E_1 > E_0$ and so on. Initially, the nodes of $tier_0$ will be considered as potential candidate gateways. Based on their energy level, some of these nodes will advertise themselves as gateways. They will act as gateways until their residual energy drops below a threshold value E_{th}. Then new gateways will be selected from the nodes of $tier_1$. The selected nodes will act as gateways until their residual energy drops below E_{th} and so on. When all tiers are considered and no more nodes can be selected as gateways based on the current E_{th}, a new cycle will start, in this cycle new gateways will be selected from $tier_0$ using smaller value of E_{th} and so on. To select the gateways, the sink broadcasts an *ADV* message. The *ADV* message contains a field for E_{th}. Initially *ADV* message is broadcasted with energy E_0 such that it reaches the nodes of $tier_0$ only. When a node receives the *ADV* message, it compares its residual energy with E_{th}, and then it responds with a *JOIN* message. A *JOIN* message contains a confirmation field. *Confirmation* is set to 1, if the node's residual energy is greater than E_{th}, i.e. the node can be a gateway and it selects the sink as its parent, otherwise *confirmation* is set to 0. After the node sends its *JOIN* message, it will act as gateway in the current round. Assuming reliable channel, it does not need a confirmation from the sink to be a gateway. All nodes send JOIN message with *confirmation field=1* will be considered gateways. If the sink receives *JOIN* messages from all nodes in the target tier and the *confirmation field =0* in all the received *JOIN* messages, then no node from the target tier can be a gateway, since we assume that all nodes can reach the sink, the sink will broadcast a new *ADV* message with higher transmission energy E_1 using the same E_{th} to select a gateway from the next tier. The nodes of the next tier will respond with *JOIN* messages according to their energy. The process will continue until all tiers are considered and no node has energy greater than E_{th}; no node can be a gateway. A new cycle will start from $tier_0$ with new E_{th}, $E_{th}(new)=eE_{th}(current)$, where $0<e<1$. Following the same procedure as above, new gateway nodes will be selected from $tier_0$. For each cycle, a fixed E_{th} will be used, and at the beginning of each new cycle, E_{th} will be reduced by the factor e. The sink and nodes will exchange messages using the CSMA mechanism. The node has to be ON until it receives the *ADV* message from the sink and then it sends the *JOIN* message. Since the node does not need confirmation from the sink, it will go to sleep immediately after sending the *JOIN* message.

A Generalized Energy-Efficient Time-Based Communication Protocol for Wireless Sensor Networks (GET)

In GET, it is assumed that each node has the ability to transmit its data for long distance, i.e. its transmission can reach the sink (Al-Khdour & Baroudi, 2009). Each node has power control capability such that the transmission energy depends on the distance to the destination node. When a node sends data to its nearest neighbor, the transmission energy will be small compared to the transmission energy required to transmit data to the sink. they assumed that all nodes are synchronized. Regarding the application of the network, they assume that the event that is being monitored is periodic, so data transmission from sensor nodes to the sink will start at specific time, and it will be repeated periodically. They assume also that all the nodes that are located close to each other and have correlated data. Hence, data aggregation will be used and it will reduce data redundancy. In GET, time is divided into rounds. Each round consists of four phases: Selecting the Gateways (SG), Building the Tree (BT), Building the Schedule (BS), and Data Transmission (DT). In the first phase, gateways are selected; the gateway is selected using the algorithm proposed in (Al-Khdour & Baroudi, 2007). In the second phase, a tree rooted at the sink is built. The tree is built using building tree algorthim proposed by(Boukerche, Cheng, & Linus, 2005). They modify the buiding tree algorithm such that building tree process will be initiated by the gatewyas not by the sink Based on this tree, a TDMA schedule is built in a distributed manner in phase-3. The schedule will be built assuming that in the data transmission period, all nodes connected to the sink through the same gateway will use the same frequency to transmit their data.. For each node, they identify two time constants: Time Ready to Receive (*TRR*) and Time Ready to Transmit (*TRT*). For a node v, TRR_v represents the time slot when the node is ready to receive data from its children, while TRT_v represents the time slot when a node can transmit data to its parent. Assuming t_0 represents the time at which the periodic sensing event occurred and the data is already collected from the monitored environment. For a leaf node, $TRT_v = t_0$. TRR_v is not valid since it does not have children. On the other hand, for a non-leaf node v:

$$TRR_v = Max(TRT_i) \qquad i = 1,2,3,\ldots n_v^c$$
$$TRT_v = TRR_v + n_v^c T_t$$

Where i represent an index for the child of node v,, n_v^c represents the count of v's children, and T_t represents the time needed to transmit one data packet. To build the schedule, initially, each leaf node will transmit its *TRT* value to its parent. When a parent receives *TRT* values from all its children, it calculates its *TRR* and *TRT* using (1) and builds the schedule for its children. Then it transmits its *TRT* to its parent and broadcasts the schedule to its children. The process will continue until all nodes receive their assigned time slot from their parents. Both leaf and non-leaf nodes use *CSMA/CA* protocol to exchange data (*TRT* and the Schedule). Eventually, we have a TDMA schedule for the whole sensor network.

In the fourth phase, data is transmitted from sensor nodes to the sink following the schedule prepared in phase-3. Data transmission period represents the time needed to forward all data packets in a single round. Data transmission period may be repeated many times in a single round.

An Energy-Efficient Distributed Schedule-Based Communication Protocol for Periodic Wireless Sensor Networks (EEDS)

In EEDS protocol (Al-Khdour & Baroudi, 2010), The energy consumption is minimized by minimizing the amount of time a sensor node is in idle listening state. EEDS is intended for applications with periodic data traffic where event reporting is initiated every specific time interval from the sensing node. In EEDS, the time is divided into rounds. Each round is composed of three phases: building the tree, building the schedule, and data transmission. In building the tree phase, an energy-aware tree is built. A TDMA schedule is constructed in a distributed manner during the building the schedule phase. This schedule will be used for data transmission in the data transmission phase. To ensure a reliable communication tree for the whole sensor network, at the beginning of each round the tree is rebuilt, and a new TDMA schedule is reconstructed.

TinyDB

Another alternative in the same direction is the work presented in (Madden et al., 2005). A distributed query processor for smart sensor devices (TinyDB) is proposed. In TinyDB, to disseminate queries and collecting results, a routing tree rooted at the base station is built. The routing tree is formed by forwarding a routing request (a query in TinyDB) from every node in the network. The root sends a request then all child nodes that hear this request process it and forward it on to their children, and so on, until the entire network has heard the request. Each node picks a parent node that is one level closer to the root. This parent will be responsible for forwarding the node's query results to the base station. To limit the scope of queries, a Semantic Rooting Tree (SRT) is built. This tree is built based on the routing tree. If a node knows that none of its children currently satisfies the query, it will not forward the query down the routing tree. Therefore, each node must have information about child attribute values.

Power-Efficient Gathering in Sensor Information System (PEGASIS)

PEGASIS protocol is an improvement of the LEACH protocol, instead of forming clusters, PEGASIS forms chains from sensor nodes (Lindsey & Raghavendra, 2002). Each node will transmit and receive from a neighbor. One node of the chain will transmit directly to the sink. Data will move from node to node until it reaches the sink. At each intermediate node, data aggregation is performed. The chain is constructed in a greedy way.

Hierarchical-PEGASIS is an extension to PEGASIS (Lindsey, Raghavendra, & Sivalingam, 2001). It is proposed to decrease the delay of packets delivered to the base station. It also proposes a solution for the data gathering problem by considering the *(energy x delay)* metric. To reduce delay, simultaneous transmissions of data packets are performed. Two approaches are used; CDMA or allowing the spatially separated nodes to transmits at the same time. In CDMA approach, the chain is constructed as a tree. Each node in a particular level will transmit data to the node in the upper level of the tree. The parallel data transmission will reduce the packet delay significantly. In the second approach, a three-level tree is constructed. Then simultaneous transmissions are scheduled carefully to reduce the interference effects.

Threshold Sensitive Energy Efficient Sensor Network Protocol (TEEN)

TEEN can be considered as data centric and hierarchical protocol (Manjeshwar & Agrawal, 2001). It is proposed for event driven applications such as sudden changes in the temperature. In

TEEN protocol, clusters are formed by the base station. Cluster head can communicate with the sink directly or through another cluster head. After cluster formulation, the cluster head broadcast two thresholds to the nodes, hard and soft threshold for sensed attribute. Hard threshold is the minimum value of the attribute at which the sensor node will turn its transmitter ON and it will transmit the corresponding data to the cluster head. The node will transmit only when the sensed attribute in the range of interest, which reduces the number of transmissions. Furthermore, when the sensed value is near the hard threshold, the sensor node will transmit data only if the attribute value changed by an amount equal to or greater than the soft threshold, which will further reduce the number of the transmission. If the attribute value does not reach the hard threshold then the node will not transmit at all, therefore the TEEN is not suitable for applications with periodic data traffic. For this kind of applications, Adaptive TEEN (APTEEN) is proposed (Manjeshwar & Agrawal, 2002). In APTEEN, after clusters formulation, the cluster head will broadcast the attributes, the threshold values and the transmission schedule to the all nodes.

Unequal Cluster Based Routing (UCR)

In UCR protocol, clusters with different size are constructed (Chen et al., 2007). Cluster heads closer to the sink will have smaller cluster sizes than those farther from the sink. Thus they can preserve some energy for the inter-cluster data forwarding. A greedy geographic and energy-aware routing protocol is designed for the inter cluster communication which considers the tradeoff between the energy cost of relay paths and the residual energy of relay nodes. The UCR protocol consists of two parts: an energy-efficient unequal clustering algorithm called EEUC and an intercluster greedy geographic and energy-aware routing protocol. Initially, the base station broadcasts a beacon signal to all sensors at a fixed power level. Based on the received signal strength, each sensor node can compute the approximate distance to the base station. It not only helps nodes to select the proper power level to communicate with the base station, but also helps us to produce clusters of unequal sizes. In EEUC algorithm, heads will be identified randomly. As in LEACH protocol, the task of being a cluster head is rotated among sensors in each round to distribute the energy consumption across the network. After cluster heads have been selected, each cluster head broadcasts a CH_ADV_MSG across the network field. Each ordinary node chooses its closest cluster head, the head with the largest received signal strength, and then informs it by sending a JOIN_CLUSTER_MSG. After forming clusters, data will be transmitted from the cluster heads to the base station. Each cluster head first aggregates the data from its cluster members, and then sends the packet to the base station via a multi-hop path through other intermediate cluster heads. Before selecting the next hop node, each cluster head broadcasts a short beacon message across the network at a fixed power which consists of its node ID, residual energy, and distance to the base station. A threshold TD_MAX in the multi-hop routing protocol is proposed. If a node's distance to the base station is smaller than TD_MAX, it transmits its data to the base station directly; otherwise, it is better to find a relay node that can forward its data to the base station.

Self-Organizing Protocol

Subramanian et al proposed architectural and infrastructural components to build sensor applications (Subramanian & Katz, 2000). In the proposed architecture, the sensor nodes can be either stationary or mobile node, they sense the environment, and they forward data to a set of nodes that act as routers. Routers are stationary nodes, and they form the backbone of the network. To be a part of network, a node must be able to reach the router. A routing architecture that requires addressing of sensor node has been proposed. Sensors are identified through the address of the router node that it is connected to. The protocol for self-organizing the router nodes and creating the routing tables consists of four phases; discovery, Organization, Maintenance, and Self-reorganizing. In the discovery phase, the nodes in the neighborhood of each sensor are discovered. In the Organization phase, groups are formed and merged to form a hierarchy. Each node assigned an address based on its position in the hierarchy. Routing tables and energy levels of nodes are updated in the Maintenance phase. In the Self-Reorganizing phase, group reorganization is performed. The proposed protocol utilizes the router nodes to keep all the sensors connected by forming a dominating set.

Energy-Aware Routing for Cluster-Based Sensor Networks

Younis et al. proposed a hierarchical routing algorithm based on a three-tier architecture (Younis, Youssef, & Arisha, 2002). In the proposed protocol, sensors are grouped into clusters. The cluster heads (gateways) are less energy constrained than normal sensors. It is assumed that cluster heads knows the location of the sensor nodes. Gateways maintain the states of the sensors and sets up multi-hop routes for collecting sensors data. Each gateway informs each node within its clusters the time slots in which it can transmit and in which it have to listen to other nodes transmission. The sensor nodes in the cluster can be in one of four states: sensing only, relaying only, sensing-relaying and inactive. In sensing state the sensor node senses the environment and generates the corresponding data. In the relaying only state, the node does not sense the environment but it forwards data from other active nodes. In sensing-relaying state, the node not only senses the environment but also forwards the data from other active nodes. In inactive state, the node neither senses the environment nor forwards data. The link cost is defined as the energy consumption to transmit data between two nodes, the delay optimization and the other performance cost. A least-cost path is found between sensor nodes and the gateway. The gateway monitors the available energy level at every sensor that is active. Rerouting is triggered by an application-related event requiring different set of sensors to probe the environment or the depletion of the battery of an active node.

Base-Station Controlled Dynamic Clustering Protocol (BCDCP)

Muruganathan et al. proposes a clustering-based routing protocol called Base Station Controlled dynamic Clustering protocol (BCDCP) (Morcos, Matta, & Bestavros, 2004). In BCDCP, the base station sets up clusters and routing paths, performs randomized rotation of cluster heads, and carries other energy intensive tasks. The key ideas in BCDCP are: formulation of balanced clusters where each cluster head serves an approximately equal number of member nodes, uniform placement of cluster heads throughout the entire sensor field, and the utilization of cluster-head-to-cluster-head(CH-to-CH) routing to transfer the data to the base station. Class-based addressing of the form <Location ID, Node Type ID> is used in BCDCP. The Location ID identifies the location of a node. It is assumed that the base station keeps up-to-date information on the location of all the nodes in the network. A Node Type ID describes the functionality of the sensor such as seismic sensing, and thermal sensing. BCDCP operates in two major phases: setup and data communication. In setup phase, clusters are formed, clusters' heads are selected, CH-to-CH routing paths are formed, and schedule is created for each cluster. During each setup phase, the base station receives information on the current energy status from all the nodes in the network. Based on this information, the base station computes the average energy level and then chooses a set of nodes, denoted S, whose energy levels are above the average value. Cluster heads for the current round will be chosen from the set S. To identify the cluster heads from the set and to from clusters, iterative cluster splitting algorithm is used. This simple algorithm first splits the network into two sub-clusters, and proceeds further by splitting the sub-clusters into smaller clusters. The base station repeats the cluster splitting process until the desired number of clusters is attained. Once the clusters and the cluster head nodes have been identified, the base station chooses the lowest-energy routing path and forwards this information to the sensor nodes along with the details on cluster groupings and selected cluster heads. The routing paths are selected by connecting all the cluster head nodes using the minimum spanning tree approach that minimizes the energy consumption and then a head is randomly selected to transmit data to the base station. The last step in this phase is building a TDMA Schedule for each cluster. In The data communication phase, Data gathering, Data fusion, and Data routing is performed using the TDMA schedule created in setup phase.

Location-Based Protocols

Information Location can be utilized to forward data with minimum energy consumption. If the region to be monitored is known, the query can be forwarded to that region. Many location-based routing protocols for WSN were proposed. In the successive subsections, I will survey many of these protocols.

Geographic Adaptive Fidelity (GAF)

GAF is energy-aware location-based routing protocol designed for mobile ad hoc protocols, but it can be applicable to sensor networks (Xu, Heidemann, & Estrin, 2001). In GAF a virtual grid for the monitored area is formed. Each node uses its GPS-indicated location to associate itself with a point in the virtual grid. Nodes associated with same point in the grid are equivalent. Some of them can be in the sleeping state to save energy while others will be in active state. Therefore, the network lifetime will increase. To balance load among nodes, equivalent nodes change their state from active to sleeping in turn. Three states are defined in GAF, discovery, sleep, and active. In the discovery state a node will determine its neighbors. While it is in sleep state, a node will turn OFF its radio. The active node will participate in data routing. A node will be in each state for particular time period which is application dependent. On the other hand, determining which nodes that will be in sleep state is application dependent. GAF is implemented for non-mobility (GAF-basic) and mobility (GAF-mobility adaptation) of nodes. To keep the network connected, a representative node must be always active for each region on its virtual grid.

Minimum Energy Communication Network (MECN)

In MECN protocol, low power GPS is utilized to find a minimum power topology for stationary nodes including the sink (Rodoplu & Ming, 1999). For each node, a relay region is identified. The relay region consists of the neighboring nodes where transmitting through those nodes is more energy efficient than direct transmission. The enclosure of a node i is the union of all relay regions that node i can reach. The protocol has two phases; in the first phase, the enclosure graph is constructed. The enclosure graph consists of all enclosures of each transmit node, and it contains globally minimum energy links which will be found in the second phase.

Geographic and Energy Aware Routing (GEAR)

In GEAR protocol, energy aware and geographical-informed neighbor selection heuristic is used to route packets towards the destination region (Younis, Youssef, & Arisha, 2002). The key idea is to restrict the number of interests in Directed Diffusion to certain regions rather than sending interest to the whole network. Each node keeps an estimated cost and a learning cost of reaching the destination through its neighbors. The estimated cost is a combination of residual energy and distance to destination. The learned cost is a refinement of the estimated cost. A hole exists in the network when a node does not have any closer neighbor to the target region. With no holes in the network, the estimated cost is equal to the learned cost. When a packet reaches the destination, the learned cost is propagated one hop back so that route setup for next packet will be adjusted. The GEAR protocol consists of two phases; in the first phase, the packets are forwarded towards the target region, when a node receives a packet, it checks its neighbors to see if there is a neighbor that is closer to the target region. The closes neighbor to the target region is selected as the next hop. When all neighbors are further than node itself, a hole exists, one of them will be selected based on the learned cost function. This selection will be updated according to the convergence of the learned cost. In the second phase, packets will be forwarded within the region; the packets are forwarded in the region by either recursive geographic forwarding or restricted flooding.

The Greedy Other Adaptive Face Routing (GOAFR)

GOAFR is a geometric ad-hoc routing algorithm combining greedy and face routing. The greedy algorithm of GOAFR always picks the closest neighbor to destination to be the next hop (Kuhn, Wattenhofer, & Zollinger, 2003). However, it can stuck at some local minimum, no neighbors closer than the current node. Other face routing is a variant of Face Routing (FR)(Stojmenovic & Lin, 1999). Other Face Routing utilizes the face structure of planer graphs such that the message is routed from node to node by traversing a series of face boundaries. The aim is to find the best node on the boundary; the closest node to the destination. It was shown that GOAFR algorithm can achieve both worst-case optimality and average-case efficiency.

SPAN

In SPAN protocol, some nodes are selected as coordinators based on their positions (Chen et al., 2002). The coordinators form a network backbone that is used to forward messages. a node should become a coordinator if two neighbors of a non-coordinator node cannot reach each other directly or via one or two coordinators.

A Mesh-Based Routing Protocol for Wireless Ad-Hoc Sensor Network (MBR)

In MBR protocol, the area of the sensor network is portioned into regions; mesh topology (Lotfifar & Shahhoseini, 2006). The nodes can communicate to their neighbor nodes through virtual channels. Forming the mesh topology is performed in three phases. In the first phase, the base node for zoning is selected. Two setup sensors are determined. One of them is located at the largest diameter and in the boundary of the area and the second sensor is located on the boundary of other orthogonal diameter of the region. In phase two, the network is divided into regions. In phase three, each sensor nodes is assigned ID. Each sensor will be known with two features: its region coordinate (X,Y) and its ID. To transmit data between source nodes and sink a path is reserved between them firstly. To reserve a path, the source node sends a reserve message, called RAP, to the sensors in its target (X,Y). Upon receiving the RAP message, each node generates a priority number and returns it to the source node using ACK message. Sensors have higher energy will have higher priority. The source sensor will select sensors to form the path among the sensors that sends ACK message. Then data will be sent based on the path determined. After transmitting data, path must be released. This is done by sending a CRP message.

Energy-Efficient Geographic Multicast Routing

Sanchez et al. proposes a novel energy-efficient multicast routing protocol called GMREE (Sanchez, Ruiz, & Stojmenovic, 2007). It aims to preserve energy and network bandwidth. GMREE protocol builds multicast trees based on a greedy algorithm using local information. GMREE protocol is based in the concept of cost over progress metric and it is specially designed to minimize the total energy used by the multicast tree. The cost is defined as the energy needed to reach the furthest neighbor in the selected set of relays plus the energy that such amount of nodes will need to process the message. GMREE incorporates a relay selection function which selects nodes from a node's neighborhood taking into account not only the minimization of the energy but also the number of relays selected. Nodes only select relays based on a locally built and energy-efficient underlying graph reduction such as Gabriel graph, enclosure graph or a local shortest path tree. Thus, the topology of the resulting multicast trees really takes advantage of the benefit of sending a single message to multiple destinations through the relays which provide best energy paths.

Energy-Aware Geographic Routing for Sensor Networks with Randomly Shifted Anchors

Anchor-based geographic routing aims at finding a small number of intermediate nodes acting as anchors so that the path length (i.e. number of hops) between the source and destination can be reduced. However, some nodes (e.g., nodes near the boundary of the network) tend to be used as anchors repeatedly by multiple flows. As a result, their energy drains quickly and the lifetime of the network is reduced. Moreover, the intermediate nodes between source and destination change very little once the anchor list is set. This also contributes to the quick depletion of the energy for some nodes. To overcome these shortcomings, Zhao et al. introduces a random shift to the location of each anchor in the routing process (Yu, Estrin, & Govindan, 2001). Each new packet will then be routed to a different anchor determined by the location of the original anchor plus the random shift. Because the shift is generated randomly, different packets will likely be routed through a different list of anchors. This allows more nodes to be involved in the routing process and the energy consumption is better distributed among nodes in the network.

Projection Distance-Based Anchor Protocol (PDA)

Zhao et al. proposes a Projection Distance-based Anchor scheme, which is called PDA, to obtain the anchor list based on the projection distance of nodes in detouring mode (Zhao, Liu, & Sun, 2007). The projection is with respect to the virtual line linking the source and destination nodes. To obtain an anchor list adaptively, the first packet of a burst is routed from the source to the destination using an existing geographic routing algorithm. During the routing of the first packet, an anchor list is built. After the first packet is delivered, the anchor list is sent back to the source from the destination, and the list is embedded into subsequently packets. A packet is then routed from the source to the first anchor node, then to the second anchor node, and so on, until it reaches the destination.

On Optimal Geographic Routing in Wireless Networks with Holes and Non-Uniform Traffic

Subramanian et al. propose a randomized geographic routing scheme that can achieve a throughput capacity of $\Theta(1/\sqrt{n})$ (within a poly-logarithmic factor) even in networks with routing holes (Subramanian, Shakkottai, & Gupta, 2007). They show that the proposed scheme is throughput optimal (up to a poly-logarithmic factor) while preserving the inherent advantages of geographic routing. They also show that the routing delay incurred by the proposed scheme is within a poly-logarithmic factor of the optimal throughput-delay trade-off curve. On the other hand, Subramanian et al. construct a geographic forwarding based routing scheme that can support wide variations in the traffic requirements as much as $\Theta(1)$ rates for some nodes, while supporting $\Theta(1/\sqrt{n})$ for others. They show that the above two schemes can be combined to support non-uniform traffic demands in networks with holes.

The randomized algorithm takes as input the number of nodes in the network, the packet to be sent, as well as the number of holes. Considering the first packet in all the source nodes, The source node for every traffic flow creates Rlog(n) copies of its packet to send. It chooses Rlog(n) independent and uniformly distributed points from the unit region and sets the NEXT-DEST field in the packet to the randomly generated location in each of these copies. The Rlog(n) packets are routed from the source in a greedy geographic manner to the location in NEXTDEST. Upon receiving a packet, a node checks if it is the NEXTDEST location. If it is not the NEXT-DEST location, it searches within its neighboring nodes for the node that is closest to the NEXT-DEST location, and forwards the packet to that node. If none of its neighbor nodes is closer to the NEXT-DEST than itself, the node drops the packet. If it is the NEXT-DEST location, it checks whether it is the final destination or not. If it is the final destination, then the packet is received. Otherwise, If the final destination is one hop away from the current node, the node forwards the packet greedily to the final destination. If the final destination is more than one hop away from the current node, the current node makes Rlog(n) copies of the packet and again generates uniform and randomly chosen locations for the NEXT-DEST in each of the packet copies, and forwards them greedily.

QoS-Aware Protocols

QoS-aware protocols consider end-to-end QOS requirement while setting up the paths in the sensor network. Many QoS-aware routing protocols for WSN were proposed. In the successive subsections, I will survey many of these protocols.

Maximum Lifetime Energy Routing

Chang et al presents a routing protocol for sensor networks based on a network flow approach (chang & Tassiulas, 2004). The protocol aims to maximize the network lifetime by defining link cost as a function of node remaining energy and the required transmission energy using that link. Finding traffic distribution is a possible solution to the routing problem. The solution to this problem maximize the network lifetime. Two maximum residual energy path algorithms were proposed to find the best link metric for the maximization problem. The two algorithms differ in their definition of link costs and the incorporation of nodes' residual energy. The link costs that are used in the two algorithms are:

$$c_{ij} = \frac{1}{E_i - e_{ij}} \text{ and } c_{ij} = \frac{e_{ij}}{E_i}$$

Where: E_i is the residual energy at node i

e_{ij} is the energy consumed when a packet transmitted over link i-j.

The least cost paths to destination are found using Bellman-Ford shortest path algorithm. The least cost path is the path whose residual energy is largest among all paths.

Maximum Life Time Data Gathering

Kalpakis et al. models the data routes setup in sensor network as the maximum lifetime data-gathering problem (Kalpakis, Dasgupta, & Namjoshi, 2002). A polynomial time algorithm to solve this problem is proposed. The data-gathering schedule specifies for each round how to get and route data to sink. For each round, a schedule has one tree rooted at the sink and spans all the nodes of the network. The network lifetime depends on the duration for which the schedule remains valid. The Maximum Lifetime Data Aggregation (MLDA) protocol is proposed to set up maximum lifetime routes taking into account data aggregation. If a schedule "S" with "T" rounds is considered, it induces a flow network G. the flow network with maximum lifetime subject to the energy constraints of sensor nodes is called an optimal admissible flow network. A schedule will be constructed by using this admissible flow network. For application with no data aggregation such as video sensors, a new scenario is presented, which is called Maximum Lifetime Data Routing (MLDR). It is modeled as a network flow problem with energy constraints on sensors.

Minimum Cost Forwarding

The objective of Minimum Cost Forwarding protocol is to find the minimum cost path in a large sensor network (Chu, Haussecker, & Zhao, 2002). The cost function for the protocols captures the effect of delay, throughput and energy consumption from any node to the sink. The protocol consists of two phases; setup phase and data transmission phase. In setup phase, starting from the sink, the cost value in all nodes is set up. The sink set its cost as zero and broadcast a message for all its neighbors. Upon receiving the message, each neighbor of the sink will set its cost as the cost of the link to sink and it broadcast its cost. And so on. Every node adjusts its cost value by adding the cost of the node it received the message from and the cost of the link. Cost adjustment is done using a back-off based algorithm. The forward of messages is deferred to allow the message with minimum cost to arrive. Therefore, optimal cost for all nodes to the sink is found. In the second phase, the source node broadcasts the data to its neighbor. Upon receiving the broadcast message, the node adds its transmission cost to sink to the cost of packet, then the node checks the remaining cost in the packet. If it is not sufficient to reach the sink, the packet is dropped. Otherwise, the node forwards the packet to its neighbors.

Sequential Assignment Routing (SAR)

SAR is a table driven multi-path protocol aiming to achieve energy efficiency and fault tolerant. In the SAR protocol, trees rooted at one-hop neighbors of the sink is created by taking QOS metric, energy resources on each path and priority level of each packet into account. By using created trees, multiple paths from sink to sensors are formed. One of these paths is selected according to energy resources and QoS in the path. Failure recovery is done by enforcing routing table consistency between upstream and downstream nodes on each path. Any local failure causes an automatic path restoration procedure locally.

BImodal Power-Aware Routing Protocol (BIPAR)

Morcos et al. proposes BImodal Power-Aware Routing Protocol (BIPAR) (Morcos, Matta, & Bestavros, 2004). BIPAR has two modes of operation; min-power and max-power routing. Min-power routing is a routing scheme that delivers packets over the minimum-power path from the source to the destination. The other mode is max power routing. Max-power routing uses more power to route packets and it favors paths of physically longer hops to those of shorter hops. The operation of BIPAR has two phases: cost establishment phase and data forwarding phase. In cost Establishment Phase, the routing status in the forwarding sensor nodes is set up. The sink sends Advertisement packet (ADV). The ADV packet is used to assign costs to each node. A node's cost is the least amount of power needed to transmit packets from this node to the sink. The ADV packet has a cost field. When the sink first broadcasts it, the ADV packet has a cost of 0. Upon receiving an ADV packet from node Y, Node X sets its own cost as the sum of the cost field in the ADV packet and the amount of power needed to transmit packets from Y to X. Then X sets the cost of the ADV to its own packet and rebroadcasts the packet. In addition, each node can utilize the ADV packet to build its list of neighbors toward the sink. This list is considered as the routing table of each node. The list of neighbors for a node X contains any node that has cost less than node X. in data forwarding phase, sensor nodes sense the environment and send their measured data back to the sink. The source assigns a power budget to each data packet it sends. This budget is the total amount of power to be used to forward this packet from the source to the sink. Along with the budget, the source node sends sender's cost and consumed power so far. Upon receiving any data packet from node Y, node X compares its own cost to the cost of the sender. Node X can only rebroadcast the packet; if its cost is less than that of Y. otherwise X drops the packet. If X decides to rebroadcast the packet, it calculates the power needed to send the packet from Y to itself and update the consumed power so far field of the packet. The latter is checked against the budget allowed for this packet. If the packet has exceeded its budget, X drops it. X then consults its neighbors' list and picks its closest neighbor to forward this packet to it. X then waits for an acknowledgement (ACK) for a predefined timeout interval. If X gets an ACK for its packet during this timeout interval, then X's job is done concerning this packet. Otherwise, X would consult its neighbors' list again, this time picking its furthest neighbor to forward the same packet to it.

SPEED

SPEED is a real-time communication protocol for sensor networks (Tian et al., 2003). It provides three types of real-time communication services; real-time unicast, real-time area-multicast and real-time area-anycast. SPEED is a stateless, localized algorithm with minimal control overhead. End-to-end soft real-time communication is achieved by maintaining a desired delivery speed across the sensor network through a novel combination of feedback control and non-deterministic geographic forwarding. SPEED is a highly efficient and scalable protocol for sensor networks where the resources of each node are scarce. In SPEED protocol, each node should maintain information about its neighbors. Geographic forwarding is used to find the paths. SPEED protocol strives to ensure end-to-end delay for the packets in the network such that each application can estimate the end-to-end delay for the packets. SPEED protocol consists of the following components: A neighbor beacon exchange scheme, a delay estimation scheme, The Stateless Non-deterministic Geographic Forwarding algorithm (SNGF), A Neighborhood Feedback Loop (NFL), Backpressure Rerouting, and Last mile processing. SNGF is the routing module responsible for choosing the next hop candidate that can support the desired delivery speed. NFL and Backpressure Rerouting are two modules to reduce or divert traffic when congestion occurs, so that SNGF has available candidates to choose from. The last mile process is provided to support the three communication semantics mentioned before. Delay estimation is the mechanism by which a node determines whether or not congestion has occurred. And beacon exchange provides geographic location of the neighbors so that SNGF can do geographic based routing.

Figure 13 shows a classification of routing protocols based on the application.

LITERATURE REVIEW OF CROSS LAYER DESIGN IN WSN

Many researchers studied the necessity and possibility of taking advantages of cross layer design to improve the power efficiency and system throughput of Wireless sensor network.

Ahmed Safwat et al proposed Optimal Cross-Layer Designs for Energy-efficient Wireless Ad hoc and Sensor Networks (Safwati, Hassanein, & Mouftah, 2003). They propose Energy-Constrained Path Selection (ECPS) scheme and Energy-Efficient Load Assignment (E2LA). ECPS is a novel energy-efficient scheme for wireless ad hoc and sensor networks. it utilizes cross-layer interactions between the network layer and MAC sublayer. The main objective of the ECPS is to maximize the probability of sending a packet to its destination in at most n transmissions. To achieve this objective, ECPS employs probabilistic dynamic programming (PDP) techniques assigning a unit reward if the favorable event (reaching the destination in n or less transmissions) occurs, and assigns no reward otherwise. Maximizing the expected reward is equivalent to maximizing the probability that the packet reaches the destination in at most n transmissions. Ahmed Safwat et. al, find the probability of success at an intermediate node *i* right before the tth transmission $f_t(i)$:

$$f_t(i) = \begin{cases} 1 & i = D \\ \max_j \sum_k p_k f_{t+k}(j) & otherwise \end{cases} \tag{10}$$

Where

D: Destination node
j: The next hop towards the destination D

Any energy-aware route that contains D and the distance between D and the source node is less or equal to n can be used as input to ECPS. The MAC sub-layer provides the network layer with the information pertaining to successfully receiving CTS or an ACK frame, or failure to receive one. Then ECPS chooses the route that will minimize the probability of error

The objective of the E2LA scheme is to distribute the routing load among a set Z of Energy-aware routes. Packets are allotted to routes based on their willing to save energy. Similar to ECPS, E2LA employs probabilistic dynamic programming techniques and utilize cross-layer interactions between the network and MAC layers. At the MAC layer, each node computes the probability of successfully transmitting packets in α attempt. E2LA assign loads according to four distinct reward schemes (Safwati, Hassanein, & Mouftah, 2003).

Parvathinathan Venkitasubrananiam et al propose a novel distribution medium access control scheme called opportunistic ALOHA (O-ALOHA) for reachback in sensor network with mobile agent (IEEE, 1999). The proposed scheme based on the principle of cross layer design that integrates physical layer characteristics with medium access control. In the O-ALOHA scheme,

Figure 13. Classification of routing protocols based on the applications

Protocol	Application		
	Query Based	Event Driven	Periodic
SPIN	v		
Directed Diffusion	v		
Shah et al.			v
Rumor Routing	v		
CADR	v		
COUGAR	v		
ACQIRE	v		
GBR	v		
O(1)-Reception Routing Protocol		v	
EMPR	v		
LEACH		v	
EAD		v	
TinyDB	v		
PEGASIS		v	
TEEN		v	
APTEEN			v
UCR		v	
BCDCP		v	
GAF		v	v
MECN		v	
GEAR	v	v	
GOAFR		v	
MBR		v	
GMREE		v	
Zhao et al. Randomly Shifted Anchors:		v	
Chang et al			v
Kalpakis et al.		v	v
Minimum Cost Forwarding		v	
SAR		v	v
Energy-Aware QoS Routing Protocol		v	v
BIPAR		v	
SPEED	v		

each sensor node transmits its information with a probability that is a function of its channel state (propagation channel gain). This function called transmission control is then designed assuming that orthogonal CDMA is employed to transmit information. In designing the O-ALOHA scheme they consider a network with n sensors communicate with a mobile agent over a common channel. It is assumed that all the sensor nodes have data to transmit when the mobile agent is in the vicinity of the network. Time is slotted into intervals with equal length equal to the time required to transmit a packet. The network is assumed to operate in time division duplex (TDD) mode. At the beginning of each slot, the collection agent transmits a beacon. The beacon is used by each sensor to estimate the propagation channel gain from the collection agent to itself which is the same as the channel gain from the sensor to the collection agent. It is assumed that the channel estimation is perfect. The propagation channel gain from sensor i to the collection agent during slot t which is denoted as $\gamma_i^{(t)}$ is modeled as:

$$\gamma_i^{(t)} = \frac{P_T R_{it}^2}{r_i^2 + d^2} \tag{11}$$

where

R_{it}^2: is Rayleigh Distribution
P_T: The transmission power of each sensor.
ri: radial distance of sensor i
d: distance from collecting agent and sensor node.

During the data transmission period, each sensor transmits its information with a probability $s(\gamma_i^{(t)})$ where $s(.)$ is a function that maps the channel state to a probability. Two transmission controls are proposed to map from the channel gain to the probability; Location independent transmission control (LIT) and Location aware transmission control (LAT). In LIT, the decision to transmit a packet is made by observing channel state γ alone, while in LAT, every sensor makes an estimate of its radial distance and the decision to transmit is a function of both the channel state γ and the location of sensor.

Mihail L. proposed a deterministic schedule based energy conservation scheme (Sichitiu, 2004). In the proposed approach, time synchronized sensors form on-off schedules that enable the sensors to be awake only when necessary. The energy conservation is achieved by making the sensor node go to sleeping mode. The proposed approach is suitable for periodic applications only, where data are generated periodically at deterministic time. The proposed approach requires the cooperation of both the routing and MAC layers. The on-off schedule is built according to the route determined by routing protocol. The proposed approach consists of two phases; the Setup and reconfiguration phase and the steady state phase. In the setup and reconfiguration phase, a route is selected from the node originating the flow to the base station then the schedules are setup along the chosen route. In the steady phase, the nodes use the schedule established in the setup and configuration phase to forward the data to the base station. In this phase, there will be three types of actions at each node; Sample action which is taking data sample from environment, Transmit action to transmit data, and Receive action to receive data. The actions at each node along with the time when each action will take place are stored in the schedule table of each node. The node can be awake at the time of each action and go to sleep otherwise.

Li-Chun Wang and Chung-Wei Wang proposed Cross layer Design of Clustering architecture for wireless Sensor Networks. The proposed scheme is called Power On With Elected Rotation (POWER) (Venkitasubramaniam, Adireddy, & Lang, 2003). The objective of the POWER is to determine the optimal number of clusters from the cross-layer aspects of power saving and coverage performance simultaneously. The basic concept of the POWER is to select a representation sensor node in each cluster to transmit the sensing information in the coverage area of the sensor node. The representative sensor node in a cluster rotated from all the sensor nodes in each cluster. In the POWER scheme, the scheduling procedure is rotated many rounds. In each round, there are two phases; the construction table phase (CTP), to construct the rotation table and the rotational representative phase (RRP) to transmit data. In CTP, all sensor nodes employ the MAC protocol and the first sensor node accessing the channel become the initiator node, then the initiator node detects other neighboring node and form s the cluster. RRP starts after constructing the rotation table. RRP is divided into many sRPs (Sub-Rotated Period). In each sRP, one node will be a representative node and all other nodes in the cluster will be in sleeping mode.

Rick W. Ha et al proposes a cross-layer sleep-scheduling-based organization approach, called Sense-Sleep Trees (SS-trees)(Rick, Pin-Han, & Shen, 2006). The proposed approach aims to harmonize the various engineering issues and provides a method of increasing monitoring coverage and operational lifetime of mesh-based WSNs engaged in wide-area surveillance applications. An iterative algorithm is suggested to determine the feasible SS-tree structure. All the SS trees are rooted at the sink. Based on the computed SS-trees, optimal sleep schedules and traffic engineering measures can be devised to balance sensing requirements, network communication constraints, and energy efficiency. For channel access a simple single-channel CSMA MAC with implicit acknowledgements (IACKs) is selected. In SS-trees approach, the WSN's life cycle goes through many stages. After the initial deployment of nodes, the WSN will enter the network initialization stage, in which the sink gathers network connectivity information from sensor nodes, compute the SS-trees, and disseminate the sleep schedules to every sensor node. Then the WSN will enter the operation stage, in which the nodes will alternate between Active and sleep stages. During long periods when sensing services are not needed the entire WSN will enter the Hibernation mode to conserve energy. The SS-trees must be computed with minimizing number of shared nodes (nodes belonging to multiple SS-trees), minimizing co-SS tree neighbors of each node, and minimizing the cost of forwarding messages between the data sink and each node. Rick W. Ha et al proposes a greedy algorithm to compute the SS-trees. The proposed algorithm follows a greedy depth-first approach that constructs the SS-trees from the bottom up on a branch-by-branch basis. After computing the SS-trees, an optimal sleep schedule that maximizes energy efficiency must be determined. The length of the active and sleep period will increase the data delay. The proposed SS-Tree design streamlines the routing procedures by restricting individual sensor nodes to only maintain local connectivity information of its immediate 1-hop neighbors.

Shuguang Cui et al, emphasize that the energy efficiency must be supported across all layers of the protocol stack through a cross-layer design(Cui et al., 2005). They analyze energy-efficient joint routing, scheduling, and link adaptation strategies that maximize the network lifetime. They propose variable-length TDMA schemes where the slot length is optimally assigned according to the routing requirement while minimizing the energy consumption across the network. They show that the optimization problems can be transferred into or approximated by convex problems that can be solved using known techniques. They show that link adaptation be able to further improve the energy efficiency when jointly designed with MAC and routing. In addition to reduce energy consumption, Link adaptation may reduce transmission time in relay nodes by using higher constellation sizes such as the extra circuit energy consumption is reduced.

Weilian Su and Tat L. Lim propose a cross layer design and optimization framework, and the concept of using an optimization agent (OA) to provide the exchange and control of information between the various protocol layers to improve performance in wireless sensor network (Su & Lim, 2006). The architecture of the proposed framework, as shown in Figure 14, which is redrawn from(Su & Lim, 2006), consists of a proposed optimization agent (OA) which facilitates interaction between various protocol layers by serving as a database where essential information such as node identification number, hop count, energy level, and link status are maintained.

Weilian Su and Tat L. conduct the performance measurements to study the effects of interference

Figure 14. A proposed cross-layer optimization framework

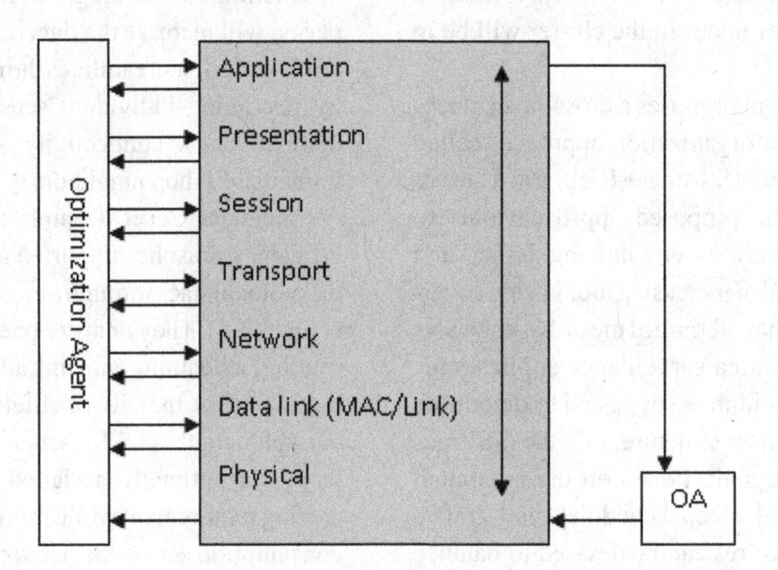

and transmission range for a group of wireless sensors. The results of their performance measurements help to facilitate the design and development of the OA. The OA can be used to trigger an increase in transmit power to overcome the effects of mobility or channel impairments due to fading when it detects a degradation due in BER. Alternatively, it can reduce the transmit power to conserve energy to prolong its lifetime operations in the absence of mobility or channel fading. The OA can also be used to provide QoS provisioning for different types of traffic. This can be done by tagging different priority traffic with different transmit power levels.

Changsu Suli et al proposed an energy efficient cross-layer MAC protocol for WSN. It is named MAC-CROSS (Changsu, Young-Bae, & Dong-Min, 2006). In the proposed protocol, the routing information at the network layer is utilized for the MAC layer such that it can maximize sleep duration of each node. in MAC-CROSS protocol the nodes are categorized into three types: Communicating Parties (CP) which refers to any node currently participating in the actual data transmission, Upcoming Communicating Parties (UP) which refers to any node to be involved in the actual data transmission, and Third Parties (TP) which refers to any node are not included on a routing path. The UP nodes are asked to wake up while other TP nodes can remain in their sleep modes. The RTS/CTS control frames are modified in the MAC-CROSS protocol. The modification is needed to inform a node that its state is changed to UP or TP in the corresponding listen/sleep period. a new field; Final_destination_Addr, is added to the RTS. On the other hand, a new field; UP_Addr is added to the CTS and it informs which node is UP to its neighbors. When a node B receives an RTS from another node A including the final destination address of the sink, B's routing agent refers to the routing table for getting the UP (node C) and informs back to its own MAC. The MAC agent of Node B then transmits CTS packet including the UP information. After receiving the CTS packets from node B, C changes its state to UP and another neighbor nodes change their states to TP and will go to sleep.

The work in (Zhang & Shen, 2009) is based on maximizing the network lifetime by balancing energy consumption among all nodes. To achieve this purpose, the network is divided into coronas centered at the sink with equal width, and all nodes in the same corona use the same probability for direct transmission and the same probability for hop-by-hop transmission. Now, the energy consumption balancing problem has two parts. The first is the intra-corona energy consumption which is solved by optimally dividing each corona to evenly distribute the amount of data received by nodes in each corona. The second part is inter-corona energy consumption balancing that is solved by optimally allocating the amount of data for direct transmission and hop by-hop transmission. Finally, the network lifetime maximization problem is solved by calculating the optimal number of coronas that the network should be divided into.

To maximize the network lifetime, the work in (He, Lee, & Guan, 2009) takes a new direction by jointly optimizing the source rate and the encoding power at each video sensor, and the link rates of each session, subject to the requirement of the collected video quality. To ease the development of an efficient distributed algorithm, the objective function as well as the inequality constraint functions is formatted as strictly convex; therefore, there exists a strictly feasible solution that satisfies all the inequality constraints

Another effort in the same direction is considered in is taken in (Kim, Wang, & Madihian, 2006). A cross-layer approach is proposed for lifetime maximization of distributed wireless sensor networks. The routing and medium access control (MAC) constraints are jointly formulated into a linear program (LP) using the flow contention graph model. The resulting formulation is a separable structure, which can be solved distributively using dual decomposition. Moreover, the MAC layer constraints are relaxed in the form of a penalty function, which facilitates distributed optimization.

A cross-layer model involving the link layer (Madan et al., 2007), the medium access control (MAC) layer, and the routing layer is considered. To maximize the network lifetime using these models, the problem is formulated to optimize the transmission schemes. The problem is solved sequentially; optimization of a single layer is considered at a time, while keeping the other layers fixed. The objective is to select the transmission rate for each link to minimize the power consumption on the links and hence maximize the network lifetime. The authors solved the optimization problems exactly for TDMA networks, while for networks with interference, approximation approaches were proposed.

A cluster formation algorithm based on the ILP is proposed in(Dai, Li, & Xu, 2006). The energy consumption is considered as an optimization parameter while clustering is imperative. A comparison between the algorithm and the cluster formation method using in LEACH is conducted.

Chamam(Chamam & Pierre, 2007) addresses the problem of maximizing sensor networks lifetime under area coverage constraint. They propose a scheduling mechanism that calculates, for every time slot of the network operating period, an optimal covering subset of sensors that will be activated while all other sensors will go on Sleep. These mechanisms aim at balancing energy dissipation over sensors, thus maximizing network lifetime. They model this problem as an Integer Linear Programming (ILP) problem, which is resolved using ILOG CPLEX Furthermore, a greedy heuristic approach is proposed to tackle the exponentially increasing processing time of CPLEX.

Friderikos et al (Papadaki & Vasilis, 2008) propose a family of mathematical programs for both the uncapacitate and capacitated joint gateway selection and routing (U/C-GSR) problem in wireless mesh networks. They formulate the problem using the shortest path cost matrix (SPM) and prove that it gives the optimal solution when applied to the uncapacitated case but can lead to an arbitrary large optimality gap in the capacitated case. Furthermore, an augmented mathematical program is developed where link capacities are allowed to take values from a discrete set depending on the link distance. In this case, the multi-rate capabilities of WMNs (via, for example, adaptive modulation and coding) can be modeled. Evidence from numerical investigations shows that using the SPM formulation realistic network sizes of WMNs can be solved.

Raja and baljai (Jothi & Raghavachari, 2005) presents a formulation to the Capacitated Minimal Directed Tree Problem. The formulation is amenable to several relaxation procedures. The proposed formulation is proven to be loop free. The objective is to minimize the total capacity at all links of the tree given the maximum capacity at each link is known.

In (Gatzianas & Georgiadis, 2008), a mobile sink problem roaming the monitored area is formulated as a linear programming problem. The study focuses on maximizing the network survivability, which is defined as the summation of time intervals during which the sink resides at location l.

Here is another work studying the placement of mobile sinks (Saad & Tourancheau, 2009) in wireless sensor networks inside buildings, to avoid the energy hole problem and to extend the network lifetime. Given all sensors are stationary and located in a grid of same size cells, find the optimal positioning of a fixed number of sinks moving in the grid from feasible site to another one until the first sensor dies. The problem is formulated as an ILP, which directly maximizes the network lifetime instead of minimizing the energy consumption or maximizing the residual energy.

In (Osais, St-Hilaire, & Yu, 2009), the authors explored the optimal placement for directional wireless sensors. Integer linear programming model is used to minimize the number of directional sensors that need to be deployed to monitor a set of discrete targets in a sensor field. The formulation considers coverage as well as connectivity problem.

The authors in (Chamam & Pierre, 2009) assume that the network is dense and the position of each sensor is fixed and known to the processing node (PN). To save network energy and increase its lifetime, a selective set of sensors will be on while all other sensors are turned off. In addition, the sensors are forming clusters and all cluster heads have to belong to a single connected graph. To maximize network lifetime while ensuring simultaneously full area coverage and sensor connectivity to cluster heads, the problem is formulated as a linear programming model such that sensors will be selected essentially according to their residual energies. The model favors the activation of sensors having relatively high residual energy and when the residual energy is relatively high, the optimal solution will tend to activate as less sensors as possible.

In (Bari, Teng, & Jaekl, 2009), a wireless sensor network composed of randomly distributed sensors is formed in a hierarchical structure to increase the network connectivity, coverage and lifetime. In addition, mobile data collectors are roaming the network along a fixed trajectory to collect data and ensure coverage. The potential positions of the relay nodes are computed based on the grid approach. Furthermore, integer linear program (ILP) formulation is presented to find the minimum number of relay nodes along with their locations taking into consideration the sensor data rates, the relay nodes buffer size and the relay node speed. The model ensures that there is no data loss due to relay node buffer overflow and the energy dissipation does not exceed a specified level.

Figure 15 shows summary of cross-layer design protocols for WSN.

FUTURE RESEARCH DIRECTIONS

Energy consumption is one of the critical issues in WSN. The sensor nodes usually have limited energy source (battery). It is usually difficult to replace these energy sources. Therefore, the need of energy efficient protocols is still one of the hot research issues in WSN. Cross layer optimization one of the techniques that can be followed to minimize the energy consumption in WSN. In cross layer optimization different factors from different layers are considered to optimize the performance of the network.

CONCLUSION

In this chapter, we present literature review for MAC, routing, and cross layer design protocols that proposed for WSN. We discuss in depth the most well-know MAC protocols for WSN. A comparison among theses protocols is presented.

Figure 15. Summary of cross layer protocols for WSN

Protocol	Layers	Approach	Evaluation method	Application	Network Topology	Cross layer Objective	Performance metrics
ECPS	MAC, Network	Mathematical Model: probabilistic dynamic programming	Experiment		Random (Static)	Maximization of probability of sending packet to its Datn transmission	Energy
E2LA	MAC, Network	Mathematical Model: probabilistic dynamic programming	Experiment		Random (Static)	Minimize Energy :- Multiple simultaneous routes Load distribution	Energy
MAC CROSS	MAC, Network	Heuristic	Simulation Hardware Implementation (MICAZ)		Random (Static)	Maximize Sleep Duration	Energy
O-Aloha	Physical, MAC	Heuristic	Simulation	SENMA	Random	Maximize throughput	Throughput
POWER	Physical, MAC, Network	Heuristic			Uniform (Static)	Optimize number of cluster	Energy
Weilian Su	ALL layers	Framework (optimization Agent)	Experimental (MICAZ)		Random	Optimize performance of WSN	Link Quality Packet Received
Shunguang Cui	Routing, MAC, Link layer	Modeling as optimization problem	Analytical		Random	Maximize network lifetime	Network lifetime
Sense-Sleep Trees (SS-Trees)	MAC, Network	Heuristic	Simulation	Surveillance	Mesh-based	Maximizing Network lifetime, and monitoring coverage	Network lifetime Energy consumed
Game Theoretic Approach	Application, Physical	Game Theory	Analytical		Random	Minimize total distortion	Distortion coverage
In Yeup Kong	Physical, MAC, Network	Mathematical	Analytical		Random	Maximize Network lifetime	
Cross Layer Scheduling	MAC, Network	Heuristic	Simulation	Periodic	Random	Maximize network lifetime	Network lifetime
Cross Layer design for cluster formulation	MAC, Physical, Network	Heuristic	Simulation	Periodic	Uniform distribution	Maximize network lifetime	Network lifetime

Moreover, the major advantages and disadvantages of each protocol are discussed. Moreover, we discuss the routing protocols for WSN. The discussed protocols are classified into data centric routing protocols, Hierarchical routing protocols, location based protocols, and QoS aware routing protocols. Moreover, A Classification of Routing Protocols based on the Application is presented. Finally, some cross layer design protocols are discussed. A comparison among the discussed protocols according to layers integrated, intended applications, cross-layer objectives, and the evaluation approach is presented.

REFERENCES

Akkaya, K., & Younis, M. (2005). A survey of routing protocols in wireless sensor networks. *The Elsevier Ad Hoc Network Journal*, *3*(3), 325–349. doi:10.1016/j.adhoc.2003.09.010

Akylidiz, F., Su, W., Sankarasubramaniam Y., & Cayirci, E. (2002 August). A survey on sensor networks. *IEEE Personal Communications Magazine,* 102-114

Al-Khdour, T., & Baroudi, U. (2007). A generalized energy-aware data centric routing for wireless sensor network. *The 2007 IEEE International Conference on Signal Processing and Communications (ICSPC 2007),* (pp. 117–120). Dubai, United Arab of Emirates (UAE).

Al-Khdour, T., & Baroudi, U. (2009). A generalized energy-efficient time-based communication protocol for wireless sensor networks. *Special Issue of International Journal of Internet Protocols*, *4*(2), 134–146. doi:10.1504/IJIPT.2009.027338

Al-Khdour, T., & Baroudi, U. (2010). An energy-efficient distributed schedule-based communication protocol for periodic wireless sensor networks. *The Arabian Journal of Science and Engineering*, *35*(2B), 153–168.

Bachir, A., Barthel, D., Heusse, M., & Duda, A. (2007). O(1)-Reception routing for sensor networks. *Computer Communications*, *30*, 2603–2614. doi:10.1016/j.comcom.2007.05.054

Bari, A., Teng, D., & Jaekel, A. (2009). Optimal relay node placement in hierarchical sensor networks with mobile data collector. *The 18th International Conference on Computer Communications and Networks* (pp. 1-6) San Francisco, CA, USA.

Biaz, S., & Barowski, Y. (2004, April). GANGS: An energy efficient MAC protocol for sensor networks. *The 42nd Annual Southeast Regional Conference.* (pp. 82-87). Huntsville, AL, USA.

Boukerche, A., Cheng, X., & Linus, J. (2005). A performance evaluation of a novel energy-aware data-centric routing algorithm in wireless sensor networks. *Wireless Networks*, *11*, 619–635. doi:10.1007/s11276-005-3517-6

Braginsky, D., & Estrin, D. (2002, October). Rumor routing algorithm for sensor networks. *The First Workshop on Sensor Networks and Applications (WSNA),* (pp. 22-31). Atlanta, GA.

Chamam, A., & Pierre, S. (2007). Energy-efficient state scheduling for maximizing sensor network lifetime under coverage constraint. *The Third IEEE International Conference on Wireless and Mobile Computing, Networking and Communications*, (p. 63). Washington, DC, USA.

Chamam, A., & Pierre, S. (2009). On the planning of wireless sensor networks: Energy-efficient clustering under the joint routing and coverage constraint. *IEEE Transactions on Mobile Computing*, *8*(8), 1077–1086. doi:10.1109/TMC.2009.16

Chang, J., & Tassiulas, L. (2004, August). Maximum lifetime routing in wireless sensor networks. *IEEE/ACM Transactions on Networking*, 609–619. doi:10.1109/TNET.2004.833122

Changsu, S., Young-Bae, K., & Dong-Min, S. (2006). An energy efficient cross-layer MAC protocol for wireless sensor networks. *The Eighth Asia Pacific Web Conference,* (pp. 410–419), Harbin, China.

Chen, B., Jamieson, K., Balakrishnan, H., & Morris, R. (2002). SPAN: An energy-efficient coordination algorithm for topology maintenance in ad hoc wireless networks. *Wireless Networks*, *8*(5), 481–494. doi:10.1023/A:1016542229220

Chen, G., Li, C., Ye, M., & Wu, J. (2007, April). An unequal cluster-based routing strategy in wireless sensor networks. *Wireless Networks*, *15*(2), 193–207. doi:10.1007/s11276-007-0035-8

Chen, Y., & Nasser, N. (2006, August). *Energy-balancing multipath routing protocol for wireless sensor networks*. The Third International Conference on Quality of Service in Heterogeneous Wired/Wireless Network, Waterloo, Ontario, Canada.

Chu, M., Haussecker, H., & Zhao, F. (2002, August). Scalable information-driven sensor querying and routing for ad hoc heterogeneous sensor networks. *International Journal of High Performance Computing Applications*, *16*(3), 293–313. doi:10.1177/10943420020160030901

Cui, S., Madan, R., Goldsmith, A., & Lall, S. (2005, May). Joint routing, MAC, and link layer optimization in sensor networks with energy constraints. *IEEE International Conference on Communications, ICC 2005, Vol. 2*, (pp. 725–729). Seoul Korea.

Dai, S., Li, L., & Xu, D. (2006, October). A novel cluster formation approach based on the ILP for wireless sensor networks. *The First International Conference on Communications and Networking in China*, (pp. 1-5, 25-27). China.

Dam, T., & Langendoen, K. (2003, Nov.). An adaptive energy-efficient MAC protocol for wireless sensor networks. *The 1st International Conference on Embedded Networked Sensor System* (pp. 171-180). Los Angeles, California, USA.

Gatzianas, M., & Georgiadis, L. (2008). A distributed algorithm for maximum lifetime routing in sensor networks with mobile sink. *IEEE Transactions on Wireless Communications*, *7*(3), 984–994. doi:10.1109/TWC.2008.060727

He, Y., Lee, I., & Guan, L. (2009). Distributed algorithms for network lifetime maximization in wireless visual sensor networks. *IEEE Transactions on Circuits and Systems for Video Technologies*, *19*(5), 704–718. doi:10.1109/TCSVT.2009.2017411

Hedetniemi, S., & Liestman, A. (1988). A survey of gossiping and broadcasting in communication networks. *NETWORKS: Networks: An International Journal*, *18*(4), 319–349. doi:10.1002/net.3230180406

Heinzelman, W., Chandrakasan, A., & Balakrishnan, H. (2002, Oct.). An application-specific protocol architecture for wireless microsensor networks. *IEEE Transactions on Wireless Communications*, *1*(4), 660–670. doi:10.1109/TWC.2002.804190

Heinzelman, W., Kulik, J., & Balakrishnan, H. (1999, August). Adaptive protocols for information dissemination in wireless sensor networks. *The 5th Annual ACM/IEEE International Conference on Mobile Computing and Networking (MobiCom'99)* (pp. 174-185). Seattle, WA.

IEEE. (1999). *IEEE 802.11 standards, part 11: Wireless medium access control (MAC) and physical layer (PHY) specifications. Technical report.* The Working Group for WLAN Standards.

Intanagonwiwat, C., Govindan, R., & Estrin, D. (2000, August). Directed diffusion: A scalable and robust communication paradigm for sensor networks. *The 6th Annual ACM/IEEE International Conference on Mobile Computing and Networking (MobiCom'00)*, (pp. 56-67). Boston, MA.

Jothi, R., & Raghavachari, B. (2005). Approximation algorithms for the capacitated minimum spanning tree problem and its variants in network design. *ACM Transactions on Algorithms*, *1*(2), 265–282. doi:10.1145/1103963.1103967

Kalpakis, K., Dasgupta, K., & Namjoshi, P. (2002). *Maximum lifetime data gathering and aggregation in wireless sensor networks*. Presented at The IEEE International Conference on Networking (NETWORKS '02), Atlanta, GA.

Kim, S.-J., Wang, X., & Madihian, M. (2006). Joint routing and medium access control for lifetime maximization of distributed wireless sensor networks. *IEEE International Conference on Communications* (pp. 3467-3472). Istanbul, Turkey.

Kuhn, F., Wattenhofer, R., & Zollinger, A. (2003). Worst-case optimal and average-case efficient geometric ad-hoc routing. *The 4th ACM International Conference on Mobile Computing and Networking*, (pp. 267-278). Dallas, TX, USA.

Li, J., & Lazarou, G. (2000, January). A bit-map-assisted energy-efficient MAX scheme for wireless sensor networks. *Hawaii International Conference on Systems Sciences*, (pp. 3005–3014). Maui, Hawaii.

Lin, P., Qiao, C., & Wang, X. (2004, March). Medium access control with a dynamic duty cycle for sensor networks. *The IEEE Wireless Communications and Networking Conference (WCNC)*, (pp. 1534-1539). Atlanta, Georgia, USA.

Lindsey, S., & Raghavendra, C. (2002, March). PEGASIS: Power efficient gathering in sensor information systems. *The IEEE Aerospace Conference*, (pp. 1125 - 1130). Big Sky, Montana.

Lindsey, S., Raghavendra, C., & Sivalingam, K. (2001, April). Data gathering in sensor networks using the energy*delay metric. *The IPDPS Workshop on Issues in Wireless Networks and Mobile Computing*, (pp. 2001-2008). San Francisco, CA.

Lotfifar, F., & Shahhoseini, H. (2006). A mesh-based routing protocol for wireless ad-hoc sensor networks. *The International Wireless Communication and Mobile Computing Conference (IWCMC'06)*, (pp. 115-120). Vancouver, British Columbia, Canada.

Madan, R., Member, S., Cui, S. L., & Goldsmith, A. J. (2007). Modeling and optimization of transmission schemes in energy-constrained wireless sensor networks. *IEEE/ACM Transactions on Networking*, *15*(6), 1359–1362. doi:10.1109/TNET.2007.897945

Madden, S., Franklin, M., Hellerstein, J., & Hong, W. (2005, March). TinyDB: An acquisitional query processing system for sensor networks. *ACM Transactions on Database Systems*, *30*(1), 122–173. doi:10.1145/1061318.1061322

Manjeshwar, A., & Agrawal, D. (2001, April). TEEN: A protocol for enhanced efficiency in wireless sensor networks. *The 1st International Workshop on Parallel and Distributed Computing Issues in Wireless Networks and Mobile Computing*, (pp. 2009-2015). San Francisco, CA.

Manjeshwar, A., & Agrawal, D. (2002, April). APTEEN: A hybrid protocol for efficient routing and comprehensive information retrieval in wireless sensor networks. *The 2nd International Workshop on Parallel and Distributed Computing Issues in Wireless Networks and Mobile computing*, (pp. 195 – 202). Ft. Lauderdale, FL.

Mishar, S., & Nasipuri, A. (2004). An adaptive low power reservation based MAC protocol for wireless sensor networks. *The IEEE International Conference on Performance, Computing, and Communications*, (pp. 731–736). Phoenix, Arizona.

Modules, S. (n.d.). *Manual data sheet for MICAz*. Retrieved from http://www.xbow.com

Morcos, H., Matta, I., & Bestavros, A. (2004). *BIPAR: Bimodal power-aware routing protocol for wireless sensor networks*. Presented at the 1st International Computer Engineering Conference New Technologies for the Information Society (ICENCO). Cairo, EGYPT.

Osais, Y., St-Hilaire, M., & Yu, F.-R. (2009). On sensor placement for directional wireless sensor networks. *IEEE International Conference on Communications*, (pp. 1-5). Dresden, Germany.

Papadaki, K., & Vasilis, F. (2008, March). Joint routing and gateway selection in wireless mesh networks. *The IEEE Conference of Wireless Communications and Networking, (WCNC)* (pp. 2325-2330). Las Vegas, USA.

Rick, H., Pin-Han, H., & Shen, X. (2006). Cross-layer application-specific wireless sensor network design with single-channel CSMA MAC over sense-sleep trees. *Computer Communications, 29*(17), 3425–3444. doi:10.1016/j.comcom.2006.01.019

Rodoplu, V., & Ming, T. H. (1999). Minimum energy mobile wireless networks. *IEEE Journal on Selected Areas in Communications, 17*(8), 1333–1344. doi:10.1109/49.779917

Saad, L., & Tourancheau, B. (2009). Multiple mobile sinks positioning in wireless sensor networks for buildings. *The Third IEEE International Conference on Sensor Technologies and Applications (SENSORCOMM)*, (pp. 264-270). Athens, Greece.

Sadagopan, N., Krishnamachari, B., & Helmy, A. (2003, May). The ACQUIRE mechanism for efficient querying in sensor networks. *The First International Workshop on Sensor Network Protocol and Applications*, (pp. 149-155). Anchorage, Alaska.

Safwati, A., Hassanein, H., & Mouftah, H. (2003, April). Optimal cross-layer designs for energy-efficient wireless ad hoc and sensor networks. *The IEEE International Conference of Performance, Computing, and Communications* (pp. 123 – 128).

Sanchez, J., Ruiz, P., & Stojmenovic, I. (2007). Energy-efficient geographic multicast routing for sensor and actuator networks. *Computer Communications, 30*, 2519–2531. doi:10.1016/j.comcom.2007.05.032

Schurgers, C., & Srivastava, M. (2001). Energy efficient routing in wireless sensor networks. *The MILCOM Proceedings on Communications for Network-Centric Operations: Creating the Information Force*, (pp. 357 - 361). McLean, VA.

Shah, R., & Rabaey, J. (2002, March). Energy aware routing for low energy ad hoc sensor networks. *The IEEE Wireless Communications and Networking Conference (WCNC)*, vol. 1 (pp. 350 - 355). Orlando, FL.

Sichitiu, M. L. (2004). Cross-layer scheduling for power efficiency in wireless sensor networks. *INFOCOM 2004, Twenty-third Annual Joint Conference of the IEEE Computer and Communications Societies*, vol. 3, (pp. 1740-1750). Hong Kong, China.

Singh, S., & Raghavendra, C. (1998, July). PAMAS: Power aware multi-access protocol with signalling for ad hoc networks. *ACM Computer Communications Review, 28*(3), 5–26. doi:10.1145/293927.293928

Stojmenovic, I., & Lin, X. (1999, November). *GEDIR: Loop-free location based routing in wireless networks*. Paper presented at In International Conference on Parallel and Distributed Computing and Systems, Boston, MA, USA.

Su, W., & Lim, T. L. (2006, June). Cross-layer design and optimization for wireless sensor networks. *The Seventh ACIS International Conference on Software Engineering, Artificial Intelligence, Networking, and Parallel/Distributed Computing* (pp. 278 – 284). Las Vegas, Nevada, USA.

Subramanian, L., & Katz, R. (2000, August). An architecture for building self configurable systems. The *IEEE/ACM Workshop on Mobile Ad Hoc Networking and Computing*, (pp. 63 – 73). Boston, MA.

Subramanian, S., Shakkottai, S., & Gupta, P. (2007, May). On optimal geographic routing in wireless networks with holes and non-uniform traffic. *The 26th IEEE International Conference on Computer Communications. INFOCOM2007*, (pp. 1019-1027). Anchorage, Alaska, USA.

Suh, C., & Ko, Y. (2005, May). A traffic aware, energy efficient mac protocol for wireless sensor networks. *IEEE International Symposium on Circuits and Systems ISCAS 2005,* vol. 3, (pp. 2975 – 2978). Kobe, Japan.

Tian, H., Stankovic, J. A., Chenyang, L., & Abdelzaher, T. (2003, May). SPEED: A stateless protocol for real-time communication in sensor networks. *The International Conference on Distributed Computing Systems*, (pp. 46-55). Providence, RI.

Venkitasubramaniam, P., Adireddy, S., & Lang, T. (2003, Oct.). Opportunistic ALOHA and cross layer design for sensor networks. *Military Communications Conference IEEE,* Vol. 1, (pp. 705 – 710). Boston, MA, USA.

Xia, L., & Chen, X. (2005). *Embedded software and systems*. Berlin, Germany: Springer.

Xu, Y., Heidemann, J., & Estrin, D. (2001, July). Geography-informed energy conservation for ad hoc routing. *The 7th Annual ACM/IEEE International Conference on Mobile Computing and Networking (MobiCom'01)*, (pp. 70-84). Rome, Italy.

Yao, Y., & Gehrke, J. (2002). The cougar approach to in-network query processing in sensor networks. *SIGMOD Record, 31*(3), 9–18. doi:10.1145/601858.601861

Ye, W., Heidemann, J., & Estrin, D. (2004, June). Medium access control with coordinated adaptive sleeping for wireless sensor networks. *IEEE/ACM Transactions on Networking, 12*(3), 493–506. doi:10.1109/TNET.2004.828953

Younis, M., Youssef, M., & Arisha, K. (2002, October). Energy-aware routing in cluster-based sensor networks. *The 10th IEEE/ACM International Symposium on Modeling, Analysis and Simulation of Computer and Telecommunication Systems (MASCOTS2002),* (pp. 129–136). Fort Worth, TX.

Yu, Y., Estrin, D., & Govindan, R. (2001, May). *Geographical and energy-aware routing: A recursive data dissemination protocol for wireless sensor networks*. UCLA Computer Science Department Technical Report, UCLA-CSD TR-01-0023.

Zhang, H., & Shen, H. (2009). Balancing energy consumption to maximize network lifetime in data-gathering sensor networks. *IEEE Transactions on Parallel and Distributed Systems, 20*(10), 1526–1539. doi:10.1109/TPDS.2008.252

Zhao, G., Liu, X., & Sun, M. (2007, Feb). *Anchor based geographic routing for sensor networks using projection distance.* Paper presented at the IEEE International Symposium on Wireless Pervasive Computing (ISWPC), San Juan, Puerto Rico.

ADDITIONAL READING

Akkaya, K., & Younis, M. (2003, May). An Energy-Aware QoS Routing Protocol for Wireless Sensor Networks. *The Proceedings of the IEEE Workshop on Mobile and Wireless Networks (MWN2003)*, (pp. 710 - 715) Providence, Rhode Island.

Baldus, H., Klabunde, K., & Musch, G. (2004). Reliable Set- Up of Medical Body-Sensor Network, EWSN 2004. *LNCS*, *2920*, 353–363.

Buczak, A., & Jamalabad, V. (1998). "Self-organization of a Heterogeneous Sensor Network by Genetic Algorithms," *Intelligent Engineering Systems Through Artificial Neural Networks*, C.H. Dagli, et. (eds.), Vol. 8, pp. 259-264, ASME Press, New York, 1998.

D. B Johnson et al. (1996). Dynamic Source Routing in Ad Hoc Wireless Networks. In Imielinski, T., & Korth, H. (Eds.), *Mobile Computing* (pp. 153–181). Kluwer Academic Publishers.

Dasgupta, K., et al. (2003). "An efficient clustering-based heuristic for data gathering and aggregation in sensor networks," in the *Proceedings of the IEEE Wireless Communications and Networking Conference (WCNC'03)*, New Orleans, Louisiana, March 2003.

Estrin, D., et al. (1999). "Next century challenges: Scalable Coordination in Sensor Networks," in the *Proceedings of the 5th annual ACM/IEEE international conference on Mobile Computing and Networking (MobiCom'99)*, Seattle, WA, August 1999.

Ganesan, D., et al. (2002). "Highly Resilient, Energy Efficient Multipath Routing in wireless Sensor Networks," in *Mobile Computing and Communications Review (MC2R)*, Vol. 1., No. 2. 2002

Gutierrez, J. A., Naeve, M., Callaway, E., Bourgeois, M., Mitter, V., & Heile, B. (2001, September/October). IEEE 802.15.4: A Developing Standard for Low-Power Low-Cost Wireless Personal Area Networks. *IEEE Network*, 12–19. doi:10.1109/65.953229

Karp, B., & Kung, H. T. (2000). "GPSR: Greedy perimeter stateless routing for wireless sensor networks," in the *Proceedings of the 6th Annual ACM/IEEE International Conference on Mobile Computing and Networking (MobiCom '00)*, Boston, MA, August 2000.

Katz, R. H., Kahn, J. M., & Pister, K. S. J. (1999). "Mobile Networking for Smart Dust," in the *Proceedings of the 5th Annual ACM/IEEE International Conference on Mobile Computing and Networking (MobiCom'99)*, Seattle, WA, August 1999.

Krishnamachari, B., Estrin, D., & Wicker, S. (2002). "Modeling Data Centric Routing in Wireless Sensor Networks," in the *Proceedings of IEEE INFOCOM*, New York, NY, June 2002.

Kulik, J., Heinzelman, W. R., & Balakrishnan, H. (2002). Negotiation-Based Protocols for Disseminating Information in Wireless Sensor Networks. *Wireless Networks*, *8*, 169–185. doi:10.1023/A:1013715909417

Lin, C. R., & Gerla, M. (1997, September). Adaptive Clustering for Mobile Wireless Networks. *IEEE Journal on Selected Areas in Communications*, *15*(7). doi:10.1109/49.622910

A. Mainwaring, J, Polastre: Wireless sensor networks for habitat Monitoring, in Proc. Internacional workshop on WSNs and applications, Atlanta, Ga, USA (Sep. performance analysis_formatted----12002).

Min, R., et al. (2001) "Low Power Wireless Sensor Networks", in the *Proceedings of Internation Conference on VLSI Design*, Bangalore, India, January 2001.

Muruganathan, S., Ma, D., Bhasin, R., & Fapojuwo, A. (2005, March). A centralized energy-efficient routing protocol for wireless sensor networks. *IEEE Radio Communication*, S8-S13.

Rabaey, J. M. (2000, July). PicoRadio supports ad hoc ultra low power wireless networking. *IEEE Computer*, *33*, 42–48. doi:10.1109/2.869369

Kay Romer, Friedemann Mattern: The Design Space of Wireless Sensor Networks, IEEE Wireless Communications, pp. 54-61 (December 2004).

Schwiebert, L., Gupta, S. K. S., & Weinmann, J. Reserach challenges in wirelesss networks of th biomedical sensors, in Proc. 7 ACM International Conference on Mobile Computing and Networking (MobiCom '01), pp. 151-165, Rome, Italy (July 2001).

Chien-Chung Shen, Chavalit Srisathapornphat, Chaiporn Jaikaeo: Sensor Information Networking Architecture and Applications, IEEE Personal Communications, pp. 52-59 (August 2001).

Elaine Shi, Adrian Perrig: Designing Secure Sensor Networks IEEE Wireless Communications, pp. 38-43 (December 2004).

Sohrabi, K. (2000, October). Protocols for self-organization of a wireless sensor network. *IEEE Personal Communications*, *7*(5), 16–27. doi:10.1109/98.878532

Tilak, S. (2002). A taxonomy of wireless microsensor network models. *Mobile Computing and Communications Review*, *6*, 28–36. doi:10.1145/565702.565708

Wang, L., & Wang, C. (2004, March). A cross-layer design of clustering architecture for wireless sensor networks. *The IEEE International Conference on Networking, Sensing & Control*, (pp. 547-552). Tapel, Taiwan.

Wendi, B. (2004, January/February). Heinzelman, Amy L. Murphy, Hervaldo S. Carvalho, Mark A. Perillo: Middleware to Support Sensor Network Applications. *IEEE Network*, 6–14.

Younis, M., Munshi, P., & Al-Shaer, E. "Architecture for Efficient Monitoring and Management of Sensor Networks," in the Proceedings of the IFIP/IEEE Workshop on End-to-End Monitoring *Techniques and Services (E2EMON'03)*, Belfast, Northern Ireland, September 2003 (to appear).

Youssef, M., Younis, M., & Arisha, K. "A constrained shortest-path energy-aware routing algorithm for wireless sensor networks," in the *Proceedings of the IEEE Wireless Communication* and Networks Conference (WCNC 2002), Orlando, FL, March 2002.

Zhao, G., Liu, X., & Sun, M. (2007, March). Energy-aware geographic routing for sensor networks with randomly shifted anchors. *The Wireless Communications and Networking Conference (WCNC 2007)*, (pp. 3454-3459). Hong Kong.

KEY TERMS AND DEFINITIONS

Cross Layer Optimization: One of the optimization techniques, in which different factors from different layers are considered to optimize the performance of the network.

Energy Efficient Protocols: Protocols whose main objective is to minimize energy consumption.

MAC Protocols: Protocols to control the access of the shared medium.

Routing Protocols: Protocols to forward data from original source to final destination.

TDMA: Time division multiple access, one of the MAC protocols which the time is multiplexed among several nodes.

Wireless Sensor Network: A set of wireless sensors which are deployed in a specific region. The sensor nodes cooperate together to forward data to the final destination.

Chapter 6
A Taxonomy of Routing Techniques in Underwater Wireless Sensor Networks

Muhammad Ayaz
Universiti Teknologi PETRONAS, Malaysia

Azween Abdullah
Universiti Teknologi PETRONAS, Malaysia

Ibrahima Faye
Universiti Teknologi PETRONAS, Malaysia

ABSTRACT

Underwater Wireless Sensor Networks (UWSNs) are finding different applications for offshore exploration and ocean monitoring. In most of these applications, the network consists of a significant number of sensor nodes deployed at different depth levels throughout the area of interest. Sensor nodes on the sea bed cannot communicate directly with the nodes near the surface level, so they require multihop communication assisted by an appropriate routing scheme. However, this appropriateness not only depends on network resources and application requirements, but environment constraints are involved as well. These factors all provide a platform where a resource aware routing strategy plays a vital role in fulfilling different application requirements with dynamic environment conditions. Realizing this fact, much of the attention has been given to construct a reliable scheme, and many routing protocols have been proposed in order to provide efficient route discoveries between the source and sink. In this chapter, the authors present a review and comparison of different algorithms proposed recently for underwater sensor networks. Later on, all of these have been classified into different groups according to their characteristics and functionalities.

DOI: 10.4018/978-1-4666-0101-7.ch006

INTRODUCTION

The ocean is vast as it covers around 140 million square miles; this is more than 70% of the earth's surface. Not only has it been a major source of nourishment production, but with the passage of time it has also taken a vital role for transportation, presence of natural resources, defense and for purposes of entertainment. With the increasing role of oceans in human life, discovering these largely unexplored areas has gained more importance during the last decades. On one side, traditional approaches used for the underwater monitoring missions have several drawbacks. At the same time, these inhospitable environments are not feasible for human presence as unpredictable underwater activities, high water pressure and vast areas are major reasons for un-manned explorations. As a result of these reasons, Underwater Wireless Sensor Networks (UWSNs) are attracting the interest of researchers, especially those who belong to terrestrial sensor networks.

Sensor networks used for underwater communication are different in many aspects from traditional wired or even terrestrial sensor networks. Firstly, energy consumption is different because some important applications require a large amount of data, but very infrequently (Heidemann, Wei, Wills, Syed & Yuan, 2006). Secondly, these networks are usually working on a common task instead of representing independent users. The ultimate goal is to maximize the

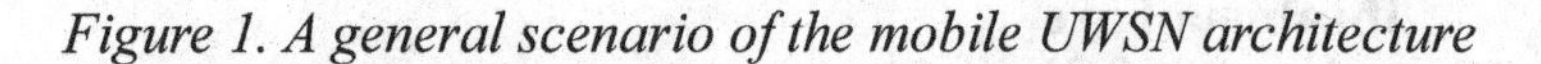

Figure 1. A general scenario of the mobile UWSN architecture

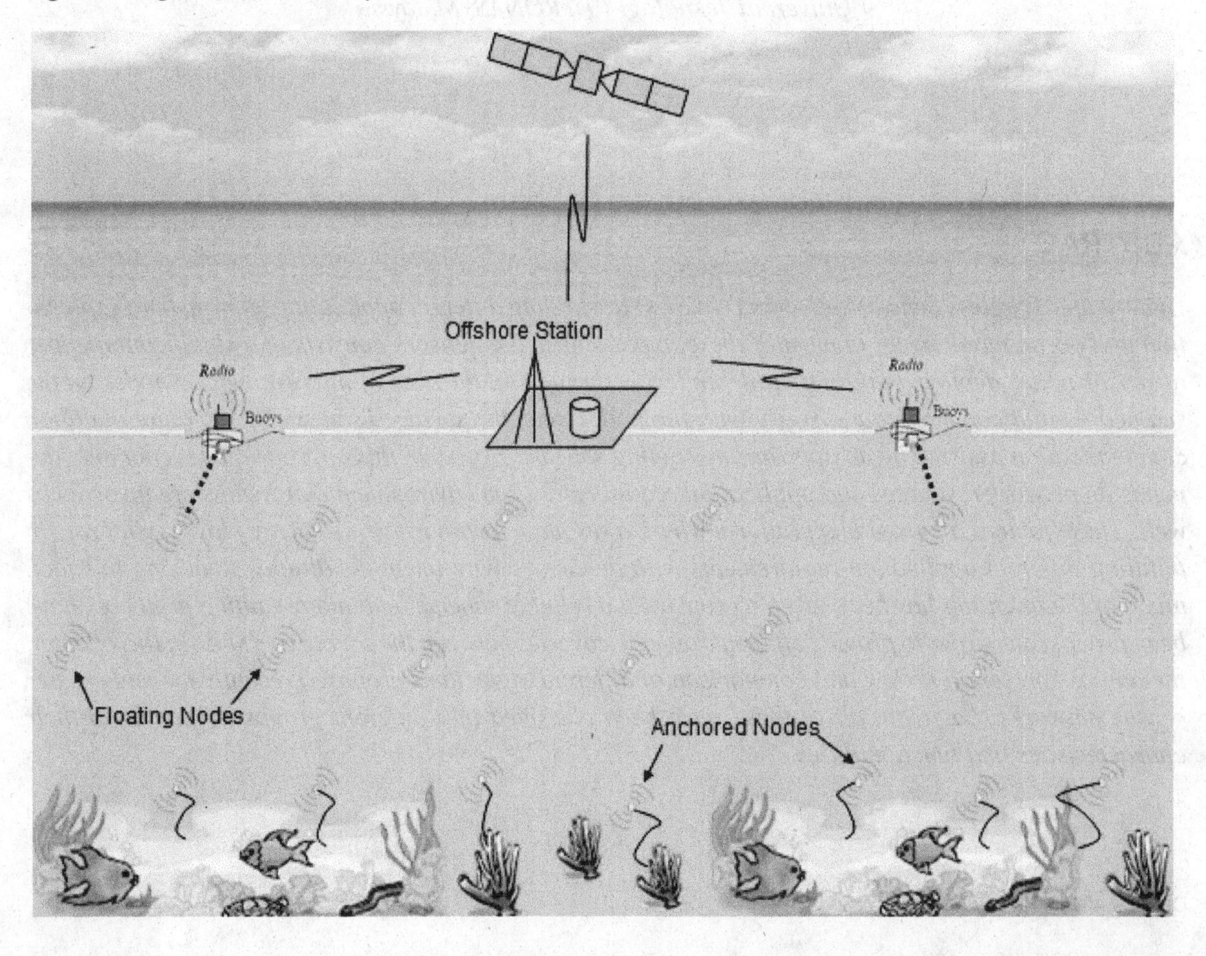

throughput rather than fairness among the nodes. Thirdly, for these networks, there is an important relationship between the link distance, number of hops and reliability. For energy concerns, packets sent over multiple short hops are preferred instead of long links, as multi-hop data deliveries are proven more energy efficient for underwater networks than the single hop (Jiang, 2008). At the same time, it is observed that packet routing over more number of hops ultimately degrades the end-to-end reliability function especially in the fragile underwater environment. Finally, most of the time, such networks are deployed by a single organization with economical hardware so, strict interoperability with the existing standards is not required. Due to these reasons, UWSNs provide a platform that supports a review of the existing structure of traditional communication protocols.

Current research in UWSNs aims to meet the above criterion by introducing new design concepts, developing or improving existing protocols, and building new applications. This chapter examines different underwater routing protocols and algorithms proposed in recent years. The main purpose of this study is to address the issues like data forwarding, coverage and localization in UWSNs under different conditions. We present a survey of more than twenty routing protocols and algorithms of different types proposed for different applications.

Problems in Existing Terrestrial Routing Protocols

The existing routing protocols, proposed for terrestrial mobile and ad hoc networks usually fall in two categories: *proactive* an *reactive*. Unfortunately, protocols belonging to both of these extremes are not suitable for underwater sensor networks. *Proactive* or *Table driven* protocols require large signalling overhead in order to establish end-to-end routes, especially for the first time and every time when any change occurs in the topology. For underwater sensor networks, we already know that continuous node movement results in continuous topology changes. On the other hand,

Figure 2. A possible system design for an UWSN (Shi, 2006)

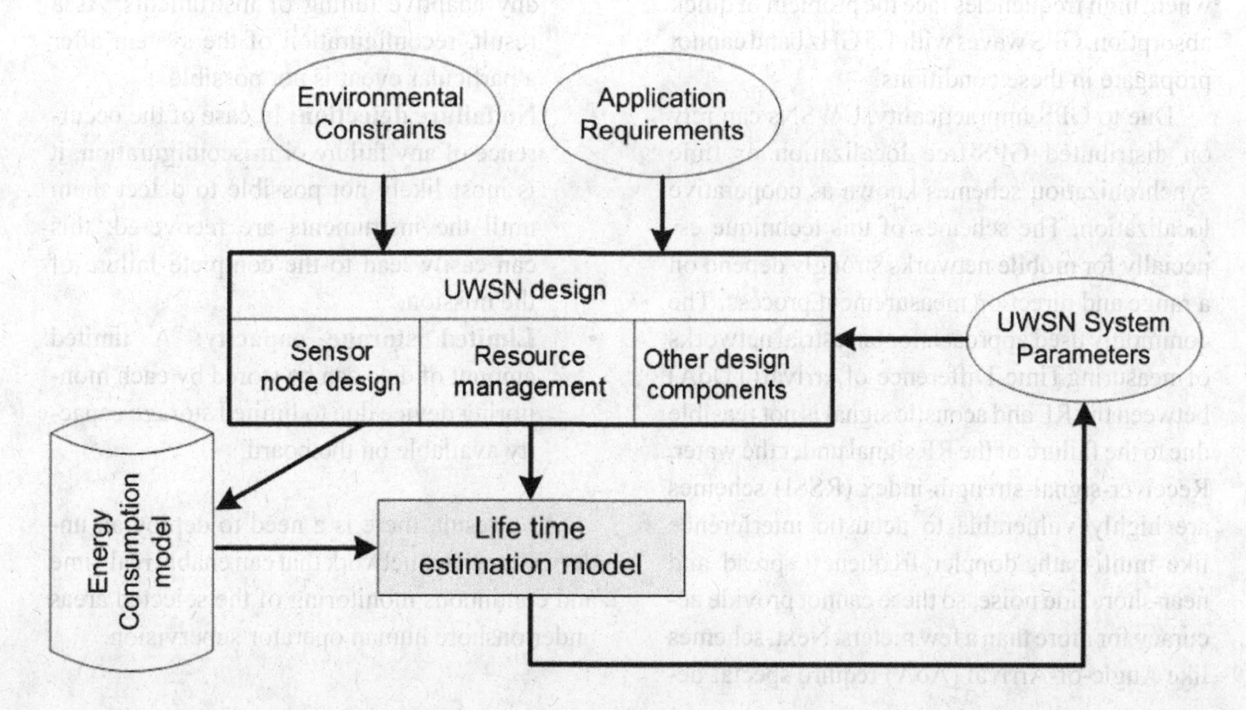

when we talk about the *Reactive* or *On Demand* routing, the protocols belonging to this category are suitable for dynamic environments but they face large delays as they require source initiated flooding of control packets for the route discovery process. Not only this, but also the experimental results show that reactive protocols provide better results with the symmetrical links in the network. For the underwater environment, the propagation delays are already high with asymmetric links, so the protocols of this type also seem not suitable for these environments.

Without any proactive neighbor information and with a small flooding option, it is a challenging task to construct a multi-hop data delivery routing scheme for a continuous mobile network (Jun-Hong, Jiejun, Gerla & Shengli, 2006). Geographical routing can be a possible solution for these situations. The protocols belonging to this type forward the data packets using the location information of their neighbors and the location of the destination. This technique has immense potential but only for terrestrial networks, where facilities like the Global Positioning System (GPS) are available. While, for underwater environments where high frequencies face the problem of quick absorption, GPS waves with 1.5 GHz band cannot propagate in these conditions.

Due to GPS impracticality, UWSNs can rely on distributed GPS-free localization or time synchronization schemes known as cooperative localization. The schemes of this technique especially for mobile networks strongly depend on a range and direction measurement process. The commonly used approach for terrestrial networks of measuring Time-Difference-of Arrival (TDoA) between the RF and acoustic signal is not feasible due to the failure of the RF signal under the water. Receiver-signal-strength-index (RSSI) schemes are highly vulnerable to acoustic interference like multi-path, doppler frequency spread and near-shore tide noise, so these cannot provide accuracy for more than a few meters. Next, schemes like Angle-of-Arrival (AoA) require special devices for directional transmission and reception. Finally, approaches like Time-of-Arrival (ToA) seem promising even though they only provide accuracy at short ranges due to the acoustic mode of communication.

Problems in Traditional Approaches

Underwater Wireless Sensor Networks (UWSNs) are becoming more and more popular for monitoring the vast oceans, as traditional approaches face the following limitations.

- **No real time monitoring:** Traditional approaches do not support interactive monitoring so all the recorded data can only be retrieved at the end of the operation, which can take several months. This is critical especially for delay sensitive applications like environmental monitoring (Akyildiz, Pompili & Melodia, 2005).
- **No on-line system reconfiguration:** Interaction between the control system and monitoring instruments is not possible, and such limitations prove to be an obstacle for any adaptive tuning of instruments. As a result, reconfiguration of the system after a particular event is not possible.
- **No failure detection:** In case of the occurrence of any failure or misconfiguration, it is most likely not possible to detect them until the instruments are recovered; this can easily lead to the complete failure of the mission.
- **Limited storage capacity:** A limited amount of data can be stored by each monitoring device due to limited storage capacity available on the board.

As a result, there is a need to deploy an underwater sensor network that can enable real-time and continuous monitoring of the selected areas under onshore human operator supervision.

Difference between UANs and UWSNs

Mobile underwater wireless sensor networks are considered as a next step with respect to existing small scale Underwater Acoustic Networks (UANs) (Jun-Hong, et al., 2006). UANs are a combination of nodes that collect information using remote telemetry or assuming point to point communication. Current UWSNs has many differences compared with traditional UANs, some of them are as follows.

- **Scalability:** A UWSN is a scalable sensor network which depends on coordinated networking among a large number of sensor nodes in order to complete its task of localized sensing and delivering data. While, an existing UAN is a small scale network which depends on data collecting strategies like remote telemetry or long-range signals to remotely collect data. UANs are considered not only less precise due to the effect of environmental conditions but also very expensive when we use them for high precision required applications. For the UAN, due to sparse deployment, multi-access techniques are not required as point to point communication is assumed; while in mobile sensor networks, nodes are densely deployed in order to achieve a better spatial coverage which requires a well designed multi-access and routing protocol.
- **Self-Organization:** Usually, in underwater acoustic networks nodes are fixed; whereas underwater sensor networks are considered as self organizing networks as the nodes can move and continue to redistribute according to the needs of the underwater activities. Thus, these nodes should not only be able to adjust their buoyancy but also move up and down according to the measured data density. This is the reason the protocols used for UANs, usually borrowed from terrestrial wireless sensor networks, cannot be used directly for mobile UWSNs.
- **Localization:** In UANs, localization is not required because nodes are fixed in most of the cases, either anchored on the sea bottom or attached to a surface buoy. While, for underwater sensor networks, some sort of localization is required as nodes can move continuously due to water currents. Now, determining the location information of mobile sensor nodes in an aquatic environment is a challenging task. On one side, we have to face the limited communication capabilities of the acoustic channel. At the same time, we have to consider immature localization accuracy.

Major Challenges for UWSNs

When we are talking about underwater wireless sensor networks, then we have to consider the following challenges, specially the unique characteristics of the underwater acoustic channel (Pompili, 2007).

- Available bandwidth is severely limited
- High propagation delay, as five orders of magnitude higher than the radio communications (.67s/Km) (Sozer, Stojanovic & Proakis, 2000).
- Underwater acoustic channel is impaired due to multi-path and fading
- High bit error rate and temporary losses of connectivity
- Ambient noise due to water movement including tides, storm, wind, etc.
- Higher power consumption as compared to radio signal and usually batteries cannot be recharged; moreover solar energy cannot be exploited

- Sensor nodes are prone to failure due to fouling and corrosion
- High cost for sensor nodes because of small relative number of manufacturers

Other than these, we have to consider some more challenges that we have to face in these environments. Some of the underwater applications including detection or rescue missions can require not only deploying the network in a short time and also without any proper planning. In such circumstances, the routing protocols should be able to determine the node locations without any prior knowledge of the network. Not only this, but the network should also be capable of reconfiguring itself with dynamic conditions in order to provide an efficient environment for the communication.

Moreover, a significant issue in selecting a system is establishing a relationship between the communication range and data rate under specific conditions. A system designed for deep water may not be suitable for shallow water or even when configured for higher data rates when reverberation is present in the environment (Chitre, Shahabudeen, Freitag & Stojanovic, 2008). Manufacturer's specifications of maximum data rates mostly are only useful for establishing the upper performance bound but are not achievable under specific conditions. Users who are well funded have resorted to purchasing multiple systems and testing them in a particular environment to determine if they will meet their needs. An international effort for standardizing the tests for acoustic communication is required, but it is not so simple, as private organizations or even government institutes which perform such comprehensive tests do not tend to publish them.

ROUTING PROTOCOLS

Routing is a fundamental issue for any network and routing protocols are considered to be in charge of discovering and maintaining the routes. When we talk about underwater senor networks, most of the research is performed on the issues related to the physical layer while issues related to the network layer like routing techniques is relatively a new area, so providing an efficient routing algorithm becomes an important task. Although, underwater acoustics has been studied for decades, underwater networking and routing protocols are still in the infant stage as a research field. In this section, we discuss the major routing protocols proposed to date for UWSNs and highlight the advantages and performance issues of each routing scheme. A comparison of the routing protocols presented in this chapter based on their characteristics can be found in Appendix at the end of the chapter, as well as some miscellaneous classifications.

Vector Based Forwarding (VBF)

High error probability is a major problem for the dense networks and in order to handle this issue, authors proposed a position based routing approach called VBF (Xie, Cui & Lao, 2006). For this, state information of the sensor nodes is not required and only a small number of nodes are involved during the packet forwarding. Data packets are forwarded along redundant and interleaved paths from the source to the sink which helps to handle the problem of packet losses and node failures. It is assumed that every node already knows its location, and each packet carries the location of all the involving nodes including the source, forwarding nodes and final destination. Here an idea of a vector, like a virtual routing pipe is proposed and all the packets are forwarded through this pipe from the source to the destination. Only the nodes closer to this pipe or "vector" from the source to the destination can forward the messages. By using this idea, not only can the network traffic be reduced significantly but it is also easy to manage the dynamic topology.

VBF, however, does have some serious problems. Firstly, here they used a virtual routing pipe from the source to the destination and creation of such a pipe can affect the routing efficiency of the network with different node densities. In some areas, if nodes are too sparsely deployed or become sparser due to some movements then it is possible that very few nodes if any lie within that virtual pipe which is responsible for the data forwarding even though it is possible that paths may exist outside the pipe. Ultimately, this results in small data deliveries in sparse areas. Secondly, VBF is very sensitive about the routing pipe radius threshold and this threshold can affect the routing performance significantly; such features may not be desirable in real protocol developments. Moreover, some nodes along the routing pipe are used again and again in order to forward the data packets from concrete sources to the destination which can exhaust their battery power. Other than these issues, VBF has a lot of communication overhead due to its 3-way handshake nature, during this process; moreover, it doesn't consider the link quality.

In order to increase robustness and overcome these problems, an enhanced version of VBF was presented in what is called Hop-by-Hop Vector-Based Forwarding (X. Peng, et al.). They use the same concept of virtual routing pipe as used by VBF, but instead of using a single pipe from the source to the destination, HH-VBF defines a per hop virtual pipe for each forwarder. In such a way, every intermediate node makes a decision about the pipe direction based on its current location. By doing so, even a small number of nodes are available to their neighbors; HH-VBF can find a data delivery path as long as a single node is available in the forwarding path within the communication range. Although, simulation results shows that, HH-VBF significantly produces better results for packet delivery ratio especially in sparse areas compared with VBF, it still has the inherent problem of the routing pipe radius threshold, which can affect its performance. Not only this, but also due to its hop-by-hop nature, HH-VBF produced much more signalling overhead as compared to VBF.

Focused Beam Routing (FBR)

Without any prior location information of the nodes, a large number of broadcast queries can be a burden on the network which can result in reducing the overall expected throughput. In order to reduce such unnecessary flooding, (Jornet, Stojanovic & Zorzi, 2008) presented the Focused Bream Routing (FBR) protocol for acoustic networks. Their routing technique assumes that, every node in the network has its own location information and every source node knows about the location of the final destination. Other than these two, there is no need to know about the location of intermediate nodes. Routes are established dynamically during the traversing of a data packet for its destination, and the decision about the next hop is made at each step on the path after appropriate nodes have proposed themselves.

Figure 3 explains the data forwarding method used in FBR. Node A has a data packet to send to destination node D. To do so, node A multicasts a request to send an (RTS) packet to its neighboring nodes. This RTS packet contains the location of source (A) and final destination (D). Initially, this multicast action will be performed at the lowest power level and later can be increased if there is no node found as a next hop in this communication range. For this, they define a finite number of power levels, P_1 through P_N, which can be increased only if necessary. Now, all the nodes that receive this multicast RTS, will calculate their current location relative to the line AD. After calculating, nodes considered as a next hop candidate are those which lie within a cone of angle $\pm$ θ/2 emanating from the transmitter towards the final destination. After calculating this angle, if a node determines that it is within the transmitting cone, it will reply to the RTS.

However, the approach followed by FBR might have some performance problems. First of all, if nodes become sparse due to water movement then it is possible that no node lies within that forwarding cone of angle where nodes are considered as

Figure 3. Illustration of the routing protocol: nodes within the transmitter's cone θ are candidate relays

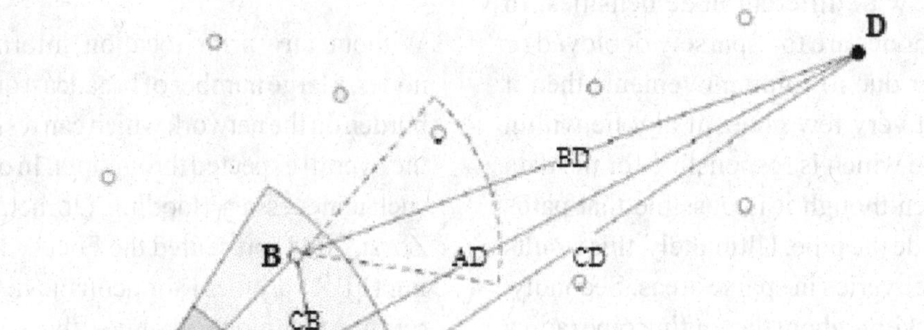

a next hop. Even so, it might possible that some nodes are available as a next hop but they exist outside this forwarding area. In such cases, when it cannot find the next relay node within this transmitting cone, it needs to rebroadcast the RTS, sometimes repeatedly, which ultimately increases the communication overhead, this in turn can affect the data deliveries in sparse areas. Secondly, it assumes that the sink is fixed and its location is already known, which also reduces the flexibility of the network.

A Reliable and Energy Balanced Routing Algorithm (REBAR)

It is common analysis that, water movement makes the underwater environment more dynamic but (Jinming, Xiaobing & Guihai, 2008) consider node mobility as a positive factor which can be helpful to balance the energy depletion in the network. The provided reason is that, when nodes move, they start to alternate around the sink which brings an effect of the balance in energy consumption in the whole network. They tried to solve the problem of network partitioning by altering the node positions as nodes near the sink are prone to die much earlier due to their frequent involvement in the routing process. Their proposed idea looks similar to VBF and HH-VBF but they designed an adaptive scheme by defining the data propagation range in order to balance the energy consumption throughout the network. Since, a network wide broadcast results in high energy consumption, here nodes broadcast in a specific domain between the source and the sink by using geographic information. Particularly, different sensor nodes have a different communication radius depending on the distance between the nodes and the sink. Nodes nearer the sink are set to a smaller value in order to reduce the chance of being involved in the routing process, ultimately, helping to balance the energy consumption among all the sensor nodes.

According to their network model, all the sensor nodes are randomly deployed in an underwater hemisphere as shown in Figure 4. The sink is stationary and fixed at the center of the surface. All the sensor nodes are assigned a unique ID and have a fixed range. It is assumed that every node knows the location of itself and the location of the sink through multi-hop routing. They also assume a data logging application where sensed data is sent towards the sink at a certain rate.

However, the idea of altering the node positions used in REBAR has a serious problem. On

Figure 4. The sphere energy depletion model

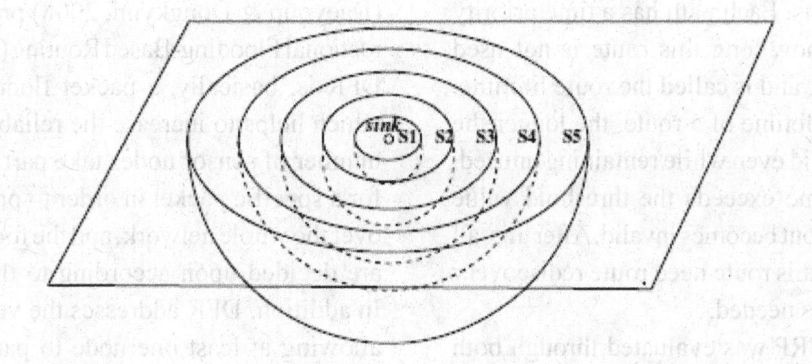

one side, they advocate node movement as a positive sign, as simulation results show that with static nodes, delivery ratios are smaller and these start to increase with the increase in node movement. It is due to some of these assumptions that they made at the start, like nodes have their current location information and the location of the final destination. In simulation results, they consider the node movements from 0 to 4 m/sec and according to this phenomenon these delivery ratios should continue to increase when movement is more than 4 m/sec. In reality however, though these node movements can helpful but they can create problems as well. Large movements can result in making the network sparser as well as requiring nodes to update their location more frequently which ultimately affects the network performance. Not only this, but it also assumes that these movements are completely dynamic in terms of direction both in the vertical and horizontal. According to this, a bottom node will reach the surface and then it can come back to the bottom. Again, in a real scenario, it might not be possible as only horizontal movements are common from 2-3 m/sec while only small fluctuations are shown vertically (Jun-Hong, et al., 2006). Moreover, the provided simulation results focus only on delivery ratios and energy consumption with different node speeds but have not provided any information about the end to end delays; those can vary according to different node movements.

Information Carrying Based Routing Protocol (ICRP)

Most of the routing protocols even for terrestrial or underwater sensor networks use separate packets for *control information* and *data transmission*. While, (Wei, Haibin, Lin, Bangxiang & Chang, 2007) proposed a novel reactive protocol called the Information-Carrying based Routing Protocol (ICRP), in order to address the routing problem for underwater communication. ICRP is used for energy efficient, real-time and scalable routing, where *control packets* used for information sharing are carried by the *data packets*. Most importantly, it doesn't require state or location information of the nodes; as well, only a small fraction of the nodes are involved in the routing process.

In ICRP, the route establishment process is initiated by the source node. When a node has a data packet to send, first it will check the existing route for the destination. If no route exists then it will broadcast the data packet which carries the route discovery message. All the nodes, which receive this packet, will in turn broadcast it while maintaining the reverse path through which this packet passes. Finally, when the destination node receives this data packet then it gets the complete reverse path from the source to the destination. Now, the destination node can use this path in order to send the acknowledgment. The path will remain valid for data packet transmission for as long as the source node continues to receive acknowledgments. Each path has a time priority which denotes how long this route is not used for transmission, and is called the route lifetime. The larger the lifetime of a route, the longer the time it can be valid even while remaining unused. When the lifetime exceeds the threshold value TIMEOUT, the rout becomes invalid. After this, all the nodes using this route need route rediscovery when the route is needed.

Although, ICRP was evaluated through both simulations and experimental implementation under a testbed; the testbed only consists of three sensor nodes which does not reflect the traffic of most of the real life UWSN scenarios. When we talk about the basic routing mechanism, it has some performance problems. Firstly, when a node doesn't have route information for a specified destination then it will broadcast the packet. More broadcasts will result in wastage of node energy; due to this, that node can die early which decreases the life of whole network. Secondly, every route has an expiry time which can be very sensitive for delivery ratios. If it is very long, then nodes can move and this route can create complexity; if too short, then it will help to increase more and more broadcasts. Moreover, routing decisions are totally based on the cached route information. For the UWSN, where nodes move continuously at 2-3 m/sec with the water currents, in such situations any intermediate node of the route can be unavailable.

Directional Flooding-Based Routing (DFR)

For UWSNs, the path establishment requires a lot of overhead in the form of control messages. Moreover, the dynamic conditions and high packet loss degrades reliability which results in more retransmissions. Existing routing protocols proposed to improve the reliability did not consider the link quality. That's why there is no guarantee about the data packet delivery especially when a link is error prone. In order to increase the reliability, (Daeyoup & Dongkyun, 2008) proposed the Directional Flooding-Based Routing (DFR) protocol. DFR is, basically, a packet flooding technique which helps to increase the reliability. A limited number of sensor nodes take part in this process for a specific packet in order to prevent flooding over the whole network, and the forwarding nodes are decided upon according to the link quality. In addition, DFR addresses the void problem by allowing at least one node to participate in the data forwarding process.

As shown in Figure 5, the flooding zone is decided by the angle between FS and FD, where F is the packet receiving node while S and D represent the source and destination node respectively. After receiving a data packet, F determines dynamically about the packet forwarding by comparing ∟SFD with a criterion angle for flooding called BASE_ANGLE which is included in the received packet. In order to handle the high and dynamic packet error rate, BASE_ANGLE is adjusted in a hop-by-hop fashion according to the link quality which helps to find a flooding zone dynamically. That is, the better the link quality is, the smaller the flooding zone is.

The performance of DFR depends on the number of nodes chosen as the next hop after flooding the data packet. Although, the problem of a void region is addressed by making sure that at least one node must participate in this process; in areas where link quality is not good then, multiple nodes can forward the same data packet so more and more nodes will join the flooding of the same data packet which ultimately increases the consumption of critical network resources. Secondly, they have controlled the void problem by selecting at least one node in order to forward the data packet towards the sink. However, when a sending node can not find a next hop closer to the sink, DFR still encounters the void problem as no mechanism is available to send the data packet in a backward direction.

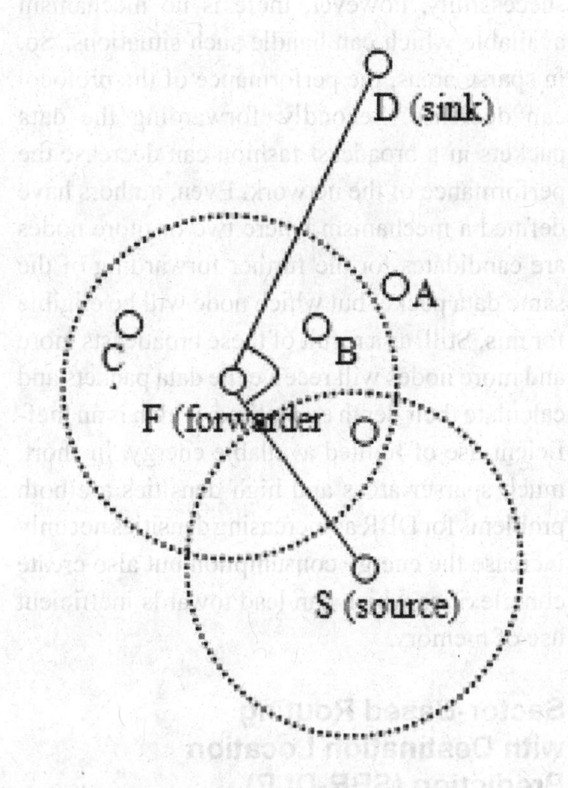

Figure 5. An example of a packet transmission in DFR

Distributed Underwater Clustering Scheme (DUCS)

Energy efficiency is a major concern for UWSNs because sensor nodes have batteries of limited power which are hard to replace or recharge in such environments. It is a fundamental problem to design a scalable and energy efficient routing protocol for these networks. (Domingo & Prior, 2007) presented a distributed energy aware and random node mobility supported routing protocol called the Distributed Underwater Clustering Scheme (DUCS) for long term but non-time critical applications.

DUCS is an adaptive self-organizing protocol where whole network is divided into clusters using a distributed algorithm. Sensor nodes are organized into local clusters where one node is selected as a cluster head for each cluster. All the remaining nodes (non-cluster heads) transmit the data packets to the respective cluster heads. This transmission must be single hop. After receiving the data packets from all the cluster members, the cluster head performs a signal processing function like aggregation on the received data and transmits it towards the sink using multi-hop routing through other cluster heads. Cluster heads are responsible for coordination among members of their clusters (intra-cluster coordination) and communication between each other (inter-cluster communication). Selection of the cluster head is completed through a randomized rotation among different nodes of the cluster in order to avoid fast draining of the battery from the specific sensor node. DUCS completes its operation in two rounds. The first round, called *set-up*, is where the network is divided into clusters and in the second round, which is called *network operation*, transfer of data packets is completed. During the second phase, several frames are transmitted to each cluster head where every frame is composed of a series of data messages that the ordinary sensor nodes send to the cluster head with a schedule. Simulation results show that DUCS not only achieves a high packet delivery ratio but also considerably reduces the network overhead and continues to increase throughput consequently.

Although, DUCS is simple and energy efficient but it also has a couple of performance issues. Firstly, node movements due to water currents can affect the structure of the clusters which ultimately decreases the cluster life. Frequent division of sectors can be a burden on the network as again and again the *set-up* phase is repeated. Secondly, during the *network operation* phase, a cluster head can transmit its collected data towards another cluster head only. Again, water currents can take two cluster head nodes away where they cannot communicate directly even if a couple of non-cluster head nodes are available between them.

Depth Based Routing (DBR)

For location based routing schemes, most of the protocols require and manage full-dimensional location information of the sensor nodes in the network which itself is a challenge left to be solved for UWSNs. Instead of requiring complete localized information, DBR (Yan, Shi & Cui, 2008) needs only the depth information of the sensor node. In order to obtain the depth of the current node, authors suggest equipping every sensor node with an inexpensive depth sensor. In their architecture, multiple data sinks placed on the water surface are used to collect the data packets from the sensor nodes. DBR makes a decision based on the depth information and forwards the data packets from the higher to the lower depth sensor nodes. When a node has a data packet to send, it will sense its current depth position as compared to the surface and place this value in the packet header field "Depth" and then broadcast it. The receiving node will calculate its current depth position and can forward only this packet if its depth is smaller than the value embedded in the packet otherwise it will simply discard it. Similarly, this process will continue until the data packet reaches any of the data sinks. Packets received at any of the data sinks are considered as delivered successfully at the final destination as these data sinks can communicate efficiently with a much higher bandwidth through a radio channel.

Although DBR has many advantages as it does not require full dimensional location information, node movement with the water currents is handled efficiently and it also takes advantage of the multiple sink architecture but it still has some serious problems. Firstly, DBR has only a greedy mode which alone is not able to achieve high delivery ratios in sparse areas. In such areas, it is possible that no node can be eligible as a forwarding node due to a greater depth as compared to the sending node; the current node will continue to make more and more attempts. Though some nodes may be available here at higher depths which can forward packets towards the data sink successfully, however, there is no mechanism available which can handle such situations. So, in sparse areas, the performance of the protocol can decrease. Secondly, forwarding the data packets in a broadcast fashion can decrease the performance of the network. Even, authors have defined a mechanism where two or more nodes are candidates for the further forwarding of the same data packet but which node will be eligible for this. Still, as a result of these broadcasts more and more nodes will receive the data packets and calculate their depth every time, which is an inefficient use of limited available energy. In short, much sparser areas and high densities are both problems for DBR as increasing densities not only increase the energy consumption but also create complexities which can lead towards inefficient use of memory.

Sector-Based Routing with Destination Location Prediction (SBR-DLP)

Recently, several location based routing techniques have been proposed like VBF, and they could achieve energy efficiency by decreasing the network overhead. Most of them assume that the destination is fixed and its location is already known to all the nodes throughout the network. This assumption may not be suitable for fully mobile networks. (Chirdchoo, Wee-Seng & Kee Chaing, 2009) proposed a routing algorithm called SBR-DLP which helps to route a data packet in a fully mobile underwater acoustic network where not only intermediate nodes but destination can be mobile as well.

SBR-DLP is a location based routing algorithm where sensor nodes do not need to carry neighbor information or network topology. However, it is assumed that every node knows its own location information and the pre-planned movement of the destination node. Data packets are forwarded to the destination in a hop-by-hop fashion instead

Figure 6. Forwarder selection at the sender

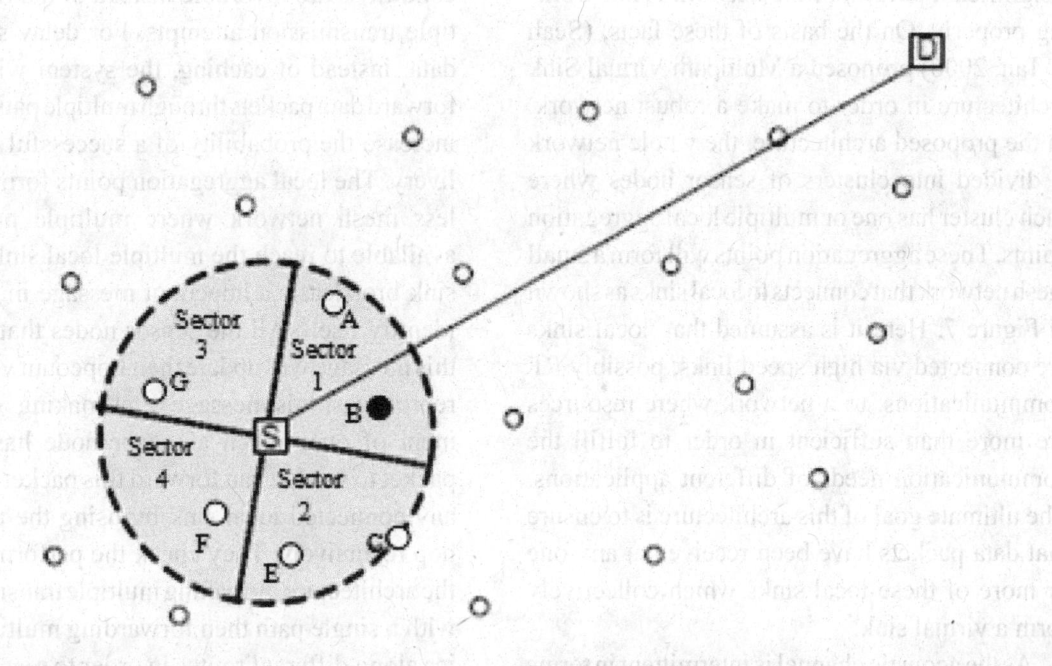

of finding an end-to-end path in order to avoid the flooding. As shown in Figure 6, a node S has a data packet to send to destination D. It will try to find its next hop by broadcasting a *Chk_Ngb* packet which includes its current position and packet ID. The neighbor node that receives the *Chk_Ngb* will check whether it is nearer to destination node D than the distance between nodes S and D as shown in Table 1. The nodes that meet this condition will reply to node S by sending a *Chk_Ngb_Reply* packet. SBR-DLP is a location based routing protocol like VBF and HH-VBF but it is different in many aspects from both of them. Firstly, instead of allowing all the candidate nodes to decide about the packet forwarding, in SBR-DLP, the sender node decides about the next hop using the information received from candidate nodes. This solves the problem of having multiple nodes acting as relay nodes.

Table 1. How node S picks its next relay node

Sector	Candidates	Distance to D	After Filtering
1	A,B	500,480	A,B
2	C	550	
3	-	-	
4	-	-	
		Next relay node	B

SBR-DLP handles the issue of destination mobility by assuming that pre-planned movements are completely known to all the sensor nodes before deploying them. However, this assumption has two issues. First, it reduces the flexibility of the network as after launching the network it is not possible to change the position or location of the destination nodes. Secondly, it is important to note that water currents can result in the destination node deviating from its scheduled movements.

Multipath Virtual Sink Architecture

The network topology is important for determining the network reliability, capacity and energy consumption. A sufficient robustness and redundancy must be available in the network in order to ensure that it will continue to work even when a significant portion of the network is not working properly. On the basis of these facts, (Seah & Tan, 2006) proposed a Multipath Virtual Sink architecture in order to make a robust network. In the proposed architecture, the whole network is divided into clusters of sensor nodes where each cluster has one or multiple local aggregation points. These aggregation points will form a small mesh network that connects to local sinks as shown in Figure 7. Here it is assumed that, local sinks are connected via high speed links, possibly RF communications, to a network where resources are more than sufficient in order to fulfill the communication needs of different applications. The ultimate goal of this architecture is to ensure that data packets have been received at any one or more of these local sinks which collectively form a virtual sink.

As the acoustic channel is intermittent in terms of connectivity and available bandwidths are very small, it can be better for sensor nodes to cache the sensed data and transmit when the channel conditions are favorable instead of making multiple transmission attempts. For delay sensitive data, instead of caching, the system will try to forward data packets through multiple paths which increase the probability of a successful data delivery. The local aggregation points form a wireless mesh network where multiple paths are available to reach the multiple local sinks. Each sink broadcasts a hopcount message in order to identify itself. All the sensor nodes that receive this message will update their hopcount value and rebroadcast this message after making an increment of one. When a sensor node has a data packet to send, it can forward this packet towards any connected local sink by using the previous hop recursively. They check the performance of the architecture by making multiple transmissions with a single path then forwarding multiple copies along different routes in order to reach different sinks.

In the proposed scheme, reliability is improved as duplicate packets are delivered towards mul-

Figure 7. Proposed underwater network topology for multipath virtual sink architecture

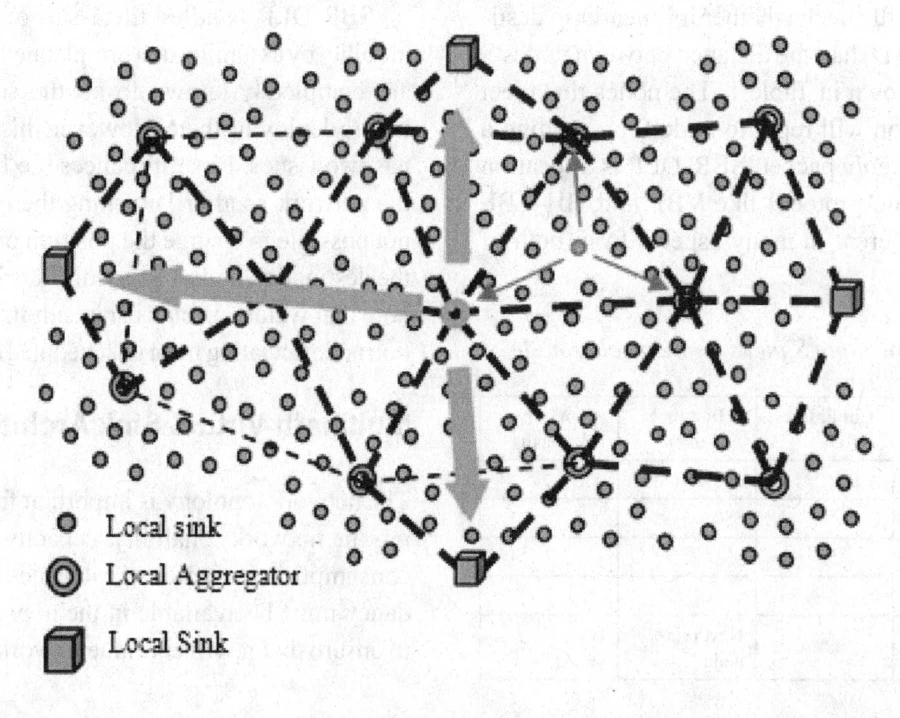

tiple sinks through multiple paths. However, the problem of redundant transmissions exists which can consume critical underwater resources.

Hop-by-Hop Dynamic Addressing Based (H2-DAB)

Most of the routing protocols proposed for UWSNs require some special network setups. Many of them make assumptions like about the full dimensional location information of a whole network being available, which is not simple. Providing the complete dimensional location information for underwater environments is a separate research issue which remains to be solved. While, the remaining ask for special hardware like every node being equipped with depth or pressure sensors; this not only increases the cost of the network, but is also a burden on the critical node energy. By considering these issues, the authors proposed a dynamic addressing based routing protocol H2-DAB (Ayaz & Abdullah, 2009), which does not make any assumption as most of the remaining schemes do.

The purpose of H2-DAB is to solve the problem of continuous node movement. Dynamic addresses are used for sensor nodes in order to solve the problem of water currents, so that sensor nodes will get new addresses according to their new positions at different intervals. In their architecture, multiple surface buoys are used to collect the data at the surface and some nodes anchored at the bottom. The remaining nodes are deployed at different depth levels from the surface to the bottom. Nodes near the surface sinks have smaller addresses and these addresses continue to increase as the nodes go down towards the bottom (Figure 8). H2-DAB completes its task in two phases; first, by assigning the dynamic addresses and second, by using these addresses in the process of data forwarding. Dynamic addresses will be assigned with the help of Hello packets; those are generated by the surface sinks. Any node which generates or receives data packets will try to deliver them towards the upper layer nodes in a greedily fashion. Packets that reach any one of the sinks will be considered as delivered successfully to the final destination as these buoys have the luxury of radio communications where they can communicate with each other at higher bandwidths and with lower propagation delays.

Although H2-DAB has many advantages like, it doesn't require any specialized hardware, no dimensional location information is required and node movement can be handled easily without maintaining complex routing tables. However, as it is based on a multi-hop architecture the problem of multi-hop routing still exists, where nodes near the sinks drain more energy because they are being used frequently.

A Mobile Delay-Tolerant Approach (DDD)

Acoustic channel impose higher energy consumption than radio signal. Due to higher power usage of acoustic modems, energy savings for underwater sensor networks becomes even more critical than in traditional sensor networks. In order to increase the energy efficiency in a resource constrained underwater environment, (Magistretti, et al., 2007) proposed a Delay-tolerant Data Dolphin (DDD) scheme for delay tolerant applications. DDD exploits the mobility of collector nodes called dolphins to harvest information sensed by stationary sensor nodes. The proposed scheme avoids energy expensive multi-hop communication and each sensor node is only required to transmit its collected data, directly to the nearest dolphin when its reach in its communication range.

In their architecture, stationary sensor nodes are deployed on the seabed in the whole area of interest. These nodes collect the information from the environment and this sensed data is stored locally after processing. These sensors periodically wake up for sensing and event generation. The acoustic modem is based on two components; the first is used for acoustic communication with the nearest

Figure 8. Assigning Hop ID's with the help of Hello packets

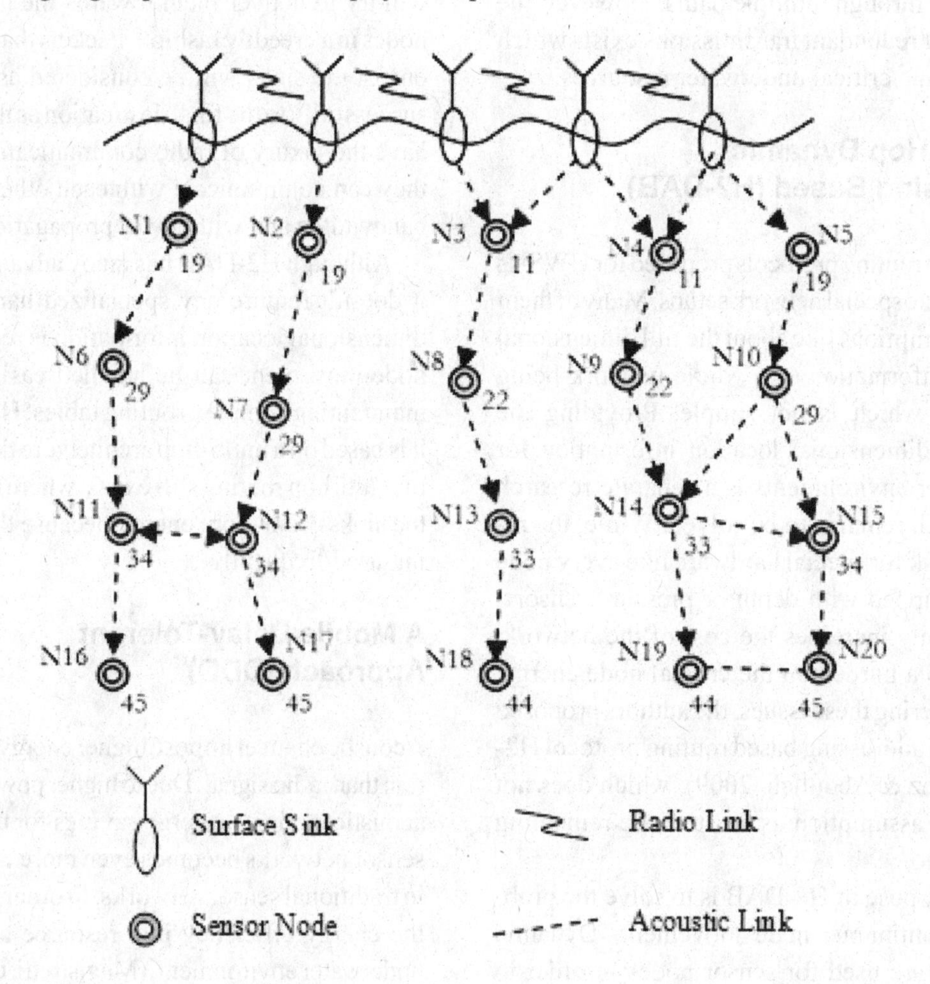

dolphin and the other is a low power transceiver used to determine the presence of dolphin nodes (by a special signal transmitted from the dolphin) and trigger the first one. Besides sensor nodes, a number of dolphin nodes are used to collect the data packets when they move within the one-hop range of scattered sensor nodes. These can move with random or controlled mobility according to network conditions. A dolphin node broadcasts beacons to advertise their presence. Beacons are transmitted at such acoustic frequencies as those compatible with the low-power sensor modem. Advertising period t is adjusted according to deployment and communication range r of sensor nodes and to the speed of dolphin v. Finally, dolphins deliver gathered data packets as soon as they reach a base station on the surface.

The quantity of dolphin nodes is the most important parameter in order to check the performance of DDD. If the number of dolphin nodes is not enough, they will not able to gather all the data packets from the sensor nodes. Since dolphins move randomly, it is possible that they may not visit some sensors directly which results in a loss of existing data packets by removing them from the limited memory of the sensor node when there is no space left. If we increase the number of dolphin nodes like 7 dolphins for 25 sensor nodes as in the simulation results, then cost becomes a major issue.

Efficient Data Delivery with Packet Cloning

In mobile sensor networks, possibly multiple paths can exist from a sensor node to the destination and these paths may or may not be disjoint. It has been shown that, routing over these multiple paths not only helps to increase the data delivery ratios but also achieves timeliness of delivery. As these paths start to converge at the destination, the possibility of contention starts to increase as well. The contention that arises among nodes in close proximity can be viewed positively. In order to get benefits from the proximity of nodes, (S. Peng, Seah & Lee, 2007) proposed a Packet Cloning technique which helps to enhance the data delivery ratios. The proposed scheme utilizes this idea to selectively clone data packets during the forwarding process to the destination. Different from the controlled broadcast or conventional multi-path routing where duplicate packets are indistinguishable because the nodes involved have no idea how many duplicates have been introduced, the current technique has the ability to control the number of packet clones according to the link quality and channel conditions in order to minimize the contention and energy expenditures.

During the packet cloning process, a relay node will not resend an incoming packet if it has already received one copy. This will help to prevent excessive network traffic. However, authors want to exploit the advantage of having two distinct copies of the same data packet along two disjoint paths. For this, distinct copies of the original packet are created while the number of distinct copies is a parameter that can be adjusted according to the conditions. A source node will first determine how many distinct copies it wants and then start to send each copy sequentially with some interval between them. In the packet header, it mentions how many copies it has produced and which copy this packet is. When a clone packet is received by an intermediate relaying node, it then can derive some information from the incoming packet. This extracted information is useful for detecting the duplicates and packet losses. For duplicate packets received, simply discard them; for new packet clones, relay them; for missed or lost packet clones, generate and transmit them. When a source node does the packet cloning, it then sends out each clone after selecting a proper value of interval which depends on the physical channel parameters. By doing so, it will help to reduce the chances of clones contending and interfering with each other.

Although, multipath routing schemes increase the network robustness not only by increasing the delivery ratios but also by helping to decrease end-to-end delays; the acoustic channel is power-hungry as compared to RF based. In order to increase the delivery ratio, more and more paths are suggested and these multiple paths continue to produce duplicates if the channel quality is not good. In short, RF based communications can support these schemes but for a highly power consuming acoustic environment, techniques like packet cloning are not easily affordable.

A Resilient Routing Algorithm for Long-Term Applications

For underwater communications, different problems are addressed at different layers, e.g. most of the impairments of an acoustic channel belong to the physical layer while characteristics like limited bandwidth, temporary losses of connectivity and node failures need to be addressed at higher layers. By considering this phenomenon (Dario Pompili, 2006) proposed a resilient routing algorithm for long term underwater monitoring applications which completes its task in two phases. In the first phase, optimal node-disjoint primary and backup multihop data paths are discovered in order to minimize the energy consumption. This is required because different from the terrestrial sensor networks where nodes are redundantly deployed, the underwater networks require a minimum number of nodes. In the second phase, an online distributed scheme observes the network and only if required, then switches to the backup paths. It is a fact that underwater monitoring missions can be highly expensive, so it is essential that the deployed network be highly reliable in order to avoid the failure of missions due to the failure of single or multiple devices.

The communication architecture used for resilient routing algorithms requires winch-based sensor devices; those are anchored to the ocean bottom. Each sensor device is equipped with a floating buoy that can be adjusted by a pump. The buoy helps the sensor device to move towards the ocean surface. The depth of the device can be regulated by adjusting the length of wire with which that node is anchored, by means of an electronically controlled engine that resides on the same device.

The proposed architecture has some strengths including, the sensor nodes are not vulnerable to weather and tampering, and the nodes are less affected by the water currents. However, this scheme is limited to long-term applications and with the proposed architecture; if we are interested in large areas then cost will become a major issue.

Pressure Routing for Underwater Sensor Networks (HydroCast)

For UWSNs, geographic routing is preferable due to its stateless nature. However, geographic routing requires distributed localization of mobile sensor nodes which can not only be expensive in terms of energy but can also take a long time to converge. In order to provide an alternative for geographic routing, (Uichin, et al.) presented HydroCast, a hydraulic pressure based routing protocol. HydroCast uses anycast routing by exploiting the measured pressure levels in order to forward the data packets towards surface buoys. The proposed hydraulic pressure based protocol is stateless and completes its task without requiring expensive distributed localization.

The basic idea of HydroCast is similar to DBR where routing decisions are made after comparing the local pressure or depth information such that data packets are greedily forwarded towards the node with the lowest pressure level among the neighbor nodes. DBR faces a serious problem of local maximum when a data forwarding node cannot find a next hop with a lower depth among its neighbor nodes. In such void regions, it does not provide any solution to handle such a situation. While in the HydroCast scheme, each local maximum node maintains a recovery route towards a neighboring node with higher depth than itself. After one or several tries of forwarding through the local maxima, a data packet can be rerouted, out of the void region and can switch back to the greedy mode.

The problem of void regions which exist in DBR is successfully solved by the HydroCast. Authors consider the quality of a wireless channel for simultaneous packet reception among the neighbor nodes. These simultaneous receptions enable the opportunistic forwarding by a subset of the neighbors that have received the data packet correctly, which ultimately increase the delivery ratios. At the same time, due to this opportunistic routing, multiple copies of the same data packet can be received at a sink, which will be a burden on the network resources. Although, simulation results show that, HydroCast provides high delivery ratios with small end-to-end delays, still, no information is available about the energy usage the pressure sensor consumes in order to find its depth.

Energy-Efficient Routing Protocol (EUROP)

Underwater sensor nodes are battery powered and as these batteries cannot be replaced easily, power efficiency is a critical issue for these environments. Additionally, extremely long delays for acoustic communications could lead to the collapse of the traditional terrestrial routing protocols due to limited response waiting time. In order to handle these issues, (Chun-Hao & Kuo-Feng, 2008) designed an energy efficient routing protocol called EUROP, where they tried to reduce a large amount of energy consumption by reducing broadcast hello messages.

In the proposed architecture, they suggest using a pressure sensor as a significant indicator for every sensor node to get its depth position. This depth sensor will eliminate the requirement of hello messages for control purposes which can be helpful in increasing the energy efficiency. These sensor nodes are deployed at different depths in order to observe the events occurring at different locations of the network. Furthermore, every node is anchored to the bottom of the ocean and equipped with a floating module that can be by a pump. This electronic module, which resides on the node, helps to push the node towards the surface and then back into position. The depth of the sensor node can be regulated by adjusting the length of the wire that connects the sensor to the anchor. All the sensor nodes at different depths will form layers, while the amount of layers depends on the depth. The sink on the surface can communicate only with the sensors belonging to shallow water. Sensor nodes on all the layers communicate through an acoustic channel after deciding to which layer it belongs by detecting the value of the pressure. Sensor nodes use RREQ and RREP packets in order to communicate with each other, and the next-hop can be determined by the rule of from deep to shallow and so on.

EUROP seems simple in terms of communication, as many of the control packets are eliminated by introducing a depth sensor inside the sensor node. Not only a depth sensor, but an electronic module also is required for every node in order to push it towards the upper layer and then back into position. Using the depth sensor and electronic module is not so simple; on one hand, cost per node will increase. On the other hand, both of these will be a burden on the critical node energy, which will ultimately decrease the life of the sensor node.

Localization and Routing Framework

In some applications, sensed data become meaningless without time and location information. Localization is essential for data labeling while some time critical applications require timely information. (M. Erol & Oktug, 2008) combine both of these tasks in a localization framework called "*catch up or pass*", where these mutually help each other. It benefits from the uncontrolled motion of underwater sensor nodes, where these nodes use the position and velocity information to help them to decide on whether to carry the data packet until they *catch up* with a sink or *pass* it to a faster or slower relay node.

The proposed framework uses a limited number of special nodes called Mobile Beacon and Sink (MBS). These MBS nodes have the ability to dive in and then return back vertically by modifying their density. The rest of the ordinary nodes stay under the water at different locations and can move with the water currents. MBSs periodically visit different depths in order to localize underwater sensor nodes and collect data packets from them. These MBSs receive the coordinates from the GPS while they are floating on the surface and upload the collected data to a ground station.

At the first stage, localization is done iteratively. Initially, MBSs get their location information via GPS. Then periodically broadcast their coordinates while diving to the deepest position of the network. After receiving from several beacons (at least four in this scenario), an ordinary node gets its location information. A localized node considered as an *active* node and can help in the localization process. It acts as a beacon and further distributes self-coordinates. Every localization phase has a fixed duration which is announced in the localization message. Duration of the interval can be adjusted according to the depth of the network and speed of the MBS nodes. Alternatively, after each dive, the duration of this interval can be updated via satellite to the MBS nodes. The location and velocity of the MBS nodes and the neighbors are learnt during the localization phase with the help of MBS message. This MBS message also includes a time stamp field which helps to determine the distance via Time of Arrival (ToA). After completing the localization round, next starts the routing phase. Sensor nodes that have data packets to send can select an MBS and forward these packets towards the sink. The routing algorithm performs the best data forwarding according to the position and relative motion of the MBS and ordinary sensor nodes.

During this localization and routing framework, authors assume that all the nodes are clock synchronized throughout the network. Such assumptions can be made for short term applications but for the long term missions we require some additional mechanism in order to achieve synchronization. Moreover, they use the ToA method when determining the distance between two nodes. Although, ToA is considered more promising than the techniques of same type like AoA (Angle-of-Arrival) and TDoA (Time-Difference-of-Arrival) it still cannot provide accuracy at long ranges and is only feasible for short ranges.

Localization Scheme for UWSNs

Location information can be used to design network architecture and routing protocols. (Melike Erol, Vieira & Gerla, 2007) proposed an idea of Dive and Rise (DNR) for the positioning system. They used mobile DNR beacons to replace static anchor nodes. The major drawback of this DNR scheme is that it requires a large number of expensive DNR beacons. (Kai Chen, 2009) proposed a hierarchical localization scheme in order to overcome this drawback of the DNR scheme. They try to decrease the requirement of mobile beacons by replacing them with four types of nodes, which are surface buoys, Detachable Elevator Transceivers (DETs), anchor nodes and ordinary sensor nodes. Surface buoys are assumed to be equipped with a GPS facility. DETs are attached to the surface buoys, mainly composed of an elevator and an acoustic transceiver. The elevator helps the DET to dive vertically in the water and then rise back up towards the water surface. An acoustic transceiver is used to communicate with the anchored nodes specifically in order to broadcast coordinate messages. Furthermore, many nodes are anchored at different positions and depth levels throughout the area of interest. These are special nodes as they have more energy and help to locate the ordinary nodes by communicating with DETs with the help of the acoustic transceiver. The fourth type of nodes is the ordinary sensor nodes; these are used for the sensing task. These ordinary nodes will listen to coordinate messages broadcast by the anchored nodes. When one receives more than 3 messages from different anchored nodes, it will then start to calculate its own position in the network.

After such specialized hardware deployments, this localization scheme has some assumptions. First of all, they assume that all the sensor nodes are equipped with a pressure sensor in order to provide its depth position or z-coordinate information. Then, after acquiring this entire infrastructure, they assume that the network is static. Although, it can be enhanced for a mobile network, during their simulation results, mobility is still not considered. However, all these arrangements are not easily possible for long term applications, plus if we are interested in large areas, then cost will become a major issue.

Underwater Wireless Hybrid Sensor Networks (UW-HSN)

In underwater wireless sensor networks, an acoustic channel is considered the only feasible means of communication. In practical use, an acoustic channel presents many key challenges specifically in shallow water; these include large propagation delays, high signal attenuation and transmission energy consumption as well as a low bandwidth. In order to handle this situation where we have only the choice of an acoustic channel, (Ali &

Hassanein, 2008) introduced a hybrid architecture called Underwater Wireless Hybrid Sensor Networks (UW-HSN).

UW-HSN is a hybrid of both, acoustic and radio communications. The basic idea here is to use the radio communication for large and continuous traffic and the acoustic for the small amount of data. Every node supports both types of communication, so they will use acoustic for underwater communication with the neighboring nodes and radio is used when nodes are on the surface in order to communicate directly with the base station. By doing so, the over-water network is a high speed, short range multi-hop with the help of a radio channel. For this purpose, any existing link layer, routing, networking and localization protocol can be used from the WSN literature with minor changes for surface communication. Every node should be equipped with both a radio and acoustic modem in order to support both types of communication. In addition to an acoustic and radio interface, every node is also equipped with a mechanical module which allows the node to swim to the surface and then dive back to different levels in the water. The philosophy is to incorporate the mobility of underwater sensor nodes in order to increase overall throughput of the network. They introduce TurtleNet, based on the hybrid concept, where nodes use piston based negative and positive bouncy for the vertical movements in order to reach the water surface and then back to the ocean bottom or to a pre-configured depth. For this architecture, they provide an algorithm called the Turtle Distance Vector (TDV), based on the distance vector approach. According to the current state of the node, TDV decides about the communication channel in order to minimize the event average delay. The event delay is defined as the time duration between its creation at the source and successful reception at the base station.

Simulation results, in order to check the performance, show that UW-HSN provides high goodput and smaller delays as compared to all-acoustic approaches. However, no information is available about the energy consumption which is an important metric in order to check the performance of TurtleNet due to its special network setup requirements. These extra hardware requirements, not only drain the crucial energy which can decrease the life of the network but also increase the cost of the network as well.

Temporary Cluster Based Routing (TCBR)

Many of the multihop routing protocols have been proposed for underwater sensor networks but most of them face the problem of multihop routing where nodes around the sink drain more energy and are expected to die early. In order to solve this problem and make equal energy consumption throughout the network, (Ayaz, Abdullah & Low Tang) proposed a Temporary Cluster Based Routing (TCBR) algorithm.

In the TCBR architecture, multiple sinks are deployed on the water surface and data packets received at any sink are considered as delivered successfully because they can communicate at a higher bandwidth and small propagation delay with the help of radio communication. Two types of nodes are used: ordinary nodes and some special nodes called Courier nodes. Ordinary sensor nodes are used to sense the event happening, collect information and try to forward these data packets to a nearer courier node. Small numbers of courier nodes (2 to 4% of the total sensor nodes) are used and these can sense as well as receive the data packets from the other ordinary sensor nodes and deliver them to a surface sink. These Courier nodes are equipped with a mechanical module which helps to push the node inside water at different defined depths and then pull it back to the ocean surface. An equipped piston can do this by creating the positive and negative buoyancy. These Courier nodes will reach different depth levels and stop for a specified amount of time. After reaching any specified position, these will broadcast hello packets so that ordinary nodes around them can know about their presence. These hello packets can be forwarded only 4 hops and if an ordinary node receives them from more than one Courier node then it will forward the data packet to the nearest one within a specified amount of time, which is defined in the hello packet.

TCBR completes its task of equal energy consumption throughout the network by only requiring a small number of Courier nodes, instead of equipping the mechanical module with every sensor node. However, data can be collected when a courier node reaches the communication range of every sensor node. Due to this, on one hand, all the sensor nodes will hold their generated data packets in a limited buffer until a Courier node visits them; on the other hand, TCBR cannot be used for time critical applications.

Multi-Sink Opportunistic Routing Protocol

(Tonghong, 2008) proposed a Multi-Sink Opportunistic routing protocol for underwater mesh networks. They defined a tiered architecture for deploying the underwater sensor nodes, where an acoustic mesh network is located between the underwater network and the central monitoring system which acts like a backbone network for sensor nodes. A quasi-stationary 2-dimensional UWSN architecture is considered for shallow-water coastal areas. This architecture is composed of five types of elements including an ordinary sensor node, a mesh node, a UW-sink, a surface buoy and a monitoring center. Among these, three of them, the sensor node, mesh node and UW-sink are anchored to the sea bed and the surface buoy is placed on the ocean surface. Furthermore, both the UW-sink and the surface sink are connected by a wire. An onshore central monitoring system is used which is connected to the internet. Compared with an ordinary sensor node, a mesh node is more sophisticated as it has more memory, a longer transmission range and better processing power. In order to help the network survive for a longer period, an underwater man controlled vehicle is used for recharging these mesh nodes.

After observing the occurred phenomena, each senor node transmits its sensed data to the nearest mesh node. Mesh nodes first aggregate the received data and then send it to the UW-sinks via a multi-hop acoustic channel. Finally, the aggregated packets are delivered to the surface sinks and from there surface buoys send them to the onshore monitoring system. The proposed scheme is a best effort protocol where data packets are forwarded along redundant and interleaved paths. The source node transmits the data packets simultaneously but not sequentially over multiple UW-sinks located at different locations. Different from the opportunistic routing, this protocol exploits the packet duplications to increase the packet delivery ratio.

However, the proposed routing protocol has some serious performance issues. First of all, it is assumed that each mesh node has information not only about its adjacencies but also about all the UW-sinks, like node IDs and their geographic positions. Secondly, authors considered a quasi-stationary network but not completely mobile that's why it is assumed that the mesh nodes and their neighbors are relatively static, which can be different in practicality. Moreover, packets are forwarded along redundant and interleaved paths so multiple copies of the same packet can be generated and these duplications will continue to increase as the number of hops along the path starts to increase.

Location-Based Clustering Algorithm for Data Gathering (LCAD)

The data transmission phase is the main source of energy consumption for a sensor node. Dissipation of energy during the data transmission is proportional to the distance between the sender and receiver. As we have already discussed, another problem with the multi-hop approach is that sensor nodes around the sink process a large number of data packets which rapidly drain their energy. In order to solve both of these problems, (Anupama, Sasidharan & Vadlamani, 2008) suggest a cluster based architecture for 3-dimensional underwater sensor networks. Here, senor nodes are deployed in the whole area of interest at fixed relative depths from each other. These sensor nodes at each tier are organized in clusters with multiple cluster heads. They suggest an algorithm for the cluster head selection at each cluster according to the node position in the network. Horizontal acoustic links are used for intra cluster communication. For energy concerns, the length of this horizontal acoustic link is restricted to the maximum of 500 m as it has been shown that the performance of the acoustic link can be optimal at this communication distance.

In the proposed architecture, the entire network is divided into 3-dimensional grids where each grid is set to approximately 30m × 40m × 500m. The entire communication process is completed in three phases: (i) Set-up phase, where the cluster head is selected. (ii) Data gathering phase, where data is sent by the nodes in the same cluster to the cluster head. (iii) Transmission phase, where data gathered by cluster heads is delivered to the base station with the help of Autonomous Underwater Vehicles (AUVs). About the cluster head (ch-node), some of the sensor nodes in every cluster have additional resources like memory and energy and such nodes can qualify as ch-node. Having multiple ch-nodes increase not only the reliability but also load balancing in the network. These ch-nodes are located approximately at the centre of the grid which helps to communicate with the maximum number of ordinary sensor nodes. These grids are organized just like the cells in the cellular network.

AUVs are used as data mules for collecting data packets from these cluster heads instead of from every sensor node in the network. As it has been proven that an acoustic link is not suggested for use with distances of more than 500m, the required number of tiers depends on the average depths of the oceans. For the best results, they advocate a dense deployment of sensor nodes at the lower tiers and a sparser distribution at the higher tiers.

The proposed protocol seems to have some serious performance issues. The performance of LCAD depends on the grid structure, especially the position of the ch-node inside it. For terrestrial sensor networks, considering such type of structure is easily possible. For underwater environments where node movement is frequent, assumption of such a grid structure is not so simple, as nodes can come to and leave different grids frequently. For performance analysis, they check the performance of LCAD in terms of network lifetime but have not provided any information about the node movement.

Location-Aware Source Routing (LASR)

Underwater sensor networks are different in many aspects from the terrestrial sensor technology as discussed already. First, radio communications are not suitable for deep water, so they have to be replaced with acoustic communication. Compared to a radio channel, acoustic communication in water has a very low data rate with high latency. These routing protocols require higher bandwidths and result in large end-to-end delays and are not suitable for such environments. Second, most of the time sensor nodes are considered as static, but underwater sensor nodes can move up to 1-3 m/sec due to different underwater activities (Nicolaou, See, Peng, Jun-Hong, & Maggiorini, 2007). These major differences between both environments lead to questions about the acoustic network performance with the protocols designed for terrestrial networks.

Dynamic Source Routing (Johnson, Maltz & Broch, 2001) is a well known routing protocol originally proposed for the MANET but suffers from high latency in the underwater acoustic environment. In these conditions, the topology rate of change is very high compared with acoustic latency so, topology continues to change more quickly that DSR can adapt. In order to solve this problem without losing the experience of DSR, (Carlson, Beaujean & An, 2006) proposed LASR, a modification of DSR. LASR uses two techniques in order to handle the high latency of the acoustic channel; the first is a link quality metric and the second is location awareness. DSR depends on only the shortest path metric which results in poor performance in highly mobile networks. LASR replaces this shortest path metric with an expected transmission count (ETX) where the link quality metric ultimately provides more-informed decisions which give better routes through the network. Location awareness can be achieved from the incoming transmissions as an aid to estimating the local network topology. Topology prediction uses a tracking system to predict the current location of other vehicles in the network based on one way and range only measurements. While all the explicit information of the network including the routes and topology information is passed in the protocol header.

After all these modifications, LASR still depends on the source routing technique inherited from DSR. Therefore, as the hop count between the source and the destination increases, the packet header continues to increase as well. This increasing header size leads to overhead for acoustic communication with a narrow bandwidth.

Adaptive Routing

An underwater sensor network can be easily partitioned due to continuous node mobility and sparse deployment. This results in the unavailability of a persistent route from a source to a destination. Therefore, an underwater sensor network can be viewed as an Intermittently Connected Network (ICN) or Delay/Disruption Tolerant Network (DTN). Traditional routing techniques are not usually suitable for ICN or DTN, since data packets will be dropped when routes are not available. Furthermore, an USN is frequently required to provide distinguished packet deliveries according to different application requirements. Therefore, it is desirable to design a smart routing technique that could manage different application requirements adaptively.

For this purpose, (Zheng, et al., 2008) proposed a novel routing technique called Adaptive routing for underwater Delay/Disruption Tolerant Sensor Networks. Here, routing decisions are made according to the characteristics of data packets and the network conditions. The purpose of this protocol is not only to satisfy different application requirements but also achieve a good trade-off among delivery ratios, end-to-end delays and energy consumption for all data packets. The packet priorities are calculated from the packet emergency level, packet age, density of the neighbors around a node and the battery level of the node. The novelty of their work is that here, different numbers of message copies are created according to the characteristics of the data packets and the network. In order to make the protocol flexible according to the conditions, all the elements in the information are variable except the emergency level. They divide the whole routing spectrum into four states, and the routing is conducted according to calculated results. Their simulation results show that such a strategy can satisfy different application requirements like delivery ratio, average end-to-end delay and energy consumption. However, the proposed scheme calculates these priorities separately for each data packet after receiving them. Such calculations require high frequent communication with the neighbor nodes; which not only can be burden on node energy but also it can help to increase end-to-end delays.

CONCLUSION

In this chapter, we presented an overview of state of the art routing protocols in underwater wireless sensor networks. Routing for the UWSN is an important issue which is attracting significant attention of the researchers. The design of any routing protocol depends on the goals and requirements of the application as well as the appropriateness, which also depends on the availability of network resources. We discussed the unique characteristics of the UWSN; protocols proposed for these environments and highlighted the advantages and performance issues of each scheme. At the end, we compared and classified these techniques according to their attributes and functionalities. The ultimate objective of this chapter is to encourage the new researchers of the area by providing a foundation about the routing protocols proposed to date.

REFERENCES

Akyildiz, I. F., Pompili, D., & Melodia, T. (2005). Underwater acoustic sensor networks: Research challenges. *Ad Hoc Networks*, *3*(3), 257–279. doi:10.1016/j.adhoc.2005.01.004

Ali, K., & Hassanein, H. (2008, 6-9 July 2008). *Underwater wireless hybrid sensor networks.* Paper presented at the IEEE Symposium on Computers and Communications, 2008.

Anupama, K. R., Sasidharan, A., & Vadlamani, S. (2008, 27-28 August). *A location-based clustering algorithm for data gathering in 3D underwater wireless sensor networks.* Paper presented at the International Symposium on Telecommunications, 2008.

Ayaz, M., & Abdullah, A. (2009, 16-18 Dec. 2009). *Hop-by-hop dynamic addressing based (H2-DAB) routing protocol for underwater wireless sensor networks.* Paper presented at the International Conference on Information and Multimedia Technology, 2009.

Ayaz, M., Abdullah, A., & Low Tang, J. (15-17 June 2010). *Temporary cluster based routing for underwater wireless sensor networks.* Paper presented at the 2010 International Symposium in Information Technology (ITSim).

Carlson, E. A., Beaujean, P. P., & An, E. (2006, 18-21 Sept. 2006). *Location-aware routing protocol for underwater acoustic networks.* Paper presented at the OCEANS 2006.

Chirdchoo, N., Wee-Seng, S., & Kee Chaing, C. (2009, 26-29 May 2009). *Sector-based routing with destination location prediction for underwater mobile networks.* Paper presented at the International Conference on Advanced Information Networking and Applications Workshops, WAINA '09.

Chitre, M., Shahabudeen, S., Freitag, L., & Stojanovic, M. (2008, 15-18 September). *Recent advances in underwater acoustic communications & networking.* Paper presented at the OCEANS 2008.

Chun-Hao, Y., & Kuo-Feng, S. (2008, November 30-December 3). *An energy-efficient routing protocol in underwater sensor networks.* Paper presented at the 3rd International Conference on Sensing Technology, ICST 2008.

Daeyoup, H., & Dongkyun, K. (2008, 15-18 September). *DFR: Directional flooding-based routing protocol for underwater sensor networks.* Paper presented at the OCEANS 2008.

Dario Pompili, T. M., & Akyildiz, I. F. (2006). *A resilient routing algorithm for long-term applications in underwater sensor networks.* Paper presented at the MedHocNet.

Domingo, M. C., & Prior, R. (2007, 3-7 Sept. 2007). *A distributed clustering scheme for underwater wireless sensor networks.* Paper presented at the IEEE 18th International Symposium on Personal, Indoor and Mobile Radio Communications, PIMRC 2007.

Erol, M., & Oktug, S. (2008, 13-18 April). *A localization and routing framework for mobile underwater sensor networks.* Paper presented at the INFOCOM Workshops 2008, IEEE.

Erol, M., Vieira, L. F. M., & Gerla, M. (2007). *Localization with Dive'N'Rise (DNR) beacons for underwater acoustic sensor networks.* Paper presented at the Second Workshop on Underwater Networks.

Heidemann, J., Wei, Y., Wills, J., Syed, A., & Yuan, L. (2006, 3-6 April). *Research challenges and applications for underwater sensor networking.* Paper presented at the Wireless Communications and Networking Conference, WCNC 2006. IEEE.

Jiang, Z. (2008). Underwater acoustic networks-Issues and solutions. *International Journal of Intelligent Control and Systems, 13*, 152–161.

Jinming, C., Xiaobing, W., & Guihai, C. (2008, 24-26 Oct. 2008). *REBAR: A reliable and energy balanced routing algorithm for UWSNs.* Paper presented at Seventh International Conference on the Grid and Cooperative Computing, GCC '08.

Johnson, D. B., Maltz, D. A., & Broch, J. (2001). *DSR: The dynamic source routing protocol for multihop wireless ad hoc networks. Ad Hoc Networking* (pp. 139–172). Addison-Wesley Longman Publishing Co., Inc.

Jornet, J. M., Stojanovic, M., & Zorzi, M. (2008). *Focused beam routing protocol for underwater acoustic networks.* Paper presented at the Third ACM International Workshop on Underwater Networks.

Jun-Hong, C., Jiejun, K., Gerla, M., & Shengli, Z. (2006). The challenges of building mobile underwater wireless networks for aquatic applications. *Network, 20*(3), 12–18.

Kai Chen, Y. Z., & He, J. (2009). A localization scheme for underwater wireless sensor networks. *International Journal of Advanced Science and Technology, 4.*

Magistretti, E., Jiejun, K., Uichin, L., Gerla, M., Bellavista, P., & Corradi, A. (2007, 11-15 March). *A mobile delay-tolerant approach to long-term energy-efficient underwater sensor networking.* Paper presented at the Wireless Communications and Networking Conference, WCNC 2007. IEEE.

Nicolaou, N., See, A., Peng, X., Jun-Hong, C., & Maggiorini, D. (2007, 18-21 June). *Improving the robustness of location-based routing for underwater sensor networks.* Paper presented at the OCEANS 2007 - Europe.

Peng, S., Seah, W. K. G., & Lee, P. W. Q. (2007, 17-20 April). *Efficient data delivery with packet cloning for underwater sensor networks.* Paper presented at the Underwater Technology and Workshop on Scientific Use of Submarine Cables and Related Technologies, 2007.

Peng, X., Zhong, Z., Nicolas, N., Andrew, S., Jun-Hong, C., & Zhijie, S. (2010). *Efficient vector-based forwarding for underwater sensor networks.* Hindawi Publishing Corporation.

Pompili, D. (2007). *Efficient communication protocols for underwater acoustic sensor networks.* Georgia Institute of Technology.

Seah, W. K. G., & Tan, H. P. (2006). *Multipath virtual sink architecture for wireless sensor networks in harsh environments.* Paper presented at the First International Conference on Integrated Internet Ad Hoc and Sensor Networks.

Shi, Z. J. (2006). *Architectural challenges in underwater wireless sensor networks.* Department of Computer Science and Engineering University of Connecticut.

Sozer, E. M., Stojanovic, M., & Proakis, J. G. (2000). Underwater acoustic networks. *IEEE Journal of Oceanic Engineering*, *25*(1), 72–83. doi:10.1109/48.820738

Tonghong, L. (2008, 25-27 May). *Multi-sink opportunistic routing protocol for underwater mesh network.* Paper presented at the International Conference on Communications, Circuits and Systems, ICCCAS 2008.

Uichin, L., Wang, P., Youngtae, N., Vieira, L., Gerla, M., & Jun-Hong, C. (2010, 14-19 March). *Pressure routing for underwater sensor networks.* Proceedings IEEE INFOCOM.

Wei, L., Haibin, Y., Lin, L., Bangxiang, L., & Chang, C. (2007, 5-8 August). *Information-carrying based routing protocol for underwater acoustic sensor network.* Paper presented at the International Conference on Mechatronics and Automation, ICMA 2007.

Xie, P., Cui, J.-H., & Lao, L. (2006). *VBF: Vector-based forwarding protocol for underwater sensor networks. Networking 2006. Networking Technologies, Services, and Protocols; Performance of Computer and Communication Networks; Mobile and Wireless Communications Systems* (*Vol. 3976*, pp. 1216–1221). Berlin, Germany: Springer.

Yan, H., Shi, Z. J., & Cui, J.-H. (2008). *DBR: Depth-based routing for underwater sensor networks*. Paper presented at the 7th International IFIP-TC6 Networking Conference on Ad Hoc and Sensor Networks, Wireless Networks, Next Generation Internet.

Zheng, G., Colombo, G., Bing, W., Jun-Hong, C., Maggiorini, D., & Rossi, G. P. (2008, 23-25 January). *Adaptive routing in underwater delay/disruption tolerant sensor networks.* Paper presented at the Fifth Annual Conference on Wireless on Demand Network Systems and Services, WONS 2008.

APPENDIX

Comparison and Classification

Table 2. Comparison of routing protocols presented in this chapter based on their characteristics

Protocol/ Architecture	Single/ Multiple Copies	Clustered/ Single-entity	Hop-by-hop/ End-to-end	Multi-paths/ Single-path	Hello or Control Packets	Single/ Multi Sink
VBF	Single-copy	Single-entity	End-to-end	Multi-path	No	Single-sink
HH-VBF	Single-copy	Single-entity	Hop-by-hop	Multi-path	No	Single-sink
FBR	Single-copy	Single-entity	Hop-by-hop	Multi-path	Yes	Multi-sink
DFR	Multiple	Single-entity	Hop-by-hop	Multi-path	No	Single-sink
REBAR	Single-copy	Single-entity	Hop-by-hop	n/a	No	Single-sink
ICRP	Multiple	Single-entity	End-to-end	Single-path	No	Single-sink
DUCS	Single-copy	Clustered	Hop-by-hop	n/a	Yes	Single-sink
Packet Cloning	Multiple	Single-entity	Hop-by-hop	Multi-path	No	Multi-sink
SBR-DLP	Single-copy	Single-entity	Hop-by-hop		Yes	Single-sink
Multipath Virtual Sink	Multiple	Clustered	Hop-by-hop	Multi-path	Yes	Multi-sink
DDD	Single-copy	n/a	Single-hop	Single-path	Yes	n/a
DBR	Multiple	Single-entity	Hop-by-hop	Multi-path	No	Multi-sink
Hydrocast	Multiple	Clustered	Hop-by-hop	Multi-path	No	Multi-sink
EUROP	Single-copy	Single-entity	Hop-by-hop	Multi-path	Yes	Single-sink
UW-HSN	Single-copy	Single-entity	Hop-by-hop	Single-path	Yes	Single-sink
TCBR	Single-copy	Clustered	Hop-by-hop	Multi-path	Yes	Multi-sink
Resilient Routing	Single-copy	Single-entity	End-to-end	Single-path	No	Single-sink
Mulit-Sink Opportunistic	Multiple	n/a	Hop-by-hop	Multi-path	No	Multi-sink
H2-DAB	Single-copy	Single-entity	Hop-by-hop	Multi-path	Yes	Multi-sink
LCAD	Single-copy	Clustered	Hop-by-hop	Multi-path	Yes	Single-sink
LASR	Single-copy	Single-entity	End-to-end	Single-path	Yes	Single-sink
Adaptive Routing	Multiple	Single-entity	Hop-by-hop	Multi-path	Yes	Single-sink

Figure 9. Classification based on assumptions and requirements

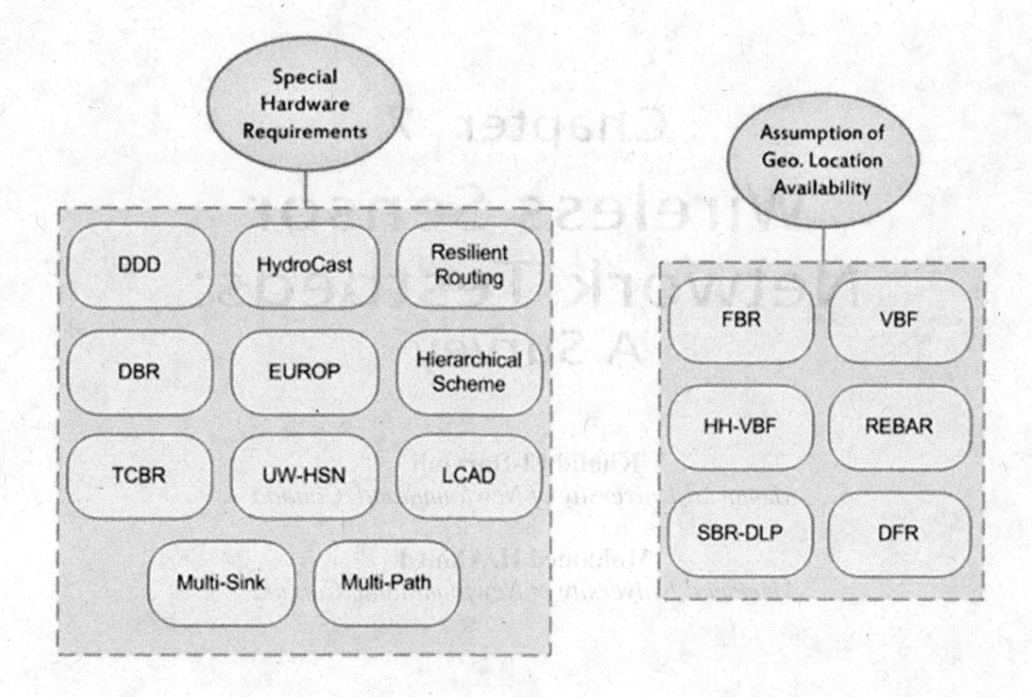

Figure 10. Some miscellaneous classifications

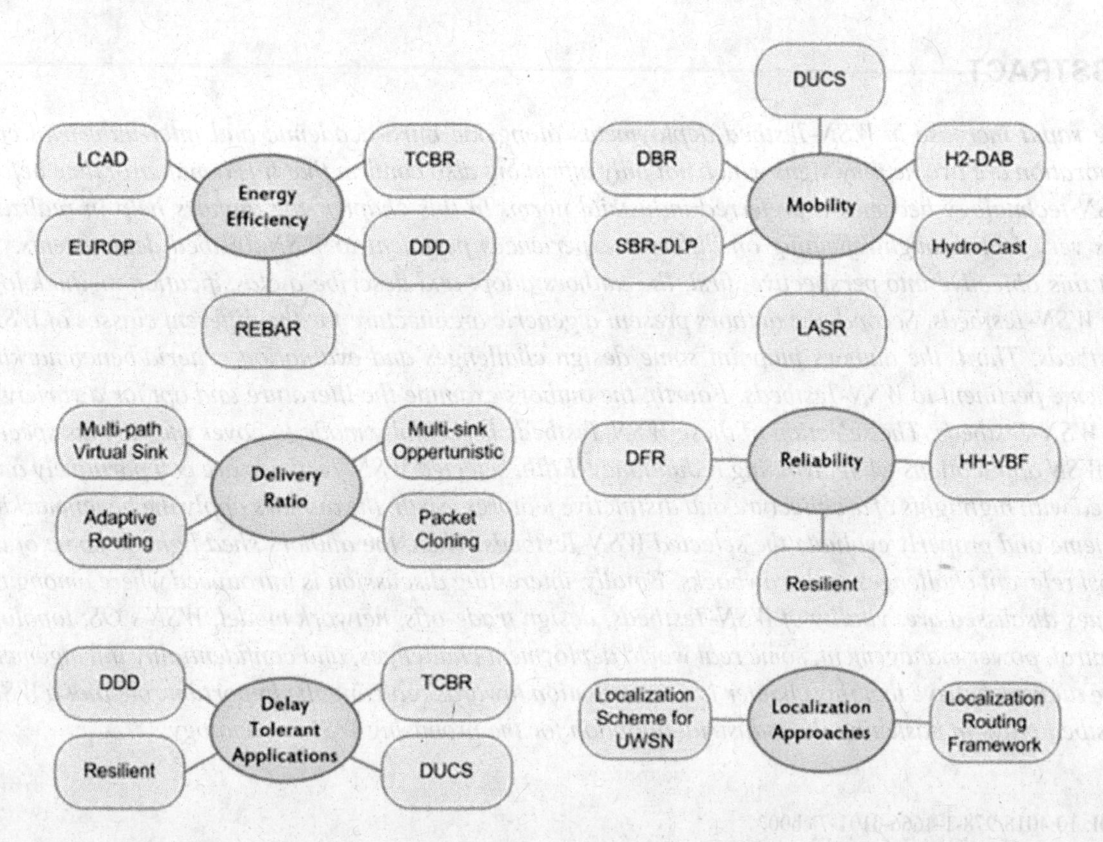

Chapter 7
Wireless Sensor Network Testbeds: A Survey

Khalid El-Darymli
Memorial University of Newfoundland, Canada

Mohamed H. Ahmed
Memorial University of Newfoundland, Canada

ABSTRACT

The rapid increase in WSN-Testbed deployments alongside intra-academic and inter-industrial collaboration are two healthy signs which not only affirm but also confirm that it is a matter of time before WSN technology becomes a preferred industrial norm. In this chapter, the authors help in realizing this very fact through reflecting on different experiences pertinent to WSN-Testbed deployments. To put this objective into perspective, first, the authors adopt and describe a classification methodology for WSN-Testbeds. Second, the authors present a generic architecture for the different classes of WSN-Testbeds. Third, the authors pinpoint some design challenges and evaluation criteria/benchmarking scheme pertinent to WSN-Testbeds. Fourth, the authors examine the literature and opt for a variety of 30 WSN-Testbeds. The selection of these WSN-Testbeds is carefully made to cover the various spectra of WSN applications while avoiding redundancy. Fifth, selected WSN-Testbeds are comparatively analyzed with highlights of architecture and distinctive features. Sixth, the authors apply the benchmarking scheme and properly evaluate the selected WSN-Testbeds. Then, the authors shed light on some of the most relevant challenges and drawbacks. Finally, interesting discussion is introduced where among the issues discussed are: vitality of WSN-Testbeds, design trade-offs, network model, WSN's OS, topology control, power management, some real world deployment challenges, and confidentiality infringement. The authors believe that this chapter is a contribution towards realizing the important role that a WSN-Testbed plays in hastening the industrial adoption for the promising WSN technology.

DOI: 10.4018/978-1-4666-0101-7.ch007

1. INTRODUCTION

A Wireless Sensor Network Testbed (WSN-Testbed) is a platform for experimentation of development projects. It allows rigorous, flexible, transparent and replicable testing of theories, computational tools and innovations. When compared to WSN simulators, WSN-Testbed enables more realistic and reliable experimentation in capturing the subtleties of the underlying hardware, software, and dynamics of the wireless sensor network. A quick look at the literature will reveal an overwhelming rapid increase in the deployments of such WSN-Testbeds. The momentum of WSN-Testbed deployment is further enhanced through an increasing collaboration between academia and industry.

Wireless Sensor Networks (WSNs) are rapidly gaining increasing attention on the experimentation level as well as the application-deployment level. Affordability, ease of deployment, and ability to monitor phenomena that were impossible to monitor using other solutions are just few among many other reasons that make WSN such a preferred choice. Thanks to recent advancements in wireless communications and MEMS technology, tiny low-cost and low-power devices known as sensor nodes or motes were developed. Motes can be spatially distributed and they are equipped with sensors that communicate over a wireless network and cooperate to monitor a physical or environmental phenomenon such as humidity, pressure, temperate, light, vibration, motion, sound, etc. (Haensel, ; Romer & Mattern, 2004).

The wide emergence of wireless sensor networks deployment was originally motivated by the military applications. Now, WSNs are deployed and being a fertile and active field of research for many military, industrial and civilian applications such as fire detection, habitat monitoring, health care, space, process monitoring, control, environmental, surveillance, security, etc. (Hadim & Mohamed, 2006).

As a result, this rapid growth enthuses universities and research institutes around the globe to set-up their own wireless sensor network testbeds (WSN-Testbeds). WSN-Testbeds enable researchers to gain hands-on experience and to investigate different kinds of scenarios. Additionally, WSN-Testbeds provide researchers with hands-on opportunity to experimentally investigate own innovation, and to test and evaluate its adaptability to real-world scenarios.

Accordingly, WSN-Testbeds are the basis for experimentation with wireless sensor networks in real-world settings; and they are also used by many researchers to evaluate specific applications pertaining to specific areas. A WSN-Testbed typically consists of sensor nodes deployed in a controlled environment. WSN-Testbeds provide researchers with an efficient way to examine and evaluate their algorithms, protocols, applications, etc. WSN-Testbed can be designed to support different features depending on the objective of the testbed. Among the important features of a WSN-Testbed is that it can be designed to remotely configure, run and monitor experiments. Another interesting feature is that the WSN-Testbed can be used for repeating experiments to produce similar results for analysis (Yick, Mukherjee, & Ghosal, 2008).

Since WSN simulators are available, why a researcher needs to use a WSN-Testbed which undeniably is a relatively costly solution? In fact, WSN simulators could be used for testing, evaluation and initial validation. For instance, they are used to test new protocols and to evaluate them. WSN simulator is based on mathematical models that attempt to model the underlying characteristics of its physical system probably taking current and potential ambient conditions into consideration. However, the fidelity of simulator is always a concern (J. Heidemann, N. Bulusu, J. Elson, C. Intanagonwiwat, K. Lan, Y. Xu, W. Ye, D. Estrin, and R. Govindan).

Selecting the appropriate level of abstraction in simulation model is a complex problem. Thus, it is obvious that the accuracy of a simulator will solely depend on its mathematical model. Accordingly, there is a trade-off between simulator's accuracy and computational complexity. The more complex the simulation model is the more computational resources and time are required to execute it. This makes the designers of such simulation models tend to make them as simple as possible.

Due to such complexity, most simulator designs focus on the specific higher layer protocols and ignore interaction between layers. Moreover, it is impossible to take all the various aspects of the wireless channel into consideration when designing a simulation model (Takai, Martin, & Bagrodia, 2001). Nonetheless, simulation tools are essential in providing affordable environment for the initial design and tuning of wireless sensor networks. Such inherent difficulty in faithful modeling motivates many researches to build their own WSN-Testbeds. Among the advantages of a real WSN-Testbed over a simulator is that it provides a realistic testing environment and allows users to get more precise testing results (De, Raniwala, Sharma, & Chiueh, 2005).

To further appreciate the important role of such WSN-Testbeds, one needs to reflect on the life cycle of a WSN product development from its inception (theory) to its final stage (early stage product). It should be noted that the role of simulation is not conflicting but rather complementary in the chain of the WSN product development. Figure 1 below depicts the essential role of WSN-Testbed in this cycle (National Science Foundation).

This chapter presents a survey of existing WSN-Testbeds in different applications and classes. The rest of this chapter is organized as follows. We present taxonomy of WSN-Testbeds in Section 2. In Section 3, we describe the architecture of WSN-Testbed taxonomies presented earlier. In Section 4, we present and discuss essential *design challenges and evaluation criteria* pertaining to wireless sensor network testbeds (WSN-Testbeds). Based on the discussed architecture and evaluation criteria, we examine the literature; carefully opt for different 30 WSN-Testbeds, and present relative comparison between the varieties of selected WSN-Testbeds in Section 5. In Section 6, we apply the *benchmarking scheme*, introduced in Section 4, to the selected WSN-Testbeds and relatively evaluate them accordingly. We then highlight some shortcomings and drawbacks for the considered WSN-Testbeds. We discuss in Section 7 some important issues relevant to WSN-Testbed design, deployment and operation. Finally, concluding remarks are presented in Section 8.

2. TAXONOMY AND ARCHITECTURE OF WSN-TESTBEDS

An analysis and perspective on the experimentation in the wireless networking research was the outcome of a workshop held in November of 2002 by the National Science Foundation (NSF) (National Science Foundation). In this work, we benefit from NSF's analysis and we tailor it to the case of WSN-Testbeds. Generally, there are two perspectives on classification of WSN-Testbeds. The first perspective is based on the objective of the WSN-Tesstbed. The second classification is based on the underlying structure. In this section we present description on these taxonomies then a generic architecture for each class is presented.

Figure 1. Development stages of a WSN system

2.1 Objective-Based Classification

WSN-Testbeds can be classified into two broad classes with distinctive objectives; Multi-user Experimental Testbeds (MXTs) and Proof-of-Concept Testbeds (PCTs) (National Science Foundation). These are depicted in Figure 2.

2.1.1 Multi-User, Experimental Testbeds (WSN-MXT)

WSN-MXTs are designed and operated to provide a service to a user community of WSN researchers. WSN-MXT provides well-supported infrastructure and tools to enable researchers to experiment with new network technologies, protocols, architectures and services (National Science Foundation).

2.1.2 Proof-of-Concept Testbeds (WSN-PCT)

WSN-PCTs are designed to advance a particular research objective, with an emphasis on demonstrating research ideas in a compelling way, in order to accelerate technology transfer and commercialization. WSN-PCTs can play a crucial role in transitioning research ideas from the laboratory to commercial practice, and are an important component of WSN networking research, which places a high value on its continuing contributions to society at large. Once the concept is proofed and the technology is commercialized, in most cases the WSN-Testbed is no more of use (National Science Foundation).

It should be understood that the missions of WSN-MXTs and WSN-PCTs are quite distinct. WSN-MXT is dedicated to providing a service to a larger group of research users with many different research objectives, while WSN-PCT seeks to advance a specific research objective. Evaluation of WSN-Testbeds should be done with a conscious recognition of which type they are, so that appropriate criteria for evaluation can be applied.

2.2 Structure-Based Classification

In terms of WSN-Testbed structure, WSN-Testbeds can be further classified into four distinctive categories that can be quite interrelated. These are depicted in Figure 3. WSN Research Kits (WSN-RK) form the most basic structure. Cluster WSN-Testbeds (WSN-CT) set a broader class that can utilize WSN Research Kits. Overlay WSN-Testbeds (WSN-OT) are typically overlaid on an existing testbed that may not be originally designed to experiment with wireless sensor networks. Finally, Federated WSN-Testbeds (WSN-FT) are the broadest class of all which can accommodate

Figure 2. Classification of WSN-testbeds into MXT and PCT

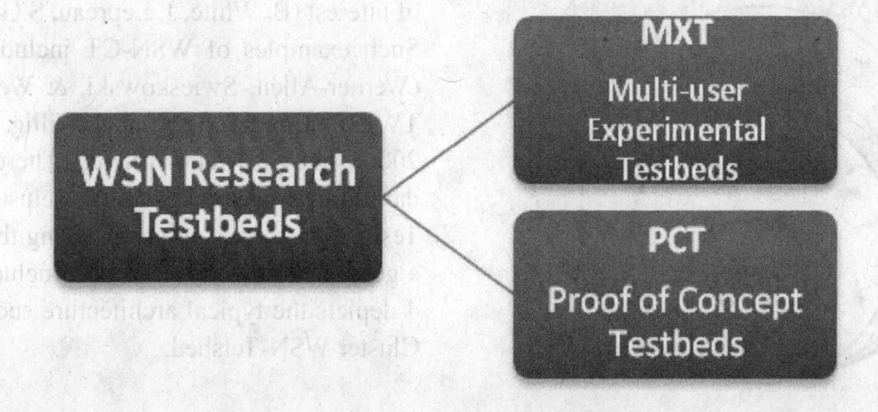

all other classes (National Science Foundation). More detailed description on each of the above-mentioned classes follows.

2.2.1 WSN Research Kits (WSN-RK)

WSN-RKs are collections of out-of-box WSN software and hardware components that are developed by a third party vendor. WSN-RK can be purchased and used by researchers to set up their own local network laboratories. Such kit can be automatically installed in minutes typically without the need for manual programming. Examples of WSN-RK include, Crossbow's Development Kits (Crossbow), Senswiz Research & Educational Kit (SensWiz), and Dust Networks' SmartMesh Kits (Dust Networks)

The constituents of a WSN research kit vary according to the specifications of the vendor. However, a general structure can be described as follows. A WSN-RK is comprised usually of a certain number of sensor nodes with various sensor options such as light, temperature, pressure, humidity, etc.; in addition to some base station(s). A mesh network firmware is typically provided and programmed for the accompanied sensor nodes

Figure 3. Further classification of WSN-testbeds based on structure

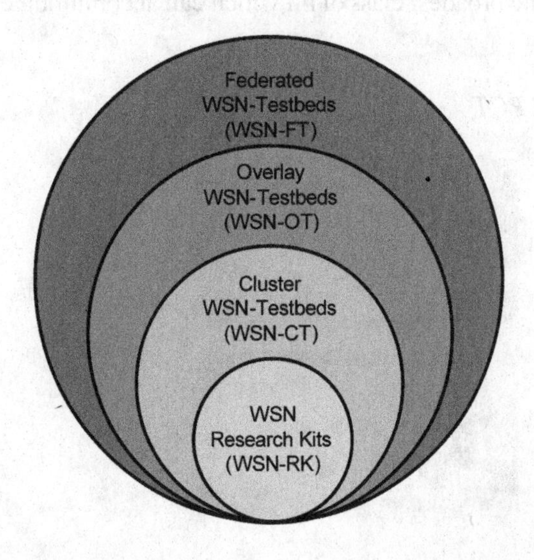

and base station(s). Additionally, WSN-RK normally comes with a pre-programmed visualization and analysis tool such as MoteView (Crossbow). Moreover, a WSN-RK may come with a variety of options. For instance, a user can select between battery-powered and USB-powered motes. The option of extending a certain WSN-RK is typically supported. The vendor makes such scalability feasible through offering some software platforms such as MoteWorks (Crossbow).

2.2.2 Cluster WSN-Testbeds (WSN-CT)

WSN-CTs are experimental research facilities that are based on the concept of emulation. Majority of existing WSN-Testbeds fall under this class. Such WSN-Testbeds bring together the elements of the WSN network (sensors, motes, links, switches, PCs, routers, etc.) in a common facility. It could be remotely accessed by users, e.g. through a web interface. Cluster WSN-Testbeds can be designed to support flexible configuration, so that users may perform experiments on a variety of distinct network topologies with programmable features to control certain aspects such as power supply, active motes, etc. Many of cluster WSN-Testbeds use open, extensible platforms so that researchers can study the behavior of individual components in detail, and modify those components to experiment with new algorithms and protocols. Cluster WSN-Testbeds centralize the infrastructure support required so that researches can focus on experimentation with the application of interest (B. White, J. Lepreau, S.Guruprasad). Such examples of WSN-CT include, Motelab (Werner-Allen, Swieskowski, & Welsh, 2005), TWIST (Handziski, Kopke, Willig, & Wolisz, 2006), and VigilNet (VigilNet). There is no standard framework for designing a clustered WSN-Testbed; however, upon surveying the literature a general architecture can be concluded. Figure 4 depicts the typical architecture scenarios of a Cluster WSN-Tetsbed.

The functionalities of the WSN-Testbed's underlying elements depicted in Figure 4 can be briefly introduced in a bottom-up approach as follows:

A. Sensor Nodes

Sensor nodes are micro-electronic wireless low-cost devices produce a measurable response to a change in a physical condition like pressure, temperature, etc. Sensor node is equipped with a transceiver which enables it to communicate with its peers. Figure 5 depicts the internal structure of a typical sensor node. The sensors continuously produce an analog signal as they listen to an input of interest. The sensed signal is digitized by an Analog to Digital Converter (ADC), and then it is sent to a micro-controller for further processing. Additionally, the micro-controller performs tasks and controls the functionality of the elements in the sensor node. Sensor nodes are equipped with limited low-power source. The power source can be battery, wall-power, energy scavenge, etc. Power consumption is an important design constraint needs to be taken into consideration.

Figure 4. Typical architecture scenarios of cluster WSN-testbed

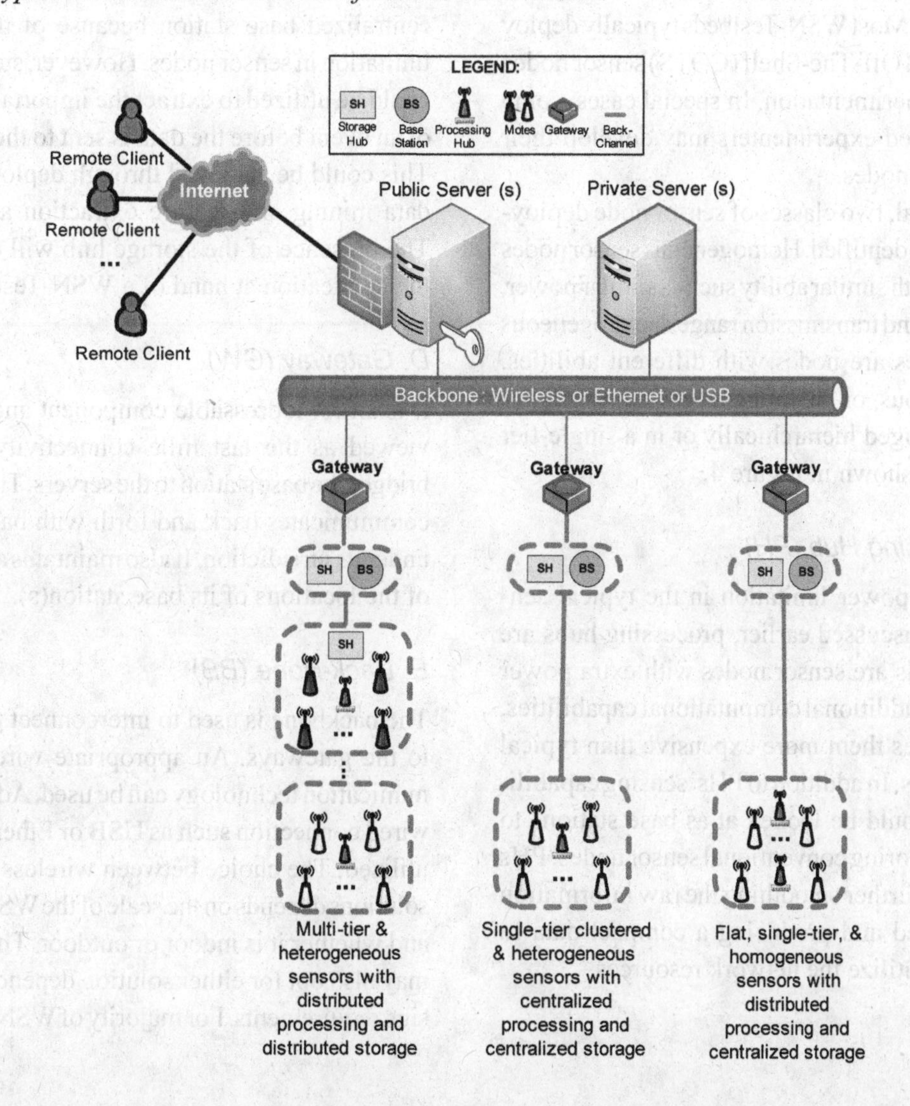

Typically, communication circuitry and antenna are the primary elements of power consumption in the sensor node (Xiaoyan Hong et al., 2001).

A typical sensor node has an on-chip micro-controller memory as well as flash memory. The memory is used to store the data required to program the node, and to store the application related data. Sensors can be passive or active. Passive sensors do not manipulate the environment by active probing, whereas active sensors actively probe the environment. Typical example of active sensor is radar sensor (Liang, 2007). Passive sensors are considered in most current WSN-Testbed deployments. Commercially, there is a wide variety of WSN sensor nodes with variant capabilities. Most WSN-Testbeds typically deploy Commercial Off-The-Shelf (COTS) sensor nodes for their experimentation. In special cases, some WSN-Testbed experimenters may develop their own sensor nodes.

In general, two classes of sensor node deployment can be identified. Homogeneous sensor nodes are nodes with similar ability such as similar power, computing and transmission range. Heterogeneous sensor nodes are nodes with different abilities. Homogeneous or heterogeneous sensor nodes can be arranged hierarchically or in a single-tier network as shown in Figure 4.

B. Processing Hub (PH)

Due to the power limitation in the typical sensor nodes discussed earlier, processing hubs are utilized. PHs are sensor nodes with extra power supply and additional computational capabilities, which makes them more expensive than typical sensor nodes. In addition to PHs' sensing capabilities, they could be looked at as base stations to their neighboring conventional sensor nodes. PHs are vital in further smoothing the raw information being sensed and producing a compact data to efficiently utilize the network resources.

Figure 5. Simple sensor node structure

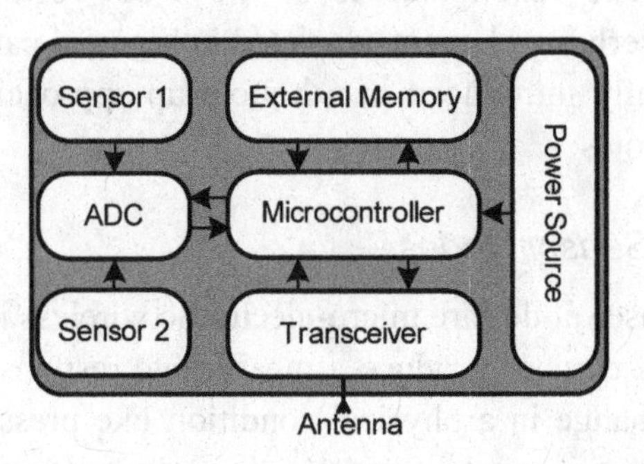

C. Storage Hub (SH)

Conventional approaches in WSNs require that data can be transferred from sensor nodes to a centralized base station because of the storage limitation in sensor nodes. However, storage hubs could be utilized to extract the important features of an event before the data is sent to the end-user. This could be achieved through deploying some data mining and feature extraction algorithms. The presence of the storage hub will depend on the application at hand of a WSN-Testbed.

D. Gateway (GW)

It is an IP-addressable component and it can be viewed as the last mile connectivity where it bridges the base station to the servers. The gateway communicates back and forth with base stations under its jurisdiction. It also maintains an estimate of the locations of its base station(s).

E. Back-Bone (BB)

The backbone is used to interconnect the servers to the gateways. An appropriate wireless communication technology can be used. Additionally, wired connection such as USB or Ethernet can be utilized. The choice between wireless and wired solutions depends on the scale of the WSN-Testbed and whether it is indoor or outdoor. The designer may also opt for either solution depending on design requirements. For majority of WSN-Testbeds, indoor deployments typically utilize wired solution while outdoor WSN-Testbeds use wireless solution. Some WSN-Testbeds use the back-bone for both communications with the testbed as well as back-channel.

F. Back-Channel (BC)

The back-channel is essential in any WSN-Testbed and it is used for sensor nodes' reprogramming, data logging and network monitoring. There are two available solutions; wired back-channel and wireless back-channel. The wired back-channel is more commonly used in most current indoor WSN-Testbeds where it utilizes an Ethernet or USB channel to perform tasks such as nodes' reprogramming, network health monitoring, etc. WSN-Testbed developers justify using wired back-channel to avoid imposing an overhead in the wireless channel so that it can be used for pure application purposes (Handziski et al., 2006; Werner-Allen et al., 2005). Some other WSN-Testbed developers argue that reliance on such solution makes such WSN-Testbeds impractical. Subsequently, they utilize the wireless link for back-channel besides its main role for communication (Dimitriou, Kolokouris, & Zarokostas, 2007). It is worth mentioning that an ongoing research investigates the use of two wireless channels, literally, one for communication and the second for back-channel (ETH Zurich).

G. Private and Public Servers (PPS)

The server hosts the WSN-Testbed's database. This database minimally provides persistent storage for debug and application data, contains information about the sensor motes, connections that have been established between the sensor motes, base stations and gateways, current geographical locations of motes, etc. The database acts as a repository where the public server places information regarding user tasks which the private server retrieves and exploits appropriately. Furthermore, the database holds the most recent status of the sensor network and the sensor nodes, which is used both by the application and the web interface. Additionally, the database provides means to achieve communication between the application and sensor network. The server machine is also used to provide a workable interface between the WSN-Testbed and end-users. Users may log onto the server to gain access to the information contained in the WSN-Testbed database and exchange messages with sensor nodes contained in the WSN-Testbed. The server interacts with the base stations through the gateways using the WSN-Testbed's backbone which supports similar communication technology.

2.2.3 Overlay WSN-Testbeds (WSN-OT)

The idea of WSN-OT is similar to the early Internet being overlaid on the telephone network. It is widely recognized in the field of wireless communications as the most effective general method for experimenting with new services and protocols (National Science Foundation). Overlay testbed offers viable means to explore unforeseen network services and architectures. It can be successfully deployed using overlay networks without changing the underlying networks. In the context of WSN-Testbeds, overlay testbeds are a class of testbeds that may not be originally designed for experimentation with wireless sensor networks. In fact, overlay testbeds are usually designed with a generic view to experiment with wireless technology at large. We refer to an overlay testbed that supports overlay of wireless sensor networks as an Overlay WSN-Testbed (WSN-OT).

The structure of an Overlay WSN-Testbed is comprised of the structure of the underlying overlay testbed as well as the wireless sensor network. The structure of the overlaid wireless sensor network is quite similar to the Cluster WSN-Testbed explained earlier. Figure 6 depicts a generic view of an Overlay WSN-Testbed

(National Science Foundation). An overlay WSN-Testbed is constrained to the limitations of the underlying resources. Examples of overlay WSN-Testbed include, PlanetLab (Drupal, 2007; Mgillis, 2009), Emulab (Emulab), and SWOON (Huang et al., 2008).

2.2.4 Federated WSN-Testbeds (WSN-FT)

WSN-FT can be created by interconnecting several independent, locally managed WSN-Testbeds. *Federation* allows researchers to carry out larger scale experiments than they could do otherwise, using the Internet to provide connectivity among the participating sites. For example, a set of institutions engaged in WSN research might link their local WSN-Testbeds to explore broader research issues than would be possible if working with a single WSN-Testbed. Users can test their network protocols and applications on remote WSN-Testbeds and can use *virtualization* to effectively merge individual WSN-Testbeds (often employing heterogeneous technologies) into an apparently seamless whole to permit larger scale experiments. The interconnection of WSN-Testbeds can also provide a wide variety of hardware to choose from, allowing researchers to match their needs much more closely than would be feasible if limited to an individual facility. Federated WSN-Testbeds also promote cost efficiency; some WSN-Testbeds can sometimes be idle, while others will not have adequate resources to meet the needs of their users. Examples of WSN-FT include, WISEBED (EU Project), WiSense (WiSense), and SensorMap (Nath, Liu, Miller, Zhao, & Santanche, 2006; SensorMap).

A simple architecture of a federated WSN-Testbed is depicted in Figure 7.

The WSN-Testbeds' federation is comprised mainly of the participating member *WSN-Testbeds* and the *federal overlay network*. Participating member WSN-Testbeds can be a mixture of variety of earlier introduced WSN-Testbed classes run by different institutions, and spatially located in different locations or even in different continents. The architecture of these individual WSN-Testbeds is similar to the structure of the testbeds introduced earlier. The owner of an individual WSN-Testbed has full access and control over their testbed. An owner can enable a local access to their WSN-Testbed through a web browser using some portal server. The overlay network acts as a federation for its member WSN-Testbeds. It interfaces with each member through the member's portal server mainly using *the Internet*. Virtually, the overlay network could be located anywhere as long as it has access to the underlying network, i.e., the Internet. This flexibility permits the overlay network to spontaneously federate multiple member WSN-Testbeds into a virtual distributed testbed and expose their services as a single unifying virtual WSN-Testbed. There are many advantages of such innovative solution. Users accessing the federated WSN-Testbed will instantly have access to member WSN-Testbeds under one umbrella. This gives users the option to experiment with a wide variety of heterogeneous solutions at their fingertips. An interesting feature of federated WSN-Testbed can give users the option to partition specific nodes for their liking from the participant WSN-Testbeds at large. In fact, we believe that such cooperation innovatively advances the experimentation with

Figure 6. An overlay WSN-testbed

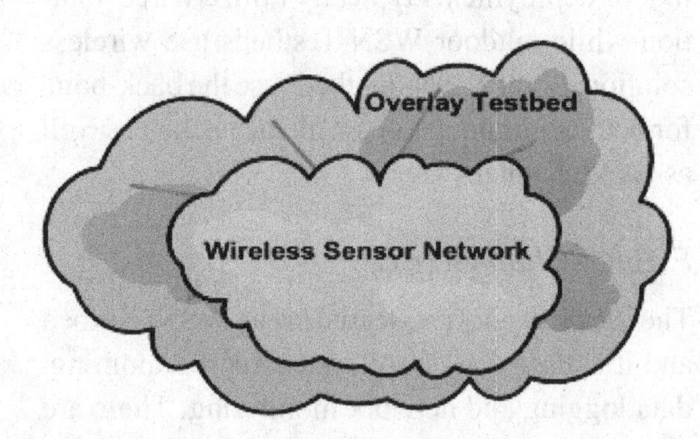

Figure 7. Federated WSN-testbed (User A; connects to a member WSN-Testbed #7 through an interlinking overlay NW using standard web-service, User B; connects to the WSN-FT using a standard web-service or portal server)

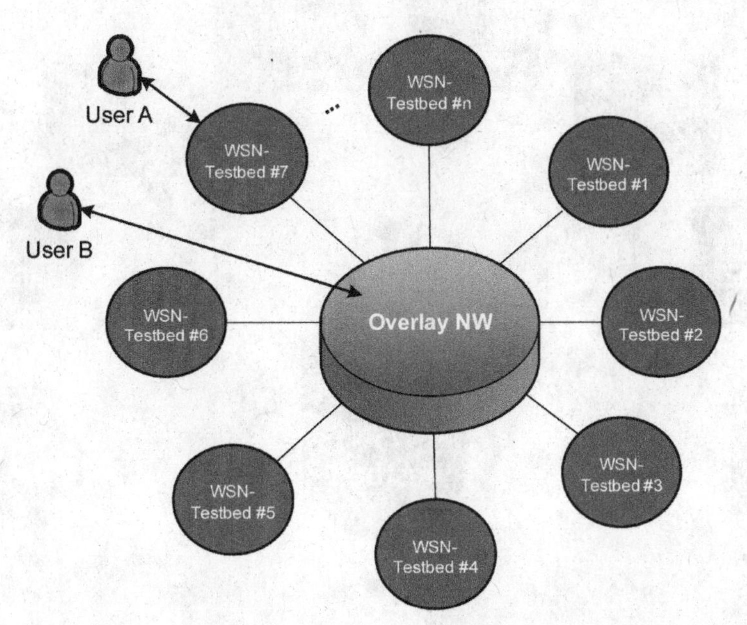

WSN technology and opens the horizon for endless opportunities. Detailed information on the architecture and the advantages of federated WSN-Testbeds can be found in (EU Project).

3. DESIGN CHALLENGES AND EVALUATION CRITERIA FOR A WSN-TESTBED

In this section we pinpoint and explain some important design and evaluation factors pertinent to WSN-Testbeds. These factors are two-fold – from the design perspective; these are design challenges that depend on the objective of the WSN-Testbed design whereby some or all of them have to be fulfilled; and from the evaluation perspective; these factors form a benchmarking scheme that can be applied in the evaluation of a WSN-Testbed. These factors are summarized in Figure 8, and introduced hereinafter.

3.1 Cost

Many of WSN-Testbed developers utilize COTS hardware for building-up their testbeds. A testbed could also utilize custom design or a hybrid of custom design and COTS technology. The choice of a certain technology including hardware is a trade-off between cost and performance. Undeniably, budget constraints and cost play one of the most important roles in setting-up the WSN-Testbed. Some of the main components required to build a WSN-Testbed were approached earlier. These components can be commercial off-the-shelf or custom made. Flexibility of operation versus *per sensor node* cost is usually a guiding factor in this selection. For example, a commercial WSN sensor node, although inexpensive, it may not fulfill the communication range as specified by the vendor. Usually, vendors of low cost COTS WSN node devices typically specify transmission distances of over 150 meters assuming a given transmission power, antenna configuration, and

Figure 8. WSN-testbed's design challenges and evaluations criteria

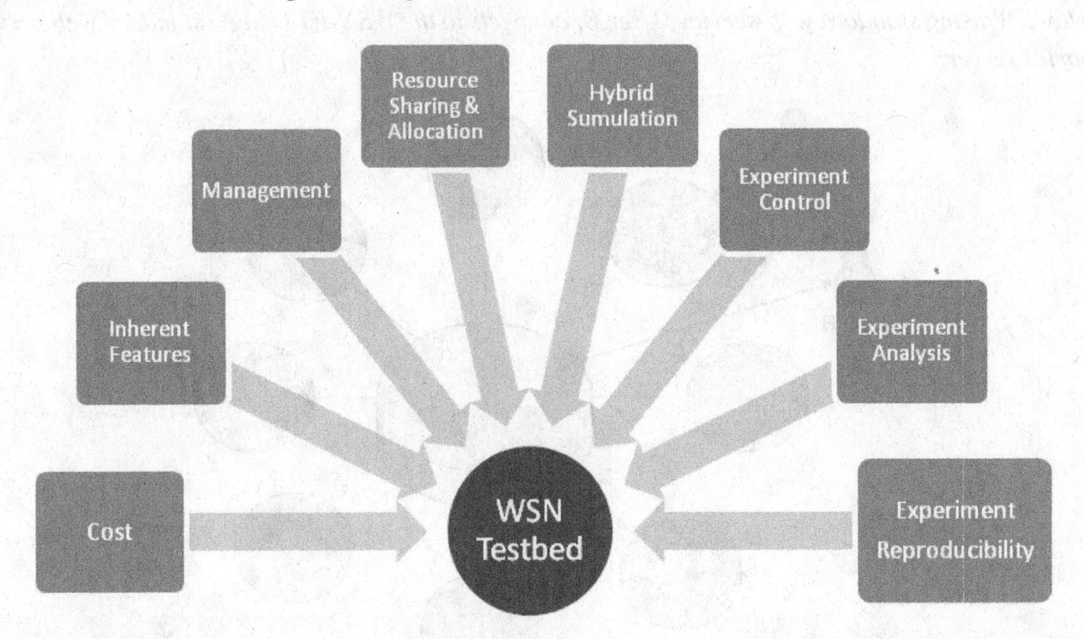

a clean line of sight. Yet, real field deployments observed a maximum transmission distance of closer to 20 meters (Evans, 2007; Nirupama Bulusu, Vladimir Bychkovskiy, Deborah Estrin and John Heidemann). It is therefore a challenging task to choose the right hardware for putting together the WSN-Testbed elements.

3.2 Inherent Design Features

A. Openness

The design of a WSN-Testbed could utilize open-source technology. Such openness will give full flexibility in the use of the testbed. Among the main characteristics inherited by an open-source testbed are: affordability, flexibility, scalability, heterogeneity and replicability. Two open-source categories can be distinguished; open-source hardware and open-source software. A testbed design could encompass either one or both of these categories. Open-source hardware means that every part of the hardware such as a sensor node (usually referred to as an open mote), from the schematic circuit to the source code of the programs running inside the hardware, is accessible and can be studied, changed and personalized. Examples of such open source hardware include; SquidBee (LibeLium), Epic (P. Dutta & Culler, 2008; P. Dutta & Culler, 2008; Prabal Dutta) and open source WSNs (hardware and software) (Leo Selavo). Open-source software refers to the openness of the operating system, components, tools and modules assisting in the design of WSN. Additionally, it can also be used to refer to the system's software of the WSN-Testbed being open for public access, replication, modification and utilization. A typical example of open-source operating system is TinyOS (UC Berkely, 2007). An example of open-source system is CodeBlue (Leo Selavo). Our review of the literature suggests that open source technology, including both open-source hardware and software, have a promising future in the field of wireless sensor networks.

B. Heterogeneity

Heterogeneity refers to the ability of the WSN-Testbed framework to accommodate and support heterogeneous mixture of WSN-Testbed hardware and software such as sensor nodes from different vendors with varying capabilities including different sensing ranges, varying sensor types, and different communication and computation capabilities. For instance, a heterogeneous WSN-Testbed may support sensor nodes with different standards. Such standards include ZigBee (ZigBee Alliance), WirelessHART (HART Communication Foundation), 6LoWPAN (Internet Engineering Task Force), and ISA100 (ISA) which all utilize the same underlying radio standard, IEEE 802.15.4. Additionally, other sensor nodes may utilize Bluetooth technology as the communication standard (ETH Zurich, ; Jens Eliasson).

Heterogeneity inherently adds flexibility to the WSN-Testbed design. A heterogeneous and flexible WSN-Testbed can distinctively achieve design equilibrium through introducing a balance between cost and performance.

C. Scalability

From a design perspective, it is often advised to keep the focus of the system design on *hardware and software* scalability rather than on capacity (Bondi, 2000). This is absolutely true because partaking in performance tuning to improve the capacity of the network is more expensive than adding new hardware resources to the WSN-Testbed's scalable framework. Scalability is a desirable property of a WSN-Testbed design. It provides the WSN-Tetsbed with the readiness required to handle expansion or growth in a graceful manner. For instance, a scalable WSN-Testbed will have the ability to cover a bigger area for environmental monitoring through integrating the required additional hardware to the scalable framework. In order for a system to be a scalable, it should inherently implement scalability into its design. Scalable WSN-Testbed should implant scalability at the heart of its protocols. For example, a scalable routing protocol will perform well as the network load increases and/or as the network expand (Alazzawi, Elkateeb, & Ramesh, 2008). Subsequently, a scalability-management trade-off is always present and it has to be wisely balanced.

3.3 Management

Design, deployment, operation and maintenance of a WSN-Testbed is an inherently complex task. Early integration of support into WSN-Testbed's design process is of a prime importance to fulfill the objective of the testbed's reliable operation and maintenance (Winnie Louis Lee).

Optimum management of WSN-Testbed is essential in maintaining the testbed's performance. This is achieved through controlling WSN nodes without direct human intervention. Unlike other kinds of network applications where network failure is exceptional, wireless sensor netwrk failures are common (Dataman Papers). Accordingly, network management systems for WSN-Testbeds are majorly concerned with monitoring and controlling node communication.

The management functions provided by WSN management system should integrate configuration, operation, administration, security, and maintenance of all elements and services of the testbed. WSN-Testbed monitoring is concerned with collecting information about a spectrum of parameters including: node states (e.g. on/off, battery level, communication power), network topology, wireless bandwidth, link state, coverage bounds, and exposure bounds. Based on the collected network states, a variety of management control tasks can be performed. For instance, power management entails controlling sampling frequency, switching node on/off; traffic management is based on controlling wireless bandwidth usage; fault management entails performing network reconfiguration in order to recover from node and communication faults.

Authors in (Dataman Papers) discuss three ways to classify the management systems for WSN networks. One way of classification is based on network architecture where management systems are classified into centralized, distributed and hierarchal. The second way is based on management functionality and it classifies the management systems into, traditional, fault detection, power management, traffic management, and application specific. Finally, the management systems can be further classified into four categories based on the approach employed for monitoring and control. These are: Passive, fault detection, reactive, and proactive monitoring. A survey of available WSN management systems for WSN is also provided in the same reference.

3.3.1 A Backbone at the Core of the Management System

A reliable backbone network is an important design requirement in all WSN-Testbeds. It enables continuous remote access to all the WSN nodes in the network. This guarantees access to the nodes even in cases of failure, and temporal or permanent disruptions. The network backbone makes it possible for the management system to perform programming, automatic control features such as reconfigurations, self healings and self repairs. Additionally, a reliable backbone enables configuration, management and code distribution in the WSN-Testbed. Moreover, WSN network downtime and on-site repairs are highly minimized through deploying this solution.

The backbone could be a *wired* or *wireless* backchannel. Many of the current WSN-Testbeds deploy a separate wired back-channel, where an Ethernet or USB channel is usually used for this purpose. Such method leaves the wireless channel for pure communication purposes and utilizes the wired-channel for managerial issues as explained earlier without any interference with the wireless channel. Some WSN-Testbed developers utilize the wireless channel for both communication and backbone and they justify that deploying a wired backchannel is not as efficient in terms of cost, maintenance and scalability (Dimitriou et al., 2007). Some other WSN-Testbed developers propose using a hybrid of wireless and wired mesh network backbone. This can be co-deployed in a tiered architecture of mesh nodes along with the sensor nodes (Bastian Blywis, Felix Juraschek, Mesut Günes, Jochen Schiller; Wagenknecht et al., 2008). Mesh nodes allow a cost efficient connectivity in local area networks. Wired connection and/or IEEE 802.3 and/or IEEE 802.11 links can be used to interconnect the mesh nodes (EAAST).

3.4 Resource Sharing and Allocation

Designing WSN-Testbeds that could be shared by multiple users is a hot topic in the WSN research community. Resource sharing involves allocating a subset of sensor nodes from a pool to individual experiment or multiplexing experiments on each node. This mechanism could be implemented through some *reservation system*. In a shared WSN-Testbed, multiple users may simultaneously have access to the testbed resources. Through some authentication mechanism, users could be granted or denied such access rights. Depending on the user's identity, some users may have priority and/or extra access rights compared to other users. To fulfill this objective, a WSN-Testbed should include an authentication, scheduling, and resource allocation system. Usually such a system is implemented using a web interface. This guarantees fair allocation of time slots and resources; and in turn facilitates the manageability of the WSN-Testbed.

Once the user granted access to the WSN-Testbed, the testbed should enable them to configure their allocated resources. The *configurator* provides user with a description of the allocated resources including sensor nodes. It also enables users to upload a certain operating system (e.g.,

Tiny OS) to run on the sensor nodes. Additionally, it enables users to upload their own application upon checking it for compatibility.

3.5 Hybrid Simulation

Unlike emulation where real network traffic is used inside the network (Fall, 1999), hybrid simulation refers to the interaction of a simulator with the external entities of the WSN-Testbed such as real network nodes, user applications, or some other external source of I/O. One of the interesting features one needs to consider when designing or evaluating a WSN-Testbed is its capability to run diverse set of experiments. From this perspective, the more diverse the set of experiments a WSN-Testbed supports, the better it is. To add to the diversity of a WSN-Testbed, one could enable a hybrid simulation along with the live testing (Arora, Ertin, Ramnath, Leal, & Nesterenko, 2006; Breslau et al., 2000; Girod et al., 2007; J. Heidemann, N. Bulusu, J. Elson, C. Intanagonwiwat, K. Lan, Y. Xu, W. Ye, D. Estrin, and R. Govindan, ; Park, Savvides, & Srivastava, 2000).

Typical simulation could be hybridized through replacing some of the protocol stack layers with real hardware. For example, the physical and MAC layers in the simulator could be replaced with real hardware while keeping the rest of the simulation layers intact. Among the advantages of the hybrid simulation is that it diversifies the WSN-Testbed's capability and helps to run experiments that may be impossible to execute without using hybrid simulation.

3.6 Experimental Control

A. Topology Control

In the field of Wireless Sensor Networks, *topology control* is a technique used to save energy and save the lifetime of the network. This technique is receiving an overwhelming attention in the research community. The technique could be defined as reorganization and management of node parameters and modes of operation throughout the lifetime of the WSN network to modify the topology of the network with the goal of extending its lifetime while preserving important characteristics, such as network and sensing connectivity and coverage. Topology control is an iterative process that could be divided into three distinctive phases. *Initialization*: Common to all WSN networks where nodes discover themselves and use their maximum transmission power to build their topology. *Topology Construction*: This phase builds a new reduced topology runs for a certain amount of time. *Topology maintenance:* This phase starts as soon as the reduced topology is built. It places a new monitoring algorithm throughout the lifetime of the network. The algorithm monitors the new construction and triggers a new topology construction when appropriate. The literature is rich in many algorithms that can be used in the topology construction and maintenance. These algorithms are normally classified according to the way deployed to perform the topology construction and maintenance. For in-depth information on this topic the reader is referred to (Miguel A. Labrador, Pedro M. Wightman, 2009).

B. Application Configuration

This step involves the way in which a user can interact with the WSN-Testbed to setup their application. First, a user can write their own applications. Alternatively, a library of applications could be provided by the WSN-Testbed developer from which a user can choose the appropriate application. Moreover, there can be two ways for interaction with the WSN-Testbed. First, the user can schedule a certain number of jobs to be run unattended in a batch fashion. Second, a user can set-up a job and they can interact directly with their running job in real time. The real time access feature may also allow the user to interact lively or proactively with the running job (Werner-Allen et al., 2005). This will allow researchers to do experiments such as injection of dataset for the sake of simulation or experiment reproducibility (Arora et al., 2006).

C. Experiment Execution

The user of a WSN-Testbed has the ability to initiate an experiment and then they should have the ability to stop the experiment at any point of time they wish. Additionally, throughout experiment execution a user can observe the results and could be provided with certain ways to fine-tune the experiment. This could be further enhanced by providing the user with means to pause the experiment, modify certain parameters and resume the experiment (De et al., 2006). Such features can be time-saving and will add to the advantages of the WSN-Testbed.

D. Network Debugging

Being constituted of embedded and distributed systems, a WSN-Testbed inherits the debugging problems and difficulties for both of its constituents. In fact, the literature has studies that deal with the debugging problem. Just to mention a few; authors in (M. Lodder, G.P. Halkes, K.G. Langendoen) designed a tool called Monitored External Global State (MEGS) based on a global state perspective in handling the debugging problem. Authors in (IDEALS) proposed a debugging tool using mobile actors. Mobile actors refer to allowing the computation to move to the individual nodes where the data is located. Additionally, a facility for introducing controlled faults to uncover bugs in the network could be an added advantage to the WSN-Testbed.

3.7 Experiment Analysis

A. Data Collection

The first step towards analyzing the data delivered by the WSN-Testbed is to deploy an efficient protocol to collect this data in the first place. In the literature, there are numerous data collection algorithms using different strategies. An important aspect towards selecting the most appropriate protocol is to thoroughly understand the operating environment as well as other contributing factors such as the characteristics of the radio transceiver(s) and link loss statistics from a long-term on-site radio survey. Depending on the operating environment, a WSN-Testbed could have more than one data collection protocol where it could choose the most appropriate protocol for the task at hand. The choice of a certain protocol is mainly dependent on the reliability requirement as well as the link quality. One phase pull Directed Diffusion (DD) (Intanagonwiwat, Govindan, Estrin, Heidemann, & Silva, 2003), Expected Number of Transmissions (ETX) (Heidemann, Silva, & Estrin, 2003), ETX with explicit acknowledgment (ETX-eAck) (Woo, Tong, & Culler, 2003), and ETX with implicit acknowledgment (ETX-iAck) (Le, Hu, Corke, & Jha, 2009) are four different data collection protocols discussed in (Rothery, Hu, & Corke, 2008). DD and EXT are simple data collection protocols with low routing overhead that can be used in case of loose end-to-end reliability requirement and good link quality. ETX-eAck and ETX-iAck achieve higher reliability for a good link quality (Rothery et al., 2008).

B. Data Fusion

In wireless sensor networks, *data fusion*, also called *information fusion*, refers to a set of techniques used to process the data gathered by sensor nodes and benefit from their processing capability. Data fusion techniques include filters, Bayesian and Dempster-Shafer inference, aggregation functions, interval combination functions, and classification methods. The main advantages of deploying data fusion include: reducing data traffic in the network, filtering of noisy measurements, and obtaining accurate predictions based on fused data. The inclusion of data fusion techniques in a WSN-Testbed should be guided by existing data fusion frameworks, such as the Joint Directors of Laboratories (JDL) model (The Data Fusion Server). Data fusion techniques are thoroughly surveyed in (Nakamura, Loureiro, & Frery, 2007). Reduction in the network traffic yields energy saving. Several simple aggregation techniques have been developed to fulfill this goal (Nakamura et al., 2009).

C. Data Storage and Visualization

Usually a WSN-Testbed numerically stores the gathered data in a central database server. The standards for metadata encoding and interoperability interfaces are specified by the Open Geospatial Consortium (OGC) (OGC,). The OGC sets the standards enabling real time integration of heterogeneous sensor-webs in the Internet. This allows a WSN-Testbed owner to make the gathered data visually available in a web browser at the fingertips of an end-user or an administrator. This highly facilitates the process of monitoring and controlling the WSN-Testbed. Also, it enables further interaction with the gathered data. In fact, there are several techniques to retrieve and gather data from the nodes. Some techniques map the data to nodes through using the concept of Distributed Hash Tables (DHT), while other techniques utilize protocols using flooding mechanisms (Awad, Sommer, German, & Dressler, 2008; Ratnasamy et al., 2003). There are many programs that can be used to facilitate the visualization of the gathered data. Such programs include; GSN (GSN), TosGUI(Miyashita, Nesterenko, Shah, & Vora, 2005), MonSense(MonSense), SpyGlass (Buschmann, Pfisterer, Fischer, Fekete, & Kr\oller, 2005), TinyViz (UC Berkely, 2007) and Crossbow's Surge Network Viewer as well as MoteView (Buschmann, Pfisterer, Fischer, Fekete, & Kroller, 2004).

3.8 Experiment Reproducibility

Reproducibility of an experiment refers to the ability of the WSN-Testbed to re-produce similar results for a certain input under the same external conditions. There are many advantages that can be attained from reproducible experiments. Such advantages include the possibility to evaluate the performance of wireless sensor network applications, and the possibility to achieve a realistic parameter tuning and protocol comparison. However, experimental testing of distributed event-driven wireless sensor network applications is highly unrepeatable because of the fact that the state of an event-driven system can change depending on the particular sequence of events received and their timing. Additionally, external factors such as attenuation, fading, and interference are always present and variable.

One interesting tool designed to enable and enhance experiment reproducibility in WSN-Testbeds is EnviroLog (Luo et al., 2006). EnviroLog is an event *record and replay* service to improve reproducibility for environmental events. Among the advantages that could be realized through this tool are: performance tuning without physically generating events, runtime status collection without extra hardware, virtual velocity simulation, remote replay (recording events in environment *A* while replaying them in environment *B*), and offsite replay (recording event on sensor devices while replaying them in simulator).

Another methodology for achieving reproducibility without sacrificing realism is to create a controlled environment that mimics external disruptions such as noise and interference. This could be achieved through placing the WSN-Testbed in a chamber(s) equipped with noise generators to simulate ambient conditions and disruptions (J. Slipp et al., 2008a). A WSN-Testbed supports repeatability will have an advantage over other WSN-Testbed of being a reference platform for various experiments. At this point it is obvious that there is always a trade-off between WSN-Testbed's realism and experiment reproducibility.

4. COMPARISON OF SELECTED WSN-TESTBEDS

Comprehensive literature review reveals that it is rich with variety of WSN-Testbed solutions. Accordingly, we aim at informing the reader about the broad spectrum of such WSN-Testbeds without being redundant. In order to make our survey constrained and focused, we opt for the WSN-Testbeds that serve our abovementioned aim. This section includes two subsections. In the first subsection, we taxonomize the WSN-Testbeds considered in this survey following the taxonomy described earlier in Section 2. This is followed by a brief description for each WSN-Testbed considered in the survey along with an architectural comparison. In the second subsection, we evaluate the WSN-Testbeds considered from different perspectives. Our evaluation methodology is mainly based on the *design challenges and evaluation criteria* introduced earlier in Section 4. This is followed by a description of the drawbacks/shortcomings pertinent to each WSN-Testbed.

4.1 Taxonomy and Architecture of Selected WSN-Testbeds

The WSN-Testbeds considered in our survey are provided in this section. The tables provided hereinafter depict a taxonomization as well as a comparative description for the constituents of each WSN-Testbed considered in our survey. A brief description highlights the most important features for each WSN-Testbed follows each table.

4.1.1 Survey of WSN Research Kits (WSN-RK)

The most basic class of testbeds pertinent to experimentation with wireless sensor networks is the WSN research kits. We consider four different WSN research kits. Two of these testbeds can be classified as a multiuser experimental, while the rest are proof-of-concept testbeds. Table 1 provides a comparison between these research kits. This is followed by an informative description of each kit.

Crossbow's Development Kits

Crossbow (Xbow) is a leading supplier of wireless sensor technology and inertial MEMS sensors for navigation and control. Varieties of kits are offered by Xbow to provide users with a pre-configured set of hardware to get started with Crossbow's mote hardware and platforms. The kits design varies from starter kits to professional packaged outdoor deployments kits. The variety of available kits include: Imote2.Builder Kit, Starter Kits, Professional Kits, OEM Design Kits, and Classroom Kits.

Imote2.Builder Kit is an out-of-the-box development kit for Imote2-based wireless sensor networks. The kit enables, for the first time, developers to design wireless sensor applications using the Visual Studio development environment and create proof-of-concept in hours or days, not weeks or months. Also, it is the industry's first set of tools for accelerating the development on the Marvell PXA hardware platform on which the Imote2 is built. Imote2.NET Edition is Crossbow's next generation hardware platform for wireless sensor networks licensed from Intel Corporation and is built on the Marvell PXA271 XScale ARM-based microprocessor. Imote2.NET is aimed at applications involving data-rich problems, where there is a need for both high performance and high bandwidth, which requires greater processing capability and low-power operation with a low duty cycle to achieve longer

Table 1. Some selected WSN kits

Kit	Current Price	Motes	Sensors	No. of Motes	Distinctive Features
MXT					
Crossbow's Development Kits (Crossbow)	$914 -$ 4,559	Imote, IRIS, MICA	Various	3-10	Imote2 supports for the first time the MS VS Development Environment
Senswiz Research & Educational Kit (SensWiz)	$1,000	SensWiz	Temperature, light, & tri-axis accelerometer	3	Motes are equipped with built-in health monitoring facility
PCT					
Dust Networks' SmartMesh Kits (Dust Networks)	$4500	Dust Networks	Various	10-12	Specifically designed for RHE environments with built-in support for WirelessHART
NI Wireless Sensor Starter Kit	CAD $2420	NI WSN-32xx Nodes	Temperature	2	Designed for RHE environments and provided with LabVIEW software

battery-life. The Imote2.Builder kit includes 3 Imote2.NET Edition modules (IPR2410CA which comes preloaded with .NET Micro Framework), 2 sensor boards (ITS400CA), the Microsoft.NET Micro Framework SDK, and everything required to allow engineers and developers to quickly and seamlessly create robust wireless sensor applications (Crossbow).

WSN Starter Kit is an out-of-the box starter kit provides an easy and cost-effective solution to get first-hand experience with wireless sensor networks either in the 2.4 GHz or 868/916 MHz ISM bands. This entry-level kit provides all the components needed for rapidly deploying a basic wireless sensor network. The sensor nodes and gateway are preconfigured with Crossbow's reliable, self-forming, self-healing mesh networking software (XMesh). Among other things, the kit includes 2 IRIS/Mica Sensor Nodes, Base Station and 1 WSN Kit CD. The Base Station is comprised of a processor/radio board which is the IRIS/MICA module functioning as a base station when connected to the USB PC interface; and USB PC Interface Board which is MIB520 Gateway that provides a USB Interface for data communications. The CD includes a MoteView client, a monitoring software for historical and real-time charting. MoteView also provides topology map, data export capability, mote programming and a command interface to sensor networks (Crossbow).

The Crossbow Professional Kit is another out-of-the-box evaluation and development kit. The kit includes 6 IRIS, MICAz or MICA2 motes, 1 Base Station, 1 Processor/Radio Board, 1 Data Acquisition Board, 1 USB Programming Board (MIB520), and 1 WSN Kit CD. The kit is designed specifically for deeply embedded sensor networks; these modules enable low power wireless sensor networks. The capability and utilization of high data rate and direct sequence spread spectrum radio that is resistant to RF interference provides inherent data security. The Development series kits have all of the components needed to develop, test and implement a wireless sensor network (Crossbow).

WSN OEM Design Kit is an out-of-the-box professional kit. The OEM Design Kit is targeted at applications using the 2.4 GHz frequency band. The kit provides users with OEM edition modules, a module programmer, preprogrammed reference design boards, sensor/data acquisition boards, a debug pod and a base station with Ethernet connectivity. The kit includes 5 IRIS/

MICAz OEM edition modules, 5 IRIS/MICAz OEM edition reference boards, 3 MDA300 sensor/data acquisition boards, 1 MDA100 Sensor/Data acquisition board, 1 MIB600 Ethernet gateway, 1 programming/debugging pod, and 1 OEM module programmer. The Crossbow OEM Design Kit supports fast development of wireless sensor network systems (Crossbow).

WSN Classroom Kit is another research kit specifically designed for the typical teaching lab or sensor class project. It provides hands-on training involving all aspects of both the hardware and software applications. The kit includes a set of 10 lab stations. The lab stations are comprised of 30 processor/radio boards, 20 sensor boards, 10 gateways, and 10 seats MoteWorks license (Crossbow).

SensWiz Research and Educational Kit

SensWiz is a product of Dreamajax Technologies Private Ltd. in India. It offers a research and educational kit that can be used for variety of sensing, measurement and control applications. The kit comes with 3 SensWiz wireless sensor nodes, and 1 SensWiz USB coordinator node. The on-board sensors are: temperature, light, and tri-axis accelerometer sensors. The nodes are equipped with a built-in node health monitoring system based on node's temperature and battery level. The operating frequency is 2.4 GHz, with 802.15.4 compliant RF transceiver. The kit comes with test and configuration as well as demonstration programs to enable the evaluation of the functionality of the on-board sensors. Additionally, SensWiz provides a simple Application Programming Interface for users to write their own programs. Provided example programs along with the source code can help the user to rapidly develop application utilizing wireless sensor networks. *SensWiz Accelerometer Development Kit; and SensWiz Wireless Temperature and Humidity Monitoring System* are two more kits provided by SensWiz (SensWiz).

Dust Networks' Kits

Dust Networks offers two categories of wireless sensor network kits including: *SmartMesh IA-510™ SmartStart Kit* and *SmartMesh-XT Research Development Kit.* The *SmartMesh IA-510™ SmartStart Kit* is specifically designed for development and testing with WirelessHART products. The kit is comprised of two sub-kits; *Evaluation Kit* and *Development Kit*. It supplies all the needs to design, develop, test, and prototype a new WirelessHART-complaint sensor network solution. The two sub-kits are available for purchase together or separately.

The Evaluation Kit is based on the industry-proven SmartMesh IA-510 products which are being deployed in harsh environments and provide high reliability, ultra-low power and easy to integrate. It provides all the tools required to enable a developer to evaluate the network performance against the needs of the application including; bandwidth, reliability, and ease of integration. Among other things, the Evaluation Kit includes; 1 Packaged Manager (D2510), 10 evaluation motes (EV2510), evaluation GUI (includes Mesh Topology Viewer and Data Transport Utility).

The *Development Kit* is also based on the industry proven SmartMesh IA-510. It provides the components needed for device integration and application development. Among other things, the kit is comprised of the following; 1 Packaged Manager (D2510), 5 DV2510, modules (signal breakout modules), LVTTL voltage converter, WirelessHART validation support, and IA-510 application development libraries.

Finally, the *SmartMesh-XT Research Development Kit* is part of the Research Innovation Program which provides access to the Research Support Community. The kit is based on Dust Networks' industry-proven SmartMesh-XT products. It provides all the tools required for application development and prototyping. The kit includes the following; *SmartMesh-XT Evaluation Kit*, which includes a demonstration network to evaluate the wireless sensor networking technology. It contains 1 packaged manager, 12 evaluation motes, SmartMesh Console—an example control application; serial cable for connecting a mote to a PC via RS232; *documentation*, including serial and XML API guides, datasheets, and application notes for detailed instructions to set up a prototype network, and transmit and receive data through the wireless mesh network; *software tools and utilities* to enable integrating the mote into the design and use of the APIs for full configuration of the mesh network; and *sample code* including API libraries and usage examples (Dust Networks).

National Instruments Wireless Sensor Starter Kit

National Instruments is an industrial leader with academic links and interests. Since more than 20 years, NI invented the NI LabVIEW graphical programming which has revolutionized the development of test, measurement, and control applications. NI offers a WSN Starter Kit includes the hardware and software needed to evaluate NI WSNs. The kit contains an NI WSN-9791 Ethernet gateway, a programmable NI WSN-3202 ±10 V analog input node, and a programmable NI WSN-3212 thermocouple input node. In addition, NI ships the kit with thermocouples, a potentiometer, NI-WSN software, LabVIEW WSN Module Pioneer evaluation software, and LabVIEW 2009 evaluation software. NI's LabVIEW enables users to rapidly and cost-effectively interface with measurement and control hardware, analyze data, share results, and distribute systems. The programmable NI WSN measurement nodes included with the WSN Starter Kit can be targeted and programmed with the LabVIEW WSN Module Pioneer. With this capability, user can customize the node's behavior, adding intelligence to extend battery life; increase analog and digital input performance; and interface with sensors (*NI wireless sensor network (WSN) starter kit - national instruments*). It should be noted that besides NI Wireless Sensor Starter Kit, motes from other vendors such as Crossbow may also be compatible with NI LabVIEW. Information pertinent to compatibility of LabVIEW with WSN from variety of vendors can be found in (*LabVIEW drivers for wireless sensor networks - developer zone - national instruments*).

Additional WSN Research Kits

It is worth mentioning that there are many other WSN research kits that are not highlighted in this report; to avoid redundancy. Such interesting WSN kits include those offered by Libelium (*Libelium - waspmote - wireless sensor networks mote*), Accsense (*Welcome to accsense.com.*), MeshNetics (*MeshNetics / development tools*), J- MicroStrain (*MicroStrain 2.4 GHz G-LINK wireless accelerometer node*), and Techkor Instrumentation (*Wireless accelerometers - digital accelerometers - vibration monitoring - wireless CBM - wireless sensor*).

4.1.2 Survey of Cluster WSN-Testbeds (WSN-CT)

Cluster WSN-Testbeds are the most common class for experimentation and development of WSN solutions. In realizing this importance, we carefully select 18 different WSN-Testbeds and appropriately survey them. The considered WSN-CTs are introduced hereinafter after categorizing them further into multi-user experimental testbeds (MXTs), and Proof of Concept Testbeds (PCTs).

4.1.2.a Multiuser Experimental Cluster WSN-Testbeds (Cluster WSN-MXT)

A total of 10 cluster multiuser experimental WSN-Testbeds are considered. In Table 2, we comparatively list the basic constituents for each of them. Additionally, we shed light on some distinctive features pertinent to each WSN-Testbed. This is followed by a brief description features individual cluster multiuser experimental WSN-Testbeds considered in Table 2.

TKN Wireless Indoor Sensor Network Testbed (TWIST)

TWIST is an open and scalable framework available for researchers to deploy their own instance. An instance of TWIST is installed in 2005 at the Technical University of Berlin. It is the largest European academic cluster WSN-Testbed for indoor deployment research. It utilizes cheap off-the-shelf hardware and open-source software. Top-down view would reveal that TWIST is hierarchically organized into the following 3 tiers; servers, super nodes and sensor nodes. TWIST's instance uses 37 NSLU2 (Network Storage Link for USB 2.0 Disk Drives) (NSLU2,) as *super nodes*. Super nodes connect the servers with the sensor nodes via an Ethernet backbone; moreover, *super nodes* are meant to get around the 127 USB limitation problems. Around 200 sensor nodes comprised of 100 Tmote Sky and 100 EyesIFxv2 motes are installed in three floors. Sensor nodes are both USB-powered and battery-powered. TWIST uses USB backchannel for programming. Among the interesting features of TWIST are (Handziski et al., 2006):

- It supports dynamic topology control, and power control,
- It enables injection of node failures in the system,
- It resolves the addressing issues through enabling super node to create a mapping between the manufacturer's serial number, the nodeID and the socket,
- It supports batch-based and interactive experiment control,
- It provides flexible hierarchical architecture with two node classes,
- It enables users to request access to interact with the TWIST instance through registering on TWIST's web page.

MoteLab

MoteLab is a set of software tools for managing Ethernet-connected wireless sensor nodes. It is developed at Harvard University where an instance of it is deployed in Maxwell Dworkin Laboratory in the Electrical Engineering and Computer Science building. The nodes of MoteLab are designed as a *fixed array* of wireless sensor nodes with Ethernet interface backchannel for every single node in the network which is deployed using dedicated Stargate boards for remote reprogramming and data logging. It was originally designed to handle 26 Mica2 motes attached to 26 Ethernet interface boards where 6 EPRB's were developed by Intel research and 20 Crossbow MIB-600 *emotes*. MoteLab was later upgraded to handle 30 MicaZ motes. Presently, 190 TMote Sky sensor motes are deployed. Each node includes a sensor for light, temperature and humidity. A central server handles scheduling, reprogramming nodes, logging data, and provide a web interface for users to set-up or schedule jobs and download data. Motes are not battery-powered but they are wall-powered instead, and they are connected to an in-place Ethernet which is used for debugging and reprogramming only. Among MoteLab features are (Werner-Allen et al., 2005):

- It deploys a *user quota* system for limiting the outstanding jobs that the user can post at once to make available resources for other users to complete their jobs,
- It provides users with *direct access to each node* over a TCP/IP connection. This enables users to use custom programs for injecting and monitoring data into the running applications,

Table 2. Survey of selected cluster MXT WSN-testbeds

Testbed	Motes	OS	Sensors	Base Station/ Gateway	Backbone/ Backchannel	Power Source	Some Distinctive Features
MXT							
TWIST (Handziski et al., 2006)	100 Tmote Sky & 100 EyesIFxv2 & Telos families	TinyOS	Light	Supernodes: Linksys' NSLU2, 37 NSLUs	Ethernet/ USB	USB and Battery	Remote power control for each mote; use of socket naming to enable automatic detection of mote locations
Motelab (Werner-Allen et al., 2005)	30 MicaZ & 190 TMote Sky	TinyOS	Light, Temperature and Humidity	Crossbow MIB600 gateways	Ethernet	USB	Digital millimeter for node; and power measurement
TutorNet (Tutornet)	91 tmoteSky & 13 MicaZ	TinyOS	No info	13 Stargates with 802.11 interface	Ethernet/USB	USB	Restriction on usage quota per user is 1 node hour; default 100 node hours/user
Kansei (Arora et al., 2006)	210 XSM nodes 50 Trio nodes Mobile XSM nodes	TinyOS	Magnetomete, IR, temperature, photocell, and mic	More than 210 XSM Stargates	Ethernet/ USB/802.11b	USB, Battery and Solar	Supports Hybrid Simulation; and focusing on sensing
NESTbed (Dalton & Hallstrom, 2009b)	80 tMote Sky	TinyOS	Temperature, light and humidity	Gateways	Ethernet/USB	USB	Interactive, open, and source-centric
SenseNeT (Dimitriou et al., 2007)	8 Mica2	TinyOS	No info available	1 Base-station	Wireless Channel	Battery	No wired backbone; utilizes wireless comm. channel for reprogramming and NW management
Kontest (KonTest)	TelosB (10 Moteiv and 50 Crossbow)	TinyOS	Temperature, humidity, light, and IR	NA	USB	USB	Motes are directly connected to PCs (6 rooms, 1 PC per room)
Trio (P. Dutta et al., 2006)	557 nodes Telos, XSM and Prometheus	TinyOS	Infrared, magnetometer and microphone	7 gateways	Mainly 802.11b	Solar and Battery	Evolutionary approach to platform design; Large scale outdoor deployment; Wireless Backbone
IBM WSN TB (Furrer, Schott, Truong, & Weiss, 2006b; IBM)	Crossbow MicaZ motes and others	No info	Compass, accelerometer, gyroscopes, and a thermometer	Gateways No info	Ethernet, an IEEE 802.11b WLAN & Bluetooth	No info available	FPGA (Field Programmable Gate Array) attached to sensor united to filter data before sending it to the NW; use of an advanced version of the MQtt protocol
Mobile Emulab (Johnson, Stack, Fish, Flickinger, Stoller, Ricci, & Lepreau, 2006a)	25 fixed, 6 Mica2 on Acroname Garcia Robots	TinyOS	Magnetometer	MIB500CA/ XScalebased Stargate	USB (for fixed) /802.11b (for mobile)	USB (for fixed) and Battery (for mobile)	First mobile and remotely accessible WSN-Testbed; overlay on Emulab

- It provides an in-situ power measurement option where one node on the network is connected to a networked Keithley Digital Multimeter (Keithley 2701) (Keithley Instruments Inc.). Users have the option to use this device through selecting it from the job creation page,
- It provides a public and permanent test-bed for development and testing of WSN applications.

TutorNet

Tutronet is a tiered WSN-Testbed installed in Ronald Tutor Hall at University of Southern California. The testbed is comprised of 13 clusters where each cluster is made of a top-down three tiers as follows: server, Stargates, and sensor nodes. Per each cluster, 7 sensor nodes are wired into a 7-port USB hub via USB cable. A total of 91 tmoteSky and 13 MicaZ sensor nodes are used. The USB hubs are plugged into the Stargates. Stargates are kind of a master (micro server). They act as a base station and they communicate back and forth with the server over 802.11b standard. End-users can interact with TutorNet through a website designed for this purpose. The main features of TutorNet can be summarized as follows (Tutornet):

- It provides wireless multi-hop routing between the testbed server and the Stargate, and among Stargates,
- It provides Support for remote and parallel re-programming for sensor nodes,
- It enables pre-authorized external users to reserve the testbed through a web interface at a granularity of 1 node-hour, default 100 node-hours/user.

Kansei

Kansei is a distinctive framework designed at the Ohio State University for deploying large-scale WSN-Testbeds. The framework was originally developed for the *outdoor* ExScal project (Arora) at the same university. An instance of Kansei is being deployed indoor at the Ohio State University. Kansei's instance is comprised of *three arrays* of sensor nodes. *Stationary Array* placed in 15'×14' rectangular grid with three-foot spacing which comprised of 210 Extreme Scale Motes (XSM) (Bibyk) and 210 Extreme Scale Stargates (XSS) (Arora). XSM motes were originally used by the University of California Berkeley's prototype sensor nodes and they were specially designed by Crossbow to be used for the ExScal project. *Portable Array* is comprised of 50 Trio motes (P. Dutta et al., 2006) which are designed by UC Berkeley. Trio mote is an integration of XSM sensor board (includes acoustic, PIR *Passive Infrared*, two-axis magnetometer, and temperature sensors), TMote Sky nodes and solar-power charging system. *Mobile Array* is comprised of robotic mobile nodes, operates on top of a transparent Plexiglas plane mounted over the *stationary array*. Each robot contains an XSM node communicates with the stationary array. Kansei makes use of Ethernet and USB for its backbone. This is achieved in conjunction with WLAN 802.11b. Among the main features of Kansei are (Arora et al., 2006; Ertin et al., 2006):

- It focuses on sensing through providing support for static, portable and mobile sensor systems,
- It supports health monitoring through deploying the *Chowkidar* (Bapat, Leal, Kwon, Wei, & Arora, 2009) tool,
- Through a simple Web interface, it gives users the options to define network topology, job scheduling, health monitoring, visualization as well as other sophisticated features,
- It provides support for hybrid simulation.

Network Embedded Sensor Testbed (NESTbed)

NESTbed is a WSN-Testbed framework developed through cooperation between the Western Carolina University and Clemson University. The main physical deployment is installed on the Clemson campus. NESTbed is tailored to support system debugging, profiling, and experimentation. In addition to using it by WSN network researchers; it is also used as part of a graduate course in embedded sensor network design. The current implementation targets applications developed using nesC and TinyOS. The system supports multiple physical deployments. The architecture is composed of three layers: physical network deployments; a centralized application and database server, and client interfaces for remote users, who may optionally connect one or more remote sensor subnets. Current physical deployment consisting of 80 Tmote Sky devices arranged in a dense grid. Small web cameras mounted overhead provide streaming video feeds that show the actuation state of the network. Each mote is attached to the server through a USB connection. Sensor support includes integrated temperature, light, and humidity sensors, and can be configured to support a range of additional sensors. The total equipment cost for the prototype installation, including the application and database server, is less than $10,000. NESTbed's software tools, as well as instructions for installation and use, are available for download (NESTbed). The main features of the NESTbed framework are (Dalton & Hallstrom, 2009b):

- It supports interactiveness; users can profile source and network level components across a network in real time, and they can inject transient state faults and external network traffic,
- It supports openness; developers can extend the set of exposed inter faces as appropriate to particular projects without modifying the underlying middleware,
- It is designed to be source-centric; it enables automated source code analysis, instrumentation, and compilation.

SenseNet

SenseNet is a framework for WSN-Testbed deployment developed at Athens Information Technology Peania in Greece. An instance of SenseNet is being deployed in the above-mentioned university. The structure of SenseNet is quite simple; however, it is presented as an efficient and low-power WSN-Testbed. SenseNet's instance is comprised of a base station and 8 Mica2 sensor nodes connected to a central computer. Among other things, SenseNet is used to experiment with three topology configurations; tree topology, simple star topology, and relay motes. The Oscilloscope application (WU WSN Wiki) is used for collecting motes' readings as well as for performance measurements. Features of SenseNet include (Dimitriou et al., 2007):

- It does not make assumption on sensor nodes capability,
- It does not rely on wired backchannel to enhance scalability,
- It utilizes the wireless channel for both reprogramming and network management.

Konrad Testbed (KonTest)

KonTest is another indoor wireless sensor network testbed deployed at Vrije Universiteit Amsterdam. KonTest is comprised of 60 TelosB nodes distributed between six rooms. Ten of these nodes are manufactured by Moteiv and they contain full sensor suites. The remainder 50 nodes are from Crossbow where 20 of them are equipped with full sensor suite and 30 are without. Each room contains a Pentium III PC where nodes in that room are connected to the PC using USB cabling. There is a total of 16 USB hubs. There is no information if the six PCs connected to some main server or linked together. Simple TinyOS 2.0 was written to obtain basic network connectivity. Among the main features of KonTest are (KonTest):

- It uses the wired USB connection only for power supply as well as a reliable transport backbone for protocol statistics and control commands,
- Through separation between statistic gathering and protocol operations, it aims at minimizing the interference between these two operations.

Trio

Trio is one of the biggest solar-powered, open experimental and outdoor platforms. Trio WSB-Testbed is based on the Trio mote platform (P. Dutta et al., 2006) where it borrows its name from there. It is the resultant of cooperation between University of California Berkeley's Electrical Engineering and Computer Science Dept., Arched Rock Corporation, and Moteiv Corporation. The Trio platform is composed of the Telos, XSM, and Prometheus platforms. Trio directly integrates Telos into its design. From the XSM, Trio borrows and improves upon the hardware grenade timer for fail-safe flexibility and sensing circuitry for acoustic, magnetic, and passive infrared. Prometheus provides the basis of the photovoltaic energy scavenging system in Trio but the Trio implementation improves upon the original design by adding support for fail-safe flexibility. Trio developers designed and integrated an application called Multi-Target Tracking (MTT) to help evaluate the testbed.

Trio is deployed in an area of approximately 50,000 square meters. The overall architecture of Trio could be functionally distinguished into four different tiers. Mote Tier; consists of Trio nodes responsible for sensing, local processing and communication using 802.15.4 specification. The mote tier is comprised of 557 solar-powered motes. Gateway Tier; consists of gateways, repeaters, and access points. Gateways are physically distributed throughout the mote tier and they forward traffic between the 802.15.4 mote network and the 802.11 backbone network. Repeaters simply rebroadcast traffic and serve to extend the range of the 802.11 backbone network. The access point bridges the 802.11 backbone network with an 802.3 Ethernet network that is connected to a single root server. Server Tier; consists of a root server which runs network monitoring processes, gathers statistics on network behavior, multiplexes traffic from multiple gateways, and provides information on the network condition to system users. Client Tier; consists of one or more desktop computers that run client-side applications. These applications access the network via the server tier which forwards traffic to and from the gateway tier. Trio has been used to evaluate various aspects of sensor network operation. At the system programming level, Trio was used to develop and evaluate a suite of programming tools. At the management level, it was used to evaluate network management tools. At the application level, intrusion detection, distributed target tracking, and multi-target tracking algorithms have been evaluated. At the middleware services level, the testbed has been used to evaluate collection, dissemination, and network programming algorithms. At the operating system and device driver level, it has been used to explore several battery charging algorithms and many questions surrounding the firmware, kernel, and application boundaries (P. Dutta et al., 2006).

IBM WSN-Testbed

IBM WSN-Testbed is built at the IBM Zurich Research Laboratory. It has been built to evaluate the performance of short-range wireless communication technologies such as IEEE 802.15.4/ZigBee networks, Bluetooth WPANs, and IEEE 802.11b WLANs. It also enables researchers to test new light-weight messaging protocols for asynchronous communication between sensors and the application server, and to develop applications such as remote metering and location sensing. A hierarchal architecture of IBM-WSN Testbed can be described as follows: *radio modules* with a sensor unit equipped with several types of sensors including a compass, accelerometers, gyroscopes, and a thermometer; *a sensor gateway* connecting the wireless sensor world with the enterprise computing environment; *middleware components* for distributing the sensor data to sensor applications; and *sensor application software*. What distinguishes the IBM WSN-Testbed from most other WSN-Testbeds is that the sensor units are equipped with a Field Programmable Gate Array (FPGA) (Betz,) which through filtering sensor readings helps to reduce the amount of traffic sent in the network. Some technical information such as the number of sensor units, information on sensor module vendors, detailed description of network topology and software design, etc. are not provided. Though it is not explicitly mentioned, it seems that the IBM WSN-Testbed is not open for outside researchers (Furrer, Schott, Truong, & Weiss, 2006b; IBM).

Mobile Emulab (Also Known as TrueMobile)

Mobile Emulab is presented as being the first robotic-based mobile WSN-Testbed. It is deployed in-door at the University of Utah for a range of experiments in mobile and wireless networking. It is a major extension of Emulab network testbed software (Emulab) which designed to provide unified access to a variety of experimental environments. Being based on Emulab, it can be viewed as comprised of three hierarchies. First, a web interface enables users to initiate experiments and manage them. Second, a core consists of a database and a wide variety of programs that allocate, configure, and operate testbed equipment. Finally, back-ends include interfaces to locally-managed clusters of nodes, virtual and simulated *nodes*, and a Planet-Lab (Drupal, 2007) interface. The sensor nodes of Mobile Emulab are two-fold; *fixed* and *robotic*. *Fixed sensor nodes*; a total of 25 sensor nodes (brand is not specified) where 15 sensor nodes are installed on the ceiling in a roughly 2-meter grid and 10 sensor nodes installed on the walls near the floor. Sensor nodes are attached to MIB500CA serial programming boards to allow reprogramming and communication. The 10 near-floor sensor nodes feature an MTS310 full multi-sensor board with magnetometers that can be used to detect the robot as it approaches. *Robotic sensor nodes*; a total of six Mica2 motes installed on a six two-wheeled Acroname Garcia robots (Acroname). Robots come with an onboard XScalebased Stargate; moreover, they operate wirelessly using 802.11b and a rechargeable battery. Among the main features of Mobile Emulab are (Johnson, Stack, Fish, Flickinger, Stoller, Ricci, & Lepreau, 2006b):

- It uses COTS affordable hardware and open-source software,
- It is the first mobile and remotely accessible WSN-Testbed,
- It offers a valuable complement to simulation for exploring many wireless and sensor protocols and applications,
- Its software is characterized as a scalable being an extension of Emulab.

4.1.2.b Proof of Concept Cluster WSN-Testbeds (Cluster WSN-PCT)

Eight different cluster proof-of-concept WSN-Testbeds are considered in our survey. The architectures of these WSN-Testbeds alongside some distinctive features are highlighted in Table 3. This is followed by a brief description of each testbed.

Wireless Industrial Sensor Network Testbed for Radio Harsh Environments (WINTeR)

WINTeR is designed by the Petroleum Application of Wireless Systems (PAWS) of Cape Breton University in Canada. WINTeR is an open access multi-user experimental WSN-Testbed that supports the development and evaluation of WSNs for radio-harsh environments (RHEs). It is designed to be used for investigating a variety of applications for emerging WSN technologies, including physical layer developments, routing protocols, security, power consumption modeling, the validation of wireless solutions for industrial processes, and cross-layer optimization. While we are in the process of writing this survey a beta release of WINTeR was publicly made available to the research community (*Wireless industrial sensor network testbed for radio-harsh environments (WINTeR)*). Through a simple web interface, researchers can test the performance and integrity of wireless sensor network solutions for radio harsh environments. Once connected, user can run tests with 32 wireless devices, create customized scenarios while maintaining full control of data, attenuation, power supplies, interference, and noise. Tests can be scheduled to run at a specific time and online graphing of the selected data can be generated in real-time. Among the main features of WINTeR are (*Wireless industrial sensor network testbed for radio-harsh environments (WINTeR);* J. Slipp et al., 2008b):

- It incorporates controlled Electromagnetic Interference (EMI) to enable dynamic signal strength control, and process control with wireless in-the-loop,
- It incorporates a micro-power meter in the mote platform,
- The online beta release of WINTeR enables researchers to test hardware and software in a realistic radio harsh environment,
- It also enables web users to create customized scenarios.

VigilNet

VigilNet is an integrated outdoor WSN-Testbed for energy-efficient surveillance developed through cooperation between the University of Virginia and Carnegie-Melon University. VigilNet is an energy aware design to ensure the longevity of surveillance missions. The project investigates an important military application of WSN pertaining to information acquisition and target detection. VigilNet currently consists of about 40,000 lines of NesC and Java code, running on XSM, Mica2 and Mica2dot platforms. The complete system is designed to scale to at least 1000 XSM motes and cover minimal 100×1000 square meters to ensure operational applicability. An instance of VigilNet was deployed along a 280 feet long perimeter in a grassy field that would typically represent a critical choke point or passageway to be monitored. VigilNet's instance is comprised of a group of 70 MICA2 cooperating sensor motes and cameras to detect and track the positions of moving vehicles in an energy-efficient and stealthy manner. In this deployment, a mote attached to a portable device such as a laptop is used as the base station. Camera devices are also controlled by the laptop to provide the next level of surveillance information, when triggered by the sensor field. Each mote is equipped with a sensor board that has magnetic, acoustic, motion and photo sensors on it. Among the main features of VigilNet are (VigilNet):

- It is characterized as being an integrated system with energy-awareness as the main design principle across a whole set of middleware services,
- It provides mechanisms for dynamic control, which allow trade-offs control between energy-efficiency and system performance by adjusting the sensitivity of the system;
- It features a physical implementation and extensive field evaluation that reveal the practical issues that are hard to capture in simulation.

FireSense

FireSense is a wireless sensor network testbed for forest fire detection developed by Yeditepe University in Turkey. FireSense project started with introducing and simulating an algorithm for fire detection then it was later developed to an outdoor ad hoc WSN-Testbed. FireSense was

Table 3. Survey of selected cluster PCT WSN-testbeds

Testbed	Motes	OS	Sensors	Base Station/ Gateway	Backbone/ Backchannel	Power Source	Some Distinctive Features
				PCT			
WINTeR (J. Slipp et al., 2008b)	30 COTS motes (brand not Specified)	TinyOS	RHE Sensors (Not specified)	4 Base Stations *SW Based on MoteLab*	Ethernet/USB	Single DC power source	Tailored for Radio Harish Environments (RHE) in the Oil and Gas industry
VigilNet (VigilNet)	70 MICA2 motes	TinyOS	Magnetic, acoustic, motion and photo	Motes attached to laptops act as base-stations	Wireless	Battery	Adaptable surveillance strategy achieves a significant extension of network lifetime
FireSenseTB (Kosucu, Irgan, Kucuk, & Baydere, 2009)	44 TmoteSky	TinyOS	Temperature, humidity	Base Stations	Wireless	Battery	Ad hoc WSNs for fire detection
Common Sense Net (CSN) (COMMON-Sense Net)	Mica2 motes	TinyOS	Various	Base Stations	Wireless	Battery/Solar	Affordable tool for helping poor farmers to improve farming strategies
Alarm Net (Wood et al., 2008)	MicaZ, Telos Sky, X10	TinyOS	Various	Cross Bow's Stargate	Wireless	Heterogeneous; power supply, batteries and scavenge energy	Context-aware power management system (CAPM)
Pavement Condition Monitoring (PCM TB) (Pei et al., 2009)	4 Mica2 motes	TinyOS	Temperature, moisture and infrared	1 Base Station	Wireless	Battery	COTS-based design of a noble solution for pavement condition monitoring called Sensor-Road Bottom (SRB)
SSST Testbed (Gurkan, Yuan, Benhaddou, Figueroa, & Morris, 2007)	No info	No info	Various	Gateways	Wireless and wired	Various options	Plug-and-play sensors with IEEE 1451 compatibility
MSRLab6 (Hongwei et al., 2007)	20 MSR6680	Tiny IPv6	Temperature	1 Base Station (MSR 8680)	Ethernet	Power-supply	IPv6-based

designed for large forest areas with nodes being ad hoc deployed from an aircraft in the unattained region. Currently, FireSense is being used to investigate various issues including node failures, link failures, topology configuration, ignition point positions, and sampling period variations. FireSense is comprised mainly of three software components. *Fire detection algorithm* run on the base station as a java application. *Scalable Querying Sensor protocol (SQS)* (Yeditepe University) running on TmoteSky sensor nodes as WSN network construction and querying tool. And *FARSITE* (Missoula Fire Sciences Laboratory), a forest fire simulator. An instance of FireSense is deployed in forest region of 4000m^2 nearby the university's campus. A number of 13 to 44

TmoteSky nodes were deployed in the tests. The outcome of the experiments is mainly concluded in two factors pertain to the success of fire detection algorithm. First, position of the fire ignition point to the base station is vital to determine the success rate of the fire detection algorithm, i.e., the closer the fire origin to the base station the less probable the fire detection algorithm is successful. Second, low packet reception rate, due to congestion for instance, will make the fire detection algorithm fail in detecting fires (Kosucu et al., 2009).

COMMON-Sense Net (CSN)

CSN project is a community-oriented management and monitoring system of natural resources through wireless sensor networks. It aims at the design and development of an integrated solution for agricultural management in the rural semi-arid areas of developing countries. This aids the poor farmers of that region with e-tool for decision support based on an improved knowledge of the conditions prevailing in the field, and on well-established models of crop prediction. CSN is achieved through using a network of various wireless ground sensors to periodically monitor the state of the soil such as salinity and humidity, temperature, volume of precipitations, etc. Moreover, subterranean sensors are used to monitor the level and quality of a ground-water. The project is collaboration between four partners among others. Mainly, the Swiss Agency for Development and Cooperation, Lausanne University School of Economics (HEC), by the Swiss Federal Institute of Technology, Lausanne (EPFL), and Center for Electronic Design and Technology (CEDT) and Center for Atmospheric and Oceanic Studies (CAOS), at the Indian Institute of Science, Bangalore. A testbed is deployed in Pavagada, Bangladesh comprised of geographical clusters of various wireless sensor networks and corresponding base stations. The wireless sensor network is linked to a centralized server via 802.11 backbone. The sensors are organized in groups corresponding to three applications; water conservation measures assessment, crop modeling, and deficit irrigation management. Sensors from different group clusters collaborate to relay the periodic data to the centralized server. The goals of the field trail include: assessment of the wireless sensor network to calibrate and refine existing crop models, assessment of the soil moisture probes to quantify different water conservation measures, and finally to test a simple deficit-irrigation management system (COMMON-Sense Net).

AlarmNet

AlarmNet is a novel system with a proof of concept WSN-Testbed developed by researchers from the University of Virginia. It is a context-aware wireless sensor network for assisted living and residential monitoring. AlramNet's scalability and heterogeneity is achieved through integration of physiological, environmental and activity sensors. AlarmNet is equipped with analysis programs designed to automatically collect data; and it supports ad hoc queries and the addition of new analysis programs overtime. Among the tools specifically developed for AlarmNet are: *SenQ*, a query system supports reconfigurable sensing and processing, dynamic query origination and high-level abstraction; *CAR*, an analysis program measures the rhythmic behavioral activity; and *CAPM*, a context-aware power management system. A testbed for AlramNet was developed to help the researchers perform ongoing evaluation for the system and its capabilities. The AlarmNet testbed is comprised of heterogeneous hardware of emplaced sensors, mobile body networks, gateways, and user devices. The sensor platform is comprised of MicaZ, Telos Sky motes and X10 devices. Sensors are connected to a Crossbow Stargate SBC. Users can access AlarmNet using iPAQ PDAs or PCs wirelessly through LCD-enabled motes or over the Internet. The residents' environmental and physiological conditions are measured through the mobile body networks. Such measurements include: heart rate, pulse oximetry, heart electrical activity, weight, systolic and diastolic blood pressure, motion, dust, light, optical tripwires, and magnetic reed switches (Wood et al., 2008).

Pavement Condition-Monitoring Testbed (PCM)

PCM testbed is developed by the School of Civil Engineering and Environmental Science, University of Oklahoma. It presents an affordable and vital safety solution to detect dangerous road conditions in real time. It enables the application of sensor technologies in intelligent transportation systems (ITS). Mica2 motes were deployed to collect data from three temperature and moisture external analog sensors. *Surge time synchronization* is explored in this specific application to enable the wireless sensor network to operate in a low power consumption mode. A novel solution called the *Sensor-Road Button* (SRB) is developed and validated experimentally. The testbed is used to examine domain-specific as well as platform-specific applications (Pei et al., 2009).

Smart Sensor Networking System Testbed (SSST)

SSST testbed is a smart sensor networking system testbed with IEEE 1451 compatibility. The testbed build-up process is still underway through collaboration between NASA Stennis Space Center and the Dept. of Engineering Technology in the University of Houston. SSST architecture is planned to be comprised of; High-end Internet backbone; COTS IEEE 1451 enabled smart components; wireless sensor system development kit; gateways, remote data acquisition system for interaction and monitoring of the network; automated TEDS (Transducer Electronic Data Sheets) management software such as Esensors (Esensors); and interface with indigent decision support system such as G2 (GenSym). The objectives of designing SSST include (Gurkan et al., 2007):

- It is designed for providing testing, verification and validation of different COTS plug-and-play smart sensors with IEEE 1451 compatibility,
- It provides open framework for simulating and modeling behaviors of heterogeneous and distributed control and monitoring systems based on both wireless and wired smart sensor network with different network capabilities,
- It complements the functionalities of Integrated System Health Management (ISHM),
- It offers network performance monitoring through implementing metrics and requirements in Systems of Systems (SoS).

MSRLab6

MSRLab6 is an IPv6 WSN-Testbed developed by School of Electronics and Information Engineering at Beijing Jiaotong University. It is a simple testbed designed to experiment with wireless sensor networks in combination with TCP/IPv6 based low-power wireless personal area networks (LoWPAN). LoWPANs are designed to conform to the IEEE 802.15.4 standard. MSRLab6 is comprised of PC terminals and few own developed items including: 20 sensor network nodes (MSR6680), a sink (MSR8680), an IPv6 server with database, a tiny IPv6 protocol stack and a dynamic route protocol named MSNRP6. MSRLab6 provides an IPv6-based Ethernet access. A web interface provides means for network management and remote parameter configuration. Future plans of developers include extending the number of wireless sensor nodes, and implementing the IPv6 packet header compression & segmentation, localization function and topology control. The ultimate goal of developers is to design a testbed similar to MoteLab but based on IPv6 (Hongwei et al., 2007).

4.1.3 Survey of Overlay WSN-Testbeds (WSN-OT)

We consider four different overlay testbeds that support experimentation with wireless sensor networks. Three of the considered testbeds are multiuser experimental testbeds (MXT) and the fourth is a proof-of-concept testbed (PCT). Table 4 presents comparison between these testbeds with highlights of some distinctive features. This is followed by a brief description of each testbed.

PlanetLab

PlanetLab is a testbed developed by the department of computer science at Princeton University. Since its inception in 2003, it currently consists of 1038 nodes at 485 sites. PlanetLab is designed to be a global research network for supporting the development of new network services. PlanetLab has been used by more than 1,000 industrial and academic researchers from all over the globe to develop new technologies for distributed storage, distributed hash tables, query processing, peer-to-peer systems, and network mapping. Among its main objectives, PlanetLab serves as a testbed of overlay networks. Researchers can request an online account through which they can experiment with many services including network-embedded storage, file sharing, routing and multicast overlay, QoS overlay, scalable event propagation, and network measurement tools. At the moment, the total number of active projects run on PlanetLab is more than 600. Being a large-scale overlay testbed, PlanetLab enables researchers to heterogeneously and remotely run real experiments under real conditions. To extend PlanetLab, developers created software called GENIWrapper which among other things provides support for wireless sensor networks. An early version of GENIWrapper and documentation was released in late 2008. The software is being tested between six universities for active deployment (Drupal, 2007; GENI).

AEOLUS

AEOLUS is an overlay computing platform led by the University of Patras in Greece with a total of 22 European universities and research institutes contributing to this project. AEOLUS aims at investigating the principles and developing algorithmic methods for building an overlay network to enable efficient and transparent access to the resources of an Internet-based global computer. The AEOLUS WSN-Testbed is developed under the sub-project (SP) number 6 of the AEOLUS framework, this is under a deliverable titled; "*Integration of mobile and wireless extension into the Overlay Computer Platform (OCP) testbed*" (AEOLUS). AELOUS WSN-Testbed interconnects multiple wireless sensor networks and overlay them on the AEOLUS global computing platform. The WSN-Testbed is comprised of a total of 108 wireless sensor nodes that are deployed in 10 sites located in 7 different locations across Europe. Hereinafter provided a list of the participant research institutes and the corresponding sensor nodes:

- Università degli Studi di Roma *La Sapienza*, UDRLS network at Rome, Italy (11 MOTEIV TmoteSky nodes),
- Research Academic Computer Technology Institute, CTI network at Patras, Greece (10 MICA nodes and 3 camera sensors),
- CTI network at Patras, Greece (8 MICA2 nodes and 1 camera sensor),
- CTI network at Patras, Greece (15 TelosB nodes and 1 camera sensor),
- CTI network at Patras, Greece (16 Sun SPOT motes),
- University of Patras, UOP network at Patras, Greece (5 MICA2 nodes and 1 camera sensor),
- Université de Genève, UNIGE network at Geneva, Switzerland (5 MICAz nodes),
- University of Ioannina, UOI network at Ioannina, Greece (10 MICA2 nodes),

Table 4. Survey of selected overlay WSN-testbeds

Testbed	Owner/ Developer	Number of Current Active Projects	WSN Support	Motes	Current No. of Motes (Max)	Distinctive Features
MXT						
PlanetLab (Drupal, 2007; Mgillis, 2009)	Princeton University	More than 600	Integrated in 2008 and still going through further development	No information available	No info available	GENIWrapper is a SW built on top of PlanetLab to provide support for WSN
AEOLUS (AEOLUS)	University of Patras	No info available	AELOUS WSN-Testbed	MOTEIV TmoteSky, MICA, MICA2, TELOSB, SUN SPOT and Camera sensors	108	In the process of implementing smart data center and cooling application; and IEEE 1451
Emulab (Emulab)	University of Utah	More than 500	Implemented through Mobile EmuLab	See Mobile Emulab	See Mobile Emulab	Provides interface to PlanetLab
PCT						
SWOON (Huang et al., 2008)	University of California	No info available	In progress	NA	NA	In the prosess of implementing support for ZigBee

- University of Paderborn, UPB network at Barcelona, Spain (18 Sun SPOT nodes),
- Institut National De Recherche en Informatique et en Automatique, INRIA network at Sophia-Antipolis, France (3 TelosB nodes).

WebDust framework is used for managing the testbed. WebDust is a software framework developed by CTI. It is based on the AEOLUS peer-to-peer notion to unify WSNs through providing ease of integration with advanced end-user interface capabilities. Other work in progress includes adding actuator extensions to interface with HVAC system for controlling light and air-condition units. Other planned extensions include the implementation of a smart data center cooling application; and the addition of 1-wire sensors (IEEE 1451) in order to monitor meteorological parameters (AEOLUS).

Emulab

Emulab is a testbed developed by the Flux Group at the school of computing at the University of Utah. Emulab is a common framework and a public facility that provides worldwide researchers with flexible environments to develop, debug, and evaluate their systems. Currently, Emulab software is installed in more than 24 sites around the world. Additionally, Emulab is used by professors in many universities to support education in the field of networking and distributed systems. Among the main features of Emulab are:

- It is designed to be web-based and script driven,
- It is designed to be GUI driven and supports programmatic interaction through an SML-RPC interface,
- It provides support for very large-scale experiments,

- It provides means for management and support for resource allocation, configuration and operation,
- It provides interfaces for locally managed clusters nodes as well as virtual and simulated nodes,
- It provides an interface to Planet-Lab.

Mobile Emulab is a WSN-Testbed built on top of Emulab. It consists of a set of fixed motes and mobile motes. Mobility is achieved by remotely controllable robots. This gives the user the ability to experiment with dynamic topologies and to modify the WSN topology of the network on-the-fly. Through overlaying Mobile Emulab on the Emulab testbed, many useful features were inherited. This includes: flexibility, scalability, management features, heterogeneity, etc (Emulab).

Secure Wireless Overlay Observation Network (SWOON)

SWOON is a testbed developed by a team of twelve researchers from the University of California at Berkeley and the National Chiao-Tung University. SWOON is a testbed based on emulation and designed to support scalable tests over an overlay network of 802.11 a/b/g, wireless sensor networks, WiMax, and 3G. Currently, the support for 802.11 networks has been completely implemented, whereas the implementation of wireless sensor networks is still under process. The main objective of this overlay testbed is to provide researchers with means to construct topologies and execute experiments without going through the hurdle of expensive cost and physical installation of the underlying hardware and software. This will enable researchers to investigate techniques, examine mechanisms and evaluate protocols for secure wireless communication. In terms of architecture, SWOON is built on top of the *Defense Technology Experimental Research* (DETER) testbed. DETER is a testbed built on top of Emulab for investigating security issues in wired networks. SWOON increases the scale of DETER and incorporates support for wireless networks. The developers of SWOON are in the process of adding support for wireless sensor networks including the Zigbee standard. Once the support for such sensor nodes is implemented, researchers will be able to run a variety of physical experiments on these nodes such as broadcasting and secure aggregation (Huang et al., 2008).

4.1.4 Survey of Federated WSN-Testbeds (WSN-FT)

Federated WSN-Testbeds is another important class of wireless sensor networks. Hereinafter, we consider 4 different federated WSN-Testbeds. Three of which are multiuser experimental testbeds and the fourth is a proof-of-concept. Table 5 below summarizes some important information pertinent to these testbeds. This is followed by a description on each WSN-Testbed.

Wireless Sensor Network Testbeds (WISEBED)

WISEBED is part of FIRE (Future Internet Research and Experimentation) initiative of the European Commission. It aims at establishing a pan-European federated WSN-Testbed through providing a multi-level infrastructure of interconnected WSN-Testbeds of large-scale wireless sensor networks for research purposes, pursuing an interdisciplinary approach that integrates the aspects of hardware, software, algorithms, and data. Nine academic and research institutes across Europe are partnering for this project. It was initiated in June 2008 and scheduled to be completed in May 2011. It is planned to be accessible by European researchers of all fields for experiments and to also serve as a showcase for European industries. They plan to achieve this goal through bringing together and extending different existing testbeds across Europe and forming a federation

Table 5. Survey of selected federated WSN-testbeds

Testbed	Owner/ Leader	No. of partners/ Testbeds	Progress of Deployment	Type of Nodes	Type of Sensors	Distinctive Features
			MXT			
WISEBED (EU Project)	EU	9 European academic and research institutes	Started June 2008, and expected to complete by May 2011	TelosB, Tmote Sky and MicaZ	Various	Pan-Europe, flexible and heterogeneous
Global WSN (SANE)	FOKUS	12 academic and research institutes	Deployed	FOKUS	Various	Inter-continental; where Local NWs are connected via IPv6/6loWPAN
WiSense (WiSense)	University of Ottawa	4 universities and 11 industrial partners	Started in 2008 and planned to finish in 2013	Heterogeneous	TBD	First federated WSN-Testbed in Canada
			PCT			
SensorMap (Nath et al., 2006; SensorMap)	Microsoft	Various	Deployed	Various	Various	Geographical visualization of sensors and data

of distributed test laboratories. The project seems to be well-planned and its workload is split over six work packages (WP). WP1 addresses the hardware design and implementation issues. WP2 deals with the software issues. WP3 handles the algorithms of the project. The objective of WP3 is to design a scalable and heterogeneous library that contains different algorithms for a number of purposes, ranging from standard algorithms to the latest research developments. WP4 deals with data issues pertaining to simulations, experiments and possible benchmarks. WP5 concerned with dissemination and joint research Activities. And finally, WP6 deals with management, review and self-assessment issues. Among the features implanted in WISEBED are: heterogeneity, scalability, flexibility, interdisciplinarity, hierarchicality and well-planned manageability (EU Project).

Global Wireless Sensor Network (Global WSN)

Global WSN is a federated WSN-Testbed administered by Fraunhofer Institute for Open Communication Systems (FOKUS) and the National Information Society (NIA) of South Korea. Additionally, 10 more partners from other countries are involved. In fact, a total of 16 institutes from 3 different continents are partnering for this project. Each of them installed a set of about 6 sensor nodes and a router device. The local sensor networks are connected via IPv6/6LoWPAN to a central server at NIA in Korea. All sensor data is collected by this server and visualized by a Java web-start application. Global WSN is still in the experimental phase. Global Wireless Sensor Network is designed to be a Research Network. Among other issues under investigation are: the interconnection of separate worldwide distributed network segments related to different application domains like, ambient assisted living, safety and security, or energy efficiency; management and maintenance of such Global WSN; implementation of algorithms, automatic and autonomous configuration mechanisms, etc (SANE).

Wireless Heterogeneous Sensor Networks in the e-Society (WiSense)

WiSense is a research and development project aims at building a heterogeneous wireless sensor network testbed. WiSense commenced in 2008 and it is planned to take five years span to finish. The project is mainly funded by the Ontario Ministry of Research and Innovation, and led by the University of Ottawa. Participants in the project include three more Ontario universities and eleven leading industrial partners including IBM, Research in Motion, Alcatel-Lucent, Google Canada, and Syncrude. The project is split between partners into six different themes. These include architecture, system design, gateway access, network security, services, and testbed development. The major research activities planned for the project include (WiSense):

- Investigation and development of network architecture models enable heterogeneous wireless sensor networks,
- Development of power-efficient protocols and algorithms for internetworking with wireless sensor networks as well as with cellular systems including WiFi, and WiMax,
- Development of service software for tasks such as intelligent vehicular systems, telemedicine, intelligent building and home security, and emergency preparedness,
- The ultimate goal is to design a testbed supports selected applications.

SenseWeb

SenseWeb project at Microsoft Research provides a common platform and a set of tools for data owners (e.g., WSN-Testbed owners) to easily publish their data, and users to make useful queries over the live data sources. It utilizes some existing solutions including Geo-centric web interface such as MSN Virtual Earth (MSN) and Google Maps (Google) to visualize spatially and geographically related data. Additional useful information can be augmented in one interface using the SenseWeb. The SenseWeb platform transparently provides mechanisms to archive and index data and to process queries so that results can be aggregated and presented on geo-centric web interfaces such as MSN Virtual Earth. SenosrMap is an instance of SenseWeb deployment (SensorMap).

5. EVALUATION OF SURVEYED WSN-TESTBEDS

In this section we apply the *benchmarking scheme* introduced earlier in Section 4 to the WSN-Testbeds surveyed in this chapter. The result of the evaluation in terms of cost, inherent features, management, resource sharing and allocation, hybrid simulation, experiment control, experiment analysis, and experiment reproducibility is provided in Table 6.

In Table 6, it is interesting to note that while cost and budget constraints play an important role when building a certain WSN-Testbed, many testbed developers tend to overlook discussing this issue when presenting their WSN-Testbed. Moreover, while resource sharing and allocation is an important feature enable multiuser experimentation with WSN-Testbeds; all proof-of-concept (PCT) testbeds considered in this chapter do not incorporate this feature. The reason for this is that such PCT testbeds are designed to be application-focused where only a little number of experimenters has pre-specified controlled access to the testbed. Hybrid simulation and experiment reproducibility are two interesting features that add a new dimension to the experience of WSN experimentation.

As Table 6 shows, only few WSN-Testbed developers consider these two features. It is interesting to note that thanks to the diversity and comprehensiveness offered by Federated as well as Overlay WSN-Testbeds, these two classes do support most of the features considered in our benchmarking scheme. More reflections on these features as well as other interesting issues are presented in the Discussion of Section 6.

Table 6. Comparison of surveyed WSN-testbeds when applying the benchmarking scheme

	Testbed	Cost	Inherent Features	Management	Resource Sharing & Allocation	Hybrid Simulation	Expt. Control	Expt. Analysis	Expt. Reproducibility
Research Kits	**MXT**								
	Crossbow's Kits (Crossbow)	√	×	√	×	×	√	√	×
	Senswiz Kit (SensWiz)	√	×	√	×	×	√	√	×
	PCT								
	Dust Networks' Kits (Dust Networks)	√	×	√	×	×	√	√	×
	NI Wireless Sensor Starter Kit (*NI wireless sensor network (WSN) starter kit - national instruments*)	√	×	√	×	×	√	√	×
Cluster WSN-Testbed	**MXT**								
	TWIST(Handziski et al., 2006)	√	√	√	√	×	√	√	×
	Motelab (Werner-Allen et al., 2005)	--	√	√	√	√	√	√	×
	TutorNet (Tutornet)	--	--	√	--	×	--	√	×
	Kansei (Arora et al., 2006)	--	√	√	√	√	√	√	×
	NESTbed (Dalton & Hallstrom, 2009a)	√	√	√	√	√	√	√	×
	SenseNet (Dimitriou et al., 2007)	√	√	√	√	×	×	√	×
	KonTest (KonTest)	--	×	√	×	×	√	√	√
	Trio (P. Dutta et al., 2006)	--	√	√	×	×	√	√	×
	IBM WSN TB (Furrer, Schott, Truong, & Weiss, 2006a)	--	√	√	--	--	√	√	--
	Mobile Emulab (Johnson, Stack, Fish, Flickinger, Stoller, Ricci, & Lepreau, 2006a)		√	√	×	×	√	√	×
	PCT								
	Winter (J. Slipp et al., 2008a)	√	×	√	×	×	√	√	√
	VigilNet (VigilNet)	×	×	√	×	×	√	√	×
	FireSense TB (Kosucu et al., 2009)	×	×	√	×	√	×	√	×
	COMMON Sense Net (COMMON-Sense Net)	√	√	√	×	×	×	√	×
	Alarm Net (Wood et al., 2008)	√	√	√	×	×	×	√	×
	PCM TB (Pei et al., 2009)	×	√	√	×	×	×	√	×
	SSST TB (Gurkan et al., 2007)	×	×	√	×	×	×	√	×
	MSRLab6 (Hongwei et al., 2007)	√	×	√	×	×	×	√	×
Overlay WSN-TB	**MXT**								
	Planetlab (Drupal, 2007; Mgillis, 2009)	√	√	√	√	√	√	√	√
	AEOLUS (AEOLUS)	√	--	√	--	--	--	√	--
	Emulab (Emulab)	√	√	√	√	√	√	√	√
	PCT								
	SWOON (Huang et al., 2008)	√	√	√	--	--	--	√	√

continued on following page

Table 6. Continued

	Testbed	Cost	Inherent Features	Management	Resource Sharing & Allocation	Hybrid Simulation	Expt. Control	Expt. Analysis	Expt. Reproducibility
Federated WSN-TB	**MXT**								
	WISEBED (EU Project)	√	√	√	√	√	√	√	√
	Global WSN (SANE)	--	√	√	√	--	--	√	--
	WiSense (WiSense)	--	√	--	√	--	--	--	--
	PCT								
	SenseWeb (SensorMap)	--	√	√	√	×	√	√	×

√ denotes that the WSN-Testbed addresses the highlighted feature either fully or partially
× denotes that the WSN-Testbed has inadequate support for the highlighted feature
-- denotes that presence/absence of the highlighted feature is not known/not addressed

Table 7. Some shortcomings/drawbacks of WSN kits

WSN-Testbed	Shortcomings/Drawbacks
MXT	• Relatively expensive when compared to experimentation with other classes of WSN-Tetsbeds, • Limited support for inherent features such as; scalability, heterogeneity, and openness, • Insufficient support for resource sharing and allocation, • Security issues pertaining to transactional confidentiality possess a real challenge to real-world deployment of this class.
Crossbow's Kits (Crossbow)	
Sensewiz (SensWiz)	
PCT	
Dust Networks Kits (Dust Networks)	
NI Wireless Sensor Starter Kit (NI wireless sensor network (WSN) starter kit - national instruments.)	

While Table 6 provides a quick and interesting view into the features and shortcomings of each WSN-Testbed, it leaves some more specific shortcoming/drawbacks go un-captured. Below we summarize the major shortcomings and/or drawbacks pertaining to each WSN-Testbed considered in our survey. Our view is focally based on the information available by the WSN-Testbed developers. Tables 7-11 list the shortcomings/drawbacks for MXT, PCT WSN Research Kits, Cluster WSN-Testbeds, Overlay WSN-Testbeds, and Federated WSN-Testbeds respectively.

It is obvious that MXT and PCT WSN-Testbeds differ in that MXT is meant to be more generic while PCT is designed to be application-oriented. However, the advantages and drawbacks of such WSN-Testbeds are pretty much dependent on the design goals and objectives. For instance, while WINTeR (J. Slipp et al., 2008a) and MSRLab6 (Hongwei et al., 2007) are being classified as PCT WSN-Testbeds, (i.e. both of them fall under the same class being application-oriented), there is a dramatic difference between two of them in terms of structure, features and drawbacks as evidenced in Table 9.

Table 8. Some shortcomings/drawbacks of MXT cluster WSN-testbeds

WSN-Testbed	Shortcoming(s)/Drawback(s)
TWIST (Handziski et al., 2006)	• Software's script-based approach lacks usability, • Lack of a power meter, • Complete reliance on Ethernet/USB as a backchannel for reprogramming, debugging and power supply.
MoteLab (Werner-Allen et al., 2005)	• Lack of hierarchy which leads to the problem of un-scalability, • Lack of node heterogeneity, • No real-time profiling or fault injection support, • Users have no control on a test run once it is scheduled.
TutorNet (Tutor-net)	• Programming: Slow speed because the Stargate used has USB 1.1 port, • Routing over 802.11b: They tried Kernel AODV, Roofnet, and OLSR, • Network Partition: Interference from 802.11 – *sflisten* data collection uses; 802.11 – worse 802.15.4 reliability than Motelab, • Manual frequency allocation does not always work because other groups and several students may have motes on their desk, • Multi-hop wireless routing (OLSR, Roofnet, even static routing) is fragile, • Sub-optimal connectivity across multiple Stargate hops makes quick interaction with motes difficult, • Up to 20 minutes to program 40 KB image on all the motes due to old FTDI driver + slow Stargate CPU, • Supports TinyOS 1.x and TinyOS 2.x but by running separate serial forwarder binaries on the Stargates.
Kansei (Arora et al., 2006)	• No real-time profiling or fault injection support, • Limited support for logging network traffic, • No clear information on the degree of controllability provided to external applications by API, • Unexpected failure of certain hardware components, • Issues with the physical construction - e.g. bending of the Plexiglas surfaces, blocking of certain sensing capabilities due to the materials used for the construction, • Issues with the facilities - external noise, interference from external 802.11 networks, minor water leakage, labor involved in cleaning and maintenance, dealing with fire codes, • Image-centrism (pre-compiled application must be uploaded by the developers) leads to preclusion of source-level analysis and instrumentation.
NESTbed (Dalton & Hallstrom, 2009a; Dalton & Hallstrom, 2009b)	• Solely dependent on wired connections for powering as well as back-channel, • Lack of means to enable users to monitor the status of distributed state conditions relevant to system correctness and performance.
SenseNet (Dimitriou et al., 2007)	• No information provided on reprogramming speed over the air, • No information about the source of power supply for motes, • The network topology used in the experiment is quite simple and straightforward. It is not clear if similar results could be obtained for complex network with bigger number of nodes, • Authors refer to their testbed as scalable, however, experiment run does not proof that, • Utilization of wireless channel is stressed over and over but no detailed comparison is provided to depict how advantageous to do so in such experimental environment, • Lack of heterogeneity.
KonTest (KonTest)	• Lack of network hierarchy, • It does not support automatic testbed reprogramming and data collection, • It does not support experiment or job scheduling, • No web interface.
Trio (P. Dutta et al., 2006)	• Insufficient support for low-power operation in the system software including the radio stack of TinyOS 802.15.4, • Middle services are not properly designed to efficiently handle intermittent connectivity and fluctuating power, • • The power management algorithm is not appropriately designed to handle a broad set of practical scenarios.
IBM WSN TB (Furrer, Schott, Truong, & Weiss, 2006a)	• No information provided on the means used for powering, • Lack of information pertinent to back-channel and reprogramming, • Support of multiple users is not clearly discussed, • No information provided on the operating system used, • While Matlab is used to visualize the sensor data, it is not clear if the testbed supports hybrid simulation through Matlab.

continued on following page

Table 8. Continued

WSN-Testbed	Shortcoming(s)/Drawback(s)
Mobile Emulab (Johnson, Stack, Fish, Flickinger, Stoller, Ricci, & Lepreau, 2006a)	• One can conclude that there is a wireless back-channel; however, nothing mentioned about it and if it has any negative impact on over-all communication, • The WSN network used is quite simple; it is merely comprised of few motes, • Only a single power meter is designed and installed on a single fixed mote • Lack of a per-experiment MySQL database and a simple reservation system, • Inability to transparently inject simulated sensor data into the *environment,* • Restriction on the classes of experiments due to using waypoint-based motion model.

Table 9. Some shortcomings/drawbacks of PCT cluster WSN-testbeds

WSN-Testbed	Shortcoming(s)/Drawback(s)
WINTeR (J. Slipp et al., 2008a)	• Testbed supports only flat hierarchy of WSN. No support for tiered-structure, • Only batteries are used to power the platform motes, although it is an indoor WSN-Testbed, • Though it is mentioned that COTS technology was deployed, no information provided on the type of nodes being used in the mote platform.
VigilNet (VigilNet)	• An important aspect of such surveillance testbed is the *network security*. Specifically, the use of MICA2 motes on the ISM frequency band might enable an enemy to intrude the network and jeopardize the whole mission. This issue is not discussed, • The process of event-based tracking and reporting arise serious concerns from the perspective of network security. It is not clear how this issue is handled • Increasing the network scale yields degradation in the operational performance requirements.
FireSense (Kosucu et al., 2009)	• Packet-loss pertinent to link failure can jeopardize the operability of the system. Although this challenge is identified, it is not clear how this substantial issue is handled, • Node failure and network congestion are two additional factors not handled that can highly degrade the system's ability in detecting ignition points, • The fire detection algorithm is negatively impacted when the ignition points are close to top-level of the tree topology of the network (base station), • The fire detection algorithms used are not adaptive.
COMMON Sense Net (CSN) (COMMON-Sense Net)	• Being crop models rely on an indirect assessment of soil moisture based on rain fall and soil characteristics rather than in-field data, make such models inaccurate, • The project is being deployed in a poor village in a third-world country. This makes it posses few challenges. Specifically, though the integration of *web access* is being identified as a future work, it is not clear how the farmers will benefit from this feature being in a village most probably lacks the Internet's infrastructure.
AlarmNet (Wood et al., 2008)	• The use of open system architecture based on COTS technology makes it susceptible to carrier-frequency-based transactional confidentiality attacks, • The routing algorithm used does not offer enough support in terms of transactional confidentiality protection, • Message transmission of event-driven body sensors is based on the rate at which a particular event is generated. This leaves AlarmNet susceptible to adversarial confidentiality attacks (Pai et al., 2008).
PCM TB (Pei et al., 2009)	• The use of dry ice to test the testbed is not as efficient if compared to a real-world icy road deployment, • Simplicity of the algorithm used makes it unsuitable for deployments other than for the deployed limited capability onboard processing power motes.
SSST (Gurkan et al., 2007)	• IEEE-1451-compatible COTS technology is being promoted as the preferred choice for implementation of the testbed; however, no information is provided on selection of the actual COTS hardware such as sensor nodes, gateways, etc, • Important issues such as powering, network structure, routing, etc. are not clearly addressed, • Practical issues pertaining to the actual testbed will not be apparent before physical deployment, though generic design aspects are presented.
MSRLab6 (Hongwei et al., 2007)	• Simplicity of the structure and topology make it barely suitable for performing basic functions, • Incomplete configuration of the IPv6 Protocol Stack due to hardware limitation, • Although the testbed is low-power oriented, no detailed information is provided on power supply, • No information is provided on back-channel.

Table 10. Some shortcomings/drawbacks of overlay WSN-testbeds

WSN-Testbed	Shortcoming(s)/Drawback(s)
	MXT
PlanetLab (Drupal, 2007; Mgillis, 2009)	• Network security issues are not handled by PlanetLab Central but rather handled by the local organization using it, • Dedicating an CIDR address block boundaries increases the potentiality of unused scarce address space.
AEOLUS (AEOLUS)	• The use of COTS technology raises security issues pertaining to transactional confidentiality. This possess a real challenge to AEOLUS.
Emulab (Emulab)	• Popularity of the network leads to increase in the number of users. This yields a critical limitation in the resource sharing and allocation capabilities.
	PCT
SWOON (Huang et al., 2008)	• As of yet, support for WSN is not implemented. It is planned for future work.

Table 11. Some shortcomings/drawbacks of federated WSN-testbeds

WSN-Testbed	Shortcoming(s)/Drawback(s)
	MXT
WISEBED (EU Project)	• The project is scheduled to be fully implemented by May 2011. Accordingly, real-world pros and cons of the testbed will not be apparent before full implementation, though some information on the design is available, • The issue of network security and confidentiality with respect to the COTS technology used does not seem to be thoroughly addressed.
Global WSN (SANE)	• The few number of *per-partner* sensor nodes (6 nodes on average) along with the simple topology of the network makes it unreliable for in-depth experimentation, • It is not clear if the partners deploy similar kind of sensor nodes. If so, an important aspect pertaining to network heterogeneity is overlooked.
WiSense (WiSense)	• No information is available yet as the project is in its early deployment stages and it is still underway.
	PCT
SenseWeb (Sensor-Map)	• Live data search is not supported, though SensorMap enables an interesting feature to search on the metadata, • Sensed data is shown as points on the map, it was to be more beneficial if it provides support for overlaying graphs such as a gradient map, temperature contour map on top on sensor map, • Security concerns pertaining to public access to SensorMap and availability of the actual locations of sensors are not addressed, • Confidentiality concerns pertaining to the use of COTS technology are not addressed.

While an PCT WSN-Testbed has the advantage of being tailored to a certain application, a designer may be overtaken by this fact while overlooking some other important aspects. To highlight one example, many WSN-Testbed designers considered in this survey such as VigilNet, AlarmNet, etc. are stressing the advantage of COTS hardware being deployed in their testbeds while overlooking serious security concerns associated with COTS technology and pertaining to transactional confidentiality. This observation is witnessed for most of the WSN-Testbeds highlighted in Tables 7 - 11.

Finally, as highlighted in Table 7, though WSN research kits are relatively expensive and they lack some interesting features when compared to other classes of WSN-Testbeds, WSN-RKs provide WSN apprentices with an opportunity for a head start to jump into the race and it gives them a chance to contribute to the world of WSN.

Figure 9. Taxonomic view of the surveyed 30 WSN-testbeds

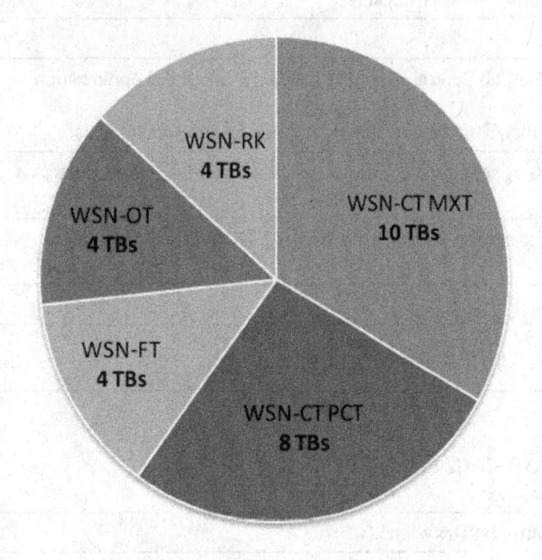

Figure 10. A perspective based on the number of collaborators

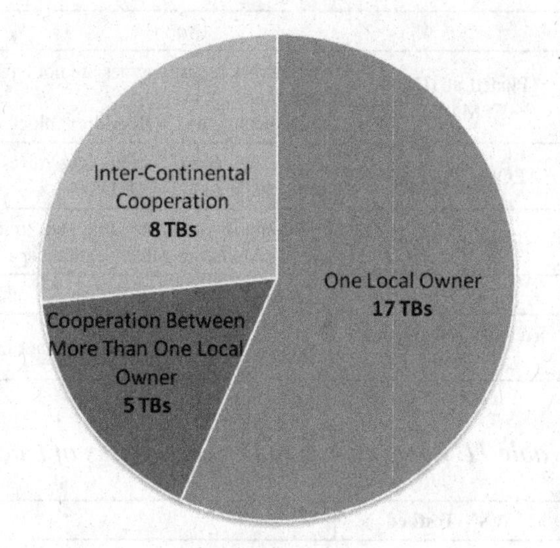

6. DISCUSSION

In this work, we surveyed a total of 30 WSN-Testbeds. Figure 9 depicts a pie chart for the WSN-Testbeds surveyed in this chapter with relation to the classification and the number of testbeds considered. Cluster WSN-Testbeds (WSN-CT) including both Multiuser Experimental Testbeds (MXT) and Proof-of-Concept Testbeds (PCT) form the basic structure for the rest of WSN-Testbeds and they are the most common.

6.1 Collaborator-Based View

An interesting perspective on WSN-Testbeds can be concluded through looking at the number of collaborators in building and maintaining a certain WSN-Testbed. The pie chart below is for the 30 WSN-Testbeds considered in this survey.

Figure 10 depicts that a total of 8 WSN-Testbeds considered in this survey are in fact an inter-continental cooperation between academia and industry. Additionally, 5 out of the surveyed WSN-Testbeds are also result of collaborative work between two or more owners. This simple depiction affirms the essential role of WSN-Testbeds and their vitality in driving the research fast towards commercialization.

6.2 Application-Based View

To further realize the rapid emersion of the WSN technology into various industrial applications, we list in Figure 11 the applications of the PCT WSN-Testbeds considered in this chapter.

The rapid increase in the deployment of WSN-Testbeds, the increasing level of intra-academia and/or inter-industrial collaboration, along with the endless possibilities of WSN applications are tangible facts apprehend that it is just a matter of time before WSN technology becomes an industrial norm (M2M Magazine).

6.3 Design Trade-Offs

Cost exemplified in budget constraints is one of the foremost factors influence the design of a certain WSN-Testbed. As it is shown in Table 6, many WSN-Testbed designers address the issue of cost throughout the process of design and implementation of their WSN-Testbed. Being

Figure 11. Variety of WSN applications pertinent to surveyed PCT WSN-testbeds

WINTeR
•Petroleum Industry
•RHE

VigilNet
•Surlverlaince Missions

FireSense
•Forest Fire Detection

COMMON Sense
•Management of Natural Resources; Agriculture

AlarmNet
•Assisted Living
•Residential Monitoring

PCM TB
•Road Safety
•Pavement Condition Monitoring

SSST TB
•Validation of IEEE 1451 for Industrail Applications

MSRLab6
•Deployment of IPv6 LoWPAN

SWOON
•Secure wireless communication over WSN

SenseWeb
•Map & Visualize Sensors Using Google Earth;
•SensorMap; Weather, & Traffic

Dust Networks' WSN-RK
•WSN Solutions for WirelessHART

the issue of cost addressed does not necessarily mean that the provided solution is cost-efficient. In fact, the relation between cost and performance is proportional. It is a trade-off that needs to be carefully studied, balanced and tailored to fit the anticipated design goals and objectives of the WSN-Testbed.

Research Kits (WSN-RK), Cluster (WSN-CT), Overlay (WSN-OT), and Federated (WSN-FT) are four architectural-based classes for WSN-Testbeds introduced earlier. Depending on the supported application, each of these classes can be further classified into Multiuser Experimental Testbed (MXT) or Proof-of-Concept Testbed (PCT). When deciding to build a WSN-Testbed, the right choice depends on the design challenges described in Section 4. One of these challenges is the *Cost*. If the budget of a researcher is very limited, she/he may opt for experimentation with WSN-OT or WSN-FT. Usually, the owners of such testbeds may provide some authorized external users with an online access. However, the user in this case is absolutely confined to the rules of testbed's owner and she/he may face many limitations. Additionally, as the WSN-OT/FT gets popular, more users might be using it. This increase in the number of users may lead to a serious problem pertaining to availability of the testbed as well as resource sharing and allocation.

For a WSN-RK and WSN-CT, the developers need to purchase their own sensor nodes and required tools. Finally, upon owning a certain WSN-CT a developer may wish to join a federation, i.e. WSN-FT, to enable experimentation with larger-scale, more scalable and heterogeneous WSN-Testbed.

Typically, any design process involves many trade-offs besides the cost-performance. Openness, scalability, flexibility and heterogeneity are such inherent design features pertinent to the design of WSN-Testbed. These features have to be appropriately addressed and relatively defined when designing a WSN-Testbed. It should be noted that these four factors are proportional to the management of the WSN system. In other words, management vs. inherent design features poses an important trade-off require proper balancing since early conception based on the design goals and objectives.

Integration of tools into WSN-Testbed design to enable experiment reproducibility is an interesting feature. As discussed earlier, among other advantages experiment reproducibility enables more accurate and controlled experimentation

Figure 12. Summary of trade-offs pertinent to the design of a WSN-testbed

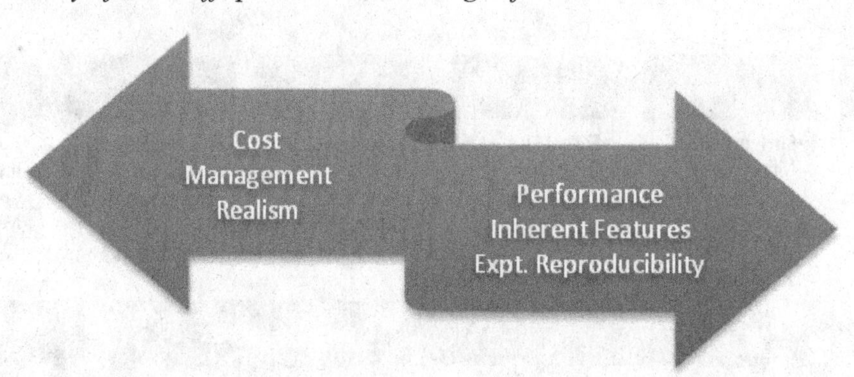

with WSN-Testbed. However, WSN-Testbed designer should bear in mind that the aim of the testbed is real-world experimentation rather than simulation. This will give rise to another important trade-off between experiment reproducibility and realism. The WSN-Testbeds that take these trade-offs into design consideration are highlighted in Table 6. Summary of the trade-offs pertinent to WSN-Testbed design is depicted in Figure 12.

6.4 Locality-Based View

Based on our earlier survey, categorization of a WSN-Testbed based on locality, i.e., indoor vs. outdoor is discussed in this subsection. While some indoor WSN-Testbeds are specifically designed for indoor applications, some others are designed to mimic some outdoor applications. Such mimicked indoor WSN-Testbeds seem to be a preferred option by many WSN-Testbed developers. The reason behind such preference is that a mimicked indoor WSN-Testbed offers an affordable, quite generic, and controllable environment for experimentation with wireless sensor networks. However, the fidelity of such a mimicked indoor deployment is in fact questionable when compared to real-world outdoor deployments. Outdoor deployments handle more realistic and firsthand experimentation. Outdoor deployments also face many challenges go unseen by most mimicked indoor deployments such as remote over-the-air programming and efficient energy usage (S. Chen, Huang, & Zhang, 2008). It should be noted that the level of attainable realism in a certain WSN-Testbed whether it is indoor or outdoor depends on how well that testbed is designed to handle the goals and objectives of interest. A WSN-Testbed can be designed to be flexible *locality-wise* in the sense that it supports both indoor and outdoor deployments; indoor-outdoor WSN-Testbed. A list of the WSN-Testbeds considered in this survey is categorized based on locality is provided in Figure 13.

6.5 Time-Based View

The process of building-up a WSN-Testbed can be cumbersome and time consuming. In the development cycle of a certain WSN application shown in Figure 1, it is obvious that such process entails proper conceptualization, simulation, and local prototyping. Developing a well-defined design goals and objectives forms the cornerstone towards a focused design. This yields a proper handling of the WSN project's time frame and adds to hastening the design process. Emanating from this important aspect, we contacted most of the cluster WSN-Testbeds' developers considered in this survey, and we inquired about the time frame entailed to build-up their WSN-Testbed beginning from conceptualization to successful deployment.

Figure 13. Categorization of surveyed WSN-testbeds into indoor, outdoor, and indoor-outdoor

Indoor TBs	Indoor-Outdoor TBs	Outdoor TBs
• TWIST	• Kansei	• Trio
• MoteLab	• AlarmNet	• VigilNet
• TutorNet	• SSST Testbed	• FireSense
• NESTbed	• WISEBED	• COMMON Sense
• SenseNet	• WiSense	• Pavement Condition Monitoring
• KonTest	• PlanetLab	• SensorMap
• IBM WSN TB	• AEOLUS	
• Mobile Emulab	• SWOON	
• WINTeR	• Dust Networks' Kits	
• MSRLab6	• Emulab	
• Global WSN	• SenseWeb	
	• CrossBow's Developmet Kits	
	• SensWiz' Kits	

The answer of those responded to our inquiry is depicted in Figure 14 below.

Figure 14 depicts that it took about 6 months to deploy MoteLab in 2003. It entailed Kansei 9 months and VigilNet about 1 year for a successful deployment in 2004. It took TutorNet 1 year, MSRLab6 about 30 months, SenseNet about 6 months, and TWIST about 1 year. All four WSN-Testbeds were successfully deployed in 2006. In 2007, FireSense was successfully deployed after about 8 months of design and implementation. KonTest was successfully deployed in June of 2008 with a time frame of 5 months. Finally, NESTbed's design process entailed 6 months for successful deployment in 2009.

From Figure 14 one can conclude that the average time frame pertaining to a WSN-Testbed's successful design and deployment is about 10.6 months. However, it should be noted that this time frame may fluctuate up and down depending on some factors such as, goals and objectives of the WSN-Testbed design, scale and size of the WSN-Testbed, whether the sensor nodes are COTS or self-designed, etc. This fluctuation in time frame deployment is witnessed in Figure 14. While 9 of the WSN-Testbeds fluctuate within close vicinity, MSRLab6 seems to be an anomalous. The reason for this anomaly is that MSRLab6 differs from the other 8 WSN-Testbeds considered above in that it does not deploy COTS elements, i.e., sensor nodes and gateways are self-designed.

It should be noted that the process of WSN-Testbed deployment is progressive in the sense that it constantly evolves and organically grows. This statement is applicable to a WSN-Testbed designed for long-term experimentation. For example, a look at TWIST will reveal that after its first successful deployment in 2005, TWIST has undergone major upgrade where its current deployment is achieved in late 2007. Another interesting example would be MoteLab. When it was initially deployed in the summer of 2003, it had 20 nodes. A year later, MoteLab was updated to enable integrating MicaZ nodes. In early 2006, 200 TelosB nodes were deployed. Additionally, MoteLab moved from MIB600 backchannels to TMote Connects.

6.6 Node-Level View

Being the basis of all WSN-Testbed classes, the physical architecture of a certain Cluster WSN-

Figure 14. Time-domain view for some surveyed cluster WSN-testbeds

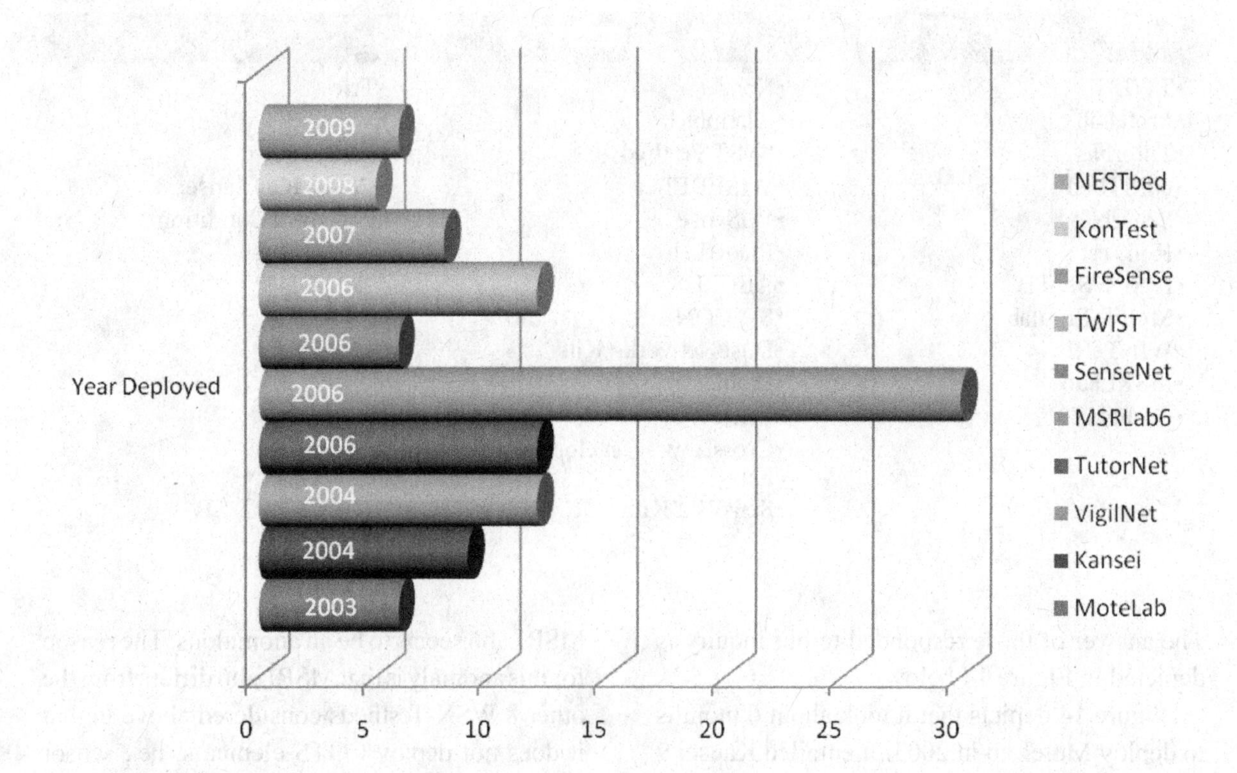

Testbed is comprised of different kinds of nodes. A top-down hierarchical architecture based on nodal view and functionality can be summarized in Figure 15. Typical operation characteristics of the four classes are discussed in detail in (Hill, Horton, Kling, & Krishnamurthy, 2004).

A more holistic approach for such nodal classification should also consider the openness of the hardware as well as the software. The topic of open-source hardware and software pertaining to wireless sensor networks was approached earlier under *Openness* in the *Design Challenges and Evaluation Criteria*.

Plug and play sensors are an interesting class of sensors for industrial applications. IEEE 1451 is the standard for this class enabling all complying sensors, instruments and systems work together with relative ease. This standard defines a wireless interface for sensors. It specifies radio-specific protocols for this wireless interface. It defines communication modules that connect the wireless transducer interface module (WTIM) and the network-capable applications processor (NCAP) using the radio-specific protocols. It also defines the Transducer Electronic Data Sheets (TEDS) for the radio-specific protocols (IEEE, 2007; IEEE,).

6.7 Operating System (OS)

OS is essential in supporting the operation of WSN nodes at the core of a WSN-Testbed. In the case of WSN, such reliable OS is expected to be open-source with specific design for wireless sensor networks being constrained in power and memory resources. WSN's OS utilizes a component-based architecture for enabling rapid implementation while minimizing code size. TinyOS is such a de facto standard implemented by most of the WSN-Testbeds considered in this survey, but it is not the only one. Among the features of TinyOS is that it

Figure 15. Cluster WSN-TB top-down network model based on nodal-view

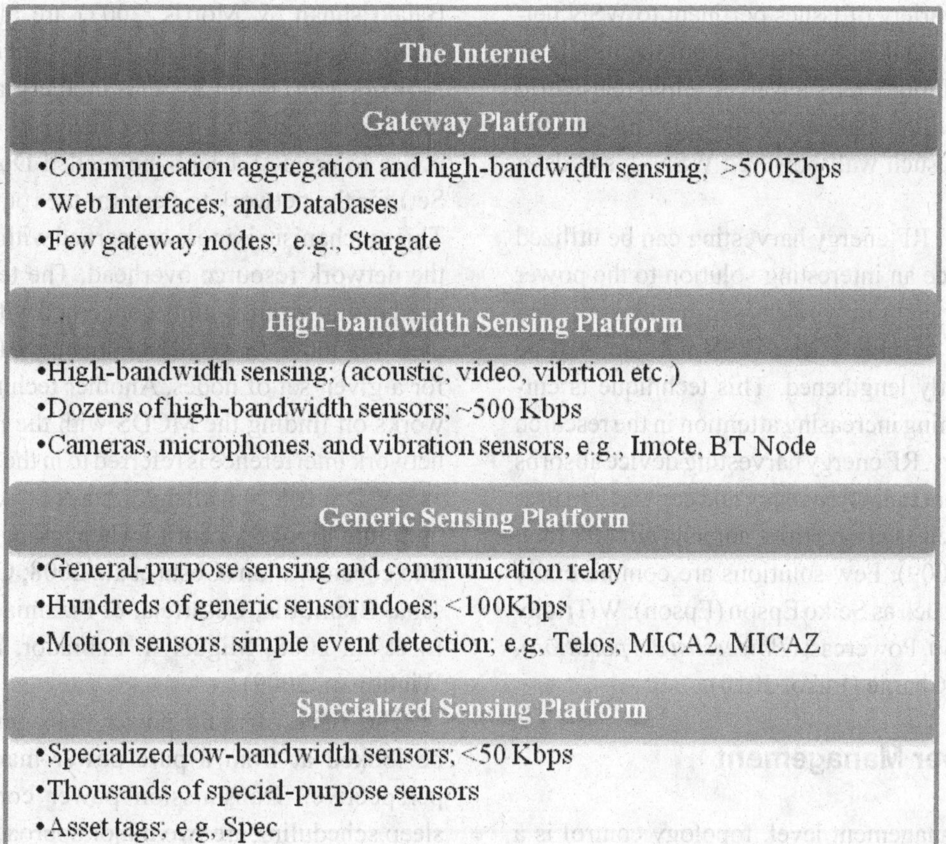

The Internet

Gateway Platform

- Communication aggregation and high-bandwidth sensing; >500Kbps
- Web Interfaces; and Databases
- Few gateway nodes; e.g., Stargate

High-bandwidth Sensing Platform

- High-bandwidth sensing; (acoustic, video, vibrtion etc.)
- Dozens of high-bandwidth sensors; ~500 Kbps
- Cameras, microphones, and vibration sensors, e.g., Imote, BT Node

Generic Sensing Platform

- General-purpose sensing and communication relay
- Hundreds of generic sensor ndoes; <100Kbps
- Motion sensors, simple event detection; e.g, Telos, MICA2, MICAZ

Specialized Sensing Platform

- Specialized low-bandwidth sensors; <50 Kbps
- Thousands of special-purpose sensors
- Asset tags; e.g, Spec

is an open-source OS with an event-driven execution model. TinyOS's component library includes network protocols, distributed services, sensor drivers; and applications. These tools can be used as-is or they can be flexibly tailored to the task at hand (Sohraby, Minoli, Znati, & John, 2007; UC Berkely, 2007). This explains the popularity of TinyOS. Besides TinyOS, many other Operating Systems for WSN are available. Such Operating Systems include: MagnetOS, SOS, EYES, Bertha, Contiki, MantisOS, CORMOS and others. A designer of a WSN-Testbed needs to carefully select the suitable operating system based on few factors such as design requirements, design objective, and hardware and software specifications. The authors in (Reddy, Kumar, Janakiram, & Kumar, 2009) introduce a recent survey on Operating Systems for WSN. The survey introduces an interesting comparison methodology aids the designer in selecting the most suitable operating system for the WSN-Testbed and application of interest.

6.8 Power Source

As it is shown throughout the surveyed WSN-Testbeds, nodes can be either wall-powered, battery-powered or utilize some energy scavenging module such as solar, thermal, vibration, etc. In many real-world scenarios, wall-power may not be possible. However, some WSN testbeds including mimicked indoor WSN-Testbeds and outdoor WSN-Testbeds may choose to utilize the wall power to power-up the testbed's nodes. While the utilization of wall-power for experimentation purposes in such scenarios enables researchers to focus on variety of issues pertinent to WSN network without being worried about the hurdle of batteries limitations, such an essential real-world problem will in fact be overlooked. This puts the fidelity of such wall-powered WSN-Testbeds at stake.

Finally, RF energy harvesting can be utilized to introduce an interesting solution to the power supply problem in wireless sensor network, and hence the lifetime of the WSN network can be significantly lengthened. This technique is currently gaining increasing attention in the research community. RF energy harvesting device absorbs energy from radio frequency and convert it to useful electricity via a rectifier antenna circuit (Tang & Guy, 2009). Few solutions are commercially available such as Seiko Epson (Epson), WiTricity (WiTricity), Powercast (*Powercast corporation*), and Wild Charge (Iniko, 2010).

6.9 Power Management

At the management level, topology control is a technique used to reduce power consumption as well as channel contention in WSN-Testbed by adjusting the transmission power of each node in a wireless sensor network. The ultimate goal is efficient resource utilization. The authors in (Li, Mao, & Liang, 2008) classify the topology control mechanisms into three distinctive classes based on the produced network topology. In *Flat Network* mechanism, nodes are treated as peers in forwarding their data. Such mechanisms include Enclosure Graph (EG), Voronoi Graph, Delaunay Graph, and Steiner Tree. In *Hierarchal Network with Clustering*, topology control mechanisms assume that the network nodes are heterogeneous with respect to functionality and/or hardware structure. The topology control protocol dynamically decides and divides the network into cluster heads, gateway nodes and cluster nodes. LEACH (Heinzelman, Chandrakasan, & Balakrishnan, 2000), HEED (Younis & Fahmy, 2004), GAF (Xu, Heidemann, & Estrin, 2001), and SPAN (B. Chen, Jamieson, Balakrishnan, & Morris, 2002) are algorithms that utilize this mechanism. Finally, *Hierarchical Network with Dominant Set* is another class where the topology control protocol chooses subset of nodes to create a backbone (called Dominating Set) for the network to perform the routing task. This mechanism is more concerned with reducing the network resource overhead. The techniques used in this regard mainly concerned with finding the Minimum Connected Dominating Set (MCDS) for a given set of nodes. Another technique that works on finding the MCDS with the minimum network interference is referred to in the literature as, I-CDS. It is a challenging task to achieve an optimum topology control when taking mobility into consideration (Costa et al., 2008; Costa, Cesana, Brambilla, Casartela, & Pizziniaco, 2008; Li et al., 2008; Miguel A. Labrador, Pedro M. Wightman, 2009).

The issue of radio power management can be looked at from a pure power management perspective. Transmission power control and sleep scheduling are two major approaches typically used to reduce power consumption in the transmission state and the idle state respectively. Additionally, the authors in (Xing et al., 2009) discuss a Unified Radio Power Management framework for the design and implementation of holistic radio power management solution in wireless sensor networks. The main idea of the framework is that it integrates transmission power control and sleep scheduling into a unified framework that is dynamically coordinated according to the network load. The framework is comprised of two key components: Minimum Power Configuration (MPC) and Unified Power Management Architecture (UPMA). The major advantage of this framework is that it enables cross-layer coordination and joint optimization of different (and/or independent) power management strategies.

6.10 Traffic Management

Efficient management of data traffic in WSN-Testbed is an important issue discussed in (Furrer, Schott, Truong, & Weiss, 2006a). Basically, two solutions are proposed. First, rather than overwhelming the enterprise network with raw sensor data, the traffic in the network can be highly reduced through placement of an edge server (computing device) between the sensor nodes and the enterprise network. The edge server will aggregate, abstract, and filter the sensor data before it is injected in the network's backbone. Second, an asynchronous publish/subscribe-based messaging system can be deployed in order to reduce the degree of coupling between the sensor network components, edge server and application server (Furrer, Schott, Truong, & Weiss, 2006a; Rooney, Bauer, & Scotton, 2005). For WSN networks deployed in remote locations, remote system management is required for configuration, control and support for software updates (Furrer, Schott, Truong, & Weiss, 2006a).

6.11 Real-World Reflections

Interesting observations pertaining to practical challenges encountered while experimenting with a WSN-Testbed are pinpointed in (He et al., 2006; VigilNet). These observations were to go unseen if experimenting with a WSN simulation. Once again, this attests to the importance of experimentation with WSN-Testbed. These challenges include; Application-specific reliability, false-alarms, race conditions, asymmetry, software calibration, TinyOS and drift in timers and sophisticated network debugging (He et al., 2006). We summarize these challenges along with their proposed solution in Table 12; then we further elaborate on them below.

Application-specific Reliability addresses the issue of packet loss in relation to the reliability of the application at hand. For instance, in the experiment provided in (He et al., 2006), a packet loss of up to %20 was encountered for MICA2 platform. To counterattack this drawback, the authors propose deploying a mechanism for multiple and selective message retransmissions based on application-specific knowledge. Accordingly, implementation of application-specific reliability guarantees in the application layer is proposed. *False Alarms* consume and waste the system's energy. They can be classified into transient and consistent false alarms. Exponential-weighted moving average (EWMA) can be used to handle transient false alarms. In-network aggregation techniques and faulty node detection algorithms surveyed in (Fasolo, Rossi, Widmer, & Zorzi, 2007) can be used to handle persistent false alarms. *The Race Condition* refers to the contention occurs when different motes try to transmit simultaneously, and when different software components on the same mote initiate transmissions simultaneously through split-phase operations. TinyOS can also lead to such race conditions due to the limited support it offers. Application-level synchronization, such as packet scheduling, can be properly deployed to coordinate the operations and avoid race conditions. *Asymmetry* is another real-world issue cannot be seen by simulators. It refers to the asymmetric nature pertinent to communication channel in low power devices such as mote due to differences in hardware, signal attenuation and battery capacity (Zhou, He, Krishnamurthy, & Stankovic, 2004). Certain symmetry detection techniques can be deployed to overcome this problem.

Unlike WSN simulators, it is common for real-world sensor devices to generate quite different readings under identical conditions. This discrepancy is due to the heterogeneity of the devices in the network. A continuous software calibration for the sensors can be used to counter-attack this problem. Another practical challenge is the drift in software timers in TinyOS especially when motes transit into sleep mode. Design of proper strategies to control the sleep-awake cycle is essential in

Table 12. Summary of some real-world WSN-testbed challenges and proposed solutions (He et al., 2006)

Real-World Challenges	Proposed Solution
Application-Specific Reliability	Implement application-specific reliability guarantees in the Application Layer
False Alarms (Transient & Persistent)	Use EWMA for Transient Alarms; and In-network aggregation techniques and faulty node detection for Persistent Alarms
Race Condition	Use application-level synchronization, such as packet scheduling
Asymmetry	Deploy symmetry detection techniques
Sensor Reading Discrepancy	Implement a continuous software calibration
Drift in SW Timers	Design proper strategies to control sleep-awake cycles
Lack of Sophisticated Debugging Tools	Handle the debugging issue appropriately

handling this challenge. The lack of proper tools for sophisticated network debugging is another practical challenge that needs to be appropriately handled. (He et al., 2006).

6.12 Confidentiality Infringements

Finally, we end our discussion with critical observations pertaining to network security in WSN. Confidentiality in wireless sensor networks can be categorized into two categories; content confidentiality (privacy) and transactional confidentiality. Content confidentiality deals with protecting the content of the data being transmitted through deploying proper authentication and encryption techniques. Transactional confidentiality is pertinent to transactional data such as message size, sensor node frequency, message routing, and message rate. The authors in (Pai et al., 2008) argue that such transactional data can be used by an intruder to reveal confidential information. In fact, being many WSN-Testbed developers utilize COTS technology makes such transactional data available to the public. Hence, this presents a serious security/confidentiality infringement problem that most WSN-Testbed developers considered in this survey and elsewhere tend to overlook. Specifically, this issue raises serious concerns from the perspective of security and/or confidentiality to the WSN-Testbeds designed for handling sensitive applications such as surveillance (e.g. VigilNet) and health monitoring (e.g., AlarmNet). Such issues have to be properly considered and handled when designing a WSN-Testbed for similar kind of applications.

7. CONCLUSION

In this chapter, we surveyed some of the most interesting WSN-Testbeds available in the literature today. Our choice of the surveyed WSN-Testbeds was carefully made to enable the reader to gain broad insight into different issues pertinent to various spectra of WSN-Testbeds. We adopted a methodology for classification, and we classified the surveyed WSN-Testbeds accordingly. We also touched on a generic architecture for each class. We then pinpointed some *design challenges and evaluation criteria/benchmarking scheme* to apply in the design of a new WSN-Testbed or the evaluation of an existing WSN-Testbed. Afterwards, we presented a comparison between 30 different WSN-Testbeds. In the comparison, we highlighted the architectures as well as some features distinct to each WSN-Testbed. We then applied the introduced *benchmarking scheme* to each WSN-Testbed and comparatively evaluated them accordingly. Then, we highlighted some of the drawbacks/shortcomings of the WSN-Testbeds considered in this survey. In the discussion, we shed light on variety of issues including: vitality of

WSN-Testbeds, applications of WSNs with respect to surveyed WSN-Testbeds, trade-offs relevant to the design of a WSN-Testbed, a perspective on surveyed WSN-Testbeds based on locality, time-based view, cluster WSN-Testbed network model based on nodal view, open-source, WSN's OS, topology control, power management, some real-world deployment challenges along with proposed solutions, and security/confidentiality infringement issues. Being the real-world enablers of WSN technology, this survey showed the important role a WSN-Testbed plays in rapidly driving the industrial adoption and normalization of wireless sensor network technology. However, as it is depicted, this role is associated with challenges and trade-offs that if properly studied and understood time and money will be saved and consequently a focused design will be realized.

REFERENCES

Acroname, I. (n.d.). *Garcia custom robot*. Retrieved October 25, 2009, from http://www.acroname.com/garcia/garcia.html

AEOLUS. (n.d.a). *SP6 — AEOLUS*. Retrieved September 27, 2009, from http://aeolus.ceid.upatras.gr/sub-projects/sp6

AEOLUS. (n.d.b). *SP6/WSNtestbed - AEOLUS IST project*. Retrieved September 27, 2009, from http://ru1.cti.gr/aeolus/SP6/WSNtestbed

AEOLUS. (n.d.c). *Welcome to AEOLUS — AEOLUS*. Retrieved September 27, 2009, from http://aeolus.ceid.upatras.gr/

Alazzawi, L. K., Elkateeb, A. M., & Ramesh, A. (2008). Scalability analysis for wireless sensor networks routing protocols. *22nd International Conference on Advanced Information Networking and Applications - Workshops, AINAW 2008*, (pp. 139-144).

Arora, A. (n.d.). *Exscal research group*. Retrieved October 25, 2009, from http://ceti.cse.ohio-state.edu/exscal/

Arora, A., Ertin, E., Ramnath, R., Leal, W., & Nesterenko, M. (2006). Kansei: A high-fidelity sensing test bed. *IEEE Internet Computing, 10*(2), 35–47. doi:10.1109/MIC.2006.37

Awad, A., Sommer, C., German, R., & Dressler, F. (2008). Virtual cord protocol (VCP): A flexible DHT-like routing service for sensor networks. *2008 5th IEEE International Conference on Mobile Ad-Hoc and Sensor Systems, MASS 2008, September 29, 2008 - October 2,* (pp. 133-142).

Bapat, S., Leal, W., Kwon, T., Wei, P., & Arora, A. (2009). Chowkidar: Reliable and scalable health monitoring for wireless sensor network testbeds. *ACM Transactions on Autonomous and Adaptive Systems, 4*(1).

Berkely, U. C. (2007). *TinyOS: Tiki RSS feed for directory sites*. Retrieved from http://www.bwsn.net/tiki-directories_rss.php?ver=2

Betz, V. (n.d.). *FPGA architecture for the challenge*. Retrieved October 25, 2009, from http://www.eecg.toronto.edu/~vaughn/challenge/fpga_arch.html

Bibyk, S. (n.d.). *ECE 582 class homepage: XScale mote*. Retrieved October 25, 2009, from http://www.ece.osu.edu/~bibyk/ee582/ee582.htm

Blywis, B., Juraschek, F., Günes, M., & Schiller, J. (n.d.). *Design concepts of a persistent WSN testbed*. AG Computer Systems & Telematics - Freie Universität Berlin. Retrieved September 7, 2009, from http://cst.mi.fu-berlin.de/publications/

Bondi, A. B. (2000). Characteristics of scalability and their impact on performance. *Proceedings Second International Workshop on Software and Performance WOSP2000, September 17, 2000 - September 20,* (pp. 195-203).

Breslau, L., Estrin, D., Fall, K., Floyd, S., Heidemann, J., & Helmy, A. (2000). Advances in network simulation. *Computer, 33*(5), 59–67. doi:10.1109/2.841785

Bulusu, N., Bychkovskiy, V., Estrin, D., & Heidemann, J. (n.d.). *Scalable, ad hoc deployable RF-based localization.* Retrieved September 8, 2009, from http://lecs.cs.ucla.edu/~bulusu/papers/Bulusu02a.html

Buschmann, C., Pfisterer, D., Fischer, S., Fekete, S. P., & Kr\oller, A. (2005). SpyGlass: A wireless sensor network visualizer. *SIGBED Review, 2*(1), 1-6.

Buschmann, C., Pfisterer, D., Fischer, S., Fekete, S. P., & Kroller, A. (2004). SpyGlass: Taking a closer look at sensor networks. *Proceedings of the Second International Conference on Embedded Networked Sensor Systems, November 3, 2004 - November 5,* (pp. 301-302).

Chen, B., Jamieson, K., Balakrishnan, H., & Morris, R. (2002). Span: An energy-efficient coordination algorithm for topology maintenance in ad hoc wireless networks. *Wireless Networks, 8*(5), 481–494. doi:10.1023/A:1016542229220

Chen, S., Huang, Y., & Zhang, C. (2008). Toward a real and remote wireless sensor network testbed. *WASA '08: Proceedings of the Third International Conference on Wireless Algorithms, Systems, and Applications,* Dallas, Texas, (pp. 385-396).

COMMON-Sense Net. (n.d.). *COMMON-sense net home.* Retrieved August 29, from http://commonsense.epfl.ch

Costa, P., Cesana, M., Brambilla, S., Casartela, L., & Pizziniaco, L. (2008). *A cooperative approach for topology control in wireless sensor networks: Experimental and simulation analysis*. 9th IEEE International Symposium on Wireless, Mobile and Multimedia Networks, WoWMoM 2008, June 23, 2008 - June 26. IEEE Computer Society.

Crossbow. (n.d.a). *Crossbow technology: Development kits.* Retrieved September 3, 2009, from http://www.xbow.com/Products/productdetails.aspx?sid=160

Crossbow. (n.d.b). *Crossbow wireless sensor networks interface - Crossbow technology.* Retrieved October 10, 2009, from http://www.xbow.com/Technology/UserInterface.aspx

Crossbow. (n.d.c). *Technical support - Software downloads - Crossbow technology.* Retrieved October 10, 20069, from http://www.xbow.com/support/wSoftwareDownloads.aspx

Dalton, A. R., & Hallstrom, J. O. (2009a). Carolina WSN testbed: An interactive, source-centric, open testbed for developing and profiling wireless sensor systems. *International Journal of Distributed Sensor Networks, 5*(2), 105–138. doi:10.1080/15501320701863403

Dalton, A. R., & Hallstrom, J. O. (2009b). An interactive, source-centric, open testbed for developing and profiling wireless sensor systems. *International Journal of Distributed Sensor Networks, 5*(2), 105–138. doi:10.1080/15501320701863403

Dataman Papers. (n.d.). *TopDisc NW management index of dataman papers.* Retrieved September 7, 2009, from http://www.cs.rutgers.edu/dataman/papers/

De, P., Raniwala, A., Krishnan, R., Tatavarthi, K., Modi, J., Syed, N. A., et al. (2006). MiNT-m: An autonomous mobile wireless experimentation platform. *MobiSys '06: Proceedings of the 4th International Conference on Mobile Systems, Applications and Services,* Uppsala, Sweden (pp. 124-137).

De, P., Raniwala, A., Sharma, S., & Chiueh, T. (2005). MiNT: A miniaturized network testbed for mobile wireless research. *IEEE INFOCOM 2005, March 13, 2005 - March 17,* (pp. 2731-2742).

Dimitriou, T., Kolokouris, J., & Zarokostas, N. (2007). SenseNeT: A wireless sensor network testbed. *MSWiM'07: 10th ACM Symposium on Modeling, Analysis, and Simulation of Wireless and Mobile Systems, October 22, 2007 - October 26,* (pp. 143-150).

Drupal. (2007). *PlanetLab - An open platform for developing, deploying, and accessing planetary-scale services.* Retrieved October 07, 2009, from http://www.planet-lab.org

Dust Networks. (n.d.). *Products: Dust networks.* Retrieved September 29, 2009, from http://www.dustnetworks.com/products

Dutta, P. (n.d.). *Epic: An open mote platform.* Retrieved September 12, 2009, from http://www.cs.berkeley.edu/~prabal/projects/epic/

Dutta, P., & Culler, D. (2008b). Epic: An open mote platform for application-driven design. *International Conference on Information Processing in Sensor Networks, IPSN '08,* (pp. 547-548).

Dutta, P., Hui, J., Jeong, J., Kim, S., Sharp, C., Taneja, J., et al. (2006). Trio: Enabling sustainable and scalable outdoor wireless sensor network deployments. *Fifth International Conference on Information Processing in Sensor Networks, IPSN '06, April 19, 2006 - April 21,* (pp. 407-415).

EAAST. (2009). *EAAST workshop volume 17: Kommunikation in verteilten sytemen 2009. Retrieved* September 6, from http://eceasst.cs.tu-berlin.de/index.php/eceasst/issue/view/24

Eliasson, J. (n.d.). *The Mulle - Bluetooth & Zigbee wireless networked sensor node.* Retrieved October 24, 2009, from http://www.csee.ltu.se/~jench/mulle.html

Emulab. (n.d.). *Emulab.net - Emulab - Network emulation testbed home.* Retrieved September 27, 2009, from http://www.emulab.net/index.php3?stayhome=1

Epson. (2008, September 28). *Epson develops non-contact power transmission module capable of transmitting 2.5 W with a 0.8mm coil.* Newsroom/Seiko Epson Corp. Retrieved April 26, 2010, from http://global.epson.com/newsroom/2008/news_20080922.htm

Ertin, E., Arora, A., Ramnath, R., Nesterenko, M., Naikt, V., Bapat, S., et al. (2006). Kansei: A testbed for sensing at scale. *Fifth International Conference on Information Processing in Sensor Networks, IPSN '06, April 19, 2006 - April 21,* (pp. 399-406).

Esensors, Inc. (n.d.). *Esensors, Inc. - Networked sensors.* Retrieved September, 30, 2009, from http://www.eesensors.com/

Evans, J. J. (2007). Undergraduate research experiences with wireless sensor networks. *Frontiers in Education Conference - Global Engineering: Knowledge without Borders, Opportunities without Passports, 2007. FIE '07. 37th Annual,* (pp. S4B-7-S4B-12).

Fall, K. (1999). Network emulation in the VINT/NS simulator. *IEEE Symposium on Computers and Communications Proceedings,* (pp. 244-250).

Fasolo, E., Rossi, M., Widmer, J., & Zorzi, M. (2007). In-network aggregation techniques for wireless sensor networks: A survey. *Wireless Communications, 14*(2), 70–87. doi:10.1109/MWC.2007.358967

Furrer, S., Schott, W., Truong, H. L., & Weiss, B. (2006a). The IBM wireless sensor networking testbed. *2nd International Conference on Testbeds and Research Infrastructures for the Development of Networks and Communities, TRIDENTCOM 2006, March 1, 2006 - March 3,* (pp. 42-46).

Furrer, S., Schott, W., Truong, H. L., & Weiss, B. (2006b). The IBM wireless sensor networking testbed. *2nd International Conference on Testbeds and Research Infrastructures for the Development of Networks and Communities, TRIDENTCOM 2006, March 1, 2006 - March 3,* (pp. 42-46).

GENI. (n.d.). *GENI: Geni - Trac.* Retrieved September 26, 2009, from http://groups.geni.net/geni

GenSym. (n.d.). *Gensym: Business rule management.* Retrieved September 30, 2009, from http://www.gensym.com/

Girod, L., Ramanathan, N., Elson, J., Stathopoulos, T., Lukac, M., & Estrin, D. (2007). Emstar: A software environment for developing and deploying heterogeneous sensor-actuator networks. *ACM Transactions in Sensor Networks*, *3*(3), 13. doi:10.1145/1267060.1267061

Google. (n.d.). *Google Maps.* Retrieved October 25, 2009, from http://maps.google.com/

GSN. (n.d.). *GSN.* Retrieved September 9, 2009, from http://sourceforge.net/apps/trac/gsn/

Gurkan, D., Yuan, X., Benhaddou, D., Figueroa, F., & Morris, J. (2007). *Sensor networking testbed with IEEE 1451 compatibility and network performance monitoring.* Paper presented at the 2007 IEEE Sensors Applications Symposium, SAS, Febrary 6, 2007 - Febrary 8, IEEE Instrumentation and Measurement Society.

Hadim, S., & Mohamed, N. (2006). Middleware: Middleware challenges and approaches for wireless sensor networks. *IEEE Distributed Systems Online*, *7*(3), 1–23. doi:10.1109/MDSO.2006.19

Haensel. (n.d.). *An FDL'ed textbook on sensor networks.* Retrieved September 2, 2009, from http://www.informatik.uni-mannheim.de/~haensel/sn_book/

Handziski, V., Kopke, A., Willig, A., & Wolisz, A. (2006). TWIST: A scalable and reconfigurable testbed for wireless indoor experiments with sensor networks. *REALMAN 2006 - 2nd International Workshop on Multi-Hop Ad Hoc Networks: From Theory to Reality, May 26, 2006 - May 26,* (pp. 63-70).

HART Communication Foundation. (n.d.). *HART communication protocol - Wireless HART technology.* Retrieved October 24, 2009, from http://www.hartcomm.org/protocol/wihart/wireless_technology.html

He, T., Krishnamurthy, S., Luo, L., Yan, T., Gu, L., & Stoleru, R. (2006). VigilNet: An integrated sensor network system for energy-efficient surveillance. *ACM Transactions in Sensor Networks*, *2*(1), 1–38. doi:10.1145/1138127.1138128

Heidemann, J., Bulusu, N., Elson, J., Intanagonwiwat, C., Lan, K., & Xu, Y. … Govindan, R. (n.d.) *Effects of detail in wireless network simulation.* Retrieved September 14, 2009, from http://www.isi.edu/~johnh/PAPERS/Heidemann01a.html

Heidemann, J., Silva, F., & Estrin, D. (2003). Matching data dissemination algorithms to application requirements. *SenSys '03: Proceedings of the 1st International Conference on Embedded Networked Sensor Systems,* Los Angeles, California, USA, (pp. 218-229).

Heinzelman, W. R., Chandrakasan, A., & Balakrishnan, H. (2000). Energy-efficient communication protocol for wireless microsensor networks. *Proceedings of the 33rd Hawaii International Conference on System Sciences,* volume 8.

Hill, J., Horton, M., Kling, R., & Krishnamurthy, L. (2004). The platforms enabling wireless sensor networks. *Communications of the ACM*, *47*(6), 41–46. doi:10.1145/990680.990705

Hong, X., Gerla, M., Bagrodia, R., Kwon, T. J., Estabrook, P., & Pei, G. (2001). *The Mars sensor network: Efficient, energy aware communications*

Hongwei, H., Hongke, Z., Yanchao, N., Shuai, G., Zhaohua, L., & Sidong, Z. (2007). *MSRLab6: An IPv6 wireless sensor networks testbed.* Paper presented at the 8th International Conference on Signal Processing, ICSP 2006, November 16, 2006 - November 20.

Huang, Y. L., Tygar, J. D., Lin, H. Y., Yeh, L. Y., Tsai, H. Y., Sklower, K., et al. (2008). SWOON: A testbed for secure wireless overlay networks. *CSET'08: Proceedings of the Conference on Cyber Security Experimentation and Test,* San Jose, CA, (pp. 1-6).

IBM. (n.d.). *IBM research: Wireless sensor networking.* Retrieved August 29, 2009, from http://domino.watson.ibm.com/comm/research.nsf/pages/r.communications.innovation2.html

IDEALS. (n.d.). *Debugging wireless sensor networks using mobile actors: IDEALS @ illinois.* Retrieved September 10, 2009, from http://www.ideals.uiuc.edu/handle/2142/4607

IEEE. (2007). *IEEE standard for a smart transducer interface for sensors and actuators - Wireless communication protocols and transducer electronic data sheet (TEDS) formats*

IEEE. (n.d.). *Welcome to IEEE xplore 2.0: IEEE standard for a smart transducer interface for sensors and actuators - Common functions, communication protocols, and transducer electronic data sheet (TEDS) formats.*

Iniko. (2010). *PureEnergy solutions takes battery charging to a whole new level.* Shop Pure Energy. Retrieved September 15, 2009, from http://www.shoppureenergy.com

Intanagonwiwat, C., Govindan, R., Estrin, D., Heidemann, J., & Silva, F. (2003). Directed diffusion for wireless sensor networking. *IEEE/ACM Transactions on Networking, 11*(1), 2–16. doi:10.1109/TNET.2002.808417

Internet Engineering Task Force. (n.d.). *IPv6 over low power WPAN (6LoWPAN).* Retrieved October 24, 2009, from http://www.ietf.org/dyn/wg/charter/6lowpan-charter.html

ISA. (n.d.). *ISA 100, wireless systems for automation.* Retrieved October 24, 2009, from http://www.isa.org/MSTemplate.cfm?MicrositeID=1134&CommitteeID=6891

Johnson, D., Stack, T., Fish, R., Flickinger, D. M., Stoller, L., Ricci, R., et al. (2006). *Mobile Emulab: A robotic wireless and sensor network testbed.* Paper presented at the INFOCOM 2006: 25th IEEE International Conference on Computer Communications, April 23, 2006 - April 29,

Keithley Instruments Inc. (n.d.). *2701 - Integrated DMM/Switch.* Keithley Instruments Inc. Retrieved October 25, 2009, from http://www.keithley.com/products/switch/dmmswitch/?mn=2701

KonTest. (n.d.). *KonTest: WSN testbed.* Retrieved September 19, 2009, from http://www.few.vu.nl/~agaba/publications/

Kosucu, B., Irgan, K., Kucuk, G., & Baydere, S. (2009). FireSenseTB: A wireless sensor networks testbed for forest fire detection. *IWCMC '09: Proceedings of the 2009 International Conference on Wireless Communications and Mobile Computing,* Leipzig, Germany, (pp. 1173-1177).

Labrador, M. A., & Wightman, P. M. (2009). *Topology control in WSN.* Netherlands: Springer.

LabVIEW drivers for wireless sensor networks - developer zone - national instruments. (n.d.). Retrieved December 17, 2009, from http://zone.ni.com/devzone/cda/tut/p/id/5435

Le, T., Hu, W., Corke, P., & Jha, S. (2009). ERTP: Energy-efficient and reliable transport protocol for data streaming in wireless sensor networks. *Computer Communications, 32*(7-10), 1154–1171. doi:10.1016/j.comcom.2008.12.045

Lee, W. L. (n.d.). *WSN network management survey.* Winnie Louis Lee's Homepage. Retrieved September 7, 2009, from http://www.csse.uwa.edu.au/~winnie/

Li, X., Mao, Y., & Liang, Y. (2008). A survey on topology control in wireless sensor networks. *2008 10th International Conference on Control, Automation, Robotics and Vision, ICARCV 2008, December 17, 2008 - December 20,* (pp. 251-255).

Liang, Q. (2007). Radar sensor networks: Algorithms for waveform design and diversity with application to ATR with delay-doppler uncertainty. *EURASIP Journal on Wireless Communications and Networking*, (1): 18–18.

LibeLium. (n.d.). *Main page - SquidBee.* Retrieved September 12, 2009, from http://www.libelium.com/squidbee/index.php?title=Main_Page

LibeLium Waspmote. (n.d.). *Wireless sensor networks mote.* Retrieved September 17, 2009, from http://www.libelium.com/products/waspmote

Lodder, M., Halkes, G. H., & Langendoen, K. G. (n.d.). *A global-state perspective on sensor network debugging.* Retrieved September 9, 2009, from http://www.st.ewi.tudelft.nl/~koen/papers/globalstate.pdf

Luo, L., Hei, T., Zhou, G., Gu, L., Abdelzaher, T. F., & Stankovic, J. A. (2006). *Achieving repeatability of asynchronous events in wireless sensor networks with EnviroLog*. Paper presented at the INFOCOM 2006: 25th IEEE International Conference on Computer Communications, April 23, 2006 - April 29, M2M Magazine. (n.d.). *The business value of sensors.* Retrieved October 6, 2009, from http://www.m2mmag.com/issue_archives/story.aspx?ID=7890

MeshNetics. (n.d.). *Development tools.* Retrieved December 17, 2009, from http://www.meshnetics.com/dev-tools/

Mgillis. (2009). *PlanetLab team brings collaboration and policy expertise to GENI.* GENI. Retrieved September 17, 2009 from http://www.geni.net

MicroStrain. (n.d.). *2.4 GHz G-LINK wireless accelerometer node.* Retrieved December 17, 2009, from http://www.microstrain.com/g-link.aspx

Missoula Fire Sciences Laboratory. (n.d.). *FireModels.org - FARSITE.* Retrieved October 25, 2009, from http://www.firemodels.org/content/view/112/143

Miyashita, M., Nesterenko, M., Shah, R. D., & Vora, A. (2005). TOSGUI visualizing wireless sensor networks: An experience report. *2005 International Conference on Wireless Networks, ICWN'05, June 27, 2005 - June 30,* (pp. 412-419).

MonSense. (n.d.). *MonSense - FEUP WSN group.* Retrieved September 12, 2009, from http://whale.fe.up.pt/wsnwiki/index.php/MonSense

MSN. (n.d.). *MSN virtual earth.* Retrieved October 25, 2009, from http://local.live.com.qe2a-proxy.mun.ca

Nakamura, E. F., Loureiro, A. A. F., & Frery, A. C. (2007). Information fusion for wireless sensor networks: Methods, models, and classifications. *ACM Computing Surveys, 39*(3).

Nakamura, E. F., Ramos, H. S., Villas, L. A., de Oliveira, H. A. B. F., de Aquino, A. L. L., & Loureiro, A. A. F. (2009). A reactive role assignment for data routing in event-based wireless sensor networks. *Computer Networks, 53*(12), 1980–1996. doi:10.1016/j.comnet.2009.03.009

Nath, S., Liu, J., Miller, J., Zhao, F., & Santanche, A. (2006). SensorMap: A web site for sensors world-wide. *SenSys '06: 4th International Conference on Embedded Networked Sensor Systems,* (pp. 373-374).

National Science Foundation. (n.d.). *NSF network research testbed workshop report: Executive summary.* Retrieved September 2, 2009, from http://www-net.cs.umass.edu/testbed_workshop/exec_summary_html.htm

NESTbed. (n.d.). *NESTbed- Project hosting on Google Code.* Retrieved October 25, 2009, from http://code.google.com/p/nestbed/

NI. (n.d.). *Wireless sensor network (WSN) starter kit - National Instruments.* Retrieved December 17, 2009, from http://sine.ni.com/nips/cds/view/p/lang/en/nid/206916

NSLU2. (n.d.). *NSLU2-Linux - Main home page browse.* Retrieved October 25, 2009, from http://www.nslu2-linux.org/wiki/Main/HomePage

OGC. (n.d.). *Welcome to the OGC website OGC®.* Retrieved September 17, 2009, from http://www.opengeospatial.org

Pai, S., Bermudez, S., Wicker, S. B., Meingast, M., Roosta, T., & Sastry, S. (2008). Transactional confidentiality in sensor networks. *IEEE Security and Privacy, 6*(4), 28–35. doi:10.1109/MSP.2008.107

Park, S., Savvides, A., & Srivastava, M. B. (2000). SensorSim: A simulation framework for sensor networks. *MSWIM '00: Proceedings of the 3rd ACM International Workshop on Modeling, Analysis and Simulation of Wireless and Mobile Systems,* Boston, Massachusetts, United States, (pp. 104-111).

Pei, J., Ivey, R. A., Lin, H., Landrum, A. R., Sandburg, C. J., & Ferzli, N. A. (2009). An experimental investigation of applying Mica2 motes in pavement condition monitoring. *Journal of Intelligent Material Systems and Structures, 20*(1), 63–85. doi:10.1177/1045389X08088785

Powercast corporation. (n.d.). Retrieved April 26, 2010, from http://www.powercastco.com/

Project, E. U. (n.d.). *WISEBED - Wireless sensor network testbeds - EU project.* Retrieved August 29, 2009, from http://www.wisebed.eu/

Ratnasamy, S., Karp, B., Shenker, S., Estrin, D., Govindan, R., & Yin, L. (2003). Data-centric storage in sensornets with GHT, a geographic hash table. *Mobile Networks and Applications, 8*(4), 427–442. doi:10.1023/A:1024591915518

Reddy, A. M. V., Kumar, A. V. U. P., Janakiram, D., & Kumar, G. A. (2009). Wireless sensor network operating systems: A survey. *International Journal of Sensor Networks, 5*(4), 236–255. doi:10.1504/IJSNET.2009.027631

Romer, K., & Mattern, F. (2004). The design space of wireless sensor networks. *IEEE Wireless Communications, 11*(6), 54–61. doi:10.1109/MWC.2004.1368897

Rooney, S., Bauer, D., & Scotton, P. (2005). *Edge server software architecture for sensor applications.*

Rothery, S., Hu, W., & Corke, P. (2008). An empirical study of data collection protocols for wireless sensor networks. *3rd Workshop on Real-World Wireless Sensor Networks, REALWSN 2008, April 1, 2008 - April 1,* (pp. 16-20). Retrieved from http://dx.doi.org/10.1145/1435473.1435479

SANE. (2009). *Global WSN, SANE, Fraunhofer FOKUS.* Retrieved from http://www.fokus.fraunhofer.de/en/sane/projekte/laufende_projekte/global_wsn/index.html

Selavo, L. (n.d.). *Open source wireless sensor networks.* Retrieved September 12, 2009, from http://www.openwsn.com/

SensorMap. (n.d.). *SensorMap home.* Retrieved from http://atom.research.microsoft.com/sensewebv3/sensormap/

SensWiz. (n.d.). *SensWiz: Dreamajax products*. Retrieved from http://www.senswiz.com/index.php?page=shop.browse&category_id=9&option=com_virtuemart&Itemid=54&vmcchk=1&Itemid=54

Slipp, J., Changning Ma, Polu, N., Nicholson, J., Murillo, M., & Hussain, S. (2008a). *WINTeR: Architecture and applications of a wireless industrial sensor network testbed for radio-harsh environments*.

Slipp, J., Ma, C., Polu, N., Nicholson, J., Murillo, M., & Hussain, S. (2008b). WINTeR: Architecture and applications of a wireless industrial sensor network testbed for radio-harsh environments. *6th Annual Communication Networks and Services Research Conference, CNSR 2008,* May 5, 2008 - May 8, (pp. 422-431). Retrieved from http://dx.doi.org/10.1109/CNSR.2008.42

Sohraby, K., Minoli, D., Znati, T., & John. (2007). *Wireless sensor networks: Technology, protocols, and applications.* Retrieved from http://www.wiley.com/WileyCDA/WileyTitle/productCd-0471743003.html

Takai, M., Martin, J., & Bagrodia, R. (2001). Effects of wireless physical layer modeling in mobile ad hoc networks. *Proceedings of the 2001 ACM International Symposium on Mobile Ad Hoc Networking and Computing: MobiHoc 2001,* October 4, 2001 - October 5, (pp. 87-94). Retrieved from http://dx.doi.org/10.1145/501416.501429

Tang, L., & Guy, C. (2009). Radio frequency energy harvesting in wireless sensor networks. *IWCMC '09: Proceedings of the 2009 International Conference on Wireless Communications and Mobile Computing,* Leipzig, Germany (pp. 644-648). Retrieved from http://doi.acm.org/10.1145/1582379.1582519

The Data Fusion Server. (n.d.). *Workshop on the theory on belief functions.* Retrieved from http://www.data-fusion.org/article.php?sid=70

Tutornet. (n.d.). *Tutornet.* Retrieved from http://enl.usc.edu/projects/tutornet/

VigilNet. (n.d.). *UVA VigilNet: Home.* Retrieved from http://www.cs.virginia.edu/wsn/vigilnet/

Wagenknecht, G., Anwander, M., Braun, T., Staub, T., Matheka, J., & Morgenthaler, S. (2008). MARWIS: A management architecture for heterogeneous wireless sensor networks. *6th International Conference on Wired/Wireless Internet Communications, WWIC 2008, May 28, 2008 - May 30, 5031 LNCS* (pp. 177-188). Retrieved from http://dx.doi.org/10.1007/978-3-540-68807-5_15

Welcome to Accsense.com. (n.d.). Retrieved from http://www.accsense.com/

Werner-Allen, G., Swieskowski, P., & Welsh, M. (2005). MoteLab: A wireless sensor network testbed. Paper presented at the *4th International Symposium on Information Processing in Sensor Networks, IPSN 2005, April 25, 2005 - April 27,* (pp. 483-488). Retrieved from http://dx.doi.org/10.1109/IPSN.2005.1440979

White, B., Lepreau, J., & Guruprasad, S. (n.d.). *Lowering the barrier to wireless and mobile experimentation.* Retrieved May 9, 2009, from http://www.cs.utah.edu/flux/papers/barrier-hotnets1-base.html

Wireless Accelerometers - Digital accelerometers - Vibration monitoring - Wireless CBM - Wireless sensor. (n.d.). Retrieved from http://www.techkor.com/industrial/wireless.htm

Wireless industrial sensor network testbed for radio-harsh environments (WINTeR). (n.d.). Retrieved from http://winter.cbu.ca/

WiSense. (n.d.). *WiSense.* Retrieved from http://wisense.ca/

Witricity. (n.d.). *Wireless power Transfer| Electricity Transmission |Nikola Tesla.* Retrieved from http://www.witricitynet.com/

Woo, A., Tong, T., & Culler, D. (2003). Taming the underlying challenges of reliable multihop routing in sensor networks. *SenSys '03: Proceedings of the 1st International Conference on Embedded Networked Sensor Systems,* Los Angeles, California, USA (pp. 14-27). Retrieved from http://doi.acm.org.qe2a-proxy.mun.ca/10.1145/958491.958494

Wood, A. D., Stankovic, J. A., Virone, G., Selavo, L., He, Z., & Cao, Q. (2008). AlarmNet context-aware wireless sensor networks for assisted living and residential monitoring. *IEEE Network*, *22*(4), 26–33. Retrieved from http://dx.doi.org.qe2a-proxy.mun.ca/10.1109/MNET.2008.4579768doi:10.1109/MNET.2008.4579768

WU WSN Wiki. (n.d.). *TinyOS 1.x installation on Windows XP - WSN.* Retrieved from http://www.cs.wustl.edu/wsn/index.php?title=TinyOS_1.x_Installation_on_Windows_XP

Xing, G., Sha, M., Hackmann, G., Klues, K., Chipara, O., & Lu, C. (2009). Towards unified radio power management for wireless sensor networks. *Wireless Communications and Mobile Computing, 9*(3), 313-323. Retrieved from http://dx.doi.org/10.1002/wcm.622

Xu, Y., Heidemann, J., & Estrin, D. (2001). Geography-informed energy conservation for ad hoc routing. *MobiCom '01: Proceedings of the 7th Annual International Conference on Mobile Computing and Networking,* Rome, Italy (pp. 70-84). Retrieved from http://doi.acm.org/10.1145/381677.381685

Yeditepe University. (n.d.). *Scalable querying sensor protocol (SQS).* Retrieved from http://cse.yeditepe.edu.tr/tnl/sqs.php?lang=en

Yick, J., Mukherjee, B., & Ghosal, D. (2008). Wireless sensor network survey. *Computer Networks, 52*(12), 2292–2330. Retrieved from http://dx.doi.org/10.1016/j.comnet.2008.04.002doi:10.1016/j.comnet.2008.04.002

Younis, O., & Fahmy, S. (2004). HEED: A hybrid, energy-efficient, distributed clustering approach for ad hoc sensor networks. *IEEE Transactions on Mobile Computing, 3*(4), 366–379. Retrieved from http://doi.ieeecomputersociety.org/10.1109/TMC.2004.41doi:10.1109/TMC.2004.41

Zhou, G., He, T., Krishnamurthy, S., & Stankovic, J. A. (2004). Impact of radio irregularity on wireless sensor networks. *MobiSys 2004 - Second International Conference on Mobile Systems, Applications and Services,* June 6, 2004 - June 9, (pp. 125-138). Retrieved from http://dx.doi.org/10.1145/990064.990081

ZigBee Alliance. (n.d.). *Our mission.* Retrieved from http://www.zigbee.org/About/OurMission/tabid/217/Default.aspx

Zurich, E. T. H. (n.d.). *BTnodes - A distributed environment for prototyping ad hoc networks: Main - overview browse.* Retrieved October 10, 2009, from http://www.btnode.ethz.ch/

Chapter 8
Mitigation of Hot Spots on Wireless Sensor Networks:
Techniques, Approaches and Future Directions

Fernando Gielow
NR2 – Federal University of Paraná, Brazil

Michele Nogueira
NR2 – Federal University of Paraná, Brazil

Aldri Santos
NR2 – Federal University of Paraná, Brazil

ABSTRACT

The use of Wireless Sensor Networks (WSNs) has increased over the past years, supporting applications such as environmental monitoring, security systems, and multimedia streaming. These networks are characterized by a many-to-one traffic pattern. Hence, sensor nodes near to the sink have higher energy consumption, being prone to earlier deaths and failures. Those areas overloaded with high traffic rates are called Hot Spots, and their emergence creates and expands energy holes that compromise network lifetime and data delivery rates, and may result in disconnected areas. This chapter provides an overview of techniques to mitigate Hot Spot impacts, such as the uneven distribution of sensors, routes that balance energy consumption, sink mobility, and the use of unequal clustering. Further, it depicts the approach for achieving mitigation of sink centered Hot Spots. Finally, this chapter presents conclusions and future research perspectives.

DOI: 10.4018/978-1-4666-0101-7.ch008

INTRODUCTION

Improvements on wireless networking have highlighted the importance of distributed systems in our current society. Wireless Sensor Networks (WSNs) have been envisaged to support different applications, such as environmental monitoring, security systems, military systems, industrial control and others (Potdar, 2009). WSNs consist of tiny devices with the capability of communication and sensing. Batteries power these devices, limiting the network lifetime and their processing power due to not only size but also energy constraints. In order to obtain effective communication among sensor nodes and satisfy application requirements, the establishment and maintenance of routes from a data source to a base station (or sink) is necessary.

Energy consumption is different on each sensor node of a WSN. Nodes distributed in a homogeneous way suffer a funneling effect due to the many-to-one traffic pattern. This traffic pattern characterizes most of the data gathering applications. As sensors get close to the sink, the number of routes decreases, overloading some areas with data traffic and triggering a gradual process that creates and expands an energy hole around the sink (Bi, 2007). These areas, overburdened with high traffic rates, are called Hot Spots. Their emergence is common, but countermeasures must be taken in order to mitigate their impacts, or else, network performance can be severely harmed.

Balancing network-wide energy consumption is crucial for routing protocols on WSN. The emergence of Hot Spots can compromise the network lifetime, data delivery rate, and may even disconnect sensors of the network. These aspects are destructive for the network performance, impacting the quality of services and applications, which should be preserved. Hot Spots have their nature bounded with routing and, therefore, they must be addressed at the routing layer.

Several works have analyzed the impact of Hot Spots in WSNs. In (Perillo, 2005), authors evaluated strategies proposed to mitigate Hot Spots. The use of the sink mobility approach to manage energy at nodes was examined in (Thanigaivelu, 2009). In (Sesay, 2006), authors analyzed the use of different parameters to detect Hot Spots, such as buffer occupation, packet loss and link layer contention. In order to mitigate Hot Spots, different approaches have been used. In (Liu, 2008), energy balance was modeled as a particle swarm optimization problem by redefining the particles fly rules for the routing optimization. In (Wu, 2006), Wu and Chen proposed the uneven distribution of sensor nodes in the network, deploying more nodes close to the sink. DAR (Bi, 2007) uses a slightly different technique, which tries to establish longer routes, enforcing a more balanced participation of nodes. Over the last years, unequal sized clustering has been extensively employed to mitigate Hot Spots. For instance, UCR (Chen, 2009) and EECRP (Xi-Rong, 2009) decrease the cluster size close to the sink, creating more possible routes to reach it.

Albeit there are many protocols intending to mitigate Hot Spots, they all have distinct disadvantages. The suitability of them is compromised due to unrealistic assumptions and operations and, the analysis of performance was not accurate enough to prove their efficiency. Also, backbone maintenance is hardly concerned. This creates energy-efficient routes that do not relay packets correctly. This chapter presents a general discussion about different techniques to mitigate Hot Spots on WSN. Its goal lies in presenting an overview of existing solutions, emphasizing open issues. Finally, we describe the proposed energy-efficient architecture capable of mitigating Hot Spot effects without performance degradation of network operations, and a new routing protocol called RRUCR, based on the proposed architecture.

BACKGROUND

Wireless Sensor Networks

Wireless Sensor Network (WSN) is an increasing area that has attracted intense research. With progresses on MEMS (*Microelectromechanical Systems*) and NEMS (*Nanoelectromechanical Systems*), tiny hardware pieces have become cheaper. Those advances promote the miniaturization of devices that own the capability of processing, sensing and communicating, called sensor nodes. WSNs are composed of sensor nodes, scaling from tens to thousands of them. The use of WSNs relies on the cooperation among these nodes, which can be employed by several applications in different fields, such as automation, control, healthcare, military, tracking and monitoring.

Despite of advantages, WSNs impose constraints. Firstly, all difficulties from the wired networks are inherited, and new ones are created. Due to the open nature of the medium, guaranteeing security is a demanding task. In this new context, collisions are more difficult to handle, the range of transmissions is limited, and the concept of interferences arises. Further, nodes have limited capabilities. Their processing power is not high, their available memory is much lower and they are energy-constrained. In the majority of cases, batteries on sensor nodes cannot be replaced.

Communication on a WSN follows either a single-hop or a multi-hop approach. On the single-hop approach, sensor nodes communicate directly with the destiny of the message. They need to use higher transmission powers in order to reach distant sensor nodes. With higher transmission powers, more energy is consumed, and more interference is produced in a bigger area. That results in more collisions, retransmissions and difficult spatial reuse of channels. On the multi-hop approach, sensor nodes relay their messages through several sensor nodes, forming a route up to the destiny. By using the multi-hop approach, nodes can communicate using lower transmission powers and, therefore, the energy consumption is distributed among nodes and interferences are produced on smaller regions. Multi-hop approach yields an easier spatial reuse of channels.

A common approach on WSN is to organize the network in groups of sensor nodes, or clusters. In this approach, a leader node, called cluster-head, represents a cluster. On this kind of approach, routing itself occurs through cluster-heads that relay data from sensor nodes in their clusters. Supposing a data gathering application, if a single-hop approach is employed, cluster-heads relay data directly to the sink. Considering a multi-hop approach, cluster-heads relay their cluster's node's data to other cluster-head and so on, up to the sink. Clustering approaches demonstrate several advantages, such as aggregation of the cluster's data on the cluster-head, fewer transmissions and easy spatial reuse of channels.

Hot Spots

Hot Spots comprehend areas that are overloaded with high data traffic rates. If their impacts are not mitigated, sensors covered by them are prone to premature death. Hot Spots can be generated by network topology and traffic pattern. Both of these causes can result from applications requirements.

A precise manual deployment of sensor nodes is almost impossible, since these devices may monitor dangerous or inhospitable areas. Therefore, there is no control in their deployment, and all is made at random. This allows unfortunate distributions where two regions of the network have only a few nodes connecting them. The traffic that occurs between those regions burdens the nodes connecting and, as consequence, the Hot Spot emerges resulted from the topology. Such behavior is illustrated in Figure 1, in which the dotted circle delimits a Hot Spot area.

In some applications, some sensor nodes have more importance and may be more commonly targeted as the final hop of messages (Figure 2). There are applications that rely on accessing

Figure 1. Hot Spot emergence due to topology

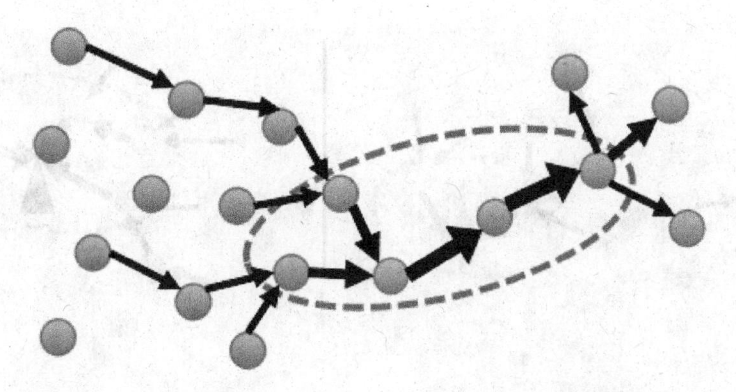

databases spread on the network, resulting in a more intense traffic on the nodes close to those databases. Also, on data gathering applications, the node with the greatest importance is the sink. Hence, all traffic is sent to it. For this reason, the network suffers from a funneling effect in which all routes converge to single points in the network. In multi-hop approaches, the total energy used by the nodes to relay data is inversely proportional to their distance to these important nodes. Hot Spots result in the death of nodes close to important ones. Such consequence makes things worse, since the distance that a message needs to travel increases with network nodes death.

ROUTING APPROACHES TO MITIGATE HOT SPOTS

Several works have analyzed the impact of Hot Spots in WSNs. In (Perillo, 2005), authors evaluated strategies based on transmission power control. They analyzed the politics of farther nodes using increased transmission powers and proposed a clustering approach in which common nodes may send their traffic to whichever cluster-head they choose in order to balance energy. The conclusion was that the studied transmission power control politics lead to ineffective use of energy, due to the inability of farther nodes to make good use of energy. Also, although the energy consumption is slightly balanced by forcing farther nodes to spend more energy, Hot Spots are not mitigated.

The use of the sink mobility approach to manage energy balance at nodes was examined in (Thanigaivelu, 2009). Authors showed that sink mobility prevent Hot Spots, avoiding to have always the same nodes close to the sink, where there is greater consume of energy due to the funneling effect. Mobile nodes keep alternating themselves throughout the sink route. However, the use of mobility in real scenarios is constrained or even impractical, due to space, trajectory or energy restrictions (Vlajic, 2009).

In (Sesay, 2006), authors say that different parameters, such as buffer occupation, packet loss and link layer contention can be combined and analyzed in order to detect Hot Spots. However, monitoring these parameters independently and combining them to accurately detect a Hot Spot would result in a complex scheme. Thus, they say that measuring a sensor node's throughput only is enough. Also, they prove that node's throughput incorporate inversely the other mentioned parameters. Their work, however, focuses on MANET multimedia applications.

In order to mitigate Hot Spots, different protocols have been developed. In (Liu, 2008), routing was modeled as a linear programming problem. The energy balance of the network topology was modeled as a particle swarm optimization problem, redefining the particles flying rules and their opera-

Figure 2. Hot Spot emergence due to traffic pattern

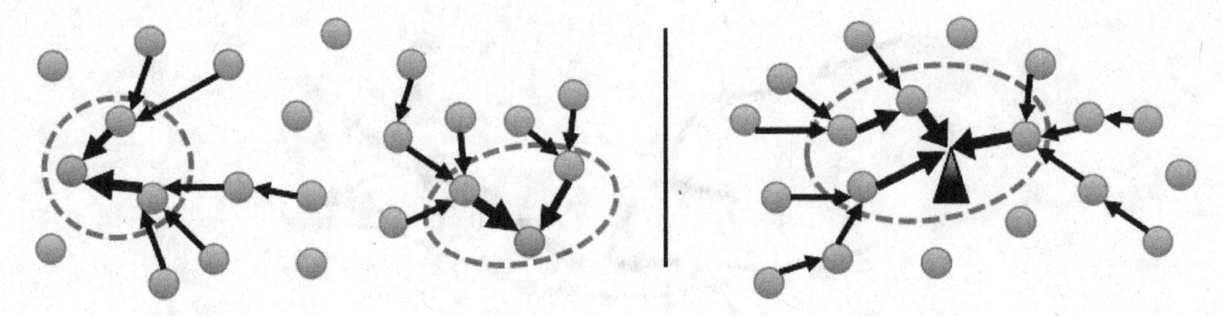

tions. However, its suitability for real applications is not proved, as authors do not measure overhead and delay added by the proposal.

In (Wu, 2006), Wu and Chen proposed the uneven distribution of sensor nodes in the network, deploying more nodes close to the sink, increasing from the outer to the inner regions in a geometric proportion. Although this approach has managed to balance the energy consumption and mitigate Hot Spots, its use in real applications is impractical as it depends on a biased deployment of sensors.

DAR (Bi, 2007) uses a slightly different technique. In order to manage energy consumption, it forces the participation of all nodes (though not at the same time) on the routing activity. The network is divided in rings, beginning on the sink, and in a division of turns, different rings are responsible for data aggregation and direct transmission of it to the sink. That way, it may establish longer routes, enforcing a more balanced participation of nodes. However, the approach results in higher energy consumption, because inner rings may have their data forwarded to outer rings, so that the latter will send the data to the sink. Further, delay is added by this approach.

In the last years, unequal sized clustering has been extensively employed to mitigate Hot Spots, such as in LUCA (Lee, 2009), UCR (Chen, 2009), EBUCP (Yang, 2009) or EECRP (Xi-Rong, 2009). In contrast to EECS (Ye, 2005), which decreases the cluster size as nodes get far from the sink, in order to preserve energy for long-haul direct transmissions to the sink, UCR and EECRP decrease the cluster size close to the sink, creating more possible routes. However, these protocols are not concerned about backbone maintenance. Thus, as cluster-heads suffer rotations, the communication link between two of them can break when more distant nodes in relation to the other cluster are selected as new cluster-heads. Such behavior damages data delivery rate due to the distance between cluster-heads being superior to the transmission range. Further, these works are not particularly focused on the energy management problem.

AN ENERGY EFFICIENT ARCHITECTURE FOR HOT SPOT MITIGATION ON WSN

In order to achieve energy balance, we proposed a Cluster-based Energy Architecture (CEA) that considers Hot Spot issues. The CEA architecture is illustrated in Figure 3, where dark gray boxes represent main modules, arrows represent interactions, and lighter boxes represent components of modules. CEA consists of four basic modules: intra-cluster energy manager, inter-cluster energy manager, data gathering and route management. Since CEA is based on clusters, energy management occurs essentially by their control, formation and maintenance. The two modules inside the dotted box have functionalities related to energy management.

Figure 3. The CEA architecture

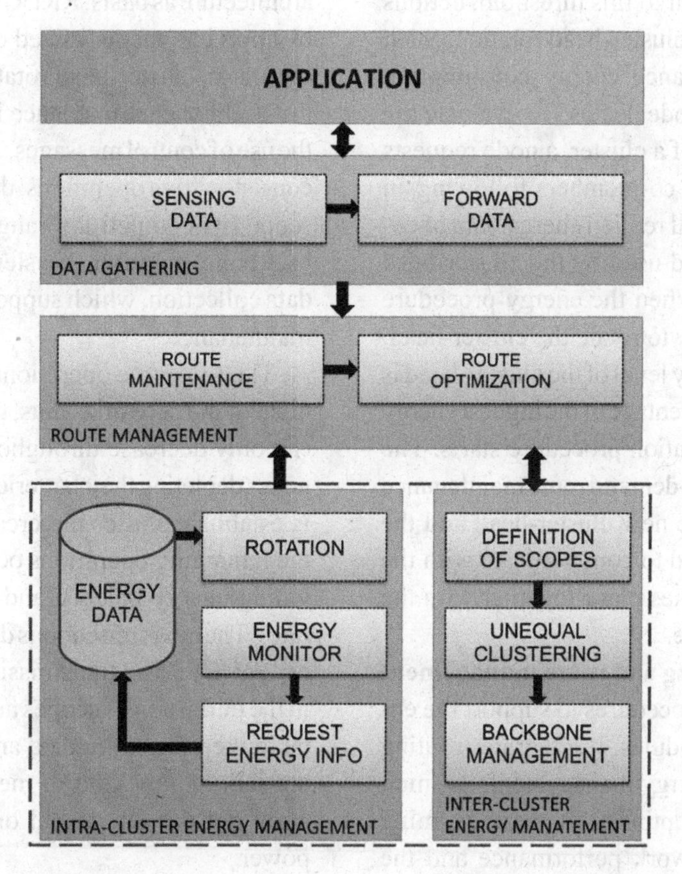

The **inter-cluster energy management** module consists of procedures responsible for managing energy control among clusters of the network. Procedures in this module intend to reduce energy consumption of nodes by efficiently defining the scope of nodes, determining unequal clusters and creating a backbone for the network. Those procedures are executed in a distributed way, in which each node runs those procedures cooperatively. This module represents a passive way to manage energy consumption across the network. Since each node executes its procedures without global information, it reduces complexity.

In the inter-cluster energy module, competition range of nodes (also called scope) is the first step towards energy management. Criteria, such as transmission power, physical distance, RSSI measurements, can be employed to determine the competition range of nodes. A previous analysis of those criteria must be done, in order to choose one. As example, in (Parameswaran, 2009), we can find analyses about the use of RSSI measurements. Considering the scopes and energy information, clusters are then formed, aiming to create more clusters close to the sink. At the end of cluster formation, the network backbone should be determined.

The **intra-cluster energy management** module is composed of procedures to monitor energy of nodes, request energy information to cluster co-members and participate in the rotation process of cluster-heads, for instance. The energy of a node is monitored periodically, to verify its level, and compared to a threshold, defined considering the highest energy level among all members of a cluster. Depending on the node energy level in relation to this threshold, actions are triggered, such as cluster-head rotation. Such action intends to balance energy consumption avoiding premature node deaths. To evaluate the highest energy level of a cluster, a node requests this information to its co-members following an efficient procedure. All replied energy data of co-members is stored and used by the cluster-head rotation procedure. When the energy procedure indicates the necessity to rotate the cluster-head, that is, when the energy level of the cluster-head is below a stipulated percentage of the highest energy on the cluster, the rotation procedure starts. The rotation triggers an on-demand route maintenance operation for both the new cluster-head and the cluster-heads that used to communicate with the previous one - this takes place together with the data gathering module.

The **data gathering** and **route management** modules consist of procedures to support the energy management modules. It integrates routing management with energy management, defining how to maintain and optimize routes and aiming to improve both network performance and the efficient use of energy. It innovates in relation to existing energy management architectures (Jiang, 2007), (Ruiz, 2003), by considering together energy and performance. This module owns also a component to sense data, being data humidity, temperature, light or others, which are determined strictly by the application needs. Data forward procedures should follow the network backbone created until reaching the sink. The backbone aims to optimize the path to the sink in terms of energy consumption.

A ROTATION REACTIVE UNEQUAL CLUSTER-BASED ROUTING PROTOCOL

This section details our proposed protocol, called Rotation Reactive Unequal Cluster-based Routing protocol (RRUCR), developed by using the CEA architecture as basis. RRUCR mitigates Hot Spots by applying unequal sized clusters (as shown on Figure 4), cluster-head rotations, and integrating the backbone maintenance in data flows without the use of control messages. The RRUCR protocol consists of five operations: definition of each node scope (its competition range), clustering, initial backbone creation, cluster-head rotation, and data collection, which supports also the backbone maintenance.

The first three operations occur at the deployment of the network. Thus, the number of clusters can only decrease throughout time as a result of nodes depleting their batteries. An initial backbone is established after the creation of clusters, and the remaining operations occur many times, providing energy balance and higher data delivery rates. The next subsections describe the operations of RRUCR. The transmission powers employed in the definition of scopes are indexed in a ranked table, kept by all nodes, and only their indexes are sent on any kind of message. Thus, indexes point out the referenced or stored transmission power.

Scope Definition

The sink initially broadcasts *INCR_POT* (increment potency) messages containing transmission power indexes that can be used by nodes. Figure 5 illustrates coverage areas determined by different transmission powers. When nodes receive INCR_POT on the first time, they store the transmission power used by the sink on a variable *RBase*. These nodes reply with an acknowledgment message. Hence, the sink knows the lowest transmission power, *RFMin*, and the highest transmission power, *RFMax*, to reach a node. Next, the sink broadcasts a *SETUP_CONFIG* message in the maximum transmission power, containing the values of the variables *RFMin* and *RFMax*. Since some nodes may not receive this message, farther nodes (which *RBase* = *RFMax*) will retransmit the *SETUP_CONFIG* message, and the remaining

Figure 4. Network organization in unequal sized clusters

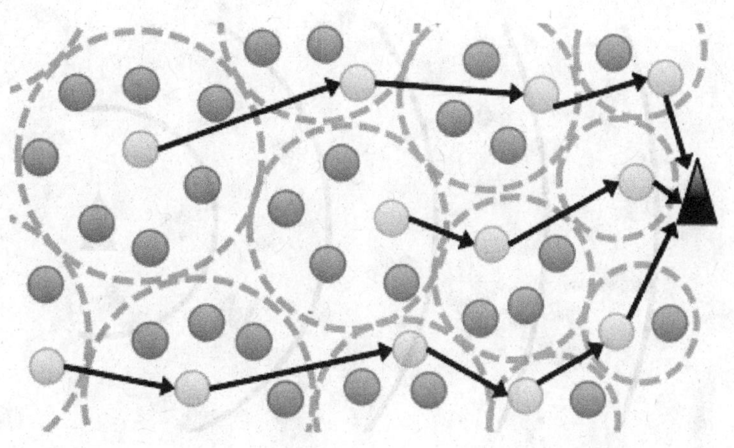

nodes, which had not been reached, retransmit the message until the entire network is covered in this flooding process. This message also contains a counter *cont* that informs how many hops it has been forwarded until reaching the current node.

The clustering operation pre-stipulates two limits for the transmission powers: *pot_limit*, the index of the maximum power that can be used by the nodes reached on the first wave of messages *SETUP_CONFIG*, and *pot_max_global*, the index of the maximum transmission power that can be used by the other nodes. These limits exist because with the cluster-head (CH) rotations nodes can move, getting far from each other and not being covered by the inter-clusters transmission power. With these limits, clusters will usually have a diameter inferior to the range of the transmission power used on inter-clusters communication, avoiding breaking links.

Figure 6 illustrates the coverage scope of some nodes in the network. For nodes which are not reached by the first wave of messages SETUP_CONFIG, coverage scope index is calculated by *Scope = (pot_limit + cont)*, being the maximum value of *Scope* determined by *pot_max_global*. Whereas, bodes reached by the first wave of the message *SETUP_CONFIG* use the transmission power of the following index on clustering operations:

$$Scope = \lfloor ((1 - \frac{(RFMax - RBase)}{(RFMax - RFMin)}) * pot\_limit) \rfloor$$

Clustering

Having the competition scope of each node defined, the clustering operation starts. Figure 7 illustrates its flowchart. Based on a pre-stipulated probability, nodes are randomly selected as candidates for CHs. Then, they send a competition message informing their energy to everyone on scope, defined on the last section. Nodes that do not receive a competition message will also candidate to CH; this measure avoids areas without any CH.

After receiving those competition messages, nodes verify if their energy is higher than the energy of their neighbors in order to become definitive CHs, then they broadcast a *FINAL_HEAD* message. All nodes count the number of received *FINAL_HEAD* messages and store it in the *finals* variable. After the time dedicated to this phase, nodes that did not receive any of these messages also become a candidate, sending a *FINAL_HEAD* message. At the end, if *finals* > 2, the candidate node gives up the election. Hence, it increases the network coverage, without creating too many clusters.

Figure 5. Transmission powers coverage

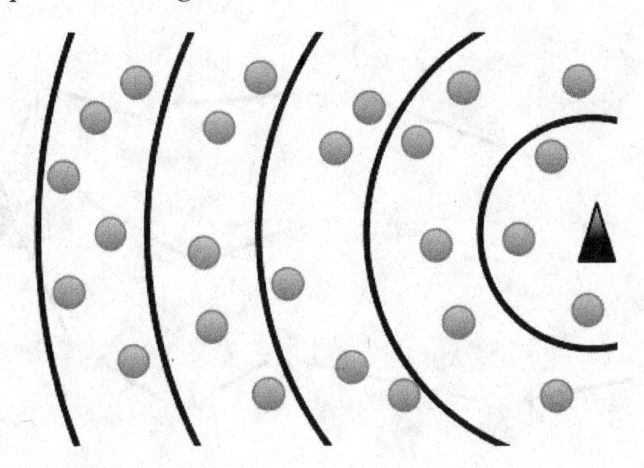

Nodes definitively established as CHs broadcast an announcement message. Thus, common nodes will be able to select a CH based on its RSSI (Received Signal Strength Indicator). Each node keeps the identifier (ID) of the selected CH, and sends a *JOIN_CLUSTER* message, informing its energy. On receiving those messages, the CH keeps the highest energy value for future rotation operations.

Initial Backbone Creation

This operation consists on the establishment of a valid initial backbone that allows nodes to reach the sink in few hops. The sink broadcasts a *BEACON_ROUTE* message to all nodes near the sink, being that area defined by a transmission power index. CHs in this area will forward received data directly to the sink, and this is usually where the Hot Spot takes place.

The *BEACON_ROUTE* message carries a *counter* field that informs how many hops the message has travelled, enabling nodes to know how far they are from the sink. Nodes keep such data in a variable *wave*. Thus, when a node receives a *BEACON_ROUTE* message, its *wave* is updated if *counter* is lower than its current *wave*, and *next_hop* is set with the ID of the node that sent

Figure 6. Unequal scopes

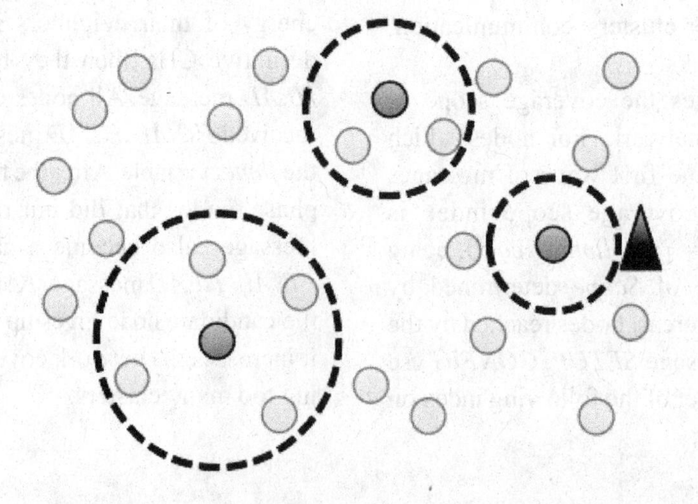

Figure 7. Flowchart of the clustering operation

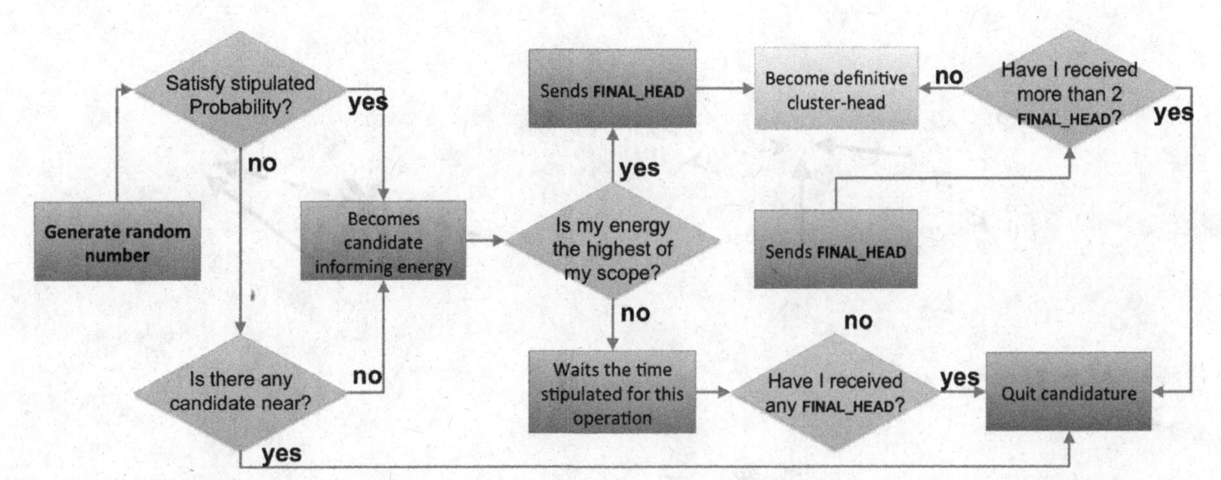

the message. *next_hop* is also updated if *counter* = *wave* and the message's RSSI is higher than the RSSI of the message that caused the last *next_hop* update. The *BEACON_ROUTE* message is then retransmitted, increasing its *counter* field.

Cluster-Heads Rotations

An energy percentage threshold, called *pRotate*, is pre-stipulated for the CHs rotation. When the CH's energy gets lower than the pRotate percentage of the highest node's energy in its cluster (originally obtained from the *JOIN CLUSTER* messages), this CH broadcasts a rotation request, informing its energy. Nodes will answer informing their energy if they belong to the cluster and have a higher energy.

The CH that requested a rotation selects as new CH the node with the highest energy, also adopting his ID as *nex_hop* (Figure 8-I and 8-II). Then the requester CH will broadcast a *DENOMINATE_CH* message, containing its previous *next_hop* and the new CH's ID. On receiving this message, the new CH will update its *next_hop* to the selected on the received message and will consider the highest energy on the cluster as being its own. The remaining nodes in the cluster will now communicate with the new CH.

As broken links may appear due to rotations (dotted link in Figure 8-II, marked with a 'X'), on receiving a *DENOMINATE_CH* message, both the new CH and the CHs that used to communicate with the old CH will be forced to update their routes as soon as a valid route is detected.

Data Gathering and Routes Maintenance

In order to send collected data to the sink, nodes broadcast a *DATA_GATHERED* message with both the data and the value of *next_hop* (thin grey line of Figure 8-III). The CH whose ID equals the received message's next hop will update the *next_hop*'s field to its own *next_hop* and then forward the message in broadcast (thick line of Figure 8-III). In such message, the CH will also send a value $wr = 100 * wave + RBase$ that identifies its distance to the base, and will be used for the backbone maintenance.

Whenever a CH receives a *DATA_GATHERED* message, it may also update its route. When the maintenance is obligatory due to rotation, the CH will only verify if the CH that sent the message is between it and the sink (its *wr* is lower), and if it is, it will be adopted as being the *next_hop* (Figure 4-IV). If the maintenance is not obligatory,

Figure 8. Repairing a broken link generated in a rotation of cluster-heads

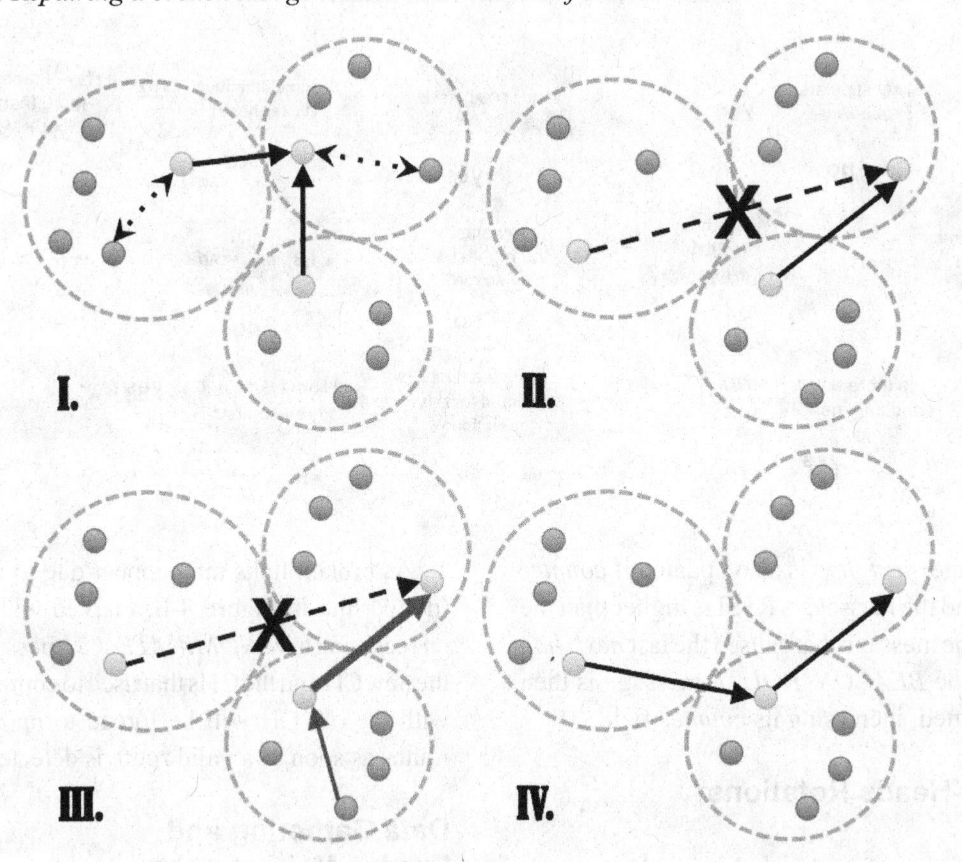

the route will be updated if the new possibility of route has a lower cost (*wr*) than the current. The cost of the selected route is always stored for future comparisons.

The cost *wr* mentioned is calculated through the variables *RBase*, *Scope* and *wave* – all of which are good estimations of distance to the sink. Nodes have these values proportional to their distance to the sink, being *wave* itself the number of hops necessary to reach it on the initial backbone.

PERFORMANCE EVALUATION

Although there are several cluster-based routing protocols, we only compared ours with the UCR (Chen, 2009) protocol, since both mitigate Hot Spots by employing unequal sized clusters, even though with different clustering algorithms. Node mobility was not considered because the mobility itself is an alternative way to mitigate Hot Spots, due to the dynamic routes (Thanigaivelu, 2009).

We implemented both protocols on the NS-2.30 environment and simulated a homogeneous WSN, based on Mica2 sensor motes. The radio parameters were set according to the CC1000 radio used by the Mica2 architecture. Each node has a 0.1% probability per second to generate data, being roughly transmitted to the sink, i.e., without any aggregation technique.

The WSN operates for 5000 seconds and consists of 700 sensors distributed in a square area, measuring 1000m at each side. The location of all nodes and sink is random in each simulation. The initial energy of each node comprehends values between 0.9 and 1.1 joules. For both protocols the probability of a node to participate in the cluster-head election was 35%, there was

32 bytes of collected data, and the inter-cluster transmission power is 3.16227mW (the highest power supported by the Mica2 motes). The area where all nodes communicate directly with the sink has 149m centered at it. In UCR, the maximum cluster radius limit was 140m. In RRUCR, it was used *pot_limit* = 0.25118mW, *pot_max_global* = 0.63095mW and *pRotate* = 65%. The power values used for scope definition were obtained from the CC1000 datasheet, and they were all crescently indexed, as previously detailed.

RRUCR had its performance evaluated under three types of simulation scenarios: operation without failures (indicated by "none" in Figures), with failures "close" to the sink and with failures "far" from the sink (the distance is quantified in hops). In the scenario with failures close to the sink, 8 nodes that take from 0 to 2 hops to reach the sink, and 8 nodes that take from 1 to 5 hops are randomly turned off. In the scenario with failures far from the sink, 25 nodes are turned off, being them 12 that take from 2 to 5 hops, and 13 that take from 3 to 6 hops. In both situations failures occur at 400s of simulation.

The metrics used for the evaluation of resilience, that is, the capability of saving energy and keeping network performance, are *number of hops from each cluster-head to the sink, total amount of energy, lifetime, number of dead nodes in relation to their distance to the sink, data delivery rate* (which considers the percentage of arrival of the last 30 data packets sent) and *number of rotations*. Those metrics assess the Hot Spot mitigation, the efficiency of the created routes, their maintenance and the energy balance of the network. We ran 35 simulations for each protocol and each kind of described network, obtaining in a 95% confidence interval. The NS-2.30 RRUCR code is available under the terms of the GLPL license and can be found at the website www.nr2.ufpr.br/~fernando/rrucr/.

Cluster Distribution and Energy Consumption

The number of clusters formed in the network is an important factor for WSNs. With too many clusters, there is more energy consumption resulted from the increased number of messages exchanged. However, a small quantity of clusters causes more overhead and higher energy consumption due to the necessity of higher transmission powers and possible retransmissions. Because of its characteristics, UCR's clusters must be smaller, in order to decrease the probability of choosing a cluster-head much far.

In Figure 9, we observe that although RRUCR owns clusters with more hops to reach the sink, it creates fewer clusters. On average, RRUCR created 43 clusters, while UCR created 67. This difference of 35.82% more clusters in UCR resulted in higher energy consumption on it, as shown in Figure 10. This happens because that with an increased number of clusters, more messages will be sent due to the higher number of rotations. Also, with more but too small clusters, the efficacy of rotations is compromised. The increase of the rotation amount is proved in the next subsection.

Network Lifetime

Lifetime is the time elapsed until the first node death. A way to extend this time on cluster-organized WSN is using cluster-head rotations to distribute cluster energy consumption. Compared to UCR, RRUCR managed better energy consumption and increased the network lifetime in 21.36% on scenarios of network without failures, as presented in Figure 11, column "none". On scenarios with failures "far" from the sink there is an increase of 17.16%, whereas under failures "close" to the sink, it is 13.55%, as observed in Figure 11.

Thus, the creation of more routes and their maintenance balance the energy consumption of

Figure 9. Cluster distribution

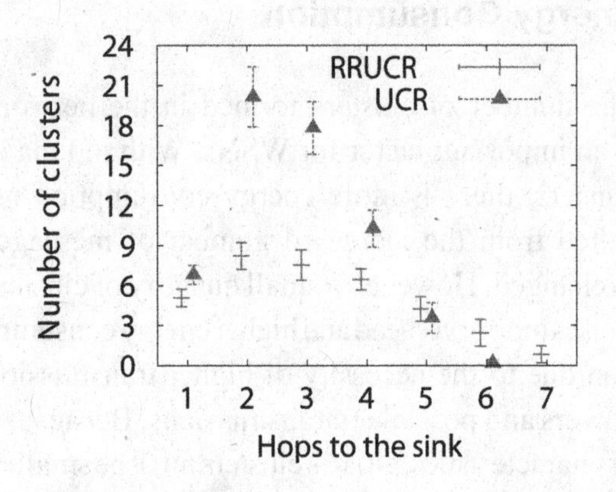

Figure 10. Energy consumption

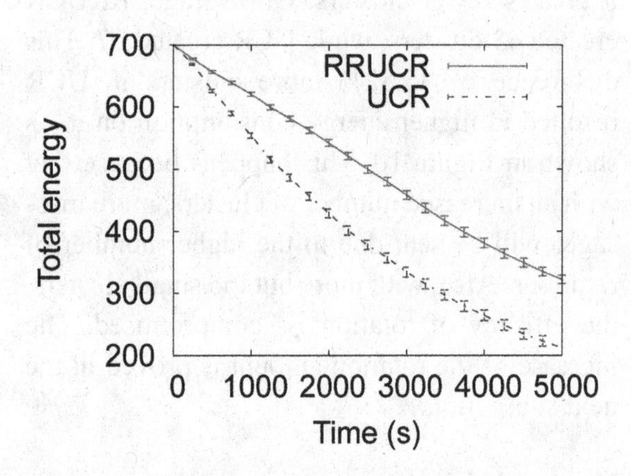

Figure 11. Network lifetime

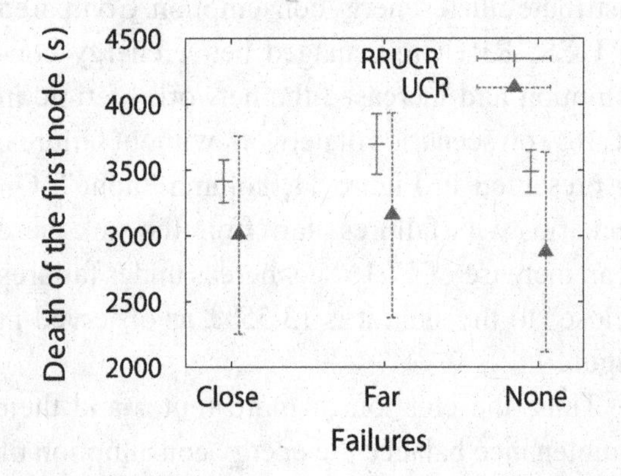

the network as a whole. On both protocols the best lifetime is reached on scenarios of networks with failures far from the sink, because there are less packets needing to travel longer distances.

To evaluate the performance of both protocols on the mitigation of Hot Spots, the number of dead nodes across the distance in meters to the sink (DS) was measured, as shown in Figure 12. In this figure, we consider distances inferior to 40m, between 40m and 80m, between 80m and 160m and superior to 160m. The number of dead nodes was measured in the simulation times of 3000, 4000 and 5000 seconds. Both protocols minimized Hot Spot effects by decreasing the number of deaths near the sink. Hence, more nodes close to the sink can be used for last hops in communication. But more nodes die on UCR, due to the increased number of clusters and the consequent cluster-head rotations.

Figure 13 compares the number of rotations per protocol and time. For each second and each protocol, the number of rotations that took place up to that time is plotted. We observe that RRUCR resulted in a lower number of rotations than UCR. In the end, RRUCR presents only half of the number of rotations UCR performed.

Data Delivery Rate

Only energy distribution and the existence of a valid initial backbone do not guarantee a satisfactory delivery rate. Figure 14 shows that RRUCR presented higher data delivery rates, and UCR matched them only in the beginning of the simulations, before any cluster-head rotation or failure takes place. When rotations start, at approximately 700s, data delivery starts to drop due to broken links.

Repairing routes is essential as broken links can happen. Our dynamic maintenance algorithm enabled RRUCR to have higher data delivery rates also when the number of cluster-head rotations is regarded, as shown in Figure 15. In general, UCR is the most compromised. It needs to

Figure 12. Dead nodes vs. distance

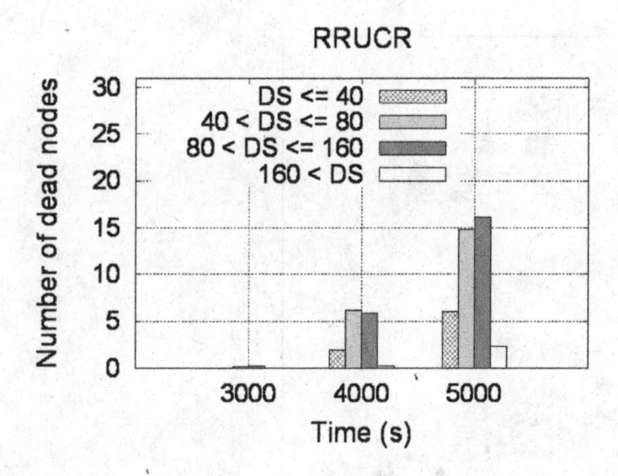

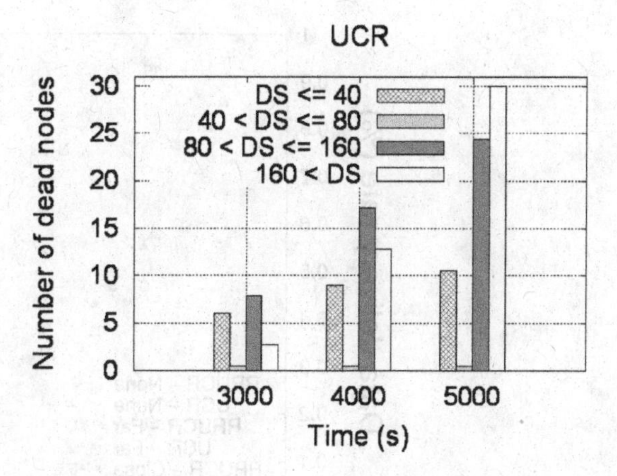

Figure 13. Cluster-head rotations

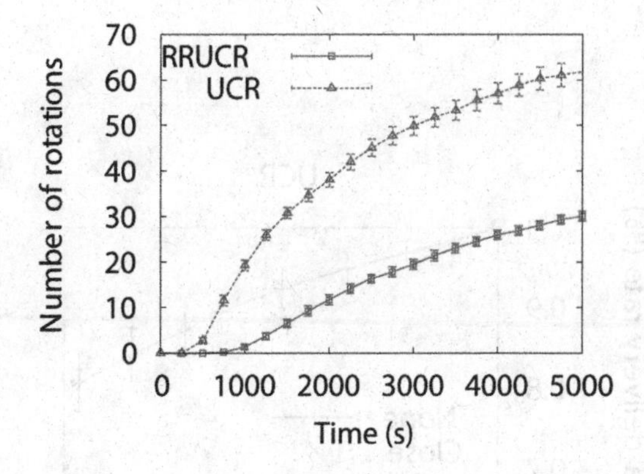

perform more cluster-head rotations, as presented in Figure 13, due to the higher quantity of clusters, as observed in Figure 9, that generates more broken links between clusters.

It is proved that a route maintenance operation is required to keep data delivery rate unharmed. By these results, we observe that RRUCR keeps better network performance when compared to UCR in terms of packet delivery rate, which can lead to better energy efficiency. If retransmissions were considered in the case of data loss, the absence of an efficient backbone would burden nodes with even more intense traffic, demanding much more energy with retransmissions.

CONCLUSION

Wireless Sensor Networks (WSNs) has become each time more important in the last few years, being applied in different domains in our current society. Despite of advantages and improvements reached by this technology, WSNs are prone to Hot Spots due to their characteristics, such as the usual many-to-one traffic pattern. Hot spots are network areas overburdened with high traffic rates. Those areas concentrate nodes close to the sink, an important node that receives almost all transmissions of data.

This chapter presented an overview of some techniques proposed to mitigate Hot Spots in WSNs, highlighting their advantages and disadvantages. Based on this analysis, we propose an architecture called CEA (Cluster-based Energy management Architecture) to mitigate WSN Hot Spots, balancing energy consumption and increasing network performance. The architecture consists of four basic modules: an intra-cluster energy manager, inter-cluster energy manager, data gathering and route management. Since it is based on clusters, energy management occurs essentially by their control, formation and maintenance. CEA innovates in relation to existing energy management architectures by considering together energy and performance.

Figure 14. Delivery rates with and without failures

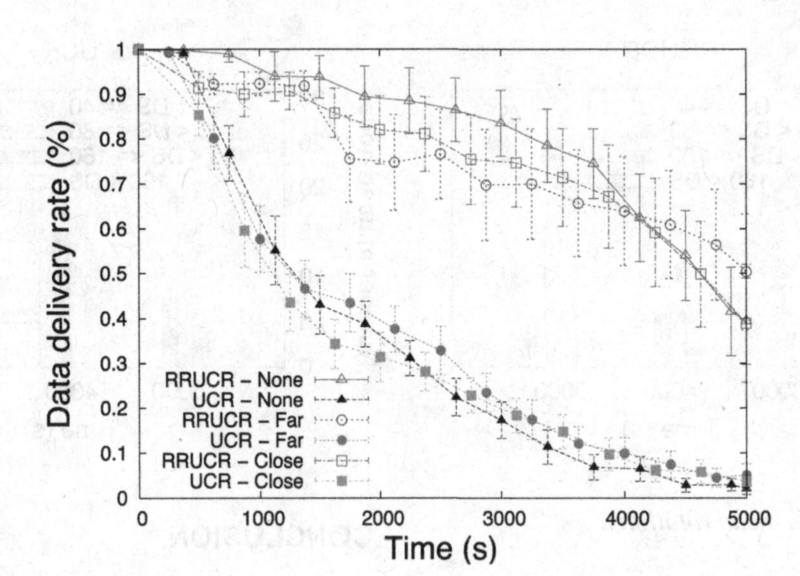

Figure 15. Delivery rates vs. cluster-head rotations

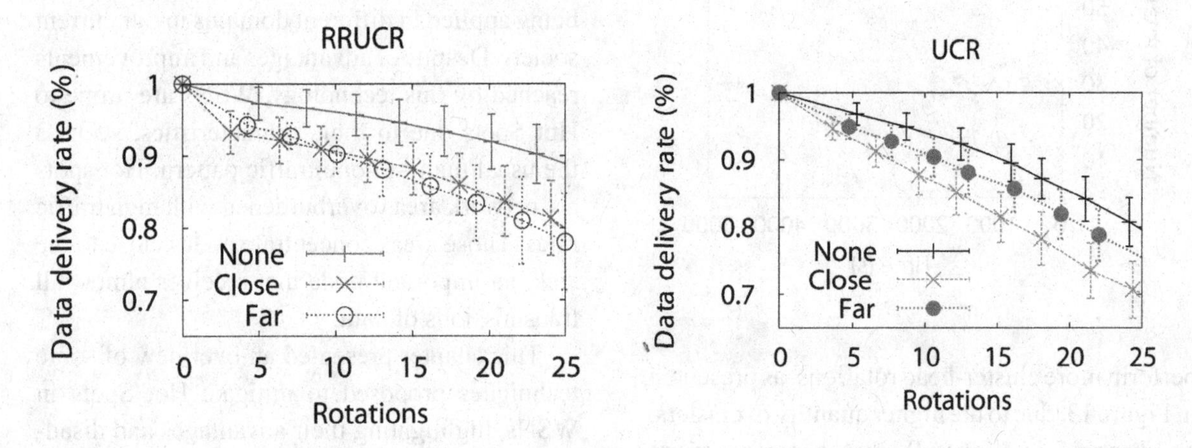

We showcase the architecture by a cluster-based routing protocol, called RRUCR. It makes dynamic route maintenance without the use of control packets, saving energy. The protocol has five operations, from which, rotation of cluster-heads and data gathering offer better energy balance and do not compromise network performance, and the clustering scheme, that employs different transmission powers and creates unequal sized clusters efficiently and in balanced quantity.

Simulation results showed that RRUCR increased the network lifetime by around 21.36% in relation to UCR. Further, the number of created clusters was 35.82% lower than UCR, spending less energy on cluster-head rotations. The RRUCR resilience was also evaluated and, although the routes maintenance of RRUCR is simple, it showed efficacy when compared to the UCR, keeping an acceptable level of network performance. Future work includes operations that check the integrity on WSN links, carrying out more complex repairs.

REFERENCES

Bi, Y., Li, N., & Sun, L. (2007). DAR: An energy-balanced data-gathering scheme for wireless sensor networks. *Computer Communications*, *30*(14-15), 2812–2825. doi:10.1016/j.comcom.2007.05.021

Chen, G., Li, C., Ye, M., & Wu, J. (2009). An unequal cluster-based routing protocol in wireless sensor networks. *Wireless Networks*, *15*(2), 193–207. doi:10.1007/s11276-007-0035-8

Jiang, X., Taneja, J., Ortiz, J., Tavakoli, A., Dutta, P., & Jeong, J. (2007). An architecture for energy management in wireless sensor networks. *SIGBED Review*, *4*(3), 31–36. doi:10.1145/1317103.1317109

Jin, W., de Dieu, I. J., Jose, A. D. L. D., Lee, S., & Lee, Y.-K. (2010). *Prolonging the lifetime of wireless sensor networks via hotspot analysis*. In International Workshop on Computing Technologies and Business Strategies for u-Healthcare (CTBuH2010). Seoul, Korea.

Lee, S., Choe, H., Park, B., Song, Y., & Kim, C.-K. (2009). Luca: An energy-efficient unequal clustering algorithm using location information for wireless sensor networks. *Wireless Personal Communications*, (pp. 1–17). DOI:10.1007/s11277-009- 9842-9

Liu, A.-F., Ma, M., Chen, Z.-G., & hua Gui, W. (2008). Energy-hole avoidance routing algorithm for WSN. *International Conference on Natural Computation, 1*, 76–80.

Parameswaran, A., Husain, M. I., & Upadhyaya, S. (2009). *Is RSSI a reliable parameter in sensor localization algorithms? An experimental study.* In Field Failure Data Analysis Workshop (F2DA'09).

Perillo, M., Cheng, Z., & Heinzelman, W. (2005). An analysis of strategies for mitigating the sensor network hot spot problem. In *Proceedings of the Second International Conference on Mobile and Ubiquitous Systems*, (pp. 474–478).

Potdar, V., Sharif, A., & Chang, E. (2009). Wireless sensor networks: A survey. In *WAINA '09: Proceedings of the 2009 International Conference on Advanced Information Networking and Applications Workshops*, (pp. 636–641). Washington, DC: IEEE Computer Society.

Ruiz, L., Nogueira, J., & Loureiro, A. (2003). MANNA: A management architecture for wireless sensor networks. *Communications Magazine, IEEE, 41*(2), 116–125. doi:10.1109/MCOM.2003.1179560

Sesay, S., Xiang, J., He, J., Yang, Z., & Cheng, W. (2006). Hotspot mitigation with measured node throughput in mobile ad hoc networks. In L. Li (Ed.), *The 6th* IEEE *International Conference on ITS Telecommunications (ITST 2006)*, (pp. 749–752).

Thanigaivelu, K., & Murugan, K. (2009). Impact of sink mobility on network performance in wireless sensor networks. *International Conference on Networks and Communications*, (pp. 7–11).

Vlajic, N., & Stevanovic, D. (2009). *Sink mobility in wireless sensor networks: a (mis)match between theory and practice* (pp. 386–393). IWCMC.

Wu, X., & Chen, G. (2006). On the energy hole problem of nonuniform node distribution in wireless sensor networks. In *Proceedings of The IEEE International Conference on Mobile Adhoc and Sensor Systems (MASS)*, (pp. 180–187).

Xi-Rong, B., Zhi-Tao, Q., Xue-Feng, Z., & Shi, Z. (2009). An efficient energy cluster-based routing protocol for wireless sensor networks. In *CCDC'09: Proceedings of the 21st Annual International Conference on Chinese Control and Decision*, (pp. 4752–4757). Piscataway, NJ, USA: IEEE Press.

Yang, J., & Zhang, D. (2009). *An energy-balancing unequal clustering protocol for wireless sensor networks*. Retrieved from http://www.scialert.net/pdfs/itj/2009/57-63.pdf

Ye, M., Li, C., Chen, G., Wu, J., & Al, M. Y. E. (2005). Eecs: An energy efficient clustering scheme in wireless sensor networks. In *Proceedings of the IEEE International Performance Computing and Communications Conference* (pp. 535–540). IEEE Press.

KEY TERMS AND DEFINITIONS

Energy Balance: Efficiently distribute energy consumption among sensor nodes of the network.

Energy Management Architecture: Architecture that specifies the operations behavior and relation with each other when energy is considered.

Hot Spot: An area overburdened with intense traffic.

Mitigation: To alleviate the intense traffic by distributing it.

Route Maintenance: Operation that fixes broken links in the formed backbone, achieving better network performance.

Routing: Determines how to establish and maintain routes between traffic sources and possible destinations, such as the sink.

Unequal Clustering: Creation of unequal sized clusters, usually intending to optimize somehow the use of energy.

Wireless Sensor Network: A network composed of tiny devices with capabilities of sensing and of wireless communication.

Chapter 9
Node Localization: Issues, Challenges and Future Perspectives in Wireless Sensor Networks (WSNs)

Noor Zaman
King Faisal University, Saudi Arabia

Azween Abdullah
University Technology PETRONAS, Malaysia

Muneer Ahmed
King Faisal University (KFU), Saudi Arabia

ABSTRACT

Wireless sensor networks (WSNs) are taking a major share with almost all types of different applications and especially, it is most suited in very harsh and tough environments, where it is too hard to deploy conventional network applications, for example in the forest fire area, battlefields during the war, chemical and thermal sites, and also for few underwater applications. WSNs are now becoming part of almost all applications because of their ease in deployment and cheaper cost. These networks are resource constraints, very small in size, computation, and with much less communication capabilities. Nodes are normally deployed in random fashion, and it's too hard to find their location because there is no any predefined way like conventional networks to discern location. Location is highly important to know the data correlation: for example its target tracking, and to know actual vicinity of the any event occurrence. This chapter describes the current available approaches, issues, and challenges with current approaches and future directions for node localization, one by one. Node localization is highly important for large sensor networks where users desire to know about the exact location of the nodes to know the data location.

DOI: 10.4018/978-1-4666-0101-7.ch009

INTRODUCTION

Wireless Sensor Networks, are rapidly growing with a high scale with a good number of good applications. Its adoptability is very high because of its cost, working and feasibility of deployment with a range of applications. This network is highly different than the conventional networks with all aspects. Conventional networks have high power capabilities with very high computation, storage and communication strengths along with their global ID to be locate the conventional node and easily can find its location. While with sensor networks are almost different almost with all aspects, because these type of networks are small in small in size with holding small nodes without and centralized administrative control, there distribution is random in nature and nodes are normally not aware about their location when they are collecting the data and passing it to their neighboring nodes. researchers are trying to find out different ways to find out its location by using different approaches like GPS (Zaman & Abdullah, 2011), but it is still little hard to reach up to a bench mark. As these networks are resource constraints to most of the approaches are valid up to assumption base only.

The aim of the localization in the sensor network is to define the exact identification of the sensor node or about its exact position. The correct information is highly required in case of localization, currently researchers defined different approaches which can be classified further into classes on the range base, such as.

1. Range Based Technique

Under this mechanisms, extra hardware and extra computation is required to know about the location of the sensor nodes exact location. The various range based techniques are Radio Interferometric Measurement (RIM) (Merico & Bisiani, 2006), Multidimensional Scaling (MDS), 3D - Landscape (Zhou, Zhang, & Cheng, 2006), DV-distance, DV-hop, Euclidean distance (Chraibi, 2005). These techniques works with the help of the distance or angle metrics, and those are normally linked with the operational timings, such as Time of Arrival (ToA), Time difference of arrival (TDoA) and also Angle of arrival (AoA), and that can be measured through (RSSI) Received Signal Strength Indicator. This mechanism become costly for the WSN, as the WSN type network is very low in cost, but when the additional hardware is required it needs more cost.

2. Range Free Technique

With range free technique, the location of the node can be identified by its neighboring nodes, based on hop count. The popular range free technique are such as APIT (Blum, Stankovic, Abdelzaher, He & Huang, 2005), chord selection approach (Ou & Ssu, 2008), three dimensional multilateration approach (Jin, Wang, Tian, Liu, & Mo, 2007), SerLOC (Lazos & Poovendran, 2005), centroid scheme (Martins, So, Chen, Huang, & Sezaki, 2008) etc. Many more techniques are discussed in (Mao, Fidan & Anderson, 2007)(Yu, Shun, & Mei, 2006)(He, Stoleru, Stankovic, 2006)(Chraibi, 2005)(Tseng & Wang, 2008a)(Tseng & Wang, 2008b). the range free techniques are normally providing good solution but they contain almost around 10-12% error rate too. These techniques are normally very cheaper if we compare it with range based techniques where extra hardware and computation is required, but less in accuracy comparable with range based techniques.

Furthermore more, node localization in wireless sensor network can be grouped in Centralized and Distributed locations. While distributed localization will be further sub divided into nine different groups such as following, and as shown in Figure 1.

Figure 1. Classification of localization in WSN. (Lloret, Tomas, Garcia, & Canovas, 2009)

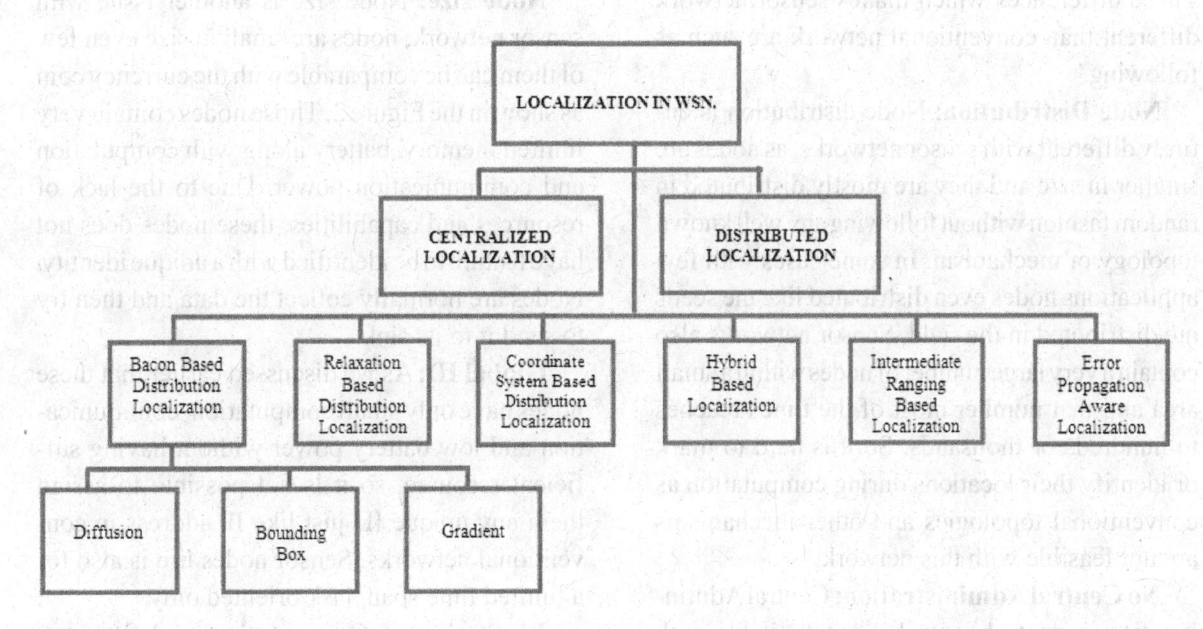

- Beacon Based Distribution Localization
  - Diffusion
  - Bounding Box
  - Gradient
- Relaxation Based Distribution Localization
- Coordinate System Based Distribution Localization
- Hybrid Based Localization
- Intermediate Ranging Based Localization
- Error Propagation Aware Localization

RELATED WORK

Node localization is a key requirement of wireless sensor network same as like its energy efficiency is concerns, a wide range of research work already has been done by the research community about it. Some extensive existing literature surveys (http://www.cse.ust.hk/~yangzh/yang_pqe.pdf, n.d), (Lloret, Tomas, Garcia, & Canovas, 2009) are already conducted on this topic, and in these surveys different common localization techniques and analysis of those techniques were present with its comparison with others, such as multidimensional scaling (MDS) and proximity based map (PDM)(Cheng, Lui, & Tam, 2007) or MDS. These techniques have high accuracy in low communication and in WSN localization. Interferometric ranging based localization has been proposed in (Maroti, Kusy, Balogh, Volgyesi, Nadas, Molnar, Dora, & Ledeczi, 2005), (Patwari & Hero, 2006), (Huang, Zaruba, & Huber, 2007). Channel fading and noise corruption error propagation issues were suppress by a localization scheme has been propose in (Alsindi, Pahlavan, Alavi, 2006). Literature gives comprehensive summary of existing localization schemes. We research will try to focus about all existing node localization techniques and its challenges along with its future directions to solve these issue.

PROBLEM STATEMENT

Nature of sensor networks is different than conventional networks, and those differences become the reason of problem to identify the node locations.

Those differences which makes sensor network different than conventional network are such as following.

Node Distribution: Node distribution is entirely different with sensor networks, as nodes are smaller in size and they are mostly distributed in random fashion without following any well known topology or mechanism. In some cases with few applications nodes even distributed like the seeds are distributed in the field. Sensor networks also contain a very large number of nodes within a small area and that number most of the times reaches to hundreds or thousands. So it is hard to mark or identify their locations during computation as conventional topologies and other mechanisms are not feasible with this network.

No Central Administration: Central Administration or central control is not available with sensor networks, as nodes are randomly distributed here without any identification, so keeping their central administration is not possible with sensor networks. The lack of central administration is also becomes mostly main reason for nodes to available for open threats. Central administration is also not be implemented in the absence of the node identification.

Node size: Node size is another issue with sensor network, nodes are small in size even few of them can be comparable with the currency coin as show in the Figure 2. Those nodes contain very limited memory, battery along with computation and communication power. Due to the lack of resources and capabilities, these nodes does not have feature to be identified with a unique identity. Nodes are normally collect the data and then try to send it to its sink.

Global ID: As we discussed earlier that these nodes have only small computation, communication and low battery power without having sufficient resource, so it is not possible to assign them any unique ID just like IP address in conventional networks. Sensor nodes life is also for a limited time span, task oriented only.

Limitation of Computation and Storage: Size of the sensor node is very small hence its battery and other capabilities is also very low, its memory often in Mega Bytes only with very small size of processing capabilities. Then main functionality is to just record the data resulting to any movement and then just to send that data to the Sink, while small nodes are distributed randomly with a rich number as shown in Figure 3.

Figure 2. Wireless sensor node

Figure 3. Randomly distributed sensor nodes with sink in a center

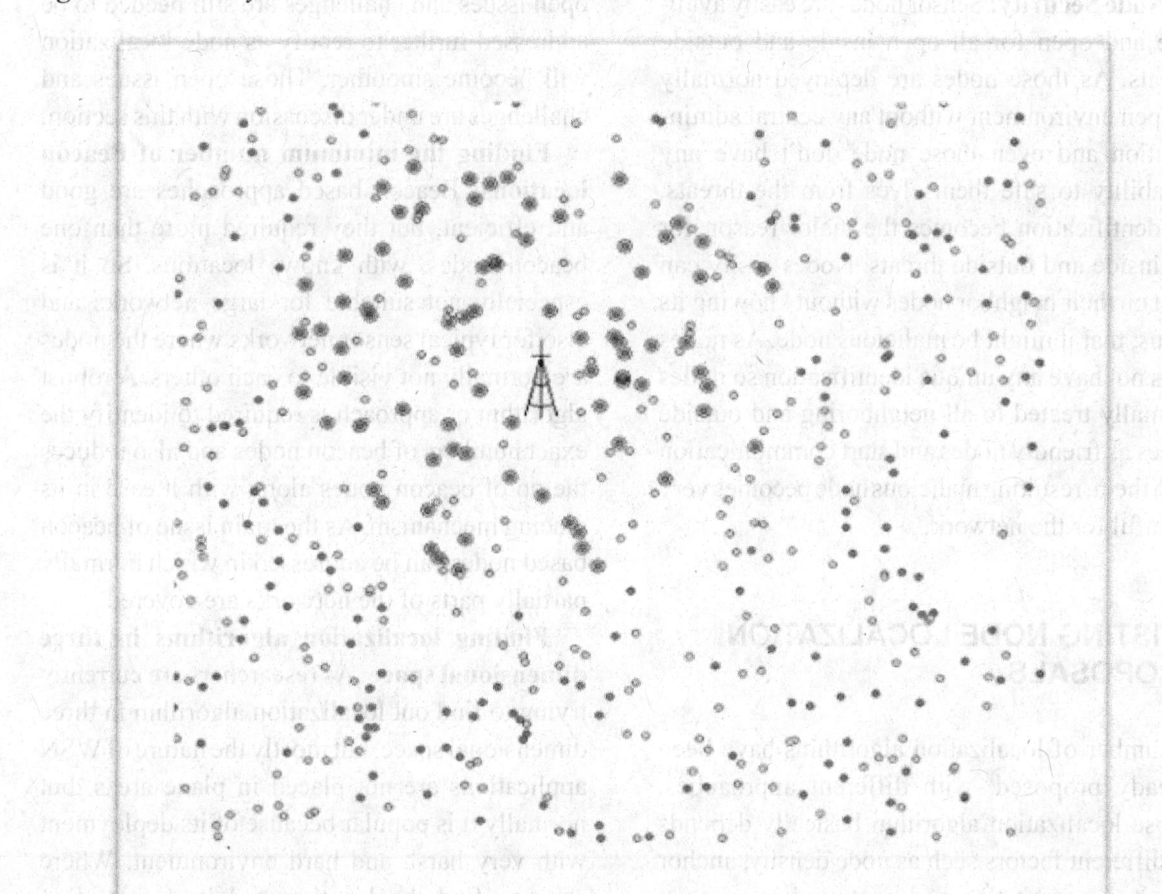

Working Mechanism: Working mechanism is also different than conventional networks. In wireless sensor networks nodes are normally not responsible for more computational tasks and also node are not liable for any safety mechanism, or even nodes not under any central administration control mechanism. The primary tasks for the sensor nodes is to simply record any physical event, if it is occurred resulting of any action, and then simply transfer that collected data to its sink.

Node Deployment: Node deployment is also different with wireless sensor network. sensor networks contain a huge number of nodes as those nodes are small in size and they are responsible to collect the data simply and then transfer it to the sink. Normally node deployment is random in fashion without following any mechanism or topology which we normally used with conventional networks. This type of deployment is also becoming reason for not available of node identification.

Sensor Node Applications: Wireless sensor network is rich with respect to its applications, it can be easily deployed with almost any kind of application. This is the most specialty of sensor network that it can be deployed even with very harsh and hard applications, such as battle field, high temperature thermal fields, for surveillance of fire in forests. Application property gives high importance to this type of network, this all become easy because its ease in deployment and its size along its cost.

Node Security: Sensor nodes are easily available and open for all open inside and outside threats. As those nodes are deployed normally in open environment without any central administration and even those node don't have any capability to safe themselves from the threats. Unidentification becomes the major reason for the inside and outside threats. Nodes easily can trust on their neighbor nodes without knowing its status, that it might be malicious node. As nodes does not have any unique identification so nodes normally treated to all neighboring and outside nodes as friendly nodes and start communication with them, resulting malicious node becomes very harmful for the network.

EXISTING NODE LOCALIZATION PROPOSALS

A number of localization algorithms have been already proposed, with different approaches. Those localization algorithm basically depends on different factors such as node density, anchor density, computation and communication cost. All node localization approaches have different pros and cons. All algorithms are related to centralized and distributed mechanism, few of them are very productive and efficient with low cost, while others have limited benefits. The detailed descriptions of all node localization algorithms are shown in Table 1. Moreover some schemes perform well in high anchor density while some need only few anchors. As shown in (http://www.cse.ust.hk/~yangzh/yang_pqe.pdf, n.d).

OPEN ISSUE AND CHALLENGES

Even though a large amount of research has been already conducted related to the node localization in WSN, but still there is a huge gap, as most of the research results are based on the assumption bases not on implementation basis. Especially few open issues and challenges are still needed to be addressed further to rectify as node localization will become smoother. Those open issues and challenges are under discussion with this section.

Finding the minimum number of Beacon locations: Beacon based approaches are good and efficient, but they required more than one beacon nodes, with knows locations. So it is especially not suitable for large networks and also for typical sensor networks where the nodes are normally not visible to each others. A robust algorithm or approach is required to identify the exact numbers of beacon nodes and also reduces the no of beacon nodes along with it ease in its placing mechanism. As the main issue of beacon based nodes can be addressed in which normally partially parts of the networks are covered.

Finding localization algorithms in three dimensional space: As researchers are currently trying to find out localization algorithm in three dimensional space, but mostly the nature of WSN applications are not placed in plane areas, but normally it is popular because of its deployment with very harsh and hard environment. Where even to find the location and its information almost not possible for example deployment in hilly war battle area. So there is a sufficient gap to work with this area and can come up robust algorithm particularly for this three dimensional node localization area.

Attack the challenges of Information Asymmetry: Sensor networks are very popular because of its range of varsity application and ease in deployment. It is the only kind of application which can be deployed very easily, quickly with very hard and hard environment. Sensor networks normally used to detect the enemies movements during the war, in the battle fields. Those networks also deployed for the surveillance of the fire forest, thermal, chemical, mining and even underwater for security and surveillance. It is really hard if any beacon node algorithm provide little wrong information then the use of entire data will be useless. In case if any information is leakage is

Table 1. Different node localization algorithms

Algorithms or (Proposals)	Objective	Centralized or Distributed	Description	Accuracy	Cost in term of Messaging / Computation
Cheng, Lui, and Tam (2007)	Distributed Algorithm Composed of MDP and PDM	Distributed	In the first phase some sensors are selected as secondary anchors which are localized through MDS. In the second phase the normal sensors are localized through proximity distance mapping	High	Low
Ahmed, Shi and Shang (2005)	Proposed a location scheme by using MDS and APS.	Distributed	First MDS is used to relatively localize reference nodes. Then APS uses these nodes as anchors to localize the rest of the nodes.	High	Low
Lloret, Tomas, Garcia, and Canovas, (2009)	Proposed a location scheme using inductive and deductive approach.	Distributed Algorithm	The algorithm uses a probabilistic model using both deductive and inductive approach.	High	Low
Maroti, Kusy,Balogh, Volgyesi, Nadas, Molnar, et al. (2005)	Proposed a location scheme based on Interferometric ranging	Distributed Algorithm	The algorithm uses a genetic algorithm approach to optimize the location problem using interferometric ranging.	High	High
Patwari & Hero (2006)	Propose a location scheme based on interferometric ranging.	Distributed Algorithm	Proposed an interferometric ranging based localization which first estimates pair wise distances and then uses them to estimate coordinates via distributed weighted multidimensional scaling.	High	High
Huang, Zaruba, & Huber (2007)	Propose a localization scheme based on Interferometric ranging for larger networks	Distributed Algorithm	Propose an iterative algorithm to solve the localization problem using interferometric ranging	High	Low
Alsindi, Pahlavan, Alavi, (2006)	Present an error propagation aware routing for localization in WSNs	Distributed Algorithm	Each node gets position error variance from some anchors and formulates a weighting matrix. Next each node calculates its position by incorporating the weighting matrix into WLS.	High	Low
Shang, Rumi, Zhang, Fromherz (2003)	Present an MDS-MAP algorithm that uses connectivity information to derive the locations of the nodes in the network.	Centralized	Presented and algorithm MDS-MAP that uses and all-pairs shortest paths algorithm to roughly estimates the distance between each possible pair of needs. Then they have used MDS to derive node locations that fit those estimated distances and finally they have normalized the resulting coordinated to take into account any nodes whose positions are known	High	High

continued on following page

Table 1. Continued

Algorithms or (Proposals)	Objective	Centralized or Distributed	Description	Accuracy	Cost in term of Messaging / Computation
Kannan, Mao, & Vucetic (2006)	Implement a two phase localization method based on the simulated annenling technique which also takes account the error due to flop ambiguity	Centralized	Proposed a two-phase simulated annealing based localization algorithm. Where an initial location estimate is obtained in simulated annealing technique and the large error due to flip ambiguity is mitigated in the refinement phase using neighborhood information of nodes.	High	High
Alippi and Vanini (2006)	Present a multi-hop localization technique for WSN, exploiting acquired received signal strength indications	Centralized	The RSSI values of the packet exchange among nodes at different power levels are collected (RF mapping Phase) and processed both to build to build the ranging model to be fed into a centralized Minimum Least Square (MLS) algorithm	High	High
He, Huang, Blum, Staneovic, and Abdelzaheer, (2003)	Describe a range free algorithm to make the scheme cost effective than range based approaches.	Distributed	Test whether the anchor node is inside or outside the triangles made by three anchors and then by utilizing combination of anchor positions. The diameter of the estimated area in which the node resides can be reduced to provide good localization estimate.	Low	Low
Savvides, Park, and Srivastava (2002)	Present a collaborative multi approach which enables ad hoc deployed sensor nodes to accurately estimate their locations by using known beacon locations.	Distributed	Presented a collaborative multi approach in which all nodes obtain a bounding box of region where the node lie.	Low	Low
Simic and Sastry (2002)	Present a distributed algorithm for localization of nodes in a discrete mode of random ad hoc network	Distributed	Drive the whole region in a number of square cells. Each unknown node will send hello packets to its neighbors and based on the response the beacons, the unknown nodes can update their position estimates.	Low	Low
Nagpal, Shrobe, and Bachrach (2004)	Discover the position information of ad hoc wireless network even when the elements have literally been sprinkled over the terrain	Distributed	Each seeds produces propagating gradient that allows the sensors to estimate their distance from the seed. Next the sensors combine all the distance estimates to produce their own position.	Low	Low
Priyantha, Balakrisinan, Demaine, and Teller (2003)	Propose an anchor free solution to the problem of node localization where node starts from a random initial coordinate assignment and converge to a consistent solution using only local node interactions.	Distributed Algorithm	The algorithm models the nodes as point masses connected with springs and use-force directed relaxation method to converge towards a minimum energy configuration.	Low	Low

continued on following page

Table 1. Continued

Algorithms or (Proposals)	Objective	Centralized or Distributed	Description	Accuracy	Cost in term of Messaging / Computation
Savarese, Rabaey, & Beutel (2001)	Implement a cooperative ranging approach to get rid of the burden of using beacons in wireless sensor networks.	Distributed Algorithm	Proposed a cooperative ranging scheme which transmits the local information of each node to the neighboring nodes and by iteratively doing this results in a convergence to global solution.	Low	Low
Moore, Leonard, Rus, & Teller (2004)	Localize sensor nodes in a region by the use of robust quadrilaterals.	Distributed Algorithm	Each node measures the distances of the neighboring nodes and form a cluster in some local coordinate system. In the next phase coordinate transformation can be compared to switch the clusters in a global coordinates systems	Low	Low
Meertens and Fitzpatrick (2004)	Construct a global coordinate system in a network of static computational nodes from inter-node distances	Distributed Algorithm	Each node in the network first construct a spatial map of its own position based on inter-node distance information and then each node reconcile its own map with those of nearby nodes and forms a global coordinate system	Low	Low

occurred because of it wrong identification then the entire application will play an opponent role rather than the positive one. So a huge gap is there to address and future work is to intend to overcome this most important issue.

Mobility Supportive Sensor Network Algorithms: Sensor network applications are successfully moving towards mobility support applications. With few applications it is required to move the sensor nodes with a particular time to collect the data from the different regions. Research community proposed different approached to solve the mobility issues, but the node identification with the mobility is very hard to identify, the available algorithms are not fully supporting to sensor node mobility. So this is an open issue for sensor networks to design a robust algorithm which supports fully to mobility based applications.

Interferomatric ranging based algorithm with support of error propagation localization: Interferomatric technique is more accurate and its results are closer to perfection. This ranging based technique has been proposed recently. Error propagation is still there and not limited up to desire limit even with algorithm technique. Interferomatric ranging base algorithm is not properly supporting to large sensor networks, so it is still required to find out proper way for large networks and proper solution with error propagation.

FUTURE PERSPECTIVES

Node localization is still a main issue for the sensor networks, even though a wide range of different algorithms has been proposed. To make sensor networks more reliable, with greater QoS and more useful with all aspect, it is required to work on it and specially following points should be in consideration, which are more concern with node localization or node identification.

Figure 4. Future perspective of node localization scheme

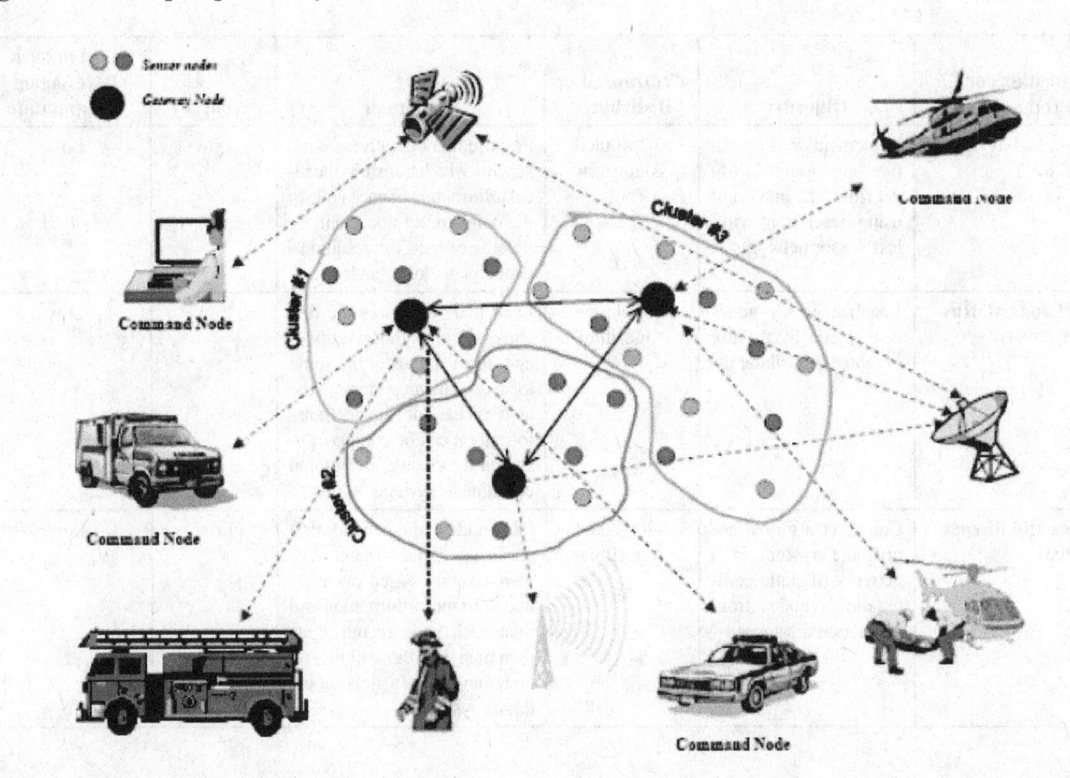

- Node should have more capable for more resources
- For node identification node should have an ID.
- Node design should be capable to estimate the distance by TOA technique
- Node formation mostly should be in cluster formation and each cluster might have one gateway node as show in Figure 4.
- Nodes might be capable with GPS capability, as involvement of GPS can enhances the range and accuracy of the sensor network.

CONCLUSION

Wireless sensor networks (WSNs) are taking a major share with almost all type of different applications and specially it is most suited in very harsh and tough environments, where even it is to hard to deploy conventional network applications, for example in the forest fire area, battle fields during the war, chemical and thermal sites and also for few underwater applications. WSN is becoming now part of almost all application because of its ease in deployment and cheaper in the cost. These networks are resource constraints and having very small in size, computation and with very less communication capabilities. Node are normally deployed in random fashion and it's too hard to find its location there is no any pet predefine way like conventional networks to find out its location. While location is highly important to know the data correlation for example its target tracking and to know exact about the actual vicinity of the any event occurrence. Our chapter describes the current available approaches, issue and challenges with current approaches and future directions for node localization, one by one. As node localization is highly important for large sensor networks where use desired to know about the exact location of the nodes to know the data location.

REFERENCES

Ahmed, A. A., Shi, H., & Shang, Y. (2005). Sharp: A new approach to relative localization in wireless sensor networks. In *Proceedings of IEEE ICDCS*, 2005.

Alippi, C., & Vanini, G. (2006). A RSSI-based and calibrated centralized localization technique for wireless sensor networks. In *Proceedings of Fourth IEEE International Conference on Pervasive Computing and Communications Workshops (PERCOMW'06),* Pisa, Italy, March 2006, (pp. 301-305).

Alsindi, N. A., Pahlavan, K., & Alavi, B. (2006). An error propagation aware algorithm for precise cooperative indoor localization. In *Proceedings of IEEE Military Communications Conference MILCOM2006*, (pp. 1-7). Washington, DC, USA, October 2006.

Bachrach, J., Nagpal, R., Salib, M., & Shrobe, H. (2004). Experimental results for and theoretical analysis of a self-organizing a global coordinate system from ad hoc sensor networks. *Telecommunications System Journal*, *26*(2-4), 213–233. doi:10.1023/B:TELS.0000029040.85449.7b

Bachrach, J., & Taylor, C. (2005). Localization in sensor networks. In Stojmenovic, I. (Ed.), *Handbook of sensor networks: Algorithms and architectures*. doi:10.1002/047174414X.ch9

Blum, B. M., Stankovic, J. A., Abdelzaher, T., He, T., & Huang, C. (2005). Range-free localization and its impact on large scale sensor networks. In *ACM Transactions on Embedded Computing Systems*, (vol. 4, pp. 877-906). November 2005.

Cheng, K.-Y., Lui, K.-S., & Tam, V. (2007). Localization in sensor networks with limited number of anchors and clustered placement. In *Proceedings of Wireless Communications and Networking Conference, 2007 (IEEE WCNC 2007)*, March 2007, (pp. 4425–4429).

Chraibi, Y. (2005). *Localization in wireless sensor network.* Master's thesis, KTH Stockholm, Sweden, 2005.

He, T., Huang, C., Blum, B., Stankovic, J., & Abdelzaher, T. (2003). Range-free localization schemes in large scale sensor networks. In *Proceedings of the Ninth Annual International Conference on Mobile Computing and Networking (MobiCom'03),* September 2003, San Diego, CA, USA, (pp. 81-95).

He, T., Stoleru, R., & Stankovic, J. A. (2006). *Range free localization.* Technical report, University of Virginia, 2006.

Huang, R., Zaruba, G. V., & Huber, M. (2007). Complexity and error propagation of localization using interferometric ranging. In *Proceedings of IEEE International Conference on Communications ICC 2007*, (pp. 3063-3069). Glasgow, Scotland, June 2007.

Jin, J., Wang, Y., Tian, C., Liu, W., & Mo, Y. (2007). Localization and synchronization for 3D underwater acoustic sensor networks. In *Ubiquitous Intelligence and Computing 2007*, (pp. 622-631). Hong Kong, China, July 2007.

Kannan, A. A., Mao, G., & Vucetic, B. (2006). Simulated annealing based wireless sensor network localization. *Journal of Computers*, *1*(2), 15–22. doi:10.4304/jcp.1.2.15-22

Lazos, L., & Poovendran, R. (2005). SeRLoc: Robust localization for wireless sensor networks. In *ACM Transactions on Sensor Networks,* (pp. 73-100). August 2005.

Lloret, J., Tomas, J., Garcia, M., & Canovas, A. (2009). A hybrid stochastic approach for self-location of wireless sensors in indoor environments. *Sensors (Basel, Switzerland)*, *9*(5), 3695–3712. doi:10.3390/s90503695

Mao, G., Fidan, B., & Anderson, B. D. O. (2007). Wireless sensor network localization techniques. *Computer Network: The International Journal of Computer and Telecommunications Networking*, *51*, 2529–2553.

Maroti, M., Kusy, B., Balogh, G., Volgyesi, P., Nadas, A., & Molnar, K..... Ledeczi, A. (2005). Radio interferometric geolocation. In *Proceedings of 3rd International Conference on Embedded Networked Sensor Systems (SenSys)*, (pp. 1-12). San Diego, California, USA, November 2005.

Martins, M., Cheung So, H., Chen, H., Huang, P., & Sezaki, K. (2008). Novel centroid localization algorithm for three-dimensional wireless sensor networks. In *IEEE Transactions on Wireless Communication, Networking and Mobile Computing*, (pp. 1-4). October 2008.

Meertens, L., & Fitzpatrick, S. (2004). *The distributed construction of a global coordinate system in a network of static computational nodes from inter-node distances.* Kestrel Institute Technical Report KES.U.04.04, Kestrel Institute, Palo Alto, 2004. Retrieved from ftp://ftp.kestrel.edu/pub/papers/fitzpatrick/LocalizationReport.pdf

Merico, D., & Bisiani, R. (2006). *Positioning, localization and tracking in wireless sensor network.* Technical report, DISCo, NOMADIS, March, 2006.

Moore, D., Leonard, J., Rus, D., & Teller, S. (2004). Robust distributed network localization with noisy range measurements. In *Proceedings of the Second ACM Conference on Embedded Networked Sensor Systems (SenSys'04)*, November 2004, Baltimore, MD, (pp. 50-61).

Ou, C.-H., & Ssu, K.-F. (2008). Sensor position determination with flying anchor in three dimensional wireless sensor networks. *IEEE Transactions on Mobile Computing*, (September): 1084–1097.

Patwari, N., & Hero, A. O. (2006). Indirect radio interferometric localization via pairwise distances. In *Proceedings of 3rd IEEE Workshop on Embedded Networked Sensors (EmNets 2006)*, (pp. 26-30). Boston, MA, May 30-31, 2006.

Priyantha, N., Balakrishnan, H., Demaine, E., & Teller, S. (2003). *Anchor-free distributed localization in sensor networks.* MIT Laboratory for Computer Science, Technical Report TR-892, April 2003. Retrieved from http://citeseer.ist.psu.edu/681068.html

Savarese, C., Rabaey, J., & Beutel, J. (2001). Locationing in distributed ad-hoc wireless sensor networks. In *Proceedings of IEEE International Conference on Acoustics, Speech, and Signal Processing (ICASSP'01)*, May 2001, Salt Lake City, Utah, USA, (vol. 4, pp. 2037-2040).

Savvides, A., Park, H., & Srivastava, M. (2002). The bits and flops of the n-hop multilateration primitive for node localization problems. In *Proceedings of the 1st ACM international Workshop on Wireless Sensor Networks and Applications (WSNA'02),* September 2002, Atlanta, Georgia, USA, (pp. 112-121).

Shang, Y., Ruml, W., Zhang, Y., & Fromherz, M. (2003). Localization from mere connectivity. In *Proceedings of ACM Symposium on Mobile Ad Hoc Networking and Computing (MobiHoc'03)*, June 2003, Annapolis, Maryland, USA, (pp. 201-212).

Simic, S., & Sastry, S. (2002). *Distributed localization in wireless ad hoc networks.* Technical report UCB/ERL M02/26, UC Berkeley, 2002. Retrieved from http://citeseer.ist.psu.edu/simic01distributed.html

Tseng, Y.-C., & Wang, Y.-C. (2008). Efficient placement and dispatch of sensors in wireless sensor network. *IEEE Transactions on Mobile Computing*, (February): 262–274.

Tseng, Y.-C., & Wang, Y.-C. (2008b). Distributed deployment scheme in mobile wireless sensor networks to ensure multi level coverage. *IEEE Transactions on Parallel and Distributed Systems*, (September): 1280–1294.

Yu, R., Sun, Z., & Mei, S. (2006). A power aware and range free localization algorithm for sensor network. *IEEE Transactions on Communications*, (August): 1–5.

Zaman, N., & Abdullah, A. B. (2011). *Position responsive routing protocol PRRP*. The 13th International Conference on Advanced Communication Technology, Phoenix. *Korea & World Affairs*, (February): 2011.

Zhou, X., Zhang, L., & Cheng, Q. (2006). Landscape-3D: A robust localization scheme for sensor networks over complex 3D terrains. In *Proceedings of the 31st IEEE International Conference Local. Computer Networks*, (November): 239–246.

Section 2
Energy Efficiency of WSN

Chapter 10
Energy Efficient Routing Protocols in Wireless Sensor Networks: A Survey

Vasaki Ponnusamy
Universiti Teknologi PETRONAS, Malaysia

Azween Abdullah
Universiti Teknologi PETRONAS, Malaysia

Alan G. Downe
Universiti Teknologi PETRONAS, Malaysia

ABSTRACT

This research presents a survey of energy efficient routing protocols in sensor network by categorizing into a main classification as architecture based routing. Architecture based routing is further classified into two main areas: flat or location based routing protocol, and hierarchical based routing protocols. Flat based routing is more suitable when a huge number of sensor nodes are deployed, and location based routing is employed when nodes are aware of their location. Hierarchical routing look into alternative approach by placing intermediate nodes in terms of cluster heads, gateway nodes, or mobile entities for efficient handling of energy. The survey is presented in order to highlight the advantage of hierarchical based routing, mainly the deployment of mobility routing. As not many surveys have been conducted in mobility based routing, this chapter can be helpful for looking into a new perspective and paradigm of energy efficient routing protocols.

DOI: 10.4018/978-1-4666-0101-7.ch010

INTRODUCTION

Sensor networks are facing many challenging issues such as routing failure, routing holes, energy depletion and network failure. Among this energy depletion should be given careful attention as this factor contributes mainly to other issues mentioned above. When sensor nodes run out of energy, the nodes go into sleep mode or dies completely. And this yield to routing hole around the network which leads to network failure. So energy efficient operation in sensor network needs careful attention and further consideration in order to provide power efficient, increased lifetime of the network. There are many researches and contributions performed in terms of energy efficient sensor network routing protocol. But there are still factors to be considered in increasing the lifetime of sensor.

Sensor networks are deployed to create "smart environments" which are useful in monitoring conditions such as temperature, sound, movement, location, light and others (Youssef et al. 2002). Wireless sensor network (WSN) differs from traditional network in terms of energy constraint of battery-backed sensors. In most of the situation, sensor nodes operate based on powered batteries and has limited capabilities as well. The design of these sensor nodes and batteries need to be compromised based on the size and cost. The higher capacity batteries can last longer but increases the size of the sensor node and the cost as well. Smaller batteries might not last longer and reduces the lifetime of the node. Sensor battery lifetime is directly related to the lifetime of the node and therefore affects the lifetime of the entire sensor network as well.

Routing in WSN differs from traditional and mobile ad hoc routing in which the latter two do not have energy as constraint in terms of routing. Whereas in WSN, since sensor nodes are energy constraint, routing path should be carefully planned to avoid having the same node to be the routing agent. But one of the potential problems with current routing protocols is that it looks for the lowest energy route and uses it for the entire communication. This leads to energy depletion at these nodes and reduces network life time (Sha et al. 2005). When considering extending the lifetime of the node and the network, energy efficiency of routing protocols is an issue (Gui & Mohapatra 2003). When energy at sensor node is depleted, the node becomes inactive or dies and this leads to routing holes in sensor network. Routing hole refers to the areas where data cannot be routed since there are no active nodes available. Most of the protocols proposed for energy efficient operation in sensor network looks into using existing architecture of fixed sensor nodes and base station. Although this architecture looks simple and efficient in terms of sensing and relaying data, there are many problems associated with it. Among those, this architecture uses multihop routing for conveying sensed data to the base station. In this case, the static base station uses more power for its receivers and more energy is utilized by other nodes to send data to the base station thus reduces network lifetime. More over when the base station is static, nodes closest to the sink and nodes which are connecting two parts of the network are utilized the most and runs out of energy first. This causes routing hole near the base station which would eventually cause network failure.

The unique characteristics of sensor network with battery-supported energy, less computing power, less prone to failure and collaborative information processing makes it different from traditional TCP/IP network. These characteristics are the basic factors for the need of a unique routing protocol in WSN to look into energy-efficiency as one of the important criteria. And most of the past research activities in WSN routing basically looks into one single criterion, i.e. energy-efficient routing, fault-tolerant routing, geographic routing etc. But there are four factors which are considered essential for good routing protocol: i) energy efficiency ii) load balancing iii) fault tolerance and iv) scalability. Based on these four important parameters, the survey in this paper is classified into two main classes as flat/location based routing and hierarchical routing. The survey conducted by looking into how these four parameters can be achieved employing such classes.

Flat/location based routing takes different paths based on the energy usage and the remaining energy of sensor nodes (Sha et al. 2005). However, one limitation of this technique is that, it might further extend the path. Hierarchical routing is achieved by employing intermediate nodes such as cluster heads, gateways or mobile entities. Employing mobile entities in WSN routing has been considered as potential solution for energy efficient operation. Mobile entities can be in the form of data mules, mobile relays, mobile agents and mobile base station. But one problem with such approach is the latency encountered. Most of the surveys conducted in WSN routing were classified into types of routing such as flat-based routing, ad hoc routing, hierarchical based routing etc. In this survey we are focusing more on the hierarchical based routing looking at the advantage of mobility in sensor network. The main focus of this category is to give authors a new thought of employing the combination of these approaches for energy efficient routing.

The rest of the paper is organized as follows: Section 2 consists of surveys on various energy efficient routing protocols discussing their protocol approach and results. A complete comparison using various metrics is given in Section 3. Section 4 concludes the paper with recommendation and future research issues.

Survey on Energy Efficient Routing Protocols in Wireless Sensor Networks

This section comprises some of the related works in the area of energy efficient routing protocols in sensor network. The survey is performed by looking into: i) flat/location based routing and ii) hierarchical routing which is further classified into cluster head/gateway and mobility based routing as shown in Figure 1. Flat/location based routing looks into multihop routing by employing some energy aware mechanisms such as shortest path, energy aware path, alternate path or additional nodes. Hierarchical routing look into employing intermediate nodes as cluster heads, mobile entities, mobile relays, mobile agents (software or hardware) and gateways. The problems, suggestions and future works discussed in this paper helps in further research and enhancement of energy efficient routing.

Flat/Location Based Routing Protocol

The flat/location based routing protocols discussed in this section mainly looks into using multihop routing. Multihop routing works by having nodes forwarding data from one node to another within the neighborhood. The closest node to the base station would eventually forward data to the base station. Researchers have been using this routing mechanism by exploiting energy efficient algorithms. The section below discusses some of the mechanisms exploited. This routing protocol can be categorized into energy aware routing and energy conserve routing.

WEAR: A Balanced, Fault-Tolerant, Energy-Aware Routing Protocol for Wireless Sensor Networks

(Sha et al. 2005) suggests that in order to deploy sensor network in a larger scale, the routing protocol in WSN has to consider four important criteria: energy-efficiency, load balancing, fault-tolerance and scalability. This protocol uses greedy geographic forwarding to achieve scalability as only local information is needed for routing purposes. It is also considered energy efficient as only the shortest path is considered for routing purposes. Load balance is achieved by forwarding messages to sensors with high remaining energy, furthest to the hole and less important. The rout-

Figure 1. Classification of energy efficient routing protocols

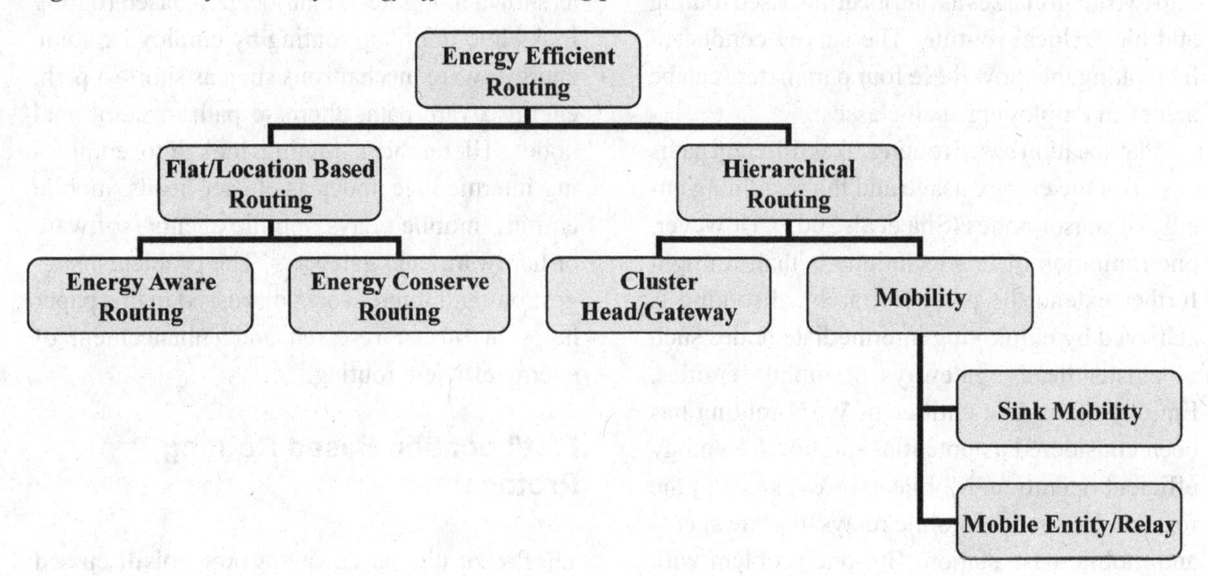

ing decision in this protocol is made based on the weighed heuristic value which is a combination of four factors described earlier: the distance to the destination, neighbor's sensor energy level, global location information and hole information of the local sensors. WEAR protocol has two different cases or ways on how it is operated. On the first case, it uses weight value to determine the next receiving node. When a sensor node receives a message, it would check whether it is the receiving node. If it is not, it would check the neighbor list to determine the next receiving node based on the smallest weight value. The process continues until the destination is reached. On the other hand the second case works by using two different modes: greedy mode and bypassing mode. The greedy mode works by forwarding the message to destination that is nearer to the destination than the current node. Then the receiving node forwards the message to the node having the smallest weight value. If no node is found closer to the destination, then the routing goes into bypassing mode. Through bypassing mode the routing is performed based on the right-hand rule in (Intanagonwiwat et al. 2000) by forwarding the message to the node that is closer to the destination than the current location where the bypassing mode resumes. This paper looks into the second alternative whereby it is more energy efficient and guaranteed delivery to the destination.

One of the main goals of WEAR is to eliminate holes and not to enlarge the hole. This can be achieved by identifying the hole and propagating the hole information to the sensors near the hole and then modifying the weight of sensors which may lead to different path. Hole information is calculated by using two processes which are hole identification and hole maintenance. Hole identification consists of hole locating, hole announcing and hole propagating. Hole identification is a process of identifying the hole using the three methods mentioned above. Hole locating would identify the hole and the information is propagated to the hole boundary nodes by hole announcing process and the information is further sent to sensor nodes within maximum hops by hole propagation process. One of advantage of this protocol is its energy efficient mechanism and guarantees delivery of data whilst avoiding routing holes.

GEAR: Geographical and Energy Aware Routing: A Recursive Data Dissemination Protocol for Wireless Sensor Networks

Like many other protocols, GEAR (Yu et al. 2001) generally focuses data centric approach where packets are forwarded to all the nodes within a target region and no location database is used for node to detect location identification. Besides that, all sensor nodes in GEAR are assumed to be static and nodes are aware of their current position. The protocol basically focuses on energy aware routing and a geographic based heuristics for neighbor selection. A recursive geographic forwarding technique is used for forwarding data. There are two phases used for forwarding packets to all nodes in the target region. The first approach is to forward packets to the target region in which a packet is always forwarded to a neighbor node which is the closest to the destination. When all nodes are further away from the destination, in which a hole exists, GEAR picks the best node which minimizes the cost value of this neighbor. The second approach is to disseminate the packet within the region by using recursive geographic forwarding algorithm for most of the cases. This protocols delivers more packets and handles routing hole efficiently.

Energy Aware Routing for Low Energy Ad Hoc Sensor Networks

(Shah & Rabaey, 2002) looks at survivability of low energy networks by maintaining a few alternative good paths for communication instead of identifying one single shortest path. The motivation for this approach is that, instead of using one single path for communication, the protocol would take different path at different times so that the energy at any single node is not depleted easily. It is also easier to respond to nodes moving in and out of network thus minimizing routing overhead. One of the potential problems with current sensor routing protocols is that, the lowest energy route is chosen for most communication. This may not be the best way to choose for network lifetime and at the same time energy depletion occurs along the nodes which lead to holes. The optimal path chosen in this approach tries to solve the problem mentioned above by ensuring the network degrades gracefully rather than causing partition in the network. In this protocol, multiple alternative paths are chosen between source and destination and each path is given a probabilistic value based on the energy metric value. Each time when a packet is sent from a source to a destination, several different paths are chosen based on the probabilistic value. In this case no single node would be used for the entire operation to ensure energy is not depleted easily. Due to the probabilistic nature of the protocol, different routes are evaluated continuously and probabilities are chosen accordingly. The protocol is categorized into three phases:

Setup phase where flooding occurs to find all the routes from source to destination and energy costs are indentified. Routing tables are built at this phase. The destination node would always initiate the connection by flooding the network towards the direction of the source node. The cost field is set to zero initially. Every intermediate node forwards the request to the neighboring node which is closer to the source rather than the destination node. Upon receiving the request, the energy metric of the neighbor that is sent the request is computed and added to the path cost. Paths that incur a very high cost would not be considered to be added into the forwarding table. A probabilistic value of each node is added into the forwarding table. The probabilistic value is inversely proportional to the cost value. In this case each node would have a number of nodes to which it can forward the packets. The average cost of reaching the destination is calculated by using the values in the forwarding table.

Communication phase where data is sent from a source to a destination is based on the probabilistic value and energy costs are identified at setup phase. Several different phases are used at this phase. Each of the nodes forwards data packets to a randomly chosen neighboring node from the forwarding table with the probability of the neighbor chosen is equal to the probability in the forwarding table. Route maintenance – flooding is performed infrequently to keep all the nodes and paths alive.

As compared to directed diffusion (Intanagonwiwat et al. 2000) routing protocol, this protocol shows that the traffic is spread over the network which results in a much 'cooler' network. The nodes in the center of the network consume energy for a longer period and the time lapse for energy conservation has increased. The average energy consumption per node has been reduced due to the low overhead of the protocol. Besides, energy differences between nodes have been reduced as well. The increase in network lifetime shows that this protocol takes a long time for the energy to be aware that network has failed as compared to directed diffusion. As a conclusion, this protocol is suitable for low energy and low bit rate networks and taking the lowest energy route is not necessarily the best solution for network lifetime. Thus, using the simple way of sending traffic through different routes using probabilistic forwarding may help in utilizing the resources more effectively and not adding much complexity at a node. The deployment of multiple paths instead of relying on single path allows energy at single node is not depleted easily.

Highly-Resilient, Energy-Efficient Multipath Routing in Wireless Sensor Networks

Earlier works that have been focusing on creating single path for dissemination of sensed data lead to a problem of flooding of data to a failed node. Such flooding effect can give impact on the lifetime of the network. So (Ganesan et al. 2001) proposes a multipath routing to increase the resilience of node failure in which, while a path is established between source and sink, an alternative path could be established to minimize the effect of flooding. There are two different multipath routing proposed in this protocol, disjoint multipath and braided multipath.

The use of multipath routing allows energy utilization to be spread throughout the nodes that leads to load balancing. Moreover, duplicate delivery of data increases the accuracy of surveillance tracking. The actual reason for the use of multipath in this protocol is to identify alternative path between the source and the sink. From these available paths, the best path (primary path) based on low latency, low loss etc. can be identified. And in situation where the primary path fails, the alternative path could replace the former. This eliminates the problem of flooding the network when the primary path fails. There are very rare occasions where both the primary and alternative paths fail at the same time.

Disjoint Multipath

Disjoint multipath is created by constructing a small number of alternative paths that are disjoint with the primary path and with each other. This node disjoint multipath is created by localized information rather than having global topology information. This is constructed by the sink node sending primary-path reinforcement to the most preferred neighbor and that neighbor then sends it to its preferred neighbor in the direction of the source. After starting to receive the data along the primary path that has been created by the method earlier, the sink then would send alternative path reinforcement to its next preferred neighbor. If the neighbor has already been on the primary path, then it would send a negative reinforcement to the sink. The next preferred neighbor then would be selected which can lead to a k-disjoint multipath by sending out k alternative path reinforcements. One of the main disadvantages of disjoint multipath is, it can be energy inefficient in which alternative disjoint paths can be longer and utilize more energy.

Braided Multipath

To overcome the energy inefficiency discussed above, braided multipath relaxes the problem to certain extent. In braided multipath, alternative path is not completely disjoint to the primary path but is partially disjoint only. There are also two types of reinforcement used for braided multipath. Sink sends primary path reinforcement to its preferred neighbor A and at the same time sends alternative path reinforcement to the next preferred neighbor B. A and the subsequent neighbor node sends out the primary path reinforcement to construct the primary path. Whereas when B sends alternative path reinforcement to neighbor C which is already on the primary path, the subsequent alternative reinforcement is not sent out by that node. This can lead to alternative path constructed may not be completely disjoint to the primary path at all.

The results presented show the use of multipath routing for energy efficient recovery from node failures in wireless sensor networks. The results also show how flooding can reduce the network lifetime and the effect of keeping small numbers of multipath alive.

Summary

Flat/location based routing protocols discussed above falls into sub classes of handling hole problems, employing shortest path, multipath routing and data centric approach. Although all these classes look into energy efficient mechanism, they basically work on the same approach using multihop communication. As discussed above, multihop communication is a traditional routing protocol used in most of the communication protocols. But this may not be the case for WSN routing, as these networks are very limited in their resources especially energy. Therefore general routing (multihop communication) may not be the ultimate solution for energy efficient communication. The following section discusses a viable solution for handling energy issues in sensor networks by employing hierarchical routing in WSN for data collection and energy provision.

Hierarchical Routing Protocol

As discussed in the previous section, most of the literature in the field of wireless sensor network assumes sensor network to be consists of densely deployed static sensor nodes that can communicate with each other through multi-hop communication. Moreover a lot of work has been conducted in energy efficient communication for static sensor network assuming these are energy efficient paradigms. However currently there has been a lot of work on the possibility of venturing into clustering and mobility as an alternative solution for energy efficient communication in sensor networks.

Cluster Head/Gateway

Cluster head approach is a mechanism by classifying the network into clusters and a single node from each cluster will be elected as a cluster head node. This node is responsible for collecting data from sensor nodes and the aggregated data is then forwarded to the base station. This approach has seen tremendous energy saving and the discussion in the following section highlights some of the protocols adopted such mechanism. Employing gateway as an intermediate node is another possible hierarchical methods for energy efficient operation.

Energy-Efficient Communication Protocol for Wireless Microsensor Networks

(Heinzelman et al. 2000) proposes a Low-Energy Adaptive Clustering Hierarchy (LEACH) that works by forming a cluster-based protocol. LEACH is hierarchical clustering, self-organizing protocol which tries to distribute the energy utilization among a randomly chosen cluster head from within the cluster. The topology is formed by a group of sensor nodes group together to form a cluster and a cluster head is randomly chosen to be the main node for certain time interval. As it can be seen in Figure 2, there are two clusters, cluster1 and cluster2 are formed by each cluster having its own cluster head. The cluster head is elected on a rotation basis so that every single sensor node can be a cluster head at certain points. The main motivation for this approach is to evenly distribute the energy among all the nodes. If only one single node is to be responsible to be the cluster head, this will eventually cause energy depletion at that particular node. The protocol also looks at data fusion at routing protocol in order to reduce the amount of information transmitted to the base station.

The operation of LEACH is organized into phases of rounds which consist of a set-up phase and a steady-state phase. Set-up phase occurs during cluster formation whereas steady-state phase occurs when data is sent to the base station. During the advertisement phase, each node makes a decision whether to become a cluster head for the current round. The decision is mainly made by the number of cluster heads to be elected which is decided in advance and the number of times a node has become cluster head. This is done by nodes choosing a number between 0 and 1 and if the number is below a threshold value then, this node becomes cluster head for the current round. The threshold value permits each node to become cluster head within certain rounds. Once a node elected itself to be cluster head, it broadcasts an advertisement message to other nodes within the cluster. Upon receiving the advertisement message, each non-cluster nodes decides which cluster it belongs to and this decision is made by the strength of advertisement signal received. The node then informs the cluster head of its decision to join that particular cluster.

Upon receiving messages from nodes, cluster head then creates a TDMA (Time Division Multiple Access) scheduling based on the number of nodes wishes to join its cluster. TDMA schedule allows nodes to know in advance its time slot to send sensed data. The TDMA schedule is propagated to all sensor nodes. Only one node at a time sends its data to the cluster head to reduce energy utilized. Other nodes can go into sleep mode by turning off the radio and comes alive when its time slot arrives. This mechanism also allows

Figure 2. Cluster formation in LEACH

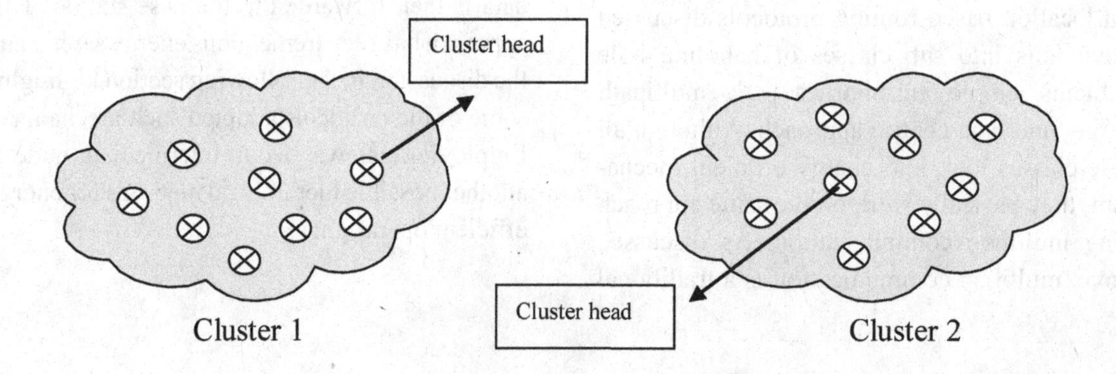

only one copy of data to be sent to the cluster head instead of sending duplicate copies. When all the data have been received from the nodes, cluster head then performs compression before sending the data to the base station.

The results obtained from LEACH shows that it reduces energy dissipation at sensor nodes by distributing energy among the nodes in the network and improves sensor network lifetime. This allows nodes to die randomly and makes sure that global energy usage is minimised since the load is distributed to all nodes at different intervals. Moreover, LEACH performs better as compared to static clustering algorithm since cluster heads that are elected are high-energy based nodes. LEACH does not require any global information from the base station and allows it to be completely distributed and no global network information is needed. Simulation results show that LEACH reduces communication energy by as much as 8 times compared with direct transmission and the first node that dies in the network occurs 8 times later than the direct transmission and the last node death happens 3 times later than in other protocols. Rotation of cluster head helps to distribute energy evenly around the sensor nodes. But one weakness of this protocol is its unsuitability in a large scale environment due to the direct communication between cluster head and base station.

A Hierarchical Routing Protocol for Survivability in Wireless Sensor Network

One of main challenges in wireless sensor network is to provide reliable network connectivity in an environment that is prone to network and node failures. The algorithm proposed in this paper (Al-Fares et al. 2009) is mainly looking at harsh environment such as forest fire and similar applications. The main concerns of this application are their wide deployment area, heterogeneous sensed data and harsh environment prone to failures. Not many researchers have been carried out in the area of unreliable link/node on the connectivity of WSN. This paper outlines a Self Organizing Network Survivability routing protocol (SONS) in WSN. It is capable of dealing with a large deployment area and should be able to handle node or link failure in WSN mainly in forest fire similar applications. This protocol is a hierarchical routing protocol in nature where the network is formed in a tree structure. One of the well known protocols, Low Energy Adaptive Clustering Hierarchy (LEACH) is a hierarchical routing protocol developed to handle network survivability and connectivity. Similar to LEACH, the protocol proposed in this paper also classifies the network into a form of cluster and the cluster head is responsible to convey messages to the base station.

SONS is developed with the motivation to handle link and node reliability problems in a hierarchical WSN. It deals with multi-hop hierarchy spanning tree to deal with the large deployment area. The fully distributed nature of SONS allows cluster head (CH) to detect the nearest parent-CH in order to forward data to the base station. In case the parent-CH is not available due to certain circumstances, then the next best CH is chosen in order to forward data. Redundant CH is introduced in this protocol to provide network fault tolerance and this may not be an issue in the high deployment area with tiny, inexpensive sensor nodes. The algorithm in this protocol is classified into start-up, setup and process phase.

Start-Up Phase

The start-up phase begins once after the sensor nodes (SNs) are fully deployed in an area of interest. This phase is basically to build the sensor network in the network layer. The initial process in this phase is where the network is formed into clusters with an elected CH. This process in similar to LEACH initially after which some additional features (remaining power source) are discussed. A parent-child tree is formed with the sink being

Figure 3. Hierarchical architecture of node communications

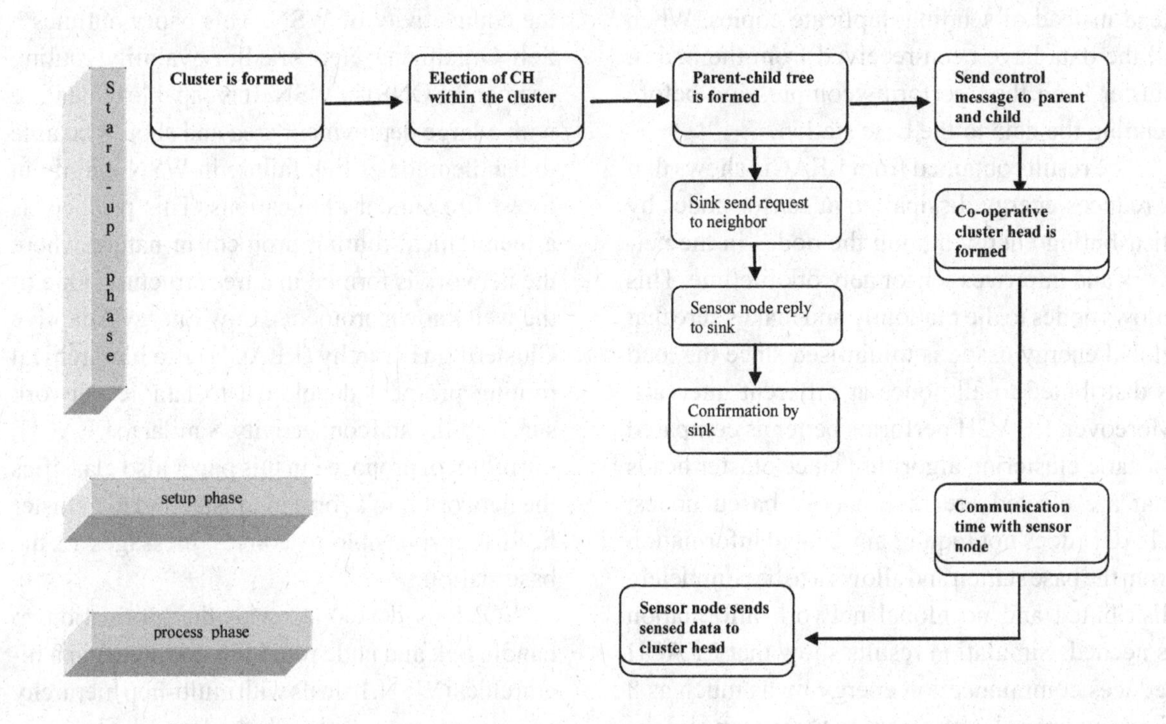

the root and all CH becomes leaves of the tree. During every election of CH within a cluster, the new CH sends a control message to the parent and child in the tree to update on the changes of the members in the tree.

Setup Phase

In this phase every CH will handle communication time with SNs within the cluster. At each cycle CH will assign a communication time slot to every cluster member based on its needs. This gives the flexibility for SNs to know when to wake up to send its data to the CH. This approach helps in terms of energy utilization whilst extending the network life time.

Process Phase

This phase takes the longest time interval compared to the other phases discussed above. This is the phase where SNs send their sensed data to the sink via CH/C-CH based on the given time slots. Besides redundant data is removed, data is compressed and data accuracy is enhanced at this phase as well (see Figure 3).

Based on the simulation results, it is noted that SONS consume less energy as compared to LEACH. Moreover, the difference in power consumption is increased with the increase in number of packets sent. The response time in SONS is improved by using redundant CH, in which the failure of one of the CHs will not terminate the communication. Redundant CH resumes the communication and based on the results, with redundancy of the communication time is 1/3 times lesser that SONS that have no redundancy. Further, when a particular CH does not receive any reply for the second time, it will update its parent table and notify the parent of the changes. This provides survivability and reliability features of the network.

A Constrained Shortest-Path Energy-Aware Routing Algorithm for Wireless Sensor Networks

This paper by (Youssef et al. 2002) explores the three important performance characteristics in a sensor network. While the majority of energy aware routing protocols in WSN attempt to extend the lifetime of the network by minimizing energy consumption, this protocol looks into two other measures as well. Besides energy, good end-to-end delay and throughput performance are also looked into as in other traditional routing protocols. This protocol introduces an energy-aware routing protocol by looking into distance as a measure for energy consumption and estimating propagation delay. The network in this protocol is organized into clusters with each cluster having a gateway node. The gateway node is responsible for mission-oriented task by identifying potential sensor nodes for sensing the environment. Gateway node becomes an interface between sensor nodes and the command node (sink). The command node is responsible for identifying the sensor nodes within the cluster and informing the sensor node ID to the gateway node.

The routing decision in this protocol can either be centralized at the gateway or distributed among sensor nodes. Routing decision refers to setting routes on the routing table that specifies the communication path of messages between nodes. Each time the network topology changes, routing table needs to be updated as in traditional routing protocols. But this process requires high usage of energy which is already limited within the sensor nodes and also creating excessive traffic on the network. Therefore, this protocol keeps routing decision to be made within the cluster in which the gateway node would be handling it. By centralizing routing decision at the gateway, resource-constrained sensor nodes are off the burden from handling such routing decision as well as data processing and communication. Since gateway node has a wider view of the entire network, it should be able to perform better routing decisions. There are two cycles in the entire operation of the network, data cycle and routing cycle. During data cycle, nodes would perform their task of sensing and collecting data. Whereas during routing cycle, gateway node determines sensor's state and inform them on how to route data.

This protocol uses a constrained shortest-path routing algorithm. The algorithm uses distance between two nodes as a measure for transmission power between two nodes. The transmission energy required is dependent on the distance, and the propagation delay is proportional to distance as well. The maximum transmission distance in constraint to a certain distance so that different network topologies can be obtained and do not limit the number of hops to just one hop direct from sensor to gateway. Even though direct transmission between sensor and gateway decreases the end-to-end delay, it incurs higher transmission energy consumption, especially nodes further away from the gateway. A smaller transmission distance creates a sparse graph with each sensor node is directly connected to its neighbor. Less energy consumed for the nearest neighbor transmission but increases end-to-end delay. So a suitable maximum transmission distance is ideal to obtain a desired performance metrics. The routing algorithm has to be dynamic in nature as in target tracking sensor network, the selected sensors vary according to the movement of target. Moreover, the gateway has to monitor the available energy level at sensors based on model-based energy consumption. The gateway makes rerouting decision based on this model using two criteria: i) reselection of active sensors takes place, ii) the battery level of any node drops. The algorithm also uses an energy-aware MAC layer protocol to increase the energy efficiency. A contention-free time division multiple access (TDMA) based MAC layer protocol is used where slot assignment is management by the gateway.

There are six important performance metrics used to analyze the performance of the protocol.

They are time for the first node to die, time for the last node to die, average node's lifetime, average delay per packet, network throughput and average energy consumed per packet. As the maximum transmission distance increases, the average energy consumed per packet increases. But end-to-end delay is reduced by increasing the throughput and the vice versa for lower transmission distance. For a moderate value of maximum transmission distance, the algorithm uses a small number of hops with a shorter distance giving better performance for all performance metrics mentioned above. This protocol maintains good end-to-end delay and shows better throughput performance.

Mobility

Mobility by means of employing mobile entities in WSN can take the routing burden away from the static sensor nodes and can possibly increase the lifetime of these static sensors. As seen in Figure 4, mobility can be seen in two perspectives: i) on function for energy conservation or scavenging and ii) on the types as mobile sink and mobile relay.

Figure 4. Classification of mobile routing protocol

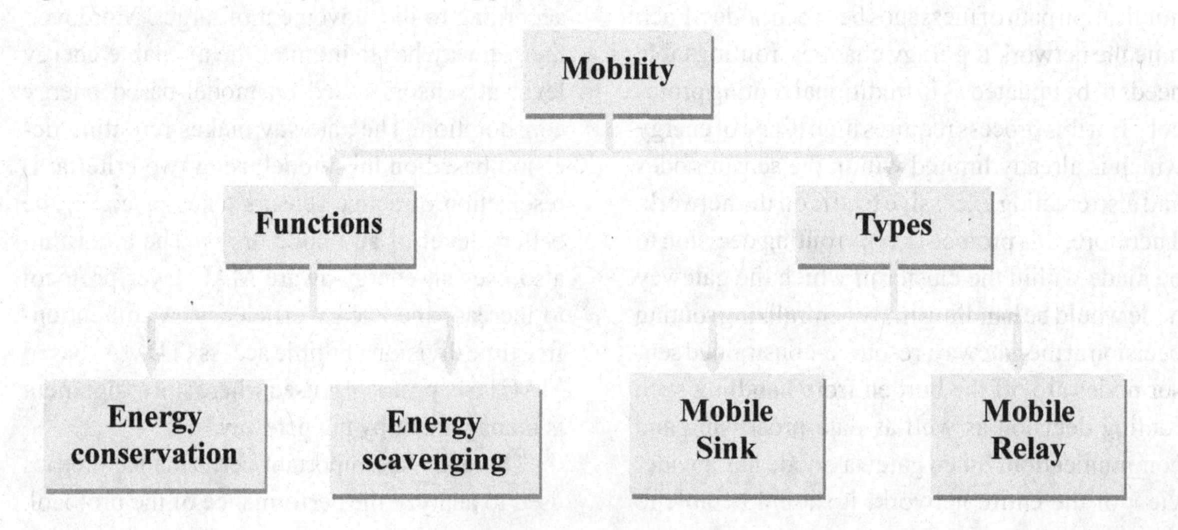

Mobility can be achieved in various ways, for example sensors themselves can be mobile by embedding mobilizers, but the weakness of such approach could be that energy consumption due to mobility may over ride over energy conservation. Moreover mobility can be supported even by having mobile sinks or electing certain nodes which are not less energy constrained. In such architecture (Figure 5) a handful of nodes known as mobile agents that have high power, memory and processing capability can be employed as data collectors. The mobile agents are placed on elements which are mobile and no additional energy consumption due to mobility itself. The following section discusses some of new ideas and suggestions presented in the literature concerning the use of mobile elements as mobile data collectors, mobile relays, mobile sinks and mobile entity as data harvesters to provide energy efficient communication rather than relying on multihop communication as performed in static sensor networks.

Figure 5. Mobile based communication

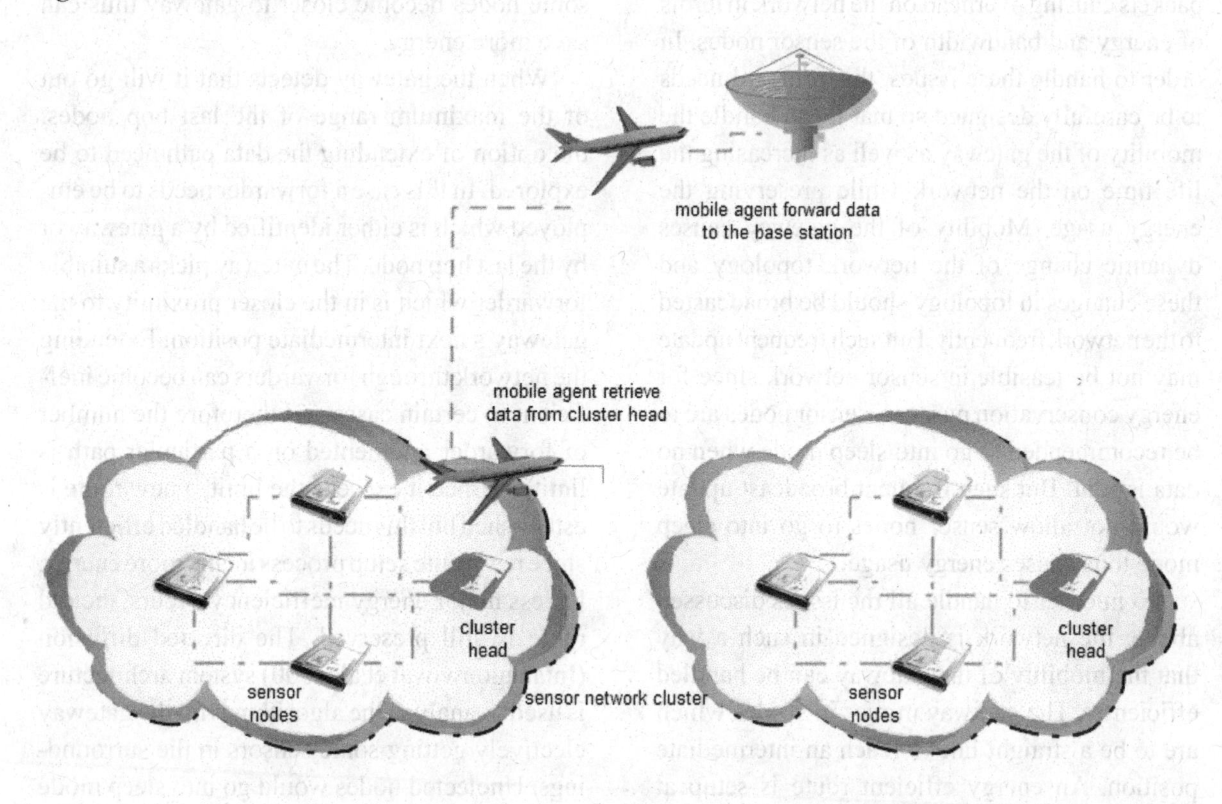

Energy-Aware Routing to a Mobile Gateway in Wireless Sensor Networks

Most of the energy-aware routing protocols that have been proposed were generally looking at optimizing network performance by forwarding data to a stationary gateway (sink node). However, the mobility of the gateway makes routes created by such protocol do not really give optimal performance. (Akkaya & Younis, 2004) looks at energy aware protocol for continuous and efficient delivery of data to a mobile gateway. In this protocol model, a set of sensors are deployed throughout an area of interest to monitor particular events. A less energy constraint gateway node is deployed within the proximity of the sensors. The gateway is responsible for organizing certain activities and collecting data from the sensor, thus the sensors should be within the communication proximity with the gateway. Many routing protocols such as have been proposed for mobility of nodes in ad hoc networks but no protocol has been specifically looking into mobility of the gateway. And these suggested protocol approaches cannot be directly applied into sensor network since these protocols were designed for IP-enabled global addressing scheme. Whereas in sensor networks, nodes are deployed in sheer numbers and no addressing is used in WSN. Besides that, sensor networks are very energy constraint and many control packets as in ad hoc networks are not suitable and in sensor network, the communication is between many sensors to one gateway which is not the case in ad hoc networks.

One of the main challenges faced in this protocol is to provide uninterruptable service to the gateway by maintaining a dynamic maintenance of multihop routes. Therefore a frequent update about the whereabouts of the gateway to the network is crucial. This requires high exchange of control packets causing overhead on the network in terms of energy and bandwidth of the sensor nodes. In order to handle these issues, this protocol needs to be carefully designed so that it can handle the mobility of the gateway as well as increasing the life time on the network while preserving the energy usage. Mobility of the gateway causes dynamic change of the network topology and these changes in topology should be broadcasted to the network frequently. But such frequent update may not be feasible in sensor network since for energy conservation purpose, sensor nodes are to be recommended to go into sleep mode when no data is sent. But such frequent broadcast update would not allow sensor nodes to go into sleep mode thus causes energy usage.

So in order to handle all the issues discussed above, the network is designed in such a way that the mobility of the gateway can be handled efficiently. The gateway moves in strides which are to be a straight line to reach an intermediate position. An energy efficient route is setup at initial gateway location to keep receiving packets uninterruptedly till reaching the next intermediate position. To overcome excessive overhead in rerouting while maintaining efficient network operation, the gateway's motion handling scheme tries to maintain continuous delivery of packets by introducing some forwarders to extend the current routes. A new route is established when gateway is unreachable due to energy consumption or the number of forwarders for the last hop node exceeds the given threshold value. When the gateway moves in strides, reassessment is performed to ensure that: i) gateway is still reachable by the last hop node, ii) the path needs to be extended and iii) rerouting is required. Even when the gateway is reachable by the last hop nodes in its current network, the nodes are instructed to readjust its transmission power to cover the next move of the gateway. Though such process increases energy consumption at such nodes, the effect is not that great since only few nodes are affected while some nodes become closer to gateway thus can save more energy.

When the gateway detects that it will go out of the maximum range of the last hop nodes, the option of extending the data path need to be explored. In this case a forwarder needs to be employed which is either identified by a gateway or by the last hop node. The gateway picks a suitable forwarder which is in the closer proximity to the gateway's next intermediate position. Extending the network through forwarders can become inefficient in certain cases and therefore the number of forwarder augmented on a particular path is limited. Once it exceeds the limit, a new route is established but this needs to be handled efficiently since new route setup process incurs more energy. Unless major energy inefficiency occurs, the old route is still preserved. The directed diffusion (Intanagonwiwat et al. 2000) system architecture is used to analyze the algorithm with the gateway electively getting some sensors in the surroundings. Unelected nodes would go into sleep mode to preserve energy. There are two different models used for gateway mobility, i) Random Waypoint Model, and ii) Target Avoidance Model (TAM). TAM was actually designed by the authors of protocol where it is suitable in battle field where motion of gateway is dynamic to escape from possible enemy or target locations. The gateway location is selected based on high volume of traffic in a certain area where possible detection of target occurs. There are three metrics used for performance results: i) network throughput, ii) average delay per packet and iii) average lifetime of a node. The effect of rerouting on energy consumption was analyzed due to frequent changing position of the gateway. The effect is quite obvious as nodes frequently changes state from sleep mode into active mode and vice versa. The total transition energy increases significantly with a smaller reroute period. Moreover, when the total number of sensors increases, the overhead of rerouting becomes significant in terms of energy usage. But this approach provides 20% increase in network lifetime with the increase in network size. Thus the protocol provides an efficient routing of sensor data to a mobile gateway. But one of the weaknesses with this protocol is its dynamic nature of topology change of the gateway node. Sensor nodes have to continuously update its topology database to know the where about of the gateway node.

Sensor Networks with Mobile Agent (SENMA)

This protocol (Tong et al. 2003) is a new mechanism of using mobile agents together with low power sensor nodes to reduce the complex tasks away from low power sensor nodes. The diagram below shows how a mobile agent can facilitate in this process by taking some of the loads away from the sensor nodes. In SENMA, mobile agents (MA) are not software units but are powerful hardware units in terms of communication and processing capability. Some of the examples of MAs are aerial vehicles, ground vehicles with terminals and power generators that can hop around the network. The proposed MA does not have to be present all the time but only needed when necessary for data collection and network maintenance (see Figure 6).

In SENMA, only MAs are responsible for data collection and therefore sensors spend minimal energy in receiving signals. All sensors listen to the neighbours in order to determine if they have to relay the packet. The communication between

Figure 6. SENMA architecture

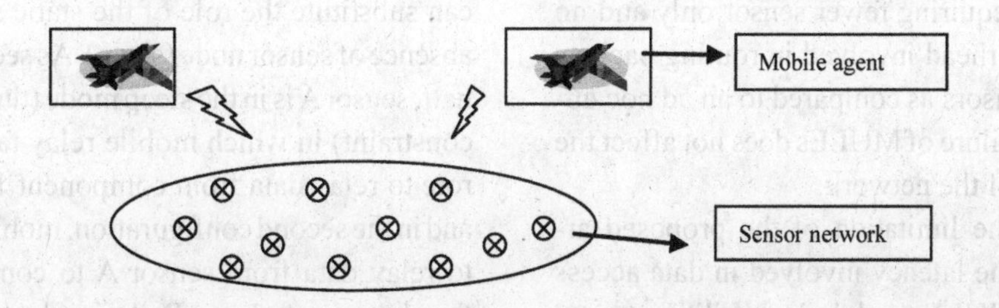

sensors and MAs is through free space medium where signal decay is only at the 2nd power distance. MAs in SENMA perform complicated signal processing and network maintenance. The fading concept introduced by MA allows sensors to only transmit if its channel is in the most favourable fading condition. Sensor nodes directly send data to a mobile agent which is an airplane flying h meters above the sensor field. This is different from the flat ad hoc architecture whereby in flat architecture sensor nodes continuously relay packet to their neighbouring node till the packet reaches the sink node. So the transmitted signal decays at the 4th power of distance whereas in SENMA signal decays at the 2nd power of distance.

The addition of mobile agents takes away the processing complexity away from sensors. Therefore, SENMA gives advantage in energy efficiency over flat ad hoc architecture. The proposed PHY/MAC layer for mobile agent exploits node redundancy.

Data MULEs: Modeling a Three-Tier Architecture for Sparse Sensor Networks

Data MULE (Shah et al. 2003; Jain et al. 2004) explores the idea of using mobile entities (MULE) to collect sensor data from sparse sensor networks. MULEs retrieve data from sensors when it reaches closer range to the sensors, buffer it and forward to the wired access point. The authors state that this can contribute to substantial power savings as sensors only have to transmit over short distance. The three-tier MULE architecture consists of three layers with the lowest layer being the sensor nodes, the middle layer is where the mobile agents are placed and the access point or base station is located at the last layer. The mobile agent proposed in this architecture can a any mobile entity such as people, animal or moving vehicles. MULEs provide scalability and flexibility features for lower cost.

The main advantage of MULEs is their huge storage capacities, renewable power and their ability to communicate with sensors and access points. The initial movement of MULEs are based on a random walk in which their movement cannot be predicted. However due to their mobility, MULEs are capable of collecting data from sensors, storing it as well as sending acknowledgements back to the sensors. Moreover MULEs can communicate with each other to improve system performance by reducing latency using multi-hop network. The latency for traffic monitoring applications can be reduced by having the MULEs equipped with an always-on connection such as a cellular or satellite phone. Failure of any MULEs does not affect the performance of any sensors hence this architecture is more robust and can be easily scalable by exploiting more mobile entities (MULEs). Scalability in terms of exploiting new sensors or MULEs does not require re-configuration of the entire network. Moreover to improve reliability, end-to-end or tier-to-tier acknowledgements can be used. Some of the key benefits if the proposed system is, less infrastructure needed than fixed base-station approach, cost can be reduced with applications requiring fewer sensor only and no additional overhead involved in routing packets from other sensors as compared to an ad hoc approach. The failure of MULEs does not affect the connectivity of the network.

Some of the limitation of the proposed architecture is the latency involved in data access speed of the MULEs and their mobility rate and guarantee in data success rate due to mobility of the MULE, the failure of MULE. Based on the simulation results, the data success rate depends on the buffer capacity of the MULE, arrival rate of MULE at sensor node and MULE density. The latency of system has to be compromised and left for future work.

Extending the Lifetime of Wireless Sensor Networks through Mobile Relays

Mobile relay (Wang et al. 2008) focuses on a heterogeneous architecture composed of a few resource rich mobile relays and large number of static sensors. The mobile relays are rich in energy that can move around the network to relieve the burden of high traffic sensors. This can help to extend the lifetime of static sensors. Unlike other mobile approach such as MULE (Shah et al. 2003; Jain et al. 2004) this wok focuses more on the use of mobile relays to deliver network resources such as energy, computational power, sensing and communication abilities instead of delivering sensor data. Mobile nodes can move to areas limited in resources such a areas where sensor deployment is with low density and requires mobile nodes to attend to sense the environment. This approach exhibits lower hardware costs instead of deploying dense sensor nodes. The mobile relays have the same communication range and sensing but with higher energy level to support energy limited sensor nodes. The use of mobile relay helps in consuming energy at sensor nodes in which mobile nodes can provide the resources required.

Figure 7 helps to explain how a mobile relay can substitute the role of the static sensor in the absence of sensor node (sleep). As seen, in the first half, sensor A is in the sleep mode (due to resource constraint) in which mobile relay takes over the role to relay data from component 1 to sensor B and in the second configuration, mobile relay jelps to relay data from sensor A to component 2 in the absence of sensor B. A simple way to do this by devising an algorithm in which mobile node switch between node A and node B and take the responsibility of the node it is substituted with for sensing and relaying purposes. With the shuttling schedule, the network lifetime can be doubled to 2T. The authors also constructed a joint mobility and routing algorithm that can improve the lifetime of the network by a factor of 4. Only nodes within certain distance to the sink will have to be aware of the location of the mobile relay.

A New Architecture for Hierarchical Sensor Networks with Mobile Data Collectors

This work by (Bari et al. 2010) presents a hierarchical architecture for heterogeneous sensor networks using mobile data collectors (MDC) at the upper tier and relay nodes at the middle tier which acts as cluster heads. Minimal number of relay nodes were used by optimally placing them with each sensor node can send data to at least one relay node. The MDC uses a fixed trajectory to visit all the relay nodes and the base station. Also the MDC is non power constrained device that collects data from the relay nodes and forwards it to the base station. Two key advantages of the architecture proposed are; relay nodes are reduced the load of routing data to the base station which results in energy savings; and the MDC does not need to visit each sensor nodes which reduces the length of the trajectory of the MDC and no frequent visit is required. Each sensor node is dedicated to one cluster and sensor nodes generate data at certain rates and send the data to the relay node corresponding to the cluster head. The data will be buffered at the relay node till the MDC comes into contact with the relay node. After forwarding data to the MDC, each relay nodes empties its buffer for the next round of data collection. The trajectory of MDC determines the buffer size and cost of the relay node, hence the speed and length of the MDC is very crucial. Through simulation results the authors have proposed: a new hierarchical architecture for heterogeneous sensor networks with mobile data collectors formulated a program to solve the node placement and the MDC visiting schedule two problems associated were solved in two separate steps

Sensor Data Collection Using Heterogeneous Mobile Devices

The authors (Prem et al. 2007) present an alternatively different approach of using mobile nodes such as mobile phones as data collectors from

Figure 7. Mobile relay

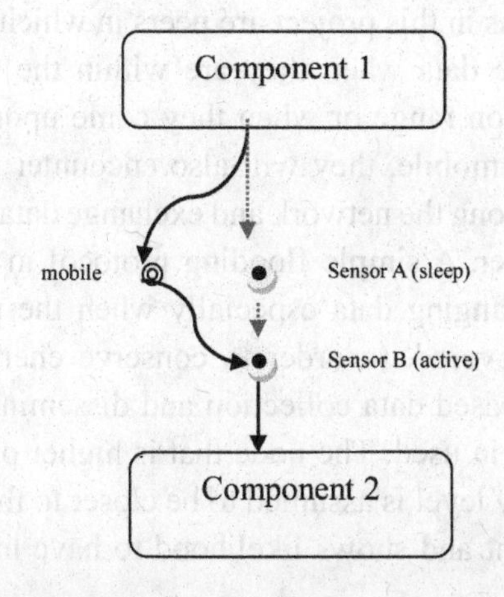

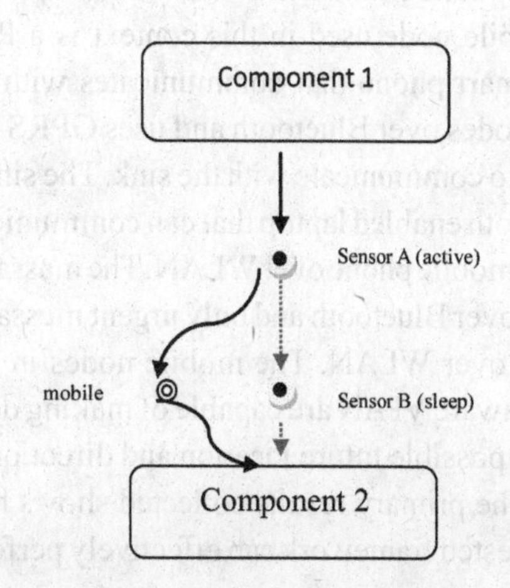

static sensor nodes. Compared to other proposals in the literatures in which mobile nodes are also equipped with sensor device with radio to communicate with the sensor nodes, this paper differs in which the mobile node communicates with sensor nodes via Bluetooth technology. Mobile devices are not equipped with sensor nodes but rather act as data carriers for sensor nodes that act as data accumulators. The idea of using mobile phones as data collectors is inspired by the notion that mobile phones have capacity for processing and communicating. The mobile phones described in this work need to be context aware in nature so that it can communicate with sensor nodes, negotiate and deliver data to the sink. These mobile nodes work within an environment which is described as virtual personal area network (VPAN). VPAN is a periphery in which nodes present in the environment are aware of their presence and their location as well.

The general architecture of this work consists of an external context server that is accessible to any mobile node as long as it is within the VPAN. The in-built context has presence and user calendar information to get user's next possible location and its presence information. The message parser is used to parse message between the mobile node and the sensor node whereas the node discovery and management is dedicated to discover nodes in the surrounding and store the location and activation schedule information of any nodes that communicate with the mobile node. The mobile node used in this context is a PDA based smart phone that communicates with the sensor nodes over Bluetooth and uses GPRS and WLAN to communicate with the sink. The sink is a Bluetooth enabled laptop that can communicate with the mobile phone over WLAN. The massages are sent over Bluetooth and only urgent messages are sent over WLAN. The mobile nodes in this context aware VPAN are capable of making decisions on possible future location and direction of travel. The primary results collected shows how the suggested framework can effectively perform data collection by reducing transmission distance while increasing network lifetime.

Energy Efficient Computing for Wildlife Tracking: Design Trade-offs and Early Experiences with Zebranet

Zebranet by (Juang et al. 2002) looks at the decision and design tradeoffs in the context of applying wireless peer-to-peer networking techniques in mobile sensor network design. This is biological based research to support wildlife tracking focused on monitoring of the zebras, carried out under large wild are. The ZebraNet is based on mobile sensor network in which all the sensors (animal) are mobile and Zebras are dedicated as mobile relays to collect data from sensors. In this project, the animals are equipped with collars embedding sensor nodes in which the collar consists of a global positioning system (GPS), flash memory, a dual band radio, wireless transceivers, a small CPU; and each node is a small, wireless computing device. The collars operate as a peer-to-peer network to deliver sensed data to researchers. One of the radio is used for communication within shorter range, for example when zebras meet near the water source. Another radio is used to communicate with the access point and for animal which are further from the others. The access point in this project is also a mobile entity in which comprise of a vehicle that traverses along the monitored area to collect data.

Zebras in this project are peers in which they exchange data when they are within the communication range or when they come upon. As they are mobile, they will also encounter other zebras along the network and exchange data with each other. A simple flooding protocol in used for exchanging data especially when the peers are discovered. In order to conserve energy, a history-based data collection and dissemination protocol in used. The node that is higher on the hierarchy level is assumed to be closer to the access point and shows likelihood to have higher success rate of data transmission. Nodes send data to their peers only after querying its hierarchy level and to which that has the highest level. The hierarchy increases when a node comes closer to the access point. Through simulation results, this mechanism is proved to be efficient in terms of energy and success rate.

BiSNET: Biologically-Inspired Middleware Architecture for self-Managing Wireless Sensor Networks

BiSNET by (Boonma & Suzuki, 2007) is a Biologically-inspired architecture for sensor networks that tries to address several issues such as autonomy, scalability, adaptability, self-healing and simplicity. BiSNET adopts certain biological properties from bee colony such as decentralization, food gathering or storage and natural selection. Besides, it analogues certain biological properties from colony such as energy exchange, pheromone emission, migration and replication to be adopted into wireless sensor network to address issues mentioned above. These issues can be tackled by biological properties of bee colony as bees are autonomic in nature by the influence of local environment conditions and local interactions with other bees.

Bee colony can be scalable in nature as their activities are non-centralized (decentralization) in which they carry out their activities without central control. They adjust themselves to dynamic changes in the environment in which when the food supply (honey) in low in the hive, they go out seeking for nectar collection (food gathering) and when there are abundant supply of honey (storage), they rest in their hive or expand the hive. The richness or deficiency of food supply allows bees to alter their behavior to adopt natural selection mechanism. For example, plenty of supply of energy indicates change in sensor readings, and agent releases pheromone to replicate itself and other neighboring agents as well. The replicated agent travels to a base station to forward sensor data. And shortage of energy causes death of agents (bees) to compromise with the demand and supply of energy over number of bees, Moreover bees have self-healing properties in which they are able to recover their pheromone traces to flowers when part of them is missing. The collective behaviors and interactions among group of bees allow them adopt desirable features such as adaptability and self-healing. Table 1 shows the mapping of bee colony analogous to BiSNET.

BiSNET is intended to be applied into oil spill detection and monitoring in the coastal environment. In BiSNET each agent is divided into attributes, body and behaviors. Attributes carry information on agent such as agent type, energy level, sensor data, timestamp and ID of sensor node. Body implements tasks of an agent, collecting and processing of sensor data. The agent collects sensor data, converts it into energy and processes it to be reported to the base station or discard it. Behavior explains all the actions performed by agents as described below:

Pheromone emission: agents secrete pheromones when there is scarce or abundant supply of energy. The pheromone secreted carries sensor data, for example temperature pheromone carries temperature data. Pheromones provoke agents to replicate themselves.

Replication: agents replicate themselves when there is abundance supply of energy and pheromones. The replicated agent exhibits same agent type of its parent's type and aggregates multiple sensor data stored. The child agent receives half of the energy stored at the parent agent and is

Table 1. Mapping of biology to BiSNET

Bee Colony	BiSNET
Bee hive	Base Station
Nectar	Sensor data
Honey	Energy
Bees	Agents
Flowers	Middleware platform

intended to report the aggregated sensor data to the base station.

Migration: agent also move from one sensor node to another when there is abundant supply of energy. Migration allows agent to report sensor data to the base station on a multi-hop or shortest path basis.

Energy exchange: agents share their energy (honey) with other agents so that their energy level remains same. Migrating agent deposits energy to other agents at the destination platform and also deposits some to the hive.

Death: when there is lack of energy, agent exhibit death behaviour to eliminate agents that carry false positive data.

Simulation results show that BiSNET is an energy efficient mechanism whereby sensor nodes (agents and platforms) autonomously adapt their sleep cycle and aggregate data from different types of nodes and also self-heal by detecting and eliminating false positive data. Several enhancements are foreseen in this project by exploiting a decentralized routing mechanism using shortest paths to the base station. A biologically-inspired routing mechanism is explored to direct agents to the base station in an effective manner.

Self-Organization in Sensor Networks Using Bio-Inspired Mechanisms

This paper by (Dressler et al. 2005) looks at autonomously working mobile nodes in the context of mobile robots in a wireless sensor networks. The motive is to teach nodes to autonomously self-organize themselves to handle events such as monitoring and tracking within the constraints such as time and energy. Researchers have been exploring methods to self-organized nodes by applying natural principles and mechanisms. So the communication methodology proposed in this paper uses bio-inspired mechanism from cell and molecular biology to teach nodes on self-organization, task allocation and energy-aware communication. The main approach of this protocol is by using mobile robots and stationary sensor networks on autonomous system as well as using bio-inspired methods for energy and application aware communication. There are three important goals to be achieved: i) self-organization without the central control, ii) energy awareness and iii) faster response from the time event occurs till it is responded.

The main motivation for this paper is by adopting the concept of feedback loop mechanism which allows a convenient operation in low-resource sensor networks. This is achieved by using the concepts of nature, more specifically from the microscopic point of view of studying the cellular feedback loop environment. There are a few jobs associated with activating a particular task. First, neighbor nodes need to be informed about available resources on the network, second routing mechanisms need to be implemented in order to send the control information. A biological diffuse communication protocol is used where a message to be sent is given priority based on the importance of the particular task. With this, the message is sent with a percentage to the direct neighbors and with a lower percentage to other remote neighbors. So a randomness factor is given to the task of dispersing the message or information.

This methodology is still under progress with simulation and real implementation using Robertino robot platform and Mica2 sensor motes running TinyOS. Feedback loop, techniques from cell biology proposed in this paper addresses some of the problems seen in traditional networks. In a resource constraint sensor networks, this methodology allows for efficient utilization of resources and achieved higher quality of global system. Bio-inspired mechanism proposed in this method would address issues in self-organization as well as adapting to environmental changes. So with the results obtained, it is evident that this protocol provides a self-organized sensor network.

Summary

Mobility in WSN can be employed in different modes such as mobile relay, mobile data collectors, data mulle and mobile base station. Based on the discussion above, mobile base station reduces the overload of nodes concentrated near the base station as it is always on the moving position. This frequent change of topology shifts the burden away from single concentrated nodes. But this still burdens the nodes near the base station when it is within the proximity as this approach still works on multihop communication and faces the same outcome as above. Mobile data collectors, data mule or mobile agents are external entities engaged to operate as routing agent or data collectors. By using such mechanism, static sensor nodes do not deplete their energy due to routing burden. In this configuration sensor nodes will only perform data collection and coordinating their tasks. So compared to traditional multihop routing, employing mobile entities would be a potential solution in handling such tasks.

COMPARISON

In section 2, the routing mechanism of two different classes of protocols has been analyzed looking into its energy efficient pattern. Based on the analysis, each of these classes can be applied into different conditions and specific to its problems. Flat and location based routing protocols are traditional approaches and a lot work has been done in this on energy aware and energy efficient routing. These protocols are more suitable for delay sensitive applications which require urgent sensor data reporting. Mobile routing is becoming popular by integrating mobile capability into sensor nodes as well as base stations. The lifetime of this approach as seen in Figure 8, can be doubled as compared to multihop routing and energy conserve/aware routing. Mobile base station also shows better results as compared to energy aware routing. So mobile platforms is discovered to be one of the potential new area in sensor networks with the mobile agent being an agent sending query to static sensors and further obtain the sensed results to forward to the base station (Chong & Kumar, 2003). An example discussed by this author is where an airborne querying device sends query to static sensors and inform the ground sensor network that it will be flying over a specific location after a while, where the response to the query should be forwarded. Due to the limitation of bandwidth in sensor network, (Qi et al. 2001) proposes mobile agents that are responsible to collect fusion based information from static sensors. Mobile agents can be a new solution to the limitation in terms of energy consumption and bandwidth limitation of sensor networks. Mobile nodes in the network can travel around the network to collect data from sensor nodes rather than utilizing higher energy for data travelling to the base station. This reduces communication between nodes thus energy is saved and the network lifetime is increased (Koutroullos & Pitisllides, 2007). Another solution for extending the network lifetime could be to have some mobile robots to travel around and refill the nodes with energy (Koutroullos & Pitisllides,

Figure 8. Normalized lifetime of various approaches (extracted from (Wang et al. 2008))

Multihop	Energy Conserve Routing	Adding energy to 25% static sensor	One mobile relay	Mobile base station
1	**3.76**	**5.4**	**7.8**	**9.67**

2007). G. Anastasi, et al, Jun et al. 2005, Somasundara et al. 2006 and Zhao et al. also looking into variants of mobility based communication for energy efficient operation. (Ponnusamy et. al. 2010) have proposed a new nature-inspired energy efficient self-healing sensor network. This work looks into bee analogy for deploying mobile agent in the context as data harvester and energy harvester/provider. The summary given in the next section highlights some of the key areas where the classes of these protocols have been looked into and issues that have been tackled. Based on the analysis performed, energy efficiency, fault tolerance, scalability and load balancing should be achieved for deploying a good routing protocol in WSN. Researchers can look into the protocols that employ such parameters and devise a better and enhanced routing protocol (see also Table 2).

CONCLUSION

A survey on energy efficient routing protocol in wireless sensor network was presented. The advantages and disadvantages of each proposed protocol have been analyzed. Also the comparison in section 3.0 is done based on various metrics useful in sensor network performance. The comparative study in terms of energy efficiency covers some available protocols that can be used to further enhance routing in sensor network. Furthermore, the static and single layer of hierarchy should be given further thought to look into the

Table 2. List of protocols

	Load Balancing	Stationary Node	Energy Efficient	Hole Problem	Shortest Path	Fault Tolerance	Scalability
Constrained Shortest path		√	√		√		
WEAR		√	√	√	√	√	
GEAR	√	√	√	√	√		√
Energy aware low energy		√	√				
LEACH		√					√
Highly resilient routing		√	√		√		
SONS		√				√	
Mobile Gateway		Mobile Base station	√				
SENMA		Partial mobile	√			√	
MULE		√	√				√
Mobile relay		√	√				
Data Collector		√	√				
Mobile device		√	√				
Zebranet		√	√				√
BiSNET	√		√				√

importance of alternative approaches such as n-tier and mobility schemes. It is worth noting that most of the works in the literature assume the energy consumption of the radio is much greater than the energy consumption due to processing of data. This is further proven by some real applications requiring more energy for processing and communication of data instead of the radio. Therefore hierarchical routings presented in this paper can help to reduce data transmission between nodes. Tremendous amount of energy can be saved by deploying resource rich nodes such as mobile entities. In the future, with the increasing need for sensor applications, the comparisons and suggestions presented in this paper could be used to further enhance the protocol.

REFERENCES

Akkaya, K., & Younis, M. (May 2003). *An energy-aware qos routing protocol for wireless sensor networks*. Paper presented at the IEEE Workshop on Mobile and Wireless Networks, Providence, Rhode Island.

Akkaya, K., & Younis, M. (Nov 2004). *Energy-aware routing to a mobile gateway in wireless sensor networks*. Paper presented at the IEEE Globecom Wireless Ad Hoc and Sensor Networks Workshop, Dallas.

Al-Fares, Z., Sun, H., & Cruickshank, H. (2009). *A hierarchical routing protocol for survivability in wireless sensor network (WSN)*. Paper presented at the International MultiConference of Engineers and Computer Scientists, IMECS 2009, Vol 1.

Anastasi, G., Conti, M., Passarella, A., & Pelusi, L. (2009). Mobile-relay forwarding in opportunistic networks. In M. Ibnkahla (Ed.), *Adaptive techniques in wireless networks.* CRC Press. Retrieved from http://info.iet.unipi.it/~anastasi/papers/mrf07.pdf

Bari, A. (2010). Lecture Notes in Computer Science: *Vol. 5935. A new architecture for hierarchical sensor networks with mobile data collectors. Distributed Computing and Networking* (pp. 116–127).

Boonma, P., & Suzuki, J. (2007). BiSNET: A biologically-inspired middleware architecture for self-managing wireless sensor networks. *Journal of Computer Networks*, *51*(16). doi:10.1016/j.comnet.2007.06.006

Chong, C., & Kumar, S. P. (2003). Sensor networks: Evolution, opportunities, and challenges. *Proceedings of the IEEE*, *91*(8), 1247–1256. doi:10.1109/JPROC.2003.814918

Dressler, F., Kuger, B., Fuchs, G., & German, R. (March 2005). Self-organization in sensor networks using bio-inspired mechanisms. *18th ACM/GI/ITG International Conference on Architecture of Computing Systems-System Aspects in Organic and Pervasive Computing (ARCS'05): Workshop Self-Organization and Emergence,* Innsbruck, Austria, (pp 139-144).

Ganesan, D., Govindan, R., Shenker, S., & Estrin, D. (2001). Highly-resilient, energy-efficient multipath routing in wireless sensor networks. *ACM SIGMOBILE Mobile Computing and Communications Review*, *5*(4), 1125. doi:10.1145/509506.509514

Gui, C., & Mohapatra, P. (2003). *SHORT: Self-healing and optimizing routing techniques for mobile adhoc networks*. Paper presented at the MobiHoc

Intanagonwiwat, C., Govindan, R., & Estrin, D. (2000). Directed diffusion: A scalable and robust communication paradigm for sensor networks. *ACM MobiCom '00*, Boston, MA, 2000, (pp. 56-67).

Jain, S., Shah, R. C., Borriello, G., Brunette, W., & Roy, S. (2006). *Exploiting mobility for energy efficient data collection in sensor networks*. Paper presented at the WiOpt.

Juang, P., Oki, H., Wang, Y., Martonosi, M., Peh, L. S., & Rubenstein, D. (2002). Energy efficient computing for wildlife tracking: Design tradeoffs and early experiences with Zebranet. In *ASPLOS-X: 10th International Conference on Architectural Support for Programming Languages and Operating Systems*, (pp. 96–107).

Jun, H., Zhao, W., Ammar, M., Zegura, E., & Lee, C. (March 2005). *Trading latency for energy in wireless ad hoc networks using message ferrying*. Paper presented at the IEEE PerCom Workshops, International Workshop on Pervasive Wireless Networking (PWN 2005).

Koutroullos, M., & Pitisllides, A. (2007) *Biological and nature inspired mechanisms for adaptive and robust self-organization in wireless sensor networks*. Paper presented at the EPL657, Projects.

Lindsey, S., & Raghavendra, C. S. (March 2002). *PEGASIS: Power efficient gathering in sensor information systems*. Paper presented at the IEEE Aerospace Conference, Big Sky, Montana.

Ponnusamy, V., Abdullah, A., & Alan, G. D. (2010). *A bio-inspired framework for wireless sensor network using mobile sensors*. Paper presented at the 5th International Conference for Internet Technology and Secured Transactions, London, November.

Ponnusamy, V., Abdullah, A., & Alan, G. D. (2010). *A biologically based wireless sensor network*. Paper presented at the Global Conference on Power Control and Optimization, Kuching, December.

Prem, P. J., Arkady, Z., & Jerker, D. (July 2007). *Sensor data collection using heterogeneous mobile devices*. Paper presented at the ICPS'07: IEEE International Conference on Pervasive Services Istanbul, Turkey.

Qi, H., Iyengar, S. S., & Chakrabarty, K. (2001, August). Multi-resolution data integration using mobile agents in distributed sensor networks. *IEEE Transactions on Systems, Man and Cybernetics. Part C, Applications and Reviews*, *31*, 383–391. doi:10.1109/5326.971666

Sha, K., Du, J., & Shi, W. (January 2005). *Wear: A balanced, fault-tolerant, energy-aware routing protocol for wireless sensor networks*. Technical Report MIST-TR-2005-001, Wayne State University.

Shah, R., Roy, S., Jain, S., & Brunette, W. (May 2003). *Data MULEs: Modeling a three-tier architecture for sparse sensor networks*. Paper presented at the IEEE SNPA Workshop.

Shah, R. C., & Rabaey, J. (March 2002). Energy aware routing for low energy ad hoc sensor networks. *IEEE Wireless Communications and Networking Conference* (WCNC), Orlando, FL, vol. 1, (pp. 350-355).

Somasundara, A. A., Kansal, A., Jea, D. D., Estrin, D., & Srivastava, M. B. (2006, August). Controllably mobile infrastructure for low energy embedded networks. *IEEE Transactions on Mobile Computing*, *5*(8), 958–973. doi:10.1109/TMC.2006.109

Tong, L., Zhao, Q., & Adireddy, S. (October 2003). *Sensor networks with mobile agent*. Paper presented at the Military Communications Intl Symp., (Boston, MA).

Wang, W., Srinivasan, V., & Chua, K.-C. (2004, March). Extending the lifetime of wireless sensor networks through mobile relays. *IEEE/ACM Transactions on Networking*, *16*(5), 1108–1120. doi:10.1109/TNET.2007.906663

Wendi, R. H., Chandrakasan, A., & Balakrishnan, H. (January 2000). *Energy-efficient communication protocols for wireless microsensor networks*. Paper presented at the Hawaii International Conference on Systems Sciences.

Younis, M., Youssef, M., & Arisha, K. (October 2002). *Energy-aware routing in cluster-based sensor networks*. Paper presented at the 10th IEEE/ACM International Symposium on Modeling, Analysis and Simulation of Computer and Telecommunication Systems, Fort Worth, TX.

Youssef, M. A., Younis, M. F., & Arisha, K. A. (2000). A constrained shortest- path energy-aware routing algorithm for wireless sensor networks. *IEEE Wireless Communications and Networking Conference* (WCNC2002), (pp. 794–799).

Yu, D., Estrin, D., & Govindan, R. (May 2001). *Geographical and energy-aware routing: A recursive data dissemination protocol for wireless sensor networks*. UCLA Computer Science Department Technical Report, UCLA-CSD TR-01-0023.

Zhao, W., Ammar, M., & Zegura, E. (May 2004). *A message ferrying approach for data delivery in sparse mobile ad hoc networks*. Paper presented at the ACM MobiHoc 2004, Tokyo (Japan).

Chapter 11
Cooperative Diversity Techniques for Energy Efficient Wireless Sensor Networks

Tommy Hult
Lund University, Sweden

Abbas Mohammed
Blekinge Institute of Technology, Sweden

ABSTRACT

Wireless Sensor Networks can be used for a multitude of applications, but they all have in common that they need to send their collected data to some central processing node. If the sensors are located at a long distance from a processing node, the transmission might need the combined transmitter power of several sensors. In this chapter, the authors investigate the use of various cooperative diversity techniques in wireless sensor networks to increase the transmission range, minimize power consumption, and maximize network lifetime.

INTRODUCTION

Wireless Sensor Networks (WSN) have been attracting great attention recently. They are relatively low cost to be deployed and to be used in many promising applications, such as biomedical sensor monitoring (e.g., cardiac patient monitoring), habitat monitoring (e.g., animal tracking), weather monitoring (temperature, humidity, etc.), seismic sensing, environment preservation and natural disaster detection/monitoring (e.g., flooding and fire) (Lewis, 2004; Tubaishat & Madria, 2003; Stankovic, Abdelzaher, Sha & Hou, 2003; Akyildiz, Sankarasubramaniam & Cayirci, 2002; Rashid-Farrokhi, Tassiulas & Liu, 1998).

An important design feature of WSN is the low energy consumption of each individual transmitting node and thereby also of the total sensor network energy consumption. The 'bottleneck' in the WSN, from an energy point of view, is when

DOI: 10.4018/978-1-4666-0101-7.ch011

the nodes need to communicate with a distant base station to forward the collected data. Since the distance to the base station usually is several magnitudes greater than the distance between the individual nodes within the network, each node would need a much more powerful transmitter than necessary considering the distances within the WSN (Singh & Prasanna, 2003). One technique for achieving this is to use cooperative space-time processing to combine the transmitting power of all the nodes into a distributed non-uniform antenna array. For example in (Cui, Goldsmith, & Bahai, 2004) the energy efficiency using cooperative multiple-input multiple-output (MIMO) system was investigated and compared with the results obtained by the reference single-input single-output (SISO) system.

The WSN applications analyzed in this chapter have a topology where a large number of wireless sensor nodes are spread out over a large or small geographic area (e.g., disaster regions, indoor factory, large sports event areas, etc.). In this topology, an inefficient use of bandwidth and transmitter power resources is resulted if each wireless sensor is transmitting its measurement data to the base station (processing central). In this case, each sensor node would have to be assigned its own frequency channel and, if the base station is located a long distance from the sensor nodes, it would also demand a higher than average sensor node transmitter power. By using a coordinating cluster head, for each cluster of wireless sensor nodes, we can instead use the combined transmitter power of the node cluster through the use of beamforming to increase the transmitter-receiver separation and/or to improve the signal-to-noise ratio (SNR) of the communication link. Another advantage of using this cooperative transmission is that we can exert power control to minimize the power consumption of each individual sensor node, and thus maximizing network lifetime. In addition, in a cooperative network the measurement data could be sent by using time division multiplexing (TDM) instead of frequency division multiplexing (FDM) which improves the overall bandwidth efficiency of the system.

The spatial properties of wireless communication channels are extremely important in determining the performance of the systems. Thus, there has been great interest in the application of classical beamforming technique and the recently popular modern multiantenna diversity techniques, using various forms of array systems, since they can offer a broad range of ways and benefits to improve wireless systems performance. For instance, diversity techniques such as multiple-input single-output (MISO), single-input multiple-output (SIMO) and multiple-input multiple-output (MIMO) can enhance the capacity, coverage, quality and energy efficiency of wireless communication systems. These spatial diversity techniques can be exploited to improve the energy efficiency since it is one of the key requirements in many WSN applications. This is particularly crucial for WSN deployed in inaccessible or disaster environments in which battery recharging and replacement is not a viable option. Thus, in this chapter we first propose to use a cooperative beamforming approach in wireless sensor networks to increase the transmission range, minimize power consumption and maximize network lifetime. This will be of particular interest for outdoor applications, especially when monitoring remote areas using aerial vehicle, such as a High Altitude Platform (HAP) or Unmanned Aerial Vehicle (UAV), as a platform for the data collecting base station. We will investigate how the required transmitter power of each sensor node is affected by the number of cooperating transmission nodes in the network. In addition, we present a comparison in the use of beamforming with the different forms of modern multiantenna diversity techniques (MISO, SIMO and MIMO) for the same purpose of achieving a longer transmission distance (or range) while maintaining a low energy consumption. The required node transmit power is calculated and comparisons are performed for different distances from the base station and various propagation environments by varying the

Rice-factor of the fading channel. Beamforming can of course be interpreted as a form of MISO system although it differs from the normal view of how a diversity system operates.

This chapter is organized as follows: we start by presenting an overview and analysis of cooperative beamforming using a large aperture random array. The MISO, SIMO and MIMO diversity schemes are later introduced and analysed using the Rician fading channel employed in the simulations. Then we present simulation results showing performance analysis and comparisons of the simulated beamformer, MISO, SIMO and MIMO diversity systems. Finally, we conclude the chapter.

COOPERATIVE DIVERSITY TECHNIQUES

Cooperative Beamforming

In this investigation we use the delay-and-sum beamforming which is the oldest and simplest diversity technique. This beamforming is done through coherent excitation/reception of amplitude and phase of the signal transmitted/received from each individual antenna element in a collection or cluster of similar antenna elements also known as an antenna array (Johnson & Dudgeon, 1993). Antenna arrays can have different configurations (e.g. linear, planar, circular, triangular, rectangular or spherical etc.). Extensive research has been done on linear array beamforming using one (linear) or two (planar) dimensional equidistant element arrays (Johnson & Dudgeon, 1993; Hansen & Woodyard, 1938; Drane, 1968), but there is also work done on beamforming using circular, triangular and rectangular arrays (Johnson & Dudgeon, 1993; Balanis, 1997).

The antenna array formed by the individual sensor node antennas is assumed to be a planar array, of randomly positioned sensor node antennas, which is parallel with the plane containing all sensor nodes plus the base station so that the sensor nodes are only extended in x and y direction and not in z direction. This is a valid assumption in most cases since the elongation of the networks in z direction in most cases is very small compared to the distance between the network cluster and the base station we want to communicate with (Jenkins, 1973).

The design of this type of cooperative array is similar to the design of large aperture arrays where we have an inter-element spacing that is random and larger than half the wavelength. There are no known techniques for synthesis of randomly spaced arrays, like Schelkunoffs polynomial method (Johnson & Dudgeon, 1993; Balanis, 1997) or the Fourier Transform method (Johnson & Dudgeon, 1993; Balanis, 1997). In the random array all properties e.g array pattern, beam width, sidelobe level and gain are stochastic variables.

In Figure 1 is shown a scenario with N=50 sensor nodes deployed inside a circular boundary in the x-y plane with the radius R. The distribution of these sensor nodes is uniform and independent. The nth sensor then has the polar coordinates (r_n, φ_n).

The signal $y_n(t)$ at the array sensor node n can then be expressed as,

Figure 1. 50 sensor nodes positioned according to an independent uniform distribution within a cluster area of radius R

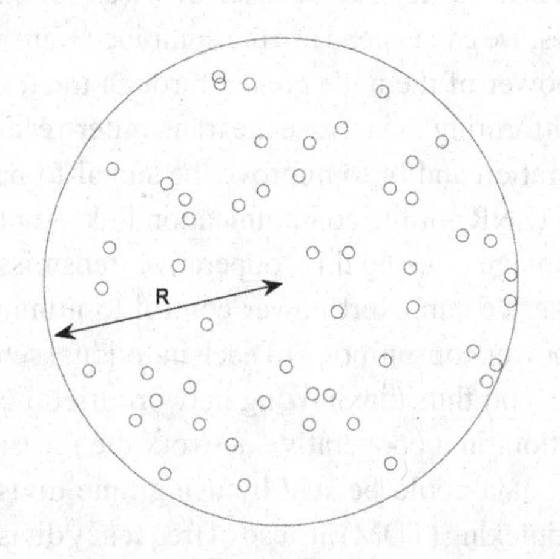

$$y_n(t) = s(t - \alpha_0 \cdot x_n), \quad (1)$$

where $s(t)$ is the signal to be transmitted/received and the nth sensor at location $\mathbf{x}_n$ transmits/receives the electromagnetic signal $y_n(t)$. The *slowness* vector α^0 is the required delay for each sensor to aim the array in a specific direction toward the signal source or target

$$\alpha_0 = \frac{\mathbf{d}_0}{c}, \quad (2)$$

where $\mathbf{d}^0$ is the direction of the wave propagation and c is the speed of light. The total output of the delay-and-sum algorithm will then be

$$z(t) = \sum_{n=0}^{N-1} w_n s(t + (\alpha - \alpha_0) \cdot x_n, \quad (3)$$

where w_n is the amplitude weights of the array tapering and α is the *slowness* vector for the direction of observation. If we assume that all the sensor nodes are approximately located in the same plane (i.e. the x-y plane) and the source/target is located at the spherical coordinates $\mathbf{d}_0=(d_0,\varphi_0,\theta_0)$ in the far-field, and if we also assume that we are transmitting a narrow band signal we can approximate Equation (3) as

$$G(\varphi,\theta) = \frac{1}{N} \sum_{n=0}^{N-1} w_n e^{j\omega\left[t - \frac{r_n}{c}(\cos(\varphi_n) \cdot u + \sin(\varphi_n) \cdot v)\right]}, \quad (4)$$

where $u = \sin(\theta)\cos(\varphi) - \sin(\theta_0)\cos(\varphi_0)$ and $v = \sin(\theta)\sin(\varphi) - \sin(\theta_0)\sin(\varphi_0)$ for the direction of the incoming/outgoing wave (φ_0,θ_0) and the direction of observation (φ,θ). The function $G(\varphi,\theta)$ is then one ensemble of the array amplitude gain function for one set of stochastic sensor locations. To find the ensemble mean of the array amplitude gain functions we assume an independent uniform distribution of the sensor locations within the radius R,

$$E\{G(\varphi,\theta)\} = \int\int G(\varphi,\theta)\, p_{R,\Phi}(r_n,\varphi_n), \quad (5)$$

where $p_{R,\Phi}(r_n,\phi_n)$ is the pdf of the sensor locations. In Figure 2 we can see the absolute squared average array gain function $|E\{G(\varphi,\theta)\}|^2$ of 250 realizations of the array amplitude gain function $G(\varphi,\theta)$. In Figure 3 is also shown the standard deviation for the distribution of the amplitude sidelobe levels.

From Figure 2 we can also estimate the mean sidelobe level will converge toward about -17 dB which is consistent with the theoretical value, N^{-1}. The amplitude of the sidelobes are Gaussian distributed.

The average SNR (signal to noise ratio) of the array is defined as $SNR_{array} = SNR_{node} G(\varphi,)$ which means that the array average SNR is $SNR_{array} = N\, SNR_{node}$ when we are aiming the array towards the incoming plane wave. The SNR_{array} is also a Gaussian distributed parameter with a mean of 17 dB, and a 95% confidence that the SNR of the array will be higher than 7 dB.

One technique to alleviate the problem of the stochastic behavior of the sidelobe level is to shape the beamformer by putting individual weights on the transmitting nodes. These weights are calculated so that the beam pattern will fit into a pre-defined desired beam pattern. This method can also be used for shaping the beam-width and the beam transmitting direction. A template of the desired beam-shape used in the simulations is shown in Figure 4.

Using the least squares method, the error between the far-field pattern $G(\varphi,\theta)$ in equation (4) and the beampattern template $G_T(\varphi,\theta)$ in Figure 4 is minimized and we attain the optimal distributed weights for the transmitting nodes,

$$\mathbf{w}_{\mathbf{opt}} = \underset{\mathbf{w}}{minimize} \; || G(\varphi,\theta) - G_T(\varphi,\theta) ||^2 \quad (6)$$

Figure 2. The absolute squared average array pattern of 250 realizations of the random sensor locations. Only a small part around the main lobe is shown in the figure.

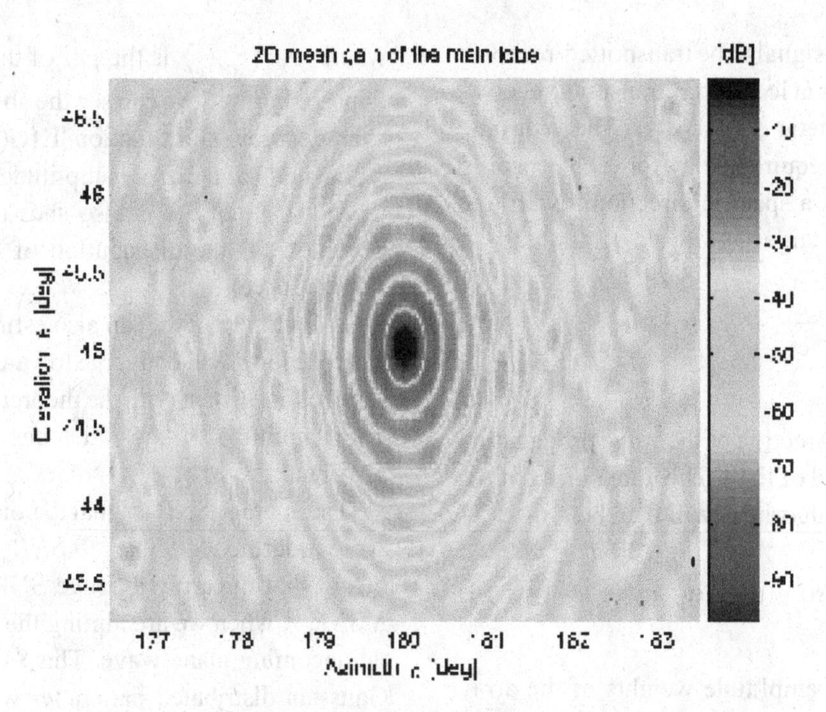

Figure 3. A cross-section of the main lobe of all 250 realizations of the array amplitude gain pattern

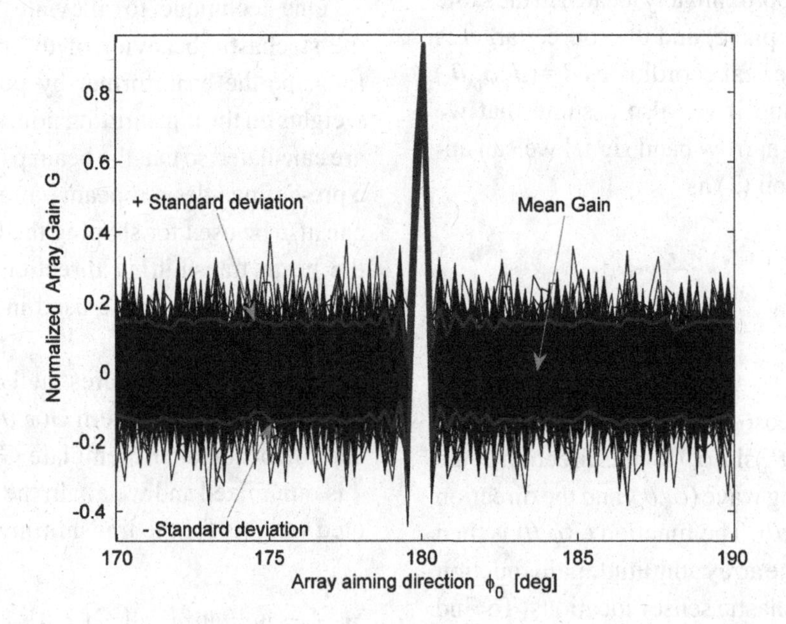

Figure 4. The template for shaping the beam pattern with an example of a sidelobe level at -20 dB and a beamwidth of 18 degrees

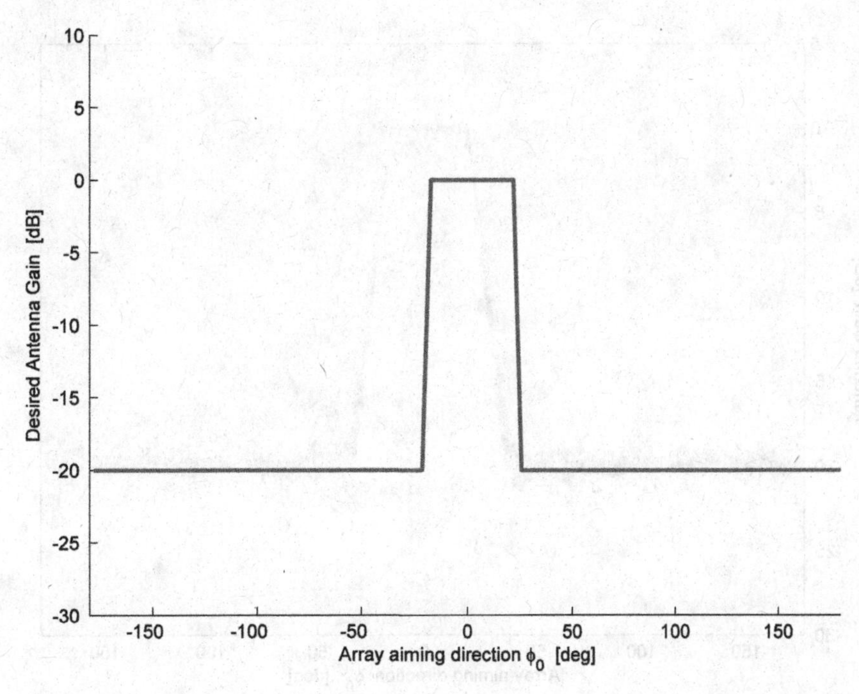

where $\mathbf{w} = [w_1, w_2, ..., w_N]$ and N is the number of nodes. Figure 5 show the generated beam patterns of an ensemble of 25 random placements of 50 and 100 distributed transmitting nodes within an area with the radius 1000 m.

Cooperative MIMO Techniques

Another recently popular technique to improve the signal to noise ratio of the long range transmission is to use some form of multiantenna diversity system. In this chapter we employ MISO, SIMO and MIMO antenna diversity systems diversity which have gained great interest in the past decade or so. Multiple transmit and receive antenna systems allow increased data rates and enhanced link reliability of wireless communication systems while reducing the transmission power requirements. In the following analysis of these diversity techniques we will assume a perfect knowledge of the propagation channel.

Cooperative MISO and SIMO

First we consider a MISO frequency flat fading propagation model with N_{tx} antenna elements at the transmitter and one antenna element at the receiver. To take full advantage of the transmit diversity system we send multiple weighed copies of the signal sample through all the transmitting antenna elements. The received baseband signal sample can then be expressed as

$$r[m] = \sqrt{\frac{E_s}{N_{tx}}} \cdot \sum_{l=0}^{L-1} h_l w_l s[m] + n[m], \qquad (7)$$

where $r[m] \in$ is the received sample, $s[m] \in$ is the transmitted sample and $n[m]$ is a noise sample with $n[m] \sim \mathcal{CN}\left(0, \sigma_n^2\right)$. The coefficient w_l is the channel weight for channel l and E_s is the transmitted average symbol energy. This can be expressed in vector notation as

Figure 5. Generated beam patterns from 25 random placements of 50 nodes (top figure) and 100 nodes (bottom figure)

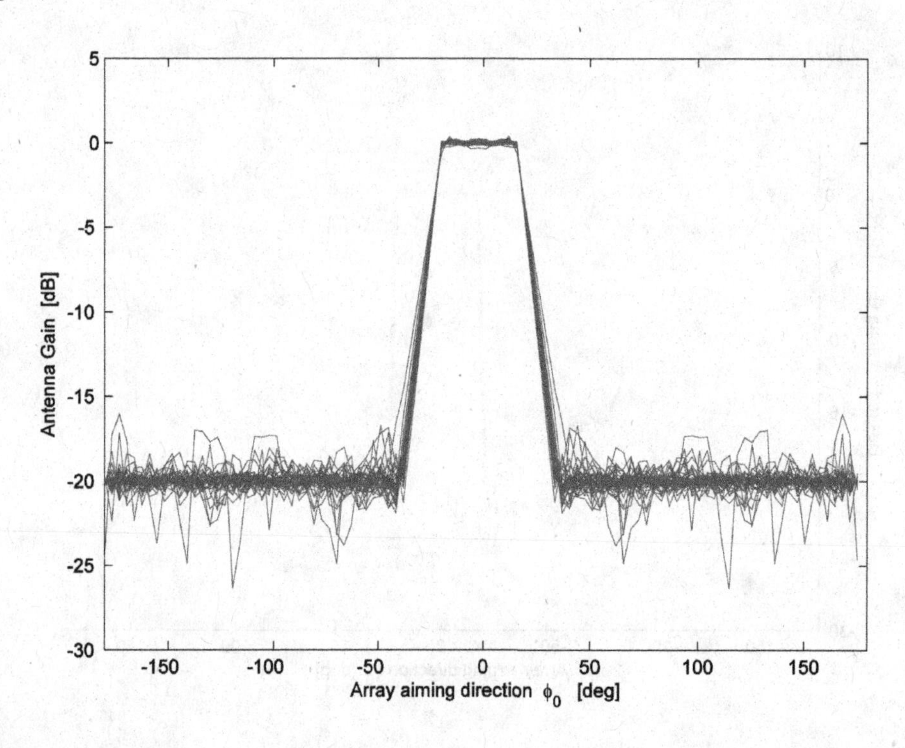

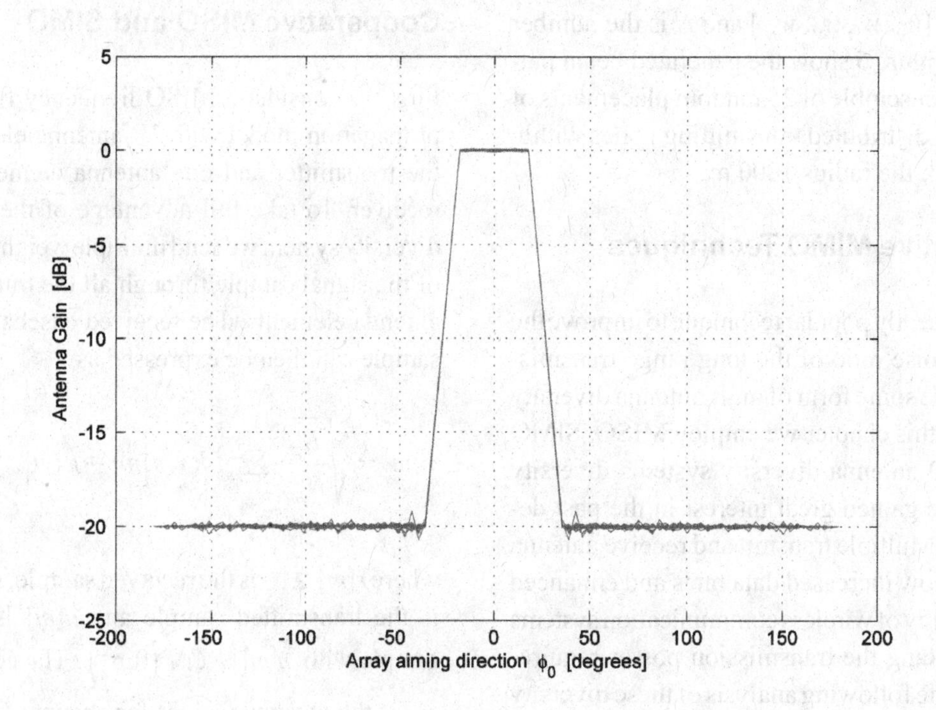

$$\mathbf{r} = \sqrt{\frac{E_s}{N_{tx}}} \cdot \mathbf{hw}s + \mathbf{n}, \quad (8)$$

where $\mathbf{h} \in \mathbb{C}^{N_{tx} \times 1}$ is the frequency flat fading channel vector with a Rice distribution. The normalized Rician channel vector $\mathbf{h}$ can, assuming the full correlation model, then be defined as (McKay & Collings, 2006).

$$\mathbf{h} \triangleq \sqrt{c_1} \cdot \mathbf{l} + \sqrt{c_2} \cdot \mathbf{R}_{tx} \mathbf{h}_w, \quad (9)$$

where $\mathbf{l}$ is the line of sight (LOS) component represented as a mean value that satisfies the condition $\|\mathbf{l}\|^2 = N_{tx}$, and $\mathbf{R}_{tx}$ is the transmit correlation vector. $\mathbf{R}_{tx}$ is assumed to be a positive definite full rank matrix. $\mathbf{h} \sim \mathcal{CN}\left(0_{N_{tx}}, 1_{N_{tx}}\right)$ is a complex valued Gaussian vector representing the non line of sight (NLOS) component. The coefficients $c_1 = K/(K+1)$ and $c_2 = 1/(K+1)$ are normalizing factors, where K is the Rice factor which represents the power ratio between the LOS and NLOS components. The weight vector $\mathbf{w}$ that maximizes the received SNR is given by

$$\mathbf{w} = \sqrt{N_{tx}} \cdot \frac{\mathbf{h}^H}{\|\mathbf{h}\|}, \quad (10)$$

which is the transmit maximum ratio combining (MRC) method and is also known as matched beamforming.

The SNR of the received signal can then be expressed as

$$\gamma_{rx} = \frac{E_s \cdot |\mathbf{h}|^2}{N_0}. \quad (11)$$

The second type of spatial diversity is receive diversity in which we are utilizing a SIMO frequency flat fading propagation channel model with N_{rx} receiving antenna elements and a single transmitting antenna element. To fully exploit the receive diversity we will receive multiple copies of the transmitted signal through all the N_{rx} receiving antenna elements. The received baseband signal sample can then be expressed as

$$r[m] = \sqrt{\frac{E_s}{N_{rx}}} \cdot \Sigma_{l=0}^{L-1} (w_l h_l) s[m] + \Sigma_{l=0}^{L-1} w_l n_l [m], \quad (12)$$

where $r_l[m] \in \mathbb{C}$ is the received sample from receiving antenna element l, $s[m] \in \mathbb{C}$ is the transmitted sample and $n_l[m]$ is a noise sample at receiving antenna element l with $n_l[m] \sim \mathcal{CN}(0, \sigma_n^2)$. the coefficient w_l is the channel weight at receiving antenna element l and E_s is the transmitted average symbol energy. This can be expressed in vector notation as

$$r = \sqrt{E_s} \cdot \mathbf{w}^H \mathbf{h}s + \mathbf{w}^H \mathbf{n}, \quad (13)$$

where $\mathbf{h} \in \mathbb{C}^{N_{rx} \times 1}$ is the frequency flat fading channel vector with a Rice distribution. The normalized channel vector $\mathbf{h}$ can then be defined as (McKay & Collings, 2006)

$$\mathbf{h} \triangleq \sqrt{c_1} \cdot \mathbf{l} + \sqrt{c_2} \cdot \mathbf{R}_{rx} \mathbf{h}_w, \quad (14)$$

where $\mathbf{l}$ is the line of sight (LOS) component represented as a mean value that satisfies the condition $\|\mathbf{l}\|^2 = N_{rx}$, and $\mathbf{R}_{rx}$ is the receive correlation vector. $\mathbf{R}_{rx}$ is assumed to be a positive definite full rank matrix. $\mathbf{h} \sim \mathcal{CN}\left(0_{N_{tx}}, 1_{N_{tx}}\right)$ is a complex valued Gaussian vector representing the non line of sight (NLOS) component. The weight vector $\mathbf{w}$ that maximize the received SNR at each antenna element is given by

$$\mathbf{w} = \sqrt{N_{rx}} \cdot \frac{\mathbf{h}^H}{\|\mathbf{h}\|}, \tag{15}$$

The SNR of the received signal after we have performed a maximum ratio combining (MRC) can then be expressed as

$$\gamma_{rx} = \frac{E_s \cdot |\mathbf{h}|^2}{N_0}. \tag{16}$$

Cooperative MIMO

By combining the MISO and SIMO diversity techniques we create a system of (N_{tx} and N_{rx}) transmitting and receiving antenna elements, respectively. If we consider a frequency flat fading $\left(N_{tx} \times N_{rx}\right)$ MIMO propagation model the received signal can be written in vector notation as

$$r = \sqrt{\frac{E_s}{N_{tx}}} \cdot \mathbf{w}_{rx}^H \mathbf{H} \mathbf{w}_{tx} s + \mathbf{w}_{rx} \mathbf{n}. \tag{17}$$

If we assume the Kronecker channel model, then the MIMO case of the Rice distributed channel matrix $\mathbf{H}$ can be derived as

$$\mathbf{H} \triangleq \sqrt{c_1} \cdot \mathbf{L} + \sqrt{c_2} \cdot \mathbf{R}_{rx}^{1/2} \mathbf{H}_w \mathbf{R}_{tx}^{1/2}, \tag{18}$$

where $\mathbf{L}$ represents the LOS component and is the arbitrary rank mean value matrix with the condition that $Tr\left(\mathbf{L}\mathbf{L}^H\right) = N_{rx} \cdot N_{tx}$, $\mathbf{R}_{rx}$ and $\mathbf{R}_{tx}$ are the correlation matrices on the transmitter and receiver side respectively.

$$H \sim \mathcal{CN}_{N_{rx} \times N_{tx}}\left(0_{N_{rx} \times N_{tx}}, \mathbf{I}_{N_{rx}} \otimes \mathbf{I}_{N_{tx}}\right).$$

To maximize the combined SNR at the receiver antenna elements we maximize

$$\gamma_{rx} = \frac{E_s}{N_0} \cdot \frac{\| \mathbf{w}_{rx}^H \mathbf{H} \mathbf{w}_{tx} \|^2}{N_{tx} \| \mathbf{w} \|^2}. \tag{19}$$

γ_{rx} is then maximized when $\mathbf{w}_{rx}$ and $\mathbf{w}_{tx}/N_{tx}$ are equal to the singular input and output vectors of the channel matrix $\mathbf{H}$ corresponding to the maximum singular value of the channel matrix $\mathbf{H}$. Equation (17) can then be written as

$$r[m] = \sqrt{E_s} \sigma_{max} s[m] + n[m]. \tag{20}$$

where $\sigma_{\max}$ is the maximum singular value of the channel matrix $\mathbf{H}$ and since σ_{max}^2 is the same as the maximum eigenvalue $\lambda_{\max}$ of $\mathbf{H}\mathbf{H}^H$ we can now express the received SNR of the MIMO diversity technique as

$$\gamma_{rx} = \frac{E_s}{N_0} \cdot \lambda_{max}. \tag{21}$$

SIMULATION RESULTS

In this section we present the simulation results and assess the performance of cooperative beamforming, SIMO, MISO and MIMO diversity techniques and compare the results together and with that of nondiversity single antenna (or SISO) system. If we consider a base station mounted on an aerial platform such as a HAP or a UAV to collect data from remote sensor networks then the amount of obstructions in the transmission path would depend on the type of environment at the sensor locations, although it can still generally be assumed that the number of obstructions will increase with a decreasing antenna elevation angle. Therefore the propagation effect of the change in elevation can be translated into a change of the Rice distribution K-factor.

In the presented simulations, the Rician K-factor was varied over an interval of

$K \in \left[1 \cdot 10^{-8}, 1 \cdot 10^{8}\right]$, where the low value represents a channel with no LOS component and very little correlation between the different signal paths and therefore resembles a Rayleigh fading channel. When the Rician K-factor is gradually increased the correlation between the signal paths will increase and the direction of departure/arrival of the signals will narrow into a smaller and smaller angular sector, until the K-factor asymptotically goes toward infinity and all signal paths will be correlated and pointing in the same direction.

In Figure 6 we compare the performance between the ordinary random array beamformer and the MISO/SIMO diversity systems. Inspecting Figure 6, we can see that the MISO/SIMO diversity system seems to maintain a constant low node transmitter power P_{tx} even in a NLOS scenario by spreading the energy over multiple paths instead of transmitting it all in one direction. Furthermore, we can see from Figure 6 that if the distance between the transmitting nodes and the basestation is increased from 1 km to 10 km, the nodes need a 100 fold increase of the total transmitted power to maintain the same capacity. This is independent of whether we are using the nodes as a beamforming array or a diversity system, which is consistent with the inverse square law of the free space loss. If we now increase the number of receiving antenna nodes to be equal to the number of transmitting antenna nodes we get a 50×50 MIMO system which will increase the array and diversity gains even further. This effect can clearly be seen in Figure 7 where the performance of the MIMO system outperforms the other systems in both LOS and NLOS scenarios. It is also clear from this figure that the nondiversity SISO system and the conventional beamformer will not function properly in this setting and in particular in NLOS conditions. These results

Figure 6. Comparison between of the array beamformer and MISO/SIMO system for different K-factor values at two different distances from the base station of 10 km and 1 km, respectively

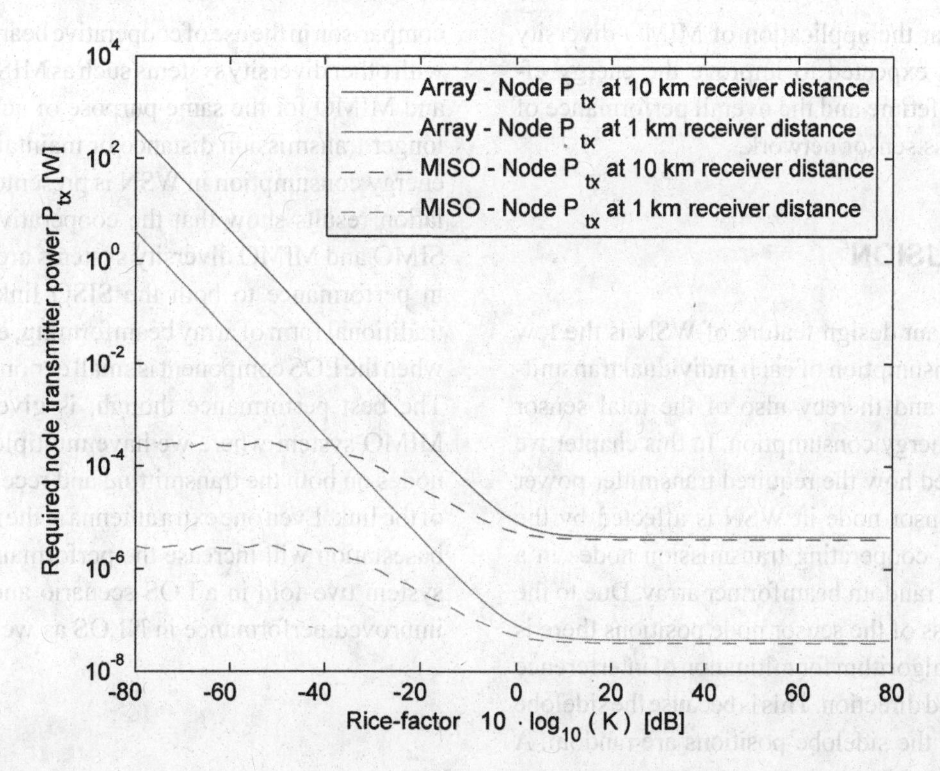

Figure 7. Performance of the array beamformer, MISO/SIMO and MIMO systems for different K-factor values and compared with a single antenna SISO system

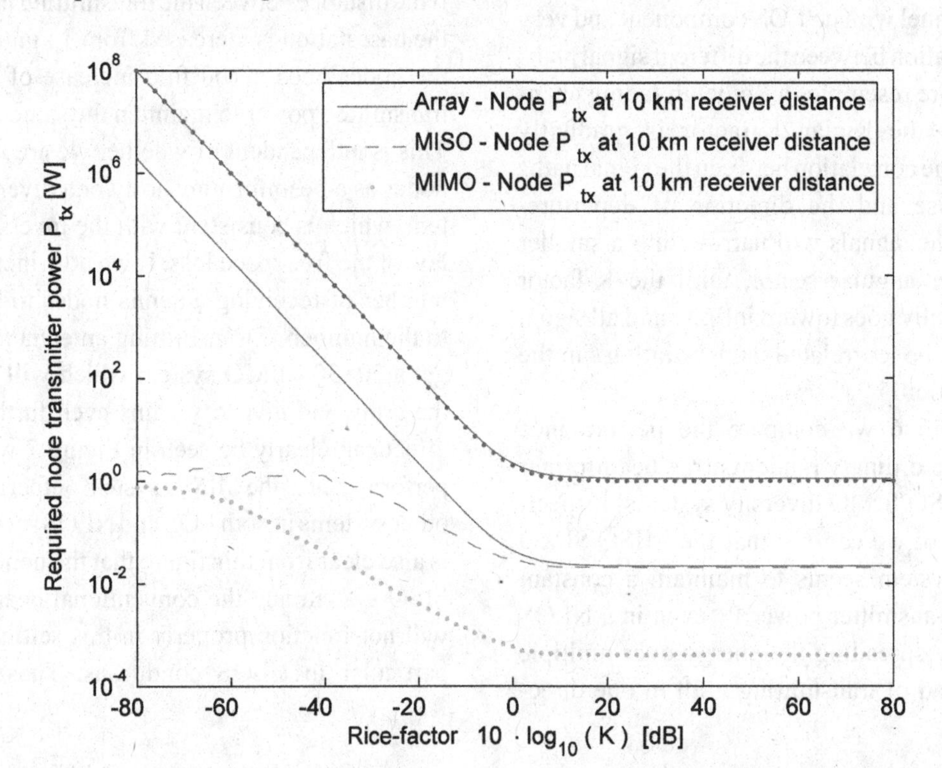

suggest that the application of MIMO diversity systems is expected to improve the energy efficiency, lifetime and the overall performance of the wireless sensor network.

CONCLUSION

An important design feature of WSN is the low energy consumption of each individual transmitting node and thereby also of the total sensor network energy consumption. In this chapter we investigated how the required transmitter power of each sensor node in WSN is affected by the number of cooperating transmission nodes in a traditional random beamformer array. Due to the randomness of the sensor node positions there is no simple algorithm for mitigation of interference from a fixed direction. This is because the sidelobe levels and the sidelobe positions are random. A comparison in the use of cooperative beamforming with other diversity systems such as MISO/SIMO and MIMO for the same purpose of achieving a longer transmission distance or maintaining low energy consumption in WSN is presented. Simulation results show that the cooperative MISO/SIMO and MIMO diversity systems are superior in performance to both the SISO link and the traditional form of array beamforming, especially when the LOS component is small or non-existent. The best performance though, is given by the MIMO system where we have multiple antenna nodes on both the transmitting and receiving end of the link. Even one extra antenna at the receiving basestation will increase the performance of the system two-fold in a LOS scenario and give an improved performance in NLOS as well.

REFERENCES

Akyildiz, I., Su, W., Sankarasubramaniam, Y., & Cayirci, E. (2002). A survey on wireless sensor networks. *IEEE Communications Magazine*, *40*(8), 102–114. doi:10.1109/MCOM.2002.1024422

Cui, S., Goldsmith, A. J., & Bahai, A. (2004). Energy-efficiency of MIMO and cooperative MIMO techniques in sensor networks. *IEEE Journal on Selected Areas in Communications*, *22*(6), 1089–1098. doi:10.1109/JSAC.2004.830916

Drane, C. J. (1968). Useful approximations for the directivity and beamwidth of large scanning Dolph-Chebyshev arrays. *Proceedings of the IEEE*.

Hansen, W. W., & Woodyard, J. R. (1938). A new principle in directional antenna design. *Proceedings IRE*, *26*(3).

Jenkins, J. (1973). Some properties and examples of random listening arrays. *IEEE Oceans, 5*.

Johnson, D. H., & Dudgeon, D. E. (1993). *Array signal processing: Concepts and techniques*. P T R Prentice-Hall Inc.

Lewis, F. L. (2004). Wireless sensor networks. In Cook, D. J., & Das, S. K. (Eds.), *Smart environments: Technologies, protocols, and applications*. New York, NY: John Wiley.

McKay, M. R., & Collings, I. B. (2006). Improved general lower bound for spatially-correlated Rician MIMO capacity. *IEEE Communications Letters*, *10*(3). doi:10.1109/LCOMM.2006.1603371

Pillutla, L. S., & Krishnamurhty, V. (2005). *Joint rate and cluster optimization in cooperative MIMO sensor networks*. IEEE 6th Workshop on Signal Processing Advances in Wireless Communications.

Rashid-Farrokhi, F., Tassiulas, L., & Liu, K. J. R. (1998). Joint power control and beamforming in wireless networks using antenna arrays. *IEEE Transactions on Communications*, *46*(10). doi:10.1109/26.725309

Singh, M., & Prasanna, V. K. (2003). *A hierarchical model for distributed collaborative computation in wireless sensor networks*. IEEE International Parallel and Distributed Processing Symposium.

Stankovic, J. A., Abdelzaher, T. F., Lu, C., Sha, L., & Hou, J. C. (2003). Realtime communication and coordination in embedded sensor networks. *Proceedings of the IEEE*, *91*(7), 1002–1022. doi:10.1109/JPROC.2003.814620

Tubaishat, M., & Madria, S. (2003). Sensor networks: An overview. *IEEE Potentials*, *22*(2), 20–23. doi:10.1109/MP.2003.1197877

Chapter 12
Energy Efficient Communication in Wireless Sensor Networks

Nauman Israr
University of Teesside, UK

ABSTRACT

Longer life time is the primary goal of interest in Wireless Sensor Networks (WSNs). Communication dominates the power consumption among all the activities in WSNs. The classical sleep and wake up scheduling scheme at the application layer is believed to be one of the best power saving schemes for dense WSNs. These schemes reduce redundant transmissions, and as a result, prolong the network life time. This chapter analyzes the effect of density on inter cluster and intra cluster communication and evaluates a hybrid cross layer scheduling schemes to enhance the life time of the WSNs. In the conventional scheduling schemes at the application layer, all the nodes whose area are covered by their neighbors are put to sleep in order to prolong the life time of the WSNs. The hybrid cross layer scheme in this chapter suggests that instead of putting the redundant nodes to sleep if they are used for some other energy intensive tasks, for example the use of redundant nodes as relay stations in inter cluster communication, will be more energy efficient compare to the conventional application layer scheduling schemes in WSNs. Performance studies in the chapter indicate that the proposed communication strategy is more energy efficient than the conventional communication strategies that employ the sleep/wake up pattern at application layer.

INTRODUCTION

Wireless Sensor Networks comprise of tiny devices with a capability of sensing, communication and computation are spread over a physical environment, perhaps for a limited period of time, with a common objective to collaborate to provide a distributed and robust sensing, storage, and communication service. These devices can be dispersed over a hostile battlefield or harsh environment to achieve a task. Wireless sensor networks have an enormous potential impact in

DOI: 10.4018/978-1-4666-0101-7.ch012

the future in countless fields such as military, civil, heath and habitat monitoring (Werner-Allen et al., 2006), (Paek, 2005), and (Cranch, 2003). The design goals of WSNs are very different from those of conventional wireless networks, such as cellular networks and wireless ad hoc networks. In these latter the primary goal of interest is usually some aspect of quality of service such as satisfying constraints on such things as signal-to-noise ratio, bandwidth, and delay or packet loss. With WSNs, because of their often remote deployment, the major constraint is in the power supply (Akyildiz, 2002), (Chong, 2003). Then if there is an imbalance in energy consumption across the network, this can lead to the premature expiration of a specific device or devices which, in turn, will shorten the lifetime of the network (Zhao, 2005),(Arias, 2006),(Shah, 2002),(Kadayif, 2004),(Lee, 2006).

The characteristics that make WSNs different from the conventional wireless networks are its dense deployment, continuously changing topology and limited resources. The dense deployment feature of the WSNs is often exploited to achieve energy efficiency. This chapter will analyze the effect of density on the life time of the WSNs and will evaluate the potential of exploiting the density at the inter cluster and intra cluster communication and will also analyze a new hybrid cross layer scheduling scheme that uses the redundant nodes at the application layer instead of putting the nodes to sleep. Performances studies in the chapter indicate that the better way to exploit density for longer life time is to use the redundancy for inter cluster communication which is the most expensive activity in WSNs.

The Chapter is organised as follow. Section II outlines the design principles that should be kept in mind while designing a communication strategy for WSNs. Section III presents a review of cluster based routing and the dominant scheduling techniques in WSNs. Section IV presents the hybrid cross layer communication strategy for cluster based WSNs. Section V presents the simulation results and finally section VI finally concludes the paper.

DESIGN PRINCIPLES OF COMMUNICATION STRATEGIES IN WSNS

The following issues should be considered while designing a communication protocol for WSNs.

Distributed

The communication strategy should be decentralized. Since nodes in WSNs are prone to failure. Centralized control with its limited resources is not feasible for WSNs. A hybrid scheme of centralized and distributed is more energy efficient for WSNs. For example, clustering in LEACH (Heinzelman et al., 2002) is distributed while centralized control is employed by sub dividing the network into smaller sub networks, resulting in an energy efficient distributed routing strategy. One of the advantages of the hybrid communication schemes is that changes in the network are localized and have no effect on rest of the network.

In-Network Computation

The communication protocol should try to perform computation on data wherever possible in order to minimize the number of transmissions in WSNs. The simplest network computation technique is data aggregation. Suppose a sink is periodically interested in the average temperature of the network, it will be more efficient to aggregate the temperature from the set of nodes and forward it rather than transmitting each and every node temperature measure.

Data Centric

The communication protocol should be data centric rather than address centric because of the maintenance cost associated with the sheer number of nodes in WSNs. This makes security hard to implement in WSNs.

Exploitation of the Activity Pattern

WSNs are sometimes deployed for real time applications which are based on event detection. In case of event detection the traffic pattern will suddenly change. In this case all nodes will try to transmit at the same time. This requires a robust communication scheme that should be able to deal with bursts as well as non real time traffic.

Single Hop or Multi Hop

The routing protocols are mostly multi hop. Since consumption of energy for large distance transmission in wireless is more than short distance transmission. But there is trade-off between multi hop and single hop. Multi hop achieves energy efficiency at the cost of delay. Multi hop transmission is also employed because the transmission range of WSNs is limited.

BACKGROUND

In order to prolong WSNs life time, scheduling at application layer and routing plays a very important role in saving energy. The aim is how to manipulate the wireless transmissions in an efficient way in order to achieve longer life time and at the same time maintain the desired quality of service. In this section we will outline some of main techniques used for scheduling and routing in WSNs for energy efficiency.

Classification of Scheduling Schemes in WSNs

The scheduling of the nodes into sleep and wake up in WSNs can be divided into three broad categories:

MAC Layer Scheduling

The sleep and wake up cycle is incorporated into the MAC layer, for example, S-Mac (Ye, 2002) in which each node follows a periodic sleep and wake up pattern and the neighboring nodes synchronize their sleep/wake up cycles. In this scheme a longer lifetime is achieved at the cost of delay because if a neighboring node is sleeping, the transmitting node will have to wait until the neighbor wakes up.

Application Layer Scheduling

In application layer scheduling if a sensor node sensing area is covered by its neighbors completely; the sensing node is put to sleep. For example, in the Tian and Geogorian algorithm (Tian, Nicolas, & Georganas, 2003) a node is put to sleep to prolong network life if its sensing area is covered by its neighbors or its sensing area is a sub set of a powerful node sensing area.

Transmission Pattern Based Scheduling

Another classification technique is the pattern of the sleep/wake up cycle. It can be either deterministic, random or on demand. For example, in LEACH the intra cluster communication follows a deterministic pattern in which each node is allotted a Time Division Multiple Access (TDMA) slot to transmit. However, the scheme is good only for data centric applications in which data is sensed most of the time.

The application layer scheduling is believed to be one of the best power saving schemes for WSNs. In this type of scheduling, if a node sensing area is covered completely by its neighbors, the node is put to sleep. The scheme reduces redundant transmissions and results in longer lifetimes for WSNs. Several schemes have been proposed in the literature for different categories of scheduling. The goal of most schemes is to extend the lifetime while manipulating the node behavior.

Tian and Geogiran present a scheduling scheme which argues that turning off some of the nodes does not affect the overall system function as long as there are enough nodes to cover the whole area. Therefore the lifetime is prolonged by exploiting the density of the WSNs. The main objective is to minimize the working nodes while maintaining full coverage. If the sensing area of one node is fully embraced by the union set of its neighbors, i.e. the neighbors can cover the current node's sensing area, this node goes to sleep. The algorithm put two types of nodes to sleep; the one whose sensing area is covered completely by its neighbors and the node whose sensing area is a sub set of a powerful node. According to the author the protocol works at the application layer and can be implemented in any energy efficient communication protocol. In the S-Mac scheme energy consumption is reduced by allowing randomly selected idle sensors to go into the sleep mode. The traffic intended for these sleeping nodes is temporarily stored by the neighboring active nodes. The sleeping sensors wake-up periodically to retrieve the stored packets from their neighboring nodes. All nodes in the network synchronize their sleep and wake up periods with their neighbors. (Jiang et al., 2007) tries to prolong the network time by turning off a selection of redundant nodes while maintaining sufficient sensing coverage and connectivity. The paper evaluates a method to figure out which nodes are redundant and put them to sleep, especially the boundary nodes. (Wang, 2007) proposes coverage based scheduling scheme for a heterogeneous network. It employs distributed coverage control strategy which is based on location information for heterogeneous WSNs. In this scheme, each node calculates the coverage degree locally, and the node scheduling scheme is based on a weak time synchronization scheme. Under the premise of guaranteeing nodes k-coverage precisely, the scheme can make as many nodes as possible inactive. Simulation results demonstrate that the scheme can reduce the number of active nodes effectively while achieving the required coverage degree, balancing energy consumption and maintaining a long network lifetime. Sichitiu (2004) presents a cross layer scheduling scheme for power efficiency in WSNs. The author argues that the single layer cannot achieve power efficiency in WSNs. He proposes a deterministic, schedule-based energy conservation scheme in which the nodes wake up only when necessary, thus maintaining a fully connected network at the same time. The author did not consider the routing factor while simulating this sleep/wake up model and at the same time also argues that power conservation is not possible without consideration of multiple layers.

It is evident from the literature that in WSNs the dominant strategy in practice to prolong the network lifetime is to put the nodes to sleep. In contrast to this argument, this chapter suggests that instead of putting the nodes to sleep, using them for multi hop intra cluster communication is more energy efficient.

Routing Strategies in WSNs

With regard to previous work in this context, Al-Karaki and Kamal (2004) classify routing in WSNs into three broad categories on the basis of network structure.

Flat Based Routing

Flat based routing comprises of a single tier network, for example, Direct Diffusion (Intanagonwiwat, Govindan, & Estrin, 2000). The main drawback of the flat based schemes is that they are not feasible for larger networks, because the nodes which are far from the base station are unable to send their data to the base station in time.

Hierarchal Routing

Hierarchical routing is based on the two tier network structure and mostly employs clustering techniques. For example LEACH (Heinzelman et al., 2002)

Clustering is an effective communication scheme in WSNs to improve the life time of the WSNs and to overcome the communication latency problem. Clustering has the following advantages compared to the flat based schemes in WSNs.

- Provides communication scalability and longer network lifetime (Younis, 2004).
- Facilitates distribution of control over the network.
- Enables reuse of resources and reduces the amount of information propagated in the network (Li, 2005).
- Reduces the amount of information that is used to store the network state.

Location Based Routing

Location based Routing is both two tier as well as single tier. For example, the Geographic Adaptive Fidelity (Xu, Heidemann, & Estrin, 2001). In location based routing the number of transmissions are reduced only by querying the region of interest, provided that the sensor network has the location information about the region of interest.

Cluster Based routing schemes are feasible for a wide variety of applications in WSNs. The work by Liyang, Neng and Xiaoqiao (2005) is an example of real time application of WSNs employing the cluster based strategy for communication. The methods of LEACH and HEED (Younis, 2004) are used for continuous data acquisition applications which involve continuous monitoring of an area for a longer period of time.

LEACH algorithm was one of the first cluster based routing algorithms for WSNs. The algorithm was developed because single-tier network can cause the gateway to overload with the increase in sensor node density. Such overload might cause latency in the communication and results in inadequate tracking of events. In addition, the single-gateway architecture is not feasible for larger WSNs (Al-Karaki, 2004). LEACH uses only two layers for communication. One is for communication within the clusters and the other is between the cluster heads and the sink. Here the cluster head selection is random and the role of cluster heads rotates to balance the energy consumption across the network. Clusters are formed depending upon the signal strength of the cluster head advertisement message. A node will join the cluster which has the strongest signal to it. The work by Heinzelman, Chandrakasan and Balakrishnan also computes the optimum number of cluster heads for WSNs. According to their work it is 5% of the entire network. Simulation results show that LEACH performs over a factor of 7 reductions in energy dissipation compared to flat based routing protocols such as Direct Diffusion (Intanagonwiwat, Govindan, & Estrin, 2000). The main problem with the LEACH protocol lies in the random selection of the cluster heads. Since the cluster head selection is random, there exists probability that the cluster heads formed are unbalanced and may be in one part of the network. This results in some parts of the network unreachable. Figure 1 depicts the formation of clusters and the data transmission in LEACH.

Younis and Fahmy proposed HEED (2004), which is a distributed cluster based routing protocol for homogeneous WSNs in which the selec-

Figure 1. Typical cluster based routing algorithm (LEACH)

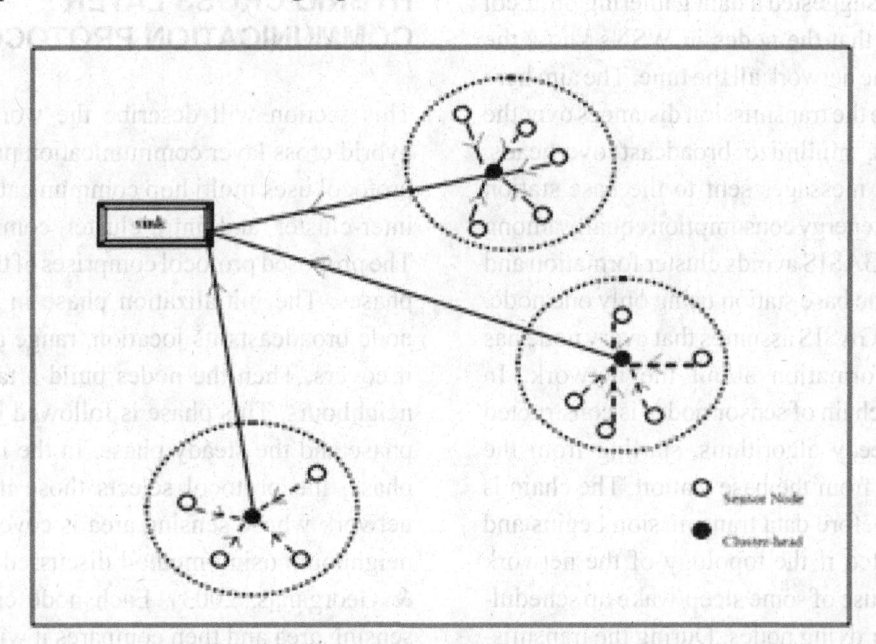

tion of the cluster head is dependent upon the residual energy of the nodes. Homogeneous WSNs after some time starts behaving as heterogeneous WSNs, there exists a probability that the lower energy nodes could have larger election probability than the higher energy nodes. Agrawal introduced a routing protocol TEEN(Threshold sensitive energy efficient sensor network protocol) (Manjeshwar, 2001) for time critical applications to respond to sudden changes in the sensed data. Here, the nodes sense the data continuously compared with the data transmission which is occasional, and only when the data is in the interest range of the user. Here the cluster head uses two value thresholds. One is a hard threshold and the other is a soft threshold. The hard threshold is the minimum value of the attribute that triggers the transmission from node to the cluster head and the soft threshold is the small change in the value of the sense attribute. The node will transmit only when the value of the attribute changes by an amount equal to or greater than the soft threshold. The soft threshold reduces transmissions further if there are no significant changes in the value of the sense attributes. The biggest advantage of this scheme is its suitability for time critical applications and also the fact that it significantly reduces the number of transmissions and gives the user control over the accuracy of the the attribute. Agrawal also proposed APTEEN (Adaptive periodic threshold sensitive energy efficient sensor network protocol) (Manjeshwar, 2002) an extension to TEEN, which is a hybrid protocol for both periodic data collection and for time critical event data collection. Here, the cluster head broadcasts four types of messages to the nodes: value of the threshold, the attributes value and a scheduling scheme for the nodes which is TDMA, allowing every node a single slot for transmission. Simulation shows that TEEN and APTEEN perform better than LEACH in terms of energy efficiency. Out of these three - LEACH, TEEN and APTEEN - TEEN performs better than the other two. The disadvantage is that since there is multilevel clustering in TEEN and APTEEN, they result in complexity and overheads. Lindsey and Raghavendra (2002) suggested a data gathering protocol that assumes that the nodes in WSNs know the topology of the network all the time. The aim here is to minimize the transmission distances over the entire WSNs, minimize broadcast overheads, minimize the messages sent to the base station and distribute energy consumption equally among all nodes. PEGASIS avoids cluster formation and transmits to the base station using only one node. However, PEGASIS assumes that every node has topology information about the network. In PEGASIS, a chain of sensor nodes is constructed using the greedy algorithms, starting from the node farthest from the base station. The chain is constructed before data transmission begins and is reconstructed if the topology of the network changes because of some sleep/wake up scheduling scheme or dying nodes. During the transmission phase the node collates the data and only one message is forwarded. For gathering data in each round, each node receives data from one neighbor, fuses it with its own data and transmits the data to the next neighbor on the chain. The node which is selected as the leader then transmits all the data to the base station. The leader in each round of communication will be at a random position in the network.

Rodoplu and Meng designed a protocol for mobile networks (1999) which is also applicable to static WSNs. The protocol identifies a relay region for every node where transmitting is more energy efficient than direct transmission. The idea behind MECN (Minimum Energy Mobile Wireless Networks) is to identify a sub network which will comprise fewer nodes and will require less power for transmission between two points in a network. So, the routing is completed while considering only the localized search.

HYBRID CROSS LAYER COMMUNICATION PROTOCOL

This section will describe the working of the hybrid cross layer communication protocol. The protocol uses multi hop communication for both inter-cluster and intra-cluster communication. The proposed protocol comprises of three distinct phases. The initialization phase in which each node broadcasts its location, range and the area it covers. Then the nodes build a table of their neighbours. This phase is followed by the setup phase and the steady phase. In the initialization phase, the protocol selects those nodes in the network whose sensing area is covered by their neighbours using method discussed in (Di Tian & Georganas, 2003). Each node calculates its sensing area and then compares it with that of its neighbours. If the sensing area of one node is fully embraced by the union set of its neighbours, that is, the neighbours can cover the current node's sensing area, this node becomes a temporary cluster head without reducing the system overall sensing coverage. All the nodes in the network

Figure 2. Random deployment of the network

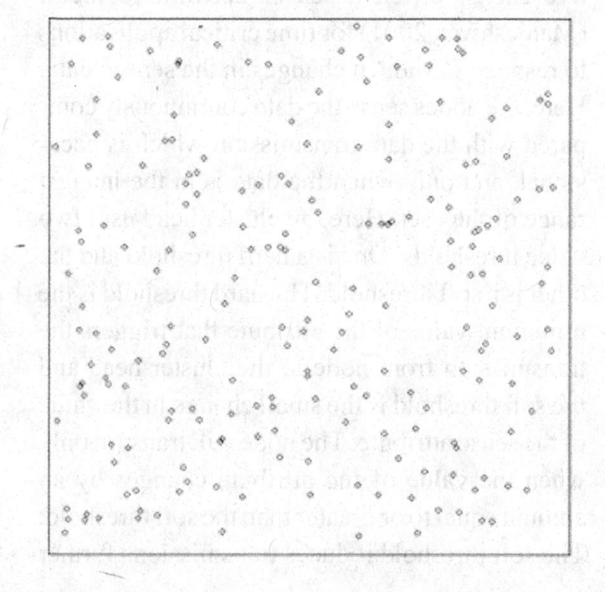

Figure 3. Formation of temporary cluster heads

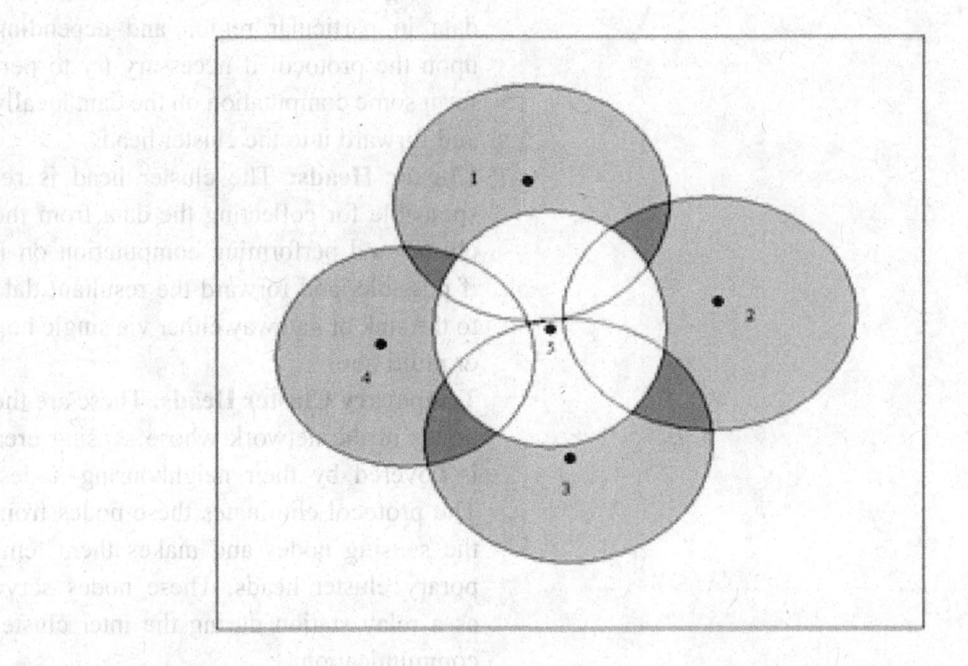

follow the same procedure. The cluster heads are selected randomly using equation (1).

At this point the algorithm then forms two layers of the network: the top layer which comprises of the temporary cluster heads and the cluster heads and the sensing layer which comprises of the sensing nodes of the clusters. After the selection of the cluster heads, the cluster heads broadcast an advertisement message. Depending on the message strength each node makes a decision to join a particular cluster. This phase uses the Code Division Multiple Access Medium Access Control (CSMAMAC) protocol. During this phase all the sensor nodes are listening. The selection of the cluster head is similar to the LEACH. During each cycle the cluster head selection is random and is dependent on the amount of energy left in the node and its probability of not being a cluster head during the last n rounds. A sensor node chooses a random number between 0 and 1. If this random number is less than the threshold T (n), the sensor node becomes a cluster-head.

$$T(n) = \begin{cases} \dfrac{P_t}{1 - P_t \cdot (r \cdot \mathrm{mod} \dfrac{1}{P_t})} & \text{if } n \in G \\ 0, & \text{otherwise} \end{cases} \quad (1)$$

Figure 4. Sensing layer of the network

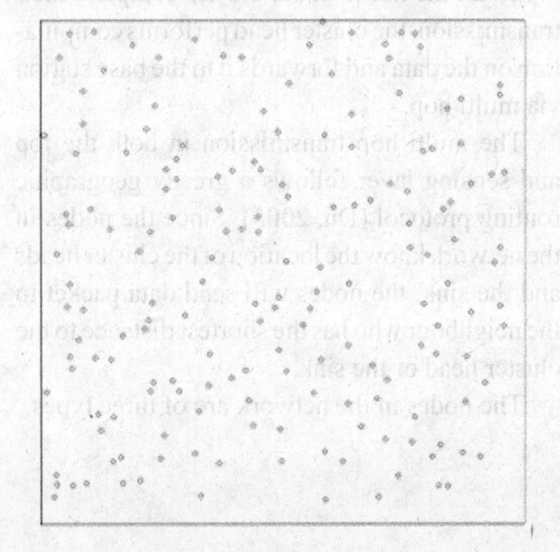

Figure 5. The top layer of the network without cluster heads

- P_t is the desired percentage of cluster head (5%).
- r is the current round.
- G is the set of nodes that have not been a cluster head in the last 1/Pt rounds.

Once the cluster heads are selected and the clusters are formed then the steady phase starts (transmission phase). In steady phase all the nodes transmit data using TDMA based scheduling. When all the nodes in the cluster complete their transmission, the cluster head performs computation on the data and forwards it to the base station via multi hop.

The multi hop transmission in both the top and sensing layer follows a greedy geographic routing protocol (Du, 2005). Since the nodes in the network know the location of the cluster heads and the sink, the nodes will send data packet to the neighbour who has the shortest distance to the cluster head or the sink.

The nodes in the network are of three types.

- **Sensing Nodes:** These nodes sense the data in particular region and depending upon the protocol if necessary try to perform some computation on the data locally and forward it to the cluster head.
- **Cluster Heads:** The cluster head is responsible for collecting the data from the cluster and performing computation on it if possible, and forward the resultant data to the sink or gateway either via single hop or multi hop.
- **Temporary Cluster Heads:** These are the nodes in the network whose sensing area is covered by their neighbouring nodes. The protocol eliminates these nodes from the sensing nodes and makes them temporary cluster heads. These nodes serve as a relay station during the inter cluster communication.

Typical sensor node energy consumption is based on equation 2, 3 and equation 4.

Total Energy = Energyt (n,d) + Energy_receive (n) (2)

Energyt (n,d) = (Energy_to_transmit*n) + (Energy_circuit*n*d*d) (3)

The energy consumption for receiving a message will be based on the following equation:

Energy_receive (n) = Energy_circuit*n (4)

SIMULATION AND DISCUSSION

This section will discuss the performance of the proposed cross layer routing protocol. The section will present simulation results for two versions of the LEACH. The first step simulates the proposed protocol by comparing it with the general

Figure 6. Energy dissipation graph of LEACH-M and proposed algorithm

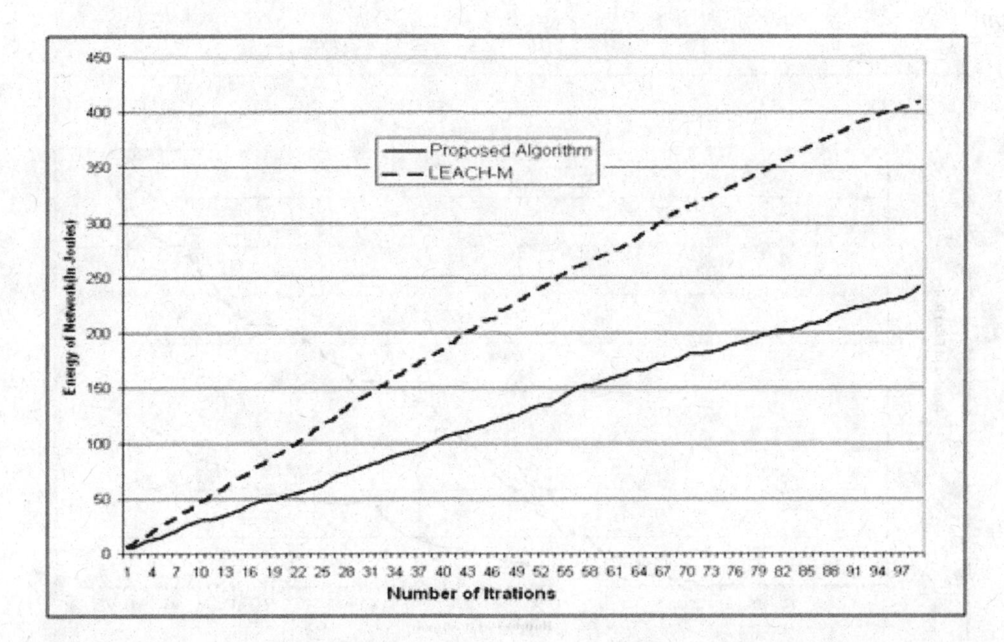

m-hop version of the LEACH which doesn't incorporate the application layer scheduling of the sleep/wakeup. The second step will compare the proposed algorithm against the modified version of the LEACH by incorporating the Tian and Georganas scheduling into the m-hop version of LEACH. The simulation will provide an insight into exploitation of density at both inter-cluster and intra-cluster layer of communication. The Simulation was based on the following assumptions.

Figure 7. Dead nodes comparison between LEACH-M and proposed algorithm

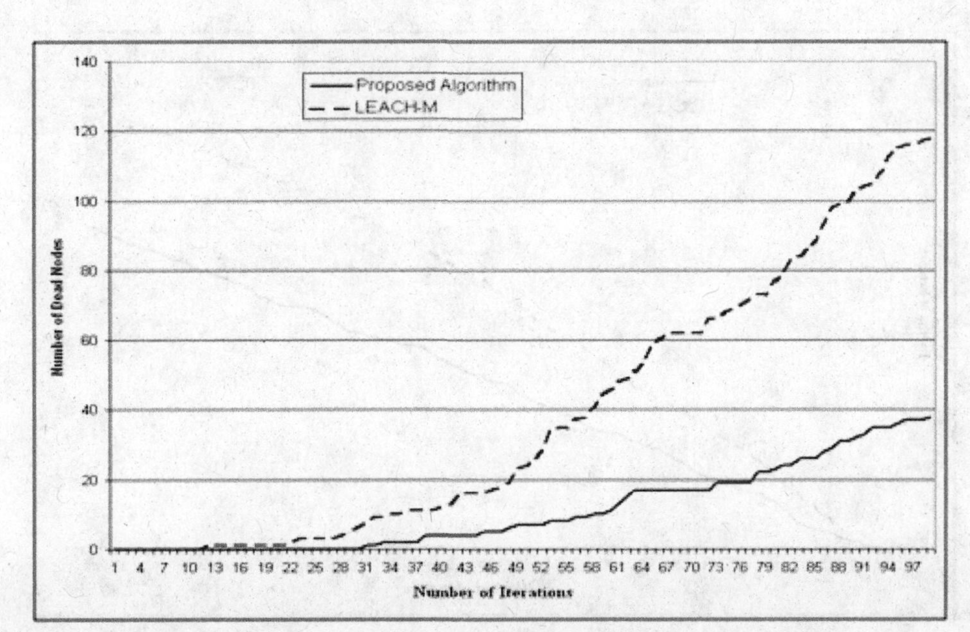

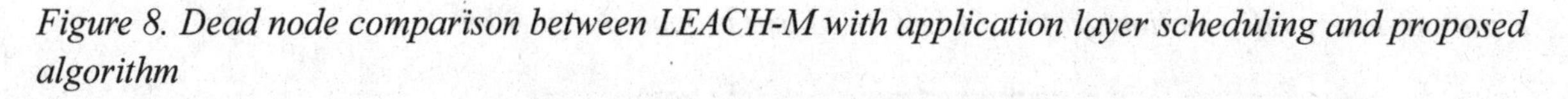

Figure 8. Dead node comparison between LEACH-M with application layer scheduling and proposed algorithm

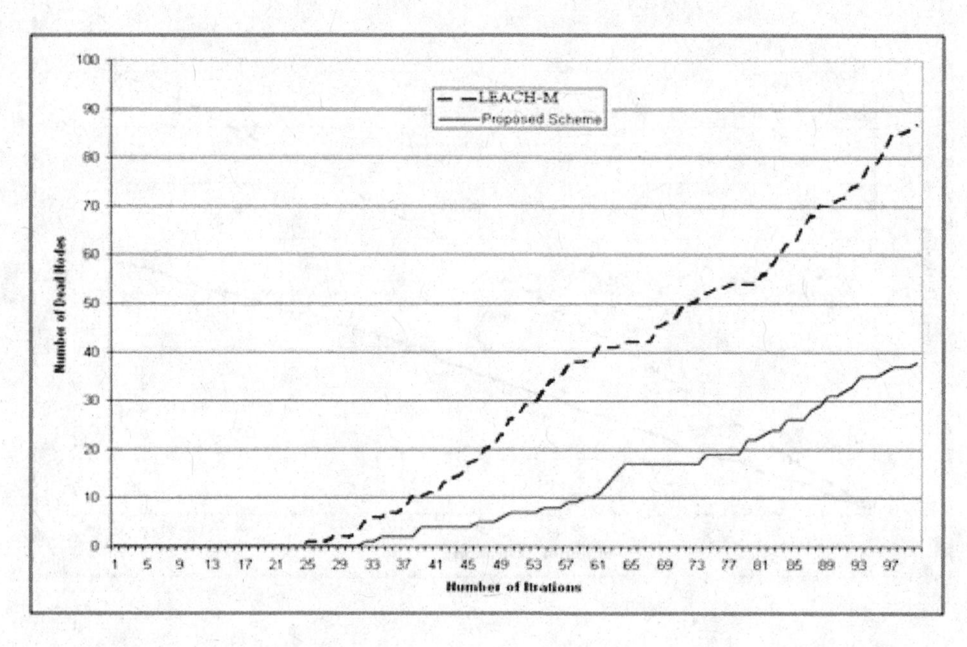

- The network comprises 200 randomly deployed sensor nodes, shown in Figure 2
- All nodes are static.
- The energy of a node is 3 joules.

Figure 9. Energy dissipation between LEACH-M with application layer scheduling and proposed algorithm

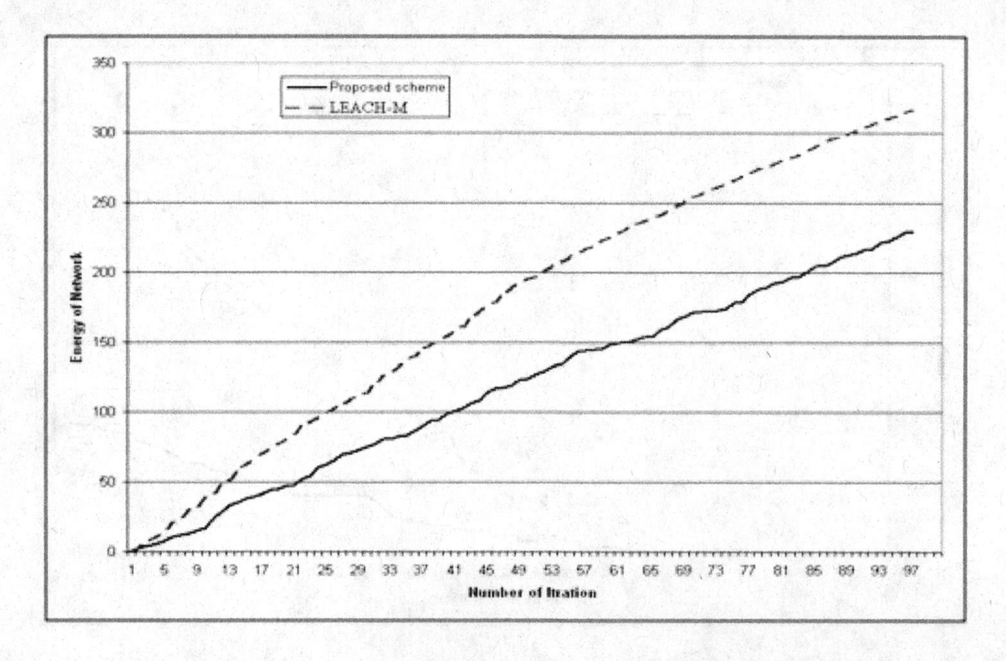

Figure 10. Number of dead nodes vs. duration of set up cycle

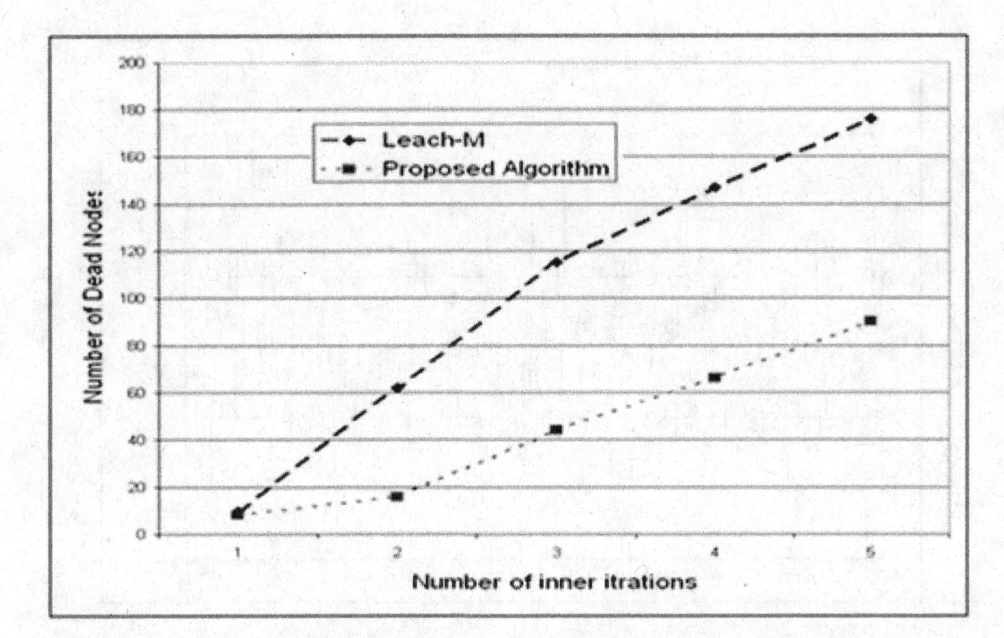

- The network is deployed randomly in an area of 500 meters square.
- The energy consumption for a node is 50 nj/bit to run the circuitry of both transmitter and receiver; and 100 Pj/bit to transmit.
- Each node knows its location via a Global Positioning System (GPS) or

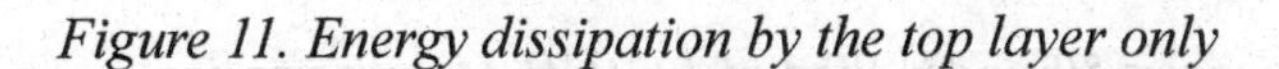

Figure 11. Energy dissipation by the top layer only

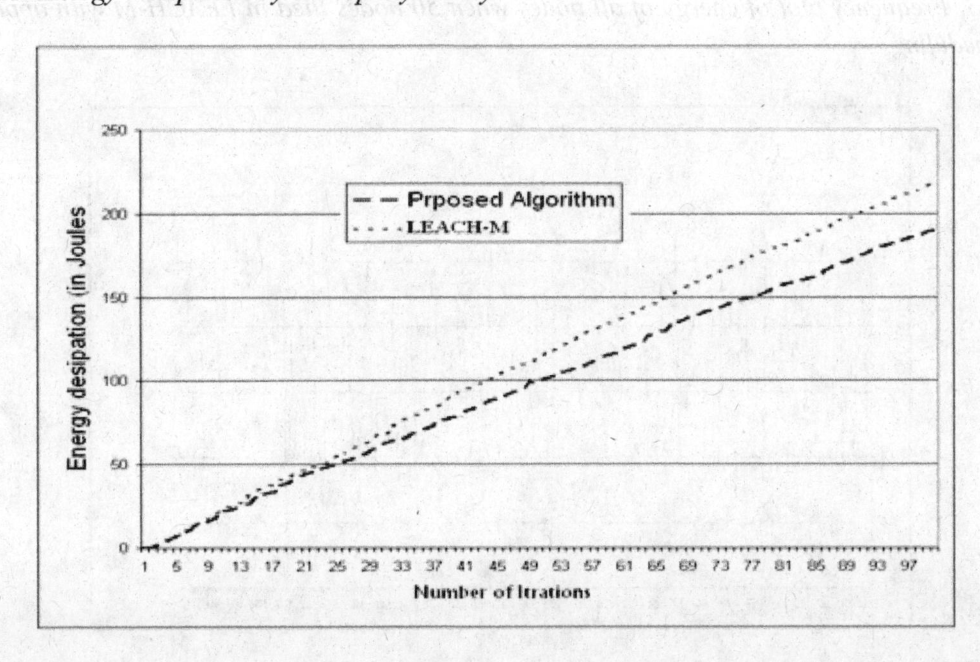

Figure 12. Frequency plot of energy of all nodes when 50 nodes died in proposed algorithm

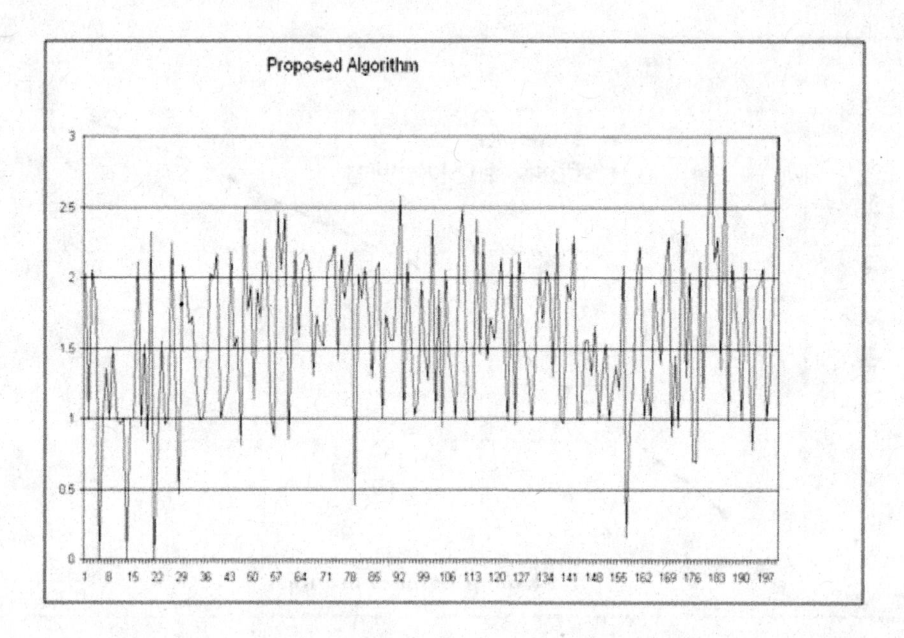

by using some localization algorithm (Bulusu, 2000),(Savvides, Han, & Strivastava, 2001), also each node holds information about its neighbors.
- The message size is 2000 bytes.

Figure 3 shows the formation of the temporary cluster heads. In the figure node 5 sensing area is covered by its neighbors so it will be marked as temporary cluster head. All the nodes in the network follow the same process. After the elimina-

Figure 13. Frequency plot of energy of all nodes when 50 nodes died in LEACH-M with application layer scheduling

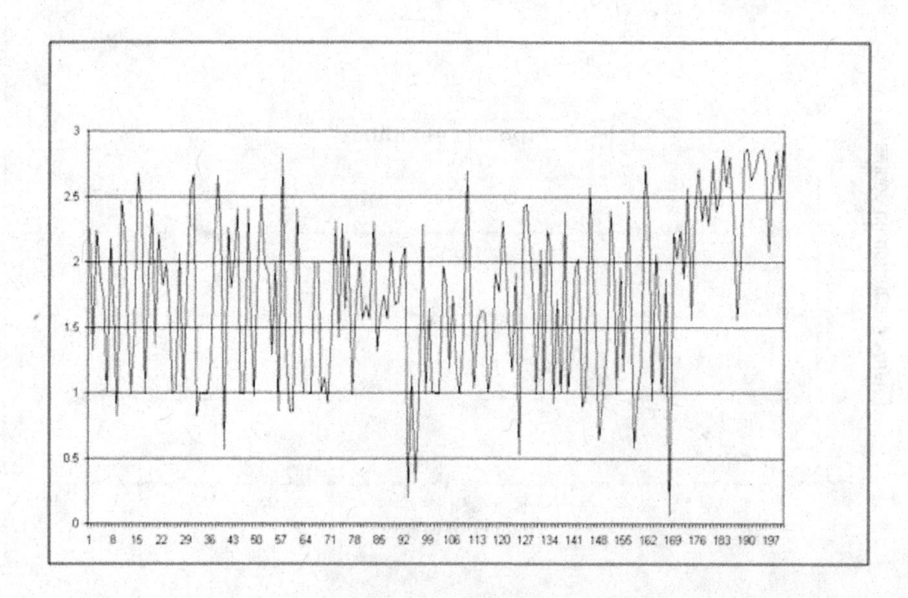

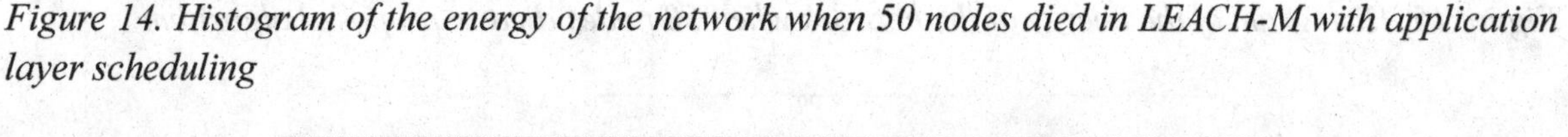

Figure 14. Histogram of the energy of the network when 50 nodes died in LEACH-M with application layer scheduling

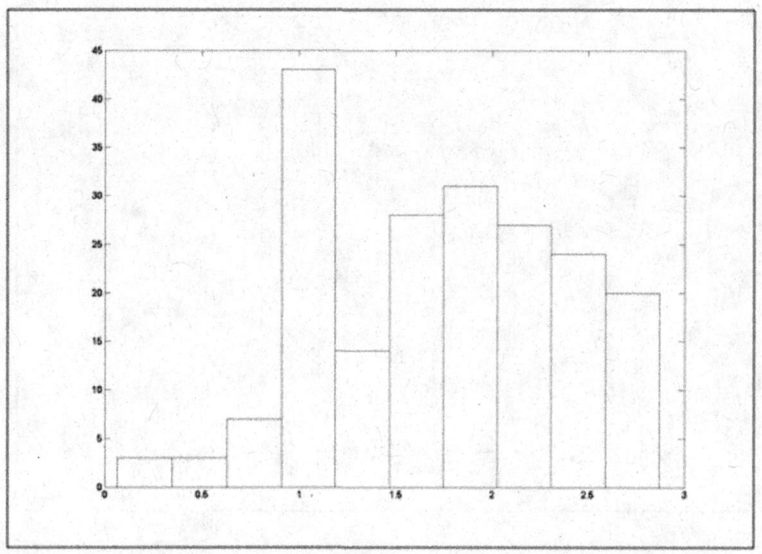

tion of temporary cluster heads, the sensing layer of the network is shown in Figure 4. Figure 5 shows the top layer of the network which comprises of the temporary cluster heads only.

Figure 6 shows the energy dissipation graph of the proposed algorithm and LEACH-M. Figure 7 shows the number of nodes that die as the simulation progresses. From both figures, it is obvious that the proposed cross layer communication strategy is more energy efficient than the conventional routing scheme. Also in the proposed scheme the density is exploited at the inter cluster communication layer compare to LEACH-M where it is exploited at the intra cluster communication layer. In LEACH-M the intra cluster communication is more dense than the inter cluster communication layer compare to proposed algorithm.

Figure 8 compares the dead nodes in the network as the simulation progresses. This figure shows simulation of the modified version of LEACH-M by incorporating the application layer scheduling in LEACH-M. From the figure, it is obvious that by incorporating the application layer scheduling the performance of LEACH-M get much better compare to Figure 7. Figure 9 compares the energy dissipation of the modified version of LEACH with application scheduling and the proposed protocol. The figure gives a clear indication that the performance of LEACH improved considerably because of the application scheduling when compared to Figure 6.

Figure 10 shows the tradeoff between the lifetime duration of a cluster head and the dead nodes at the end of the simulation. From the figure it is clear that if the duration is short then the network performs better, but if it is changed to be very short, there are more overheads for the network and also more delays. While on the other hand, if kept too long the energy in the network might not dissipate evenly. For the proposed protocol, keeping the cluster head cycle duration at 30 was decided, but for LEACH-M with application layer scheduling it is expensive. Figure 11 shows the energy dissipation of the top layer only. The figure gives a comparison of the energy dissipation of the inter cluster communication between the protocols only. From Figure 11 it is

Figure 15. Histogram of the energy of the network when 50 nodes died in proposed algorithm

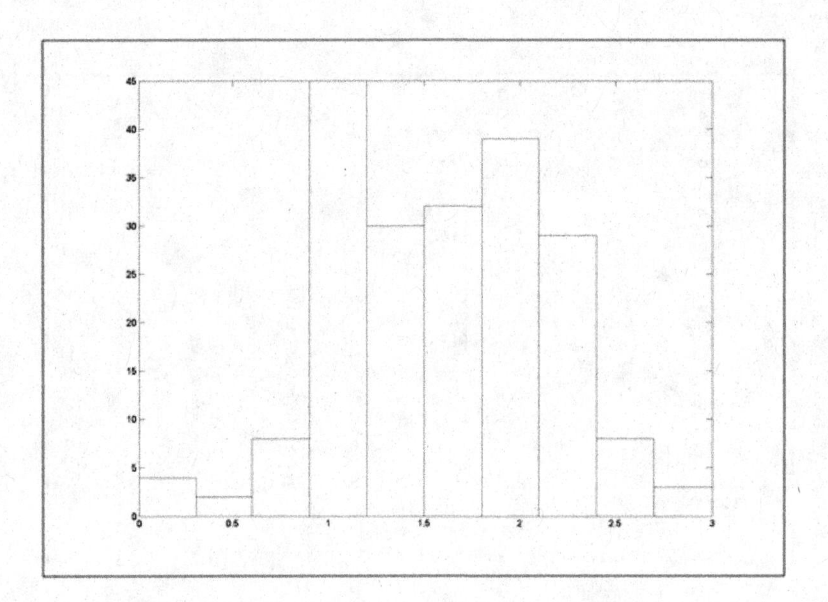

obvious that exploiting density at the inter cluster communication is more energy efficient than exploiting it at the intra cluster communication.

Figure 12 shows the frequency plot of the energy levels of all nodes in the network when 50 nodes died in the proposed protocol. Figure 13 is the frequency plot of the energy levels in LEACH-M i.e. LEACH-M with application layer scheduling. To further explain the figures, consider the energy histogram of the proposed protocol in Figure 15 and the energy histogram of the LEACH-M with application layer scheduling in Figure 14. From both the histogram, it is obvious that LEACH-M with scheduling has more different energy levels compare to the proposed protocol. This also shows that the proposed protocol is more efficient in load balancing when compared to the modified version of LEACH-M.

From simulation it is clear that the proposed cross layer communication strategy performs more efficiently compared to both LEACH-M and the enhanced version of LEACH-M with application layer scheduling.

CONCLUSION

This chapter looked at a new cross layer communication strategy for energy efficient communication in wireless sensor networks. Putting the redundant nodes to sleep is the dominant strategy in WSNs for energy efficiency. The new communication strategy proves that putting the nodes to sleep is not energy efficient in all circumstances. In fact, if the sleeping nodes are used for some energy intensive tasks it is more energy efficient compare to putting the redundant nodes to sleep. Performance analysis of the new cross layer communication strategy proves that the proposed way of using the redundant nodes is more energy efficient than putting them to sleep.

REFRENCES

Akyildiz, I. F., Su, W., Sankarasubramaniam, Y., & Cayirci, E. (2002). A survey on sensor networks. *Communications Magazine, IEEE, 40*(8), 102–114. doi:10.1109/MCOM.2002.1024422

Al-Karaki, J. N., & Kamal, A. E. (2004). Routing techniques in wireless sensor networks: A survey. *Wireless Communications, IEEE*, *11*(6), 6–28. doi:10.1109/MWC.2004.1368893

Arias, J., Santos, E., Marin, I., Jimenez, J., Lazaro, J., & Zuloaga, A. (2006). Node synchronization in wireless sensor networks. *International Conference on Wireless and Mobile Communications, ICWMC '06* (pp. 50-50).

Bulusu, N., Heidemann, J., & Estrin, D. (2000). GPS-less low-cost outdoor localization for very small devices. *Personal Communications, IEEE*, *7*(5), 28–34. doi:10.1109/98.878533

Chee-Yee, C., & Kumar, S. P. (2003). Sensor networks: Evolution, opportunities, and challenges. *Proceedings of the IEEE*, *91*(8), 1247–1256. doi:10.1109/JPROC.2003.814918

Cranch, G. A., Nash, P. J., & Kirkendall, C. K. (2003). Large-scale remotely interrogated arrays of fiber-optic interferometric sensors for underwater acoustic applications. *Sensors Journal, IEEE*, *3*(1), 19–30. doi:10.1109/JSEN.2003.810102

Di, T., & Georganas, N. D. (2003). *A node scheduling scheme for energy conservation in large wireless sensor networks. Wireless Networks and Mobile Computing* (pp. 271–290). Wiley.

Du, X., & Lin, F. (2005). *Designing efficient routing protocol for heterogeneous sensor networks*. Performance, Computing, and Communications Conference, 2005. IPCCC 2005. 24th IEEE International, 51-58.

Heinzelman, W. B., Chandrakasan, A. P., & Balakrishnan, H. (2002). An application-specific protocol architecture for wireless microsensor networks. *IEEE Transactions on Wireless Communications*, *1*, 660–670. doi:10.1109/TWC.2002.804190

Heinzelman, W. R., Chandrakasan, A., & Balakrishnan, H. (2000). Energy-efficient communication protocol for wireless microsensor networks. *Proceedings of the 33rd Annual Hawaii International Conference on Systems Sciences*, vol. 2.

Intanagonwiwat, C., Govindan, R., & Estrin, D. (2000). Directed diffusion: A scalable and robust communication paradigm for sensor networks. *Proceedings of the Sixth Annual International Conference on Mobile Computing and Networking* (pp. 56-67)

Jiang, S.-F., Yang, M.-H., Song, H.-T., & Wang, J.-M. (2007). An enhanced perimeter coverage based density control algorithm for wireless sensor network. *Third International Conference on Wireless and Mobile Communications,* ICWMC '07, (p. 79).

Kadayif, I., & Kandemir, M. (2004). Tuning in-sensor data filtering to reduce energy consumption in wireless sensor networks. Design, Automation and Test in Europe Conference and Exhibition. *Proceedings*, *2*, 852–857.

Lee, S., Kim, C., & Kim, S. (2006). Constructing energy efficient wireless sensor networks by variable transmission energy level control. *The Sixth IEEE International Conference on Computer and Information Technology,* CIT '06, (pp. 225-225).

Li, C., Ye, M., Chen, G., & Wu, J. (2005). An energy-efficient unequal clustering mechanism for wireless sensor networks. *IEEE International Conference on Mobile Ad Hoc and Sensor Systems*, (p. 604).

Lindsey, S., & Raghavendra, C. S. (2002). PEGASIS: Power-efficient gathering in sensor information systems. *Aerospace Conference Proceedings, IEEE,* (vol. 3, pp. 1125-1130).

Manjeshwar, A., & Agrawal, D. P. (2001). TEEN: A routing protocol for enhanced efficiency in wireless sensor networks. *Proceedings 15th International Parallel and Distributed Processing Symposium.*

Manjeshwar, A., & Agrawal, D. P. (2002). APTEEN: A hybrid protocol for efficient routing and comprehensive information retrieval in wireless sensor networks. *Parallel and Distributed Processing Symposium, IPDPS 2002, Abstracts and CD-ROM,* (pp. 195-202).

Paek, J., Chintalapudi, K., Govindan, R., Caffrey, J., & Masri, S. (2005). A wireless sensor network for structural health monitoring: Performance and experience. *The Second IEEE Workshop on Embedded Networked Sensors, EmNetS-II,* (pp. 1-10).

Qing, Z., & Tong, L. (2005). Energy efficiency of large-scale wireless networks: Proactive versus reactive networking. *IEEE Journal on Selected Areas in Communications, 23*(5), 1100–1112. doi:10.1109/JSAC.2005.845411

Rodoplu, V., & Meng, T. H. (1999). Minimum energy mobile wireless networks. *IEEE Journal on Selected Areas in Communications, 17*(8), 1333–1344. doi:10.1109/49.779917

Savvides, A., Han, C.-C., & Strivastava, M. B. (2001). Dynamic fine-grained localization in ad-hoc networks of sensors. *Proceedings of the 7th Annual International Conference on Mobile Computing and Networking,* Rome (pp. 166-179).

Shah, R. C., & Rabaey, J. M. (2002). Energy aware routing for low energy ad hoc sensor networks. *Wireless Communications and Networking Conference*, WCNC2002, (vol. 1, pp. 350-355).

Sichitiu, M. L. (2004). Cross-layer scheduling for power efficiency in wireless sensor networks. INFOCOM 2004. *Twenty-Third Annual Joint Conference of the IEEE Computer and Communications Societies*, (vol. 3, pp. 1740-1750).

Wang, H., Dong, B., Chen, P., Chen, Q., & Kong, J. (2007). LATEX DSL: A coverage control protocol for heterogeneous wireless sensor networks. *Second International Conference on Systems and Networks Communications,* ICSNC 2007, (p. 2).

Werner-Allen, G., Lorincz, K., Welsh, M., Marcillo, O., Johnson, J., Ruiz, M., & Lees, J. (2006). Deploying a wireless sensor network on an active volcano. *IEEE Internet Computing,* ▪▪▪, 18–25. doi:10.1109/MIC.2006.26

Xu, Y., Heidemann, J., & Estrin, D. (2001). Geography-informed energy conservation for ad hoc routing. *Proceedings of the 7th Annual International Conference on Mobile Computing and Networking*, New York, (pp. 70-84).

Ye, W., Heidemann, J., & Estrin, D. (2002). An energy-efficient MAC protocol for wireless sensor networks. *Twenty-First Annual Joint Conference of the IEEE Computer and Communications Societies Proceedings,* (vol. 3, pp. 1567-1576).

Younis, O., & Fahmy, S. (2004). HEED: A hybrid, energy-efficient, distributed clustering approach for ad hoc sensor networks. *IEEE Transactions on Mobile Computing, 3*(4), 366–379. doi:10.1109/TMC.2004.41

Younis, O., Fahmy, S., & Santi, P. (2004). *Robust communications for sensor networks in hostile environments*. Twelfth IEEE International Workshop on Quality of Service, (pp. 10-19).

Yu, L., Wang, N., & Meng, X. (2005). Real-time forest fire detection with wireless sensor networks. *Proceedings of the International Conference on Wireless Communications, Networking and Mobile Computing,* (pp. 1214-1217).

Chapter 13
Using Multi-Objective Particle Swarm Optimization for Energy-Efficient Clustering in Wireless Sensor Networks

Hamid Ali
National University of Computer and Emerging Sciences, Pakistan

Waseem Shahzad
National University of Computer and Emerging Sciences, Pakistan

Farrukh Aslam Khan
National University of Computer and Emerging Sciences, Pakistan

ABSTRACT

In this chapter, the authors propose a multi-objective solution to the problem by using multi-objective particle swarm optimization (MOPSO) algorithm to optimize the number of clusters in a sensor network in order to provide an energy-efficient solution. The proposed algorithm considers the ideal degree of nodes and battery power consumption of the sensor nodes. The main advantage of the proposed method is that it provides a set of solutions at a time. The results of the proposed approach were compared with two other well-known clustering techniques: WCA and CLPSO-based clustering. Extensive simulations were performed to show that the proposed approach is an effective approach for clustering in WSN environments and performs better than the other two approaches.

INTRODUCTION

The field of Wireless Sensor Networks (WSNs) has emerged as a very active area of research during the last few years. A WSN consists of autonomous tiny devices that cooperatively monitor physical or environmental conditions such as temperature, vibration, pressure, motion etc. These tiny devices or sensors have limited battery power, memory and processing capabilities. One of the important challenges of a WSN is the energy-efficient communication which increases the lifetime of the network. Several techniques have been

DOI: 10.4018/978-1-4666-0101-7.ch013

proposed to achieve this goal and clustering in WSNs is one of them that can help in providing an energy-efficient solution. Clustering requires the selection of cluster-heads (CHs) for each cluster. Fewer CHs result in greater energy efficiency as these nodes consume more power and energy as compared to non cluster-heads. Several techniques are available in the literature for clustering by using optimization and evolutionary techniques. The main drawback of these techniques is that they handle only one objective at a time. These techniques do not provide a freedom of choice to the user.

A wireless sensor network (WSN) which is a special type of wireless ad hoc network consists of autonomous tiny devices that are capable of cooperatively monitoring physical or environmental conditions. These nodes have limited battery, processing speed, storage, and communication capabilities. These limitations of WSNs bring new problems and challenges for the researchers. Clustering is a technique of organizing objects into meaningful groups with respect to their common characteristics. The objective of clustering in WSNs is to identify the groups of nodes in such a way that the groups are exclusive and any node in the network belongs to a single group. The nodes known as cluster-heads (CHs) are responsible for the formation of clusters, maintenance of network topology, and the allocation of resources to all nodes present in their clusters. Since the configuration of CHs can change frequently due to the mobility of sensor nodes, minimizing the number of cluster-heads becomes an essential component. Optimal selection of CHs is an NP-hard problem. The neighbourhood of a CH is a set of nodes that lie within its transmission range. Since energy-efficiency is an important requirement for a WSN that increases its lifetime, clustering can provide an energy-efficient solution as only a few nodes are involved in doing the main operations in the sensor network such as management, routing, data aggregation etc. Therefore, clustering can greatly help in achieving an energy-efficient solution for WSNs.

Optimization refers to determining one or more solutions of a given problem which correspond to extreme values of one or more objectives. It has been an active area of research as many real-world problems have become increasingly complex. Therefore, better optimization techniques are always required. Most real world problems consist of several objectives that are needed to be optimized at the same time. Such kinds of problems arise in many applications. While solving multi-objective problems (MOPs) with traditional mathematical programming techniques, a single solution is generated from a set of solutions in one run. Therefore, these techniques are not much suitable for solving multi-objective optimization problems. Evolutionary Algorithms paradigm is very suitable to solve MOPs because they are population-based and can generate a set of solutions in one run (Liang et al., 2006).

In this chapter, we introduce a Multi-objective Particle Swarm Optimization (MOPSO) based clustering algorithm for wireless sensor networks. MOPSO efficiently manages the resources of the network by finding optimal number of clusters in a multi-objective manner. Optimal number of clusters can make the WSN energy-efficient by efficiently managing the resources of the network so that the CHs can do their job in a proper manner. The proposed clustering algorithm takes into account the ideal degree and battery power for selecting the cluster-heads. MOPSO uses the evolutionary capabilities to optimize the number of clusters in a network. Instead of assigning weight to each of the parameters mentioned above, the algorithm deals directly with the multi-objective problem in order to find the Pareto-optimal solutions. The algorithm first finds the cluster-heads and then the neighbours of these cluster-heads. The neighborhood of a CH is a set of nodes that lie within its transmission range. There are some requirements for clustering in WSNs. The clustering algorithm must be distributed since each node in the network has local knowledge only and communicates outside its group only through its CH e.g., in case of cluster-based routing. The algorithm should be robust and it should be able to adapt to all the changes as the network size increases or decreases. The cluster formation should be reasonably efficient and the selected CHs should cover large numbers of nodes.

We compare the results of the proposed MOPSO-based clustering technique with two other well-known clustering algorithms i.e., Weighted Clustering Algorithm (WCA) (Chatterjee et al., 2002) and Comprehensive Learning Particle Swarm Optimization (CLPSO) based clustering (Shahzad et al., 2009). The results of our experiments show that the proposed technique covers the whole network with minimum number of clusters that can help in reducing the routing cost of the network and consuming less energy. This will also minimize the number of hops and delays of packets transmitted in a cluster-based routing environment. The proposed algorithm also has the advantage of finding multiple solutions of the problem instead of a single solution (Coello Coello et al., 2004). This diversity of solutions provides more flexibility for the experts so that they can select a solution according to their requirements. The results indicate that the proposed clustering approach is more effective and flexible as compared to the other approaches and performs better than these approaches in a WSN environment. The algorithm has the ability to optimize various parameters of nodes in a WSN for finding multi-objective solutions.

BACKGROUND

In WSNs, a sensor node sends data to the base station through intermediate sensor nodes. They also need to collaborate to send the control information from a specific sensor node to the base station. The communication between nodes is usually the main source of energy consumption in WSNs which depends on the distance of a communication link between the source and the sink. The energy consumption is calculated by the following formula:

$$e = kd^c \quad (1)$$

Where k and c are constants for a specific wireless system (usually $2 < c < 4$).

Since energy consumption is proportional to d^c where $c > 2$, the distance can be minimized by partitioning the sensor nodes into clusters which significantly minimizes the amount of energy consumption. In case of clustering the information is transmitted to the base station in a hierarchical fashion. It also decreases the load at the base station. As discussed earlier, the cluster formation is one of the main problems in sensor network applications and can affect the energy consumption in the network. In clustering each sensor node is assigned to a specific Cluster-Head (CH). All communication of the sensor node is carried out through its corresponding CH. One would like to have each sensor node to communicate with the closest CH to save its energy; however, CHs have an upper bound for their neighbouring nodes. So, if the closest CH reaches its upper bound then a node must join the second closest CH. Clustering can provide a large-scale wireless sensor network with a hierarchical network structure.

In the literature, many researchers have proposed the optimization techniques for clustering. A wireless adaptive mobile information system is presented in (Gerla & Tsai, 1995) which is multi-cluster and multi-hop packet radio network architecture. This network architecture is designed for an infrastructure-less environment. They used lowest ID-based and highest-connectivity scheme to find the suitability to a mobile network environment. In Highest-connectivity algorithm a node with maximum number of neighbouring nodes become a cluster-head. Each node broadcasts its identifier to other nodes for election procedure.

A weight-based clustering algorithm known as Weighted Clustering Algorithm (WCA) is proposed in (Chatterjee et al., 2002). This algorithm dynamically adapts itself in a dynamic topology for wireless ad hoc networks. In this approach a threshold of neighbouring node is used for the cluster-head. It uses two different parameters for selection of a cluster-head. These parameters are ideal degree and battery power of the nodes. These parameters are assigned different weight factors for the combined effect. This algorithm is executed on demand in cases when a node is isolated from the existing cluster-head. They also find the load balance factor and how it effects in the dynamic environment. The authors compare their algorithm with two other clustering techniques; Highest-Degree and the Lowest-ID heuristics.

In (Turgut et al., 2002), the authors use a genetic algorithm approach for clustering that can dynamically adapt itself with the dynamic topology of the ad hoc networks. They mapped the possible solutions generated by original WCA to the genetic algorithm technique in order to find the optimum solution from a population of solutions. The clusters-heads are stored in the chromosomes and the information given in the chromosomes is used in the selection process. In this technique each cluster-head handles maximum number of nodes. The simulation results show that the performance of GA-based WCA is better than the original WCA algorithm. This new technique also evenly balances the load between all the cluster-heads.

In (Er & Seah, 2004), a distributed clustering algorithm is proposed which can change its diameter on the basis of mobility. Authors propose two mobility metrics relative to the mobility concept: (1) variable distance between nodes over time, and (2) estimated mean distance for the cluster. These metrics are used to find the stability of a cluster. The cluster membership is also decided on the basis of these metrics. Therefore, the cluster formation depends upon the mobility pattern of nodes to guarantee the maximum cluster stability. The variable diameter of the cluster is used to achieve the desired scalability. The mobility pattern of sensors within a cluster decides the diameter of a cluster. If the sensor nodes have similar mobility pattern in the same directions then they make a cluster. It is like a group that moves towards the same destination.

A Comprehensive Learning Particle Swarm Optimization (CLPSO) based clustering algorithm is proposed for mobile ad hoc networks in (Shahzad et al., 2009). The authors use the same parameters which are used in WCA. They compare their results with WCA and divided range PSO (DRPSO). The performance of the proposed approach is better than the other two algorithms to find the optimal number of clusters. This approach is also a weight-based clustering algorithm. Each particle contains the information about the cluster-heads and their neighbours.

The basic problem with all these algorithms is that none of them include all the basic parameters of a wireless network. WCA was the first algorithm that tried to include maximum number of parameters but it failed to find the optimal number of clusters in the network. At the end it assigns weight to each objective and converts the multi-objective problem into single objective problem. The single objective algorithms find a single optimum solution. On the other hand the multi-objective evolutionary algorithms find many optimum solutions at a time. In this chapter, we introduce a Multi-objective Particle Swarm Optimization (MOPSO) based clustering algorithm for WSNs to find an array of solutions. MOPSO efficiently manages the resources of the network by finding optimal number of clusters. This gives more flexibility to the user to choose a solution which is according to its requirements.

MULTIOBJECTIVE CLUSTERING

As described earlier that multi-objective problems have many objectives that are minimized or maximized simultaneously. These objectives have trade-offs among one another. There are a number of constraints which a solution must satisfy in all the objectives. We have multi-dimensional search space in a multi-objective problem. Suppose a problem has *n* objective functions: f(s) = (f1(s), f2(s)... fn(s)). These functions are either maximized or minimized. s ∈ F is a variable which is used to evaluate the values of all objective functions. In many cases, there is a contradiction between the objective functions i.e., if we increase the value of one objective function, the value of the other objective function must be decreased. So we are interested in such a value of this variable that gives an optimum solution in terms of all the objective functions. s* is a Pareto-optimal solution which would decrease (increase) some objective function value without simultaneous increase (decrease) in at least one other objective function value. This concept almost gives a set of solutions instead of single solution called Pareto-optimal front, Pareto optimal set, or Pareto-optimal solution. These solutions included in the Pareto front are non-dominated with each other (Coello Coello et al., 2004).

In our problem, we are interested in finding such a Pareto-optimal solution that is close to true Pareto-optimal front. The Pareto fronts which are not close to the true Pareto-optimal front are not required solutions. It is also desirable to find a Pareto-optimal front in which the solutions should be diverse. Multi-objective problems have two spaces: (1) Decision variables space, and (2) Objective space. A Pareto-optimal set contains many solutions if and only if the objective functions are conflicting to each other. The Pareto-optimal front contains only a single solution if objectives are not conflicting in a given problem. Multi-objective problems have many decision and search spaces while single objective problems have only one decision space.

A solution *s1* dominates the other solution *s2* if and only if the following two conditions are true:

- The solution *s1* is no worse than *s2* in all objectives.
- The solution *s1* is strictly better than *s2* in at least one objective.

If any condition is false, the solution *s1* does not dominate the solution *s2*. The solution which dominates the other solution is a better solution.

If the solutions do not dominate each other in all objective function values then these solutions are called non-dominated solutions. The solutions exists in the Pareto-front are equally important. It is to be noted that different Pareto-fronts dominate each other. So the Pareto-front which dominates all the remaining Pareto-fronts is called Pareto-optimal front. This whole scenario can be seen in Figure 1 in which Pareto-front is achieved when we have two objective functions that are conflicting with each other (Mostaghim & Teich, 2004). Our problem is multi-objective clustering in which cluster formation is carried out according to multiple objectives. We use two objective functions from wireless sensor network environment. Since our problem is a multi-objective problem so we are interested in a set of solutions that are lying on the same optimal front.

MULTIOBJECTIVE PARTICLE SWARM OPTIMIZATION

Evolutionary algorithms have been successfully used to solve the multi-objective problems during the last decade. These multi-objective algorithms use a population of solutions at a time instead of a single solution, while the classical techniques use a single solution at a time. So, the evolutionary algorithms have the capability to find a set of optimal solutions simultaneously instead of performing a series of separate runs as in case of traditional mathematical programming techniques. Several evolutionary algorithms are available in the literatures that have different mechanisms to evolve solutions, for example, Genetic Algorithm

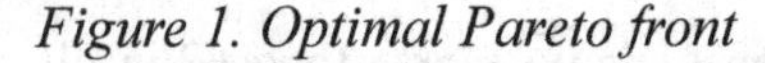

Figure 1. Optimal Pareto front

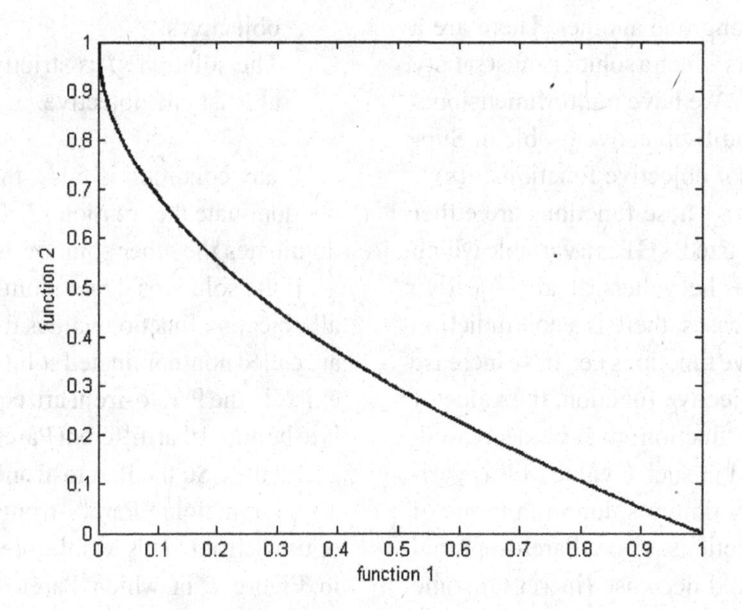

(GA), Ant Colony Optimization (ACO), Particle Swarm Optimization (PSO), Artificial Immune System (AIS), and Differential Evolution (DE) etc.

Particle swarm optimization (PSO) was proposed by Kennedy and Eberhart (Kennedy & Eberhart, 1995) inspired by the behaviour of bird flocking. In this algorithm, personal best and global best behaviours guide each individual in the flock. So, this property of the algorithm can quickly converge each individual to optimal geographical positions (AlRashidi & El-Hawary, 2009).

A complete solution is called a particle. Swarm is a group of all these solutions which search for an optimum solution or solutions. A particle *i* is defined by its position vector *Xi*. All the particles know their best position found so far. Initially, the positions and velocities of the particles are generated randomly. These positions and velocities then proceed iteratively. The velocities and positions are calculated as follows:

$$V_{id} = W \times V_{id} + c1r1\,(P_{id} - X_{id}) + c2r2(P_{gd} - X_{id}) \quad (2)$$

$$X_{id} = X_{id} + V_{id} \quad (3)$$

Where, $d = 1,2,\ldots,D$; $i = 1,2,\ldots,N$; N is the size of the population; W is the inertia weight; *c1* and *c2* are acceleration coefficients; and *r1* and *r2* are two random values in the range [0, 1]. The velocity of the ith particle is calculated using Equation (2) by using the three terms (a) the particle's previous velocity, (b) the personal best position, and (c) Global best position. The new position is updated using Equation (2). The impact of previous velocities on the current velocities is controlled by the inertia weights. We can exclude the impact of previous history by assigning zero value to inertia weights. In each generation, each particle updates its personal and global best values achieved so far (Leong & Yen, 2006). There are many variations of standard PSO available to solve the multi-objective optimization problems. PSO is very popular due to simplicity in its nature. It is computationally inexpensive in order to update the position of each individual. Most of the real world problems are dynamic i.e., they change over time. For this type of problems, the optimization algorithm has to track a moving optimum. Particle Swarm Optimization can handle both continuous as well as discrete variable types of problems.

PROPOSED TECHNIQUE

We use a multi-objective particle swarm optimization (MOPSO) algorithm to solve the problem of clustering in a WSN environment. The working of the proposed algorithm is presented in Table 1.

Multi-objective PSO starts with population *P0* of *N* randomly generated cluster-heads vector *T* which covers the whole network for communication. The objectives of each solution are calculated using the cluster-heads and their neighbours contained in the particles. First of all we calculate the distance of each sensor node to its cluster-head and then sum the distances of all the neighbours belonging to one cluster-head. Degree difference of each cluster-head is calculated using the equation $\Delta = |d - \delta|$ where Δ is degree difference, *'d'* is total neighbours of the cluster-head, and *'δ'* is a predefined threshold. In the same way we calculate the battery power consumption for all cluster-heads in a single solution in the population. After that we sum up all the distances of cluster-heads as well as the degree differences. These summations are the overall objective values of a single solution in a population. The objectives of all the population is calculated in the same way.

After finding the objective values we find the personal best cluster-heads vector. Global best vector is achieved using non-domination sorting which gives a Pareto-optimal front. Velocity of each individual is calculated by using equation (2). The new positions are achieved by the previous position and the velocities calculated in the current generation. This process is shown in Figure 2.

EXPIREMENTAL RESULTS

We implement the proposed algorithm in MATLAB 7.8.0. We conduct the experiments in a machine with 1.75GHZ dual processors with 512MB of RAM. The experiments of *M* different nodes are performed on a 100 x 100 grid. The transmission range of each node is varied from 10 to 40. In our experiments, *M* is varied between 10 and 20. The threshold for degree difference is set to 10. This restriction will ensure load balancing for the sensor network.

The parameters of MOPSO and CLPSO are chosen as follows:

- The population size is equal to 100
- The maximum generations are equal to 150
- The inertia weight *w* is 0.694
- The learning factors *c1* and *c2* are 2

Table 1. Proposed MOPSO based clustering algorithm

1. Initialize the population *POP*
2. Initialize the general parameters of MOPSO
3. Initialize the speed of each particle
4. Store the best cluster-heads vectors
5. Evaluate each of the particles in *POP*
6. Find the Pareto front with non-dominated sorting
7. Store cluster-heads vectors that represent non-dominated vectors in the repository *REP*
8. Find the global cluster-heads vector from the repository
9. WHILE maximum number of cycles not reached

   DO
   - a. Compute the velocity *VEL* of each cluster-head vector
   - b. Compute the new cluster-heads vectors of the solutions by adding the velocity produced from the previous step
   - c. Evaluate each of the particles in *POP*
   - d. Update the contents of *REP*
   - e. When the current combination of cluster-heads of the solution is better than the combination of cluster-heads contained in its memory, the particle's position is updated
   - f. Increment the loop counter
10. END WHILE

Figure 2. Flowchart of MOPSO

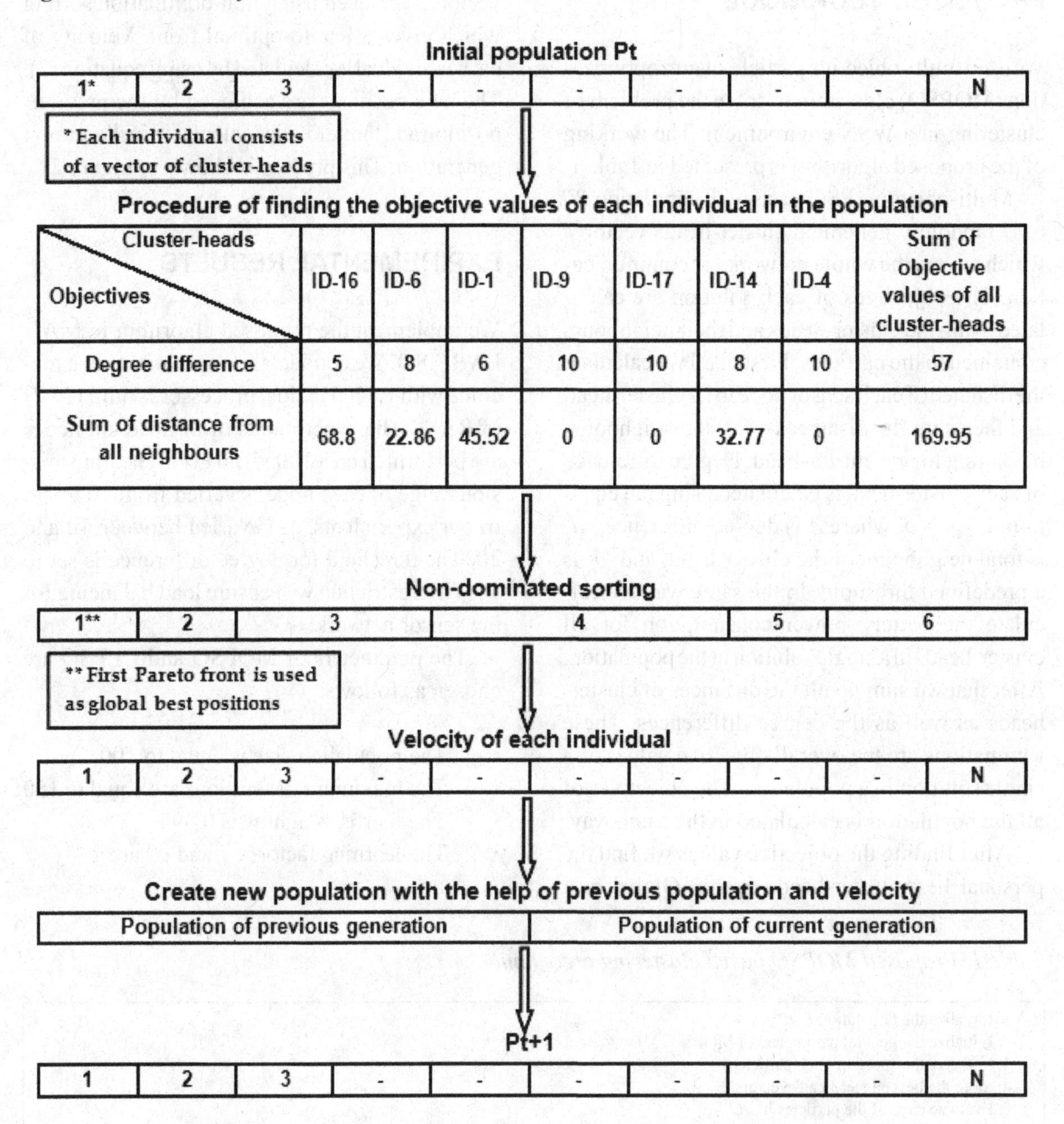

We compare the performance of multi-objective PSO based clustering technique with two other clustering algorithms using three performance metrics: (1) The number of clusters, (2) Total energy consumed in the network, (3) Load balancing factor (LBF). The results have been produced by varying different parameters such as number of nodes in the network, transmission range, grid size, and the displacement.

Each solution in a Pareto-optimal front describes the number of cluster-heads in the network. Each solution has different number of cluster-heads. The energy consumed in a cluster is calculated by adding the energy consumed within a cluster due to the communication between cluster-head and its neighbours and from cluster-head to the base station. The total energy consumed within a network is calculated by adding the energy consumed in all the clusters. The ratio between the required number of neighbours of a cluster-head and the number of neighbours achieved by the algorithm is called load balance factor.

Number of Clusters vs. Transmission Range

The experiments were conducted to find the number of clusters against different values of the transmission range. We generate the solutions by fixing the number of nodes to 20 and the grid size to 100mx100m. The initial solutions are randomly generated so we use average number of clusters as a performance metric after running the algorithms 10 times. As can be seen in Figure 3, MOPSO based algorithm produces a set of solutions against each transmission range while on the other hand, WCA and CLPSO produce a single solution in the same environment i.e., 100m x 100 m. MOPSO finds optimum numbers of clusters as compared to WCA and CLPSO and gives diverse solutions at each point.

Number of Clusters vs. Sensor Nodes in a Network

We also conduct the experiments to find the performance of the three algorithms by keeping the transmission range constant and increasing the number of nodes. Figure 4 shows that the proposed clustering algorithm performs better than other two algorithms. This shows the flexibility and robustness of the algorithm.

Figure 3. Number of clusters in case of WCA, CLPSO, and MOPSO in 100x100 m^2 area with nodes equal to 20

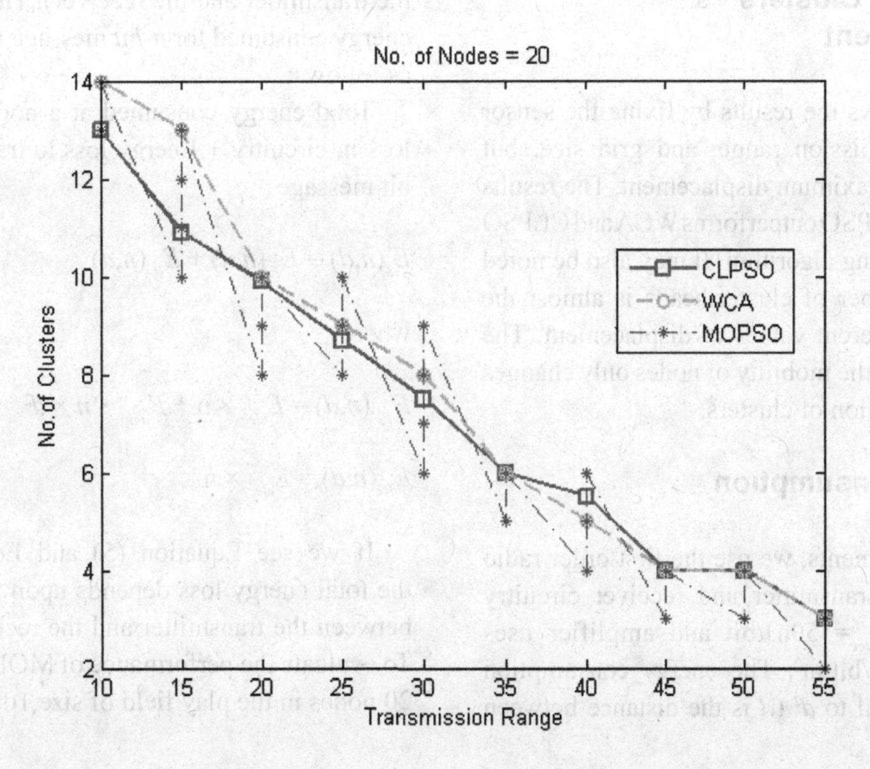

Figure 4. Number of clusters in case of WCA, CLPSO, and MOPSO in 100x100 m² area with transmission range equal to 60

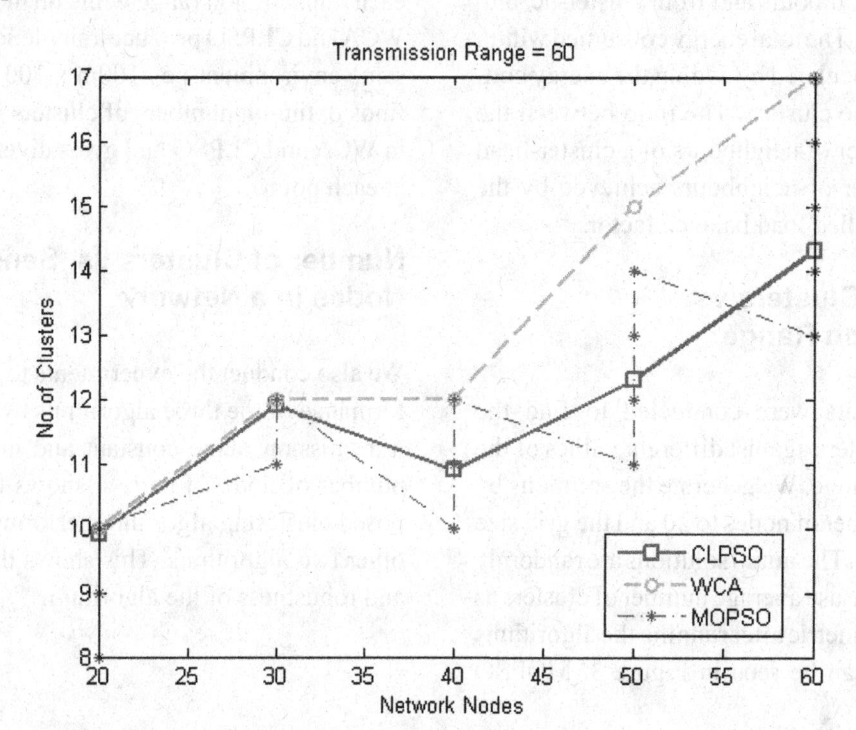

Number of Clusters vs. Displacement

Figure 5 shows the results by fixing the sensor nodes, transmission range, and grid size, but varying the maximum displacement. The results show that MOPSO outperforms WCA and CLPSO based clustering algorithm. It may also be noted that the number of cluster-heads is almost the same for different values of displacement. The reason is that the mobility of nodes only changes the configuration of clusters.

Energy Consumption

In our experiments, we use the first order radio model. The transmitter and receiver circuitry consumes E_{elec} = 50nJ/bit and amplifier uses E_{amp} = 100pJ/bit/m². The energy consumption is proportional to d^2 (*d* is the distance between the transmitter and the receiver). Thus, the total energy consumed for *n-bit* message is calculated as follows:

Total energy consumed at a node = Energy loss in circuitry + Energy loss to transmit an n-bit message

$$E_T(n,d) = E_{Tx}(n,d) + E_{Rx}(n,d) \tag{4}$$

where

$$E_{Tx}(n,d) = E_{elec} \times n + E_{amp} \times n \times d^2 \tag{5}$$

$$E_{Rx}(n,d) = E_{elec} \times n \tag{6}$$

If we see Equation (5) and Equation (6), the total energy loss depends upon the distance between the transmitter and the receiver nodes. To evaluate the performance of MOPSO, we use 20 nodes in the play field of size 100m x 100m.

Figure 5. Number of clusters vs. displacement in case of WCA, CLPSO, and MOPSO when nodes are 40 and transmission range is 20

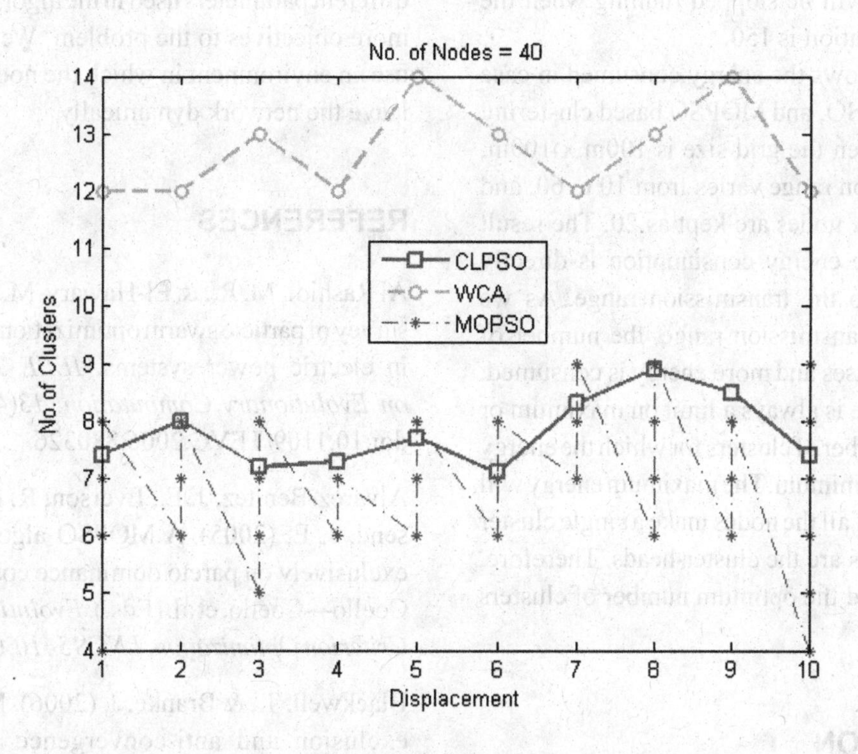

Figure 6. Energy consumed in case of WCA, CLPSO, and MOPSO when area is 100mx100m, transmission range varies from 10 to 60 and the number of nodes is 20

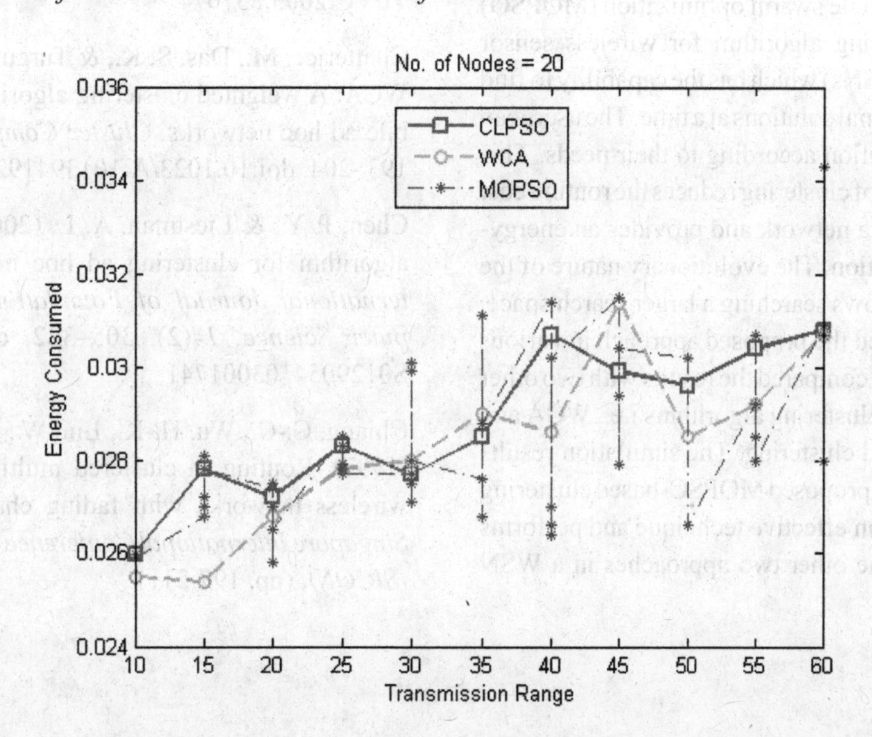

The base station is located at position (50,110). The program will be stopped running when the number of iteration is 150.

Figure 6 shows the energy consumed in case of WCA, CLPSO, and MOPSO based clustering techniques when the grid size is 100m x 100m, the transmission range varies from 10 to 60, and the numbers of nodes are kept as 20. The result shows that the energy consumption is directly proportional to the transmission range. As we increase the transmission range, the number of clusters decreases and more energy is consumed. However, there is always a limit on maximum or minimum number of clusters for which the energy consumed is minimum. The maximum energy will be consumed if all the nodes make a single cluster or all the nodes are the cluster-heads. Therefore, we want to find the optimum number of clusters in the network.

CONCLUSION

In this chapter, we have introduced a multi-objective particle swarm optimization (MOPSO) based clustering algorithm for wireless sensor networks (WSNs) which has the capability to find multiple optimal solutions at a time. The users can choose a solution according to their needs. This optimization of clustering reduces the routing cost of packets in a network and provides an energy-efficient solution. The evolutionary nature of the algorithm allows searching a larger search space. We have tested the proposed approach in various scenarios and compared the results with two other well-known clustering algorithms i.e., WCA and CLPSO based clustering. The simulation results show that our proposed MOPSO-based clustering technique is an effective technique and performs better than the other two approaches in a WSN environment.

As a future work, we would like to optimize different parameters used in the algorithm and add more objectives to the problem. We also plan to use an environment in which the nodes enter and leave the network dynamically.

REFERENCES

Al Rashidi, M. R., & El-Hawary, M. E. (2009). A survey of particle swarm optimization applications in electric power systems. *IEEE Transactions on Evolutionary Computation*, *13*(4), 913–918. doi:10.1109/TEVC.2006.880326

Alvarez-Benitez, J. E., Everson, R. M., & Fieldsend, J. E. (2005). A MOPSO algorithm based exclusively on pareto dominance concepts. In C. Coello—Coello, et al. (Eds.), *Evolutionary Multi-Criterion Optimization, LNCS 3410*, (pp. 459-73).

Blackwell, T., & Branke, J. (2006). Multiswarm, exclusion and anti-convergence in dynamic environments. *IEEE Transactions on Evolutionary Computation*, *10*(4), 459–472. doi:10.1109/TEVC.2005.857074

Chatterjee, M., Das, S. K., & Turgut, D. (2002). WCA: A weighted clustering algorithm for mobile ad hoc networks. *Cluster Computing*, *5*(2), 193–204. doi:10.1023/A:1013941929408

Chen, P. Y., & Liestman, A. L. (2003). A zonal algorithm for clustering ad hoc networks. *International Journal of Foundations of Computer Science*, *14*(2), 305–322. doi:10.1142/S0129054103001741

Chiang, C.-C., Wu, H.-K., Liu, W., & Gerla, M. (1997). Routing in clustered multihop, mobile wireless networks with fading channel. *IEEE Singapore International Conference on Networks (SICON)*, (pp. 197-211).

Coello, C. A. C., Pulido, G. T., & Lechuga, M. S. (2004). Handling multiple objectives with particle swarm optimization. *IEEE Transactions on Evolutionary Computation*, *8*(3), 256–279. doi:10.1109/TEVC.2004.826067

Deb, K., Pratap, A., Agarwal, S., & Meyarivan, T. (2000). A fast and elitist multi-objective genetic algorithm: NSGA-II. *IEEE Transactions on Evolutionary Computation*, *6*(2), 182–197. doi:10.1109/4235.996017

Donoso, Y., Fabregat, R., & Marzo, J. (2004). Multi-objective optimization algorithm for multicast routing with traffic engineering. *Proceedings of 3rd International Conference on Networking IEEE ICN'04* Pointe-à-Pitre, Guadeloupe, French Caribbean.

Er, I. I., & Seah, W. K. G. (2004). *Mobility-based d-hop clustering algorithm for mobile ad hoc networks*. IEEE Wireless Communications and Networking Conference (WCNC), Atlanta, USA.

Freshchi, F., Coello, C. A. C., & Repetto, M. (2008). Multiobjective optimization and artificial immune system: A review. In Mo, H. (Ed.), *Handbook of research on artificial immune systems and natural computing: Applying complex adaptive technologies* (pp. 1–21). Hershey, PA: IGI Global.

Gerla, M., & Tsai, J. T. C. (1995). Multicluster, mobile, multimedia radio network. *Wireless Networks*, *1*(3), 255–265. doi:10.1007/BF01200845

Hong, X., Gerla, M., Pei, G., & Chiang, C. (1999). A group mobility model for ad hoc wireless networks. In *Proceedings of ACM/IEEE MSWiM*, Seattle, WA.

Hong, X., Gerla, M., Yi, Y., Xu, K., & Kwon, T. J. (2002). Scalable ad hoc routing in large, dense wireless networks using clustering and landmarks. In *Proceedings of the IEEE International Conference on Communications. (ICC'02)*, *25*(1), 3179 –3185.

Hu, X., & Eberhart, R. C. (2002). Multiobjective optimization using dynamic neighborhood particle swarm optimization. *Proceedings of Conference on Evolutionary Computation*, Honolulu, HI, (pp. 1677-1681).

Kennedy, J. (1997). Minds and cultures: Particle swarm implications. *Socially Intelligent Agents. 1997 AAAI Fall Symposium*, (pp. 67-72). Technical Report FS-97-02. Menlo Park, CA: AAAI Press.

Kennedy, J. (1997). The particle swarm: Social adaptation of knowledge. In *Proceedings of IEEE International Conference on Evolutionary Computation*, Indianapolis, Indiana, (pp. 303-308).

Kennedy, J. (1998). The behavior of particles. In *Proceedings of 7th Annual Conference on Evolutionary Programming*, San Diego, USA.

Kennedy, J., & Eberhart, R. C. (1995). Particle swarm optimization. In *Proceedings of IEEE International Conference on Neural Networks*, Perth, Australia, (vol. 4, pp. 1942-1948).

Leong, W., & Yen, G. G. (2006). Dynamic population size in PSO-based multiobjective optimization. *IEEE Congress on Evolutionary Computing*, Vancouver, BC, Canada, (pp. 1718-1725).

Liang, J. J., Qin, A. K., Suganthan, P. N., & Baskar, S. (2006). Comprehensive learning particle swarm optimizer for global optimization of multimodal functions. *IEEE Transactions on Evolutionary Computation*, *10*(3), 281–295. doi:10.1109/TEVC.2005.857610

Mostaghim, S., & Teich, J. (2004). Covering pareto-optimal fronts by subswarms in multi-objective particle swarm optimization. In *Congress on Evolutionary Computation*, Portland, USA, (pp. 1404-1411).

Parrott, D., & Xiaodong, L. (2004). Locating and tracking multiple dynamic optima by a particle swarm model using speciation. *IEEE Transactions on Evolutionary Computation, 10*(4), 440–458. doi:10.1109/TEVC.2005.859468

Parsopoulos, K. E., Tasoulis, D. K., & Vrahatis, M. N. (2004). Multiobjective optimization using parallel vector evaluated particle swarm optimization. *IASTED International Conference on Artificial Intelligence and Applications,* Innsbruck, Austria, (pp. 823-828).

Shahzad, W., Khan, F. A., & Siddiqui, A. B. (2009). Clustering in mobile ad hoc networks using comprehensive learning particle swarm optimization (CLPSO). *Communication and Networking, CCIS, 56*, 342–349. doi:10.1007/978-3-642-10844-0_41

Sivavakeesar, S., & Pavlou, G. (2004). Stable clustering through mobility prediction for large-scale multihop intelligent ad hoc networks. In *Proceedings of the IEEE Wireless Communications and Networking Conference (WCNC'04),* Georgia, USA, (vol. 3, pp. 1488-1493).

Turgut, D., Das, S. K., Elmasri, R., & Turgut, B. (2002). Optimizing clustering algorithm in mobile ad hoc networks using genetic algorithmic approach. In *Proceedings of GLOBECOM'02*, Taipei, Taiwan, (pp. 62– 66).

Valle, Y., Venayagamoorthy, G. K., Mohagheghi, S., Hernandez, J.-C., & Harley, R. G. (2008). Particle swarm optimization: Basic concepts, variants and applications in power systems. *IEEE Transactions on Evolutionary Computation, 12*(2).

Chapter 14
Wireless Sensor Networks:
Data Packet Size Optimization

Low Tang Jung
Universiti Teknologi PETRONAS, Malaysia

Azween Abdullah
Universiti Teknologi PETRONAS, Malaysia

ABSTRACT

This chapter presents the studies and analysis on the approaches, the concepts, and the ideas on data packet size optimization for data packets transmission in underwater wireless sensor network (UWSN) and terrestrial wireless sensor network (TWSN) communications. These studies are based on the related prior works accomplished by the UWSN and TWSN research communities. It should be mentioned here that the bulk of the studies and analysis would be on the data packet size optimization techniques or approaches rather than on the communication channel modeling, but the channel model is deemed essential to support the optimization approaches. The various optimization solutions proposed in the prior arts are dealt with in depth to explore their feasibilities to accommodate the data packet size optimization algorithm proposed by the various researchers. This chapter starts off with the studies and analysis on prior arts found in UWSN and then moves on to the similar works found elsewhere in the TWSN communications counterparts. A comparison on some important issues related to data packet size optimization approaches used in UWSN and TWSN communications are summarized in a table at the end of this chapter. The findings in this chapter may be helpful to readers who are interested in the R&D of data packet size optimization techniques with the intention to formulate new data packet size optimization framework or algorithms.

INTRODUCTION

It is a known fact in WSN that the data packet size could directly affects the reliability and the quality of the communication between the wireless nodes. For instance, with a certain level of link quality, long packet sizes may be more susceptible to data bits corruption than shorter sizes thus demanding a higher frequency of data packet retransmissions. On the other hand, short packet sizes may increase data transmission reliability since the chances of bit errors over the link are less, but too short a packet size may not be efficient in the context

DOI: 10.4018/978-1-4666-0101-7.ch014

of data payload carrying capacity because of the standardized data packet overhead. The ultimate goal of communication between any nodes in WSN is to ensure a successful and efficient delivery of data packets from a source node to a sink node based on certain performance metrics of the network. Data packet size optimization is considered one of the strategies that can be used in WSNs to fulfill that goal.

With the many unique characteristics of WSN affecting the performance of the communication link, the strategies and approaches use to determine the optimal data packet size for effective and efficiency data transmission remains a fundamental problem that needs more in depth research and investigations. In general WSNs can be classified into two broad categories of UWSN (which uses acoustic waves for data transmission) and TWSN (uses radio/terrestrial waves) in which each category comes along with its own unique characteristics. The differences between these characteristics are vastly related to the type of media used to relay the data packets wirelessly among the sensor nodes. Based on these differences, presented in this chapter are several works on data packet size optimization in underwater WSN and terrestrial WSN. The data packet size optimization approaches to be discussed in this chapter are divided into two broad categories. The first category is on data packet optimization for underwater wireless sensor network communications and the second category is about data packet optimization deployable in terrestrial wireless sensor network environment. Three packet size optimization approaches would be discussed and analyzed for each of these categories.

The main objective of this chapter is to expose to readers, who are interested in WSN data transmission, on some of the important R&D works accomplished by the related research communities in data packet size optimization for WSNs data transmission. More specifically, at the end of this chapter the reader should be able to:

1. Understand the different data packet size optimization approaches that may be deployable in UWSNs and TWSNs.
2. Comprehend the performance metrics used in each of the optimization approaches.
3. Understand the important effect of optimal data packet size on the WSNs performance.
4. Know the important comparison between underwater wireless communications and terrestrial wireless communications.
5. Identify the possible research directions for packet size optimization based on other UWSN and TWSN performance metrics or the hybrid of these metrics.

Next section presents a brief description on three approaches that may be used for optimizing data packet size in UWSN and TWSN respectively. The section also highlights the optimizing parameters/metrics used in each of the approaches.

BACKGROUND

Data packet size can be optimized based on different wireless link (communication) criteria. There may not be one best solution to optimize the packet size simply because of the various unique characteristics in WSN data transmission. Therefore there exists several different ways or techniques, as proposed by different researchers, to determine optimal packet size in the context of effective and efficient data packet transmission in wireless data communications. In this section three different data packet size optimizing approaches and their related performance metrics/parameters would be briefly explained for UWSN and TWSN respectively.

1. Data Packet Optimization in UWSN Communications
   a. **Optimal data packet length qualified by maximum throughput efficiency:** In this approach the optimal data

packet length (size) is chosen based on the throughput efficiency metric. This approach is suitable for a multi-hop underwater sensor network. It is found that the optimal packet length would be highly dependent on the bit error rate and the offered load. Here, the throughput efficiency is generally defined as the ratio between the delivered bit rate (to the sink node) and the offered bit rate (from the source node). The packet length/size is chosen to offer maximum throughput efficiency.

b. **Optimal packet size qualified by Stop-and-Wait protocol efficiency:** This approach is based on the development of an efficient data link layer protocol that could be used to control the formatting of data packets together with an automatic request (ARQ) protocol. This ARQ is essentially the improved version of the simple Stop-and-Wait protocol with selective acknowledgement strategy. This approach is able to increase the throughput efficiency and therefore the throughput efficiency can be maximized by selecting the appropriate optimal packet size.

c. **Packet size optimization using generic cross-layer optimization framework:** In this approach a generic framework adapted from the terrestrial WSN is used to provide a cross-layer optimization for choosing the optimal packet size. Various performance metrics were taken into account in determining the optimal packet size which includes throughput efficiency, energy consumption per useful bit, resource utilization, etc. This framework is applicable to both shallow water and deep water environment.

2. Data Packet Optimization in TWSN Communications

a. **Adaptive data frame/packet length control:** This approach dynamically controls the data packet frame length/size based on accurate link quality estimation. The estimation technique is able to capture both physical channel conditions and other interferences imposed on the communication link. The control mechanism is implemented in the MAC layer and does include the aggregation service and fragmentation service.

b. **Packet size optimization qualified by energy efficiency:** This optimization technique is based on the energy efficiency metric for a single-hop TWSN. The energy efficiency is evaluated with respect to the number of retransmission, error control parities, and encoding/decoding energies. Data link layer packet format is the main focus in this optimization approach. A notion of energy channel is used to evaluate the energy efficiency.

c. **Packet size optimization for goodput enhancement:** In this approach optimal packet size is chosen based on the goodput or throughput in TWSN with respect to the slotted 802.15.4 network. The packet size is evaluated to maximize resource efficiency and energy efficiency. The MAC and PHY layer constraints are also taken into consideration for the optimal packet size evaluation. This approach is aimed for network in a saturation mode.

Based on the above mentioned data packet optimization approaches, the next two sections (Section D and E) shall present in detail their related concepts, ideas, techniques, etc with the intention to expose to the readers the various R&D works from the UWSN and TWSN research communities in the area of data packet size optimization. It should be mentioned here that the focus area is wireless sensor network data transmission.

DATA PACKET OPTIMIZATION IN UWSN COMMUNICATIONS

This section is subdivided into three sub-sections describing and discussing on three different packet size optimizing techniques that may be deployed in underwater wireless sensor network environment. The readers should learn to appreciate the different qualifying performance metrics/parameters used in each of the technique.

I. Optimal Data Packet Length Qualified by Maximum Throughput Efficiency

Basagni, Petrioli, Petroccia, and Stojanovic (2009) presented their findings in choosing the optimum packet size in multi-hop underwater wireless (UW) networks in a simulated environment using ns-2 simulator. Their simulation indicated that there exists an optimum packet size in underwater wireless acoustic (UWA) communications where the optimum packet size depends on the offered load and is heavily influenced by the bit error rate (BER). The main contribution in this work is in choosing the length of data packet to achieve maximal *throughput efficiency* where this efficiency is defined generally as the ratio between the effective (delivered) and the offered (attempted) bit rate.

Two realistically deployable MAC protocols were used in their work, namely, pure Carrier Sense Multiple Access (CSMA) protocol and the Distance-Aware Collision Avoidance Protocol (DACAP). Other realistic UWA channel characteristic parameters considered in their simulations include the bit rates, energy consumption models, and different BERs. An expected future UW network deployment core scenario was set up in their simulation works too which include a relatively large number of nodes randomly deployed over arbitrary shallow water. The data were generated with a rate in corresponds to different application requirements. In more specific details, the setup of their experiment using the ns2 simulator is as follow:

- An N=100 of UW static nodes are deployed randomly and uniformly over an area of 4km by 4km at a depth of 200m (shallow water).
- A centrally located common sink is placed at the sea surface to collect packets transmitted from the other nodes.
- Shortest path routing protocol.
- Nodes are equipped with an acoustic modem with a transmission range of R=1000m.
- Each packet needs an average of 2.3 hops to reach the sink.
- Transmitting power is adjusted to achieve a SNR of 20dB at a receiver which is 1000m away in the presence of ambient noise and frequency-dependent acoustic path loss.
- Receiving power and the idle power is set to 80mW.
- Carrier frequency is 24 KHz.
- Data packet payloads range from 50B to 3000B in the increments of 50B.
- The acoustic modem raw bit rates are 200bps and 2000bps.
- Two different BERs of 10^{-4} and 10^{-6} were used.

However it should be mentioned here that there is no power control in this simulation work. The total size of a packet is given by the payload bits plus the header bits added by the physical layer right through to the network layer. A minimum signal to inference ratio (SIR) of 15dB is required to correctly receive a packet.

In this research, the throughput efficiency is defined more technically as the ratio between the average bit rate delivered to the sink (correct bits) and the average bit rate offered by the network. The average bit rate offered by the network is given by $N_b\lambda$ where N_b is the packet size in bits and λ is packets per second.

Figure 1(a) and (b) below summarize the findings of their simulation. These results clearly show that there exist an optimal packet size that can be qualified by an optimal throughput efficiency. It should be noted here that that DACAP is seemed to be more suitable for larger packets while CSMA is for a smaller packets.

II. Optimal Packet Size Qualified by Stop-and-Wait Protocol Efficiency

The work by Stojanovic (2005) which focused on the design and analysis of data link protocols for UWA system, was accomplished with the aim to develop a protocol that is as efficient as possible at the data link layer for UW wireless communications. In her work the data link layer controls the formatting of data packets and implement the automatic request (ARQ) protocol.

The ARQ protocol is essentially a Stop-and-Wait (S&W) protocol which unfortunately suffers from low throughput efficiency in UWA channel. The main cause of this low efficiency is due to the inherently high BER and long propagation delay (low speed of sound propagation) in most of the UWA channels. Her work showed that the basic S&W protocol can be improved by transmitting packets in group and implementing a selective acknowledgement strategy. This improvement brings along a higher throughput efficiency which in turn can be maximized by selecting the appropriate optimal packet size. However the optimal size is found to be somehow influenced by the range of transmission, the bit rates, and the error probability.

In general, the efficiency of the S&W protocol is a function of the data packet size, the link delay, and the packet error rate. This implies that there exists an optimal packet size in obtaining a maximum efficiency Schwartz (1988). The main contribution in the work of Stojanovic (2005) is her statistical analysis of the S&W protocol efficiency that leads to the optimal packet size for a typical UWA channel. Her analysis involved three types of S&W protocols: the basic type called S&W-1, and the other two modified protocols called S&W-2 and S&W-3 respectively. These protocols are used for a group transmission of up to M packets. The efficiency of these protocols are given respectively by the expressions below. It is claimed in her work that $\eta_2 \geq \eta_3 \geq \eta_1$.

$$\text{S\&W-1: } \eta_1 = \frac{N_d T}{T_1} = (1-p)\frac{N_d T}{T(1)}$$

$$\text{S\&W-2: } \eta_2 = (1-p)\frac{MN_d T}{T(M)}$$

$$\text{S\&W-3: } \eta_3 = \frac{MN_d T}{T(M)}$$

where,

N_d is the number of data bits.
T is the bit duration ($T=1/R$, and R is bit rate).
p is the probability of packet error.

The efficiency expressions shown above indicate that they are dependable on the packet error rate p. Note that by increasing the packet size in S&W protocol it means a better utilization of the waiting time. However the chances of having more bit errors in the packet is also increased. Hence it is claimed that there exists an optimal packet size for obtaining a maximal throughput

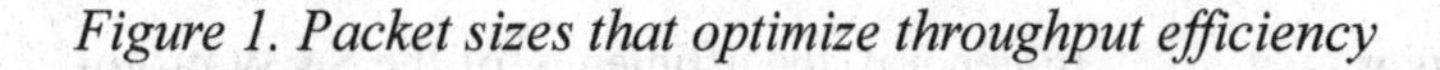

Figure 1. Packet sizes that optimize throughput efficiency

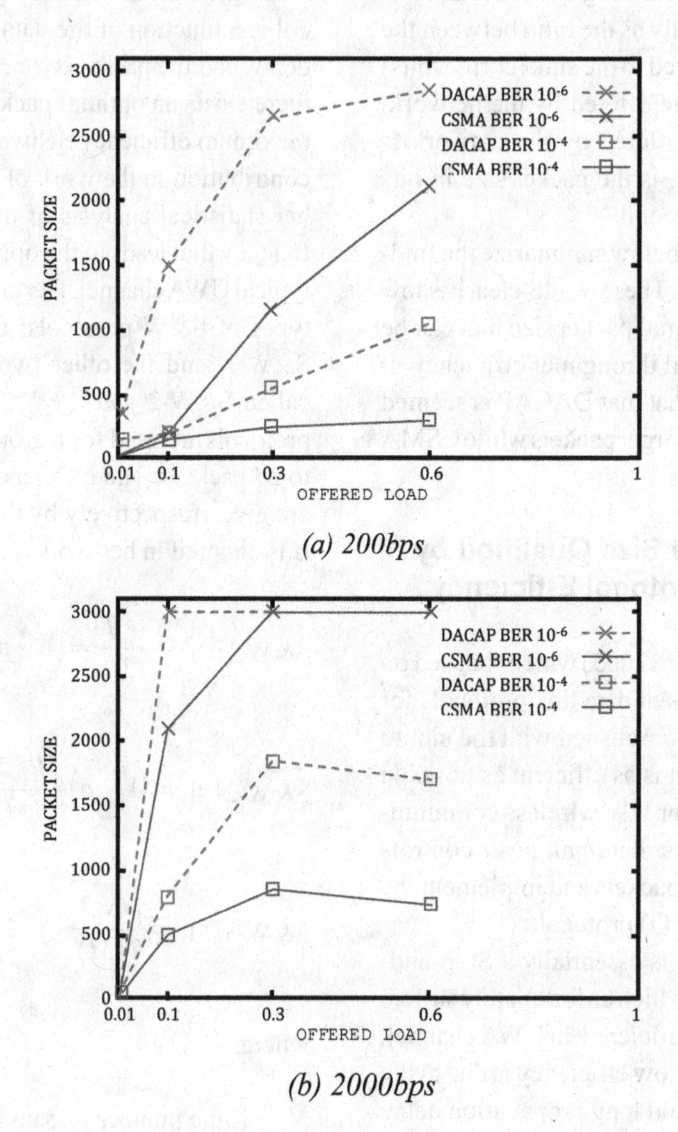

(a) 200bps

(b) 2000bps

efficiency. That is, the packet size can be varied to maximize the throughput efficiency. Further analysis on the efficiency expression of η_2 yields the expression for optimal packet size as follow,

$$N_{d,opt} = \frac{\mu}{2}\left[\sqrt{1+\frac{4}{\mu\rho}} - 1\right]$$

of which,

$$\mu = \mu_o + \frac{2}{Mc} lR$$

where,

μ_o is the packet overhead bits
M is packets transmitted
c is speed of sound in water in m/s
l is link distance in meter
R is the bit rate in bps

and

$$\rho = \ln \frac{1}{1 - P_e}$$

In which P_e is the probability of bit error in the packet. Figure 2 below shows one of the results obtained from the work of Stojanovic (2005). This graph depicts the optimal packet size plotted as a function of range-rate product (lR) for bit error probability (P_e) of 10^{-3} and 10^{-4}. The plot is for S&W-1 and S&W-2. For future system design Stojanovic (2005) suggests that an adaptive ARQ scheme should be used.

Two aspects that worth considering are: (1) adaptive adjustment of the time-out in accordance with the measured instantaneous round-trip time, and (2) adaptive adjustment of the packet size in accordance with the measured instantaneous error probability and link delay. It is the second aspect that the author of this research work is focusing on.

Figure 2. Optimal packet size $N_{d,\,opt}$ as a function of range-rate product

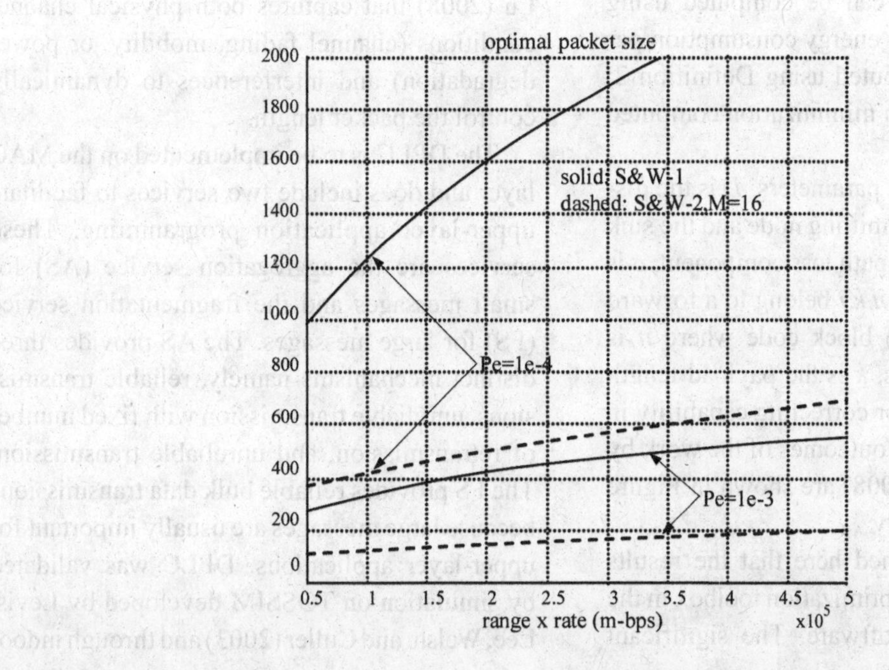

III. Packet Size Optimization Using Generic Cross-layer Optimization Framework

In the work of Vuran and Akyildiz (2008), these authors proposed a generic cross-layer optimization framework to determine the optimal packet size in terrestrial WSN. However the framework was extended to cater for the more challenging environment of UW and underground (UG) sensor networks. This work shows that an optimization solution can be formalized using three different performance metrics namely packet throughput, energy consumption per useful bit, and resource utilization. More specifically the metrics also include the latency and reliability of the multiple hop link (or path). The definition of the three performance metrics, according to Vuran and Akyildiz (2008) are given as below.

Definition 1: Packet throughput,

$$U_{tput} = \frac{l_D\left(1 - PER_{e2e}\right)}{T_{flow}}$$

where l_D is the payload length in bits, PER_{e2e} is the end-to-end packet error rate, and T_{flow} is the end-to-end latency in seconds.

Definition 2: Energy per useful bit,

$$U_{eng} = \frac{E_{flow}}{l_D\left(1 - PER_{e2e}\right)}$$

where E_{flow} is the end-to-end energy consumption in Joules to transport a packet from a source to a sink.

Definition 3: Resource utilization,

$$U_{res} = \frac{E_{flow} T_{flow}}{l_D\left(1 - PER_{e2e}\right)}$$

It is highlighted in their works that the energy consumption of a flow is mainly a function of packet size and the SNR threshold. In other words, the choice of the SNR threshold value may determines the optimum packet size. Moreover the packet size may also be affected by the routing decision.

For a given set of parameters (D,η,σ,n,k,t) maximum throughput can be computed using Definition 1; minimum energy consumption per useful bit can be computed using Definition 2; and resource utilization minimization computed by Definition 3.

For that given set of parameters: D is the distance between the transmitting node and the sink node in meters, η is the path loss component, σ is the fading component, n,k,t belong to a forward error correction (FEC) block code where n is the block length in bits, k is the payload length in bits, and k is the error correcting capability in bits. The sample of the outcomes of the work by Vuran and Akyildiz (2008) are shown in Figure 3(a) and (b) respectively.

It is worth mentioned here that the results obtained were via the optimization toolbox in the MatLab application software. The significant difference in Figure 3(a) and (b) shows that the propagation characteristic in deep water and that in shallow water would affect the optimum packet size remarkably.

DATA PACKET OPTIMIZATION IN TWSN COMMUNICATIONS

This section is again subdivided into three subsections describing and discussing on three different packet size optimizing techniques proposed by three different research groups for the terrestrial wireless sensor network environment. These subsections are intended to expose to the readers the different qualifying performance metrics/parameters used in TWSN.

I. Adaptive Data Frame/Packet Length Control

A dynamic packet length control (DPLC) scheme for WSN was proposed by Dong, Liu, Chen, He, Liu, and Bu (2010). In their work they incorporated a lightweight and accurate link quality estimation method adapted from Hackmann, Chipara and Lu (2008) that captures both physical channel conditions (channel fading, mobility, or power degradation) and interferences to dynamically control the packet length.

The DPLC is to be implemented on the MAC layer and does include two services to facilitate upper-layer application programming. These services are the aggregation service (AS) for small messages and the fragmentation service (FS) for large messages. The AS provides three distinct mechanisms namely, reliable transmissions, unreliable transmission with fixed number of retransmission, and unreliable transmission. The FS provides reliable bulk data transmissions because large messages are usually important for upper-layer applications. DPLC was validated by simulation on TOSSIM developed by Levis, Lee, Welsh, and Culler (2003) and through indoor

Figure 3. Optimum packet size for deep water and shallow water

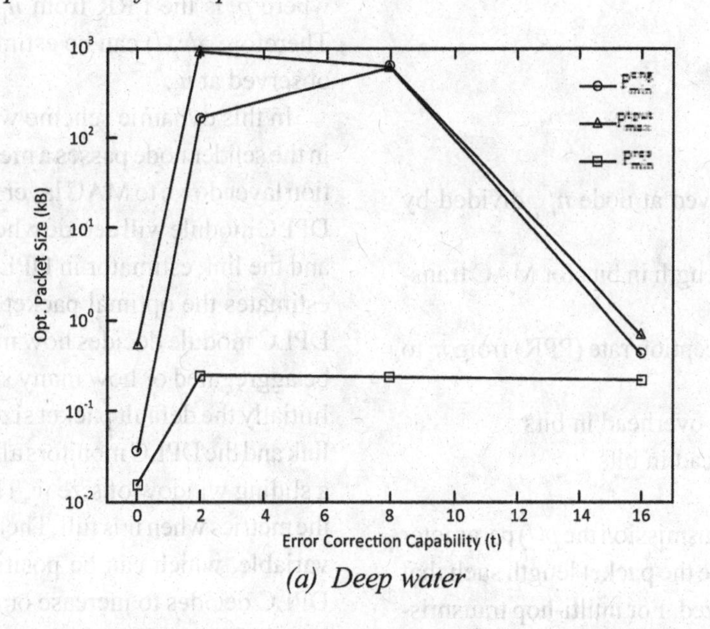

(a) Deep water

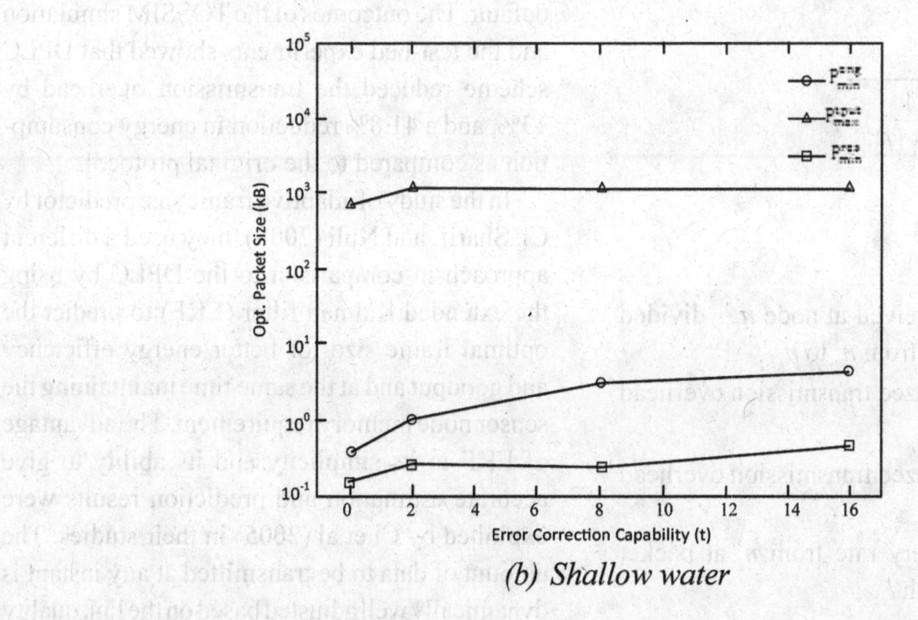

(b) Shallow water

test-bed experiments via 20 TelosB motes running the CTP protocol proposed by Gnawali, Fonseca, Jamieson, Moss, and Levis (2009).

DPLC uses the metric of transmission efficiency (ε) for dynamic adaptation. In general the transmission efficiency is defined as, $\varepsilon = U_b/T_b$ where U_b is the received useful bytes and T_b the overall transmitted bytes. Two variants of ε were used in DPLC in which one is for single-hop transmission and the other one for multi-hop transmission. For single-hop transmission the efficiency is given by,

$$\varepsilon_i = \frac{l.p(l)}{l + H + O}$$

Where,

ε_i equals U_b received at node n_{i+1} divided by T_b from node n_i

l is the payload length in bits for MAC transmission

$p(l)$ is the packet reception rate (PPR) from n_i to n_{i+1}

H is MAC header overhead in bits

O is DPLC overhead in bits

For single-hop transmission the $p(l)$ parameter is monitored to decide the packet length such that the metric is maximized. For multi-hop transmission the metric is given by,

$$\varepsilon_k = \frac{1}{\varepsilon_k^{-1}(l) + \varepsilon_{k-1}^{-1}(\frac{1}{dr_k(l)})}$$

where,

ε_k equals U_b received at node n_{k+1} divided by overall T_b from n_1 to n_{k+1}

$\varepsilon_k^{-1}(l)$ is the normalized transmission overhead from n_k to n_{k+1}

ε_{k-1}^{-1} is the normalized transmission overhead from n_1 to n_k

$dr_k(l)$ is data delivery rate from n_k at packet payload length l

It is mentioned that for reliable multi-hop transmission the strategy is equivalent to maximizing the ε_i as in single-hop transmission. However for unreliable multi-hop trans-mission with fixed number of retransmissions, the data delivery rate is related to PRR as,

$$dr_i = 1 - (1 - p_i)^{m+1}$$

where p_i is the PRR from n_i to n_{i+1} $(1 \leq i \leq k)$. Therefore, $dr_k(l)$ can be estimated from the PRR observed at n_k.

In this dynamic scheme when the application in the sender node passes a message from application layer down to MAC layer for transmission the DPLC module will decide whether to use AS or FS and the link estimator in DPLC will dynamically estimates the optimal packet length. That is, the DPLC module decides how many messages shall be aggregated or how many shall be fragmented. Initially the default packet size is sent through the link and the DPLC monitors all packets by keeping a sliding window of size w. The DPLC computes the metrics when w is full. Then based on a gradient variable, which can be positive or negative, the DPLC decides to increase or decrease the packet length. The gradient variable is set to positive by default. The outcomes of the TOSSIM simulation and the test-bed experiments showed that DPLC scheme reduced the transmission overhead by 13%, and a 41.8% reduction in energy consumption as compared to the original protocol.

In the study of adaptive frame size predictor by Ci, Sharif, and Nuli (2005), they used a different approach in comparison to the DPLC by using the extended Kalman filter (EKF) to predict the optimal frame size for better energy efficiency and goodput and at the same time maintaining the sensor node memory requirement. The advantage of EKF in its simplicity and its ability to give accurate estimation and prediction results were exploited by Ci et al (2005) in their studies. The amount of data to be transmitted at any instant is dynamically well adjusted based on the link quality estimated by the EKF. In other words, the frame size is optimized dynamically with respect to the channel quality predicted by an accurate predictor. They proposed an algorithm that is able to reduce the number of retransmissions due to frame errors and it is understood that the rate of frame errors is sensitive to frame size. More specifically, the algorithm is able to predict the optimal frame size based on the network parameters such as channel quality, frame length, protocol overheads, and collisions.

In a nutshell they used the EKF filtering characteristics coupled with the known present channel quality to keep track of the channel history and thus to predict the optimal frame size. The EKF has a unique capability of estimating the past, present, and future states of a system even without the precise knowledge of the modeled system as presented in the papers by Haykin (2002), and Julier and Uhlmann (1997). The main focus of Ci et al (2005) is to maintain the network performance and at the same time to improve the energy efficiency of WSNs at the MAC layer.

In order to use EKF this research team had developed two models namely, the *process model* and the *observation model* to fit into the EKF model. Channel throughput was chosen to be the main performance parameter in their developed models because their main goal is to maximize channel throughput ρ, at every transmission by predicting an optimal frame size taking into consideration of the collision and frame errors. The *process model* developed is given by the following equations and give the frame length L at time k and $k+1$,

$$L_{k+1} = L_{max} \quad L_{k+1} > L_{max}$$

$$= L_{opt} \quad L_{min} < L_{k+1} < L_{max}$$

$$= L_{min} \quad L_{k+1} < L_{min}$$

where,

L_{max} is the maximum frame size;
L_{min} is the minimum frame size;

and,

$$L_{opt} = \frac{-u \pm \sqrt{u^2 - 4L_k Pb_{k+1} \upsilon}}{2L_k Pb_{k+1}}$$ (the optimal frame size)

where,

Pb is the bit error rate under a known channel quality;

with,

$$u = (2L_k Pb_{k+1} H + L_k SR_k Pb_{k+1} - Pb_k L_k^2 - 2L_k Pb_k H \\ -H - Pb_k H^2 - HN_k - SR_k - SR_k Pb_k L_k \\ -SR_k Pb_k H - SR_k N_k - ACK(1 - Pb_k)^{-ACK} \\ -OR_k (1 - Pb_k)^{-ACK} - N_k C_k)$$

and,

$$v = (L_k H + L_k Pb_{k+1} H^2 + L_k HN_k + L_k SR_k \\ +L_k SR_k Pb_{k+1} H + L + kSR_k N_k \\ +L_k ACK(1 - Pb_{k+1})^{-ACK} \\ +L_k OR_k (1 - Pb_{k+1})^{-ACK} + L_k N_k C_k)$$

in which,

L is payload size of a frame
R is data transmission rate
H is MAC protocol header of a frame
N is average number of collisions occurred between two renewal points
ACK is frame length of an acknowledgment frame
O is overhead of ACK frame
C is average length of collisions
S is average number of random backoff time slots

The *observation model* is given by,

$$\rho_{k+1} = Q(L_{k+1}, Pb_{k+1})$$

where,

ρ_{k+1} is observation time at time $k+1$;
Q is the observation function.

The proposed optimal frame size predictor algorithm were tested by integrating it into the MAC implementation of the Berkerly motes for performance evaluation under different channel quality conditions by modifying the PHY layer parameters. Four network scenarios were considered with 2, 5, 10, 20, and 50 nodes respectively. It was claimed that the algorithm achieved a 15% reduction in energy consumption, the goodput was doubled, and the delay reduced by 20%.

Modiano (1999) presented a work on adaptive algorithm for optimizing data packet size with wireless data link layer ARQ protocol focusing on bit error rate (BER). His algorithm is making used of the acknowledgment history to estimate the channel BER and allows the ARQ protocol to dynamically optimize the packet size. It is claimed that this algorithm is particularly useful for wireless channel where BER tends to be high and time variable. He observed in his work that it was not necessary to have an accurate estimation of the channel BER by using the Selective Repeat Protocol (SRP), an optimal ARQ protocol, to choose a good packet size. Thus his algorithm is able to perform well with a short observation history of just 10,000 bits.

Modiano (1999) has chosen SRP to be an optimal ARQ protocol in the sense that SRP attempts to retransmit only packets containing errors as according to Bertsekas and Gallager (1987). An estimation of the channel BER can be made based on the acknowledgement history of the most recently transmitted packets. So with a given number of packets that required retransmission an optimal packet size can be chosen to maximize the expected efficiency of the data link protocol. In SRP, for a given channel BER p, the expected efficiency of the protocol using a packet size of k is given by Modiano (1994) as,

$$\eta = \left(\frac{k}{k+h}\right)\frac{1}{(1-p)^{-(k+h)}}$$

where,

k is the number of information bits;
h is the number of header bits in the packet;
p is the channel BER.

The first term of the above expression represents the ratio of information bits to total bits in a packet, while the second term represents the average number of transmission attempts per packet. If the number of retransmission request R out of the last M packet transmissions is known then the expected efficiency of the protocol can be given as,

$$\eta_R(k) = \int_p \left[\frac{k(1-p)^{k+h}}{(k+h)} \times \frac{\binom{M}{R} E^R (1-E)^{M-R}}{\int_p \binom{M}{R} E^R (1-E)^{M-R}}\right]$$

where,

E is the probability that a packet contains errors and is given by

$$E = 1-(1-p)^{k'+h}$$

Take note that the packet size used in the previous M transmissions is given by k'. In this equation, it shows that the value of frame size k, can be chosen to maximize the protocol efficiency for a given value of R out of the previous M transmissions.

Figure 5 from Modiano (1999) shows the performance of the proposed algorithm for various bit observation history. The results in Figure 5 is the evaluation of the steady state performance of the algorithm based a Markov chain in Figure 4 by Modiano (1999) which depicts the packet size of 200, 500, 1000, and 1500 bits.

Figure 4. System Markov chain with 4 states

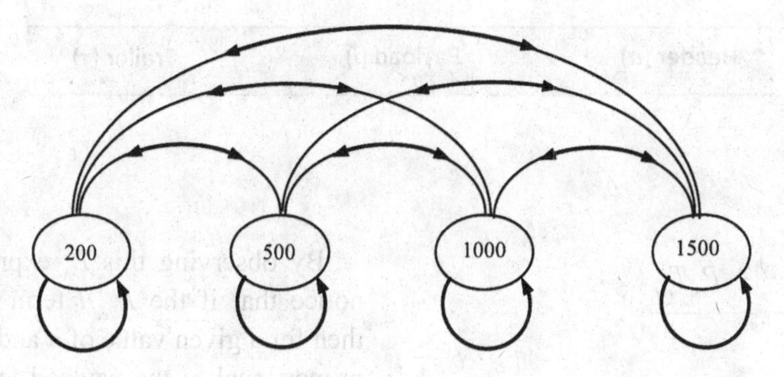

II. Packet Size Optimization Qualified by Energy Efficiency

By choosing energy efficiency as the optimization metric Sankarasubramaniam, Akyildiz, and McLaughlin (2003) aimed to determine the optimal data packet size for communication between neighboring nodes. They also examined the relationship between energy efficiency based on the effect of retransmissions, error control parities and encoding/decoding energies.

Based on the general data link layer packet format in Figure 6 and the energy model outlined in Shih et al (2001), Sankarasubramaniam et al (2003) has expressed the energy required to transmit and receive one bit of information across a single-hop as,

$$E_b = E_t + E_r + \frac{E_{dec}}{l}$$

where E_t is energy consumed in transmitter and E_r is energy consumed in receiver with E_{dec} as decoding energy per packet.

It is further highlighted that E_t and E_r are respectively given by,

$$E_t = \frac{\left(\left(P_{te} + p_o\right)\frac{(l + \alpha + \tau)}{R} + P_{tst}T_{tst}\right)}{l}$$

Figure 5. Performance of the adaptive algorithm

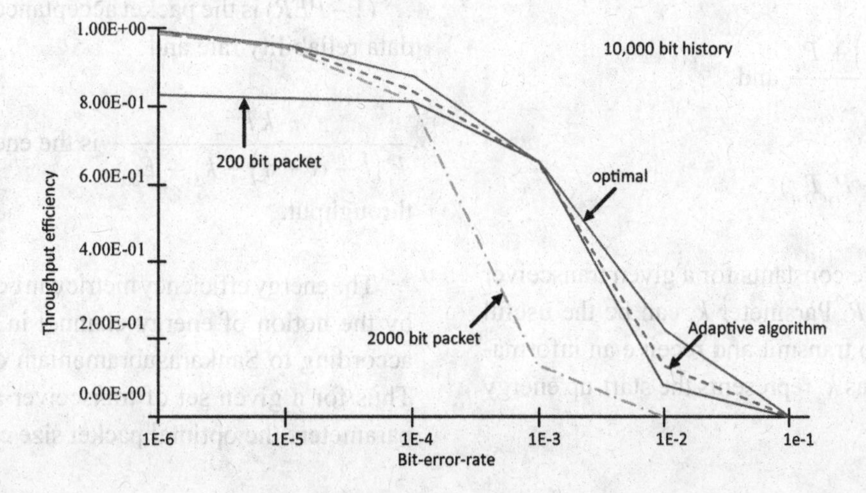

Figure 6. Data link layer packet format

Header (α)	Payload (l)	Trailer (τ)

$$E_r = \frac{P_{re}\frac{(l+\alpha+\tau)}{R} + P_{rst}T_{rst}}{l}$$

where,

$P_{te/re}$ power consumed in the transmitter/receiver electronics;
$P_{tst/rst}$ start-up power consumed in transmitter/receiver;
$T_{tst/rst}$ transmitter/receiver start-up time;
P_o output transmit power;
R data rate in bps.

The expression for the energy required to transmit and receive one bit of information across a single-hop can then be written in terms of radio parameters k_1 and k_2 as,

$$E_b = k_1 + k_1\frac{(\alpha+\tau)}{l} + \frac{k_2 + E_{dec}}{l}$$

where,

$$k_1 = \frac{\left(P_{te} + P_o\right) + P_{re}}{R} \text{ and}$$

$$k_2 = (P_{tst}T_{tst} + P_{rst}T_{rst})$$

k_1 and k_2 are constants for a given transceiver and data rate R. Parameter k_1 can be the useful energy used to transmit and receive an information bit whereas k_2 represents the start-up energy consumption.

By observing this E_b expression, it can be notice that if the E_{dec}/l term is kept constant, then for a given value of α and τ, E_b is inversely proportional to the payload length l. It implies that E_b shall become a constant k_1 if the length l is allowed to increase to a large value. However from practical experiences it is well understood that long packet sizes are associated with greater loss rates and on the other hand, shorter packets are reliable but are energy inefficient. Therefore there exists an optimal packet size/length that can be chosen to balance these conflicting interests.

Sankarasubramaniam et al (2003) are of the opinion that energy efficiency is the most suitable metric to capture the energy and reliability constraints. They thus defined the energy efficiency as follow,

$$\eta = \frac{k_1 l}{k_1\left(l+\alpha+\tau\right) + k_2 + E_{dec}}(1-PER)$$

where,

$(1-PER)$ is the packet acceptance rate i.e. the data reliability rate and

$\frac{k_1 l}{k_1\left(l+\alpha+\tau\right) + k_2 + E_{dec}}$ is the energy throughput.

The energy efficiency metric can be represented by the notion of energy channel in Figure 7 as according to Sankarasubramaniam et al (2003). Thus for a given set of transceiver and channel parameters the optimal packet size can be deter-

Figure 7. The notion of energy channel

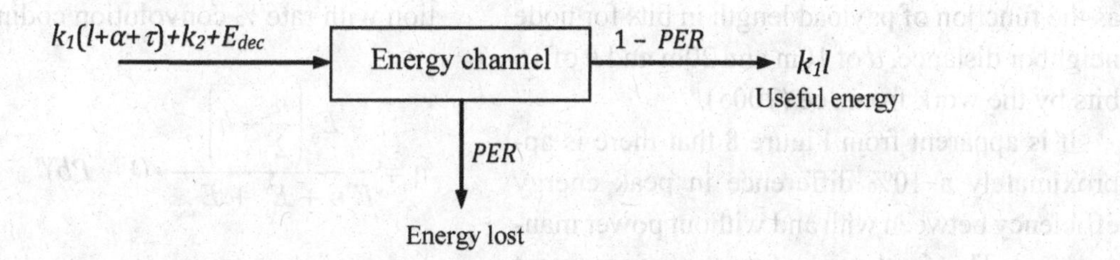

mined by maximizing the energy efficiency metric (η) as defined above.

It should be mentioned here that the proposed approach is for the optimization of fixed size packet based on the parameters estimated at the time of network design to give the maximal energy efficiency. It is thus not the same as those described in the Adaptive Frame Length Control Approach where the packet size is dynamically control based on the channel quality and other parameters. The additional computation overhead and resource management costs are the main reasons why Sankarasubramaniam et al (2003) refrained from the dynamic control approach. These researchers had shown that significant improvement in η can be achieved with forward error control (FEC) as compared to no error control i.e. with τ set to 0.

Joe (2005) proposed a method to improve energy efficiency in WSN using optimal packet length with channel coding capability but without power management. His work showed that energy efficiency can be improved by using optimal packet length at the data link layer. He also showed that energy efficiency may not be maximized via power management. In power management approach the transceiver is turned off at idle state to conserve energy. He argued that since sensor nodes normally communicate using short packets thus due to the dominance of start-up energy the energy efficiency in the nodes could actually be reduced. However he did emphasize that even though power management does not improve energy efficiency, it should be employed to minimize energy wastage. He defined the energy efficiency as,

$$\eta = E_{th} \cdot R$$

where,

E_{th} is the energy throughput i.e. the ratio of energy consumed for actual data transmission to entire packet transmission;
R is the reliability i.e. the successful packet reception rate.

In his work, Joe (2005) further expressed the energy efficiency as a function of packet length with power management and without power management as follow,

$$\eta = \frac{E_c l}{E_c(l+h)+E_s}(1-PER)$$; with power management

$$\eta = \frac{l}{l+h}(1-PER)$$; without power management

where,

E_c is communication energy consumption;
E_s is the start- up energy consumption;
l the payload length;
h the header length;
PER is packet error rate.

Figure 8 shows the plot of energy efficiency as the function of payload length in bits for node neighbor distance, d of 10m and 20m and h of 16 bits by the work from Joe (2005).

It is apparent from Figure 8 that there is approximately a 10% difference in peak energy efficiency between with and without power management. That is, the use of power management cannot improve energy efficiency. The plots do show that there exists an optimal packet length in attaining a maximal energy efficiency. For instance, with power management in place, when energy efficiency is at peak, the optimal payload length is 280 bits for 10m of neighbor distance and it is 60 bits for 20m of neighbor distance.

Joe (2005) provides further analysis in energy efficiency with optimal packet length in terms of channel coding i.e. with forward error correction (FEC). Channel coding is one of the most commonly used approach in increasing the channel reliability. He put an effort to find the relationship between FEC and energy efficiency by considering the Binary BCH coding and the rate ½ convolution coding. His analysis showed that with channel coding the energy efficiency can be improved significantly and BCH code is better than the convolution code by a factor of 2. The reason being that the redundancy bits in convolution code is almost double that used in BCH coding. The energy efficiency equation with BCH coding with t correction capability is given as below.

$$\eta = \frac{E_c\left(n-h-\tau\right)}{E_c n + E_s + E_{dec}} \cdot \sum_{j=0}^{t} \binom{n}{j} P_b^j (1-Pb)^{n-j}$$

Where, n is the BCH code length of $h+l+\tau$ of which h is the header, l is payload length, and τ is the trailer (parity bits), E_c is the encoding energy, E_{dec} is the decoder energy at the receiver side, and P_b is the probability of bit error for BCH code. For completion purpose the energy efficiency equation with rate ½ convolution coding is given as,

$$\eta = \frac{E_c\left(\frac{n}{2}-h\right)}{E_c n + E_s + E_{dec}} \cdot (1-Pb)^n$$

where,

P_b is the convolution code probability of bit error;
n is the packet length;
h is the header.

III. Packet Size Optimization for Goodput Enhancement

One of the more recent works in investigating the relationship between optimal packet size and the goodput or throughput in WSN systems can be found in Zhang and Shu (2009). They proposed a new analytical model to calculate the goodput and energy consumption with respect to packet size optimization for slotted IEEE 802.15.4 networks. Take note that the IEEE 802.15.4 Standard (2006) is released to regulate low-rate and low-cost short distance wireless personal area networks (LR-WPANs).

They investigated the issue of how to optimize the packet size in terms of maximizing resource efficiency and energy efficiency by considering both MAC and PHY layer constraints with the assumption that the system under study is in saturation mode i.e. every node in the system always has a packet to be transmitted at any moment. More specifically their research was to deduce the optimized packet size in terms of resource and energy efficiency maximization.

System level simulation was used to verify their analytical model and the outcomes of the simulation were claimed to bring significant performance enhancement to IEEE 802.15.4 networks by segmenting the data stream in a more efficient way. Technically their new model takes into ac-

Figure 8. Energy efficiency as function of payload length

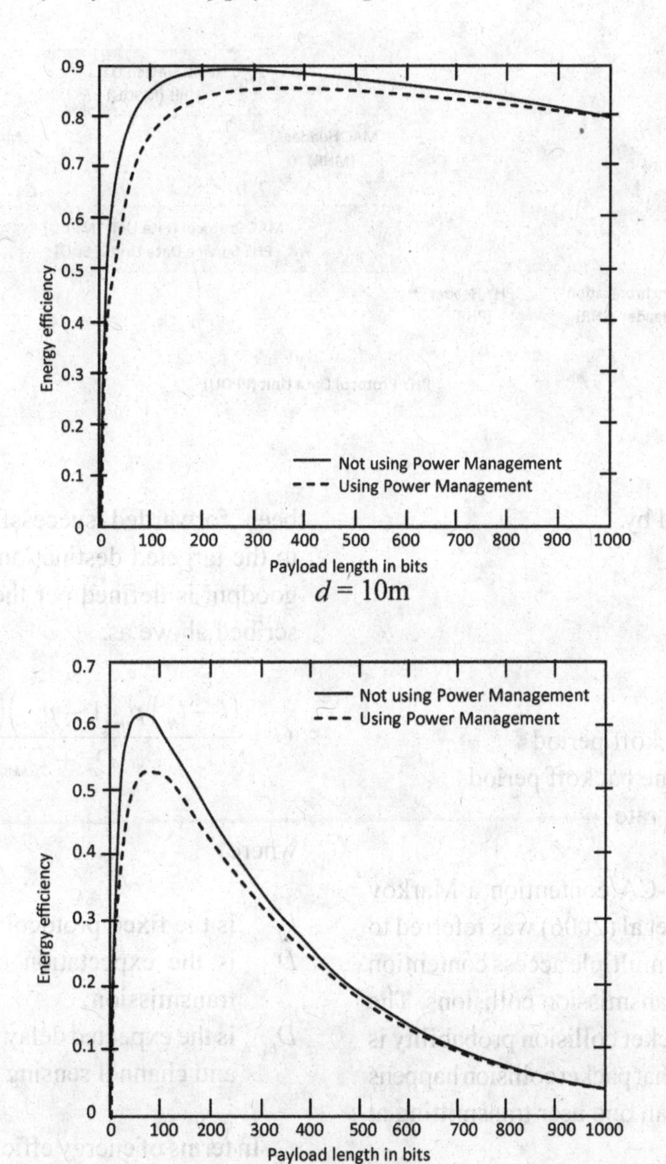

count of the protocol overhead, channel condition, CSMA-CA contention, resource efficiency, and energy efficiency. The protocol overhead can be illustrated by the standard IEEE 802.15.4 data frame format in Figure 9.

Note that in Figure 9 the only part in the data packet that carries the useful information i.e. the user data is the MAC Service Data Unit (MSDU). The rest of the fields are simply overhead needed by the protocol. For the channel condition at PHY layer, the bite error rate (BER) is used as the indicator. For a given value of BER, the packet error rate can be deduced by,

$PER = 1 - (1 - BER)^l$; where l is the packet length in bits

Figure 9. IEEE 802.15.4 data frame format

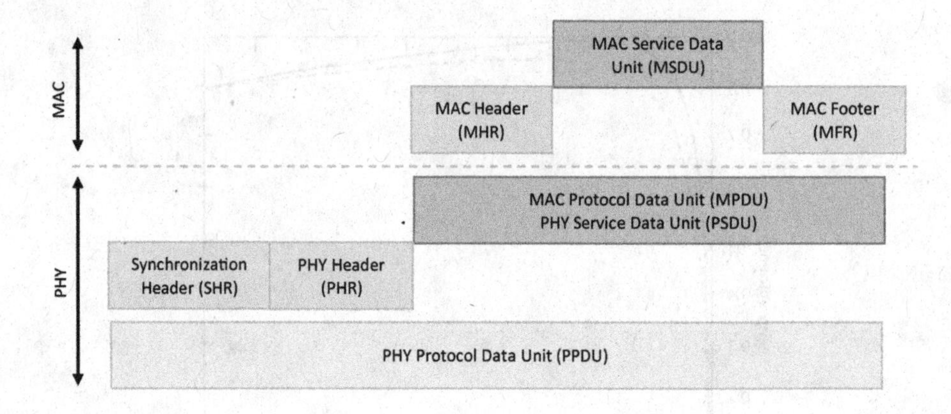

and l can be computed by,

$$l = Lt_b R$$

where,

L is the number of backoff period
t_b is time interval of one backoff period
R is the PHY layer bit rate

As for the CSMA-CA contention a Markov chain model in Pollin et al (2006) was referred to derive the behavior of multiple access contention thus to describe the transmission collisions. The new expression for packet collision probability is given below to denote that packet collision happens when there is more than one user transmitting at the same moment.

$$P_{col} = 1 - (1 - p_{tr})^N - Np_{tr}(1 - p_{tr})^{N-1}$$

where,

p_{tr} is the probability to transmission;
N is the network size (number of nodes).

In terms of resource efficiency, Y. Zhang et al (2009) used goodput as the metric. It denotes the number of information bits, excluding protocol overhead and data retransmissions, which has been forwarded successfully from the source to the targeted destination per unit of time. The goodput is defined per the three constraints described above as,

$$G = \frac{\left(l - l_H\right) p_{tr}\left(1 - p_{col}\right)\left(1 - PER\right)}{D_{tr} + D_{bk}}$$

where,

l_H is the fixed protocol overhead;
D_{tr} is the expectation of time used for data transmission;
D_{bk} is the expected delay due to random backoff and channel sensing.

In terms of energy efficiency they used energy consumption per bit of goodput to do the measurement. The energy efficiency is given as,

$$E = \frac{LP_{TX}p_{tr} + D_{bk}P_{SL} + L_{CCA}\left(P_{RX} - P_{SL}\right)}{\left(l - l_H\right) p_{tr}\left(1 - p_{col}\right)\left(1 - PER\right)}$$

where,

L is the number of backoff period;
p_{tr} is the probability to transmission;

P_{TX} is energy consumptions per backoff period in transmission state;
P_{RX} is energy consumptions per backoff period in receiving state;
P_{SL} is energy consumptions per backoff period in sleeping state;
L_{CCA} is the interval of channel sensing in both CCA1 and CCA2.

A link adaptation scheme for multi-rate wireless networks which combines adaptive modulation and coding (AMC) at physical layer and with type-II hybrid ARQ (HARQ) at data link layer to enhance channel utilization and goodput was proposed by Wu, Ci, Sharif, and Yang (2007). Then based on the goodput performance analysis of the scheme they focused on goodput enhancement in delivering messages from the transport layer by optimizing packet sizes in a cross-layer fashion. An effective and efficient algorithm for searching (golden section search) optimal packet sizes was developed. In relating the goodput to optimal packet size search D. Wu et al (2007) started off with a general goodput equation given by Qiao, Choi, and Shin (2002) as,

$$J_M = \frac{L}{T} \cdot (1 - P)$$

where,

L is the number of bits in a packet;
T is the average time for transmitting a packet;
P is the average PER of an L-bit information packet after the maximal possible N^{th} transmission attempts.

Then in general, in a transport layer session, each transport layer message is broken into MAC packets. In their work they assumed that a MAC packet consists of L information bits. Thus each message is fragmented into multiple packets of size L. It follows that the last packet will contains the left-over bits. In other words, a transport layer message of length I would be fragmented into I/L packets with the final packet contains between 1 and L bits. Their analysis showed that the goodput of transmitting a message can be given by,

$$J = \frac{I}{\left|\frac{I}{L}\right| \cdot T + t} \cdot (1 - P)^{\left|\frac{I}{L}\right|} \cdot (1 - p)$$

where t and p denote the average time and PER of transmitting a packet of leftover bits.

Based on the convexity property of the above goodput expression they have developed a one-dimension golden search algorithm to search for the optimal packet size. The search algorithm by Qiao, Choi, and Shin (2002) is listed as below.

Step 1: set $x_2 = a + 0.618(b - a)$, $J_2 = J(x_2)$, go to Step 2.
Step 2: set $x_1 = a + 0.382(b - a)$, $J_1 = J(x_1)$, go to Step 3.
Step 3: if $\left|(b - a)\right| \leq \varepsilon$, set $x^* = \frac{a+b}{2}$, stop. else, go to Step 4.
Step 4: if $J_1 < J_2$,
$a = x_1, x_1 = x_2, J_1 = J_2$,
go to Step 5.
If $J_1 = J_2$,
$a = x_1, b = x_2$,
go to Step 1.
If $J_1 > J_2$,
$b = x_2, x_2 = x_1, J_2 = J_1$,
go to Step 2.
Step 5: set $x_2 = a + 0.618(b - a)$, $J_2 = J(x_2)$, go to Step 3.

This algorithm shall search for an optimal packet size, x on an initial packet size interval $[a,b]$ for a given accuracy of ε. That is, the optimal packet size should have $x \in [a,b]$. Technically speaking, the golden section algorithm requires an initial packet size interval $[a, b]$ on

Table 1. Comparison of data packet size optimization in UWSN and TWSN

UWSN Communications	TWSN Communications
Packet size optimization in terms of energy efficiency, throughput, BER, and the types of protocol.	Packet size optimization in terms of energy efficiency, throughput, BER, and the types of protocol.
Dynamic control of packet size for each packet transmission is not very practical due to the slower speed of acoustic wave that brings along possible high propagation delay.	Dynamic control of packet size for each packet transmission is possible to be implemented at MAC layer and below especially at data link and physical layer.
Dynamic control of packet size involved link quality computation and some over-head resource utilization. Due to slower transmission rate and high propagation delay the computed link quality para-meters may be out of date to support the required response time for adequate con-trolling of the packet size. Thus the next packet send out may not be of optimal size.	Does involved link quality computation and overhead resource utilization how-ever is possible to adjust the packet size dynamically with the proper selection of protocol and adaptive channel modu-lation and coding (AMC) at the PHY layer.
If dynamic or adaptive packet length control is not a good strategy then the fixed optimal packet size shall be deter-mined at the network design stage with reference to the desired network parameters.	Has the luxury to use fixed size optimal packet or dynamically adaptive packet length. However fixed size packet trans-mission seems to be preferred in prac-tice.
A more practical approach to optimize or increase the data rate, the efficiency, the and the bandwidth of UWA trans-mission is the MIMO technique.	Not necessary to go into MIMO. New approaches are being investigated by many researchers.

which the function $J(x)$ is a convex function of packet size x. It is claimed that this scheme does not incurred much calculation complexity and computation overhead.

With reference to the six approaches described above a comparison table is created in the next section to summarize the important comparisons between the proposed optimization approaches published in the reviewed literatures.

DATA PACKET SIZE OPTIMIZATION IN UWSN VS OPTIMIZATION IN TWSN

Table 1 summarizes some of the important comparisons of data packet size optimization techniques described in Section E for UWSN and TWSN communications. It provides a quick reference for the readers to know the possible performance metrics that can be considered in finding optimal data packet size in UWSN and TWSN data packets transmission. Table 1 is by no means an exhaustive comparison but does serve as an information provider to interested readers.

FUTURE RESEARCH DIRECTIONS

With the fast technology advancement in UWSN and TWSN data packet size optimization approaches discussed in this chapter are by no means a static techniques. They are certainly to be evolved over time. It is especially so for the UWSN in which a lot of new effort have been put in place by many UWSN researchers to elevate its performance to be at least at par with the TWSN counterparts if not better. Various matured techniques/technology in the TWSN are the potential candidates to be adopted or adapted into the UWSN.

For UWSN, the effect of packet length on the collision probability may worth a further investigation to support a new optimization algorithm. Some of the unique application requirements in UWSN (such as data packet encryption/decryption, real-time data packet sampling, etc) may be taken into consideration in the effort to develop a generic or comprehensive framework for data packet size optimization. In view of the very limited number of literatures and publications in underground WSN, the authors are of the opinion that packet size optimization for underground WSN is another vast area opened for further research. This is because some of the concepts and principles, such as the usage of acoustic wave as the transmission media, are found to be similar between the underground WSN and the underwater WSN.

CONCLUSION

The studies and the relevant analysis presented in this chapter can be categorized into two broad categories, namely data packet size optimization in underwater wireless sensor network (UWSN) communications and data packet size optimization in terrestrial wireless sensor network (TWSN) communications. Three optimization methods or approaches for each of the category were discussed. Some of the main performance metrics/parameters used in the methods described include throughput efficiency, BERs, energy efficiency, and protocol efficiency. Table 1 serves as a quick reference for readers who wish to know some important issues regarding data packet size optimization in UWSN communications and TWSN communications respectively.

Various essential equations and their related parameters were given in the text as quick references. Should the reader be interested to know in depth a particular equation and /or parameters, the cited references are readily available from the referenced sources and most of them can be downloaded in full text format from the Internet by visiting the relevant websites. The authors of this chapter would like to emphasized that some of the approaches highlighted in this chapter could be adopted in future research to formulate new optimization framework or algorithm for data packet size in WSN data transmission.

REFERENCES

Basagni, S., Petrioli, C., Petroccia, R., & Stojanovic, M. (2009, January). *Choosing the packet size in multi-hop underwater networks*. Northern University, Boston, MA.

Bertsekas, D. P., & Gallager, R. (1987). *Data networks*. Englewood Cliffs, NJ: Prentice-Hall.

Ci, S., Sharif, H., & Nuli, K. (2005, February). Study of an adaptive frame size predictor to enhance energy conservation in wireless sensor networks. *IEEE Journal on Selected Areas in Communications*, *23*(2).

Dong, W., Liu, X., Chen, C., He, Y., Chen, G., Liu, Y., & Bu, J. (2010, March). Dynamic packet length control in wireless sensor networks. In *IEEE INFOCOM2010*. San Diego: DPLC. doi:10.1109/INFCOM.2010.5462063

Gnawali, O., Fonseca, R., Jamieson, K., Moss, D., & Levis, P. (2009). *Collection tree protocol*. In ACM SenSys '09, Berkeley, CA, USA.

Hackmann, G., Chipara, O., & Lu, C. (2008, November). Robust topology control for indoor wireless sensor networks. In *6th ACM Conference on Embedded Networked Sensor Systems, ACM SenSys '08*, Raleigh, NC, USA.

Haykin, S. (2002). *Adaptive filter theory* (4th ed.). New York, NY: Prentice-Hall.

IEEE. 802.15.4. (2006). *Wireless medium access control (MAC) and physical layer (PHY) specifications for low rate wireless personal area networks* (LR-WPANS). Standard, IEEE.

Joe, I. (2005, May). Optimal packet length with energy efficiency for wireless sensor networks. In *IEEE International Symposium on Circuits and Systems: Vol. 3* (pp. 2955-2957).

Julier, S. J., & Uhlmann, J. K. (1997, July). New extension of Kalman filter to nonlinear systems. In I. Kadar (Ed.), *AeroSense '97, Vol. 3068., Signal Processing, Sensor Fusion, and Target Recognition VI, Multisensor Fusion, Tracking, And Resource Management II,* (pp. 182-193). Orlando, FL, USA.

Levis, P., Lee, N., Welsh, M., & Culler, D. (2003). Accurate and scalable simulation of entire TinyOS applications. In *ACM SenSys, '03*. Los Angeles, CA, USA: TOSSIM.

Modiano, E. (1994, October). *Data link protocols for LDR MILSTAR communications*. Communications Division Internal Memorandum. MIT Lincoln Laboratory, Lexington, MA.

Modiano, E. (1999, July). An adaptive algorithm for optimizing the packet size used in wireless ARQ protocols. *Wireless Networks*, *5*(4), 279–286. doi:10.1023/A:1019111430288

Pollin, S., Ergen, M., Ergen, S., Bougard, C., Van der Perre, L., & Catthoor, F. … Varaiya, P. (2006, November). Performance analysis of slotted carrier sense IEEE 802.15.4 medium access layer. In *49th IEEE Global Telecommunications Conference, GLOBECOM 2006*. San Francisco, CA, USA.

Qiao, D., Choi, S., & Shin, K. G. (2002, October). Goodput analysis and link adaptation for IEEE 802.11a wireless LANs. *IEEE Transactions on Mobile Computing*, *1*(4), 278–292. doi:10.1109/TMC.2002.1175541

Sankarasubramaniam, Y., Akyildiz, I. F., & McLaughlin, S. W. (2003, May). Energy efficiency based packet size optimization in wireless sensor networks. In *1st IEEE Int. Workshop on Sensor Network Protocols and Applications (SNPA'03)*, Anchorage, USA.

Schwartz, M. (1988). *Telecommunication networks*. Addison Wesley.

Shih, E., Cho, S. H., Ickes, N., Min, R., Sinha, A., Wang, A., & Chandrakasan, A. (2001, July). Physical layer driven protocol and algorithm design for energy-efficient wireless sensor networks. In *ACM MobiCom'01* (pp. 272-286). Rome, Italy.

Stojanovic, M. (2005, June). Optimization of a data link protocol for an underwater acoustic channel. In *IEEE Oceans'05*. Brest, France: Conference.

Vuran, M. C., & Akyildiz, I. F. (2008, April). *Cross-layer packet size optimization for wireless terrestrial, underwater, and underground sensor networks*. In IEEE INFOCOM '08 Phoenix, Arizona.

Wu, D., Ci, S., Sharif, H., & Yang, Y. (2007). Packet size optimization for goodput enhancement of multi-rate wireless networks. In *IEEE Wireless Communications and Networking Conference, WCNC 2007* (pp. 3575-3580). Hong Kong.

Zhang, Y., & Shu, F. (2009, April). Packet size optimization for goodput and energy efficiency enhancement in slotted IEEE 802.15.4 networks. In *IEEE Wireless Communications and Networking Conference, WCNC '09* (pp. 1-6). Budapest.

ADDITIONAL READING

Abdellaoui, M. (2009). Generic Algorithm with Varying Block Sizes to Improve the Capacity of a Wireless Communications Networks. *Journal of Computer Science*, *5*(4), 323–329. doi:10.3844/jcssp.2009.323.329

Breed, G. (2003). Bit Error Rate: Fundamental Concepts and Measurement Issues. *High Frequency Electronics. Summit TechnicalMedia, LLC*, 46–48.

Dong, W., Liu, X., Chen, C., He, Y., Chen, G., Liu, Y., & Bu, J. (2010). DPLC: Dynamic Packet Length Control in Wireless Sensor Networks. *In 29th Conference on Computer Communications, INFOCOM '10* (pp. 1-9).

Foote, K. G. (2008). Underwater Acoustic Technology: Review Of Some Recent Developments. In *IEEE OCEANS '08: Vol 2008-Supplement* (pp. 1-6). Quebec City.

Hara, S., Ogino, A., Araki, M., Okada, M., & Morinaga, N. (1996). Throughput Performance of SAW-ARQ Protocol with Adaptive Packet Length in Mobile Packet Data Transmission. *IEEE Transactions on Vehicular Technology*, *45*(Issue 3), 561–569. doi:10.1109/25.533771

Holland, M. (2007). *Optimizing Physical Layer Parameters for Wireless Sensor Networks*. Unpublished Master degree dissertation, University of Rochester, Rochester, New York.

Jiang, Z. (2008). Underwater Acoustic Networks – Issues and Solutions. *International Journal of Intelligent Control and Systems*, *13*(3), 152–161.

Kim, S., Cheon, H., Seo, S., Song, S., & Park, S. (2010, March). A Hexagon Tessellation Approach for the Transmission Energy Efficiency in Underwater Wireless Sensor Networks. *Journal of Information Processing Systems*, *6*(1), 53–66. doi:10.3745/JIPS.2010.6.1.053

Lee, H. C. (2005, December). IP Packet Size to Maximize Throughput of Wireless ATM Link Layer. In *Applied Electromagnetics Asia-Pacific Conference*. Johor Baru, Malaysia.

Lettieri, P., & Srivastava, M. B. (1998). Adaptive Frame Length Control for Improving Wireless Link Throughput, Range, and Energy Efficiency. *In 17th Annual Joint Conference of the IEEE Computer and Communications Societies INFOCOM '98. Vol. 2* (pp. 564-571).

Lin, Y., & Wong, W. S. (2006). Frame Aggregation and Optimal Frame Size Adaptation for IEEE 802.11n WLANs. In *Global Telecommunications Conference, GLOBECOM '06*. San Francisco, CA.

Lin, Y. D., Yeh, J. H., Yang, T. H., Ku, C. Y., Tsao, S. L., & Lai, Y. C. (2009). Efficient Dynamic Frame Aggregation in IEEE 802.11s Mesh Networks. *International Journal of Communication Systems*, *22*(Issue 10), 1319–1338. doi:10.1002/dac.1028

Liu, B., Wen, H., Ren, F., & Lin, C. (2008). Performance Analysis of Reliable Transport Schemes Joint With the Optimal Frequency And Optimal Packet Length In Underwater Sensor Networks. In *World of Wireless, Mobile and Multimedia Networks, WoWMoM 2008 International Symposium* (1-5), Newport Beach, CA.

Luo, H., Guo, Z., Wu, K., Hong, F., & Feng, Y. (2009). Energy Balanced Strategies for Maximizing the Lifetime of Sparsely Deployed Underwater Acoustic Sensor Networks. *Open Access journal (Molecular Diversity Preservation International, Basel, Switzerland)*. *Sensors (Basel, Switzerland)*, *2009*, 6626–6651. doi:10.3390/s90906626

Mahasukhon, P., Sharif, H., Hempel, M., Zhou, T., & Wang, W. (2009). Optimal Packet Size Estimation Using Pseudo Gradient Search Based on 2-Additive Measures. In *International Symposium on Performance Evaluation of Computer & Telecommunication Systems, SPECTS '09, Vol. 41* (pp. 131-136). Istanbul.

Martins, J. A. C., & Alves, J. C. (1990). ARQ Protocols with Adaptive Block Size Perform Better Over a Wide Range of Bit Error Rates. *IEEEE Transactions On Communications*, *38*(6), 737–739. doi:10.1109/26.57462

Morash, J. P. (2008). Implementation of A Wireless Underwater Video Link. Unpublished doctoral dissertation, Department of Electrical Engineering and Computer Science, Massachusetts Institute of Technology.

Proakis, J. G., Sozer, E. M., Rice, J. A., & Stojanovic, M. (2001). Shallow Water Acoustic Networks. *IEEE Communications Magazine*, *39*(Issue: 11), 114–119. doi:10.1109/35.965368

Seo, S., Song, S., Kim, E., & Kim, S. (2008). A New Energy Efficient Data Transmission Method for Underwater Wireless Sensor Networks. In *International Conference on Soft Computing as Transdisciplinary Science and Technology (CSTST '08)*. Paris, France.

Sozer, E. M., Stojanovic, M., & Proakis, J. G. (2000). Underwater Acoustic Networks. *IEEE Journal of Oceanic Engineering, 25*(1), 72–83. doi:10.1109/48.820738

Torres, D., Friedman, J., Schmid, T., & Srivastava, M. B. (2009, November). Software-Defined Underwater Acoustic Networking Platform. *In 4th ACM International Workshop on UnderWater Networks (WUWNet) in conjunction with ACM SenSys 2009*, Berkeley, CA, USA.

Yang, K. T., Lai, W. K., & Shieh, C. S. (2009). A Cross-Layer Approach to Packet Size Adaptation for Improved Utilization in Mobile Wireless Networks. *International Journal of Innovative Computing, Information and Control (ICIC International), Vol. 5, No. 11(B)*, 4335–4346.

Yin, J., Wang, X., & Agrawal, D. P. (2004). Optimal Packet Size in Error-Prone Channel For IEEE 802.11 Distributed Coordination Function. In *IEE Wireless Communications and Networking Conference, WCNC '04. Vol. 3* (pp. 1654-1659). Atlanta, Georgia, USA.

Zhou, G., & Shim, T. (2007). Adaptive Transmission Technique in Underwater Acoustic Wireless Communication. In Kang, L., Liu, Y., & Zeng, S. (Eds.), *ICES 2007, LNCS 4684* (pp. 268–276). Springer-Verlag Berlin Heidelberg. doi:10.1007/978-3-540-74626-3_25

KEY TERMS AND DEFINITIONS

Channel Coding: Is an approach used to increase the wireless communication channel reliability by incorporating a certain form of forward error correction capability in data packet transmissions.

Data Packet Size: Interchangeable with the term “data packet length”. Is a data packet which comprises generally of three portions namely header, payload, and trailer, according to IEEE data packet transmission standard.

Data Packet Size Optimization: A process or an algorithm used to determine the optimal packet size for transmission in UWSN or TWSN. The optimization shall depend on bit-error-rate, throughput efficiency, or energy efficiency.

Energy Channel: Is a notion used to represent energy efficiency metric in energy throughput analysis.

Shallow Water: Refers to tropical waters of depth from 50m to 200m.

Throughput Efficiency: Is the ratio between the effective delivered bit rate of a sink node to the offered or attempted bit rate at a source node. Throughput may be referred as goodput in some analysis.

TWSN: Terrestrial Wireless Sensor Networks are referred to wireless sensor networks that used terrestrial or radio wave for data transmission.

UWSN: Underwater Wireless Sensor Networks are referred to wireless sensor networks deployed in underwater environment where acoustic wave is the means for wireless data transmission.

WSN: Wireless Sensor Network is a network comprises of many sensor nodes in which each node is equipped with a wireless communication device allowing each node to communicate with one another wirelessly. The nodes are designed to communicate either by means of radio frequency (terrestrial waves) or by acoustics waves.

Chapter 15
Reducing Complexity and Achieving Higher Energy Efficiency in Wireless Sensor Network Communications by Using High Altitude Platforms

Zhe Yang
Blekinge Institute of Technology, Sweden

Abbas Mohammed
Blekinge Institute of Technology, Sweden

ABSTRACT

In this chapter, a novel approach is explored to employ high-altitude platforms (HAPs) to remove the relaying burden and/or de-centralize coordination from wireless sensor networks (WSNs). The approach can reduce the complexity and achieve energy efficiency in communications of WSNs, whereby applications require a large-scale deployment of low-power and low-cost sustainable sensors. The authors review and discuss the main constraints and problems of energy consumptions and coordination in WSNs. The use of HAPs in WSNs provides favorable communication links via predominantly line of sight propagation due to their unique position and achieves benefits of reduced complexity and high energy efficiency, which are crucial for WSN operations.

1. INTRODUCTION

Recent advances in low-power and low-cost wireless sensors are revolutionizing the way we interact with the physical world. These sensors are generally equipped with data processing, communication and information collection capabilities (Chong & Kumar, 2003). They can detect the variation of ambient conditions in the environment surrounding the sensors and transform them into electric signals. Sensors, which send collected data via radio link to a sink either directly or through other nodes in a multi-hop fashion, can organize themselves in networks.

DOI: 10.4018/978-1-4666-0101-7.ch015

The research field is driven by the desire to expand current communications capability beyond the reach of conventional networks (Technology Review, 2003). A typical sensor network is shown in Figure 1, where a large number of sensor nodes with data processing and communication capabilities send collected data via radio transmitter, to a sink either directly or through other nodes in a multi-hop fashion. The sink in Figure 1 could be either a fixed or mobile node with the capability of connecting sensor networks to the outer existing communication infrastructure, e.g. Internet, cellular and satellite networks.

Challenges in Wireless Senor Networks

Using wireless sensor networks (WSNs) in environmental monitoring and control applications, such as fire detection, disaster prevention, and control of urban environments, are new trends of applications and have flourished in recent years. In general, these applications, which require a large-scale deployment of sensors over vast geographical areas and deliver a large unpredicted amount of information about the environments, face many challenging issues such as:

- **Energy constraints:** The process of data routing in WSNs can be greatly affected by energy considerations, routing path and radio link. If sensor networks consist of a large number of sensors (in the order of tens of thousands or higher) over a large area, it may not be energy efficient to gather measured data from sensors to sinks using data aggregation. Sensors are highly expected to work in a considerable long period, e.g. months or years, and be disposable, which gives more constraints to energy consumptions (Cheng, et al., 2008).
- **Dynamic networks:** Basically a WSN consists of sensor node, sink and event. If sensors are randomly deployed in remote geographical regions, inaccessible environments and disaster areas, it may be challenging or unfeasible to deploy powerful sinks or impractical to provide communications to sinks sending data externally. Moreover urgent situation in dynamic environment monitoring applications demands unpredicted reports to the sink in an unexpected period.
- **Propagation environment:** Sensor nodes deployed on the ground have a relative low effective height of antenna and a small distance to the radio horizon. Because dense environment of deployment can cause severe attenuations, non line of sight signal transmission is predominant in most directions. Signal power at a distance d away from the transmitter may be estimated as

Figure 1. General communication scenarios of a WSN (redrawn from (Akyildiz, et al., 2002))

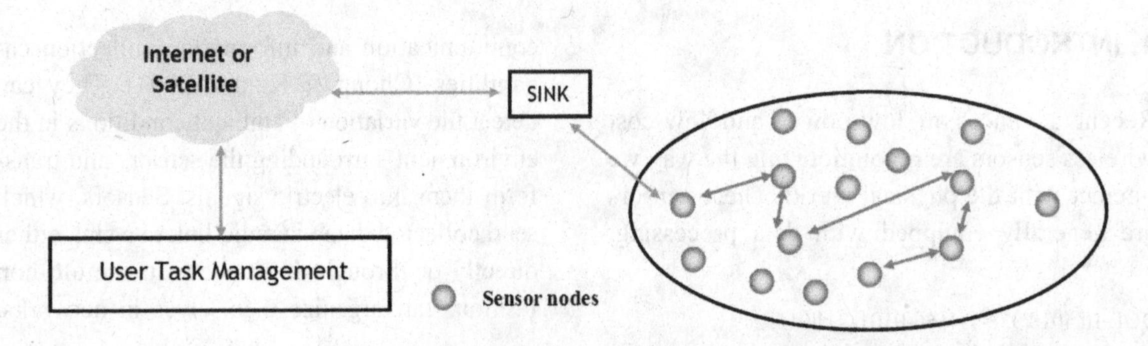

$1/d^n$, where n=2 is for propagation in free space, but n is estimated to be between 2 and 4 for low-lying antenna deployments (Vincent, et al., 2006).

Advantages of Using HAPs in WSNs

High altitude platforms (either aircrafts or airships operating at an altitude of 17 km above the ground) have been proposed as an alternative solution to challenges and constrains existing in WSN and its applications requiring a large-scale deployment and energy aware consumptions (Dovis, et al., 2001; Mohammed, et al., 2008; Mohammed & Yang, 2010). Solutions based on the platform have been recently proposed as a novel approach for the delivery of wireless broadband services to fixed and mobile users. Solutions based on HAPs is competitive in terms of greater capacities and cost-effective deployment when compared to conventional terrestrial and satellite systems. A Swiss company has started to develop three different platforms based on lighter-than-air vehicles or conventional aircraft operating from 3 km to 17 km above the ground to provide different services and applications such as mobile broadband multimedia services, local navigation and positioning, remote sensing, etc (StratXX, 2008).

We consider a novel application of HAPs employed to establish a HAP-WSN system, where HAPs can collect information from sensor nodes as a remote sink above the ground. Advantages of using HAPs are high receiver elevation angle, line of sight (LOS) transmission, large coverage area and mobile deployment etc. for WSN applications. A comparable scenario to HAP-WSN is to use unmanned aerial vehicles (UAVs) to transfer information in distributed WSNs as an energy-efficient solution (Vincent, et al., 2006). The HAP-WSN can be deployed in inaccessible or disaster environments, where sensor nodes are powered by battery. Employing unpiloted, solar-powered platforms in different altitudes can ultimately make the solution more reliable and competitive in the future. Main advantages of employing HAPs in WSNs can be summarized as:

- **Reducing complexity of multi-hop transmission and achieving energy-efficiency:** Multi-hop routing has been mostly investigated in different research topics related to WSN, because radio links in WSNs on the ground are usually constrained by obstructions. HAPs are often considered to be located a few kilometres above the ground, where it can establish a LOS link between the sensor node and the HAP sink. Therefore the platform has a potential of reducing and/or removing transmission burden in WSNs, and would manage transmission based on energy efficient protocols to reduce energy consumption in sensor networks.
- **Cost-efficient and mobile deployment:** HAPs can act as base-stations or relay nodes, which may be effectively regarded as a very tall antenna mast or a very Low-Earth-Orbit (LEO) satellite to provide a large coverage area. Therefore it is believed the cost of HAP is considerably cheaper than that of a satellite. The HAP as a sink in WSNs can be reused, repaired and replaced quickly for WSNs' applications, such as disaster and emergency surveillance where the platform has clear advantages (Yang & Mohammed, 2008). The platform can stay in the sky for periods up to a year or more, which can prolong the life of WSNs.

In this chapter, reliable communication links are analyzed between sensor nodes and HAPs to achieve LOS in most cases based on the height of the platform. Energy efficient design is considered to deploy the HAP-WSN system in inaccessible or disaster environments, where sensor nodes and HAPs can be both powered by battery, which means energy consumption is the key factor in the system design. We propose and investigate approaches by using aerial platforms to reduce complexity of networking structure, improve the energy efficiency of communications and extend the lifetime of WSNs.

2. ENERGY EFFICIENT DESIGN OF USING HAPS IN WSN APPLICATIONS

The importance of wireless sensor networks lies in their ability to continuously monitor physical conditions in remote and inaccessible locations without human interference. One of the limitations of low power sensor devices is the battery life, which is often required to monitor, detect and report on events for a long period of time without an external power supply. It is well understood that communication in a sensor network usually consumes more energy than computation, so it is important to develop more energy efficient solutions for this purpose. Due to a large number sensor nodes deployed in a region and transmitting data simultaneously over the same transmission medium, they inevitably lead to an increased collision probability, which is a common problem and most changeling issue in wireless networks. Using reliable multiple access techniques can overcome this problem and allow the accommodation of multiple users to provide effective communication.

Current research in HAPs has widely adopted two types of cell planning in HAP system. By subdividing the coverage area of the HAP into one or multiple cells, the HAP antenna payload has potential to provide a high gain in each cell planning scenario. In (Thornton, et al., 2003; Yang, et al., 2007), the coverage area has been divided into 121 and 19 cells in order to improve the capacity of HAP system. We propose two scenarios based on the geometric cell planning for the platform.

Scenario I

The platform is employed as a mobile sink, which is able to connect sensor networks to the outer existing communication infrastructure through the platform. The sensor nodes deployed inside the HAP coverage areas can transmit information directly to the platform without relaying information inside WSNs in a multi-hop fashion, therefore a considerable energy reduction associated with relaying data can be reduced. This scenario is shown in Figure 2.

Figure 2. A HAP-WSN system in a single cell configuration

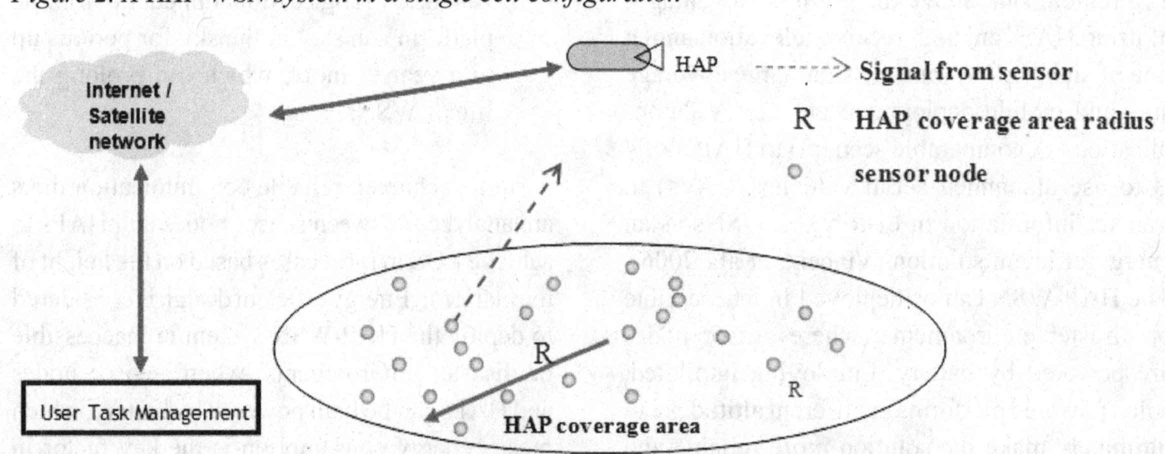

A. Medium Access Protocols

In scenario I, a HAP allows all nodes with a suitable connection to transmit data directly to the HAP, which remove the burden of a multi-hop transmission and energy-consumption on overhead with packets forwarding in the sensor network on the ground. Channel assignment and access control is managed outside the WSN by a HAP, which naturally avoids the energy consumption inside the WSN. Compared with access scheduling by a WSN on the ground, it improves the efficiency by decreasing the number of failure nodes and increasing the scalability of the sensor network. Sensor nodes wouldn't need to operate in a multitask fashion and periodically switch on. Although a longer propagation distance to the HAP may require a higher transmit power, it has to noticed that the radiated power from a sensor node represents a small portion of the overall energy consumption for transmission.

B. Multiple Access Scheme

A multiple scheme based on direct sequence provides a promising solution in scenario I, since it doesn't require a downlink channel from a HAP. Each sensor nodes can transmit data to the HAP by using its own spreading code. The received signal can then be despread by multiplying the same spreading code at the HAP. HAP can distinguish the individual signal if codes assigned to different senor nodes can keep orthogonal to others and the system is synchronized. The time accuracy to achieve a correct synchronization based on the low-cost sensor nodes could be difficult. This scheme is illustrated shown in Figure 3.

Scenario II

In the scenario II, sensor nodes inside the HAP cell can be organized into clusters, where one node with a higher energy could be selected as the cluster head. Senor nodes being members of the cluster collect information and send to the cluster head, which is responsible to transmit data to the HAP. The main aim of the scenario is to reduce the complexity of a multi-hop WSN and maintain a low average energy consumption of most sensor nodes. A high data transmission between cluster heads and the platform can be achieved depending on the requirement for certain WSN applications, such as transmission of a video stream. The key advantages of scenario II is removal of energy burden associated with access scheduling in a WSN on the ground and

Figure 3. Sensors using spread spectrum multiple access to access HAPs

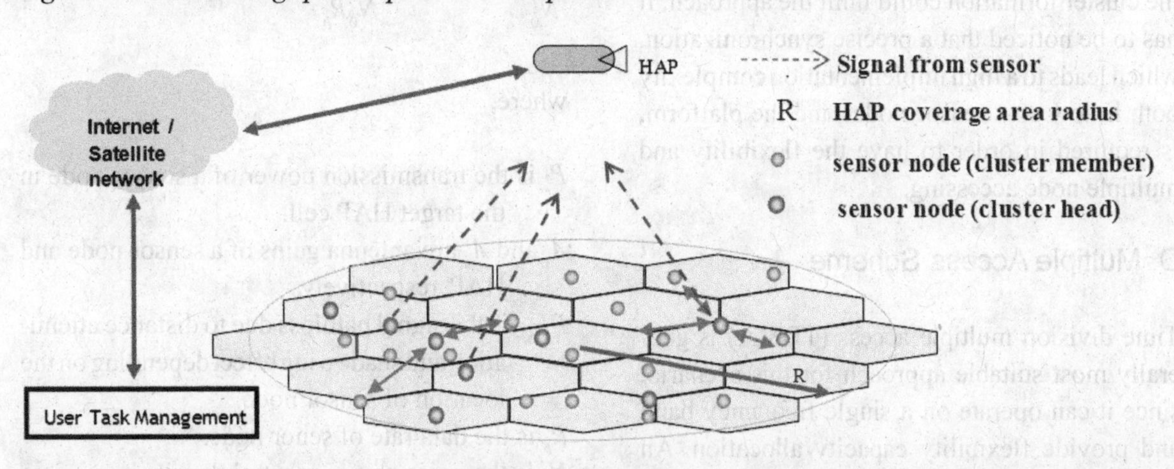

Figure 4. A HAP-WSN system in a multi-cell configuration

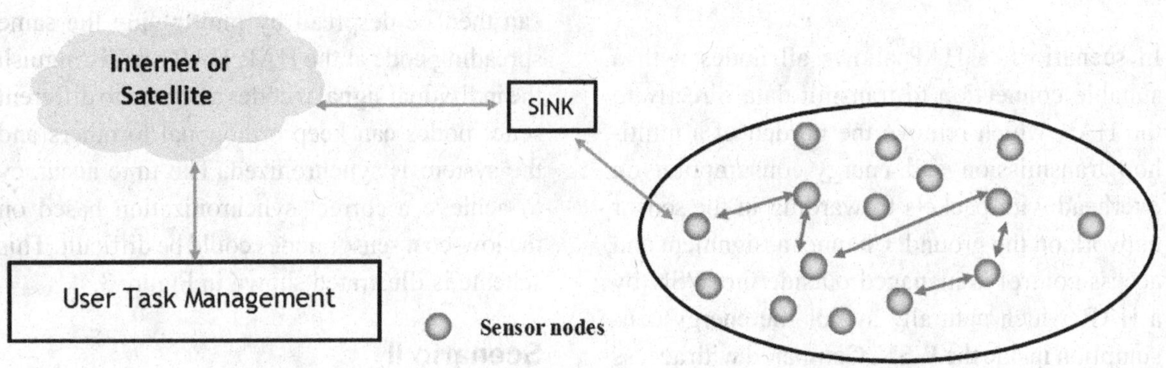

synchronization through the downlink channel. This scenario is shown in Figure4.

C. Medium Access Protocols

In this scenario, channel assignment and synchronization can be partly managed by a HAP by using a schedule-based protocol. For example, a LEACH protocol, where one node with the higher-energy is selected as the cluster head, can be implemented in each HAP cell to organize nodes into a cluster (Demirkol, et al., 2006). Local clusters can be scheduled regionally by a cluster head, which will consume a lot of energy due to being constantly switched on. The schedule burden can be distributed among associated nodes by rotating the role of the cluster head. Overheads generated due to the cluster formation could limit the approach. It has to be noticed that a precise synchronization, which leads to a high implementation complexity both in low cost sensor nodes and the platform, is required in order to have the flexibility and multiple node accessing.

D. Multiple Access Scheme

Time division multiple access (TDMA) is generally most suitable approach for this scenario, since it can operate on a single frequency band and provide flexibility capacity allocation. An orthogonal frequency division multiplexing (OFDM) transmission scheme in conjunction with TDMA is shown in Figure 5. The ultimate goal is to allow multiple sensor nodes to communicate with the platform base on orthogonal frequency-division multiple access (OFDMA), by which nodes can occupy different portions of frequency bands (k+3 time slot at in Figure 5).

3. IMPLEMENTATION ISSUES

Considering a sensor node in the location *(x,y)* on the ground to communicate with the HAP, performance can be evaluated by an energy per bit to noise per spectral density ratio in (1):

$$\frac{E_b}{N_0}(x,y) = \frac{P_S A_S A_H PL_{SH}}{N_0 R_b} \quad (1)$$

where,

P_s is the transmission power of a sensor node in the target HAP cell.

A_s and A_u are antenna gains of a sensor node and HAP, respectively.

PL_{SH} is the signal pathloss due to distance attenuation and shadowing effect depending on the location of sensor node.

R_b is the data rate of senor node.

N_0 is the noise power spectral density.

Figure 5. Sensors using time division or frequency division to access HAPs

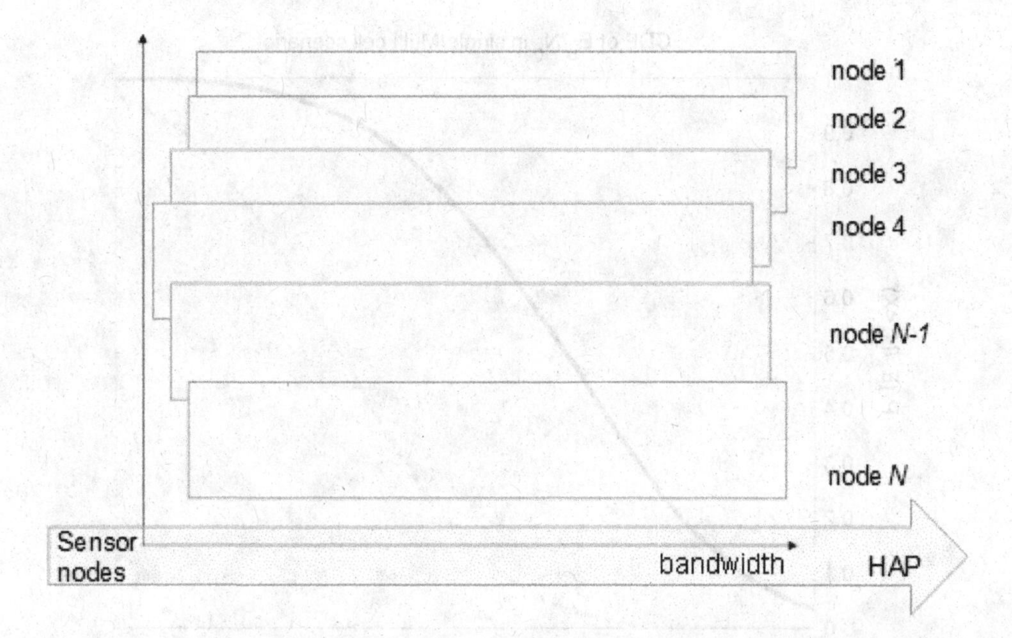

Selected parameters are shown in Table 1. Parameters, e.g. data rate, sensor node transmit power, are referred to product data sheets of the company *Crossbow®* specializing on the sensor network technology (Crossbow, 2008). Parameters of the low speed (R_b=38.4 kbps) and high speed (R_b=250 kbps) senor nodes are adopted for different WSN applications.

The cumulative distribution function (CDF) of E_b/N_0 is evaluated. The result is shown in Figure 6 shows the CDF of E_b/N_0 of the received signal in single cell and multi cell scenario with different transmission rate. According to the product data sheet in (Crossbow, 2008), industrial-scientific-medical (ISM) band at 868 MHz and 2.4 GHz is selected, respectively. BPSK is the most common modulation technique employed by sensor nodes. An uncoded BPSK signal requires an E_b/N_0 value of 10.4 dB for bit error rate (BER) at a 10^{-6}. It can be seen that transmission from sensor node to HAP at 17 km in two scenarios is possible under the coverage area of 30

Table 1. Parameters for low-speed and high-speed sensor nodes to platform communications

Parameters	Values
Data Rate (R_b) Tx Power (P_s) Tx Antenna Gain Rx (A_s)	250 kbps / 38.4 kbps 3 dBm / 5 dBm 1 (0 dBi)
HAP Antenna Boresight (G_H) HAP Height Coverage Radius (R) Cell Radius	7 dBi / 16 dBi 17 km (typical) 30 km (typical) 30 km (scenario I)/8 km (scenario II)
Pathloss Exponent (*n*) Shadowing Standard Deviation Operation Bands Noise Power Spectral Density (N_0)	2 (free space loss) 2 dB (Log-normal) 2.4 GHz /868 MHz 3.98e-21 W/Hz

Figure 6. E_b/N_0 of signal received from sensor nodes to the platform

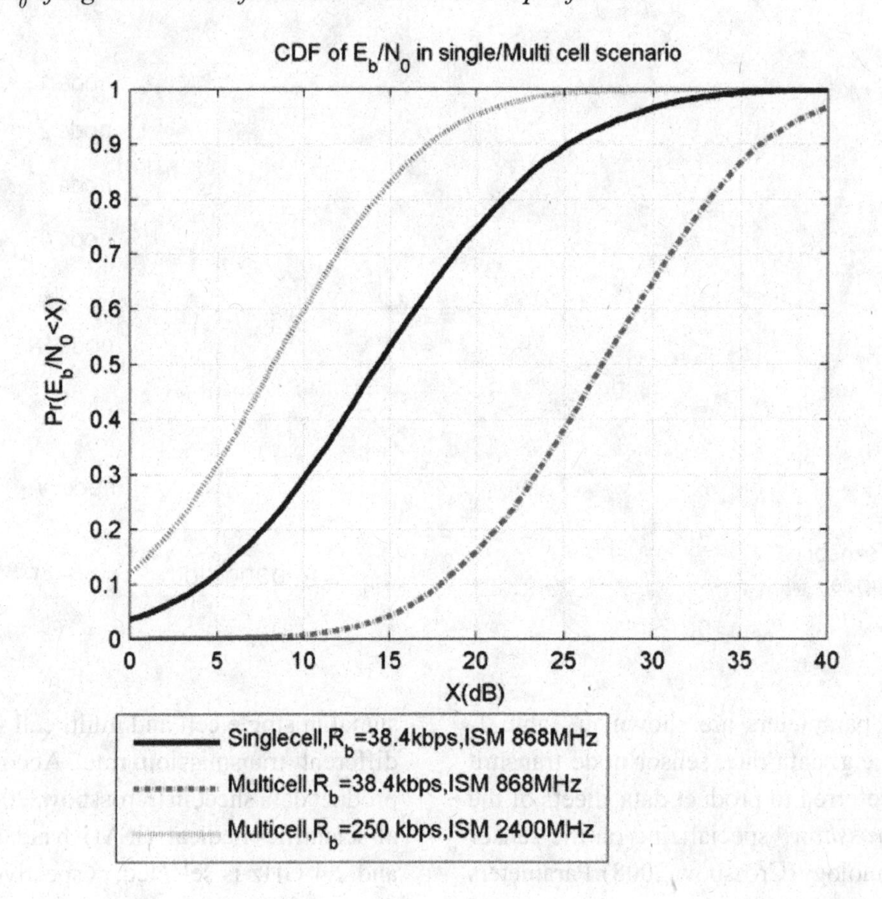

km in radius. The performance of sensors in multi cell scenario is even enhanced compared to the single cell HAP-WSN system with the same transmission rate due to a HAP cellular antenna radiation profile.

Typically, low-power sensor nodes are regularly spaced at 10 meter interval with a corresponding node density at 0.01 per m^2. A single platform with a typical coverage area radius at 30 km, which is the value adopted by most research articles, can provide access to maximum 28,000,000 nodes. Assume an individual node sends a 64-bit packet message every 10 seconds at the transmission rate 38.4 kbps over a transmission range at 500 meters in an environment monitoring application. The time to transmit is 1.67ms. If we implement the HAP-WSN application by using the first scenario allowing all sensor nodes to send data directly to a HAP based on a time-division access fashion, it takes about 13 hours for a platform to receive all messages from a WSN on the ground. In reality, not all sensor nodes need to report an event by sending packets to the sink, which reduces the interval between node reports. Furthermore, radius of the coverage area for a platform and node density would be less for applications in inaccessible areas. It is important to consider these facts for an emergency situation, (e.g. unusual temperature variation) because the packet can be transmitted earlier to reach the platform.

4. FUTURE RESEARCH DIRECTIONS

A detailed study of multiple access scheme based on a specific designed medium access protocol for the platform is promising. Protocols, which have been well studied for WSN, have limitations when being implemented for HAP-based sensor network applications. For example, schedule based protocol LEACH could be implemented in an individual cell of the platform to coordinate with in a cluster. Synchronization can be still difficult for a large number of sensors nodes within the coverage area of a platform in first scenario I or an individual cluster inside a cell of the second scenario. Frequency division multiple access techniques are not particularly suitable for WSN applications base on the platform due to complexity and cost to transceivers. Although synchronization is difficult to achieve both in code division and time division based multiple access schemes, time division based schemes could be interesting due to the operation on a single frequency band and potential flexible capacity assignment.

Research and development in the future need to be done in order to demonstrate the commercial and applicable applications based on the platform associated with WSNs. A promising research area and application would be related to an environment monitoring and surveillance in an incredible disaster scenario, where disposable sensor nodes are deployed in random fashion like dropping from a helicopter or aircraft. A HAP or UAV can be employed to collect information from sensors.

5. CONCLUSION

In this chapter, we have shown the two scenarios of using HAP as a sink in the WSN and examined the performance in terms of E_b/N_0. A HAP-WSN can be an energy-efficient solution by reducing complexity of the WSN and effectively decreasing or removing the multi-hop transmission in order to prolong the lifetime of sensor node by. The platform is potential to provide the extended coverage area of WSN due to the unique height of the HAP. A LOS free space pathloss with a log-normal shadowing model has been employed to examine the radio link between HAP and sensor nodes. It can be seen that employing HAP as a sink is a promising alternative to senor nodes deployed in a large scale area.

REFERENCES

Akyildiz, I. F., Su, W., Sankarasubramaniam, Y., & Cayirci, E. (2002). A survey on sensor network. *IEEE Communications Magazine.*

Cheng, Z., Perillo, M., & Heinzelman, W. B. (2008). General network lifetime and cost models for evaluating sensor network deployment strategies. *IEEE Transactions on Mobile Computing, 7*(4), 484–497. doi:10.1109/TMC.2007.70784

Chong, C.-Y., & Kumar, S. P. (2003). Sensor networks: Evolution, opportunities and challenges. *Proceedings of the IEEE, 91*(8), 1247–1256. doi:10.1109/JPROC.2003.814918

Crossbow (2008). *Product reference guide.* Retrieved from http://www.xbow.com/

Demirkol, I., Ersoy, C., & Alagöz, F. (2006). MAC protocols for wireless sensor networks: A survey. *IEEE Communications Magazine.*

Dovis, F., Fantini, R., Mondin, M., & Savi, P. (2001). 4G communications based on high altitude stratospheric platforms: Channel modeling and performance evaluation. *Global Telecommunications Conference, GLOBECOM '01 IEEE, 1*, (pp. 557-561).

Mohammed, A., Arnon, S., Grace, D., Mondin, M., & Miura, R. (2008). Advanced communications techniques and applications for high-altitude platforms. *Editorial for a Special Issue, EURASIP Journal on Wireless Communications and Networking.*

Mohammed, A., & Yang, Z. (2010). Next generation broadband services from high altitude platforms. In Adibi, S., Mobasher, A., & Tofighbakhsh, M. (Eds.), *Fourth-generation wireless networks: Applications and innovations* (pp. 249–267). Hershey, PA: Information Science Reference. doi:10.4018/978-1-61520-674-2.ch012

Strat, X. X. (2008). *StratXX near space technology*. Retrieved from http://www.stratxx.com/products/

TechnologyReview. (2003, February). 10 emerging technologies that will change the world. *Technology Review, 106,* 33-49.

Thornton, J., Grace, D., Capstick, M. H., & Tozer, T. C. (2003). Optimizing an array of antennas for cellular coverage from a high altitude platform. *IEEE Transactions on Wireless Communications*, *2*(3), 484–492. doi:10.1109/TWC.2003.811052

Vincent, P. J., Tummala, M., & McEachen, J. (2006, April 2006). *An energy-efficient approach for information transfer from distributed wireless sensor systems.* Paper presented at the IEEE/SMC International Conference on System of System Engineering, Los Angeles, CA, USA.

Yang, Z., & Mohammed, A. (2008). *On the cost-effective wireless broadband service delivery from high altitude platforms with an economical business model design*. IEEE 68th Vehicular Technology Conference, 2008. VTC 2008-Fall.

Yang, Z., Mohammed, A., Hult, T., & Grace, D. (2007). *Assessment of coexistence performance for WiMAX broadband in high altitude platform cellular system and multiple-operator terrestrial deployments.* 4th IEEE International Symposium on Wireless Communication Systems (ISWCS'07).

Chapter 16
Wireless Sensor Network:
Quality of Service (QoS) Issues and Challenges

Noor Zaman
King Faisal University, Saudi Arabia

Azween Abdullah
University Technology PETRONAS, Malaysia

Khalid Ragab
King Faisal University, Saudi Arabia

ABSTRACT

Wireless Sensor Networks (WSNs) are becoming common in use, with a vast diversity of applications. Due to its resource constraints, it is hard to maintain Quality of Service (QoS) with WSNs. Though they contain a vast variety of applications, at the same time they are also required to provide different levels of QoS, for various types of applications. A number of different issues and challenges still persist ahead to maintain the QoS of WSN, especially in critical applications where the accuracy of timely, guaranteed data transfer is required, such as chemical, defense, and healthcare. Hence, QoS is required to ensure the best use of sensor nodes at any time. Researchers are trying to focus on QoS issues and challenges to get maximum benefit from their applications. With this chapter, the authors focus on operational and architectural challenges of handling QoS, requirements of QoS in WSNs, and they discuss a selected survey of QoS aware routing techniques by comparing them in WSNs. Finally, the authors highlight a few open issues and future directions of research for providing QoS in WSNs.

INTRODUCTION

WSN is a wireless network consisting of small nodes with sensing, computation, and wireless communications capabilities (Al-Karaki & Kamal, 2004), operating in an unattended environment, with limited computational and sensing capabilities. Normally nodes routs data back to the base station BS (Akyildiz, Su, Sankarasubramaniam, & Cayirci, 2002a). Data transmission is usually a multi- hop, from node to node toward the base station or gateway. Sensor nodes are equipped with small, often irreplaceable batteries with limited power and computation capacity. Impor-

DOI: 10.4018/978-1-4666-0101-7.ch016

tant concern is the network lifetime and QoS as nodes run out of power, congestion may caused, the connectivity decreases and the network can finally be partitioned and become dysfunctional (Akyildiz, Su, Sankarasubramaniam, & Cayirci, 2002b),(Bhardwaj & Chandrakasan, 2002),(Ettus, 1998), which directly involved to reduce the QoS of network. The minimum energy routing problem linked to QoS has been addressed in (Rodoplu & Meng, 1999),(Estrin, Govindan, Heidemann, & Kumar, 1999), (Hac, 2003), (Wood & Stankovic, 2010),(Sadek, Su, & Liu, 2007), (Sadek, Su, & Liu, 2010). The service is bound to the data and to the application QoS vs. QoI (Quality of Information), QoSu (Jeong, Sharafkind, & Du, 2009) (Quality of Surveillance) and data delivery can be continuous, event-driven, query-driven, or hybrid. An important concern is the QoS of network. QoS model can be defined as under, what the users have required from the network. User demands for guaranteed data transfer timely, guaranteed bandwidth in case of image and video data, data accuracy timely in case of any critical applications such as defense or health application. Resulting sensor nodes' network must satisfy requirements of users as shown in figure 1 with QoS simple model.

RELATED WORK

Quality of Service (QoS) always has very key role for all types of applications in networks, including conventional, wireless ad hoc and wireless sensor network. QoS routing is performed usually through resource reservation in a connection oriented communication in order to meet the QoS requirement for each individual connection. While couple of different mechanisms have been proposed for routing QoS constrained image and video type of data in wire based network (Lee et al, 1995)(Wang & Crowcraft, 1996)(Ma & Steenkiste, 1997)(Zhang, et al. 1993)(Crowley, et al., 1998), they cannot be directly applied to wireless sensor network,

Figure 1. QoS model of WSN (Bhuyan, et al., 2009)

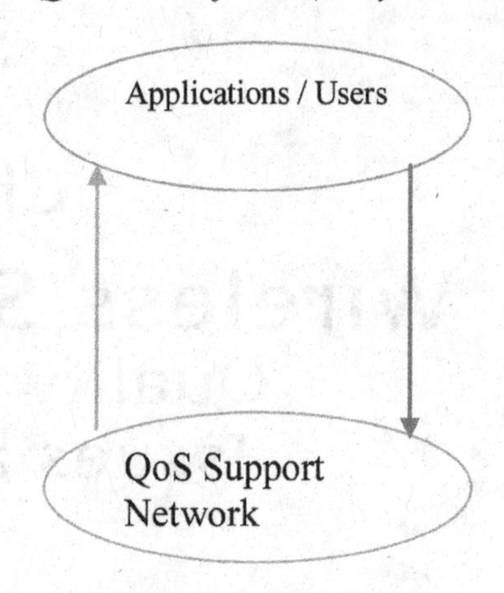

because of its different architecture, structure and resource constraints. Therefore several new protocols have been proposed for QoS routing in wireless networks taking the dynamic nature of the network into account (Querin & Orda, 1997) (Chen & Nahrstedt, 1999)(Srivakumar, et al., 1998)(Lin, 2000)(Zhu & Corson, 2002). Some of the proposed protocols consider the imprecise state information while determining the routes (Querin & Orda, 1997)(Chen & Nahrstedt, 1999).

QOS REQUIREMENT IN WSN

Requirements of QoS in wireless sensor networks is different from wired networks. e.g. traditional end-to-end QoS parameters may not be sufficient to describe them with WSN. As a result, new parameters are used to measure the QoS performance in WSN (Jaballah & Tabbane, 2009). The existing researches related to the QoS in WSN can be classified in three categories (Chen & Varshney, 2004): traditional end-to-end QoS, reliability assurance, and application-specific QoS. QoS always have very important role in all types of network, including conventional, wireless ad hoc and wireless sensor network. QoS routing is performed usually through resource reservation in a connection oriented communication in order to meet the QoS requirement for each individual connection. While specially in wireless sensor network, many QoS based routing protocols have been proposed but they normally make their primary metrics to energy consumption, and they can be grouped on the basis of the problem they solve, like:

Prioritization

Differentiate services on the basis of the definition of classes of traffic (Q-MAC (Liu, Elhanany, & Qi, 2005), SAR (Zhu, Papavassiliou, & Yang, 2006)

Timeliness

Guarantee delivery within a given time (MMSPEED (Felemban, Lee, & Ekici, 2006), SPEED (He, Stankovic, Lu, & Abdelzaher, 2003), DEED (Kim, Abdelzaher, & Kwon, 2005), Data Relaying in Hierarchical WSNs (Benkoczi, Hassanein, Akl, & Tai, 2005)

Reliability

Support probability of delivery (MMSPEED (Jeong, Sharafkandi, & Du, 2006)], REINFORM (Deb, Bhatnagar, & Nath, 2003)

In Network Processing

Improve the performance of the network by processing data along the path from the source to the destination (Q-DAP & LADCA (Zhu, Papavassiliou, & Yang, 2006))

Scheduling

Coordinate sensors in accessing channel or in sensing the environment (CoCo (Li, Shenoy, & Ramamritham, 2004), MAC Coding (Felemban, Lee, & Ekici, 2006), Scheduling with Quality of Surveillance (Jeong, Sharafkandi, & Du, 2006), EAD (Boukerche, Cheng, & Linus, 2005), QoS Reliability of Hierarchical Clustered WSNs (Xing & Shrestha, 2006)).

Node Relocation

Change node position in order to increase efficiency (Sink Repositioning (Akkaya, Younis, & Bangad, 2005), SAFER (Zhao & Tong, 2003)

Generic Metric Minimization

Improve the performance of the network with respect to some cost function (Energy-Aware QoS Routing (Akkaya & Younis, 2003), Dynamic Routing (Ouferhat & Mellouck, 2006), DAPR (Perillo & Heinzelman, 2004))

As sensor network have some specific applications like Military and health, where very high precision is required otherwise the complete application becomes useless and all the above work is not fully fulfilling its requirement. WSN networks Characteristics such as loose network state information, dynamically varying network topology unrestricted mobility of hosts, unrestricted mobility of hosts, limited availability of bandwidth, and battery power make QoS very demanding.

WSN QOS ISSUES AND CHALLENGES

WSN is a resource constrained network, in which energy efficiency is a main issue. All operations of sensor network depend on its low level battery. Routing is the process of finding a path for transferring data from a source node to the destination. Nodes broadcast their messages more than once to find out active live node around them to establish the connection with it. This is the most energy consuming task among all with sensor networks, researcher gave more attention to save its energy rather to take QoS as a primary goal. As sensor network has a different kind of network other than conventional networks (Zhu & Corson, 2002), (Akkaya, Younis, & Bangad, 2005). issues and challenges which are effecting directly or indirectly to QoS of WSNs, are coverage, data gathering, network mobility, energy consumption, topology or node deployment, reliability, scalability, transmission media, connectivity, data aggregation, security, operation environment, production cost, fault tolerance and congestion (Felemban, Lee, & Ekici, 2006). Quality of Service (QoS) has very high importance in any kind of network; QoS in WSN can be characterized by reliability, timeliness, robustness, availability, and security, among others (Estrin, Govindan, Heidemann, & Kumar, 1999). Some QoS parameters may be used to measure the degree of satisfaction of these services, such as throughput, delay, jitter, and packet loss rate. There are many other QoS parameters worth mentioning, but these four are the most fundamental (Hac, 2003), (Wood & Stankovic, 2010), (Sadek, Su, & Liu, 2007), (Sadek, Su, & Liu, 2010). Although a lot of work has been done in the last decade with WSN field and most of the things are clear now, but the researchers are mainly paying more attention to its energy efficiency and secondly about efficient routing protocols rather than QoS issue significantly. However a number of different QoS challenges are still exist, but few of them are worth mentioning here, including WSN open issues, which effects sensor networks quality of service directly or indirectly such as follow.

1. **Jitter:** Jitter can be defined as the measure of variation over time of the packet latency across a network. If a network with constant latency then it means it has no jitter, to maintain jitter is a main issue with sensor networks due to special characteristics.
2. **Delay:** Delay is the main cause within sensor networks to lower the QoS. As sensor network are deployed in un attending environment without any proper topology and its topology change with time after few time of operation. Routing and data gathering is little hard and cause of delay with sensor networks.
3. **Throughput:** Throughput can be defined as the average successful rate of data packet transfer which is directly linked with QoS of the sensor network, in sensor networks routing paths differ from conventional networks and those paths are also not persists for a long time like conventional networks. Therefore it is really hard to manage sensor networks throughput from start to end.
4. **Packet loss:** Packet loss can be defined as a packet which is unable to reach to the destination due to any reason, here in sensor networks packet loss is high due to data redundancy. Sensor nodes are normally collecting the data and then transferring its data to their cluster heads, usually all nodes within the same clusters are collecting the same data type or information at same time. Hence packet loss ratio is high comparable with conventional networks.
5. **Resource Constraints:** Sensor network is normally consist on a very large number of nodes distributed randomly in environment. Sensor nodes are very small in size with limited computation and communicational capabilities, with small often irreplaceable batteries. So researchers are focusing to power consumption rather than to its QoS.
6. **Platform Heterogeneity:** Wireless sensor nodes are platform heterogenic in nature and that causes lower the QoS.
7. **Dynamic Network Topology:** Network topology has key role for defining routing paths and its persistence later on. WSN has special characteristics in distribution and deployment, so WSN is not following any specific topology type of conventional networks, which effects the data collection performance, routing and QoS of NETWORK.

8. **Mixed Traffic:** Mixed traffic become another reason for the lower of QoS, as sensor nodes are randomly distributed in un attending environment without any central administration, so mixed and un desire data also collected by the sensor nodes which uses the resources of the network meaningless.
9. **Resources self management:** WSN are special type of network with limited resources and with un attending without central administration control network. So resource management is on self management basis with sensor network. Which again become the cause of lower the QoS and sensor networks are not intelligent to maintain their resource efficiently.
10. **Data Redundancy:** Data redundancy is common with sensor network, usually nodes sense the data resulting of any physical event occurrence and then transfer it to the sink. Nodes are not intelligent, they are collecting the same data within the same cluster and then transfer that data to the sink. Usually all nodes collected data is the same which causes high redundancy.
11. **Mobility:** Mobility in sensor network is highly required for few applications where different connected components can communicate on move. Sensor nodes don't have sufficient capability to support the mobility in better way, which causes the lower of QoS in sensor networks.
12. **Multiple Sink:** Data is gathered in sensor networks, after collecting the data by the nodes and then transfer that data to multiple sinks or cluster heads. As described earlier that nodes are normally not efficient and collecting redundant data before transferring to the cluster heads. In the same fashion multiple cluster heads or sinks are also forwarding same data to the base station in most of the cases, which causes lower in QoS of network. Besides of above factors few more also effects QoS of network such as Service Oriented Architecture, QoS- Aware Communication Protocols, QoS-Aware Power Management, Scalability and Packet criticality.
13. **QoS based Routing:** A number of QoS based routing has been already proposed, but most of them tends their primary priority to energy efficiency, while they put QoS on second priority. We show comparison of few selected QoS based routing protocols, as shown in Table 1.

QOS OPEN RESEARCH ISSUES

To maintain QoS with traditional conventional network is litter easier and we know, that QoS-enabled conventional wired networks make it sure: That applications/users have their QoS requirements almost satisfied. While at the same time also ensuring an efficient resource usage such as, efficient bandwidth utilization and the most important QoS requirements satisfied during network overload, by efficient resource usage in mean of efficient bandwidth utilization, along with a minimal usage of energy. Hence QoS control mechanism should include in WSN QoS support, which can helps to eliminate unnecessary energy consumption during data delivery. Furthermore, during network overload, the traffic should still have its QoS requirements satisfied in the presence of different types of network mobility, which may cause due to node or wireless link failure. We have mentioned the main technical challenges in our QoS issues and challenges section. Based on these challenges, we identified open research issues with QoS support in WSNs, such as:

1. **Simpler QoS models:** Interserv and Diffserv models are might not be applicable within WSNs due to the complexity of those models. Simple QoS models may required to identify the architecture for QoS support in WSNs.

Table 1. Comparison of QoS based aware routing protocols

Routing Protocol	Mobility	Energy Aware	Data Aggregation	QoS	Multipath	Query Based	Position Awareness
SAR	No	Yes	Yes	Yes	No	Yes	No
Minimum Cost Forwarding Protocol	No	No	No	Yes	No	No	No
An Energy Aware Routing Protocol	No	Yes	No	Yes	No	No	No
SPEED	No	No	No	Yes	Yes	Yes	No
MMSEED	No	No	No	Yes	Yes	Yes	No
ReInForM	No	No	No	Yes	Yes	No	No
Mobicast	Yes	Yes	No	Yes	No	Yes	Yes
DAST	No	Yes	Yes	Yes	No	No	No
Position Responsive Routing Protocol (PRRP)	No	Yes	Yes	Yes	Yes	No	Yes

As Cross layer models are more suitable instead of traditional layered design.

2. **Services:** What kind of non-end-to-end services can WSNs provide and are those services are even better than the traditional one networks.
3. **QoS-aware data dissemination protocols:** It is very highly required to analyze how proposed routing protocols support QoS-constrained traffic while minimizing energy consumption. Is there any mechanism to support the identified and then support the priority data request.
4. **Trade-offs:** Data redundancy is very common in WSNs and it can be avoided to enhance the reliability. Normally data aggregation is applied with collected data, but it is recommended to apply data fusion, it can reduce data redundancy in order to save energy, but it also introduces much delay into the network. What is an optimum trade-off among them? This optimum trade-off may be achieved analytically or by network simulations.
5. **Traditional end-to-end energy-aware QoS support:** End to end support is normally not fully available with sensor networks, it might applicable with few applications. It is important to investigate end to end QoS support with WSN and to know its limits.
6. **Adaptive QoS assurance algorithms:** It is desirable to maintain QoS throughout the network life instead of having a gradual decay of quality as time progresses.
7. **Service differentiation:** Service differentiation, should it be based on traffic types, data delivery models, sensor types, application types, or the content of packets? If we consider the memory and processing capability limitations, then we cannot afford to maintain too many flow states in a node.
8. **QoS control mechanism:** QoS control mechanism should be efficient as sensors may send excessive data (not desired) sometimes and become reason of waste of resources. In some cases it might send inadequate data, so that the quality of the application cannot be met. So it is required to come up with some novel QoS control mechanism based algorithms.
9. **The integration of QoS support:** QoS support in WSNs may be very different from that in traditional or conventional networks. However, the final data collection

from the sensor nodes can be done through a traditional network such as the Internet. Research is required for the integration of traditional and sensor networks for reducing the difference between them and increasing the QoS of network.

10. **QoS support via a middleware layer:** If QoS requirements from an application are not feasible in the network, the middleware may negotiate a new quality of service with both the application and the network. Such a middleware layer, which may be used to translate and control QoS between the applications and the networks, is of great interest.

FUTURE PERSPECTIVES

As wireless sensor network (WSN) becomes most attractive area for the researchers now day, due to the wide range of its application areas. Due to distributed nature of these networks and their deployment in remote areas, the application of sensor networks become more meaningful for even very hard and valuable applications installations. The importance and the successful use of sensor applications is directly linked with the QoS of network. It is desired to consider more about QoS while designing of routing algorithms, as routing is more resource hungry task in WSN, besides of energy efficiency.

CONCLUSION

Wireless Sensor Networks (WSN's) are becoming common in use, with vast diversity of its applications. Due to its resource constraints type of network, it is hard to maintain Quality of Service (QoS) of network with WSN. As (WSN's) has vast variety of applications, at the same time it is also required to provide different levels of QoS, for various types of applications. A number of different issues and challenges are still persists ahead to maintain the QoS of WSN, specially in critical applications. Where the accuracy, timely guaranteed data transfer is highly required such as chemical, defense, and healthcare. Hence, QoS is required to ensure the best use of sensor nodes at any time. With this chapter we try to focus on QoS requirement in WSN and then also highlighted few important QoS issues and challenges along with its open research issues. A comparative study of few QoS based routing protocols also discussed. Finally we describes few open issue of QoS in WSN and conclude that by addressing those issue, further QoS research will be initiated.

REFERENCES

Akkaya, K., & Younis, M. (2003). An energy-aware QoS routing protocol for wireless sensor networks. In *Proceedings of the 23rd International Conference on Distributed Computing Systems*, 2003.

Akkaya, K., Younis, M. F., & Bangad, M. (2005). Sink repositioning for enhanced performance in wireless sensor networks. *Computer Networks*, *49*(4).

Akyildiz, I. F., Su, W., Sankarasubramaniam, Y., & Cayirci, E. (2002). A survey on sensor network. *IEEE Communications Magazine*, *40*(8), 102–116. doi:10.1109/MCOM.2002.1024422

Al-Karaki, J. N., & Kamal, A. E. (2004). Routing techniques in wireless sensor network: A survey. *IEEE Wireless Communication*, December 2004.

Benkoczi, R., Hassanein, H., Akl, S., & Tai, S. (2005). QoS for data relaying in hierarchical wireless sensor networks. In *Proceedings of the 1st ACM International Workshop on Quality of Service & Security in Wireless and Mobile Networks*, 2005.

Bhardwaj, M., & Chandrakasan, A. (2002). Bounding the lifetime of sensor networks via optimal role, assignments. *Proceedings of the INFOCOM, 02*, 1587–1596.

Bhuyan, B. (2009). Quality of service (QoS) provisions in wireless sensor networks and related challenges. *Wireless Sensor Network, 201*(2), 861–868.

Boukerche, A., Cheng, X., & Linus, J. (2005). A performance evaluation of a novel energy-aware datacentric routing algorithm in wireless sensor networks. *Wireless Networks, 11*(5). doi:10.1007/s11276-005-3517-6

Chen, D., & Varshney, P. K. (2004). *QoS support in wireless sensor networks: A survey*. International Conference on Wireless Networks 2004.

Chen, S., & Nahrstedt, K. (1999). Distributed quality-of-service routing in ad-hoc networks. *IEEE Journal on Selected Areas in Communications, 17*(8).

Crowley, E., et al. (1998). *A framework for QoS based routing in the Internet*. Internet-draft, draft-ietf-qosrframework-06.txt, Aug. 1998.

Deb, B., Bhatnagar, S., & Nath, B. (2003). ReInForM: Reliable information forwarding using multiple paths in sensor networks. In *Proceedings of the 28th Annual IEEE International Conference on Local Computer Networks,* 2003.

Estrin, D., Govindan, R., Heidemann, J., & Kumar, S. (1999). Next century challenges: Scalable coordination in sensor networks. *Proceedings of 5th Annual IEEE/ACM International Conference on Mobile Computing and Networking*, (pp. 263-270).

Ettus, M. (1998). System capacity, latency, and power consumption in multihop-routed SS-CDMA wireless networks. *Proceedings of RAWCON, 98*, 55–58.

Felemban, E., Lee, C. G., & Ekici, E. (2006). MMSPEED: Multipath multi-speed protocol for QoS guarantee of reliability and timeliness in wireless sensor networks. *IEEE Transactions on Mobile Computing, 5*(6). doi:10.1109/TMC.2006.79

Hac, A. (2003). *Wireless sensor network design*. Wiley. doi:10.1002/0470867388

He, T., Stankovic, J. A., Lu, C., & Abdelzaher, T. (2003). Speed: A stateless protocol for real-time communication in sensor networks. In *Proceedings of the 23rd International Conference on Distributed Computing Systems.*

Jaballah, W. B., & Tabbane, N. (2009). Multi path multi speed contention window adapter. *International Journal of Computer Science and Network Security, 9*(2).

Jeong, J., Sharafkandi, S., & Du, D. H. C. (2006). Energy-aware scheduling with quality of surveillance guarantee in wireless sensor networks. In *Proceedings of the 2006 Workshop on Dependability Issues in Wireless Ad Hoc Networks and Sensor Networks*.

Kim, H. S., Abdelzaher, T. S., & Kwon, W. H. (2005). Dynamic delay-constrained minimum-energy dissemination in wireless sensor networks. *Transactions on Embedded Computing Systems, 4*(3).

Lee, W. C. (1995). *Routing subject to quality of service constraints integrated communication networks. IEEE Network*. July/August.

Li, H., Shenoy, P., & Ramamritham, K. (2004). Scheduling communication in real-time sensor applications. In *Proceedings of the 10th IEEE Real-Time and Embedded Technology and Applications Symposium*, 2004.

Lin, C. R. (2000). On demand QoS routing in multihop mobile networks. *IEICE Transactions on Communications*, July.

Liu, Y., Elhanany, I., & Qi, H. (2005). *An energy-efficient QoS-aware media access control protocol for wireless sensor networks*. In IEEE International Conference on Mobile AdHoc and Sensor Systems Conference, 2005.

Ma, Q., & Steenkiste, P. (1997). Quality-of-service routing with performance guarantees. *Proceedings of the 4th IFIP Workshop on Quality of Service*, May 1997.

Ouferhat, N., & Mellouck, A. (2006). QoS dynamic routing for wireless sensor networks. In *Proceedings of the 2nd ACM International Workshop on Quality of Service & Security for Wireless and Mobile Networks*, 2006.

Perillo, M., & Heinzelman, W. (2004). *DAPR: A protocol for wireless sensor networks utilizing an application-based routing cost.*

Querin, R., & Orda, A. (1997). QoS-based routing in networks with inaccurate information: Theory and algorithms. In *Proceedings of IEEE INFOCOM'97*, Japan, (pp. 75-83).

Rentel, C. H., & Kunz, T. (2005). Mac coding for QoS guarantees in multi-hop mobile wireless networks. In *Proceedings of the 1st ACM International Workshop on Quality of Service & Security in Wireless and Mobile Networks*, 2005.

Rodoplu, V., & Meng, T. H. (1999). Minimum energy mobile wireless networks. *IEEE Journal on Selected Areas in Communications*, *17*(8), 1333–1344. doi:10.1109/49.779917

Sadek, A. K., Su, W., & Liu, K. J. R. (2007). Multi-node cooperative communications in wireless networks. *IEEE Transactions on Signal Processing*, *55*(1), 341–355. doi:10.1109/TSP.2006.885773

Sivakumar, R., et al. (1998). *Core extraction distributed ad hoc routing (CEDAR) specification*. IETF Internet draft draft-ietf-manet-cedar-spec-00.txt, 1998.

Wang, Z., & Crowcraft, J. (1996). QoS-based routing for supporting resource reservation. *IEEE Journal on Selected Area of Communications*, September.

Wood, A. D., & Stankovic, J. A. (2010). Security of distributed, ubiquitous, and embedded computing platforms. In Voeller, J. G. (Ed.), *Wiley handbook of science and technology for homeland security*. Hoboken, NJ: John Wiley & Sons. doi:10.1002/9780470087923.hhs449

Xing, L., & Shrestha, A. (2006). *QoS reliability of hierarchical clustered wireless sensor networks*. In 25th IEEE International Performance, Computing, and Communications Conference, 2006.

Zhang, L., et al. (1993). RSVP: A new resource reservation protocol. *IEEE Network*, September.

Zhao, Q., & Tong, L. (2003). *QoS specific medium access control for wireless sensor networks with fading.*

Zhu, C., & Corson, M. S. (2002). QoS routing for mobile ad hoc networks. *Proceedings - IEEE INFOCOM*, 2002.

Zhu, J., Papavassiliou, S., & Yang, J. (2006). Adaptive localized QOS-constrained data aggregation and processing in distributed sensor networks. *IEEE Transactions on Parallel and Distributed Systems*, *17*(9).

Chapter 17
Low Complexity Processor Designs for Energy-Efficient Security and Error Correction in Wireless Sensor Networks

J. H. Kong
The University of Nottingham- Malaysia Campus, Malaysia

J. J. Ong
The University of Nottingham- Malaysia Campus, Malaysia

L.-M. Ang
The University of Nottingham- Malaysia Campus, Malaysia

K. P. Seng
The University of Nottingham- Malaysia Campus, Malaysia

ABSTRACT

This chapter presents low complexity processor designs for energy-efficient security and error correction for implementation on wireless sensor networks (WSN). WSN nodes have limited resources in terms of hardware, memory, and battery life span. Small area hardware designs for encryption and error-correction modules are the most preferred approach to meet the stringent design area requirement. This chapter describes Minimal Instruction Set Computer (MISC) processor designs with a compact architecture and simple hardware components. The MISC is able to make use of a small area of the FPGA and provides a processor platform for security and error correction operations. In this chapter, two example applications, which are the Advance Encryption Standard (AES) and Reed Solomon (RS) algorithms, were implemented onto MISC. The MISC hardware architecture for AES and RS were designed and verified using the Handel-C hardware description language and implemented on a Xilinx Spartan-3 FPGA.

DOI: 10.4018/978-1-4666-0101-7.ch017

INTRODUCTION

Secure data encryption is usually applied for data integrity and protection. Encrypted data transmission requires error correction codes to reliably recover the data during decryption. To implement security and error correction features into an independent system, a processor core has to be pre-established in order to serve as its own instruction execution mechanism. With this base mechanism, functions and programs can be run at this platform with hardware re-configurability features. The original motivation for the One Instruction Set Computer (OISC) processor, sometimes called the URISC (Ultimate Reduced Instruction Set Computer) was meant for educational ideologies of a much simpler version of computer organization to explain the sophisticated Computer Architectures (Gilreath & Laplante, 2003). Concepts of the hardwired or micro-programmed control are to consider the execution sequence of the instruction sets. However, each individual instruction sets of a complex instruction architecture lack of functionality and independency. By breaking down the complex instruction sets, a simplified model of the existing computer organization can be developed. The proposed MISC architecture serves as a new approach, to demonstrate simple arithmetic functions and conditional 'jump' capabilities for the universality of a Computing Device (Gilreath & Laplante, 2003).

The objective to design tiny WSN sensor nodes has always been a challenge. With the nature of the WSN having resource constraint environment and environment adaptation requires on-field restructuring and reprogramming in software and hardware, design approaches often go towards low-area, low-cost and low complexity system designs. Visual data processing and compression modules usually occupy most of the area of the on-board FPGA microcontroller. This leads to a design for low complexity and low area system. In this chapter, the MISC is used to provide data security and error correction, fulfilling both criteria: low-area and low-complexity designs. The Advance Encryption Standard (AES) and Reed Solomon Error Control System (RS) are implemented using the MISC processor to provided security and error correction features to the WSN.

BACKGROUND

Ultimate Reduced Instruction Set Computer (URISC)

The URISC which was first proposed by (Mavaddat & Parhami, 1988) is meant for educational purpose. It has been an inspiration and insight to the CISC (Complex Instruction Set Computer) and RISC (Reduced Instruction Set Computer). This simplified model of computer architecture is flexible with only a single instruction incorporated can be further expanded and implemented on hardware easily. The URISC uses only one instruction called the SBN instruction (Subtract and Branch If Negative). By using only the SBN instruction, the URISC is able to perform data addition and subtraction. Logical operations can be performed to execute data movement from one location to another. The URISC consists of an Adder circuit as its sole ALU. Detailed operation of the URISC can be found in (Mavaddat & Parhami, 1988). Figure 1 shows the schematic of the URISC architecture.

The 'Subtract and Branch if Negative' (SBN) processor was first proposed by Van der Poel (Gilreath & Laplante, 2003). With this primitive SBN instruction, the URISC is built from its basic processor. The basic operations of URISC are moving operands to and from the memory, with addresses corresponding to the registers. The arithmetic computation can be performed and the results are stored in the 2nd operand's memory location. Similarly, to execute URISC instructions, the Core subtracts the 1st operand from the 2nd operand, storing the results in the 2nd operand's memory location. If the subtraction results a

Figure 1. The URISC architecture

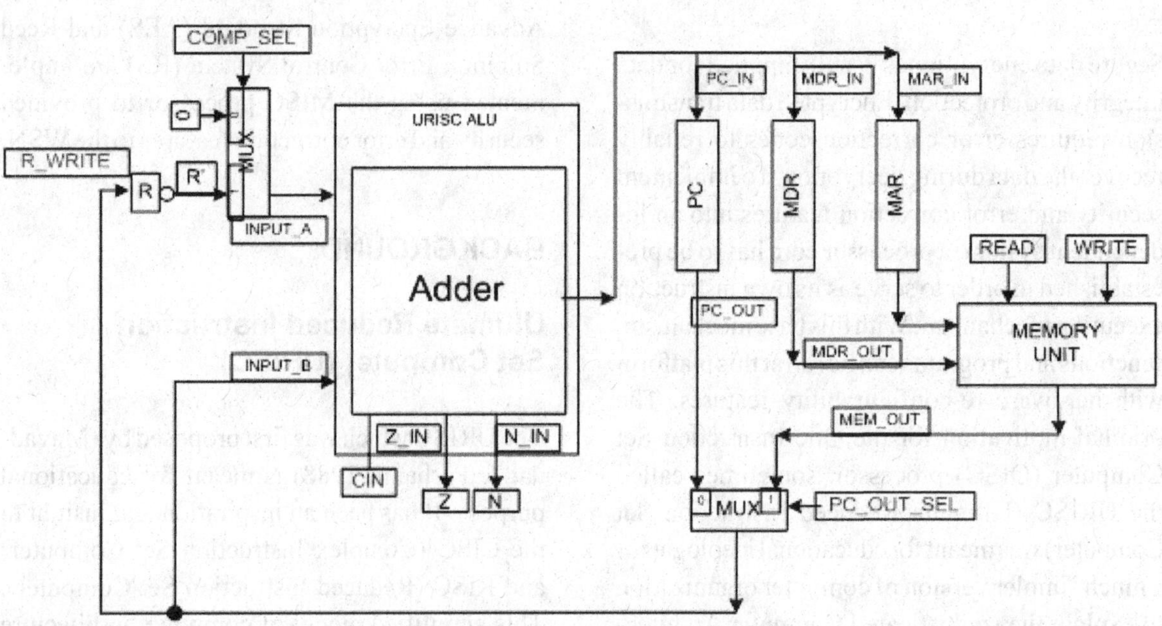

negative value, it will 'jump' to the target address, else, it proceeds to execute the next instruction in the following sequence (Gilreath & Laplante, 2003). Figure 2 shows the pseudo-code format of the SBN instruction written in programming.

System Overview of the Wireless Sensor Network

The Wireless Sensor Network (WSN) consists of many wireless nodes and a base-station. The sensor nodes will monitor and collect the information and data from sensors (depending on what kinds of sensors that it is connected to) and send it to the base-station or the sink. The base-station will receive the information from the nodes and response to the gathered data. A typical wireless visual sensor node consists of a visual sensor with image processing and compression modules. On the other hand, the node might not be able to provide any security or error protection on the data transmitted across the network due to the hardware and memory constraints. Therefore, we propose a minimalist perspective of error control system and an encryption system to be integrated into the visual sensor nodes of a WSN. In Figure 3, it can be seen that each of these proposed visual sensor nodes consists of a visual sensor (camera),

Figure 2. Pseudo-code format of the SBN instruction

```
SBN (Subtract and Branch If Negative)
Mem_B = Mem_B – Mem_A
If Mem_B < 0 Goto (PC + C)
Else Goto (PC + 1)
```

Figure 3. Overview of the error control system or encryption system in WSN

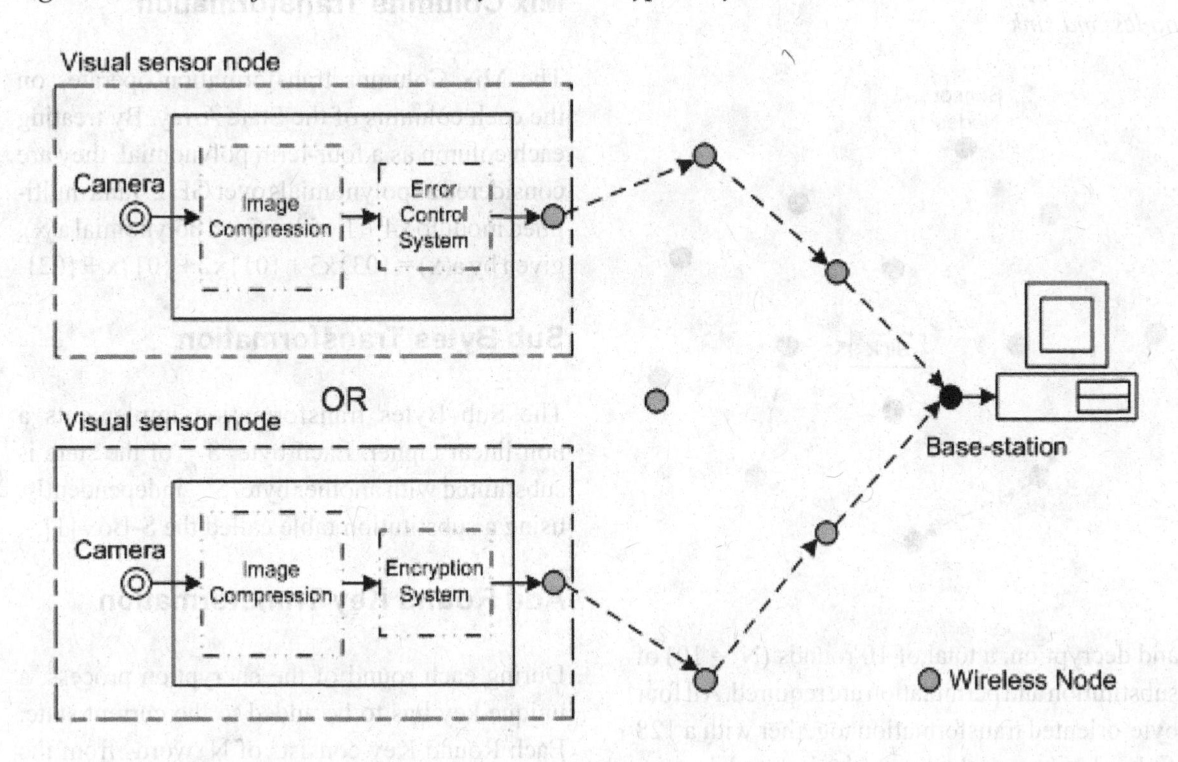

image compression module, error control system or encryption system module, and a wireless node Yick, Mukherjee, & Ghosal, 2008).

A Wireless Sensor Network (WSN) has the data transmitted from one sensor node to other sensor nodes until it reaches the sink or base-station. Typical wireless communication uses the ZigBee® technology to transmit the data across the communication channel. Figure 4 shows the sensor nodes and a sink arrangements with the data transmitted from one sensor nodes to the sink through a few sensor nodes. In normal applications, usually a WSN is deployed in a geographical region where it is meant to collect data through its sensor nodes. The applications for WSNs are varied, typically involving environmental monitoring and tracking. Some dedicated WSNs specifies in object tracking, fire detection, land slide detection and traffic and vehicle monitoring. One of the main drawbacks of WSNs is their low battery life span and limited hardware resources, especially the memory unit.

Advance Encryption Standard (AES)

The AES is a symmetric block cipher that uses data blocks of 128 bits as input and using cipher keys with lengths of 128, 192 or 256 bits. The basic unit for processing in the AES algorithm is a byte and the input, output and cipher key bit sequences are processed as array of bytes. The state array is a two-dimensional array of bytes. It consists of 4 rows of bytes. Each row contains N_b bytes where N_b is the block length (128 bits) divided by 32. The input plain text is stored in the state array at the start of the encryption process and it is executed over a number of rounds determined by the key size. The round function can be broken down into four different byte-oriented transformations: Sub Bytes, Shift Rows, Mix Columns and Add Round Key.

Figure 5 shows the round function for the AES encryption and decryption processes and the transformations in each round (Goodman & Benaissa, 2006). For a complete AES encryption

Figure 4. A typical WSN arrangement with sensor nodes and sink

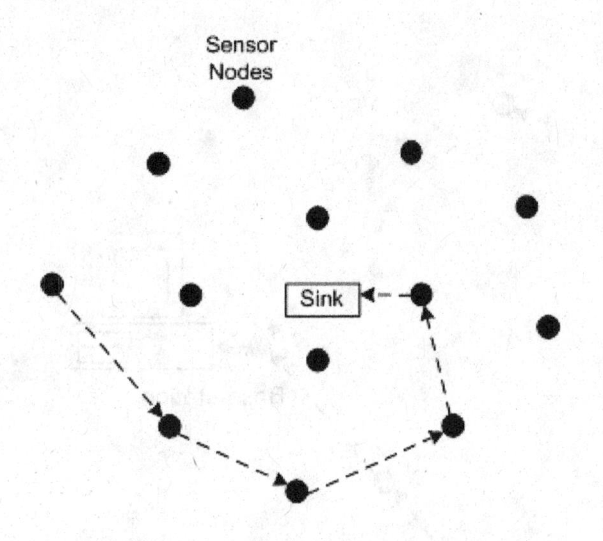

and decryption, a total of 10 rounds ($N_r = 10$) of substitution and permutation are required. All four byte-oriented transformation together with a 128 cipher key are applied onto the input plain text. For each round, another unique cipher key will be generated by the Key Expansion algorithm. In every round, the Round Key will be added to the plain text before going through rounds of S-Box Substitution (Sub Bytes), Mix Columns, and Shift Rows.

Shift Rows Transformation

In the Shift Rows transformation, it involves a cyclic shifting of the bytes in the last three rows with a different number of byte offset. The first row is not shifted. The bytes in the second row (r=1) are cyclically shifted left by one byte and the bytes of the subsequent rows are shifted by 2 and 3 bytes. For $0 \leq r < 4$ and $0 \leq c < N_b$, the Shift Rows Transformation can be expressed as: $S'r,c = Sr,c + \text{shift}(r, N_b) \bmod N_b$, where the shift value shift(r, N_b) depends on the row number r.

Mix Columns Transformation

The Mix Columns transformation operates on the each columns of the State Array. By treating each column as a four-term polynomial, they are considered as polynomials over GF(2^8) and multiplied modulo x4 + 1 with a fixed polynomial a(x), given by a(x) = {03}x3 + {01}x2 + {01}x + {02}.

Sub Bytes Transformation

The Sub Bytes transformation implements a non-linear cipher. Each byte, $S_{r,c}$ of the state is substituted with another byte, $S'_{r,c}$ independently, using a substitution table called the S-Box [1].

Add Round Key Transformation

During each round of the encryption process, a unique key has to be added to the current state. Each Round Key consists of N_b words from the key schedule of the current round, generated by the Key Expansion algorithm. Those N_b words are each added into the columns of the state, such that, [S0,c, S1,c, S2,c, S3,c] = [S'0,c, S'1,c, S'2,c, S'3,c] XOR [w_{round} * N_{b+c}] for $0 \leq c < N_b$ [1]. The first Round Key addition occurs when round = 0. The subsequent rounds are added with Round Keys generated by the Key Expansion algorithm.

Key Expansion Algorithm

The Key Expansion algorithm generates unique cipher keys for each round of the Add Round Key Transformation. The initial round key is filled by an initial cipher key. The key expansion generates a total of Nb (Nr + 1) words and the resulting key schedule consists of a linear array of 4-byte words, denoted [w_i], with i in the range $0 \leq i < N_b(Nr + 1)$.

The RotWord takes in a word of four-bytes as input, performs a permutation, and returns the cyclic shifted word, similarly to a Shift Row transformation. The SubWord function that, which is a Sub Bytes transformation is applied to each

Figure 5. The encryption and decryption process in AES

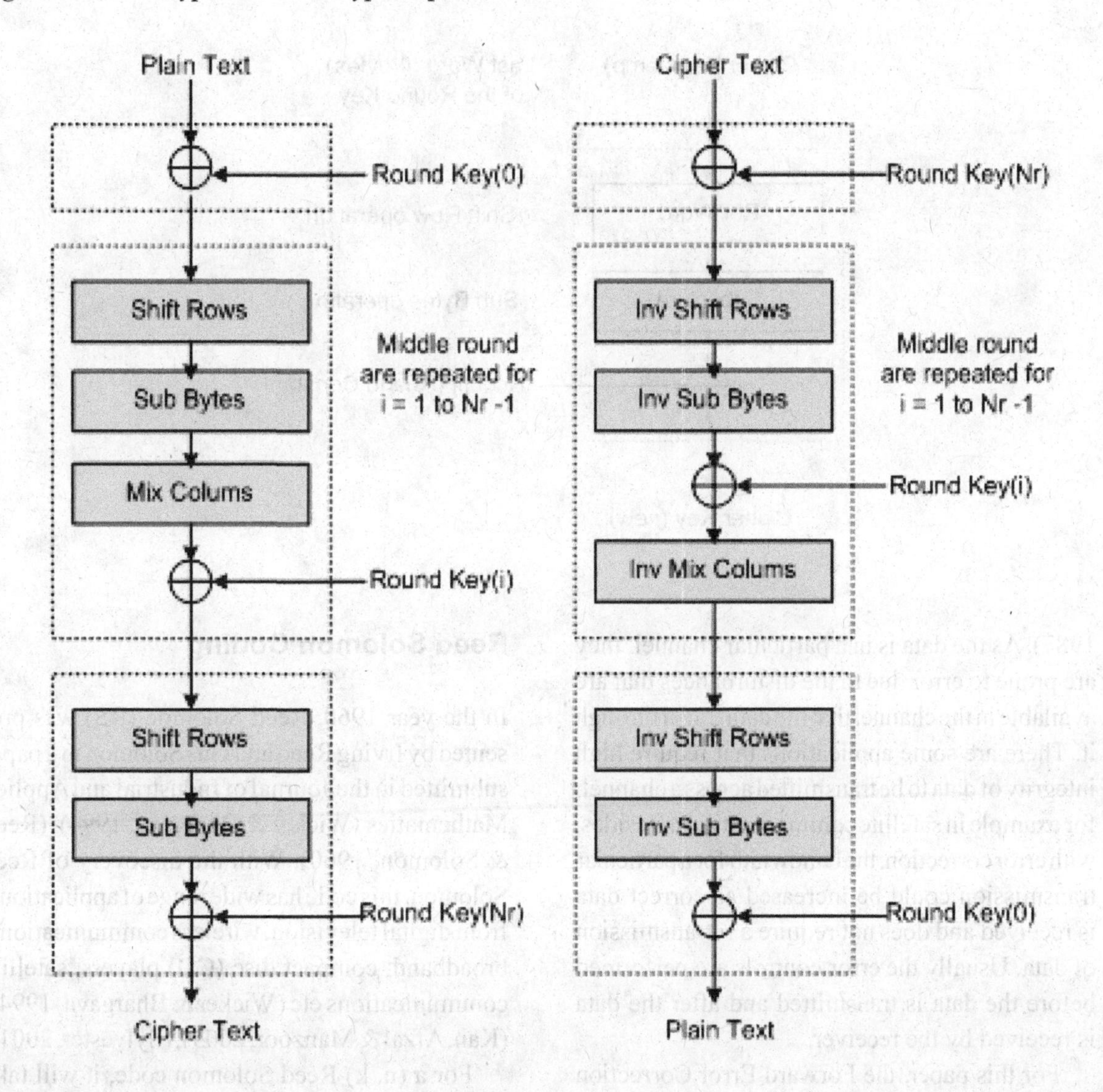

of the bytes to produce a substituted output word. The Round Constant, $R_{con}[i]$, holds the values given by $[x_{i-1}, \{00\}, \{00\}, \{00\}]$, with xi-1 being powers of x (x is denoted as {02}) in the field GF (2^8). Note that 'i' starts at 1, not 0. Figure 6 shows the key generating process of Sub Word, RotWord, and adding the $R_{con}[i]$.

For words in positions that are a multiple of N_k, a transformation is applied to w [i-1] prior to the XOR, followed by an XOR with a round constant, $R_{con}[i]$. This transformation consists of a Shift Row of the bytes in a word (RotWord), followed by the application of a byte substitution onto all four bytes of the word (SubWord).

Reed Solomon Error Control System (RS)

Error control is important in the communication area as to sustain the reliability and integrity of the data transmitted across through a channel (Lin & Costello, 2004; Berlekamp, Peile, & Pope,

Figure 6. The key expansion operation

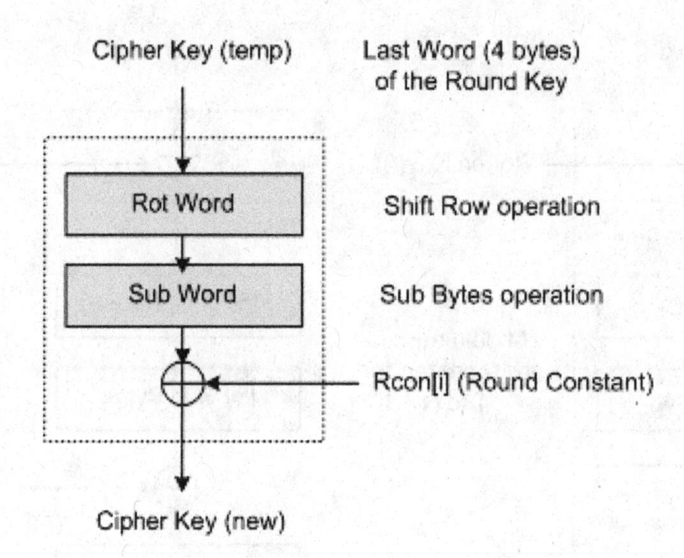

1987). As the data is in a particular channel, they are prone to error due to the disturbances that are available in the channel that the data travel through it. There are some applications that require high integrity of data to be transmitted across a channel, for example in satellite communications. Besides, with error correction, the bandwidth for a particular transmission could be increased as correct data is received and does not require a retransmission of data. Usually the error controls are performed before the data is transmitted and after the data is received by the receiver.

For this paper, the Forward Error Correction (FEC) (Forouzan, 2007) method is implemented. For the forward error correction, the transmitter will encode redundant bits onto the message and then combine it with the message to give a codeword. Then the codeword will be transmitted across the communication channel. After receiving the codeword, the receiver will detect any errors and correct any errors using the redundant bits available in the message. Figure 7 shows the forward error correction technique implemented onto a communicating system.

Reed Solomon Coding

In the year 1960, Reed Solomon (RS) was presented by Irving Reed and Gus Solomon in a paper submitted in the Journal of Industrial and Applied Mathematics (Wicker & Bhargava, 1994), (Reed & Solomon, 1960). With the discovery of Reed Solomon, this code has wide range of applications, from digital television, wireless communications, broadband, compact disc (CD) players, satellite communications etc (Wicker & Bhargava, 1994), (Kan, Afzal & Manzoor, 2003), (Sylvester, 2001).

For a (n, k) Reed Solomon code, it will take in k information symbols and generates $2t = n - k$ of redundant symbols (Sylvester, 2001), (Chio, Sahagun, & Sabido, 2001), (Sha, Yaqub, & Suleman, 2001). The redundant symbols produced by the RS encoder are also known as the parity symbols that will be used in RS decoding. With 2t redundant symbols, the encoder will combine both parity symbols and information symbols to get one block of codeword with n number of symbols. Therefore, the minimum distance for the Reed Solomon code will then be, $d_{min} = 2t$

Figure 7. Block diagram of atypical data transmission system with error control system

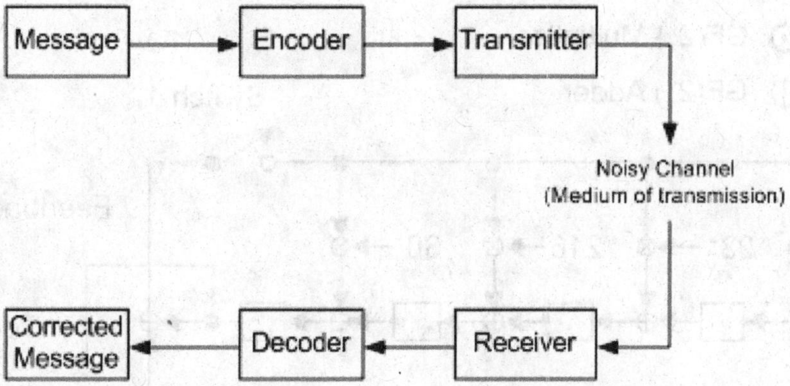

+ 1. Figure 8 shows the breakdown of one block of codeword produced by the (255, 251) Reed Solomon encoder.

Reed Solomon Encoder

By consider α being the primitive element in GF (2^8), then (255,251) Reed Solomon encoder will have the roots of the Galois field GF (2^8) as α, α2… α4. Then the generator polynomial would be shown in equation (12). With coefficients of the generator polynomial known, the Reed Solomon encoder in the Linear Feedback Shift Register (LFSR) circuit can be determined. Figure 9 shows the circuitry of the (255,251) Reed Solomon encoder in LFSR circuit arrangement.

For each block of codeword, 251 message symbols are input to the circuit and at the same time, it outputs as part of the codeword. Then after all the 251 message symbols are input to the register, the 4 redundant symbols will be shifted out from the registers to form one block of complete codeword. Figure 10 shows the generator polynomial used.

Reed Solomon Decoder

As for the RS decoder, the received codeword, r(x) will be stored in a buffer. At the same time, the received codeword is used to generate syndrome values. The generated syndrome values will determine whether there is any error occurs at the codeword. Further on, the Peterson's direct method can be used to decode and correct less than two errors (Wade, 2000). This method is straightforward in determining the error locator polynomial, σ(x), but it will be computationally inefficient as the number of error to correct is increase. Instead, using the iterative techniques – the Berlekamp-Massey frequency domain algorithm or the Euclid's algorithm would be preferred to solve for the error locator polynomial (Lin &

Figure 8. One (255, 251) Reed Solomon codeword

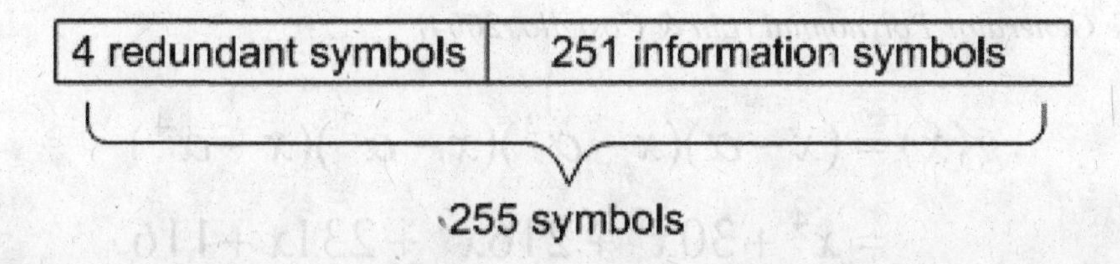

Figure 9. A (255,251) Reed Solomon encoder

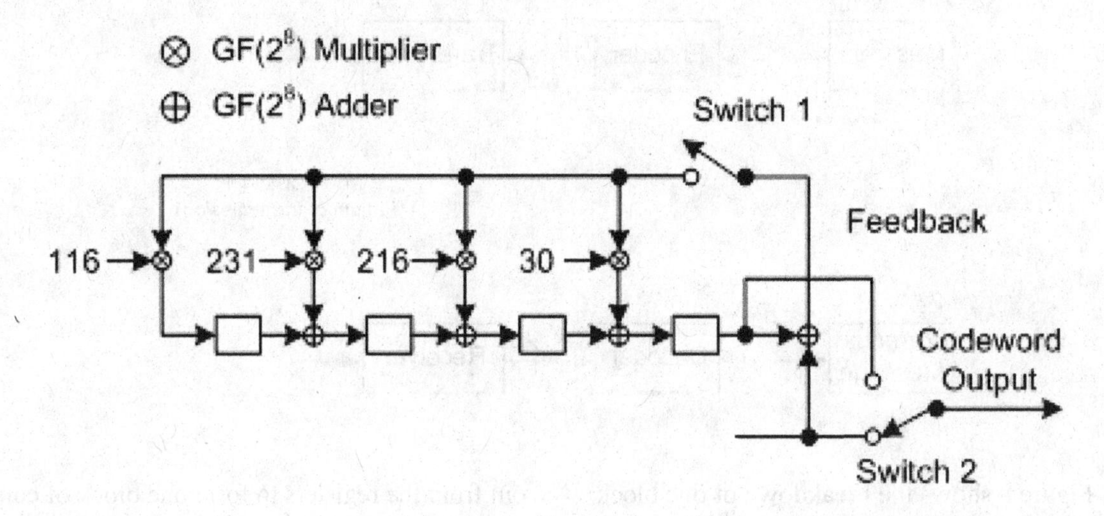

Costello, 2004), (Wicker & Bhargava, 1994), (Chio, Sahagun, & Sabido, 2001), and (Wade, 2000). Since the Wireless Sensor Network requires small amount of hardware implementation, the Peterson's direct method is considered in the implementation for the moment.

Next, apply Chien Search onto the error locator polynomial to determine the error locations. With error locations known, the error magnitude is determined that will be used to correct the errors occurred at the received codeword, r(x). The error magnitudes are determined using the Forney algorithm. Finally, the known error magnitudes will be added to the received codeword to correct the errors that occurred at a particular location in the codeword. Figure 11 shows a general block diagram of the RS decoder decoding a received codeword and produces a correct codeword.

PROPOSED MINIMAL INSTRUCTION SET COMPUTER PROCESSOR

To come up with a small footprint design, both the error control system and encryption system are implemented on the Minimal Instruction Set Computer (MISC) processor. The MISC processor is a low-complexity computer architecture that provides a minimalist ideology of how a simple computer processor can be used to process and execute reduced instructions for encryption and error correction operations. The MISC processor has only a few specific instructions that are sufficient to execute the specific encryption and error correction functions. Figure 12 shows the general MISC processor architecture implemented for the encryption system and error control system.

The MISC architecture can be broken down into several important sections, from its ALU to its memory block. The ALU blocks can be seen

Figure 10. The Generator Polynomial (Lin & Costello, 2004)

$$g(x) = (x-\alpha)(x-\alpha^2)(x-\alpha^3)(x-\alpha^4)$$
$$= x^4 + 30x^3 + 216x^2 + 231x + 116$$

Figure 11. RS decoder block diagram

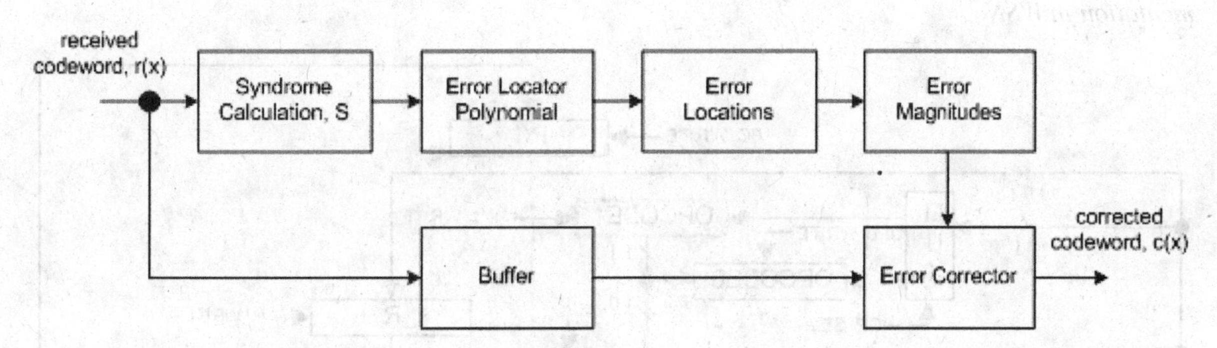

and represented by alphabets A, B, and C in figure 12. Judging on what function the MISC has to perform, the programmer himself can put in any ALU hardware blocks in any of the A, B, and C space. The memory block shown in figure 12 is represented by a square block. The MISC can also be configured to have 2 separate memory block for faster program execution, but in this chapter, we will on focus on the 1 memory architecture, which is also known as the von Neumann architecture.

Similar to the SBN OISC, the MISC branching ability is inherited from the OISC. An N register is used to indicate a resultant negative is the outcome of a single instruction output. In other words, an SBN instruction can be used to branch to any memory item of the block RAM in the FPGA.

To differentiate what instruction the computer is processing, an op code is appended into the instruction sets. Note that the MISC instruction sets are 3-tuple, holding 3 bytes for each of the instructions. In figure 12, there are 2 registers, opcode1 and opcode2, both storing 1 bit of the appended opcodes. With these values, the ALU output multiplexer will be triggered and will only store the intended data to the MDR register.

AES Minimal Instruction Set Computer Processor

The MISC AES Processor is derived from the URISC. This extended version of URISC, together with 4 customized ALU for the AES Encryption and Decryption, has the ability to perform any transformation in the AES algorithm. The MISC ALU consists of 4 basic hardware blocks as the MISC ALU: Adder, XOR, xTime and Sub Bytes. With an external 1-bit input switch, the MISC is able to switch its mode between encrypt mode and decrypt mode.

MISC AES Processor Architecture

The MISC only uses one memory unit to store both program and data for the AES. The MISC inherits the traits of the URISC. With the SBN instruction, the MISC is able to branch to any PC values in the memory unit and execute the instructions in any location of the memory unit. With 7 registers, 5 multiplexers, 1 memory unit and 4 ALU blocks, the MISC is being constructed. Similar to the structure of URISC, the MISC loads in the first memory address and subsequently loads in the first data item. This operation is repeated for the second data item. Once both data are loaded into the MISC, they are sent into the ALU for computation and the outputs will be chosen regarding to the Function Code embedded into the first address loaded. The function code is a 2-bit value, concatenated to the first data address in the memory unit. With the 2-bit MSB value, the MISC is able to determine which instruction is used for the current processor cycle and what

Figure 12. General MISC processor proposed for error control system and encryption system implementation in WSN

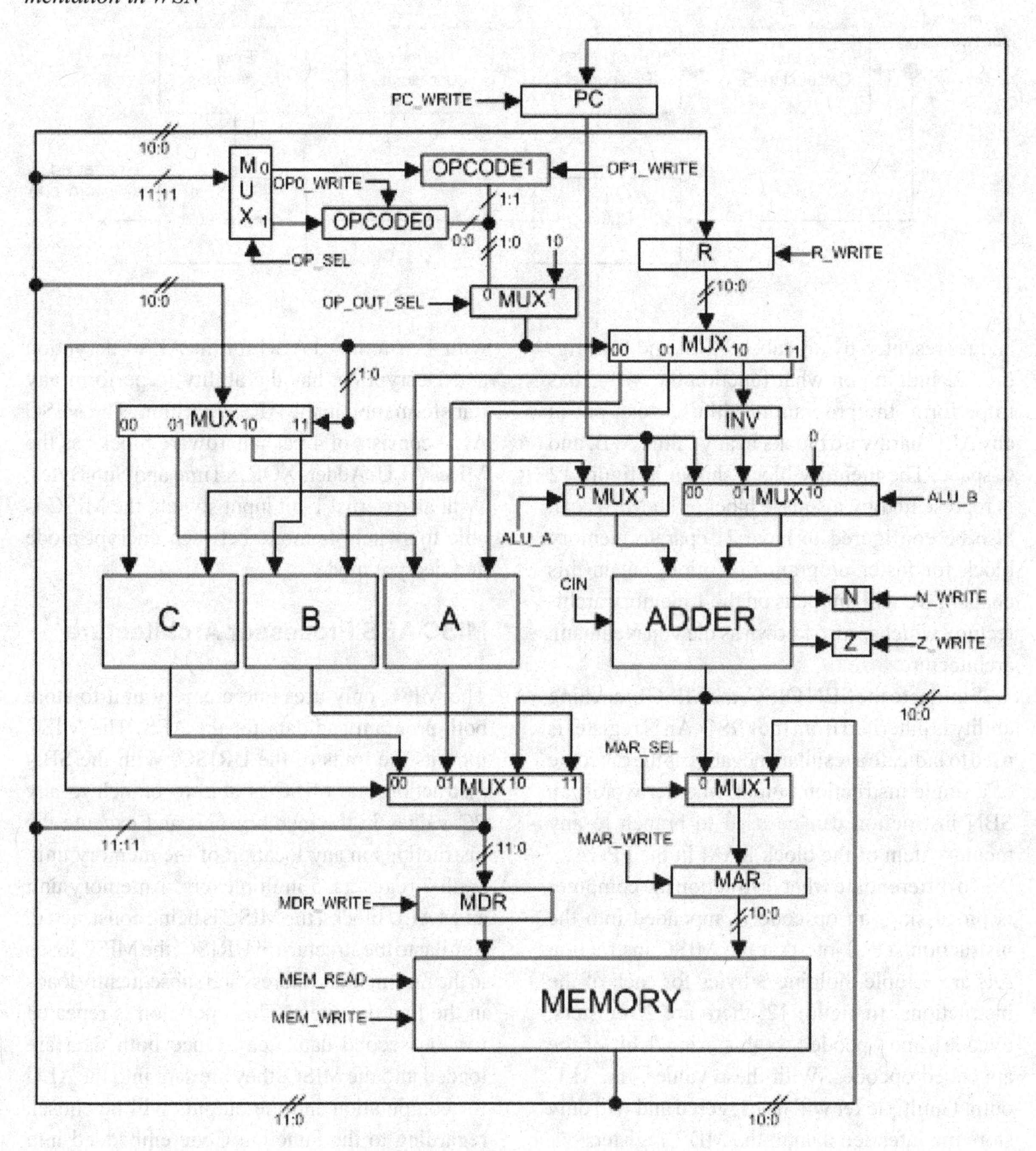

data are stored back to the memory. The MISC Data Path is shown in Figure 13.

The representation of MISC AES datapath architecture is slightly different than the MISC RS is due to the difference of application and hardware blocks in the ALU. Both MISC AES and RS are the same in terms of execution clock cycle, data-path structure and processing mechanism.

Figure 13. General MISC processor proposed for error control system and encryption system implementation in WSN

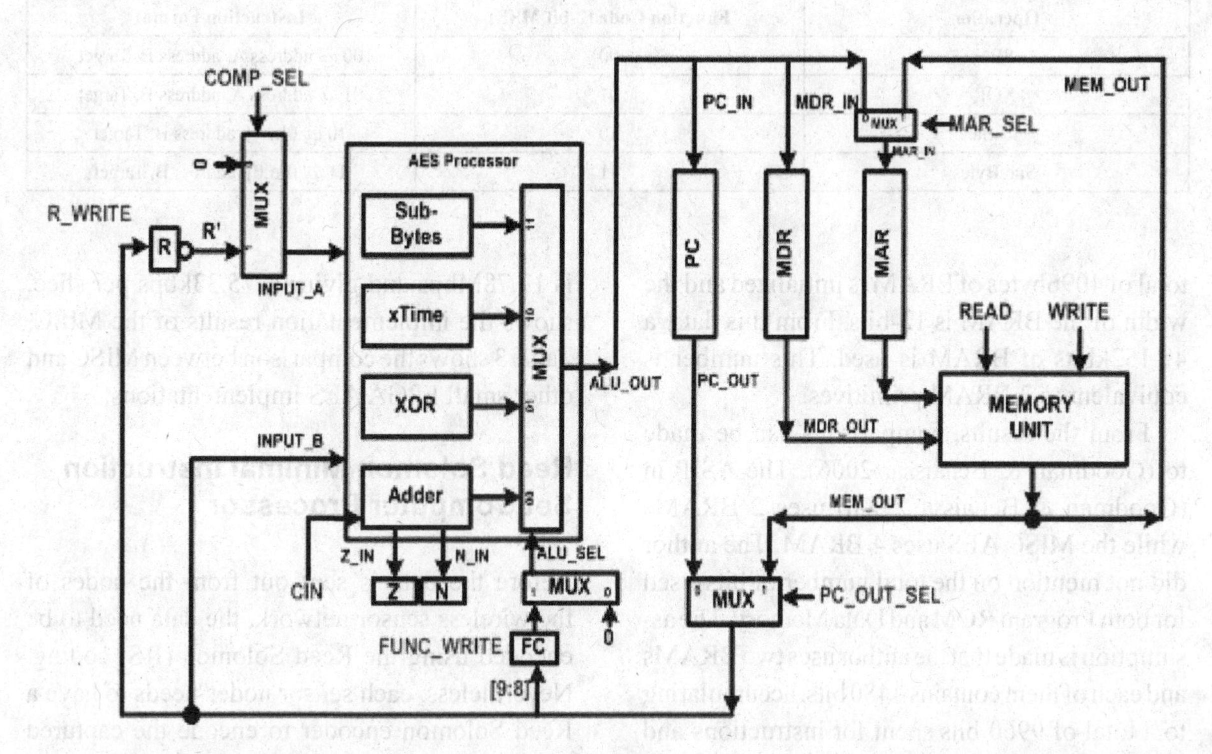

AES byte-oriented transformations are adapted. The MISC AES instructions are specific and comply only with the MISC AES Processor. Each instruction execution will result to byte-oriented processing onto the set of data addressed in the instruction's operand. Each instruction consists of three bytes: address of first operand, address of second operand, target address. Note that a 2-bit function code is used, occupying the 2-bit location of the first data address. The function codes are used to differentiate which instruction does the line of three bytes stands for. With this function code, the MISC is able to perform necessary operations in order to execute the correct command of data processing and storage. From the pseudo-code in Figure 2, the SBN and XOR instruction takes in two input data items, whereas the xTime and Sub Bytes instruction takes in only one data item. Table 1 shows the simplified format for the MISC instructions.

Implementation Results and Discussions

The MISC AES Processor is designed and tested using the DK Design Suite software environment, which provides a Handel-C Hardware Descriptive language to ease the design process. A Celoxica RC10 board which houses the Spartan 3 XCS1500L-4 FPGA is used and on-board LEDs are used to observe the data memory items (see Table 2).

The result simulated showed a successful encryption and decryption process, corresponding to its original plaint text. Only a small amount of hardware is used in the MISC AES implementations. 1% of available flip-flops and LUTs in the hardware are used. From the Xilinx datasheet (DS099 - Xilinx DS099 Spartan-3 FPGA Family data sheet, n.d.), each BRAM primitive contains a total of 18.432kbits. From the MISC design, a

Table 1. Simplified instruction format with their respective operations

Operation	Function Code (2-bit MSB)	Instruction Format
SBN	00	00 @ address A, address B, Target
XOR	01	01 @ address A, address B, Target
xTime	10	10 @ 0[n:0], address B, Target
Sub Bytes	11	11 @ 0[n:0], address B, Target

total of 4096bytes of BRAM is initialized and the width of the BRAM is 12-bits. From this data, a 49.152kbits of BRAM is used. This number is equivalent to 3 BRAM primitives.

From the results, comparisons can be made to (Goodman & Benaissa, 2006), The ASIP in (Goodman & Benaissa, 2006) uses 2 BRAMs while the MISC AES uses 4 BRAM. The author did not mention on the total number of bits-used for both Program ROM and Data Memory. The assumption is made that the author uses two BRAMs and each of them contains 4480 bits, accumulating to a total of 9960 bits spent for instructions and data locations. On the other hand, the architecture in (Goodman & Benaissa, 2006) only uses 122 slices of flip-flops, whereas MISC uses 110 slices of flip-flops, with 12 FF improvement. The throughput for (Goodman & Benaissa, 2006) is 2.18Mbps while our MISC AES has an average throughput of (8-bit / 9 clocks) x 20 MHz, which is 17.78Mbps and giving a 75.33kbps per slice. shows the implementation results of the MISC. Table 3 shows the comparison between MISC and other small FPGA AES implementations.

Reed Solomon Minimal Instruction Set Computer Processor

Before the data is sent out from the nodes of the wireless sensor network, the data need to be encoded using the Reed Solomon (RS) coding. Nevertheless, each sensor nodes needs to have a Reed Solomon encoder to encode the captured environment data. Then it will only transmit the encoded data to the sink through the communication channel. Therefore, a Reed Solomon encoder is required to be implemented onto the sensor nodes.

Table 2. Implementation results

Components	Quantity	Total	Usage
No. of Slice Flip Flops	110	13,312	1%
No. of Occupied Slices	236	13,312	1%
Total of No. of 4 Input LUTs	428	26,624	1%
No. of LUTs used a logic	405	428	95%
No. of LUTs used a route-thru	22	428	5%
No. of LUTs used a Shift Registers	1	428	~0%
No. of Bonded IOBs	28	221	12%
No. of BRAMs	3	32	9%
No. of GCLKs	4	8	50%
No. of DCMs	1	4	25%

MISC RS Processor Architecture

The (255,251) Reed Solomon (RS) Minimal Instruction Set Computer (MISC) processor is developed based on the One Instruction Set Computer architecture. This MISC processor are similar to the developed (255,223) RS MISC processor (Ong, Ang, & Seng, 2010). Figure 14 shows the architecture of the (255,251) RS MISC processor that can be programmed to function as RS encoder and decoder. With the (255,251) RS MISC processor, it is then programmed using only three specific instructions, which are the SBN, GF and XOR. The developed processor can only be programmed for processing 255 symbols in one block of codeword since the Galois field (GF) multiplier will be different for different number of symbols in one block codeword.

The (255,251) RS MISC processor shown in Figure 14 consists of 6 registers, 6 multiplexers, one memory block, one GF(2^8) multiplier block, an 8-bit XOR block, an Adder block and many control inputs. The function of each components insider this RS MISC processor are the same as the (255,223) RS MISC processor (Ong, Ang, & Seng, 2010). The difference is that the adder handles 11-bit word length of data instead of 10-bit word length in the (255,223) RS MISC processor.

Table 3. Comparison results with other small FPGA designs

Design & FPGA (device)	MISC AES ENC/ DEC (von Neumann) Spartan-III (XC3S50-4)	Tim et al (ASIP) Spartan-II (XC2S15-6)	Tim et al (Picoblaze) Spartan-II (XC2S15-6)	Chodowiec & Gaj [5] Spartan-II (XC2S30-6)	Rouroy et al [6] Spartan-III (XC3S50-4)	Zhang [1] Virtex-E (XCV1000e-8)
Encryption Algorithm	AES	AES	AES	AES	AES	AES
Max. Clock Freq. (MHz)	20	72.3	90	60	71	168.4
Data-path Bits	12	8	8	32	32	128
No. of Slices of Flip-Flop utilized	110	122	119	222	163	11022
No. of Block RAMs used	4	2	2	3	3	0
Block RAM size (kbits)	4	4	4	4	18	-
Bits of block RAM used	49152	4480	10666	9600	34176	0
Equiv. slices for Memory	126	140	333	300	1068	0
Total Equiv. Slices (Est.)	236	262	452	522	1231	11022
Max Throughput (Mbps) (excl. key expansion)	-	-	-	166	208	21556
Ave. Throughput (Mbps) Average encryption – decryption including key expansion	17.78	2.18	0.71	69	87	21556
Performance, Typical throughput per slice (kbps/ slice)	75.33	8.3	1.6	132	70	1956
Summary	Smallest	Smallest (previously known)	Software	Best speed VS. area for 32-bit	Fastest 32-bit	Loop Unrolled

Table 4. Hardware utilization of (255,251) RS MISC processor

Component	Quantity	Total	Usage
Flip Flop	291 slice	13312	2%
4 input LUTs • Logic • Route-thru • Dual Port Rams • Shift registers	520 476 42 0 2	26624 - - - -	1% - - - -
Bonded IOB	46	221	20%
Block RAMs	2	32	3%
BUFGMUXs	3	8	37%
DCMs	1	4	25%

Implementation Results and Discussions

The proposed (255,251) RS Minimal Instruction Set Computer processor was implemented onto a Celoxica RC10 board. The processor was programmed into the Xilinx Spartan 3S1500L-4 FPGA which is on the board. As for the RS MISC processor implementation, the processor requires only a small amount of hardware. From Table 4, it can be seen that only 2% of the flip-flop was utilized and 1% of Look up Tables (LUTs) was used for this implementation. Whereas for the memory of the processor, it takes up to 2 block RAMs in FPGA, for 3072 bytes of the available memory to store the program instructions and data.

By having the (255,251) Reed Solomon MISC processor developed, either a Reed Solomon encoder or decoder can be implemented onto a FPGA. This can be done by programming the (255,251) RS MISC processor to function as an RS encoder or RS decoder. Since the RS MISC processor utilized a small amount of hardware, it can be produced either as an encoder or decoder chip that is possible to be integrated into communication system.

FUTURE RESEARCH DIRECTIONS

A few identified research directions for the MISC AES and MISC RS are to develop a joint architecture system. Reconfigurable error correction system with data encryption feature is one of the promising research areas that yield a unified system to accommodate both encryption and error-correction. With the MISC engine platform established, any symmetric cipher with 8-bit or 16-bit data lengths or any hardware implementable error correction coding scheme can be applied to the MISC, as a unified module for security and error correction operations.

CONCLUSION

The size of WSN nodes varies from a shoebox down to the size of a grain of sand. Tiny hardware designs have always been the preferred solution for resource constraint devices. Given that the designs for both MISC AES and RS are tiny, the remaining resources can be used for other processing units.

In this paper, the MISC AES Processor for security and error-correction is proposed. With the simplistic MISC, a small and compact AES and RS Processor can be realized. Simple instructions are used to perform both encryption and error correction. The proposed MISC focuses on minimizing the processor area occupancy thus providing a unified platform suitable for data encryption and error correction function. Nonetheless, the MISC Architecture implements on simple hardware components, adding value to the MISC's reconfigurable, compact and flexible nature.

Figure 14. The (255,251) Reed Solomon MISC processor architecture

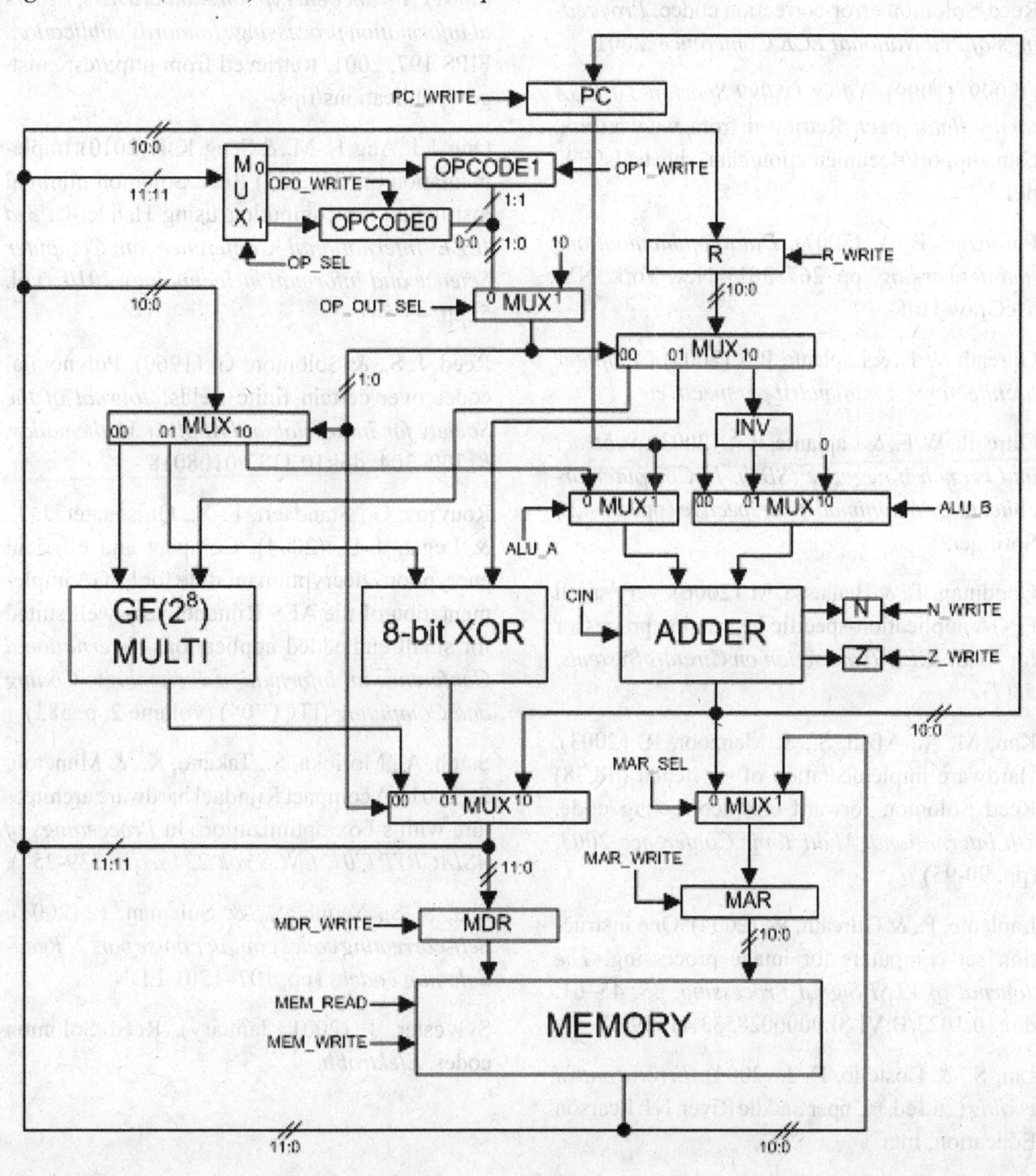

REFERENCES

Berlekamp, E. R., Peile, R. E., & Pope, S. P. (1987). The application of error control to communications. *IEEE Communications Magazine*, *25*(4), 44–57. doi:10.1109/MCOM.1987.1093590

Chio, A. M. P., Sahagun, J. A., & Sabido, I. X. D. J. M. (2001). VLSI implementation of a (255,223) Reed-Solomon error-correction codec. *Proceedings of 2nd National ECE Conference*, 2001.

DS099. (1999). *Xilinx DS099 Spartan-3 FPGA family data sheet*. Retrieved from www.xilinx.com/support/documentation/data_sheets/ds099.pdf

Forouzan, B. A. (2007). *Data communications and networking* (pp. 267–301). New York, NY: McGraw-Hill.

Gilreath, W. F., & Laplante, P. A. (2003). *Computer architecture: A minimalist perspective.*

Gilreath, W. F., & Laplante, P. A. (2003). *Subtract and branch if negative (SBN). In Computer architecture: A minimalist perspective* (pp. 41-42). Springer.

Goodman, T., & Benaissa, M. (2006). Very small FPGA application-specific instruction processor for AES. *IEEE Transaction on Circuits Systems, 53*(7).

Kan, M. A., Afzal, S., & Manzoor, R. (2003). Hardware implementation of shortened (48,38) Reed Solomon forward error correcting code. *7th International Multi Topic Conference 2003*, (pp. 90-95).

Laplante, P., & Gilreath, W. (2004). One instruction set computers for image processing. *The Journal of VLSI Signal Processing*, *38*, 45–61. doi:10.1023/B:VLSI.0000028533.41559.17

Lin, S., & Costello, D. J. (2004). *Error control coding* (2nd ed.). Upper Saddle River, NJ: Pearson Education, Inc.

Mavaddat, F., & Parhami, B. (1988). URISC: The ultimate reduced instruction set computer. *International Journal of Electrical Engineering Education*, *25*, 327–334.

Mui, E. N. C. (2007). *Practical implementation of Rijndael SBox using combinational logic.*

National Institute of Standards and Technology. (2001). *Advance encryption standard AES, Federal information processing standards publication*. FIPS 197, 2001. Retrieved from http://csrc,nist.gov/publications/fips

Ong, J. J., Ang, L.-M., & Seng, K. P. (2010). Implementation of (255,223) Reed Solomon minimal instruction set computing using Handel-C. *3rd IEEE International Conference on Computer Science and Information Technology 2010,* (vol. 5, pp. 49-54).

Reed, I. S., & Solomon, G. (1960). Polynomial codes over certain finite fields. *Journal of the Society for Industrial and Applied Mathematics*, *8*, 300–304. doi:10.1137/0108018

Rouvroy, G., Standaert, F.-X., Quisquater, J.-J., & Legat, J.-D. (2004). Compact and efficient encryption / decryption module for FPGA implementation of the AES Rijndael very well suited for small embedded applications. *International Conference on Information Technology: Coding and Computing* (ITCC'04) (volume 2, p. 583).

Satoh, A., Morioka, S., Takano, K., & Munetoh, S. (2001). A compact Rijndael hardware architecture with s-box optimization. In *Proceedings of ASIACRYPT'01, LNCS vol. 2248*, (pp. 239-254).

Sha, S. S., Yaqub, S., & Suleman, F. (2001). *Self-correcting codes conquer noise part 2: Reed-Solomon codecs* (pp. 107–120). EDN.

Sylvester, J. (2001, January). Reed Solomon codes. *Elektrobit*.

Wade, G. (2000). *Coding techniques, an introduction to compression and error control*. New York, NY: Palgrave.

Wicker, S. B., & Bhargava, B. K. (1994). *Reed-Solomon codes and their applications*. IEEE Press.

Xhang, X., & Parhi, K. K. (2004). High-speed VLSI architectures for the AES algorithm. *IEEE Transactions on Very Large Scale Integration (VLSI). Systems, 12*(9), 957–967.

Yick, J., Mukherjee, B., & Ghosal, D. (2008). Wireless sensor network survey. *Computer Networks, 52*(12), 2292–2330. doi:10.1016/j.comnet.2008.04.002

ADDITIONAL READING

Ana-Belén García-Hernando. (2008). *José-Fernán Martinez-Ortega, Juan-Manuel López-Navarro, Aggeliki Prayati, Luis Redondo-López, "Problem Solving for Wireless Sensor Networks*. Spinger.

Daemen, J. (2002). *Vincent Rijmen, "The design of Rijndael: AES--the advanced encryption standard*. Springer.

Jean-Pierre Descamps, Jose Luis Imana, and Gustavo D. Sutter, "Hardware implementation of finite-field arithmetic", The McGrawHill Companies, Inc, 2009, pp. 163-169.

Elwyn, R. (1987, April). Beriekamp, Robert E. Peile, and Stephe P. Pope, "The application of error control to communications. *IEEE Communications Magazine, 25*(4), 44–57.

George Cyril Clark, J. (1981). *Bibb Cain., "Error-correction coding for digital communications*. Springer.

Hauck, S. (2008). *André DeHon., "Reconfigurable computing: the theory and practice of FPGA-based computation*. Morgan Kaufmann.

Jonathan Katz and Yehuda Lindell. "Introduction to Modern Cryptography", 2008, Chapman & Hall/CRC Taylor & Francis Group, 6000 Broken Sound Parkway. NW, Suite 300 Boca Raton, FL 33487 - 2742.

Lin, S., & Costello, D. J. (2004). *Error Control Coding* (2nd ed.). Pearson Education, Inc.

Liu, D. (2007). *Peng Ning., "Security for wireless sensor networks*. Springer.

Mali, M., Novak, F., & Biasizzo, A. (2005)... *Journal of Electrical Engineering, 56*(9-10), 265–269.

Mastrovito, E. (1991). VLSI Architectures for Compositions in Galois Fields. In *Ph.D. dissertations*. Linkoping, Sweden: Linkoping Univ.

Maya Gokhale. (2005). *Paul S. Graham, "Reconfigurable computing: accelerating computation with field-programmable gate arrays*. Birkhäuser.

Parhami, B. (2005). *Computer Architecture: From Microprocessors to Supercomputers* (pp. 151–153). Oxford University Press.

C. S. Raghavendra, Krishna M. Sivalingam, and Taieb Znati, Wireless Sensor Networks, Springer Science+Business Media, 2006.

Rajanish K. Kamat, Santhosh A. Shinde, Vinod G. Shelake, Department of Electronics, Shivaji University, Kolhapur, India, "Handel C: A Boon for Software Professionals", *Unleash the System On Chip using FPGAs and Handel C*, Springer; 1 edition, United States of America, pp. 32-33, April 14, 2009.

Ren, K. (2008). *Wenjing Lou., "Communication Security in Wireless Sensor Networks*. VDM Verlag.

Richard, R. (2003). *Oehler, Reduced instruction set computer (RISC), Encyclopedia of Computer Science* (4th ed., pp. 1510–1511). Chichester: John Wiley and Sons Ltd.

Satoh, A., & Morioka, S. K.takano, and S.Munetoh, "A Compact Rijndael Hardware Architecture with S-Box Optimization," in Proc. LNCS ASIACRYPT'01, Dec. 2001, vol. 2248, pp. 239-254.

Schneier, B. (1996). *Applied cryptography: protocols, algorithms, and source code in C*. Wiley.

Stephen, B. (1994). *Wicker and Vijay K. Bhargava, "Reed-Solomon codes and their applications*. New Jersey: IEEE Press.

Syed Shahzad Sha, Saqib Yaqub, and Faisal Suleman, "Selfcorrecting codes conquer noise Part 2: Reed-Solomon codecs", EDN, pp. 107-120, March 2001.

Van der Poel. (1952). W. 1, "A simple electronic digital computer. *Applied Scientific Research. B, Electrophysics, Acoustics, Optics, Mathematical Methods*, *2*, 367–400. doi:10.1007/BF02919783

Xinmiao Zhang and Keshab K. Parhi. High-Speed VLSI Architectures for the AES Algorithm, IEEE transactions on very large scale integration (VLSI) systems, vol. 12, no. 9, 2004.

Yang Xiao. (2006). *Security in sensor networks*. Auerbach Publications.

Yee Wei Law, Jeroen Doumen, Pieter Hartel, "Survey and Benchmark of Block Ciphers for Wireless Sensor Network", ACM Transactions on Sensor Networks (TOSN), Volume 2, Issue 1 (February 2006), pp: 65 – 93, 2006.

KEY TERMS AND DEFINITIONS

3-Tuple: A sequence with 3 elements, in this case, 3 bytes.

Branching Instruction: An instruction that changes the value of the Program Counter.

Instruction Sets: A set of commands, understood by computer hardware.

Minimal Instruction Set Computer: A computer architecture with the minimal number of instructions, sufficient to execute the desired operations.

Round Function: Iterated function execution.

Sink: A sensor node that acts as a gateway to a server computer.

Symmetric Cipher: Cipher with identical keys on both ends.

Chapter 18
Medium Access Control Protocols for Wireless Sensor Networks: Design Space, Challenges, and Future Directions

Pardeep Kumar
Free University Berlin, Germany

Mesut Güneş
Free University Berlin, Germany

ABSTRACT

This chapter provides an overall understanding of the design aspects of Medium Access Control (MAC) protocols for Wireless Sensor Networks (WSNs). A WSN MAC protocol shares the wireless broadcast medium among sensor nodes and creates a basic network infrastructure for them to communicate with each other. The MAC protocol also has a direct influence on the network lifetime of WSNs as it controls the activities of the radio, which is the most power-consuming component of resource-scarce sensor nodes. In this chapter, the authors first discuss the basics of MAC design for WSNs and present a set of important MAC attributes. Subsequently, authors discuss the main categories of MAC protocols proposed for WSNs and highlight their strong and weak points. After briefly outlining different MAC protocols falling in each category, the authors provide a substantial comparison of these protocols for several parameters. Lastly, the chapter discusses future research directions on open issues in this field that have mostly been overlooked.

DOI: 10.4018/978-1-4666-0101-7.ch018

Figure 1. Some of the common sensor platforms used by industrial and research organizations for several WSN related applications and testbed implementations. They differ from each other in processing, storage, and communication capabilities and are suitable for an application or the other.

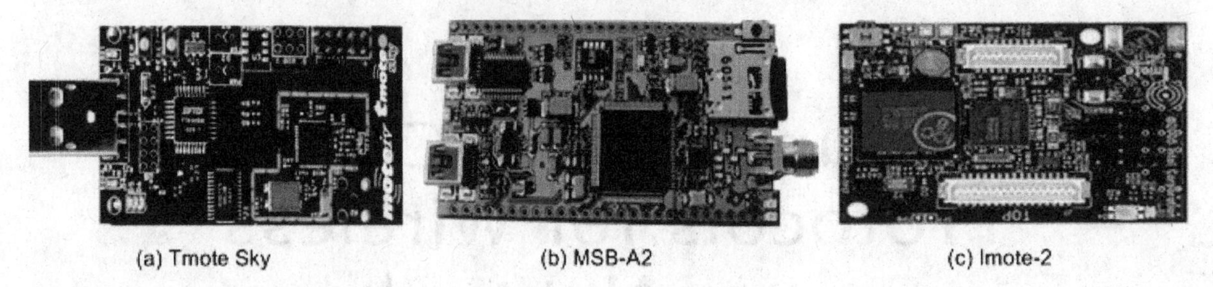

(a) Tmote Sky (b) MSB-A2 (c) Imote-2

INTRODUCTION

The pervasiveness, self-autonomy, and self-organization of low-cost, low-power, and long-lived WSNs (Karl & Willig, 2006; Ilyas & Mahgoub, 2006; Li X. Y., 2008; Misra, Woungang, & Misra, 2009; Sohraby, Minoli, & Znati, 2007; Tubaishat & Madria, 2003; Akyildiz & Varun, 2010) have brought a new perspective to the world of wireless communication. This domain is destined to play a vital role to our future ubiquitous world as it extends the reach of cyberspace into physical and biological systems. Coupled with sensing, computation, and communication into a single tiny device, WSNs are emerging as an ideal candidate for several daily-life applications, particularly in monitoring and controlling domains. Demands placed on these networks are expending exponentially with the increase in their dimensions. The development of new hardware, software, and communication technology, and continuous refinements of current approaches is also pushing this domain even further. Besides the development of new algorithms and protocols, many commercial hardware vendors are also engaged designing novel and efficient architectures for sensor nodes[1]. Figure 1 shows some of the sensor nodes used for deployment, experiment, and evaluation of different WSN related applications, whereas Table 1 gives hardware details in terms of microcontroller, radio chip, and memory available to these platforms.

However, unique characteristics along with limited resources available to sensor nodes pose

Table 1. Detailed hardware specifications of the WSN platforms shown in Figure 1

	Tmote Sky	MSB-A2	Imote-2
CPU - Speed	TI MSP430 8 MHz	NXP LPC2387 Upto 72 MHz	PXA271 XScale 13 – 416 MHz
Radio - Frequency - Data Rate - RX Current - TX Current - Modulation - Output Power	Chipcon CC2420 2.4 GHz 250 kbps 18.8 mA 17.4 mA DSSS +0 dBm	Chipcon CC1100 315/433/868/915 MHz upto 500 kbps 15.6 mA 28.8 mA 2-FSK/GFSK/MSK/ OOK/ASK +10 dBm	Chipcon CC2420 2.4 GHz 250 kbps 18.8 mA 17.4 mA DSSS +0 dBm
Memory - RAM - Flash	10 KB 48 KB	98 KB 512 KB	32 MB 32 MB

several challenges in the design of sensor networks. Integrating sensing, processing, and communication functionalities into a tiny sensor node has added a lot of complexities. Moving from sensors with only few hours of life time to one with many years of life time demands several iterations of energy efficient techniques. Shrinking size of nodes requires small size transceivers. Mapping overall system requirements down to individual device capabilities is not an easy task. Moreover, the direct interaction with the real world and the application-specific nature of WSNs require them to respond accordingly. As a result, a detailed understanding of capabilities, requirements, constraints, and limitations of WSNs is required.

MAC BASICS FOR WSNS

The MAC sublayer is a part of the data link layer specified in the communication protocol stack that is shown in Figure 2. It provides channel access mechanisms to several medium-sharing devices. On a wireless medium, which is broadcast in nature, when one device transmits, every other device in the transmission range receives its transmission. This could lead to an interference and collision of frames when a transmission from two or more devices arrives at one point simultaneously. Sensor nodes usually communicate via multi-hops over the wireless medium in a scattered, dense, and rough sensor field. A MAC protocol for WSNs manages the shared-medium and creates a basic network infrastructure for nodes to communicate with each other. Thus it provides a self-organizing capability to nodes and tries to enforce the singularity in the network by letting the sender and receiver communicate with each other in collision- and error-free fashion.

Moreover, the typical requirement to increase lifetime of a WSN without the need of any power replacement and/or human interaction has prompted the development of novel protocols in all layers of the communication stack. However, prime gains can be achieved at the data link layer, where a MAC protocol directly controls the activities of the radio, which is the most power-consuming component of resource-scarce sensor nodes. In brief, a MAC protocol for WSNs specifies how nodes employ the radio, share the channel, avoid collision in correlated environments, response the inquirer timely, and survive for a long period.

MAC Services

In general, the fundamental task of any MAC protocol is to regulate the *fair access* of nodes to a shared medium in an efficient way in order to achieve good individual throughput and better channel utilization (Labrador & Wightman, 2009). However, constrained resources, redundant deployment, and collaboration rather than competition among sensor nodes considerably change the responsibilities of the MAC protocol for WSNs. On one hand, some relaxations may be granted to such MAC protocol. For example, nodes in WSNs usually send very small frames and use the channel occasionally, either periodically or whenever an important event occurs. Therefore, fairness is not as important as in other networks. In WSNs nodes remain idle or in the sleep mode most of the time and rarely compete for the channel. Achieving good channel utilization is usually not considered as an important metric. The data flow in WSNs is usually unidirectional, i.e., from nodes to the sink, and end-users generally focus on the collective information rather than the individual throughput.

On the other hand, the WSN MAC protocol has some extra responsibilities to deal with as well. First and foremost is the issue of energy conservation. Since a distributed network of several nodes demands for long-time and maintenance-free operations, a MAC protocol - irrespective to the scheme and work space it uses - certainly must have built-in power-saving mechanism. Along

Figure 2. The communication protocol stack. This five-layered simplified model is commonly applied to network research as apposite to the seven-layered OSI model. An end-user can use application-specific software/algorithms at the application layer. The transport layer helps maintaining the sensor data flow. The network layer routes data on an appropriate path. The Logical Link Control (LLC) sub-layer of the data link layer provides framing, flow control, error control, and link management facilities, whereas the MAC sub-layer shares the wireless medium and helps in energy aware operations for nodes. The physical layer takes care of the radio, modulation, and transmission/reception of bits on a physical medium.

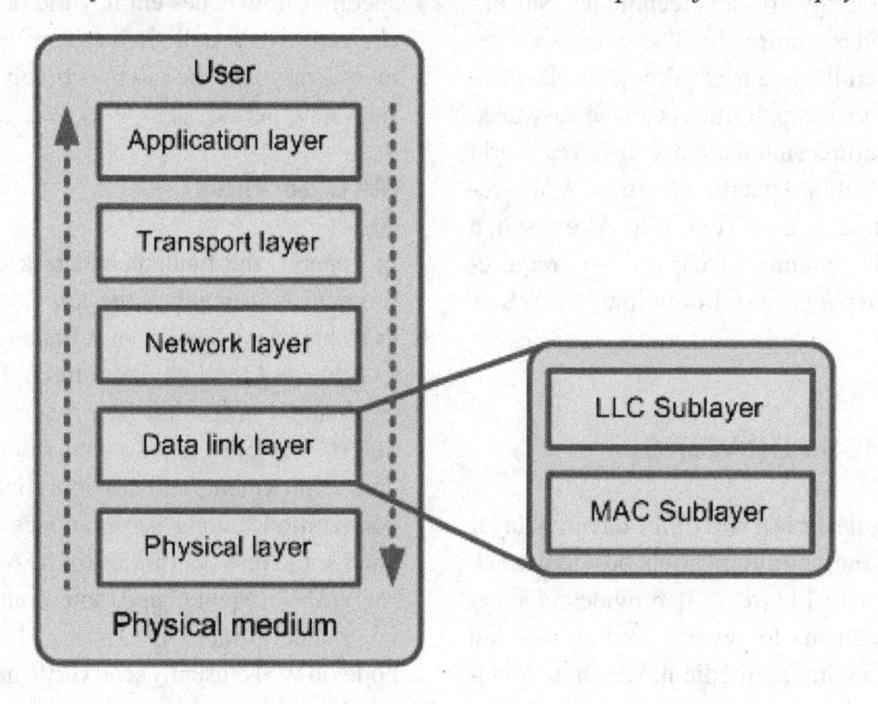

with energy efficiency and as per application requirements, provision of timeliness, adaptability to traffic and topology conditions, scalability, support for non-synchronized operations, and interaction with other layers via cross-layering may also play an important role in designing the MAC protocol for WSNs. Furthermore, due to the dense deployment and small transmission distances, the transmission power of sensor nodes can often be lesser than the receiving power. It is unlike other traditional networks that usually consider the receiving power to be negligible. Therefore, the WSN MAC protocol has to preferably consider a balanced equation between transmitting and receiving packets. Additionally, the ideal MAC protocol ensures self-stabilization, graceful adaptation, an acceptable delivery rate, low overhead, and low error rates for a WSN.

MAC Challenges

The design of the MAC protocol for WSNs is a complex task due to the energy constraints, low transmission ranges, and compact hardware design of sensor nodes. Along with these factors, the event- or task-based network behavior and application diversity of WSNs also demand for peculiar MAC schemes, which are not common with traditional wireless networks. Additionally, by virtue of the wireless broadcast medium, WSNs inherit all the well-known problems of wireless communication and radio propagation in the shape of interference, fading, path loss, attenuations, noise, and high error-rates (Tse & Viswanath, 2005; Garg, 2007). The MAC protocol is directly influenced by all these effects as it sits right above the physical (PHY) layer, thereby presuming total control over the medium and its access rate. The frequent use of the unlicensed Industrial, Scientific, and Medical (ISM) band for most sensor networks and sensing applications can also worsen these effects.

The need of energy efficient operations for WSNs usually results in trades-off among energy and other parameters such as latency, packet reception, throughput, fairness, and scalability. As a result, many existing architectures and protocols for traditional wireless networks such as IEEE 802.11 and Bluetooth are not suitable for WSNs, as they usually target higher data rates with less emphasis on energy and other WSN specific issues. Among the three basic responsibilities of WSNs, namely *sensing*, *processing*, and *communication*, each performed by the sensors, microcontroller (MCU), and radio respectively, the later is the major power consumer (Akyildiz & Varun, 2010). The communication power of the node depends on several factors that include the type of modulation scheme used, data rate, transmit power, operational modes of the radio, and the switching frequency between these modes. At the same time, a MAC protocol can be made accountable for the following sources of energy waste in WSNs, which mainly relate to the communication (Bachir, Dohler, Watteyne, & Leung, 2010; Akyildiz & Varun, 2010).

Idle listening: Since a node in a WSN usually does not know when it will receive a message, it keeps its radio in ready-to-receive mode, which consumes almost as much energy as in the receive mode. In low traffic applications, this is considered one of the major sources of energy waste.

Collisions: A collision is a wasted effort when two frames collide with each other and get discarded. A collision usually results in retransmission that drains more energy in transmitting and receiving extra packets.

Overhearing: An overhearing occurs on the shared-medium when a node receives and processes a gratuitous packet that is not addressed for it. In heavy traffic situations, this could lead to a serious energy problem for sensor nodes.

Control packet overhead: An increase in the number and size of control packets results in overhead and energy waste for WSNs, especially when only a few bytes of real data are transmitted in each message. These control signals also decrease the channel capacity.

Over-emitting: An over-emitting or deafness occurs due to the transmission of a message when the destination node is not ready to receive it.

Complexity: Computationally expensive algorithms might decrease the time the node spends in the sleep mode. They might limit the processing time available for the application and other functionalities of the protocol. An overly simple MAC algorithm can save higher energy than a complex one, but it may not be able to provide the complex functions such as adaptation to traffic and topology conditions, clustering, or data aggregation.

Duty cycling, which is discussed in the next section, is widely considered as a powerful mean to cope with most of the energy related issues in WSNs. However, this duty cycling results in high latency and low throughput, and a deep consideration is required in selecting a proper duty cycle value for nodes. Along with energy efficiency, several applications of WSNs may need delay bound operations. Unlike traditional distributed systems, the timeliness guarantee for WSNs is more challenging. They interact directly with the real world, where physical events occur in an unpredictable manner with different traffic and delay requirements. Duty cycling, dynamic topology, and limited memory and computation power also restrict the design space we could trade off.

Common MAC Approaches

There is no universal *best* MAC protocol for WSNs; the design choice mainly depends on the nature of the application (Karl & Willig, 2006). The stringent design requirements of a MAC protocol for WSNs can be met by a plethora of approaches. The most widely used approaches in designing such MAC protocols with their implications are outlined below.

Duty cycling: Though the application domain of WSNs is diverse and broad, environmental monitoring and surveillance have been the most visible applications of WSNs. The data traffic generated and processed by sensor nodes in such applications can be distinguished in two different classes; *periodic traffic* and *event-based traffic*. The periodic traffic class senses the environment usually at a regular interval and collects the information about temperature, air pressure, humidity, and/or light values of a physical object and report to the sink node (Tolle, Polastre, Szewczyk, & et-al., 2005; Goense, Thelen, & Langendoen, 2005; Mainwaring, Polastre, Szewczyk, Culler, & Anderson, 2002). In the event-based traffic class, nodes do not follow a periodic monitoring mechanism but report the sink node or sound an alarm when something significant occurs in the sensor field (Arora, Dutta, Bapat, & et-al., 2004; Bhuse & Gupta, 2005; Simon, Maroti, Ledeczi, & et-al., 2004). Nodes in this class remain idle most of the time but usually generate a burst of packets during the short time period when an event occurs.

A WSN generally generates much less data traffic and sends very small data frames as compared to traditional wireless or wired networks. Sensor nodes therefore remain idle most of the time either waiting for their periodic turn to generate data or listening the idle channel for something to occur. Since the radio consumes as much energy during idle listening as in receiving data packets, switching it into the low-power sleep mode and waking up shortly at a periodic interval can significantly conserve the energy of nodes. All these facts are sketched down in Figure 3, where a node periodically switches its radio between sleep and listen periods rather than constantly listens the idle channel. It turns to the sleep mode for a sleep period t_s and wakes up and checks the medium for a short listen period t_L. The sum of the sleep period and the listen period is called a wake-up period t_W, whereas the ratio of the listen period to the wake-up period t_L./ t_W is called duty cycle of a node.

Duty cycling significantly increases system lifetime of a dense WSN by reducing idle listening and overhearing among nodes. Importantly, nodes usually do not need any additional hardware or a complex algorithm to perform duty cycling. However, these advantages have other implications too. The transceiver is usually kept in the sleep mode most of the time, which could end up in a significant competition among neighbors at wake-up periods. This could ultimately lead to collisions, low throughput, and high network latency for a WSN, particularly in heavy load situations. The important question that arises here is to select an optimal value of duty cycle for an application. Choosing a long sleep period induces significant per-hop latency, since a sending node has to wait an average of half a sleep period before the receiver can accept packets. Too short sleep phases, i.e., more frequent switching of the radio between on and off modes also outweighs the benefits of duty cycling because the switching is not instantaneous and consumes additional energy (Akyildiz & Varun, 2010). Hence, the optimal selection of the duty cycle value is a critical step towards achieving the desired system performance.

Timeliness: While designing the MAC protocol for WSNs, the timeliness factor is often ignored by researchers. With ever increasing applications of WSNs in many diverse fields, new concepts of offering timeliness related Quality of Service (QoS) are inevitable. Generally, timeliness related applications can be categorized into Hard Real-Time (HRT) and Soft Real-Time (SRT) based applications (Li, Chen, Song, & Wang, 2007). A deterministic end-to-end delay is required in HRT applications, where a strict deadline is applied on the arrival of messages. Alternatively, a tolerable

Figure 3. Duty cycling in WSNs. Sensor nodes usually generate/process data at a very low rate as compared to other traditional networks. They use the channel occasionally, either periodically or whenever an important event occurs. Therefore, in order to reduce idle listening and overhearing, nodes perform duty cycling, remain in the sleep mode most of the time, and wake up shortly to sense the channel.

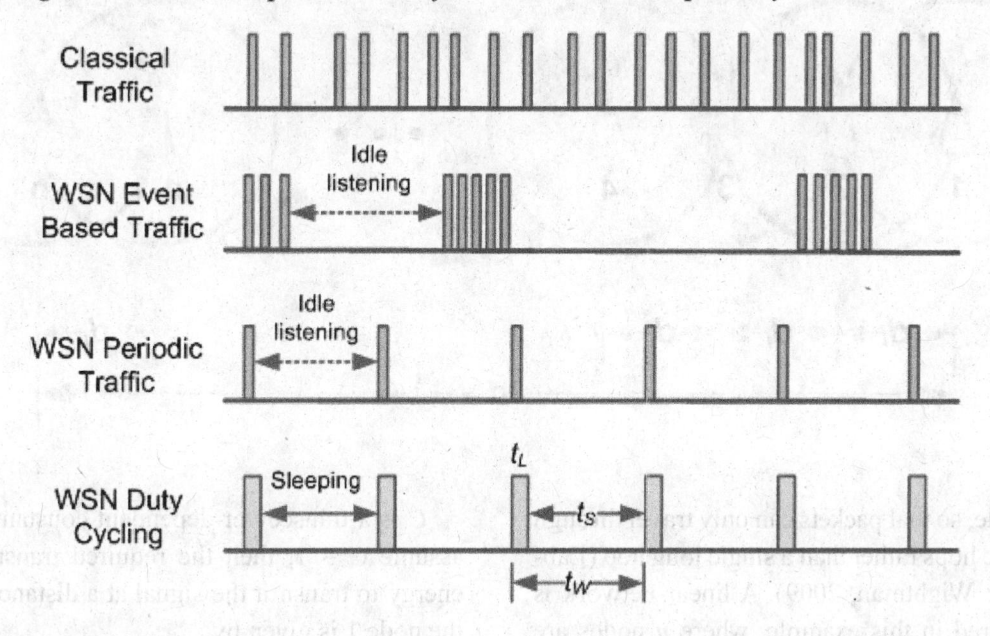

and probabilistic delay guarantee is supported for SRT applications.

Limited resources, low node reliability, dynamic network topology, and direct interaction with the physical world makes HRT very difficult in WSNs. With a time scheduling mechanism, which is discussed later, a bounded and predictable delay in WSNs can be achieved. However, even in that case, along with other implications, the average queuing and access delays are much higher as a node has to wait for its allocated slot before accessing the medium. Consequently, the probabilistic based SRT guarantee in many applications of WSNs is mostly permissible.

Channel access method: Acquiring and releasing the channel is the core in the design of any MAC protocol, and in dense and energy-limited WSNs its importance increases even more. Several channel accessing methods that have been proposed in the literature are elaborated in the next section.

Topology control: The goal of topology control is to build a reduced topology by dynamically changing transmitting range of nodes in order to save energy and preserve important network characteristics, such as connectivity and coverage (Santi, 2005). Since the transmission energy often dominates the total communication energy and grows with the increase of transmission distance, topology control can reduce this consumption by forcing packets to travel through multiple hops. The topology control function is usually located between the MAC layer and the network layer and interacts with both of them. Dynamic topology construction can also be exercised by turning unnecessary nodes off. The topology control mechanism reduces energy consumption by reducing collisions, contentions, and exposed terminal problems. However, idle listening, overall latency, complexity, and increased packet loss probability remain core issues with this mechanism.

Figure 4 illustrates that energy consumption can be reduced by minimizing transmission range

Figure 4. A simple topology control example in WSNs, where nodes dynamically change their transmission power to save energy. They transmit packets via their adjacent neighbors rather than sending over the long distance.

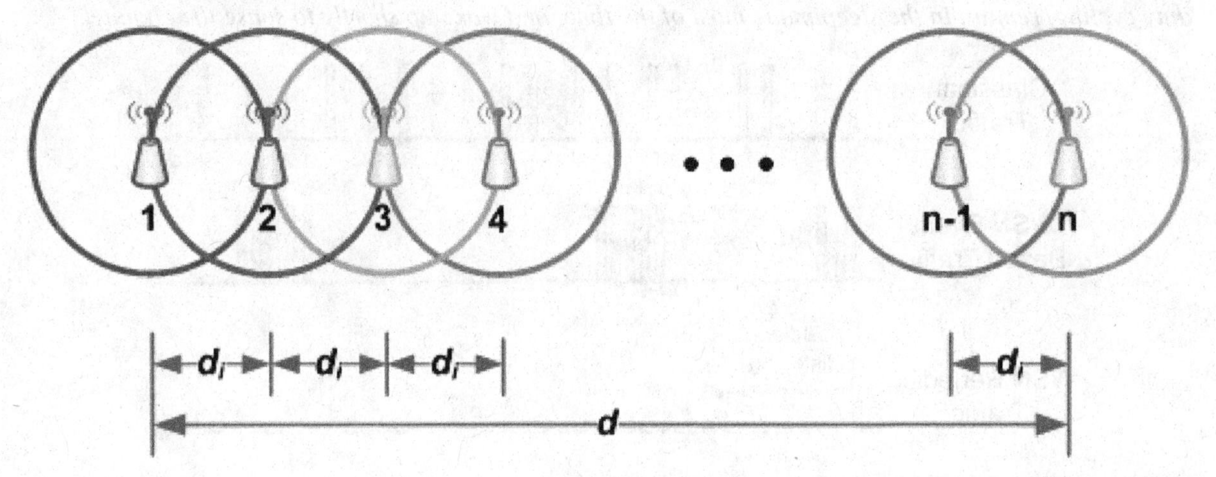

of a node, so that packets can only travel through multiple hops rather than a single long hop (Labrador & Wightman, 2009). A linear network is considered in this example, where *n* nodes are equally spaced by distance *d*. If node 1 directly communicates with node *n* over the total distance of *D*, and if path loss exponent of 2 is assumed, then the received power at a distance *D* can be given by the Friis free-space propagation model, as follows:

$$P_r = P_t \times G_t \times G_r \times \left(\frac{\lambda}{4 \times \pi \times D} \right)^2$$

Where P_t is the power at which the signal was transmitted, G_t and G_r are the antenna gains of the transmitter and receiver respectively, and λ is the wavelength. This equation can be written as:

$$P_r = C \times \frac{P_t}{D^2}$$

$$C = \frac{G_t \times G_r \times \lambda^2}{(4 \times \pi)^2}$$

C is a transceiver-dependent constant. If we assume $C = 1$, then the required transmission energy to transmit the signal at a distance *D* for the node 1 is given by:

$$P_t = P_r \times D^2$$

Alternatively, if the node 1 chooses to transmit a packet to the node *n* via its adjacent neighbors, i.e., over $n-1=h$ hops then the energy needed by each node to reach its immediate neighbor is given by:

$$P_t = P_r \times d^2$$

Therefore, the power saving with multi-hop communication is given by:

$$P_s = \frac{P_r \times D^2}{h \times P_r \times d^2} = \frac{P_r \times (h \times d)^2}{h \times P_r \times d^2} = h$$

This shows that larger the number of hops *h*, the higher the power saving for a node. However, with multi-hop communication $(n-1)$ nodes process and hence consume energy as opposite to the only two nodes with the single-hop communication. Moreover, this is a very simplistic example because in reality factors like Bit Error Rate (BER) and probability of packet loss with the varying transmission distance also need to be considered.

Scheduling and synchronization: Many MAC protocols for WSNs assume that all nodes in a network follow a fixed schedule to switch their radios between wake-up and sleep modes. They also assume that nodes are timely synchronized. However, in reality such time synchronization in dynamic and resource-limited WSNs is very difficult to achieve as it induces a lot of overhead and may need extra hardware. Collisions and retransmissions increase dramatically if all nodes wake up simultaneously. Therefore, it is wise to use random and non-synchronized wake-up and sleep schedules for WSN nodes.

Cross-layering: Most of the proposed MAC protocols for WSNs follow the traditional layered architecture, where they try to improve performance only at the respective layer. With very limited resources available to sensor nodes, a trend of the cross-layer design is emerging in order to achieve aggregate optimization among different layers. Unlike layered networks, WSNs cannot afford significant layered overhead due to their limited energy, storage, and processing capabilities. Moreover, application-aware communication and low-power radio considerations motivate for cross-layer architecture for WSNs. Recent studies in (Melodia, Vuran, & Pompili, 2005; Kawadia & Kumar, 2005; Haapola, Shelby, Pomalaza-Raez, & Mahonen, 2005; Hoesel, Nieberg, Wu, & Havinga, 2004; Kumar, Gunes, Mushtaq, & Schiller, 2010) affirm improvement in WSN performance by using cross-layering. There is, however, still much to be done to provide unified cross-layer communication architecture for dynamic and resource-constrained WSNs.

Miscellaneous techniques: Along with common methods of the MAC designing, some unconventional approaches have also been endeavored in the literature. Some researchers counsel for having two different channels with each node; the *data channel* and the *control channel* (Schurgers, Tsiatsis, Ganeriwal, & Srivastava, 2002). The data channel is always kept in the sleep mode except when transmission of data and/or ACK packets occurs. However, control packets are only exchanged on the control channel. This approach does not need synchronization but increases complexity in terms of hardware, cost, energy consumption, and handling of two transceivers at each node.

Other researcher opt for the positive aspects of more than one channel accessing approaches to form a common set of two or more characteristics of different methods. Such *hybrid schemes*, which are explained later in the chapter, may improve overall performance of WSNs, but they also increase overhead and complexity as more than one channel accessing methods need to be handled simultaneously.

The protocols proposed in (Lin, Rabaey, & Wolisz, 2004; Sun, Gurewitz, & Johnson, 2008) suggest shifting transmission initiation from the sender to receiver side. When the receiver is awake and ready to receive a frame, it sends a beacon and starts monitoring the channel for incoming frames for a while. On reception of the beacon, a sender node can send the actual payload. Such schemes are in fact similar to the classical ones with only the difference of who is going to start the communication. Therefore, they carry all the common problems of this domain in one or the other way. Moreover, such protocols cannot be used for broadcast and multicast communication and can increase latency and idle listening at the sender side.

CLASSIFICATION OF MAC PROTOCOLS

WSN MAC protocols depending on how they allow nodes to access the channel can be classified into the general categories of *contention based*, *scheduling based*, *channel polling based*, and *hybrid protocols* (Klues, Hackmann, Chipara, & Lu, 2007). Figure 5 depicts this classification.

Figure 5. Channel accessing taxonomy in WSNs. MAC protocols can be classified into four main categories depending on how nodes access the shared channel.

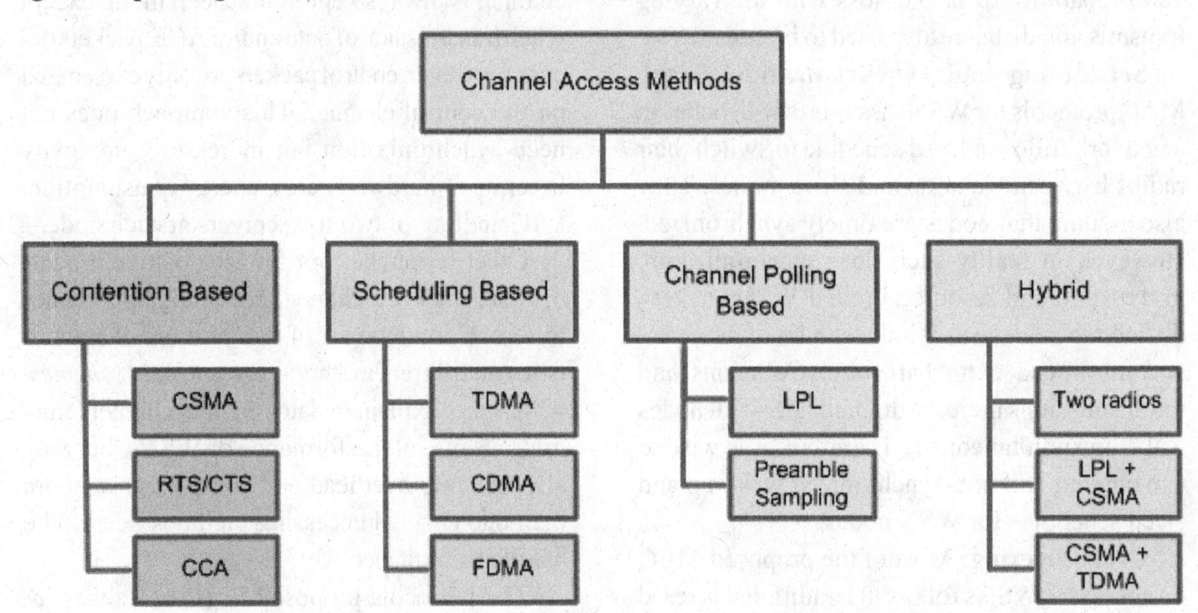

Contention Based MAC Protocols

With the contention-based Carrier Sense Multiple Access (CSMA) method, a transmitting node competes with its neighbors to acquire the channel. Before any transmission, it first senses the carrier. If the carrier is found idle, the node starts with its transmission, otherwise it defers the transmission for some random time usually determined by the back-off algorithm. Such MAC protocols consume less processing resources and are suitable for event-driven WSN applications. They are flexible to network scale and dynamics as no clustering and/or topology information is required. However transmission with this approach is purely handled by the sender, and the problem of hidden- and exposed-terminals may occur causing collisions, overhearing, idle listening, and less throughput for a WSN.

Various MAC protocols falling in this category consider that contention times of nodes are synchronized according to a schedule, i.e., at each periodic interval, all neighboring nodes wake up simultaneously to exchange packets (Ye, Heidemann, & Estrin, 2004; Dam & Langendoen, 2003; Lin, Qiao, & Wang, 2004). In that case, chances of collisions are very high as all neighboring nodes compete for the channel simultaneously. Collided packets are usually retransmitted, which results in higher energy consumption and delays. The formation and maintenance of synchronization in resource-limited WSNs lead to complexity and communication overhead and may require special hardware and/or algorithms. Moreover, the clock drift on each node can affect the schedule coordination and synchronization by causing timing errors (Zheng & Jamalipour, 2009).

The *Request To Send* and *Clear To Send* (RTS/CTS) handshake is among the several approaches which have been proposed in order to minimize the hidden- and exposed-terminal problems in CSMA. The sender and receiver first perform the RTS/CTS handshake and then communicate by transmitting and receiving the actual data packet. Unfortunately, this method is not fully able to eliminate the hidden-terminal problem, and moreover it incurs additional overhead in transmitting control packets.

An ample amount of WSN MAC protocols working on the contention based mechanism is available in literature. We depict next some of the representative protocols of this category.

S-MAC

The design of the Sensor-MAC (S-MAC) (Ye, Heidemann, & Estrin, 2004) was one of the initial attempts to significantly reduce idle listening, collisions, and overhearing in WSNs by putting nodes in *listen* and *sleep* periods. Listen periods are normally of fixed size according to some PHY and MAC layer parameters, whereas length of sleep periods depends on a predefined application based duty cycle parameter. S-MAC attempts to coordinate neighboring nodes by letting them share common listen periods according to a *schedule*. This requires formation and maintenance of synchronization among nodes. In order to reduce the hidden-terminal effect, S-MAC uses the RTS/CTS handshake scheme for unicast messages.

The listen period is further divided into two intervals; *SYNC* and *data* periods. During the SYNC period, a node tries to receive a SYNC packet from its neighbors, which consists of the sender ID and the remaining time until the sender switches to the sleep mode. At initialization, if the node does not hear the schedule from its neighbors, it immediately chooses its own schedule and broadcasts it. The data period is used to exchange data related messages, which may include RTS, CTS, DATA, or ACK messages. To minimize costly retransmissions, S-MAC fragments long messages into short frames and sent them in a burst. The RTS/CTS is only required before transmitting the first short frame.

Although S-MAC reduces idle listening and overhearing for duty cycled nodes, it has several other drawbacks. S-MAC is rigid and optimized for a predefined set of workloads as there is no mean to adapt length of listen and sleep periods with changing traffic conditions. As discussed earlier, the formation, maintenance, and compliance of synchronization has serious consequences in WSNs. Longer and fixed sleep periods of S-MAC have serious impact on system latency. It could be worsen if intermediate nodes on a route do not share a common schedule. To decrease latency, an adaptive listening mechanism is employed with S-MAC. A node that overhears the on-going transmission of its neighbors, wakes up just at the end of the transmission. The node receives and processes the packet if the packet is destined for it, otherwise it goes back to the sleep mode.

Another drawback of S-MAC is the possibility of following more than one schedule, which results in more energy consumption via idle listening and overhearing. By following the same common schedule, nodes form a virtual cluster. Thus the S-MAC network is most likely to have several virtual clusters. Border nodes need to adapt to two or more different schedules for successful formation of virtual clusters. Thus, border nodes expend more energy than non-border nodes as they need to maintain network connectivity among clusters. With fragmentation in S-MAC, overhead and retransmission can be reduced, but it comes at the expense of unfairness as a node reserves the channel for a whole burst duration. A neighboring node carrying delay bound data would have to wait longer to gain access of the channel.

Several improvements and enhancements have been proposed to overcome the weaknesses of S-MAC, here we discuss some of the popular variants of S-MAC. Most of them, by default, inherit overhead and complexity incurred in synchronizing common listen and sleep schedules for nodes.

T-MAC

The Time-out MAC (T-MAC) (Dam & Langendoen, 2003) protocol improves energy efficiency of S-MAC protocol, especially under variable traffic conditions by adaptively shorten listen periods of nodes. Unlike S-MAC, where listen periods are always rigid, the listen period of a T-MAC node ends when no activation event has occurred

Figure 6. T-MAC vs. S-MAC. T-MAC reduces energy consumption of S-MAC by adaptively terminating the listen period of a node if no activation event has occurred for a TA period. However, this early sleeping could result in packet loss, low throughput, and high latency.

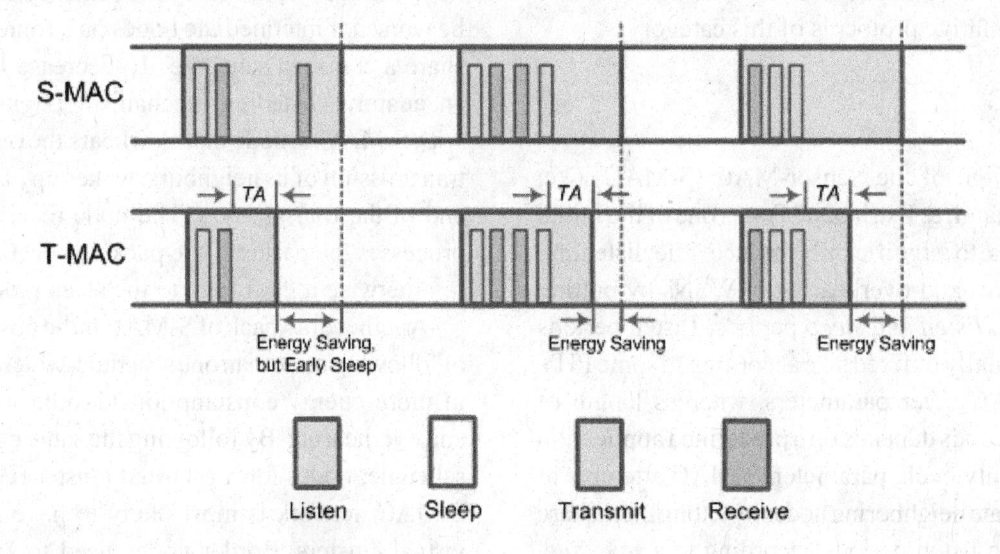

for a threshold period *TA*. This dynamic aspect frees the application from selecting an appropriate duty cycle value. The comparison between S-MAC and T-MAC shown in Figure 6 confirms this improvement. However, this improvement is heavily dependent on the *TA* value. *TA* must be long enough so that a node can sense the carrier and hear a potential CTS from a neighbor.

Although T-MAC performs better under variable loads, synchronization of the listen periods within virtual clusters may partially break down. This could lead to the *early sleep* problem for T-MAC nodes. The early sleep happens when a node, especially third hop one, goes to the sleep mode when a neighbor still has messages for it. However with T-MAC, *Future Request To Send* (FRTS) frames can be sent to the third hop nodes either to extend their TA expiration, or to let them awake by the appropriate time. T-MAC saves more power than S-MAC and minimizes collisions and redundancy, since idle nodes switch back to the sleep mode relatively earlier. However, this comes at the cost of reduced throughput and higher network latency. T-MAC also suffers from synchronization and scaling problems.

DSMAC

The Dynamic S-MAC (DSMAC) (Lin, Qiao, & Wang, 2004) improves latency of S-MAC by dynamically adjusting the duty cycle value of nodes as per traffic and energy conditions. A node using DSMAC keeps track of its energy consumption level and average latency it has experienced and tries to dynamically adjust its duty cycle accordingly. All nodes start with the same duty cycle value and share their one-hop latency values in the SYNC period. When the node notices that the traffic has been increased or low latency is required, it adds extra active periods, shortens its sleep time, and sends an updated SYNC message to its neighbors. On receiving the updated SYNC message, a neighboring node checks its queue for packets destined to that node. If there is a one, it doubles its duty cycle provided that its battery level is above a specified threshold.

Latency observed with DSMAC is better than S-MAC, but nodes consume higher average energy and use less frame duration. Thus, DSMAC achieves less throughput at high traffic. With varying duty cycle values, synchronization of a virtual cluster may get affected. A high duty cycled node can receive a SYNC message sent by a node operating at a low duty cycle. Thus, complexity in adapting duty cycle values, particularly within a virtual cluster and under high traffic loads, manifolds the pre-existing synchronization overhead.

Global Schedule and Fast Path Algorithms

As mentioned earlier, nodes with S-MAC protocol may follow more than one schedule resulting in higher energy consumption by spending more time in active periods. To minimize the number of active schedules, the number of border nodes, and latency of S-MAC, two algorithms have been proposed (Li, Ye, & Heidemann, 2005). The first algorithm called Global Schedule Algorithm (GSA) minimizes the number of active schedules of S-MAC. GSA uses age of the schedule to determine which schedule to prefer and keep. On the availability of more than one schedule for a node to choose with, it prefers the oldest schedule. Over the time, all nodes migrate toward the oldest common global schedule in the network.

The second algorithm is called Fast Path Algorithm (FPA) that provides fast data forwarding paths by adding additional wake-up periods on the nodes along paths from sources to the sink node. Given a source, sink, and the path between them, additional wake-up periods along the path are added such that they occur exactly when the previous-hop node is ready to send the packet.

Results achieved in (Li, Ye, & Heidemann, 2005) show that more than half of nodes using S-MAC have more than one active schedules, and that GSA converges to one schedule in a network of 50 Mica2Dot nodes deployed in a linear fashion quite well. However, in addition to the synchronization overhead, nodes using GSA/FPA need more processing in deciding the fate of schedules and data forwarding paths.

STEM

The Sparse Topology and Energy Management (STEM) protocol (Schurgers, Tsiatsis, Ganeriwal, & Srivastava, 2002) uses two different channels for each node, the *wake-up channel* and the *data channel,* as shown in Figure 7. The wake-up channel only monitors for control signals, whereas the data channel is solely used for data packets. On the wake-up channel, the time is divided into fixed-length *wake-up periods*, which are further subdivided into *listen* and *sleep periods*.

To gain the attention of the receiver, two different variants of STEM are used. In **STEM-B**, the transmitter tries to wake up the receiver by sending contention-free *beacons* on the wake-up channel, each containing source and destination addresses. The sender sends these beacons at least for a complete wake-up period. However, it stops sending further beacons as soon as it receives an ACK frame from the receiver. In that case, both the sender and receiver switch on their transceiver for the data channel and start communicating. The non-targeted nodes can go back to sleep mode once they recognize that the packet is not addressed to them. In **STEM-T**, the transmitter sends a *busy tone* signal on the wake-up channel, but it contains no destination address and in result, all the neighbors may sense the tone and switch on their data channels. STEM-T looks very similar to the traditional channel polling based protocols except for using two separate channels instead of one.

STEM-B cuts back the number of beacons to be sent as the transmitter does not need to always send a beacon burst of full wake-up period length. However, more than one transmitter might send a beacon simultaneously, resulting in a beacon collision. STEM-T uses a simpler and cheaper transceiver on the wake-up channel, but busy tones are sent for the maximum time. Non-targeted neighbors can receive busy tones and unnecessarily switch on transceiver for their

Figure 7. The working of STEM. It uses two channels for each node; the wake-up channel is used to transmit and receive control packets and the data channel for data packets. A sender sends a beacon on the wake-up channel with the STEM-B and a busy tone with the STEM-T variant to pull in the intended receiver. A node turns to the data channel only if a beacon or a busy tone is found on the wake-up channel.

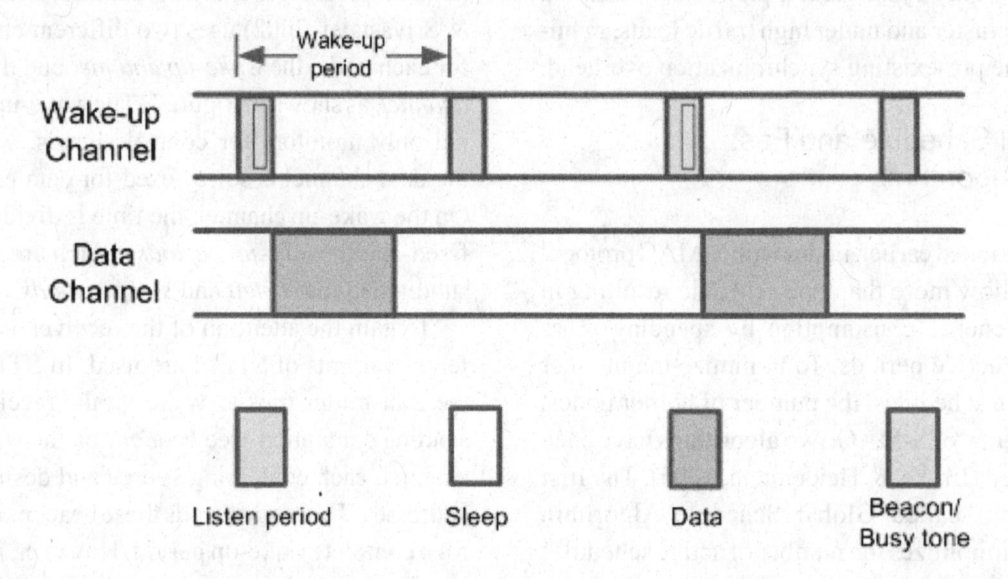

data channels. On a whole, the employment of two different channels for each node is a complex and expensive process in itself.

Scheduling Based MAC Protocols

Scheduling based schemes assign collision-free links to each node in the neighborhood usually during the initialization phase. Links may be assigned as time slots (TDMA), frequency bands (FDMA), or spread spectrum codes (CDMA). However, due to the complexities incurred with FDMA and CDMA schemes, TDMA schemes are preferred as scheduling methods for WSNs. With TDMA schemes, the system time is divided into slots and each one is allocated to a node in the neighborhood. A *schedule* in such schemes regulates which participant may use which resource at what time. The schedule can be fixed or computed on demand (or a hybrid) and is typically regulated by a central authority. A node can only access its allocated time slot and does not need any contention with its neighbors.

The cardinal advantages of scheduling based schemes include minimum collisions, less overhearing, and implicitly avoidance of idle listening. They also provide a bounded and predictable end-to-end delay. However, the average queuing delay is much higher as a node has to wait for its allocated slot before accessing the channel. Overhead and extra traffic required in setting up and maintaining synchronization among nodes, no mean to adapt with varying traffic and topology conditions, reduced scalability, and low throughput are the major concerns with these schemes. Allocating conflict-free TDMA schedules is indeed a difficult task in itself. A peer-to-peer based communication is usually not possible as nodes are normally allowed to communicate only with the central authority.

LEACH

The Low-Energy Adaptive Clustering Hierarchy (LEACH) protocol (Heinzelman, Chandrakasan, & Balakrishnan, 2002) is mostly a scheduled based protocol, which divides a dense and homogeneous WSN into several clusters. Each cluster is supervised by a cluster head. The cluster head is responsible for creating and maintaining TDMA schedules, communicating with its cluster members, and forwarding received messages to the sink node. As the cluster head is always switched on, the chances of a cluster head to die earlier are high. However, LEACH uses a randomized rotation mechanism for selecting a cluster head. Each node can independently decide to become the cluster head with preference is given to the node that has not been cluster head for a long time. Thus LEACH tries to distribute the energy among nodes in an evenly manner.

LEACH works in rounds, each further divided into the *set-up* and *steady-state* phases. The cluster formation occurs during the set-up phase, where each cluster head broadcasts an advertisement (ADV) message by using CSMA to invite its members. The cluster head then creates and broadcasts a TDMA schedule for nodes that have sent join-request (REQ) to it. In order to reduce inter-cluster interference, the cluster head chooses a random CDMA code for its members. Once the set-up phase is completed, the steady-state phase starts, where a node can transmit data to its head by using the allocated slot. Upon receiving packets from its members, the cluster head aggregates and sends them to the sink node. Since LEACH does not allow any inter-cluster communication, the cluster head directly communicates with the sink node using CSMA.

The cluster head has to perform highly computational and energy consuming tasks. It prepares and maintains the TDMA schedule, remains awake for the whole round, aggregates data, and transmits it directly to the sink node. LEACH guarantees that each member node belongs to at most one cluster. However, due to an ADV collision, it cannot guarantee that each member node belongs to the cluster. In that case, LEACH considers that all nodes are within the range of the sink node. The lack of such multi-hop communication capabilities severely limits the network scalability of LEACH. The channel under-utilization occurs with LEACH as it considers that nodes always have data to send during their allotted time. Perfect correlation among nodes is assumed, which is hardly possible in WSNs.

TRAMA

The TRaffic-Adaptive Medium Access (TRAMA) protocol (Rajendran, Obraczka, & Garcia-Luna-Aceves, 2003) is mostly a TDMA based protocol that creates schedules for time-synchronized nodes on a distributive manner based on traffic information. The system time is divided into *cycles*, each containing a *random access* and *scheduled access* periods. The random access period consists of a collection of signalling slots, whereas the scheduled access period contains collection of data transmission slots, as depicted in Figure 8.

TRAMA consists of three components: a *Neighbor Protocol (NP)*, a *Schedule Exchange Protocol (SEP)*, and an *Adaptive Election Algorithm (AEA)*. NP works during the random access period and is used to exchange one-hop neighbor information and to gather two-hop topology information for each node in the network. During SEP, a node transmits its current transmission schedule and also picks up schedules of its neighbors. To win a time slot, a node computes its own priority and the priority of all its two-hop neighbors for each time slot. AEA also works during the scheduled access and uses neighborhood and schedule information to select the transmitters and receivers for the current time slot, letting all other nodes to switch to the low-power sleep mode. For efficient channel utilization, AEA uses traffic based information that is exchanged among nodes during SEP and attempts to reuse slots that are not used by the selected transmitter.

A node using TRAMA can have one of the three states: transmit, receive, and sleep. The state of the node is determined based on its two-hop neighborhood information and the schedules

Figure 8. Time slot organization of TRAMA. During the random access period, nodes perform contention-based channel acquisition, and one-hop neighbor information is propagated by using signaling slots. Transmission slots are used for collision-free data exchange and schedule propagation among neighboring nodes.

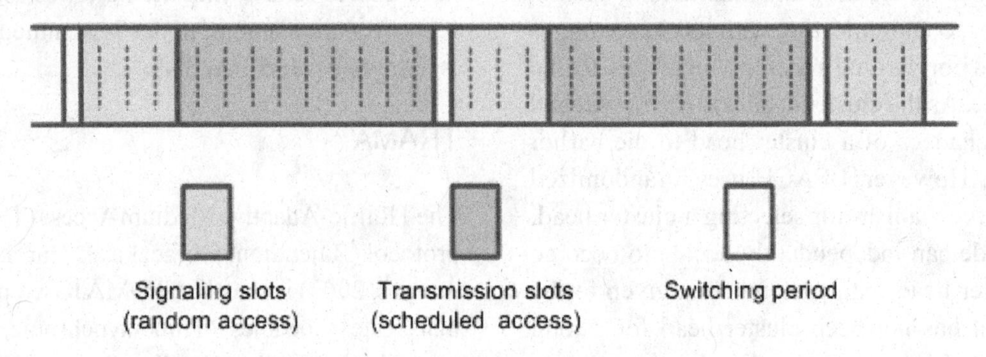

announced by its one-hop neighbors. The node switches to transmit state if it has data to send with the highest priority among its contending set. If not, the node consults the schedule sent by the current transmitter. If the transmitter has traffic destined to this node in the current slot, it stays in the receive mode, otherwise it goes to the sleep mode.

Reuse of time slots, utilization of neighborhood and traffic information, and hybrid scheme are the positive features of TRAMA. Simulation results presented in (Rajendran, Obraczka, & Garcia-Luna-Aceves, 2003) show higher percentage of sleep time, less collision probability, and better data delivery with TRAMA as compared to S-MAC and IEEE 802.11. However, all nodes in TRAMA are defined to be either in receive or transmit states to exchange schedules during the random access period. And for each time slot, every node calculates priorities for itself as well as for its two-hop neighbors. This results in significant computation, large queuing delays, and ineffective channel and memory utilization due to the fact that the two-hop neighborhood in a dense WSN could be reasonably large.

SMACS

The Self-organizing Medium Access Control for Sensor networks (SMACS) protocol (Sohrabi, Gao, Ailawadhi, & Pottie, 1999) is a distributive and infrastructure building protocol that forms a flat topology for WSNs. Nodes using SMACS discover their neighbors and establish transmission/reception schedules for communicating with them without the need of global synchronization or clustering. SMACS combines the neighbor discovery phase with the channel assignment phase and assigns the channel to a link immediately after the existence of the link is discovered instead of waiting to finish the network-wide neighbor discovery process.

To reduce collisions between adjacent links, each link operates on a random FDMA or CDMA code. Each node regularly executes a neighborhood discovery procedure and establishes a *directional* link to each discovered neighbor by assigning a time slot to that link. Nodes maintain a TDMA-like *superframe* to communicate with known neighbors. The fixed-size superframe is further divided into smaller and variable size frames. The superframe helps a node maintaining its time slot schedules with all its neighbors such that nodes are required to direct their radios to the proper FDMA or CDMA code for the successful communication.

SMACS avoids computation and communication overhead of transmitting neighborhood information to a central node by using a local scheme instead of a global assignment. Since the neighborhood discovery process is executed regularly, the protocol assumes to efficiently adapt to all the topology changes. As a densely deployed WSN usually has low traffic load, nodes using SMACS will have highly populated schedules and have to wake up quite often just to discover that there is no packet destined to them. The length of a superframe is also a decisive factor. It should be large enough to accommodate the highest node degree in the network, since a smaller superframe cannot conciliate all the neighbors. With SMACS, the link between two nodes is directional. For bidirectional communication between them, two such links are required.

Channel Polling Based MAC Protocols

With the channel polling scheme, also known as preamble sampling or Low Power Listening (LPL), a sending node prefixes data packets with extra bytes called a *preamble* and sends it over the channel to ensure that the destination node detects radio activity and wakes up before the actual payload is sent. On a wake-up, if radio activity is detected, the receiver turns on its radio to receive data packets. Otherwise, the node goes back to the sleep mode until the next polling interval. To avoid deafness, a sender prefixes the preamble at least as long as the check interval duration of the receiver to ensure that the receiver wakes up and performs channel sampling at least once while the preamble is being sent (Polastre, Hill, & Culler, 2004). Figure 9 illustrates how channel polling works in WSNs.

Since channel polling based protocols do not use common active/sleep schedules, they do not need any synchronization, scheduling, or clustering among nodes. Nodes, specially the receiving

Figure 9. Channel polling in WSNs. With the traditional channel polling scheme, a sender first sends an extended preamble that is at least as long as the check interval duration of the receiver to ensure that the receiver will be awake by the time when the data packet is sent.

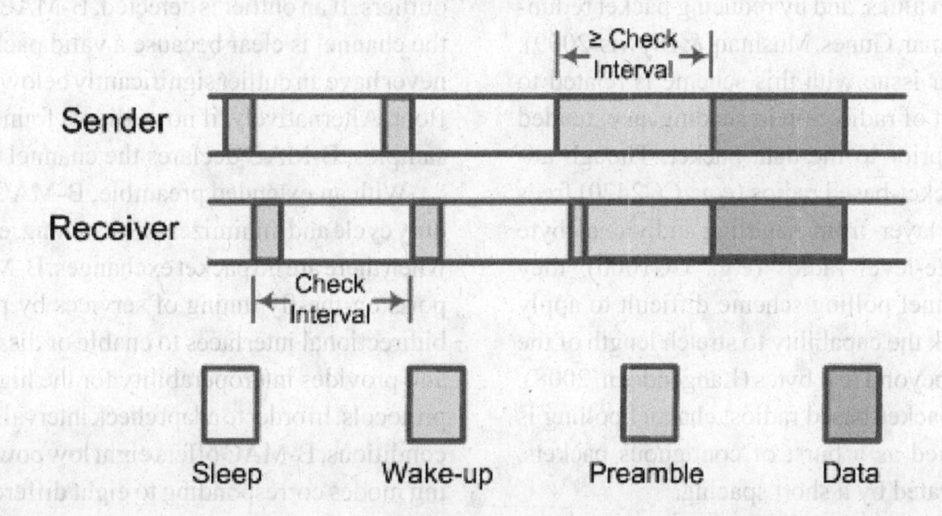

ones consume significantly less energy as they wake up for very short period of time to check the availability of the preamble on the channel. Hence, this scheme is considered to be more energy efficient, particularly under low traffic conditions than other schemes. However, the sending nodes pay the price in sending long and extended preambles, and once the receiving node senses the preamble, it stays on and continues to listen until the data packet is received. Moreover, once the non-targeted nodes overhear the preamble, they can distinguish that they are not the addressed nodes only after receiving the complete preamble. This leads to increased transmission and reception lengths and an increased collision probability, which may further increase as traffic load increases. Another drawback of channel polling is related to the limitation of the duty cycle value. Lowering the duty cycle extends the check interval. That is good from the receiver point of view, but it significantly increases the transmission cost in the shape of long and extended preambles for the sender. Consequently, extended and long preambles cause unnecessary energy consumption both at the receiver and sender ends, overhearing at non-target receivers, excessive latency at each hop, and the bandwidth waste on the broadcast medium. These issues can be tackled by using short preambles, adaptive duty cycle values, and by reducing packet redundancy (Kumar, Gunes, Mushtaq, & Blywis, 2009).

Another issue with this scheme is related to the support of radio chip in sending an extended preamble prior to the data packet. Though advanced packet-based radios (e.g., CC2420) frees the MAC layer from handling individual byte unlike byte-level radios (e.g., CC1000), they make channel polling scheme difficult to apply as they lack the capability to stretch length of the preamble beyond few bytes (Langendoen, 2008). For such packet-based radios, channel polling is implemented as a burst of contiguous packets, each separated by a short spacing.

B-MAC

The channel polling scheme has been renamed as the Low Power Listening (LPL) in the Berkeley MAC (B-MAC) protocol (Polastre, Hill, & Culler, 2004). It is one of the initial MAC protocols working on the channel polling mechanism, where each node has independent *awake* and *sleep* periods. The sum of both these periods is called *check interval* duration. While transmitting, a node precedes the data packet with a preamble that is slightly longer than the check interval duration of the receiver. Nodes wake up at each awake period and sample the medium shortly. If the node detects a preamble, it remains awake to receive the whole preamble. If the preamble is destined to this node, it further extends its wake-up time to receive the data packet; otherwise it goes back to the sleep mode. With the extended preamble, a sender is assured that at some point during the preamble sending, the receiver will wake up and detect the preamble.

B-MAC uses Clear Channel Assessment (CCA) to determine whether the channel is clear. Instead of using a threshold, which is the common method for many CSMA protocols, B-MAC improves the quality of CCA by using an outlier detection method. When a node wants to transmit, it takes a sample of the channel and searches for outliers. If an outlier is detected, B-MAC declares the channel is clear because a valid packet could never have an outlier significantly below the noise floor. Alternatively, if no outlier is found for five samples, B-MAC declares the channel is busy.

With an extended preamble, B-MAC reduces duty cycle and minimizes idle listening, especially when there are no packet exchanges. B-MAC supports on-the-fly tuning of services by providing bidirectional interfaces to enable or disable them and provides interoperability for the higher layer protocols. In order to adapt check intervals to traffic conditions, B-MAC offers eight low power listening modes corresponding to eight different check intervals. It also draws the optimal listening mode for a node based on the neighborhood size in the network. The efficient CCA helps in eliminating most of the false positives and adapting the node to its surroundings.

However, during transmissions, the preamble sent by the sender needs to be longer than check interval durations of receivers. Therefore, B-MAC carries all the stated problems of the extended preamble technique. The modified CCA leads to the increased delay, complexity, and memory usage, since each node needs to have several channel measurements before determining the status of the channel and to keep the statistical track of these measurements.

WiseMAC

The WiseMAC (El-Hoiydi & Decotignie, 2004) is among the initial protocols working on the non-persistent CSMA combined with the channel polling mechanism. The basic functionality of WiseMAC is more or less similar to B-MAC protocol. However, in order to mitigate idle listening and to reduce energy consumption incurred by the long and fixed length preamble, WiseMAC lets a node learn about the awake periods of its neighbours by piggy backing the remaining time to the next awake period in ACK messages. They also keep an updated table containing sampling time offsets of their neighbors. A sending node sends the preamble just before the receiving node wakes up, and hence keeps the preamble length at minimum. Clock drifts may make the transmitter to send a long enough preamble to cover up the estimated drift.

All nodes in a network sample the medium with the common basic cycle duration. However, their awake and sleep periods are independent and left unsynchronized. WiseMAC uses short preambles for regular traffic and switches to longer preambles for infrequent communication. For very low traffic loads, where data packets can be smaller than preambles, WiseMAC repeats the data frame instead of the extended preamble.

With WiseMAC, over-emitting can occur if the receiver is not ready at the end of the preamble due to factors such as interference or collision. This over-emitting can increase further with the increase in the preamble and data packet size. As nodes are unsynchronized, keeping wake up times for all neighbors is a memory and time consuming task for a node. In case of broadcast communication, a transmitter has to deliver the same packet many times to each neighbor. This redundant transmission leads to higher latency and energy consumption for nodes. In addition, the hidden terminal problem can spring up when one node transmits the preamble to a node that is already receiving packets from another node. WiseMAC does not provide a mechanism to adapt schedules of nodes to varying traffic patterns.

AREA-MAC

The Asynchronous Real-time Energy-efficient and Adaptive MAC (AREA-MAC) protocol (Kumar, Gunes, Mushtaq, & Blywis, 2009) provides an application-specific optimized performance in terms of energy efficiency and latency for WSNs. AREA-MAC adapts preamble sampling for packet-based radios and sends out a stream of short preambles instead of one long preamble. It incorporates the ACK mechanism by adding source and destination addresses with each preamble.

Nodes using AREA-MAC have unsynchronized *sleep* and *wake-up* periods. They remain in the sleep mode most of the time and wake up very shortly at every *check interval* duration to check the availability of a preamble on the channel. A node immediately sends a *pre-ACK* frame to the sender if the preamble is found and the destination address of the preamble matches with its address. The node then switches the radio to the receive mode to receive a data packet. A node goes back to the sleep mode immediately if a preamble is not found, or its address does not match with the destination address of the available preamble. However, for broadcast communication, a node sends a pre-ACK frame as soon as it receives the preamble without checking the destination address. A sending node transmits a burst of short preambles with a short spacing and turns its radio to receive mode in between in order to receive the pre-ACK. A data packet is sent as soon as the pre-ACK framè is received, which creates a sort of pre-established link between the sender and receiver. This minimizes the chances of the data packet being dropped as the probability of data collision decreases. However, collisions might occur with preambles, which is not the severe case as the node continues sending short preambles for full check interval duration.

Short preambles unified with the ACK mechanism can impede the problems of extended and long preambles and can reduces energy consumption both at the receiver and sender ends, overhearing at non-target receivers, and excessive latency at each hop. Moreover, the AREA-MAC preamble also contains a unique sequence number of the upcoming data packet, which helps the receiving node to check whether the upcoming data packet is already received/processed. In case of the redundant packet, the receiver sends a *pre-NACK* to the sender, which then discards that data packet in order to reduce packet redundancy. If the sender node has more packets waiting in its queue to be transmitted, it indicates the receiving node by setting the *more-to-follow* bit of the data packet. The receiving node can then adapt its next wake-up schedule in accordance with the sender, which further decreases latency and energy consumption of nodes. Results presented in (Kumar, Gunes, Mushtaq, & Blywis, 2009) show an improved performance of AREA-MAC in terms of energy, delay, data packet delivery, preamble sending/receiving, and overhearing over the protocols that use extended preambles. However, adding extra information with each preamble can increase overhead for nodes as the receiving node has to extract this information from every received preamble.

Hybrid MAC Protocols

Hybrid MAC protocols combine the strengths of two or more different MAC schemes in order to achieve a joint improvement. They usually combine a synchronized scheme with an asynchronous one. Though hybrid protocols aggregate advantages of multiple schemes, they also carry, among other, scaling and complexity problems in maintaining two or more different working modes.

IEEE 802.15.4

The IEEE 8021.5.4 (IEEE 802.15.4, 2006) (referred to as 802.15.4 hereinafter) deals with the Low-Rate Wireless Personal Area Network (LR-WPAN). A LR-WPAN is a simple, low cost, low power, and low data-rate communication network that facilitates ease of installation, reasonable battery life, and reliable data transfer within a limited range of around 10 meters. The 8021.5.4 protocol defines the PHY and MAC layers for a LR-WPAN. It supports both star and peer-to-peer operation with Fully and Reduced Function Devices (FFDs and RFDs). Though this protocol was not specifically designed for WSNs, its capability for fitting different requirement of WSNs by adequately tuning its parameters has made it a front runner for several WSN related applications. In fact, its pertinence to WSNs has already been supported by several commercial sensor vendors. Together with ZigBee[2], which provides the upper (network and application) layers, 802.15.4 defines a full protocol stack suitable for several surveillance, home automation, health care, industrial, and agricultural related applications of WSNs.

The PHY layer of 802.15.4 is responsible for activation and deactivation of the radio transceiver, Energy Detection (ED), Link Quality Indication (LQI) for received packets, channel frequency selection, Clear Channel Assessment (CCA), and transmitting as well as receiving packets across the physical medium. The radio operates at three different frequency bands of 868 MHz,

915 MHz, and 2.4 GHz. The features of the MAC layer include beacon management, channel access, Guaranteed Time Slots (GTS) management, frame validation, acknowledged frame delivery, association, and disassociation of nodes. The MAC protocol supports two operational modes selected by the coordinator; the *non beacon-enabled mode*, in which MAC is simply ruled by non-slotted CSMA/CA and the *beacon-enabled mode*, in which beacons are periodically sent by the coordinator to synchronize associated nodes.

The format of the (optional) *MAC superframe* is defined by the PAN coordinator and is depicted in Figure 10. The superframe is bounded by network beacons sent by the coordinator and is divided into 16 equally sized slots. A beacon frame is transmitted in the first slot of each superframe and is used to synchronize attached devices, to identify the PAN, and to describe the structure of the superframe. The superframe can be divided in two parts; Contention Access Period (CAP) and Contention Free Period (CFP). Any device wishing to communicate during the CAP period competes with other devices using a slotted CSMA/CA mechanism. For low-latency applications or applications requiring specific data bandwidth, the PAN coordinator may dedicate contention-free TDMA-like GTS portion of the CFP to devices. The combination of CSMA and TDMA makes 802.15.4 MAC a hybrid protocol.

Though 802.15.4 MAC protocol fulfils many of the WSN requirements, it endures various limitations, especially for timeliness, energy, and bandwidth critical applications, and its performance can certainly be improved. All these limitations and their proposed solutions, along with the detailed working of 802.15.4 are elaborated in (Kumar, Gunes, Almamou, & Hussain, 2008).

Z-MAC

The Zebra MAC (Z-MAC) (Rhee, Warrier, Aia, & Min, 2005) protocol is a hybrid scheme that combines strengths of TDMA and CSMA while offsetting their weaknesses. Z-MAC is a traffic adaptive protocol in the sense that under low contention, it switches to CSMA to achieve high channel utilization and low delays, and under high contention, it switches to TDMA to achieve high channel utilization, fairness, and less collisions. Unlike the traditional TDMA scheme, a node using Z-MAC can also utilize slots assigned to other nodes. However, the owner of the slot always has higher priority over the others, which reduces the chances of collisions.

Figure 10. MAC superframe structure. It is divided into 16 equally sized slots and is bounded by network beacons. A superframe can have an active and an inactive portion.

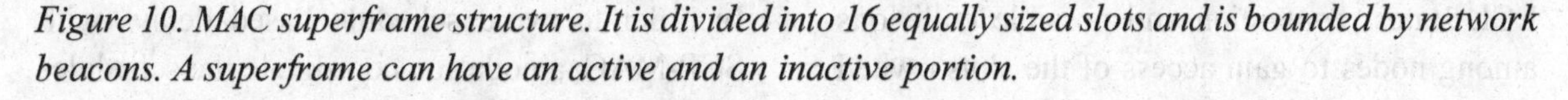

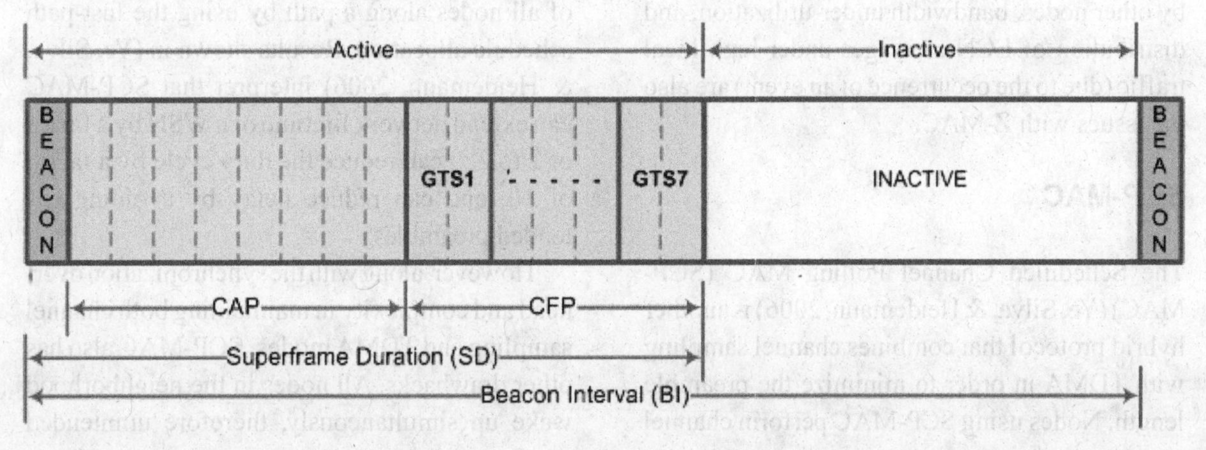

To reduce the hidden terminal effect during high contention, a node using Z-MAC sends an Explicit Congestion Notification (ECN) frame to the neighbor it has message for. Note that nodes observe the contention level by tracking the time they spend in backoff due to the unsuccessful carrier sensing. The neighbor further broadcasts the ECN to its neighbors. On reception of the ECN message, a node enters a High Contention Level (HCL) state, where it only attempts to transmit in its slot and those of its direct neighbors, thus the node helps in reducing contention between neighbors two hops apart. The node switches back to the Low Contention Level (LCL) state if it does not receive an ECN within a time period.

Z-MAC uses CSMA as the baseline MAC scheme, and in the worst case it always falls back to CSMA. Z-MAC provides a simple two-hop synchronization scheme, where each sending node adjusts its frequency based on its current data rate and resources. Hence, Z-MAC is robust to timing and slot assignment failures, channel conditions, and topology changes. However, during the start up phase, Z-MAC requires global time-synchronization, which certainly is an energy, time, and memory consuming for lightweight nodes. A highly dynamic WSN may need to perform this costly procedure more than once. Apart that, complexity in maintaining both CSMA and TDMA modes, switching between the HCL and LCL states, contention and possible collisions among nodes to gain access of the slots owned by other nodes, bandwidth under-utilization, and distribution of ECN messages under high local traffic (due to the occurrence of an event) are also the issues with Z-MAC.

SCP-MAC

The Scheduled Channel Polling MAC (SCP-MAC) (Ye, Silva, & Heidemann, 2006) is another hybrid protocol that combines channel sampling with TDMA in order to minimize the preamble length. Nodes using SCP-MAC perform channel polling periodically and switch to the sleep mode when there is no traffic available on the channel. However, unlike the channel polling scheme, sampling times of nodes are synchronized. In result of that, a very short *wake-up tone* can be sent to wake up the receiver, which largely reduces the overhead of transmitting long preambles.

The basic working of SCP-MAC is illustrated in Figure 11. Before communicating, the node silently waits in the sleep state and performs carrier sense within the first contention window (CW1) just before the polling time of the receiver. On finding the channel idle, the sender sends a short wake-up tone to activate the receiver. Otherwise, it goes back to the sleep mode and performs regular channel polling. After the sender wakes up a receiver, it enters the second contention window (CW2) and sends data if the node still detects the channel idle. The receiving node that has received a wake-up tone extends its wake-up period to receive the data packet.

SCP-MAC uses two separate contention phases to achieve low collision probability. The RTS/CTS handshake can be enabled or disabled with SCP-MAC. When the RTS/CTS is disabled, overhearing is performed by examining destination address from the packet header. Adaptive listening is also supported in SCP-MAC, where the MAC layer, after transmitting a packet, immediately polls the channel for additional traffic. In order to avoid schedule based delays with SCP-MAC, a node can coordinate the schedules of all nodes along a path by using the fast-path schedule allocation. Results shown in (Ye, Silva, & Heidemann, 2006) interpret that SCP-MAC can extend network lifetime of a WSN by a factor of 2 to 2.5, can reduce the duty cycle by a factor of 10, and can reduce delay by avoiding extended preambles.

However, along with the synchronization overhead and complexity in maintaining both channel sampling and TDMA modes, SCP-MAC also has other drawbacks. All nodes in the neighborhood wake up simultaneously, therefore unintended

Figure 11. Basic working of SCP-MAC. A sender waits in sleep mode until the wake-up time of the receiver, sends a wake-up tone to wake it up, and then sends a data packet. To minimize the collision probability, it performs carrier sense twice; one before sending a wake-up tone and the other before sending the data packet.

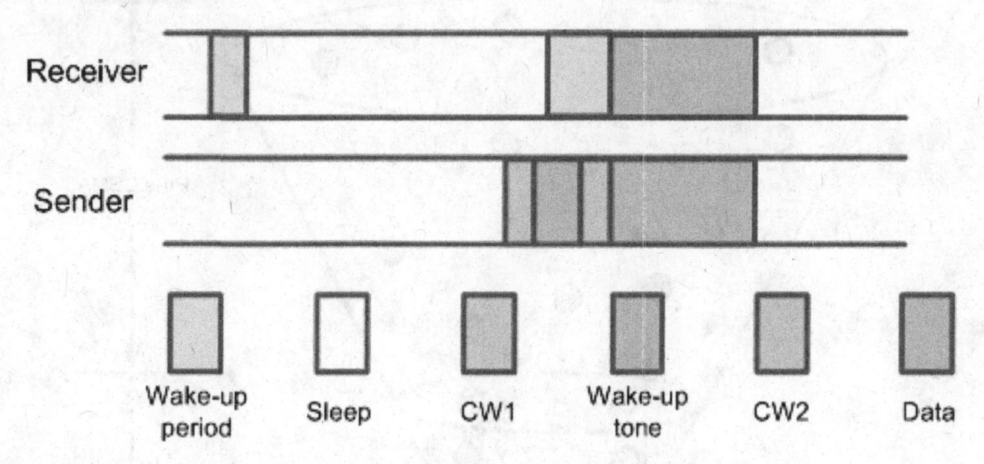

receivers cannot avoid overhearing packets (particularly preambles) and participating in higher contention at that time. Collisions may occur, which force nodes to postpone their transmission to the next synchronized wake-up time and thus inducing higher latency and energy consumption for nodes. Double contention also increases the system latency.

Funneling-MAC

In WSNs, packets generated by nodes usually travel hop-by-hop in a many-to-one pattern, i.e, from nodes to the sink node. Thus, they exhibit a unique funneling effect, which could lead to an intensified traffic, collisions, delays, and energy drain as events move closer toward the sink node. A hybrid, localized, and sink-oriented Funneling-MAC protocol (Ahn, Miluzzo, Campbell, Hong, & Cuomo, 2006) exposes this WSN phenomenon. Funneling-MAC is mainly a CSMA/CA protocol with a localized TDMA algorithm, which works only within a few hops from the sink node in the *intensity region*. The basic working of Funneling-MAC is shown in Figure 12.

The TDMA scheduling is managed by the sink node and operates locally in the intensity region. The burden of computing and maintaining the depth of the intensity region also falls on the sink node. The sink node broadcasts a beacon that triggers the TDMA scheduling. All nodes perform CSMA by default unless they receive the beacon and are then considered to be in the intensity region and called *F-nodes*. The area of the intensity region is defined by the sink node by controlling the transmission power of the beacon. The F-nodes switch back to CSMA, if they do not receive the beacon within the specified time period. The sink node calculates the TDMA schedule as per traffic conditions and then broadcasts it to all the F-nodes. Each F-node transmits its scheduled packet at the allocated time slot specified in the TDMA frame. In order to allow the transmission of data packets that have not been allocated slots yet, a CSMA frame is reserved between two consecutive TDMA frame schedules.

With Funneling MAC, F-nodes consume more energy in receiving several beacons and schedule frames. Beacon and schedule frames are sent at potentially high power and could interfere with on-going communication. Many duty cycled

Figure 12. Basic working of Funneling-MAC. Nodes that receive beacon(s) inside the intensity region are called F-nodes. The area of the intensity region is defined by the sink node by controlling the transmission power of the beacon.

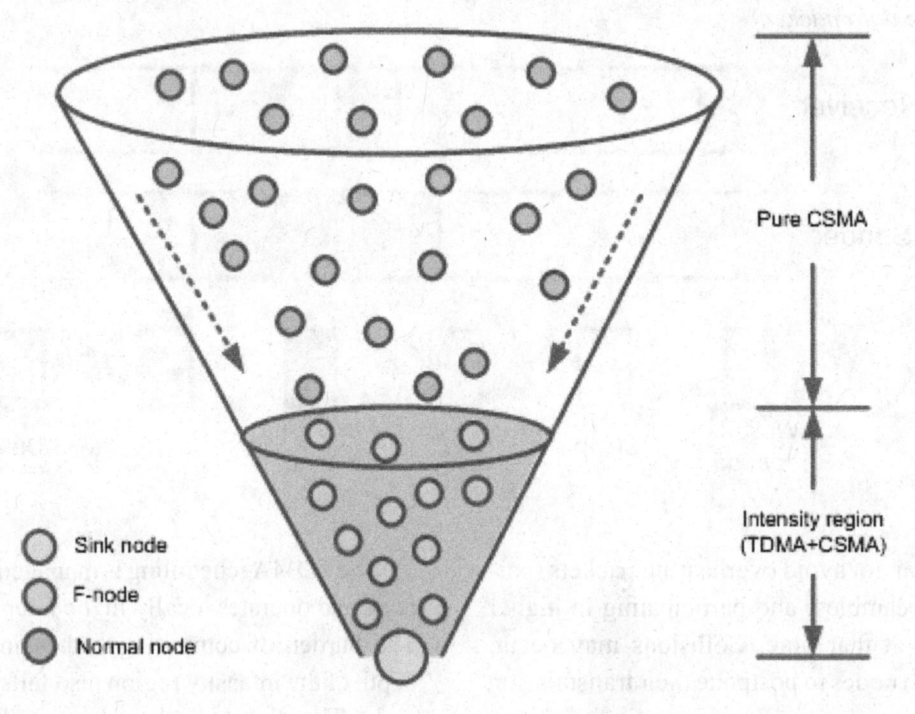

nodes within the intensity region could become interferes by missing out the beacon if their periodic wake-up times do not match with the beacon time. Funneling MAC also carries complexity in managing intensity region over the time and in performing CSMA after every second TDMA frame. The number of F-nodes in a dense WSN could be very high, and in result each F-node has to wait longer for its turn to communicate with the sink. This will increase network latency.

Comparison of Different MAC Protocols

Having discussed the detailed working, advantages, and drawbacks of the major channel accessing categories used for WSNs and several MAC protocols falling in each category, we next present a comprehensive comparison of these protocols in Table 2. Most of the time, it is hard to compare protocol architectures that use different working space, we in the table, however, try to analytically compare all the discussed MAC protocols for their support to some metrics that are vital in designing any WSN MAC protocol. The considered metrics are related to the energy efficiency, timeliness, asynchrony, adaptability, and scalability factors.

It is clear from the table that most of the protocols do not support all of these parameters. They either support a parameter partially or trade-off with another parameter, particularly in favor of energy. Several protocols are suboptimal for a given parameter and can be improved further. Significantly, most protocols do not consider delay as an important metric in their design.

Table 2. Comparison of different MAC protocols proposed for WSNs based on the support for different parameters

Protocol	Type	Energy Eff.	Delay Eff.	Asynchronous	Adaptive	Scalable
S-MAC	CSMA	yes[1]	no	yes[2]	yes[1]	yes[1]
T-MAC	CSMA	yes[1]	no	yes[2]	yes	yes[1]
DSMAC	CSMA	yes[3]	yes[1]	no	yes	no
GSA	CSMA	yes[1]	yes[1]	no	no	no
STEM	CSMA	yes[4]	no	yes	no	yes
LEACH	TDMA	yes	no	no	no	no
TRAMA	TDMA	yes[1]	no	no	yes	no
SMACS	TDMA	yes[1]	no	yes	yes[1]	yes
B-MAC	LPL	yes[1]	no	yes	yes[1]	yes
WiseMAC	LPL	yes[1]	yes[1]	yes	yes	yes
AREA-MAC	LPL	yes	yes	yes	yes	yes
802.15.4	Hybrid	yes[1]	yes[1]	yes[5]	no	no
Z-MAC	Hybrid	yes	no	no	yes	no
SCP-MAC	Hybrid	yes[1]	yes[1]	no	yes	no
Funn-MAC	Hybrid	yes[1]	yes[1]	yes[6]	yes	yes

[1] Suboptimal, can be improved
[2] Uses common listen periods
[3] Trades-off with other parameter(s)
[4] Uses two different channels
[5] Depends on the topology
[6] Supports partially

FUTURE RESEARCH DIRECTIONS

The research community has witnessed intense research related to the MAC design for WSNs over the last years. Various MAC protocols target different objectives and have different performance priorities. The most visible and prime consideration of almost all of them is the issue of energy efficiency. Other typical performance metrics like latency, adaptability to traffic and topology conditions, scalability, fairness, throughput, and bandwidth utilization are mostly overlooked or dealt as secondary objectives due to the school of thought that nodes are densely deployed and they collaborate with each other rather than compete. However, with the advent of new Micro Electro-Mechanical Systems (MEMS) technologies and energy harvesting techniques, and with the dynamic increase in several WSN related applications, the research community is set to experience many diverse directions in this domain. For numerous applications of WSNs, some of these metrics may temporarily outweigh energy efficiency. Hence, energy consumption, which no doubt is the most critical parameter for WSNs performance, should not be the *only* focal point in designing a MAC protocol.

For medical urgency, surveillance, security, terrorist attacks, home automation, flood, fire, and seismic detection applications, the provision of timeliness is as crucial as saving energy. For example, a patient wearing an e-textile integrated with sensor nodes needs an automatic but timely attention from doctors or emergency servicemen when he suffers from a severe disease. In this case, the importance of delivery ratio may also be increased as one disease could immediately result in other diseases, and body sensors deployed on or near different body parts need to urgently inform medical personnel about the relative body part.

Last but not least, selecting a proper hardware scheme has a lasting impact on the performance of a MAC protocol and ultimately on a whole WSN system. Support for multi-channel hardware, balancing an appropriate memory size, use of wake-up radios, and selection of packet- or bit-based radios are the important factors (Bachir, Dohler, Watteyne, & Leung, 2010).

CONCLUSION

Throughout this chapter, we have dwelt on different factors and techniques that need definite consideration while designing a WSN MAC protocol. Idle listening, collisions, overhearing, control packets, and over-emitting are the main sources of energy waste in WSNs. As the traffic generation in WSNs is usually low, nodes spent large amounts of time listening to an idle channel. To cope with most of the energy related issues, an eminent amount of research work supports duty cycling in WSNs, where nodes remain in the sleep mode most of the time. However, this duty cycling results in high latency and low throughput, and a deep consideration is required to select a proper duty cycle value.

MAC protocols for WSNs have been categorized into four classes in this chapter; contention based, scheduling based, channel polling based, and hybrid protocols. Along with the detailed working of each category and protocols falling in, their advantages as well as disadvantages have also been discussed in detail. Contention based MAC protocols usually follow the common active/sleep schedules, where all neighboring nodes wake up simultaneously at fixed periodic intervals and access the channel through CSMA. Such protocols are scalable, robust to network dynamics, and do not need any clustering or topology information. However, traffic rigidity, latency, collisions, overhearing, idle listening, and hidden nodes are the issues with such protocols. Many of these issues are avoided in scheduling based protocols, where nodes are forced to access their allocated slots. But, overhead and extra traffic needed to setup and maintain schedules and synchronization, higher queuing delays, adaptability to traffic and topology changes, and scalability are the major concerns with such schemes.

The channel polling based mechanism of MAC designing is getting more attention from researchers. Nodes in this scheme neither follow common active/sleep schedules nor do they need any sort of synchronization or clustering. The sender prefixes a preamble before each data frame to ensure that the receiver detects the radio activity and wakes up before the actual payload is sent. In traditional channel polling protocols, preambles sent by the sender need to be longer than the check interval duration of the potential receiver to guarantee that the receiver will be awake when the data portion arrives. Such extended preambles introduce higher overhead, latency, and energy consumption for the sender, intended receiver, and also for the non-targeted receivers. These issues can be handled by using short preambles, adaptive duty cycle values, and reduced redundancy.

After discussing several protocols from each category, we have presented a substantial comparison of these protocols for different parameters in Table 2. At the end of the chapter, future research directions in designing a MAC protocol have been envisioned, where we conclude that energy efficiency even though being the most critical metric is not sufficient to address. Factors like timeliness, scalability, asynchrony, delivery ratio, and adaptability to traffic and topology changes may also need important considerations as per requirements of the application.

REFERENCES

Ahn, G.-S., Miluzzo, E., Campbell, A. T., Hong, S. G., & Cuomo, F. (2006). *Funneling-MAC: A localized, sink-oriented MAC for boosting fidelity in sensor networks*. 4th ACM Conference on Embedded Networked Sensor Systems (SenSys'06). Boulder, Colorado, USA.

Akyildiz, I. F., & Varun, M. C. (2010). *Wireless sensor networks*. John Wiley and Sons Ltd. doi:10.1002/9780470515181

Arora, A., Dutta, P., & Bapat, S. (2004). A line in the sand: A wireless sensor network for target detection, classification, and tracking. *Computer Networks*, *46*, 605–634. doi:10.1016/j.comnet.2004.06.007

Bachir, A., Dohler, M., Watteyne, T., & Leung, K. (2010). MAC essentials for wireless sensor networks. *IEEE Communications Surveys and Tutorials*, *12*(2), 222–248. doi:10.1109/SURV.2010.020510.00058

Bhuse, V., & Gupta, A. (2005). *Anomaly intrusion detection in wireless sensor networks*. Western Michigan University, Kalamazoo, MI-49008 Report.

Dam, T. v., & Langendoen, K. (2003). An adaptive energy-efficient MAC protocol for wireless sensor networks. *1st ACM Conference on Embedded Networked Sensor Systems (SenSys)*, (pp. 171 - 180).

El-Hoiydi, A., & Decotignie, J. (2004). WiseMAC: An ultra low power MAC protocol for the downlink of infrastructure wireless sensor networks. *Ninth IEEE Symposium on Computers and Communication, ISCC04*, (pp. 244 - 251).

Garg, V. (2007). *Wireless communications and networking*. Elsevier - Morgan Kaufmann Publishers.

Goense, D., Thelen, J., & Langendoen, K. (2005). *Wireless sensor networks for precise phytophthora decision support*. 5th European Conference on Precision Agriculture (5ECPA). Uppsala, Sweden.

Haapola, J., Shelby, Z., Pomalaza-Raez, C., & Mahonen, P. (2005). Cross-layer energy analysis of multi-hop wireless sensor networks. *European Conference on Wireless Sensor Networks, EWSN'05*, (pp. 33-44). Istanbul, Turkey.

Heinzelman, W., Chandrakasan, A., & Balakrishnan, H. (2002). An application-specific protocol architecture for wireless microsensor networks. *IEEE Transactions on Wireless Communications*, *1*(4). doi:10.1109/TWC.2002.804190

Hoesel, L. V., Nieberg, T., Wu, J., & Havinga, P. J. (2004). Prolonging the lifetime of wireless sensor networks by cross-layer interaction. *IEEE Wireless Communication*, *11*(6), 78–86. doi:10.1109/MWC.2004.1368900

IEEE. 802.15.4. (2006). *MAC and PHY specifications for LR-WPANs*. Retrieved from http://ieee802.org/15/pub/TG4.html

Ilyas, M., & Mahgoub, I. (2006). *Sensor network protocols*. Taylor and Francis Group.

Karl, H., & Willig, A. (2006). *Protocols and architectures for wireless sensor networks*. John Wiley and Sons Ltd.

Kawadia, V., & Kumar, P. R. (2005). *A cautionary perspective on cross-layer design*. IEEE Wireless Communication Magazine.

Klues, K., Hackmann, G., Chipara, O., & Lu, C. (2007). *A component-based architecture for power-efficient media access control in wireless sensor networks*. 5th International Conference on Embedded Networked Sensor Systems (SenSys'07). Sydney, Australia.

Kumar, P., Gunes, M., Almamou, A. A., & Hussain, I. (2008). *Enhancing IEEE 802.15.4 for low-latency, bandwidth, and energy critical WSN applications*. 4th International Conference on Emerging Technologies (IEEE ICET 2008). Rawalpindi, Pakistan.

Kumar, P., Gunes, M., Mushtaq, Q., & Blywis, B. (2009). A real-time and energy-efficient MAC protocol for wireless sensor network. [IJUWBCS]. *International Journal of Ultra Wideband Communications and Systems*, *1*(2), 128–142. doi:10.1504/IJUWBCS.2009.029002

Kumar, P., Gunes, M., Mushtaq, Q., & Schiller, J. (2010). *Performance evaluation of AREA-MAC: A cross-layer perspective*. Fifth International Conference on Mobile Computing and Ubiquitous Networking (ICMU 2010). Seattle, USA.

Labrador, M. A., & Wightman, P. M. (2009). *Topology control in wireless sensor networks*. Springer Science and Business Media B.V.

Langendoen, K. G. (2008). Medium access control in wireless sensor networks. In Wu, H., & Pan, Y. (Eds.), *Medium access control in wireless networks* (pp. 535–560). Nova Science Publishers, Inc.

Li, X. Y. (2008). *Wireless ad hoc and sensor networks, theory and applications*. Cambridge University Press. doi:10.1017/CBO9780511754722

Li, Y., Chen, C. S., Song, Y.-Q., & Wang, Z. (2007). *Real-time QoS support in wireless sensor networks: A survey*. 7th IFAC International Conference on Fieldbuses and Networks in Industrial and Embedded Systems. Toulouse, France.

Li, Y., Ye, W., & Heidemann, J. (2005). *Energy and latency control in low duty cycle MAC protocols*. IEEE WCNC.

Lin, E.-Y., Rabaey, J., & Wolisz, A. (2004). *Power-efficient rendezvous schemes for dense wireless sensor networks* (pp. 3769–3776). IEEE ICC.

Lin, P., Qiao, C., & Wang, X. (2004). *Medium access control with a dynamic duty cycle for sensor networks* (pp. 1534–1539). IEEE WCNC.

Mainwaring, A., Polastre, J., Szewczyk, R., Culler, D., & Anderson, J. (2002). *Wireless sensor networks for habitat monitoring*. First ACM International Workshop on Wireless Sensor Networks and Application (WSNA). Atlanta, GA, USA.

Melodia, T., Vuran, M. C., & Pompili, D. (2005). Lecture Notes in Computer Science: *Vol. 3883. The state of the art in cross-layer design for wireless sensor networks. Proceedings of EuroNGI Workshops on Wireless and Mobility*. Springer.

Misra, S., Woungang, I., & Misra, S. C. (2009). *Guide to wireless sensor networks*. Springer-Verlag London Limited.

Polastre, J., Hill, J., & Culler, D. (2004). Versatile low power media access for wireless sensor networks. *2nd ACM Conference on Embedded Networked Sensor Systems (SenSys 2004)*, (pp. 95 - 107). Baltimore, MD.

Rajendran, V., Obraczka, K., & Garcia-Luna-Aceves, J. J. (2003). *Energy-efficient, collision-free medium access control for wireless sensor networks*. Los Angeles, CA: ACM SenSys.

Rhee, I., Warrier, A., Aia, M., & Min, J. (2005). *ZMAC: A hybrid MAC for wireless sensor networks*. 3rd ACM Conference on Embedded Networked Sensor Systems (SenSys'05).

Santi, P. (2005). Topology control in wireless ad hoc and sensor networks. *ACM Computing Surveys*, *37*(2), 164–194. doi:10.1145/1089733.1089736

Schurgers, C., Tsiatsis, V., Ganeriwal, S., & Srivastava, M. (2002). Optimizing sensor networks in the energy-latency-density design space. *IEEE Transactions on Mobile Computing*, *1*(1), 70–80. doi:10.1109/TMC.2002.1011060

Simon, G., Maroti, M., Ledeczi, A., & et-al. (2004). *Sensor network-based countersniper system*. 2nd ACM Conference on Embedded Networked Sensor Systems (SenSys). Baltimore, MD, USA.

Sohrabi, K., Gao, J., Ailawadhi, V., & Pottie, G. J. (1999). *Protocols for self-organization of a wireless sensor network*. 37th Allerton Conference on Communication, Computing and Control.

Sohraby, K., Minoli, D., & Znati, T. (2007). *Wireless sensor networks, technology, protocols, and applications*. Wiley Interscience Publications. doi:10.1002/047011276X

Sun, Y., Gurewitz, O., & Johnson, D. B. (2008). *RI-MAC: A receiver-initiated asynchronous duty cycle MAC protocol for dynamic traffic loads in wireless sensor networks*. 6th ACM Conference on Embedded Networked Sensor Systems, SenSys'08. Raleigh, NC, USA.

Tolle, G., Polastre, J., Szewczyk, R., et al. (2005). *A macroscope in the redwoods*. 3rd ACM Conference on Embedded Networked Sensor Systems (SenSys). San Diego, CA, USA.

Tse, D., & Viswanath, P. (2005). *Fundamentals of wireless communication*. Cambridge University Press.

Tubaishat, M., & Madria, S. (2003). Sensor networks: An overview. *IEEE Potentials*, 20-23.

Ye, W., Heidemann, J., & Estrin, D. (2004). Medium access control with coordinated adaptive sleeping for wireless sensor networks. *IEEE/ACM Transactions on Networking*, *12*(3), 493–506. doi:10.1109/TNET.2004.828953

Ye, W., Silva, F., & Heidemann, J. (2006). *Ultra-low duty cycle MAC with scheduled channel polling*. 4th ACM Conference on Embedded Networked Sensor Systems (SenSys'06).

Zheng, J., & Jamalipour, A. (2009). *Wireless sensor networks: A networking perspective*. Hoboken, NJ: John Wiley and Sons Inc. doi:10.1002/9780470443521

ENDNOTES

[1] Terms such as sensor node, node, wireless node, mote, and smart dust are used somehow interchangeably.

[2] http://www.zigbee.org

Chapter 19
Routing Optimization and Secure Target Tracking in Distributed Wireless Sensor Networks

Majdi Mansouri
University of Technology of Troyes, France

Khoukhi Lyes
University of Technology of Troyes, France

Hichem Snoussi
University of Technology of Troyes, France

Cédric Richard
Université de Nice Sophia-Antipolis, France

ABSTRACT

Due to the limited energy supplies of nodes in wireless sensor networks (WSN), optimizing their design under energy constraints, reducing their communication costs, and securing their aggregated data are of paramount importance. To this goal, and in order to efficiently solve the problem of target tracking in WSN with quantized measurements, the authors propose to jointly estimate the target position and the relay location, and select the secure sensor nodes and the best communication path. Firstly, the authors select the appropriate group in order to balance the energy dissipation and to provide the required data of the target in the WSN. This selection is also based on the transmission power between a single sensor and a cluster head (CH). Secondly, the authors detect the malicious sensor nodes based on the

DOI: 10.4018/978-1-4666-0101-7.ch019

information relevance of their measurements. Thirdly, they select the best communication path between the candidate sensor and the CH. Then, the authors estimate jointly the target position and the relay location using Quantized Variational Filtering (QVF) algorithm. The selection of candidate sensors is based on multi-criteria function, which is computed by using the predicted target position provided by the QVF algorithm. The proposed algorithm for the detection of malicious sensor nodes is based on Kullback-Leibler distance between the current target position distribution and the predicted sensor observation, while the best communication path is selected as well as the highest signal-to-noise ratio (SNR) at the CH. The efficiency of the proposed method is validated by extensive simulations in target tracking for wireless sensor networks.

I. INTRODUCTION

Wireless Sensor Network (WSN) is defined as highly distributed networks of small and light-weight wireless nodes, which are deployed in large numbers to mentor the environment or system by measuring physical parameters such as temperature, pressure, or relative humidity (Chen & Zhao, 2005). Due to their small dimension, sensor nodes are typically operated by lightweight batteries, which are difficult to be replaced or recharged. Then, WSN is a typical energy-restricted network. As observed in almost applications, the communication power consumption accounts for about 70% of the total power in WSN (Akyildiz et al., 2002), i.e. most energy of sensor nodes is spent in the exchange of routing information and data. A direct consequence is that sensor nodes will use up their energy quickly, which is uneconomical. Recently, researchers have proposed a special type of node called a "relay node" which is responsible for routing data packets. By using relay nodes, each sensor node would be able to send only its own data without a need to relay traffic from others (Wang et al., 2005). To this goal, minimizing the communication costs between sensor nodes is critical to extend the lifetime of sensor networks. Another important metric of sensor networks is the accuracy of the sensing result of the target in that several sensors in the same cluster can present redundant data. Because of physical characteristics of sensor networks, such as distance, modality, or noise model of individual sensors, the data packets genetared from different sensors can have various qualities. Hence, the accuracy depends on the selection achieved by the cluster head on sensors and communication links.

In this chapter, we address the problem of secure target tracking, relay localization and sensors selection using WSN based on quantized proximity sensors. Target tracking using quantized observations is a nonlinear estimation problem that can be solved using estimation solutions such as unscented Kalman filter (KF) (Julier & Uhlmann, 2004), particle filters (PF) (Djuric et al., 2003) or variational filtering (VF) (Snoussi & Richard, 2006).

Recently, the VF has been proposed as an efficient solution for solving the target tracking problem since: (i) it respects the communication constraints of sensors, (ii) the online update of the filtering distribution and its compression are simultaneously performed, and (iii) it has the nice property to be model-free, ensuring the robustness of data processing.

There has already been a certain amount of research in the area of relay node placement estimation, sensors selection, and secure target tracking for wireless sensor networks. Li *et al.* (Li, Kao, & Ke, 2009) have proposed a Voronoi-based relay node placement scheme to balance the energy consumption of each sensor node spent in communication. Bari et al. (Bari, Teng, & Jaekel, 2009) have studied the relay node placement with a mobile data collector. Iranly *et al.* (Iranli, Maleki, & Pedram, 2005) have addressed the joint problem of energy provisioning and relay node placement. Hou et al. (Hou et al., 2005) have formulated the issues discussed above as a mixed-integer nonlinear programming problem. In (Ergen & Varaiya, 2006), Ergen and Varaiya have proposed an approximation algorithm where the location of each relay node is restricted to a square lattice. An energy-constrained relay node placement problem has been explored by Wang *et al.* (Wang, 2008) aiming at minimizing the network cost with constraint on the network lifetime. The work in (Chen, Hou, & Sha, 2004) has explored a topic similar to tracking security by proposing a protocol that verifies securely the time of encounters in multi-hop networks. However, (Chen, Hou, & Sha, 2004) did not address the security issue in the context of Bayesian tracking. Sensors selection based on expected information gain was introduced for decentralized sensing systems in (Manyika & Durrant-Whyte, 1995). A mutual information between a predicted sensor observation and a current target location distribution was proposed in (Liu, Reich, & Zhao, 2003)–(Luo & Giannakis, 2008) to evaluate the expected information gain about the target position attributable to sensors. On the other hand, without using the information theory, Yao *et al.* (Yao et al., 1998) found that the overall localization accuracy depends not only on the accuracy of individual sensors but also on the sensor locations relative to the target position during the development of localization algorithms. In addition, sensor selection can be discussed within a framework of optimal estimation, where the error estimation is normally used as the cost function of the optimization. In previous works (Ucinski & Patan, 2007; Kaplan, 2006; Han et al., 2010), the problem of sensors selection is formulated as a constrained 0-1 integer programming problem, where the error estimation is minimized.

Most of these previous works do not consider a tradeoff between the quality of sensed data, the transmit power and the power stored in candidate nodes to select the candidate sensors. Moreover, these works have simplified or ignored the security issue in the context of Bayesian tracking and the communication costs through a sensors-cluster head path.

Our contribution lies in the following aspects: 1) we improve the use of variational filtering algorithm (VF), which perfectly fits the highly non-linear conditions and eliminates the transmission error; 2) we jointly estimate the target position and relay location by using QVF algorithm; 3) we jointly select the best group of sensors that participates in data collection and detect the malicious sensor nodes; 4) we explore the impact of the communication path selection on QVF algorithm performances and introduce an adaptive quantization algorithm.

The remainder of the chapter is organized as follows. Section II presents both the observation model and the general state evolution models. Section III is devoted to the technique aimed at adaptively estimating the target position and the relay node location. The appropriate group of sensors selection is presented in Section IV. The malicious sensor nodes detection method is presented in Section V. Then, Section VI presents the best communication path selection scheme. Section VII gives some numerical results. Finally, Section VIII concludes the chapter.

II. PROBLEM FORMULATION

The variational filtering (VF) algorithm for target tracking inherits many desirable properties from the Bayesian Inference framework. An important step in the Bayesian target tracking is the recursive estimation of the predictive distribution described as follows,

$$p(X_t|Z_{1:t-1}) = \int p(X_t|X_{t-1})\, p(X_{t-1}|Z_{1:t-1}) dX_{t-1}, \qquad (1)$$

Where $p(X_t|X_{t-1})$ is the conditional distribution used to model the prior time evolution of the target state. By incorporating the observation model $p(Z_t|X_t)$, the new estimate of the targets state X_t is updated using the following predictive distribution $p(X_t|Z_{1:t-1})$:

$$p(X_t \mid Z_{1:t}) = \frac{p(Z_t \mid X_t)p(X_t \mid Z_{1:t-1})}{p(Z_t \mid Z_{1:t-1})}$$

where

$$p(Z_t|Z_{1:t-1}) = \int p(Z_t|X_{t-1})\, p(X_{t-1}|Z_{1:t-1})dX_t. \tag{2}$$

The observation model $p(Z_t|X_t)$ depends on the sensing mode employed by the sensors, while the state evolution model $p(X_t|X_{t-1})$ is always described by a parametric model.

A. General State Evolution Model

In this chapter, we use a General State Evolution Model (GSEM) described in (Snoussi & Richard, 2006; Vermaak, Lawrence, & Perez, 2003; Teng, Snoussi, & Richard, 2007; Mansouri et al., 2009), which is more adaptive to practical situations and has no restriction on velocity or moving direction of the target. As defined above, at instant t, the joint hidden state $X_t = \{x_t, R_t\}$ to be estimated contains the target position x_t, a set of activated relays locations $R_t = \{r_t^i\}_{i=1}^{m_t}$, where m_t denotes the number of activated relays. For a relay i for example, r_t^i is assumed to be a Gaussian variable, whose expectation is its latest estimate value $\hat{r}_t$, and the precision matrix v^i indicates its position offset due to deployment error and other spatial factors. The target x_t is assumed to follow an extended Gaussian model, where the expectation μ_t and the precision matrix λ_t are both random, with a Gaussian distribution and a Wishart distribution respectively:

$$\begin{cases} r_t^i \sim N(\hat{r}^i, v^i) \\ x_t \sim N(\mu_t, \lambda_t) \\ \mu_t \sim N(\mu_{t-1}, \bar{\lambda}) \\ \lambda_t \sim W_n(\bar{V}, \bar{n}) \end{cases} \tag{3}$$

Where $\bar{\lambda}$ is the initial precision matrix which reflects the uncertainty of the target location estimation at instant t with respect to the previous one. The target state precision matrix λ_t is modelled by a d dimensional Wishart distribution ($n = 2$ in this work), with V and n denote the precision matrix and the degrees of freedom, respectively. Notice that $\bar{.}$ denotes the initial fixed parameter. Assuming a random mean and covariance for the state, x_t leads to a probability distribution covering a wide range of tail behaviors, which allows discrete jumps in the target trajectory. In fact, the marginal state distribution is obtained by integrating over the mean and precision matrix (Eq. 4):

$$p(x_t|x_t-1) = \int N(\mu_t,\lambda_t)p(\mu_t,\lambda t|x_{t-1})d\mu_t d\lambda_t, \tag{4}$$

In (4), the integration with respect to the precision matrix leads to a known class of scale mixture distributions given by Barndorff-Nielsen (Barndorff-Nielsen, 1977). The low values of degrees of freedom n reflect the heavy tails of the marginal distribution $p(x_t|x_t-1)$.

B. Quantized Proximity Observation Model QPOM

Consider a wireless sensor network in which the sensor locations, for an activated sensor i, are denoted by $s^i = (s_1^i, s_2^i), i = 1,2,..., N_s$ (the activation procedure is explained in section IV). Their observations are modelled by:

$$\gamma_t^{i,x} = C\left\|x_t - s^i\right\|^\eta + \epsilon_t, \tag{5}$$

Similarly,

$$\gamma_t^{i,r} = C\left\|r_t - s^i\right\|^\eta + \epsilon_t, \tag{6}$$

where ε_t, is a Gaussian noise with zero mean and variance σ_n^2, and η and C are known constants. The observation for target tracking problem is quantized, before being transmitted, by partitioning the observation space into N_t^i intervals $R_k = [\tau_k, \tau_{k+1}]$, $k \in \{1,..., N_t^i\}$.

Similarly, where a sensor collects information to estimate the relay location, it partitions the observation space into M_t^i regions $R_i = [\tau_i, \tau_{i+1}]$, $i \in \{1,..., M_t^i\}$. The equations (7) and (8) show the quantized signals.

$$y_t^{i,x} = d_{k,x} \text{ if } \gamma_t^{i,x} \in [\tau_k(t), \tau_{k+1}(t)], \quad (7)$$

Similarly,

$$y_t^{i,r} = d_{j,r} \text{ if } \gamma_t^{i,r} \in [\tau_j(t), \tau_{j+1}(t)], \quad (8)$$

where $d_k^{\,x}$ (resp. $d_j^{\,r}$) is the centroid of k-th cell during the target tracking problem (resp. the centroid of j-th cell during the relay localization). Then, the signals received by the cluster head from the sensor i at the sampling instant t are written as,

$$z_t^{i,x} = \beta_i y_t^{i,x} + n_t; \quad (9)$$

Similarly,

$$z_t^{i,r} = \alpha_i y_t^{i,r} + n_t; \quad (10)$$

where n_t is a random Gaussian noise sensor with a zero mean and a variance σ_n^2, and β_i is the i-th sensor channel attenuation coefficient.

Formulation of the Observation Model $p(z_t|X_t)$

As illustrated above, the Bayesian filtering involves the construction of the observation model $p(z_t|x_t,R_t)$. To track the target x_t, the available observations at the activated CH are denoted by $z_t^{\,s,x} = \{z_t^{\,i,x}\}_{i=1}^{m_t'}$, where m_t' is the number of sensors in the activated cluster. Assuming that the noise samples ε_t are independently distributed, we have,

$$p(z_t^{s,x} \mid x_t) = \prod_{i=1}^{m_t'} \sum_{j=0}^{N_t^i - 1} p(\tau_j(t) < \gamma_t^{s,x} < \tau_{j+1}(t)) N(\beta_i d_j, \sigma_\epsilon^2) \quad (11)$$

where,

$$p(\tau_j(t) < \gamma_t^{s,x} < \tau_{j+1}(t)) = \int_{\tau_j(t)}^{\tau_{j+1}(t)} N(\rho_{\gamma_t^{s,x}}(x_t), \sigma_n^2) d\gamma_t^{s,x} \quad (12)$$

is computed according to the quantization rule defined in (7), in which

$$\rho_{\gamma_t^{s,x}}(x_t) = K \left\| x_t - s \right\|^{\eta} \quad (13)$$

Similarly,

$$p(z_t^{s,r} \mid r_t) = \prod_{i=1}^{m_t} \sum_{k=0}^{n_t - 1} p(\tau_k(t) < \gamma_t^{s,r} < \tau_{k+1}(t)) N(\alpha_i d_k, \sigma_\epsilon^2) \quad (14)$$

Explicitly, $z_t^{\,i,r}$ denotes the observations collected during the relay localization phase between the i-th sensor and the neighboring relays. At a sampling instant t, the observation of the target $z_t^{\,i,x}$ is incorporated together with $z_t^{\,i,r}$ to help refining the localization of the relay i.

C. Overview of the Proposed Algorithm

Generally, in order to ensure a precise target tracking, the relay locations need to be known a priori. In our proposed algorithm, initially we suppose that relays are placed in roughly known positions. The true position r^i of the relay i is

Gaussian distributed around its initial setting value $\bar{\bar{r}}^i$ with precision v^i. This initial assignment results in a node layout that resembles an unfolded and scaled version of the actual deployment, roughly preserving the topological ordering of nodes. After the deployment, the sensors proceed to exchange information with neighboring relays in their communication range. The observations between them can thus be detected, and stored in the corresponding sensors. Then, a localization phase is launched locally by incorporating only the observations between the sensors and the relays, in the aim to improve a priori information on the relay locations. Consider the relay i for example,

$$p(\hat{r}^i \mid z^{i,r}) \propto N(\overline{r}^i, v^i) p(z^{i,r} \mid r^i). \quad (15)$$

Thus, the estimation of the relay position i is refined by incorporating the observation $z^{i,r}$, according to the prior distribution $r^i \sim N(\bar{\bar{r}}^i, v^i)$. After having localized all the relays, much more precise information on their locations is provided, which is incorporated as the prior information $\hat{r}^i$ for the adaptive scheme.

Once an intrusion in the WSN is identified, a cluster of sensors around the phenomenon of interest is activated. Note that, as mentioned above, the adaptive scheme procedure is distributively executed on a cluster head in order to minimize energy and bandwidth consumption. The sensors that have detected the presence of the target in their sensing ranges broadcast their residual energy level. The sensor having a maximum residual energy is elected as the cluster head (CH) to take charge of signal processing. The other clusters which detect sensors transfer to the CH their timely observation of the target, and the observations between their neighboring relays which are stored during the pre-localization phase. Note that, the information concerning the detection of relay locations and the estimations of the target are simultaneously updated in the CH by the adaptive scheme algorithm based on these observations. According to the Bayesian framework, the adaptive scheme algorithm works in a recursive way, where the filtering distribution is transferred to the next CH for further uses.

III. ADAPTIVE SCHEME ALGORITHM

The classical Bayesian framework is used by the adaptive scheme algorithm to estimate unknown states over time by using the incoming observations. As mentioned in section II, two distinct phases compose the Bayesian filtering framework: Prediction and Update. The prediction phase uses the state estimated from the previous sampling instant to produce an estimate of the state at the current instant according to Equation (1). In the update phase of Equation (2), measurement information at the current instant is used to refine this prediction to arrive at a new and hopefully more accurate state estimate. In a distributed context, the estimations of the detecting sensors are updated and stored locally, whereas the filtering distribution of the target needs to be transferred for future use. Concerning energy and bandwidth efficiency, the variational approach compresses the filtering distribution of the target to a single Gaussian distribution between successive clusters in a consistent manner (Snoussi & Richard, 2006). Thus, the distributed signal processing is achieved effectively. The details of the proposed adaptive scheme algorithm are illustrated in what follows.

A. Variational Filtering Approximation

A variational approach is employed here to approximate the posterior probability $p(\alpha_t|Z_{1:t})$ by a separable distribution $q(\alpha_t)$, which minimizes the Kullback-Leibler (KL) divergence error:

$$D_{KL}(q \,||\, p) = \int q(\alpha_t) \log \frac{q(\alpha_t)}{p(\alpha_t \mid Z_{1:t})} (d\alpha_t),$$

where

$$q(\alpha_t) = \prod_i q(\alpha_t^i) = q(x_t) q(\mu_t) q(R_t),$$

and

$$q(R_t) = \prod_{i=1}^{m_t} q(r_t^i). \tag{16}$$

Using a variational computation, the following approximate distribution yields (Snoussi & Richard, 2006),

$$q(\alpha_t^i) \propto \exp \left\langle \log p(Z_{1:t}, \alpha_t) \right\rangle_{\prod_{j \neq i} q(\alpha_t^j)}, \tag{17}$$

where $\langle . \rangle_{q(\alpha_t^j)}$ denotes the expectation operator relative to the distribution $q(\alpha_t^j)$. Taking into consideration the separable approximate distribution at time $t-1$, that is, $\hat{p}(\alpha_{t-1} \mid Z_{1:t-1}) = q(\alpha_{t-1})$, the filtering distribution at time t is deduced,

$$\hat{p}(\alpha_t \mid Z_{1:t}) = \frac{p(z_t \mid \alpha_t) \int p(\alpha_t \mid \alpha_{t-1}) d\alpha_{t-1}}{p(z_t \mid Z_{1:t-1})}$$
$$\propto p(z_t \mid x_t, R_t) p(x_t \mid \mu_t, \lambda_t) p(\lambda_t) p(R_t) q_p(\mu_t)$$

with

$$q_p(\mu_t) = \int p(\mu_t \mid \mu_{t-1}) q(\mu_{t-1}) d\mu_{t-1}. \tag{18}$$

Therefore, the filtering distribution $p(\alpha_t|z_{1:t})$ can be sequentially updated through a simple integral with respect to μ_{t-1}. Considering the general state evolution model proposed in (3), the evolution of μ_{t-1} is Gaussian, namely $p(\mu_t|\mu_{t-1}) = N(\mu_{t-1}, \bar{\lambda})$. Defining $q(\mu_{t-1}) = N(\mu_{t-1}^*, \lambda_{t-1}^*)$, $q_p(\mu_t)$ is also Gaussian (Vermaak, Lawrence, & Perez, 2003), with the following parameters,

$$q_p(\mu_t) = N(\mu_t^p, \lambda_t^p),$$

where

$$\mu_t^p = \mu_{t-1}^*$$

and

$$\lambda_t^p = (\lambda_{t-1}^{*\,-1} + \bar{\lambda}^{-1})^{-1}. \tag{19}$$

Hence, the temporal dependence is reduced to the incorporation of only one Gaussian component approximation $q(\mu_{t-1})$. The update and the approximation of the filtering distribution $p(\alpha_t|z_{1:t})$ are jointly performed, yielding a natural and adaptive compression (Snoussi & Richard, 2006), (Teng, Snoussi, & Richard, 2007b). According to the equation (17) and taking into account (18) and (19), the variational calculus leads to closed-form expressions of $q(\mu_t)$ and $q(\lambda t)$:

$$\begin{cases} q(\mu_t) = N(\mu_t^*, \lambda_t^*) \\ q(\lambda_t) = W_2(V_t^*, n^*) \end{cases} \tag{20}$$

where the parameters are iteratively updated until convergence, according to the following scheme:

$$\begin{cases} \mu_t^* = \lambda_t^{*-1}\left(\langle\lambda_t\rangle\langle x_t\rangle + \lambda_t^p \mu_t^p\right) \\ \lambda_t^* = \langle\lambda_t\rangle + \lambda_t^p \\ n^* = \bar{n} + 1 \\ V_t^* = \left[\langle x_t x_t^T\rangle - \langle x_t\rangle\langle\mu_t\rangle^T - \langle\mu_t\rangle\langle x_t\rangle^T + \langle\mu_t\mu_t^T\rangle + \bar{V}^{-1}\right]^{-1} \end{cases}$$
(21)

Notice that < . > designates the expectation relative to the distribution $q(\alpha_t^j)$. The mean state and the precision matrix distributions represented respectively by $q(\mu_t)$ and $q(\lambda t)$ have closed forms, such that their expectations are easily derived:

$$\begin{cases} \langle \mu_t \rangle = \mu_t^* \\ \langle \mu_t \mu_t^T \rangle = \lambda_t^{*-1} + \mu_t^* \mu_t^{*T} \\ \langle \lambda_t \rangle = n^* V_t^* \end{cases} \quad (22)$$

Nevertheless, the target state x_t and the activated relay positions R_t do not have closed forms. Combining the equations (17) and (18), $q(x_t)$ and $q(r_t^i)$ have the following expressions:

$$\begin{aligned} q(x_t) &\propto N(\langle \mu_t \rangle, \langle \lambda_t \rangle) \prod_{i=1}^{m_t} p(z_t^{i,x} \mid x_t), \\ q(r_t^i) &\propto N(\hat{r}^i, v^i) \prod_{i=1}^{m_t} p(z_t^{i,r} \mid r_t), \end{aligned} \quad (23)$$

Hence, the general state evolution model (3) and the observation model (4) and (5) are naturally incorporated to update $q(x_t)$ and $q(r_t^i)$. Their distribution forms immediately suggest an Importance Sampling (IS) procedure, where the samples are drawn from the Gaussian distributions $N(\langle \mu_t \rangle, \langle \lambda_t \rangle)$ and $N(\hat{r}^i, v^i)$ respectively, and are weighted according to their likelihoods (taking into account the observation model):

$$\begin{aligned} x_t^{(k)} &\sim N(\langle \mu_t \rangle, \langle \lambda_t \rangle), w_t^{(k)} \propto \prod_{i=1}^{m_t} p(z_t^{i,x} \mid x_t^{(k)}), \\ r_t^{i,(k)} &\sim N(\hat{r}^i, v^i), w_t^{i,(k)} \propto \prod_{i=1}^{m_t} p(z_t^{i,r} \mid r_t^{(k)}), \end{aligned} \quad (24)$$

Therefore, $q(x_t)$ and $q(r_t^i)$ can be approximated by the Monte Carlo method:

$$\begin{cases} \langle x_t \rangle = \sum_{k=1}^{N} w_t^{(k)} x_t^{(k)} \\ \langle r_t^i \rangle = \sum_{k=1}^{N} w_t^{i,(k)} r_t^{i,(k)} \end{cases} \quad (25)$$

As mentioned above, the standard adaptive scheme solution includes both prediction and update steps. Besides the update of the filtering distribution $p(\alpha_t|z_{1:t})$, the predictive distribution $p(\alpha_t|z_{1:t-1})$ can also be efficiently calculated by the variational approach. In fact, by incorporating the separable approximate distribution $q(\alpha_{t-1})$ in the place of $p(\alpha_{t-1}|z_{1:t-1})$, the recursive adaptive scheme algorithm calculates the predictive distribution $p(\alpha_{t-1}|Z_{1:t-1})$ in the following form:

$$\begin{aligned} \hat{p}(\alpha_t \mid Z_{1:t-1}) &\propto \int p(\alpha_t \mid \alpha_{t-1}) q(\alpha_{t-1}) d\alpha_{t-1} \\ &\propto p(x_t \mid \mu_t, \lambda_t) p(\lambda_t) p(R_t) q_p(\mu_t). \end{aligned}$$

(26)

The exponential form solution, which minimizes the Kullback-Leibler divergence between the predictive distribution $p(\alpha_{t-1}|Z_{1:t-1})$ and the separable approximate distribution $q_{t|t-1}(\alpha_t)$, yields Gaussian distributions for the predicted expectations, and a Wishart distribution for the target precision matrix:

$$\begin{aligned} q_{t|t-1}(x_t) &\propto N\left(\langle \mu_t \rangle_{qt|t-1}, \langle \lambda_t \rangle_{qt|t-1}\right), \\ q_{t|t-1}(\mu_t) &\propto N\left(\mu_{t|t-1}^*, \lambda_{t|t-1}^*\right) \\ q_{t|t-1}(\lambda_t) &\propto W_2(V_{t|t-1}^*, n_{t|t-1}^*), \\ q_{t|t-1}(R_t) &\propto N(\hat{R}, v), \end{aligned} \quad (27)$$

Where, the parameters are updated according to the same iterative schemes (21) and (22). The target state and the activated relays are now evaluated by the following expressions:

$$\begin{aligned} &\langle x_t \rangle_{qt|t-1} = \langle \mu_t \rangle_{qt|t-1}, \langle x_t x_t^T \rangle_{qt|t-1} = \\ &\langle \lambda_t \rangle_{qt|t-1}^{-1} + \langle \mu_t \rangle_{qt|t-1} \langle \mu_t \rangle_{qt|t-1}^T, \\ &\langle Rt \rangle_{qt|t-1} = \langle \hat{R} \rangle. \end{aligned} \quad (28)$$

In the next section, we present a technique aiming at adaptively (and jointly) selecting the appropriate group of candidate sensors that participate in data collection for tracking the target.

IV. OPTIMAL GROUP OF CANDIDATE SENSORS SELECTION BASED ON MULTI-CRITERIA FUNCTION (MCF)

Since the predicted target position $x_p(t) = \langle x_t \rangle_{q_{t+1|t}}$ at time t+1 is available, it could be used to select sensors. The sensors inside the disk centred at $x_p(t)$ with radius R_{max} are pre-selected. After the pre-selection of sensors within R_{max} range, the CH divides the pre-selected sensors into $M = \sum_{j=4}^{N_s} C_{N_s}^j$ groups G_t (at least four sensors are needed to sense the target within their range (Chen, Hou, & Sha, 2004)). Then, it computes for each group of sensors the MCF (detailed in IV-A) and activates the appropriate group which has the highest MCF value to participate in data aggregation.

In the next subsection, we detail the MCF function with the used parameters permitting to select the appropriate group of sensors that participate in data collection for tracking the target.

A. Multi-Criteria Function

The multi-criteria function for sensors selection aims is to define the main parameters that may influence the relevance of the participation in cooperation, which are: 1) $MI(x_t, z^{G_t})$: the information that can be transferred from the group of candidate sensors, 2) $G_t(D(i))$: the transmitting distance between the sensor i and the CH in group, 3) $E(i)$: the stored energy in candidate sensors.

The problem is how to formulate the criteria for the CH to select the appropriate group of sensors that provide relevant data and balances the energy level among all sensors. We define a combinative measurement for the group of candidate nodes G_t, denoted as F, which is given by:

$$F(G_t) - n_1 MI(x_t, z^{G_t}) - n_2 \sum_{i=1}^{M} R_t^2(i). \tag{29}$$

Where $R(i) = \frac{E(i)}{D(i)}$, n_1 and n_2 are the importance factor for the Mutual Information (MI) function and the ratio between the energy stored in the sensor (E) and the transmit distance (D). In order to select the best group of candidate sensors based on equation (27), the objective is to choose the appropriate group of sensors G_t so that,

$$\hat{G}_t = \arg\max_{G_t \subset C} F(G_t) \tag{30}$$

The efficiency measurement of given information is often achieved by the mutual information function. This later is a quantity measuring the amount of information that the observable variable z^{G_t} carries about the unknown parameter x_t. The mutual information between the observations z^{G_t} and the source x_t is proportional to:

$$MI(x_t, z^{G_t}) \propto p(z^{G_t} \mid x_t) \log(p(z^{G_t} \mid x_t)) \tag{31}$$

The likelihood function (L) is expressed as,

$$L(s^{G_t}) = p(z^{G_t} \mid x_t) = \prod_{i=1}^{M} \sum_{j=0}^{N_t^i - 1} p(\tau_j(t) < \gamma^i < \tau_{j+1}(t) N(d_j, \sigma_\epsilon^2) \tag{32}$$

where,

$$p\left(\tau_j(t) < \gamma_t^i < \tau_{j+1}(t)\right) = \int_{\tau_j(t)}^{\tau_{j+1}(t)} N\left(\rho_{\gamma_t^i}(s^i), \sigma_n^2\right) d\gamma_t = \frac{1}{\sqrt{\pi}} [erfc\left(\frac{\tau_j(t) - \rho_{\gamma_t^i}(x_t)}{\sqrt{2\sigma_n^2}}\right) - erfc\left(\frac{\tau_{j+1}(t) - \rho_{\gamma_t^i}(x_t)}{\sqrt{2\sigma_n^2}}\right)] \tag{33}$$

using the quantization rule defined in (7), in which

$$\rho_{\gamma_t^i}(s^i) = K \left\| x_t - s^i \right\|^{\eta}, \tag{34}$$

It is worth noting that the expression of the MI given in (31) depends on the target position x_t at the sampling instant t and on the group of candidate sensors G_t. However, as the target position is unknown, the MI is replaced by its expectation according to the predictive distribution $p(x_t \mid z_{1:t-1}^{G_t})$ of the target position:

$$< MI(s^{G_t}) >= E_{p(xt|z_{1:t-1}^{G_t})}[MI(s^{G_t})] \qquad (35)$$

The computation of the above expectation is analytically untractable. However, as the VF algorithm yields a Gaussian predictive distribution $N(x_t; \mu_{t|t-1}, \lambda_{t|t-1})$, the expectation (35) can be efficiency approximated by a Monte Carlo scheme:

$$< MI(s^{G_t}) > \simeq \frac{1}{J}\sum_{j=1}^{J} MI(\tilde{x}_t^j, s^{G_t}), \qquad (36)$$

where x_t^j is the j-th drawn sample at instant t, and J is the total number of drawn vectors x_t.

The next section presents a malicious sensor nodes detection technique based on the Kullback-Leibler distance (KLD) between the current target position distribution and the predicted sensor observation.

V. MALICIOUS SENSOR NODE DETECTION BASED ON KULLBACK-LEIBLER DISTANCE (KLD)

A. Problem Definition

The following are the assumptions made to resolve the malicious sensor detection problem:

1. We assume that malicious nodes can successfully authenticate with the sensor network, and their data can be collected with other nodes in the network. Otherwise, the CHs are assumed to well-behave and not malicious.
2. We assume a centralized scenario, in which a CH processing unit collects tracking reports from sensor nodes, determines which of them are malicious and removes them from sensor networks.
3. The purpose of the fictitious path is to allow the enemy to avoid surveillance. However, the fictitious path does not go beyond the sensing range of the nodes.
4. We assume that sensor nodes have unlimited communication bandwidth among each other. Meanwhile, an unknown number of nodes are malicious and they are injecting false tracking reports into the network. The problem is how to detect those malicious nodes, and to provide a correct target trajectory.

B. Computing the Kullback Leibler Distance (KLD)

The measurement of the distance between two statistical models is needed to resolve certain problems. For example, this distance can be used in evaluating the training algorithm or classifying the estimated models (Juang & Rabiner, 1985). The Kullback-Leibler distance or the relative entropy arises in many contexts as an appropriate measurement of the distance between two distributions. The KLD between the two probability density functions p and $\hat{p}$ is defined as (Cover & Thomas, 2006):

$$KLD(p \,\|\, \hat{p}) = \int p \log \frac{p}{\hat{p}} \qquad (37)$$

The computation of the distribution function is very complex for hidden Markov models, and practically it can be only computed via a recursive procedure; the "forward/backward" or "upward/

downward" algorithms (Ronen, Rohlicek, & Ostendorf, 1995), (Rabiner, 1989). Thus, there is no simple closed form expression for the KLD for these models. Commonly, the Monte-Carlo method is used to numerically approximate the integral in (37) as,

$$KLD(p \,||\, \hat{p}) = E_p(\log(p) - \log(\hat{p})) \quad (38)$$

Hereafter, we detail the two proposed schemes for the detection of malicious sensors.

C. Case 1: Reactive Cluster Formation

At every sampling instant, a cluster (CH and candidate secure sensors) is dynamically formed. The candidate sensors are activated in a manner explained in Section IV. The cluster head CH_t is chosen to be the nearest sensor to $\langle x_t \rangle_{q_{t|t-1}}$ i.e:

$$CH_t = \arg\min_{i=1,\ldots,|B_t|} \left\{ \left\| \langle x_t \rangle_{q_{t|t-1}} - s_t^m \right\| m \in B_t \right\} \quad (39)$$

where $|.|$ denotes the cardinality, and B_t is the set of activated sensors.

Let's assume that sensors *i*, *j* and *k* are activated and the sensors *j* and *k* can overhear the *i*'s transmission. At time step *t*, the sensors *j* and *k* compute their predictive target distributions $p(x_t \mid z^j_{t+1|t})$ and $p(x_t \mid z^k_{t+1|t})$ by executing the QVF algorithm. At time step (*t*+1), they overhear the *i*'s transmitted value, and compare the *i*-th sensor predictive distribution $p(x_t \mid z^i_{t+1|t})$ to the two predictive distributions $p(x_t \mid z^j_{t+1|t})$ and $p(x_t \mid z^k_{t+1|t})$ by computing the Kullback Leiber distances (detailed in V-B) $K_{i,j}$ and $K_{i,k}$. If the $K_{i,j}$ and $K_{i,k}$ are larger than K_0 (predefined threshold), as shown in Figure 1, then the sensors *j* and *k* estimate that the sensor *i* is malicious and send a notification to the CH. This later compares the $p(x_t \mid z^i_{t+1})$ to the $p(x_t \mid z^i_{t+1|t})$ distribution by computing the distance $K_{i,i}$, if this latter is also larger than K_0, then, it deletes the *i*-th fictitious path from the sensor network.

The major advantages of this algorithm can be summarized as follows: i) The accuracy of tracking is guaranteed by choosing the most potential sensors to dynamically form a cluster. ii) Sensors Kullback Leibler distance (KLD) K_0 (a predefined threshold). iii) This scheme is much more robust to external attacks. iv) As the lifetime of WSN is defined as the time elapsed until the first sensor depletes its energy (Misra, Dolui, & Das, 2005), it is essential to evenly distribute the energy consumption over the whole WSN. By dynamically forming the clusters, the CHs performing series of energy-intensive functions are changing frequently in order to balance the energy expenditure. However, since the signal processing task is assigned to all the slave sensors, all these advantages cited above are at the expense of homogeneous high hardware configuration.

The scheme discussed previously will be referred to as Re-QVF; its procedure is described in Algorithm 1.

D. Case 2: Proactive Cluster Formation

For this second case, we suppose that CHs are statically selected at the time of sensors deployment. At sampling instant t, the QVF provides the predicted target position $\langle x_t \rangle_{q_{t|t-1}}$. As shown in Figure 2, based on this predicted information, the cluster head CH_t at sampling instant (t−1) selects the next cluster head CH_t. If the predicted target position $\langle x_t \rangle_{q_{t|t-1}}$ remains in the vicinity of CH_{t-1}, which means that at least four of its slave sensors can detect the target, then $CH_t = CH_{t-1}$. Otherwise, if $\langle x_t \rangle_{q_{t|t-1}}$ is going beyond the sensing range of the current cluster, then a new CH_t is activated

Figure 1. A simple example for KLD computing

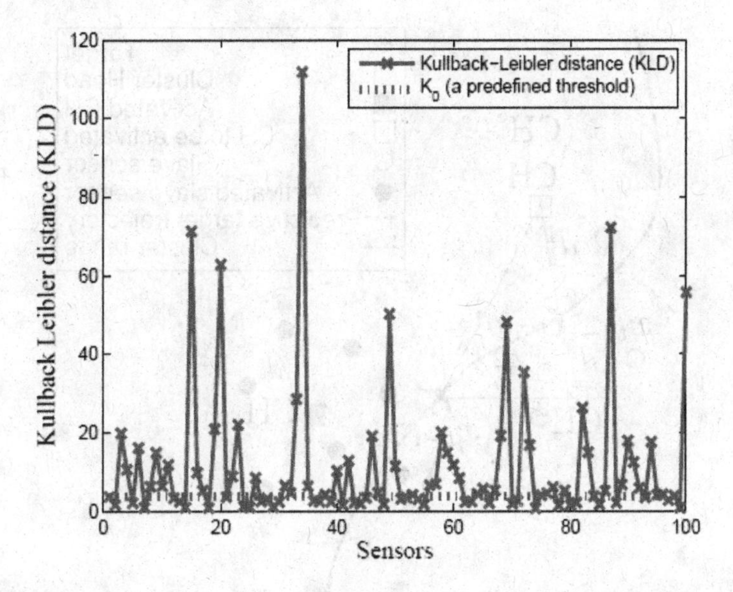

based on the target position prediction $\langle x_t \rangle_{q_{t|t-1}}$ and its future tendency.

$$CH_t = \arg\max_{k=1,\dots,K}\{\frac{\cos\theta_t^k}{d_t^k}\}$$

where

$$d_t^k = \| \langle x_t \rangle_{q_{t|t-1}} \| - L_{CH_t^k}$$

and

$$\theta_t^k = angle(\overrightarrow{\langle x_{t-1} \rangle \langle x_t \rangle_{q_{t|t-1}}}, \overrightarrow{\langle x_{t-1} \rangle L_{CH_t^k}}) \quad (40)$$

Where K is the number of CHs in the neighborhood of CH_{t-1} and $L_{CH_t^k}$ is the location of the k-th neighboring CH_t.

The initial distribution of the target position $p(x_0)$, at the instant t=0, is assumed to be known and stored in CH unit. Then, at time step t, the activated sensor i sends its quantized observation

Algorithm 1.

— Initialization:
1. Determine the CH by using the equation (39).
2. Select the appropriate group of sensors according to Section IV
3. Quantize the sensors' measurements according to Sub.II-B.
4. Execute the QVF algorithm.

— Iterations:
1. Compute the multi-criteria function by using the equation (29).
2. Select the appropriate group of sensors according to Section IV
3. Determine the CH using the equation (39).
4. Detect the malicious sensors according to Sub.V-C.
5. Delete the fictitious path according to Sub.V-C.
6. Quantize the sensors' measurements according to Sub.II-B.
7. Execute the QVF algorithm.

Figure 2. Prediction-based CH_t activation

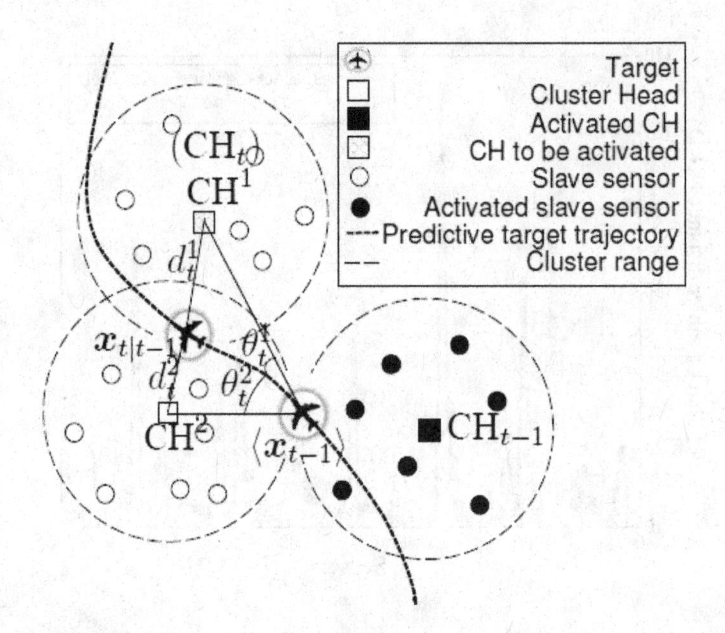

y_t^i to the CH_t. When the CH_t receives z_t^i, which is corrupted by an additive white Gaussian noise n_t, it executes the QVF algorithm, which provides in addition to the estimated target distribution $\hat{p}(z_t \mid x_t)$, the predictive target distribution for the *i*-th sensor $p(x_t \mid z^i_{t+1|t})$. The CH_t sends this predicted information to CH_{t+1}, which receives also a measurement from the *i*-th sensor. Based on this information, CH_{t+1} computes for *i*-th sensor, the Kullback Leiber distance $K_{i,i}$ between the predictive distribution $p(x_t \mid z^i_{t+1|t})$ received from CH_t and the distribution $p(x_t \mid z^i_{t+1})$. In a similar way, CH_{t+1} calculates $K_{i,j}$ and $K_{i,k}$ and compares them with the constraints, if the differences are greater than 0, then it estimates that the *i*-th sensor node is malicious. Finally, the CH deletes the fictitious path corresponding to the malicious sensor from the sensor networks.

In summary, our proposed scheme includes many several advantages. Firstly, the consumed energy and the required bandwidth in communication are considerably reduced. The tracking process is performed only by the activated CH, while the slave sensors are unable to take over this task. The slave sensors candidates are required to collect and transmit their measurements over short distances to the CH. Secondly, the cost of hardware configuration drops sharply owing to the low-cost of slave sensors. Furthermore, avoiding CH competition puts an end to unnecessary resource and energy consumption. Only when the hand-off operation occurs does the active CH need to communicate the temporal dependence information to the subsequent CH. However, the occurrence of hand-off operations is reduced by the non-myopic selective CH activation rule.

In addition, the QVF algorithm reduces the dependence of temporal informations to only statistic parameters of a Gaussian distribution. We refer to the approach described in this section by

Pro-QVF summarized by the steps presented in Algorithm 2.

The next section is devoted to the method aimed at adaptively selecting the best communication path between the candidate sensor and the cluster head.

Algorithm 2.

— Initialization:
1. Determine the cluster head by using the equation (40).
2. Select the best group of candidate sensors according to Section.IV.
3. Quantize the sensors' measurements according to Sub.II-B.
4. Execute the QVF algorithm.

— Iterations:
1. Compute the predictive target distribution.
2. Compute the multi-criteria function by using the equation (29).
3. Select the appropriate group of sensors according to Section.IV.
4. Determine the CH using the equation (40).
5. Detect the malicious sensors according to Sub.V-C.
6. Delete the fictitious path according to Sub.V-C.
7. Quantize the sensors' measurements according to Sub.II-B.
8. Execute the QVF algorithm.

VI. BEST COMMUNICATION PATH SELECTION METHOD

This section aims at selecting the best communication path as well as the highest SNR at the CH. In what follows, we assume that the relay placement is already estimated.

A. System Model

Figure 3 illustrates the idea of the proposed model. The communication is established between a sensor node and the cluster head through the selected relay including direct and indirect links. The selected relay is achieved according to the best communication procedure. The cluster head can get the best copy of the source signal transmitted by the source sensor S, via the best communication path selection scheme. The first one is from the source sensor (direct link), while the second one is from the best path as shown in Figure 3.

The parameter α_i shown in Figure 3 is the channel coefficient between the source sensor and the i-th sensor. α_i and α_j, β_i and are the flat Rayleigh fading coefficients which are mutually independent and non identical for all i and j.

The signal is simply amplified at the relay sensor i using the gain $g = 1/\sqrt{\alpha_i^2 E_s + N_\varepsilon}$, where E_s is the transmitted signal energy of the source. It is easy to prove that the source to cluster head SNR of the indirect path, $S \to i \to CH$ can be written as,

$$\gamma_{S\to i\to CH} = \frac{\gamma_{\alpha i}\gamma_{\beta i}}{\gamma_{\alpha i} + \gamma_{\beta i} + 1} \quad (41)$$

where $\gamma_{\alpha_i} = \alpha_i^2 E_s / N_\varepsilon$ is the instantaneous SNR of the source signal at the sensor i, $\gamma_{\beta_i} = \beta_i^2 E_i / E_0$ is the instantaneous SNR of the sensor signal (sensor i) measured at the cluster head, and E_i is the signal transmitted energy of the relay i.

The best communication path will be selected as the path which achieves the highest source-to-end SNR of indirect and direct paths. The SNR for the best path is given by,

$$\gamma_b = \max(\gamma_{\beta_0}, \gamma_{S\to i\to CH})_{i=1\dots M} \quad (42)$$

where $\gamma_{S\to i\to CH}$ is the instantaneous SNR between the source and the cluster head and M is the total number of links inside cluster. The upper-bound of $\gamma_{S\to i\to CH}$ is given by (Ikki & Ahmed, 2007). Computing the above expression is analytically untractable. However, (42) can be efficiency approximated (Valenzuela, 1987) by,

$$\gamma_{S\to i\to CH} \le \gamma_i = \min(\gamma_{\alpha i}\gamma_{\beta i}), \quad (43)$$

Figure 3. Illustration of the diversity network with the best communication path selection scheme

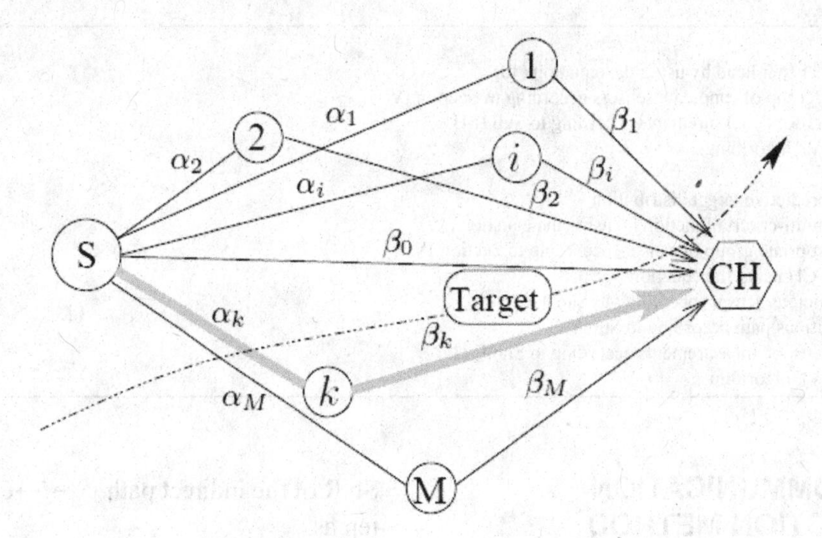

this approximation simplifies the derivation of the SNR statistics: Cumulative Distribution Function (CDF), Probability Distribution Function (PDF) and Moment Generating Function (MGF).

B. Error Performance Analysis

The error probability p_e of the best path is given as,

$$p_e(\gamma_{\beta_0}, \gamma_b) = A\,erfc\left(\sqrt{B \max(\gamma_{\beta_0}, \gamma_b)}\right) \quad (44)$$

Where $erfc(x) = \frac{2}{\sqrt{\pi}} \int_x^{\infty} \exp(-t^2)dt$, and A and B are dependent on the modulation type. Over the random variables representing the SNR values of the best communication path, the average error probability is given by,

$$P_E = \int_0^{\infty} p_e(\gamma_b) f_{\gamma_b} d\gamma_b \quad (45)$$

using the alternative definition of the $erfc(x)$ function as (Simon & Alouni, 2005)

$$erfc(x) = \frac{2}{\pi} \int_0^{\frac{\pi}{2}} \exp\left(-\frac{x^2}{\sin^2\theta}\right) d\theta \quad (46)$$

and by substituting (45) into (46) we obtain

$$P_E = \int_0^{\infty} \frac{2}{\pi} \int_0^{\frac{\pi}{2}} \exp\left(-\frac{B\gamma_b}{\sin^2\theta}\right) f_{\gamma_b}(\gamma_b) d\gamma_b$$
$$= \frac{2}{\pi} \int_0^{\frac{\pi}{2}} M_{\gamma_b}\left(\frac{B}{\sin^2\theta}\right) d\theta \quad (47)$$

where, $M_{\gamma_b} = \int_0^{\infty} f_{\gamma_b}(\gamma_b) exp(-s\gamma_b) d\gamma_b$ is the MGF of γ_b and f_{γ_b} is the PDF of γ_b.

In order to find the P_E, it is necessary to find the PDF and MGF of γ_b. We can write the CDF of γ_b as follows,

$$F_{\gamma b}(\gamma) = P(\gamma_b \leqslant \gamma) = P(\gamma_i \leqslant \gamma) P(\gamma_{\beta_0} \leqslant \gamma)_{i=1...M} =$$
$$\left[\prod_{i=1}^{M} (1 - e^{-\gamma/\gamma_i})\right] (1 - e^{-\gamma/\gamma_{\beta_0}}) \quad (48)$$

In the aim of giving a coherent analytic form, we assume $\gamma_{M+1} = \gamma_0$, the PDF can be found by taking the derivative of (48), with respect to γ, and after doing some manipulations, $f_{\gamma_b}(\gamma)$ can be written as,

$$f_{\gamma_b}(\gamma) = \sum_{n=1}^{M+1}(-1)^{(n+1)}\sum_{k_1=1}^{M-n+2}\sum_{k_2=k_1+1}^{M-n+3}\sum_{k_n=k_{n-1}+1}^{M+1}\prod_{j=1}^{n}(e^{-\gamma/\bar{\gamma}_{k_j}})\sum_{j=1}^{n}\left(\frac{1}{\bar{\gamma}_{k_j}}\right) \quad (49)$$

by using the PDF in (49), the MGF can be written as,

$$M_{\gamma_b}(s) = \int_0^{\infty} e^{-s\gamma}\sum_{n=1}^{M+1}(-1)^{(n+1)}\sum_{k_1=1}^{M-n+2}\sum_{k_2=k_1+1}^{M-n+3}\sum_{k_n=k_{n-1}+1}^{M+1}\prod_{j=1}^{n}(e^{-\gamma/\bar{\gamma}_{k_j}})\sum_{j=1}^{n}\left(\frac{1}{\bar{\gamma}_{k_j}}\right)d\gamma \quad (50)$$

and the integral can be evaluated in a closed form as,

$$M_{\gamma_b}(s) = \sum_{n=1}^{M+1}(-1)^{(n+1)}\sum_{k_1=1}^{M-n+2}\sum_{k_2=k_1+1}^{M-n+3}\sum_{k_n=k_{n-1}+1}^{M+1}\frac{\Psi_n}{s+\Psi_n} \quad (51)$$

where $\psi_n = \sum_{j=1}^{n}\frac{1}{\bar{\gamma}_{k_j}}$. Substituting (51) in (47) and finalizing the integration using (Papoulis & Pillai, 2002), we can write PE in a closed form as follows,

$$P_E = A\sum_{n=1}^{M+1}(-1)^{(n+1)}\sum_{k_1=1}^{M-n+2}\sum_{k_2=k_1+1}^{M-n+3}\sum_{k_n=k_{n-1}+1}^{M+1}\left(1-\sqrt{\frac{B/\Psi_n}{1+1/\Psi_n}}\right) \quad (52)$$

VII. RESULTS

In this section, we evaluate the proposed algorithms based on a synthetic example, which involves the mobile target tracking, the relay localization, the secure sensor node detection and the optimal communication path selection. The purpose of the synthetic example is to establish a baseline performance comparison on a relatively difficult problem.

In what follows, we proceed to compare the tracking accuracy of the adaptive quantized variational filtering (AQVF) algorithm, with the binary variational filtering (BVF) algorithm (Teng, Snoussi, & Richard, 2007b), and the centralized quantized particle filter (QPF) algorithm (Zuo, Niu, & Varshney, 2007). In the simulation, we have considered the following parameters: η =2 for free space environment, the constant characterizing the sensor range is fixed for simplicity to C=1, the cluster head noise power is $\sigma_n^2 = 10^{-3}$, the sensor noise power is $\sigma_\varepsilon^2 = 10^{-4}$, the maximum sensing range R_{max} (resp. the minimum sensing range R_{min}) is fixed to 10 m (resp. 0 m) and 200 particles were used in QVF, BVF and QPF algorithms. All sensors have equal initial battery energies of E_i = 1 *Joule*. All the simulations shown in this chapter are implemented with Matlab version 7.1, using an Intel Pentium CPU 3.4 GHz, 1.0 G of RAM PC.

The quantized proximity observation model, formulated in the equation (6), was adopted for the QPF algorithm. Figure 4 and Figure 5 show the performances comparison of AQVF and BVF. We can observe from Figure 4.a) and Figure 5.a) that even with sudden changes in the target trajectory, the desired quality is achieved by the AQVF algorithm and outperforms the BVF algorithm. Their tracking accuracies are compared in Figure 4.b) and Figure 5.b), in terms of Mean Square Error (MSE) (Eq. 53) where the sensors number varies in {200, 600}.

$$\text{MSE} = \text{E}\left((x-\hat{x})^2\right) \quad (53)$$

Where x (resp. $\hat{x}$) is the true trajectory (resp. the estimated trajectory).

The performances expressed by the proposed method demonstrate the effectiveness of the adaptive quantization and the impact of neglecting the

Figure 4. a) Tracking accuracy between BVF and AQVF algorithms b) Mean Square Error Comparison between QPF and AQVF algorithms where the sensors number=200

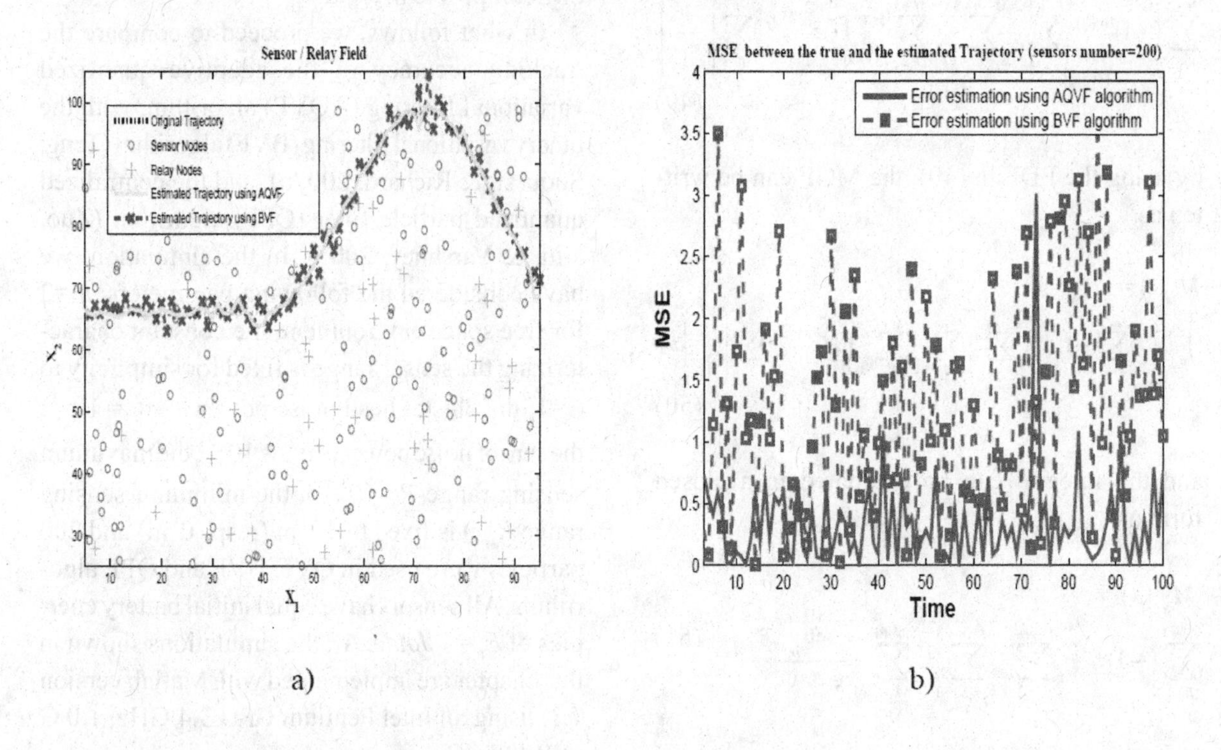

Figure 5. a) Tracking accuracy between BVF and AQVF algorithms b) Mean Square Error comparison between QPF and AQVF algorithms where the sensors number=300

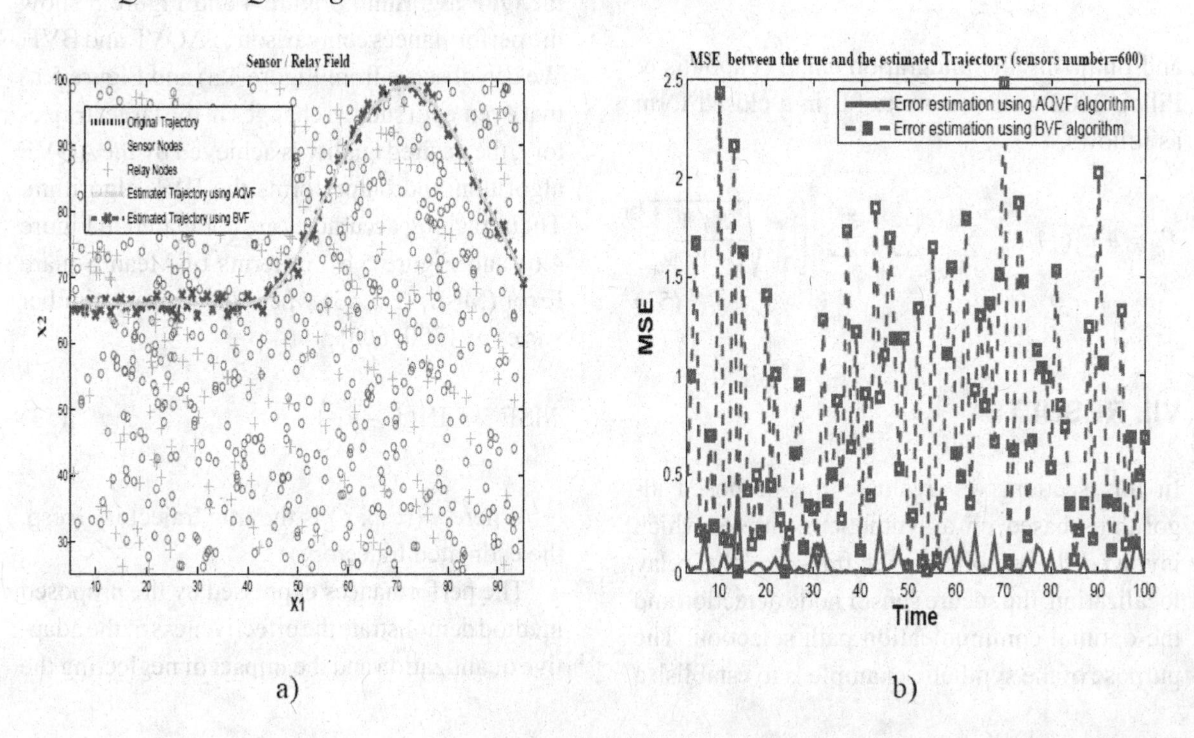

Figure 6. a) Tracking accuracy between QPF and AQVF algorithms b) Mean Square Error comparison between QPF and AQVF algorithms where the sensors number=200

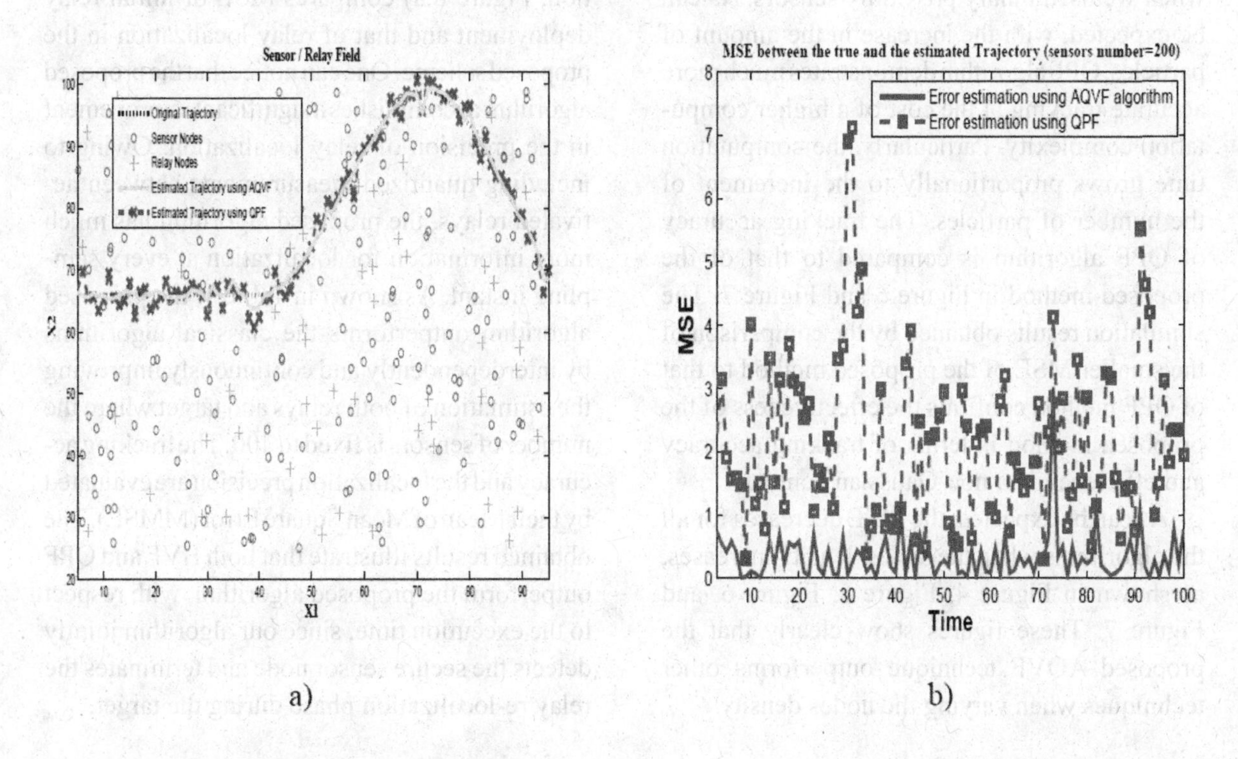

Figure 7. a) Tracking accuracy between QPF and AQVF algorithms b) Mean Square Error comparison between QPF and AQVF algorithms where the sensors number=600

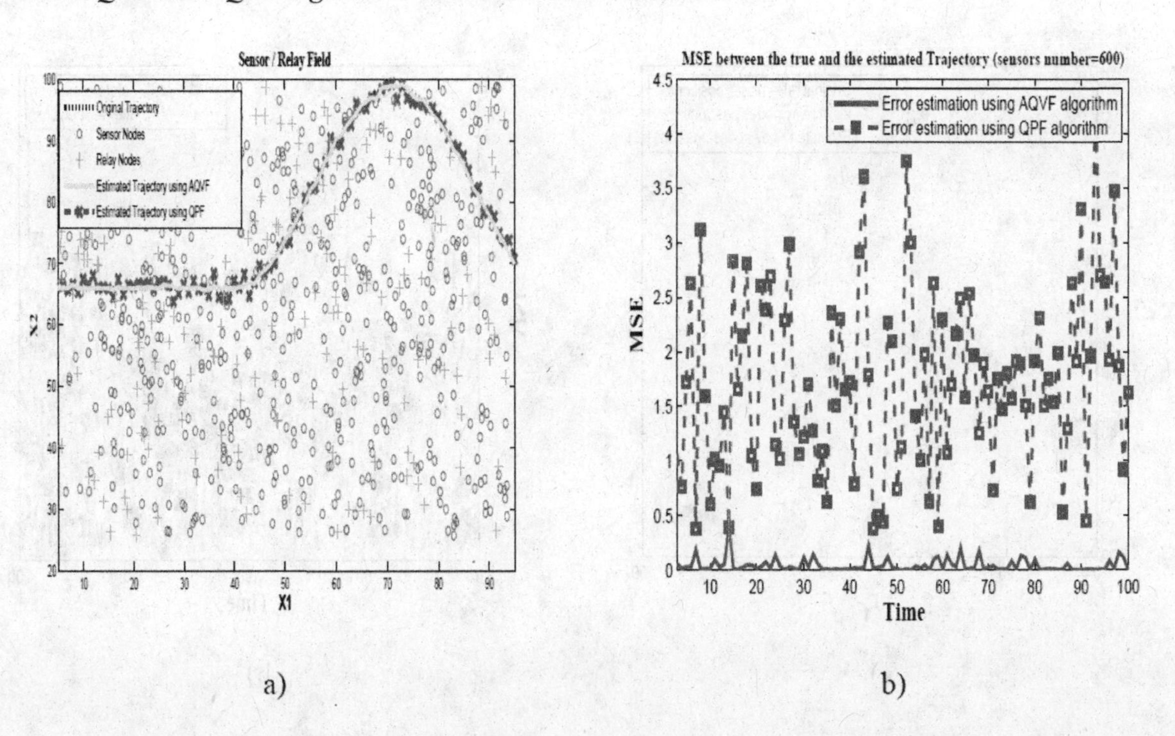

information relevance of sensor measurements, when we used binary proximity sensors. As can be expected, with the increase in the amount of particles, QPF algorithm demonstrates much more accurate tracking at the cost of a higher computation complexity. Particularly, the computation time grows proportionally to the increment of the number of particles. The tracking accuracy of QPF algorithm is compared to that of the proposed method in Figure 6 and Figure 7. The simulation results obtained by the comparison of the smaller MSE of the proposed method to that of QPF method confirms the effectiveness of the proposed method in terms of tracking accuracy and efficiency in a non-Gaussian context.

As can be expected, the MSE decreases for all the algorithms when the nodes density increases, as shown in Figure 4, Figure 5, Figure 6, and Figure 7. These figures show clearly that the proposed AQVF technique outperforms other techniques when varying the nodes density.

Concerning the precision of relay localization, Figure 8.a) compares MSE of initial relay deployment and that of relay localization in the proposed scheme. One can notice that the proposed algorithm accomplishes a significant improvement in the precision of relay localization. Owing to including quantized measurements between activated relays, the proposed algorithm has much more information for localization at every sampling instant. As shown in Table 1, the proposed algorithm outperforms the classical algorithms by interdependently and continuously improving the estimation of both relays and target where the number of sensors is fixed to 400. The tracking accuracy and the localization precision are evaluated by their Mean of Mean Square Error (MMSE). The obtained results illustrate that both BVF and QPF outperform the proposed algorithm, with respect to the execution time, since our algorithm jointly detects the secure sensor node and terminates the relay re-localization phase during the target.

Figure 8. a) Relay nodes estimation b) MSE before and after estimation

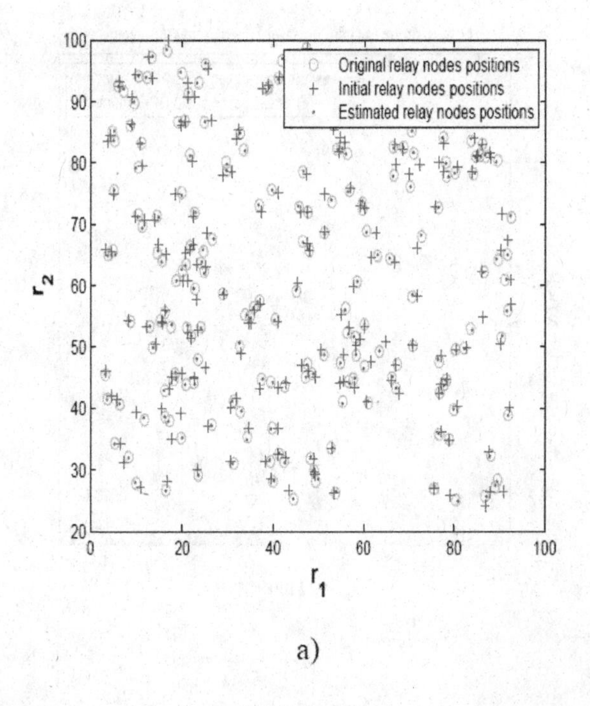

a)

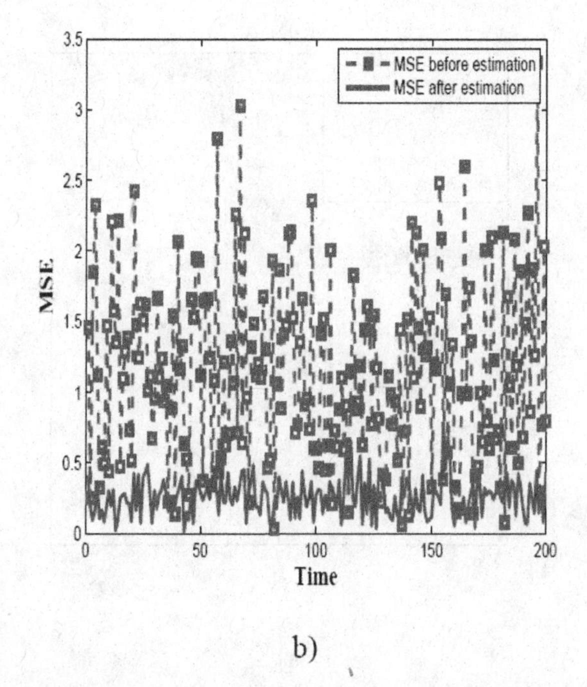

b)

Figure 9 shows the bit error rate for BPSK with different numbers of paths sensors-CH (M).

As can clearly observed in a high SNR regime, the improvement of bit error rate is proportional to the number of sensors-CH links (M).

The evaluation of the energy consumption is done following the model proposed in (Heinzelman, Chandrakasan, & Balakrishnan, 2000). We can observe from Figure 10 that our model successfully balances the trade-off between the energy consumption even with several abrupt changes in the trajectory where the bits quantization number is fixed to 3.

Hereafter, we evaluate the performances of the malicious sensor node detection schemes presented in Section V in terms of ROC curves. The evaluation of the malicious sensor detection technique for WSNs depends on whether it can satisfy the mining accuracy requirements while maintaining the resource consumptions of WSNs to a minimum (Gama & Gaber, 2007). While keeping the false alarm rate low, the techniques of malicious sensor nodes detection are required to maintain a high detection rate. The detection rate represents the percentage of vulnerable sensor that are correctly considered as malicious, and the false positive rate, represents the percentage of normal sensor that are incorrectly considered as malicious. ROC curves (Lazarevic et al., 2003) is frequently used to represent the trade-off between the detection rate and the false positive rate. The larger the area under the ROC curve, the better the performance of the corresponding techniques. An example of ROC curves is illustrated in Figure 11.

VIII. CONCLUSION AND PERSPECTIVES

In this chapter, a distributed variational filtering solution to simultaneously localize relays, track mobile target, select the secure candidate sensor and select the best communication path is proposed in the context of WSN. Without any a priori information on the target motion, the proposed adaptive scheme algorithm aims at continuously updating and improving the estimation of the activated relay locations and the target trajectory, and optimizing the routing by selecting the best communication path. As the target can travel arbitrarily and the location information of the activated relays is rather coarse, a general state evolution model is proposed in this chapter to describe the hidden state, which is more adaptable to the non-linear / non-gaussian situation than other kinematic parameter models. The adaptive scheme algorithm is executed on a fully distributed cluster scheme, in order to minimize the resources consumption in WSN. The variational method allows an implicit compression of the exchanged statistics between clusters. This method permits not only to reduce the inter-cluster communication, but also it terminates the error propagation problem, which is always unavoidable in other approximation methods. Furthermore, by incorporating the quantized proximity observation model, the energy and bandwidth consumed by the intra-cluster communication are dramatically reduced. In conclusion, as the target moves freely in WSN, a large number of quantized measurements are generated, which facilitates both the activated relays localization and the target tracking. The

Table 1. Comparison of target tracking algorithms

Comparison	Mean of MSE (MMSE)	Execution time
AQVF algorithm	0.2511 m	1.5938 s
BVF algorithm	2.69 m	1.5163 s
QPF algorithm	2.832 m	1.1273 s

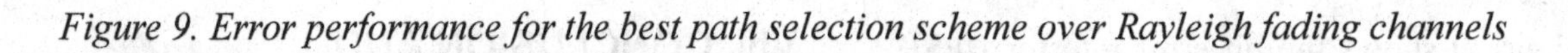

Figure 9. Error performance for the best path selection scheme over Rayleigh fading channels

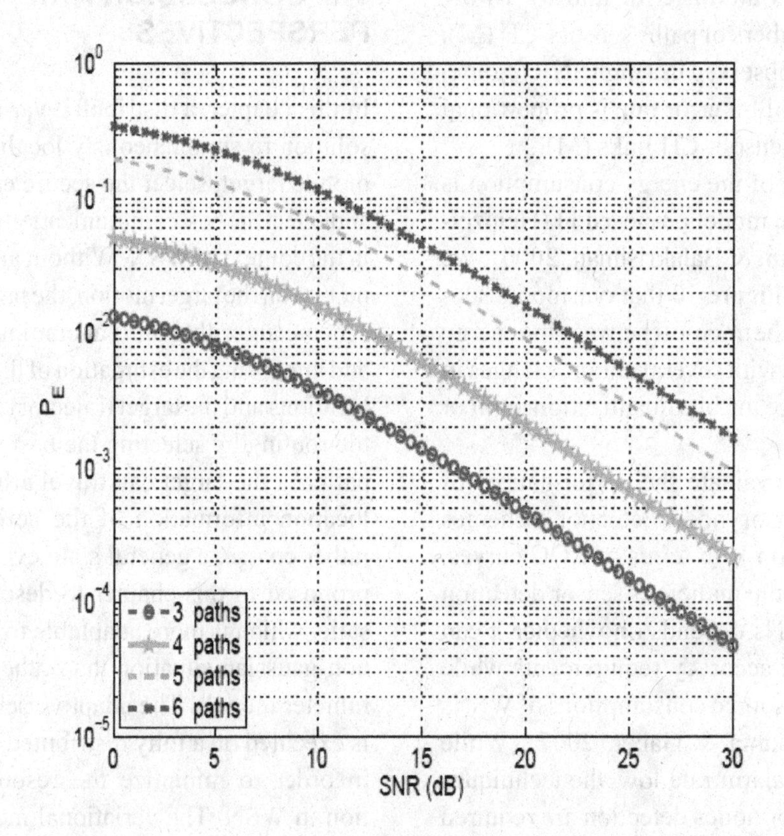

Figure 10. Energy consumption comparison where the quantization level is fixed to 3

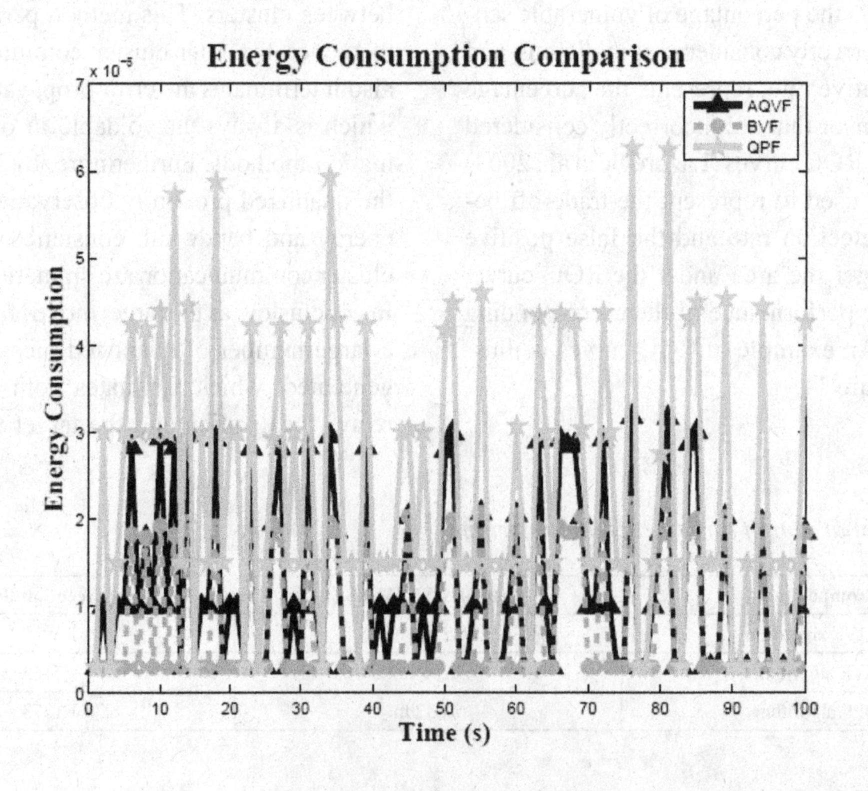

Figure 11. ROC curves for the proposed malicious sensors detection techniques

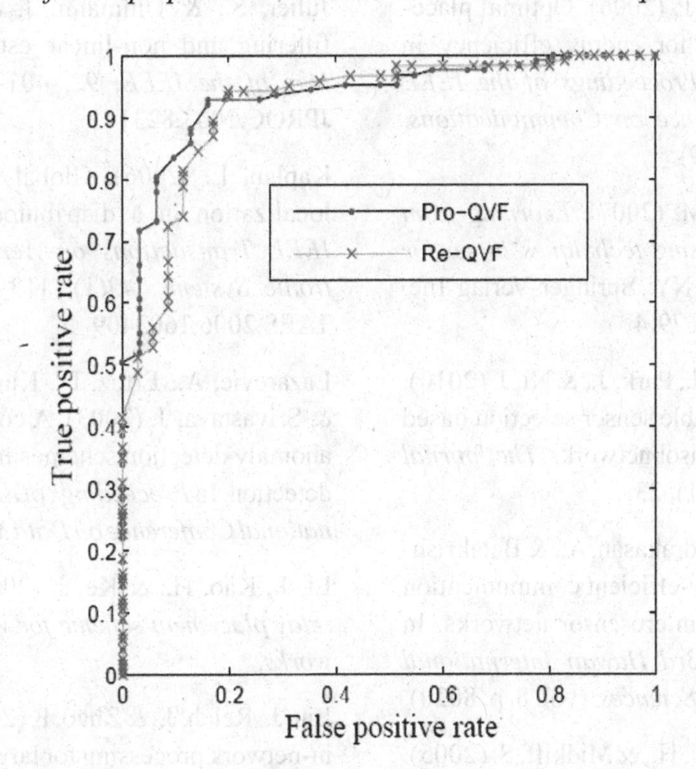

promising results obtained by simulations and presented in this chapter have shown clearly that the estimation of relay locations and that of target are interdependently and continuously improved on-line.

REFERENCES

Akyildiz, I., Su, W., Sankarasubramaniam, Y., & Cayirci, E. (2002). Wireless sensor networks: A survey. *Computer Networks*, *38*(4), 393–422. doi:10.1016/S1389-1286(01)00302-4

Bari, A., Teng, D., & Jaekel, A. (2009). Optimal relay node placement in hierarchical sensor networks with mobile data collector. *18th International Conference on Computer Communications and Networks*, (pp. 1–6). IEEE Computer Society.

Barndorff-Nielsen, O. (1977). Exponentially decreasing distributions for the logarithm of particle size. *Proceedings of the Royal Society of London*, *353*, 401–419. doi:10.1098/rspa.1977.0041

Chen, W., Hou, J., & Sha, L. (2004). Dynamic clustering for acoustic target tracking in wireless sensor networks. *IEEE Transactions on Mobile Computing*, 258–271. doi:10.1109/TMC.2004.22

Chen, Y., & Zhao, Q. (2005). On the lifetime of wireless sensor networks. *IEEE Communications Letters*, *9*(11), 976–978. doi:10.1109/LCOMM.2005.11010

Cover, T., & Thomas, J. (2006). *Elements of information theory*. Wiley-Interscience.

Djuric, P., Kotecha, J. Z. J., Huang, Y., Ghirmai, T., Bugallo, M., & Miguez, J. (2003, September). Particle filtering. *IEEE Signal Processing Magazine*, *20*, 19–38. doi:10.1109/MSP.2003.1236770

Ergen, S., & Varaiya, P. (2006). Optimal placement of relay nodes for energy efficiency in sensor networks. In *Proceedings of the IEEE International Conference on Communications,* (vol. 8, pp. 3473–3479).

Gama, J., & Gaber, M. (2007). *Learning from data streams: Processing techniques in sensor networks*. New York, NY: Springer-Verlag Inc. doi:10.1007/3-540-73679-4

Han, G., Shu, L., Ma, J., Park, J., & Ni, J. (2010). Power-aware and reliable sensor selection based on trust for wireless sensor networks. *The Journal of Communication*, *5*(1), 23.

Heinzelman, W., Chandrakasan, A., & Balakrishnan, H. (2000). Energy-efficient communication protocol for wireless microsensor networks. In *Proceedings of the 33rd Hawaii International Conference on System Sciences*, (vol. 8. p. 8020).

Hou, Y., Shi, Y., Sherali, H., & Midkiff, S. (2005). On energy provisioning and relay node placement for wireless sensor networks. *IEEE Transactions on Wireless Communications*, *4*(5), 2579. doi:10.1109/TWC.2005.853969

Ikki, S., & Ahmed, M. (2007). Performance analysis of cooperative diversity wireless networks over Nakagami-m fading channel. *IEEE Communications Letters*, *11*(4), 334. doi:10.1109/LCOM.2007.348292

Iranli, A., Maleki, M., & Pedram, M. (2005). Energy efficient strategies for deployment of a two-level wireless sensor network. In *Proceedings of the 2005 International Symposium on Low Power Electronics and Design,* (p. 238).

Juang, B., & Rabiner, L. (1985). A probabilistic distance measure for hidden Markov models. *AT & T Bell Laboratories Technical Journal*, *64*(2), 391–408.

Julier, S., & Uhlmann, J. (2004). Unscented filtering and non-linear estimation. *Proceedings of the IEEE*, *92*, 401–422. doi:10.1109/JPROC.2003.823141

Kaplan, L. (2006). Global node selection for localization in a distributed sensor network. *IEEE Transactions on Aerospace and Electronic Systems*, *42*(1), 113–135. doi:10.1109/TAES.2006.1603409

Lazarevic, A., Ertoz, L., Kumar, V., Ozgur, A., & Srivastava, J. (2003). A comparative study of anomaly detection schemes in network intrusion detection. In *Proceedings of the Third SIAM International Conference on Data Mining*, (pp. 25–36).

Li, J., Kao, H., & Ke, J. (2009). *Voronoi-based relay placement scheme for wireless sensor networks.*

Liu, J., Reich, J., & Zhao, F. (2003). Collaborative in-network processing for target tracking. *EURASIP Journal on Applied Signal Processing*, *2003*, 378–391. doi:10.1155/S111086570321204X

Luo, X., & Giannakis, G. (2008). Energy-constrained optimal quantization for wireless sensor networks. *EURASIP Journal on Advances in Signal Processing*, *2008*, 1–12. doi:10.1155/2008/462930

Mansouri, M., Ouchani, I., Snoussi, H., & Richard, C. (2009). *Cramer-Rao Bound-based adaptive quantization for target tracking in wireless sensor networks*. In IEEE/SP Workshop on Statistical Signal Processing, SSP'09.

Manyika, J., & Durrant-Whyte, H. (1995). *Data fusion and sensor management: A decentralized information-theoretic approach*. Upper Saddle River, NJ: Prentice Hall PTR.

Misra, I., Dolui, S., & Das, A. (2005). *Enhanced-efficient adaptive clustering protocol for distributed sensor networks*. ICON, 2005.

Papoulis, A., & Pillai, S. (2002). *Probability, random variables and stochastic processes. McGraw-Hill Education*. India: Pvt Ltd.

Rabiner, L. (1989). A tutorial on hidden Markov models and selected applications inspeech recognition. *Proceedings of the IEEE*, *77*(2), 257–286. doi:10.1109/5.18626

Ronen, O., Rohlicek, J., & Ostendorf, M. (1995). Parameter estimation of dependence tree models using the EM algorithm. *IEEE Signal Processing Letters*, *2*(8), 157–159. doi:10.1109/97.404132

Simon, M., & Alouini, M. (2005). *Digital communication over fading channels*. IEEE, 2005.

Snoussi, H., & Richard, C. (2006). Ensemble learning online filtering in wireless sensor networks. In *10th IEEE Singapore International Conference on Communication Systems, ICCS 2006,* (pp. 1–5).

Teng, J., Snoussi, H., & Richard, C. (2007). Prediction-based proactive cluster target tracking protocol for binary sensor networks. In *2007 IEEE International Symposium on Signal Processing and Information Technology,* (pp. 234–239).

Teng, J., Snoussi, H., & Richard, C. (2007b). Binary variational filtering for target tracking in sensor networks. In IEEE/SP 14th Workshop on Statistical Signal Processing, SSP'07, (pp. 685–689).

Ucinski, D., & Patan, M. (2007). D-optimal design of a monitoring network for parameter estimation of distributed systems. *Journal of Global Optimization*, *39*(2), 291–322. doi:10.1007/s10898-007-9139-z

Valenzuela, R. (1987). A statistical model for indoor multipath propagation. *IEEE Journal on Selected Areas in Communications*, *5*(2), 128–137. doi:10.1109/JSAC.1987.1146527

Vermaak, J., Lawrence, N., & Perez, P. (2003). Variational inference for visual tracking. In *Proceedings of the 2003 IEEE Computer Society Conference on Computer Vision and Pattern Recognition*, vol. 1.

Wang, Q., Xu, K., Takahara, G., & Hassanein, H. (2005). *Locally optimal relay node placement in heterogeneous wireless sensor networks.*

Wang, Y. (2008). *Topology control for wireless sensor networks* (pp. 113–147). Wireless Sensor Networks and Applications.

Yao, K., Hudson, R., Reed, C., Chen, D., & Lorenzelli, F. (1998). Blind beamforming on a randomly distributed sensor array system. *IEEE Journal on Selected Areas in Communications*, *16*(8), 1555–1567. doi:10.1109/49.730461

Zuo, L., Niu, R., & Varshney, P. (2007). A sensor selection approach for target tracking in sensor networks with quantized measurements. In *Proceedings of the 2007 IEEE International Conference on Acoustics, Speech, and Signal Processing*, II, (pp. 2521 – 2524).

Chapter 20
Improved Energy-Efficient Ant-Based Routing Algorithm in Wireless Sensor Networks

Adamu Murtala Zungeru
University of Nottingham-Malaysia Campus, Malaysia

Li-Minn Ang
University of Nottingham-Malaysia Campus, Malaysia

SRS. Prabaharan
University of Nottingham-Malaysia Campus, Malaysia

Kah Phooi Seng
Sunway University, Malaysia

ABSTRACT

High efficient routing is an important issue in the design of limited energy resource wireless sensor networks (WSNs). This chapter presents an Improved Energy-Efficient Ant-Based Routing Algorithm (IEEABR) in wireless sensor networks. Compared to traditional Basic Ant-Based Routing (BABR), Improved Ant-Based Routing (IABR), and Energy-Efficient Ant-Based Routing (EEABR) approaches, the proposed IEEABR approach has advantages of reduced energy usage and achieves a dynamic and adaptive routing that can effectively balance the WSN node power consumption and increase the network lifetime. This chapter covers applications and routing in WSNs, different methods for routing using ant colony optimization (ACO), a summary of routing algorithms based on ant systems, and the Improved Energy-Efficient Ant-Based Routing Algorithm approach. Simulations results were analyzed while also looking at open research problems and future work to be done. The chapter concludes with a comparative summary of results with IABR and EEABR.

DOI: 10.4018/978-1-4666-0101-7.ch020

1 INTRODUCTION

The advancement in technology has produced the availability of small and low cost sensor nodes with capability of sensing types of physical, environmental conditions, data processing, and wireless communication (Akyildiz et al., 2002; Katz et al., 1999; Min et al., 2001; Rabaey et al., 2000; Sohrabi et al., 2000). The sensing circuitry measures ambient conditions related to the environment surrounding the sensor, which transforms them into an electric signal. Processing such a signal, reveals some properties about objects located and/or events happening in the vicinity of the sensor. The sensor sends such collected data, usually via radio transmitter, to a command center (sink) either directly or through a data concentration center (a gateway). The decrease in the size and cost of sensors, resulting from such technological advances, has fueled interest in the possible use of large set of disposable unattended sensors. Such interest has motivated intensive research in the past few years addressing the potential of collaboration among sensors in data gathering and processing and the coordination and management of the sensing activity and data flow to the sink. A natural architecture for such collaborative distributed sensors is a network with wireless links that can be formed among the sensors in an ad hoc manner.

The main goal of our study is to maintain network lifetime at a maximum, while reducing Energy usage by the nodes and discover the shortest paths from the source nodes to the base node (sink) using a Improved Energy-Efficient Ant-Based Routing Algorithm (IEEABR).

The chapter is organized in the following format: The introductory part of the chapter provided in section 1, covers a general perspective and the objective of the chapter. Section 2 the WSNs and previous work on routing, deals with wireless sensor networks while looking at its applications and routing algorithms, different methods of routing using the popular ACO and generally comparing the different methods of routing. Section 3 gives detailed explanation of our proposed algorithm, the IEEABR. Section 4 looks in to analyze simulation results while also looking at the simulation environment and the trace files. Finally, section 5 concludes the chapter with an open research problems and future work to be done, and a comparative summary of our results with the IABR and EEABR. At the end of the section were provision for References and Additional readings for detail of some routing Algorithms.

2 PREVIOUS WORK ON ROUTING IN WSNS

2.1 Wireless Sensor Networks; Applications and Constraints

WSNs are collections of compact-size, relatively inexpensive computational nodes that measure local environmental conditions, or other parameters and forward such information to a central point for appropriate processing. WSN nodes can sense the environment, communicate with neighboring nodes, and in many cases perform basic computations on the data being collected. The environment can be the physical world, a biological system, or an information technology (IT) framework. However, the characteristic of wireless sensor network (WSN) require more effective methods of data forwarding and processing. Though, WSN is used in many applications such as; radiation and nuclear-threat detection systems, weapon sensors for ships, toxins and to trace the source of the contamination in public-assembly locations, structural faults (e.g., fatigue-induced cracks) in ships, Volcanic eruption, earthquake detection, aircraft, and buildings, biomedical applications, habitat sensing, and seismic monitoring. More recently, interest has focused on networked biological and chemical sensors for national security applications, physical security, air traffic

Table 1. Classifying applications of WSNs based on areas of application

APPLICATIONS OF WSNs	
Areas of Applications	**Applications**
Home	Home automation, Instrumented environment, Automated meter reading
Commercial	Environmental control in industrial and office buildings, Inventory control, Vehicle tracking and detection, Traffic flow surveillance
Environmental	Microclimates, Forest fire detection, Flood detection, Precision agriculture
Military	Targeting, Monitoring inimical forces, Monitoring friendly forces and equipment, Military theater or battle-field surveillance, Battle damage assessment, Nuclear, biological, and chemical attack detection
Health	Remote monitoring of physiological data, Tracking and monitoring doctors and patients inside a hospital, Drug administration, Elderly assistance

control, traffic surveillance, video surveillance, industrial and manufacturing automation, process control, inventory management, distributed robotics, weather sensing, environment monitoring, national border monitoring, and building and structures monitoring (Chong et al., 2003). The area of applications of the WSN is classified in to Five (5) as shown in Table 1.

Despite the wide areas of application of WSNs, it also has many restrictions such as limited energy supply, limited computation and limited communication capabilities, Limited functional capabilities, including problems of size, Power factor, Node costs, Environmental factors, Transmission channel factor, Topology management complexity and node distribution.

2.2 Routing in WSN

Researchers have designed and developed various routing protocols specifically for WSNs due to the differences between routing in WSNs and other wireless networks. Routing in sensor networks involves a lot of task due to differences in some properties between them and contemporary communication and wireless ad hoc networks (Akkaya et al., 2005). First of all, it is not possible to build a global addressing scheme for the deployment of sheer number of sensor nodes. Therefore, classical IP-based protocols cannot be applied to sensor networks. Second, in contrary to typical communication networks almost all applications of sensor networks require the flow of sensed data from multiple regions (sources) to a particular sink. Third, generated data traffic has significant redundancy in it since multiple sensors may generate same data within the vicinity of a phenomenon. Such redundancy needs to be exploited by the routing protocols to improve energy and bandwidth utilization. Fourth, sensor nodes are tightly constrained in terms of transmission power, on-board energy, processing capacity and storage and thus require careful resource management. Due to such differences, many new algorithms have been proposed for the problem of routing data in sensor networks. These routing mechanisms have considered the characteristics of sensor nodes along with the application and architecture requirements. Almost all of the routing protocols can be classified as data-centric, hierarchical or location based although there are few distinct ones based on network flow or quality of service (QoS) awareness (Uchhula et al., 2010). Some protocols also fits into other categories as such we will summarize the routing protocols categories and their differences according to some metrics as shown in Table 2.

Table 2. Classification and comparison of routing protocols in WSNs

Routing Protocols	Classification	Data Aggregation	QoS	Query based	Power usage	Position Awareness
Rumor Routing	Data-centric	Yes	No	Yes	Low	No
Sensor Protocol for Information via Negotiation	Data-centric	Yes	No	Yes	Limited	No
Directed Diffusion	Data-centric	Yes	No	Yes	Limited	No
Flooding & gossiping	Data-centric	Yes	No	Yes	Limited	No
Energy Aware Routing	Data-centric	No	No	Yes	Low	Yes
Gradient-based routing	Data-centric	Yes	No	Yes	Low	No
Constrained Anisotropic Diffusion Routing	Data-centric	Yes	No	No	Limited	No
COUGAR	Data-centric	Yes	No	Yes	Limited	No
Active Query forwarding in sensor networks	Data-centric	Yes	No	Yes	Low	No
Low-energy Adaptive Clustering Hierarchy	Hierarchical	Yes	No	No	Maximum	No
Power-efficient Gathering in Sensor Information Systems	Hierarchical	No	No	No	Maximum	No
Threshold sensitive Energy Efficient sensor Network protocol & Adaptive TEEN	Hierarchical	Yes	No	No	Maximum	No
Energy Aware Routing for Cluster-Based SNs	Hierarchical	Yes	No	No	Low	No
Self-Organizing protocol	Hierarchical	No	No	No	Low	No
Minimum energy communication network & Small MECN	Location Based	No	No	No	Limited	Yes
Geographic adaptive fidelity	Location Based	No	No	No	Limited	Yes
Geographic & energy-aware routing	Location Based	No	No	No	Limited	Yes
Maximum Lifetime Energy Routing	Network flow & QoS-Aware	No	Yes	No	Low	No
Maximum Lifetime Data Gathering	Network flow & QoS-Aware	Yes	Yes	No	Low	Yes
Minimum Cost Forwarding	Network flow & QoS-Aware	Yes	Yes	Yes	Low	No
Sequential Assignment Routing	Network flow & QoS-Aware	Yes	Yes	Yes	Low	No
Energy-Aware QoS Routing Protocol	Network flow & QoS-Aware	No	Yes	Yes	Low	No
SPEED	Network flow & QoS-Aware	No	Yes	Yes	Low	No

2.3 Ant Colony Optimization (ACO) and its Methods for Routing in WSNs

2.3.1 Ant Colony Optimization (ACO)

The optimization of network parameters for the WSN routing process for enhancement of networks' lifetime might be considered as a combinatorial optimization problem. A lot of research work has been done on the collective attitude of biological species such as ants as a natural model for combinational optimization problems (Bonabeau et al., 1999; Hussain et al., 2008; Iyengar et al., 2007, White et al., 1998). Ant colony optimization (ACO) algorithm simulating the behavior of ant colony has been successfully applied in many optimization problems such as the asymmetric travelling salesman (Liu et al., 2005), vehicle routing (Rodoplu et al., 1999), and WSN routing (Akkaya et al., 2005; Buczak et al., 1998; Camilo et al., 2006). Also, mobile agents system that is based on Ant Colony Optimization (ACO) method informally, the AntNet algorithm was also proposed by M. Dorigo and G. Di Caro, which is based on packet routing in communication networks (Dorigo et al., 1997).

ACO is a class of optimization algorithms modeled on the actions of an ant colony and a subset of Swarm Intelligence that helps in finding optimal solutions to optimization problems. In ACO, a set of software agents called artificial ants search for good path to their food which is in turn adopted as a solutions to a given optimization problem. To apply ACO, the optimization problem is transformed into the problem of finding the best path on a weighted graph. The artificial ants (hereafter ants) incrementally build solutions by moving on the graph. The solution construction process is stochastic and is biased by a pheromone model, that is, a set of parameters associated with graph components (either nodes or edges) whose values are modified at runtime by the ants (Anandamoy et al., 2006). Ant colony optimization which uses the natural metaphor of ants and stigmergy to solve problems over a physical space which is basically a simulation of the behaviors of ant swarms. The ACO mimics the way real ants find the shortest route between a food source and their nest. The ants communicate with one another by means of pheromone trails and exchange information about which path should be followed. The more the number of ants traces a given path, the more attractive this path or trail becomes and is followed by other ants by depositing their own pheromone. This auto catalytic and collective behavior results in the establishment of the shortest route.

The basic ACO Routing Algorithm is described as below:

1. Initialization of the parameters which determines the pheromone trail,
2. While (until result conditions supplied) do,
3. Generate solutions,
4. Apply Local Search,
5. Update Pheromone Trail,
6. End

Associated with the solution of optimum tour in ant colony systems, the following steps describe the process of the implementation.

a. Pheromone initialization. When the algorithm begin, m ants is placed on the sources and every path is endued the same number of pheromone, the number of iteration N begin with zero (N=0) and maximum number is N_{max}.

b. Selection strategy. An ant, say k, moves from the present node v_i to the next node v_j according to the state transition rule given by equation (1)

$$S_{k=}\begin{cases}\arg\max_{u\in J_{k(r)}}\left\{\left[\tau(r,u)\right]^{\alpha}.\left[\eta(r,u)\right]^{2}\right\}, q\leq q_0 \\ S, else\end{cases} \quad (1)$$

$$P_k(r,s) = \begin{cases} \dfrac{[\tau(r,s)]^{\alpha}.[\eta(r,s)]^{\beta}}{\sum_{u \in J_{k(r)}} [\tau(r,s)]^{\alpha}.[\eta(r,s)]^{\beta}}, s \notin J_k(r) \\ 0, else \end{cases} \quad (2)$$

Where q is a random number distributed in the space [0,1), which tends to increase the number of searching tours, and can weaken the trend of ants being trapped in a local optimum. q_o is a predefined parameter ($0 \leq q_o \leq 1$). It controls the relative importance of exploring new tour versus utilizing transcendent knowledge. This state transition rule favors transitions towards nodes connected by short edges and with a large amount of pheromone. Forward ants will get a random number q before selecting the next node. If $q \leq q_o$ then the best next node v, according to equation (1) is chosen, otherwise $q \leq q_o$, a node is chosen according to equation (2). If the vehicle capacity is met, the ant will return to the depot before selecting the next node. This selection process continues until each node is visited and the tour is complete.

c. Local pheromone update. The local updating rule in equation (3) is applied to change pheromone level of edges after an ant completes its route.

$$\tau(r,s) \leftarrow (1-\rho) * \tau(r,s) + \rho * \tau_o \quad (3)$$

$$\tau_o = (n * L_m)^{-1} \quad (4)$$

$\tau_o = (n * L_m)^{-1}$ Is the initial pheromone level of edges, where n is the number of nodes and L_m is the tour length produced by the nearest neighbor heuristic. In a case the best tour did not improve within a defined number of generations, the pheromone level of each edge is then reset to the initial pheromone level τ. The forward ants having the lifetime, if does not arrive at the destination and not be close to the lifetime, go to step **b.**

d. Global pheromone update. Forward ants bring the information to the backward ants, after all the forward ants complete the research. Whenever a node r receives a backward ant coming from a neighboring node, it updates its routing pheromone in the following manner.

e. Examine the termination condition of iteration. Ninety percent of forward ants search the same optimum tour in max iteration N_{max}. While $Nc \geq N$max, iterations terminate, else, Nc = Nc + 1, go to step **a.**

Where s_k = the next node of the forward ants to select, $J_k(r)$=the set of nodes that remain to be visited, β=the parameter that determines the relative pheromone versus distance savings, η= the savings of combining two nodes on one tour as opposed to serving them on two different tours, τ= the pheromone level on edge, (r,s,u)= node identifier, P_k = the probability with which ant k chooses to move from node to node, τ_o=the pheromone increment level on edge, ρ = evaporation coefficient of local research, L_{nn} = the tour length produced by the nearest neighbor heuristic, $\Delta\tau(r,s)$= the pheromone level increment on edge, L_{gb} = the length of global optimum tour and α=evaporation coefficient of global research.

Above complex global behaviors are the result of self-organizing dynamics driven by local interactions and communications among a number of relatively simple individuals. The simultaneous presence of these and other fascinating and unique characteristics have made ant societies an attractive and inspiring model for building new algorithms and new multi-agent systems.

Hence, the procedure iterated and the algorithm runs until a predefined time of halt is met or, a certain number of generations have been done or the average quality of the solution found by the ants of a generation has not changed for several generations.

The ant colony optimization algorithm combines such characteristics as quick problem solving like; Quadratic Assignment Problems (QAP), Job-shop Scheduling Problems (JSP), vehicle routing, graph coloring, environmental and habitat monitoring, health care application, traffic control, wild ecological survey and, global optimization as well as the high degree of self-organization and quickly routing in wireless sensor network. This caused us to explore the overall energy balance of the wireless sensor network routing protocol based on ant colony algorithm.

Quite a couple of protocols have been design to solve the optimization problems using the popular ACO. A list of few among the review ones are: AntNet Routing Algorithm (AntNet), Basic Ant Based Routing Algorithm (BABR), Improved Ant Based Routing Algorithm (IABR), Energy Efficient Ant Based Routing Algorithm (EEABR), Ant Based Control Routing (ABC), Ant Colony based Routing Algorithm Routing (ARA), Probabilistic Emergent Routing Algorithm (PERA), and Ant agents for Hybrid Multipath Routing (AntHocNet).

2.3.2 Basic Ant Based Routing for WSN

Basic Ant Routing (White et al., 1998) proposed a scheme for routing in circuit-switched networks which was also inspired by ant colonies' foraging behavior and in particular by Ant system (AS). In their work, ants are launched in time there is a connection request. A connection request can be a point-to-point or a point to multipoint request. In the point to point case, in a network of N nodes, N ants are launched to look for the best path to the destination, given a cost function associated with every node. For a point to multipoint request with m destinations, N.m ants are launched. Which is similar to Ant System, but it relies on pheromone concentration on the links (Dorigo et al., 1998).

2.3.3 Improved Ant Based Routing Algorithm

The Improved Ant Based Routing for WSN as proposed (Camilo et al., 2006). They proposed two improvements in the basic ant-based routing algorithm in order to reduce the memory used in the sensor nodes and also to consider the energy quality of the paths found by the ants. They found out that in the basic ant-based algorithm, the forward ants are sent to no specific destination node, which means that sensor nodes must communicate with each other and the routing tables of each node must contain the identification of all the sensor nodes in the neighborhood and the corresponding levels of pheromone trail. This can be problem in a large network, since nodes will need to have quite a large amounts of memory to save the neighbor nodes that are in the direction of the sink nodes, and considerably reduces the size of the routing tables, and in consequence, the memory needed by the nodes (Kulik et al., 2002).

Since sensor nodes are devices with a very limited energy capacity, which means that the quality of a given path between a sensor node and the sink node should be determine not only on the distance (number of nodes of the path), but also in terms of the energy level of the path. Hence, it will then be preferable to choose a longer path with high energy level than a shorter path with very low energy levels. In order to consider the energy quality of the paths on the basic algorithm, they proposed a new function to determine the amount of pheromone trail that the backward ant will drop during its returning journey as:

$$\Delta\tau = \frac{1}{c - [Aug(E_k) - 1 / (Min(E_k))]} \quad (5)$$

Where E_k is a new vector carried by forward ant k with the energy levels of the nodes of its path, C, the initial energy level of the nodes, $Avg(E_k)$ the average of the vector values and $Min(E_k)$ is the minimum value of the vector.

2.3.4 Energy Efficient Ant-Based Routing Algorithm

The Energy-Efficient Ant Based Routing for WSN as proposed (Camilo et al., 2006) which is an improved version of the Ant based routing in WSN which does not only considered the nodes in terms of distance, but also in terms of energy level of the path transverse by the ants. And pointed out in his proposal that in the basic algorithm the forward ants are sent to no specific destination node, which means that sensor nodes must communicate with each other and the routing tables of each node must contain the identification of all the sensor nodes in the neighborhood and the correspondent levels of pheromone trail. Which could be problem since nodes would need to have big amounts of memory to save all the information about the neighborhood. If the forward ants are sent directly to the sink-node, the routing tables only need to save the neighbor nodes that are in the direction of the sink-node. This considerably reduces the size of the routing tables and, in consequence, the memory needed by the nodes. Since one of the main concerns in WSN is to maximize the lifetime of the network, which means saving as much energy as possible, it would be preferable that the routing algorithm could perform as much processing as possible in the network nodes, than transmitting all data through the ants to the sink-node to be processed there.

2.3.5 AntNet: Ant Algorithm

AntNet is a protocol design purposely for packet routing in communication networks, which was proposed by Dorigo et al. (Dorigo et al., 1997). AntNet routing algorithm is a mobile agents system that is based on Ant Colony Optimization (ACO) method. In AntNet, a group of mobile agents (or artificial ants) build paths between pair of nodes; exploring the network concurrently and exchanging obtained information to update the routing tables. Informally, the AntNet algorithm and its main characteristics can be summarized as follows:

1. At regular intervals, and concurrently with the data traffic, from each network node mobile agents are asynchronously launched towards randomly selected destination nodes.
2. Agents act concurrently and independently, and communicate in an indirect way, through the information they read and write locally to the nodes.
3. Each agent searches for a minimum cost path joining its source and destination nodes.
4. Each agent moves step-by-step towards its destination node. At each intermediate node a greedy stochastic policy is applied to choose the next node to move to. The policy makes use of (i) local agent-generated and maintained information, (ii) local problem-dependent heuristic information, and (iii) agent-private information.
5. While moving, the agents collect information about the time length, the congestion status and the node identifiers of the followed path.
6. Once they have arrived at the destination, the agents go back to their source nodes by moving along the same path as before but in the opposite direction.
7. During this backward travel, local models of the network status and the local routing table of each visited node are modified by the agents as a function of the path they followed and of its goodness.
8. Once they have returned to their source node, the agents die.

AntNet Algorithm

AntNet was first presented in 1997 by Dorigo et al. (Dorigo et al., 1997). The algorithm was based on data network, with N nodes, and s as source node, and a destination d. The Algorithm was described with two Ants;

1. Forward Ant, denoted $F_{s\to d,}$ which will travel from the source node s to a destination d.
2. Backward Ant, denoted $B_{s\to d}$ that will be generated by a forward ant $F_{s\to d}$ in the destination d, and it will come back to s following the same path traversed by $F_{s\to d}$ with the purpose of using the information already picked up by $F_{s\to d}$ in order to update routing tables of the visited nodes.

Every ant deposits a stack $S_{s\to d}(k)$ of data, where the k index refers to the k^{st} visited node, in a journey, where $S_{s\to d}(0)$= s and $S_{s\to d}(m)$= d, being m the amount of jumps performed by $F_{s\to d}$ for arriving to d.

Let k be any network node; its routing table will have N entries, one for each possible destination.

Let j be one entry of k routing table (a possible destination).

Let N_k be set of neighboring nodes of node k.

Let P_{ki} be the probability with which an ant or data packet in k, jumps to a node i, i∈N_k, when the destination is j ($j\neq k$). Then, for each of the N entries in the node k routing table, it will be n_k values of P_{ji} subject to the condition:

$$\sum_{i\in N_k} P_{ji} = 1; j = 1,\dots, N. \qquad (6)$$

The routing table and list of trips updating methods for k are described as follows:

1. The k routing table is updated for the entries corresponding to the nodes k' between k and d inclusive. For example, the updating approaches for the d node, when $B_{s\to d}$ arrives to k, coming from f, f∈N_k.

A P_{df} probability associated with the node f when it wants to update the data corresponding to the d node is increased, according to:

$$P_{df} \leftarrow P_{df} + (1-r') * (1-P_{df}). \qquad (7)$$

Where r′ is an a dimensional measure, indicating how good (small) is the elapsed trip time T with regard to what has been observed on average until that instant. Experimentally, r' is expressed as:

$$r' = \begin{cases} \frac{T}{c\mu}, c \geq 1 \; if \; \frac{T}{c\mu} \prec 1 \\ 1, else \end{cases} \qquad (8)$$

Where: μ is the arithmetic observed trip time T average, c is a scale factor experimentally chosen like 2 (Dorigo et al., 1997)

The other neighboring nodes ($j\neq f$) P_{dj} probabilities associated with node k are diminished, in order to satisfy equation (10), through the expression:

$$P_{dj} \leftarrow P_{dj} - (1-r')P_{dj} \,.\, j\in N_k \,, j \neq f \qquad (9)$$

A list $trip_k$ (μ_i, σ_i^2) of estimate arithmetic mean values μ_i and associated variances σ_i^2 for trip times from node k to all nodes i ($i\neq k$) is also updated. This data structure represents a memory of the network state as seen by node k. The list trip is updated with information carried by $B_{s\to d}$ ants in their stack $S_{s\to d}$. For any node pair source-destination, μ after (n+1) samples (n>0) is calculated as follows:

$$\mu_{n+1} = \frac{n\mu_n + x_{n+1}}{n+1} \qquad (10)$$

Where: x_{n+1} represents the trip time T sample n+1, μ_n is the arithmetic mean after n trip time samples.

A modified version of AntNet with five main steps that performs basically the same actions, but differences on how the routing tables and the lists trips (now known in as traffic local model M_k) are updated, was later proposed in 1998 (Dorigo et al., 1998).

The two main differences are;

Suppose that a $B_{s\to d}$ arrives to a node k, in its return trip to node s. The $B_{s\to d}$ ant will update the traffic local model M_k and the neighbor nodes probabilities of k associated to node d in the routing table τ_k. Also, as in the earlier AntNet, the update is performed in the entries corresponding to every node $k' \in S_{s\to d}$, $k' \neq d$ in the sub-paths followed by $F_{s\to d}$ after visiting k. If a sub-path trip time T is statistically good (i.e.: T is smaller than $\mu_i + I(\mu, \sigma)$, where I is a μ interval confidence estimator), then *T* is used to update the statistics related and the routing table. However, if *T* is bad, it is not used, because it doesn't give a true idea about the time required to arrive to the sub-path nodes. The traffic local model M_k and the routing table τ_k are updated for a generic destination $d' \in S_{s\to d}$ in the following way:

M_k is updated with the values carried in $S_{s\to d'}$. The trip time $T_{k\to d'}$ employed by $F_{s\to d}$ to travel from *k* to *d'* is used to update μ_d, $\sigma^2_{d'}$ and the best observed value inside window W_d according to the expressions:

$$\mu_{d'} \leftarrow \mu_{d'} + \eta(T_{k\to d'} - \mu_{d'}) \quad (11)$$

$$\sigma^2_{d'} \leftarrow \sigma^2_{d'} + \eta((T_{k\to d'} - \mu_{d'})^2 - \sigma^2_{d'}) \quad (12)$$

Where η is the weight of each trip time observed, the effective number of samples will be approximately 5(1/η). Therefore, for 20 samples, η=0.25, and for 100 samples η=0.05. The $T_{k\to d'}$ mean value and its dispersion could vary strongly, depending on traffic conditions: a poor (large) time with low data traffic could be very good with relation to another measure with more traffic. The statistical model should reflect this variability and continue the traffic fluctuations in a robust way. This model plays a critical role in routing table updating.

2. The routing table for k is updated in the following way:

The value $P_{fd'}$ (the probability for selecting the neighbor node *f*, when the node destination is *d'*) is incremented by means of the expression:

$$P_{fd'} \leftarrow P_{fd'} + r(1 - P_{fd'}). \quad (13)$$

where *r* is a reinforcement factor indicating the goodness of the followed path.

The $P_{nd'}$ probabilities associated to the other nodes decreases respectively:

$$P_{nd'} \leftarrow P_{nd}' - r P_{nd'}, \; n \in N_k, \; n \neq f \quad (14)$$

The factor of reinforcement *r* is calculated considering three fundamental aspects: (i) the paths should receive an increment in their probability of selection, proportional to their goodness, (ii) the goodness is a traffic condition dependent measure that can be estimated by M_k, and (iii) they should not continue all the traffic fluctuations in order to avoid uncontrolled oscillations. It is very important to establish a commitment between stability and adaptability. Between several tested alternatives (Dorigo et al., 1997), expression (15) was chosen to calculate *r*:

$$r = c_1\left(\frac{W_{best}}{T}\right) + c_2\left(\frac{I_{sup} - I_{inf}}{(I_{sup} - I_{inf}) + (T - I_{inf})}\right) \quad (15)$$

Where W_{best} represents the best trip of an ant to node *d'*, in the last observation window $W_{d'}$, I_{inf}

$= W_{best}$ stands for lower limit of the confidence interval for μ, $I_{sup} = \mu + z^*(\sigma / \sqrt{|w|}$ represents the upper limit of the confidence interval for μ, with $Z = 1/\sqrt{1-\gamma}$, while γ = confidence level, $\gamma \in [0.75, 0.8]$, c_1 and c_2 are the weight constants, chosen experimentally as $c_1 = 0.7$ and $c_2 = 0.3$ (10).

2.3.6 Ant Based Control (ABC) Routing

ABC (Schoonderwoerd et al., 1996) proposed the ABC designed purposely for telephone network application in mind. It is a very interesting adaptive routing algorithm based on the use of ants that modify the routing policy at every node in a network by depositing a virtual pheromone trail on routing table entries. In this Algorithm, every node in the network has a pheromone table entry for every possible destination in the network, and each table has an entry for every neighbor. Initially all are assumed to have 0.5 probabilities. Ants are launched from any node in the network. Each node has random destination. Ants move from node to node, selecting the next node to move to according to the probabilities in the pheromone tables for their destination node. Arriving at a node, they update the probabilities of that node's pheromone table entries corresponding to their *source* node. They alter the table to increase the probability pointing to their previous node. When ants have reached their destination, they die. The increase in these probabilities is a decreasing function of the age of the ant, and of the original probability. The ants get delayed on parts of the system that are heavily used. Some noise can be added to avoid freezing of pheromone trails. The method (Anandamoy et al., 2006) used to update the probabilities is quite simple: when an ant arrives at a node, the entry in the pheromone table corresponding to the node from which the ant has just come is increased according to the formula:

$$P = (P_{old} + \Delta P) / (1 + \Delta P) \quad (16)$$

Here p is the new probability and is the probability increase. The other entries having the probability decreased P' in the table of this node are decreased according to:

$$P' = (P'_{old}) / (1 + \Delta P) \quad (17)$$

The probabilities are updated according to the following formula, where *age* stands for the number of time steps that passed since the launch of the ant:

$$\Delta P = (((0.08 / age) + 0.0005) \quad (18)$$

The *delay* in time steps that is given to the ant is a function of the spare capacity *s* of the node:

$$delay = \lfloor 80.e^{-0.07Ss} \rfloor \quad (19)$$

2.3.7 Ant Colony Based Routing Algorithm (ARA)

Ant Colony Based Routing Algorithm (ARA) (Dorigo et al., 1997, White et al., 1998) works in an on demand way, with ants setting up multiple paths between source and destination at the start of a data session FANTs are broadcasted by the sender to all its neighbors. Each FANT has a unique sequence number to avoid duplicates. A node receiving a FANT for the first time creates a record [destination address, next hop, pheromone value] in its routing table. The node interprets the source address of the FANT as destination address, the address of the previous node as next hop, and computes the pheromone value depending on the number of hops the FANT needed to reach the node. Then the node relays the FANT to its neighbors. When the FANT reaches destination, it is processed in a special way. The destination node extracts the information and then destroys the FANT. A BANT is created and sent towards the source node. In that way, the path is established and data packets can be sent. Data packets are used to maintain the path, so no overhead is introduced. In this method, Pheromone values are changing.

2.3.8 Probabilistic Emergent Routing Algorithm (PERA)

This algorithm is an on-demand protocol, with ants being broadcast towards the destination at the start of a data session (Baras et al., 2003). Multiple paths are set up, but only the one with the highest pheromone value is used by data and the other paths are available for backup. The route discovery and maintenance is done by flooding the network with ants. Both forward and backward ants are used to fill the routing tables with probabilities. These probabilities reflect the likelihood that a neighbor will forward a packet to the given destination. Multiple paths between source and destination are created. First of all, neighbors are discovered using HELLO messages, but entries are only inserted in the routing table after receiving a backward ant from the destination node. Each neighbor receives an equal probable value for destination. This value is increased as a backward ant comes from that node, establishing a path towards destination. As ants are flooded, the algorithm uses sequence numbers to avoid duplicate packets. Only the greater sequence number from the same previous hop is taken into account. Forward ants with a lower sequence number are dropped. This approach is similar to AODV Route Request packets, but discovers a set of routes instead of one. Data packets can be routed according to the highest probability in the routing table for the next hop.

2.3.9 Ant Agents for Hybrid Multipath Routing (AntHocNet)

AntHocNet (Di Caro et al., 2005) is a multipath routing algorithm for mobile ad-hoc networks that combines both proactive and reactive components. It maintains routes only for the open data sessions. This is done in a Reactive Route Setup phase, where reactive forward ants are sent by the source node to find multiple paths towards the destination node. Backward ants are used to actually setup the route. While the data session is open, paths are monitored, maintained and improved proactively using different agents, called proactive forward ants.

2.3.10 Comparison of the Different Ant Based Routing Algorithms

AntNet a switched wired networks protocol (Dorigo et al., 1997) uses HELLO messages initially to discover the neighbors. In PERA, HELLO messages are broadcasted whenever there is a movement of node to a different location so that node can discover its new neighbors. In ARA entry in the routing table for each node is created when a forward ant arrives at that node. Pheromone value is the number of hops required by the forward ant to reach the current node from the destination. ARA is quite similar to PERA. One difference is that both forward and backward ants leave pheromone behind: forward ants update pheromone about the path to the source, while backward ants update pheromone about the path to the destination. Another difference is that also data packets update pheromone, so that paths which are in use are also reinforced while the data session is going on. This comes down to repeated path sampling, so that ARA keeps more of the original ACO characteristics than PERA. PERA uses routing table which has the following structure: [Destination, Next hop, Probability].

Each node periodically sends forward ant to randomly chosen destination, whereas source node is chosen according to some probability of data flow. ABC performs same ants flooding as applied in AntNet, where nodes send ants continuously to a chosen destination. Each ant has an associated age, which is increased proportionally to the load of each visited node, as the move from their source to destination; ant updates the pheromone for the path backward to s, based on its age which is a main difference with AntNet. AntHocNet is most efficient in maintaining paths. It has greater chance of exploring new paths due to proactive nature with a hint of probability. This is due to the fact that proactive ants are normally unicast to sample the existing path found by reactive forward ants but also have a small probability at each node of being broadcast. Therefore even though AntHocNet maintains paths between nodes and explores new routes it is costly and requires more resources.

3 IMPROVED ENERGY-EFFICIENT ANT-BASED ROUTING ALGORITHM

IEEABR as the proposed Algorithm, consider the available power of nodes and the energy consumption of each path as the reliance of routing selection, improves memory usage, utilizes the self organization, self-adaptability and dynamic optimization capability of ant colony system to find the optimal path and multiple candidate paths from source nodes to sink nodes then avoiding using up the energy of nodes on the optimal path and prolong the network lifetime while preserving network connectivity. This is necessary since for any WSN protocol design, the important issue is the energy efficiency of the underlying algorithm due to the fact that the network under investigation has strict power requirements. It has been proposed (Kulik et al., 2002), for forward ants sent directly to the sink-node; the routing tables only need to save the neighbor nodes that are in the direction of the sink-node. This considerably reduces the size of the routing tables and, in consequence, the memory needed by the nodes. Since one of the main concerns in WSN is to maximize the lifetime of the network, which means saving as much energy as possible, it would be preferable that the routing algorithm could perform as much processing as possible in the network nodes, than transmitting all data through the ants to the sink-node to be processed there. In fact, in huge sensor networks where the number of nodes can easily reach more than Thousands of units, the memory of the ants would be so big that it would be unfeasible to send the ants through the network. To implement these ideas, the memory M_k of each ant is reduced to just two records (Camilo et al., 2006), the last two visited nodes. Since the path followed by the ants is no more in their memories, a memory must be created at each node that keeps record of each ant that was received and sent. Each memory record saves the previous node, the forward node, the ant identification and a timeout value. Whenever a forward ant is received, the node looks into its memory and searches the ant identification for a possible loop. If no record is found, the node saves the required information, restarts a timer, and forwards the ant to the next node. If a record containing the ant identification is found, the ant is eliminated. When a node receives a backward ant, it searches its memory to find the next node to where the ant must be sent. In this section, we proposed two modifications on EEABR and the Basic Ant Based Routing Algorithm to improve the Energy consumption in the nodes of WSNs and also to in turn improve the performance and efficiency of the networks. The Algorithm of our proposed method is as below.

Initialize the routing tables with a uniform probability distribution;

$$P_{ld} = \frac{1}{N_k} \tag{20}$$

Where P_{ld} is the probability of jumping from node l to node d (destination), N_k the number of nodes.

2. At regular intervals, from every network node, a forward ant k is launched with the aim to find a path until the destination. The identifier of every visited node is saved onto a memory M_k and carried by the ant.

Let k be any network node; its routing table will have N entries, one for each possible destination.

Let d be one entry of k routing table (a possible destination).

Let N_k be set of neighboring nodes of node k.

Let P_{kl} be the probability with which an ant or data packet in k, jumps to a node l, $l \in N_k$, when the destination is d ($d \neq k$). Then, for each of the N entries in the node k routing table, it will be n_k values of P_{ld} subject to the condition:

$$\sum\nolimits_{l \in N_k} P_{ld} = 1; d = 1, \ldots, N \quad (21)$$

At every visited node, a forward ant assigns a greater probability to a destination node d for which falls to be the destination among the neighbor node, $d \in N_k$. Hence, initial probability in the routing table of k is then:

$$P_{dd} = \frac{9N_k - S}{4N_k^2} \quad (22)$$

Also, for the rest neighboring nodes among the neighbors for which $m \in N_k$ will then be:

$$P_{dm} = \begin{cases} \frac{4N_k - S}{4N_k^2}, if N_k > 1 \\ 0, if N_k = 1 \end{cases} \quad (23)$$

Of course equation (22) and (23) satisfy (21). But if it falls to the case where by none among the neighbor is a destination, equation (20) applies to all the neighboring nodes.

Else,

Forward ant selects the next hop node using the same probabilistic rule proposed in the ACO metaheuristic:

$$P_k(r,s) = \begin{cases} \frac{\left[\tau(r,s)\right]^{\alpha} \cdot \left[E(s)\right]^{\beta}}{\sum\limits_{u \notin M_k} \left[\tau(r,u)\right]^{\alpha} \cdot \left[E(s)\right]^{\beta}}, & s \notin M_k \\ 0, else \end{cases} \quad (24)$$

Where $p_k(r,s)$ is the probability with which ant k chooses to move from node r to node s, τ is the routing table at each node that stores the amount of pheromone trail on connection (r,s), E is the visibility function given by $\frac{1}{(C - e_s)}$ (c is the initial energy level of the nodes and e_s is the actual energy level of node s), and α and β are parameters that control the relative importance of trail versus visibility. The selection probability is a trade-off between visibility (which says that nodes with more energy should be chosen with high probability) and actual trail intensity (that says that if on connection (r,s) there has been a lot of traffic then it is highly desirable to use that connection.

5. When a forward ant reaches the destination node, it is transformed in a backward ant which mission is now to update the pheromone trail of the path it used to reach the destination and that is stored in its memory.
6. Before backward ant k starts its return journey, the destination node computes the amount of pheromone trail that the ant will drop during its journey:

$$\Delta\tau = \frac{1}{C - \left[\frac{EMin_k - Fd_k}{EAvg_k - Fd_k}\right]} \tag{25}$$

And the equation used to update the routing tables at each node is:

$$\tau(r,s) = (1-\rho) * \tau(r,s) + \left[\frac{\Delta\tau}{\varnothing Bd_k}\right] \tag{26}$$

Where φ is a coefficient and Bd_k is the travelled distance (the number of visited nodes), by the backward ant k until node r. which the two parameters will force the ant to lose part of the pheromone strength during its way to the source node. The idea behind the behavior is to build a better pheromone distribution (nodes near the sink node will have more pheromone levels) and will force remote nodes to find better paths. Such behavior is important when the sink node is able to move, since pheromone adaptation will be much quicker (Camilo et al., 2006)).

7. When the backward ant reaches the node where it was created, its mission is finished and the ant is eliminated.

Figure 1. An IEEABR forward ant flow chart

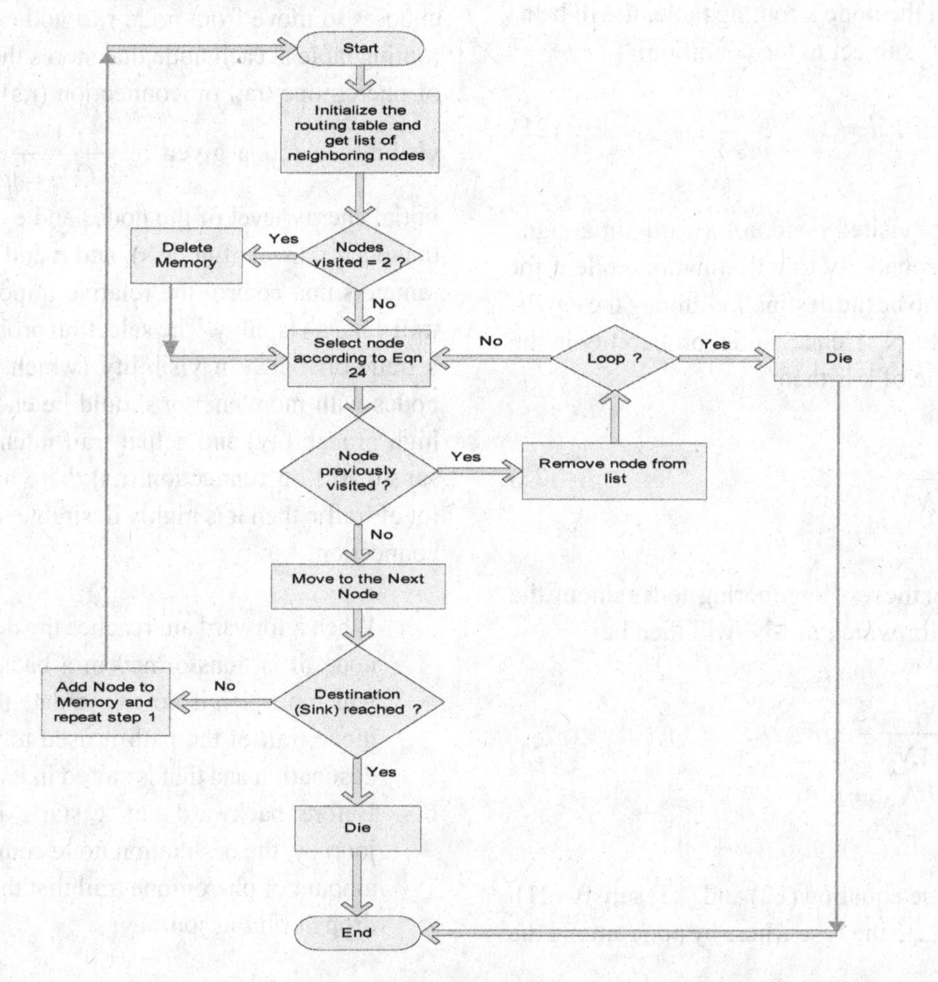

By performing this algorithm several iterations, each node will be able to know which are the best neighbors (in terms of the optimal function represented by Equation (26)) to send a packet, towards a specific destination. The flow chart describing the action of movement of forward ant for our proposed Algorithm is as shown in Figure 1. The backward ant takes the opposite direction of the flow chart.

3.1 Algorithm Operations

After the initialization of the routing table, and setting up a forward ant for hoping from node to node in search for the sink, at every point in time, a node becomes a source holding in its stack or memory information about an event around itself (neighbors). The information gathered in its memory is transferred or disseminated towards the sink node with the help of neighbor nodes behaving as repeaters. Associated raw data generated at each source (nodes) is divided in to M pieces known as data parts. An integer value M also represents the number of ant agents involving in each routing task. This raw data provided by the source node about an event contains information such as source node identification, event identification, time and the data about the event. The data size is chosen based on the sensor nodes deployed and the size of the buffer. After the splitting of the raw data, each part is associated with routing parameters to build a data packet ready to transfer. These parameters are code identification, describing the code following as data, error or acknowledge; C_{ID}. Next is the node identification to which the packet is transferred; N_{ID}. Packet number also represents the ant agent k; S_N is the sequence number, N_k which contains the number of visited nodes so far, and the kth data part as shown in Figure 2. In this figure, the group of the first four fields is named the data header. When delivery of all data packages is accomplished, the base combines them into raw data.

When a node participating in a routing received a data packet whose agent number is given, it makes decision about the next destination for that packet of data. The decision on the next node or destination for which the packet of data should be transfer, will depend on equation (22) else (24) with the highest $P_k(r,s)$. The pheromone level of the neighbors which is the first determining factor follows by the Energy levels of the neighbor nodes which are most important on the decision rule. For any of the neighbor chosen, the N_{ID} field of the node is updated and the packet is then broadcasted. The remaining neighbors among the chosen node also hear the broadcast, they check the N_{ID} field and understand that the message is not made for them; they as such quickly discard the packet immediately after only listening to the N_{ID} field of the packet. The N_k is updated with an increment by one after ensuring that S_N is not in the list of the tabus of the chosen node. The next node is determined to update the N_{ID} field by performing the same operation as perform earlier by the first node and the sequence continue till the packet gets to the sink node. The reversed operation is done for the backward ant as for the acknowledgement, which get to the source now the last bus stop for the backward ant and die off after reaching the source.

Figure 2. Data packet content

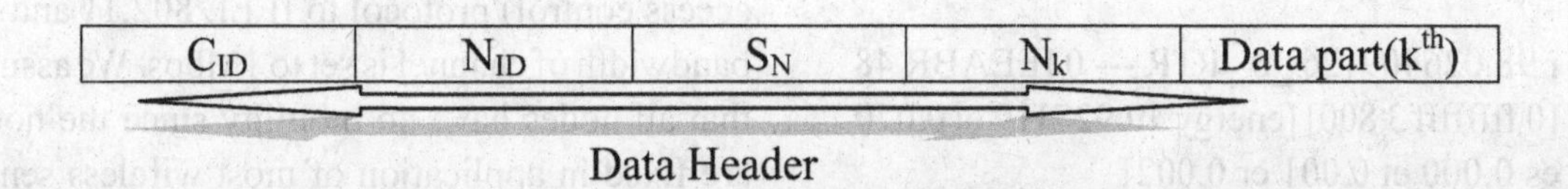

4 SIMULATIONS AND RESULTS

This section of the chapter discusses the simulation of the proposed routing algorithm (IEEABR) and evaluates its performance on the wireless sensor networks platform, while comparing it with some standard already existing routing algorithm. The simulation environment in visual form normally seen with the Network Animator (NAM) of which its file ends with '.nam' and sample of the trace files also ends with the '.tr' for analysis are presented below.

4.1 The Trace File

The trace files created by a simulation in NS-2 are always incredibly large. They are so detailed that it is almost impossible to find out any high level statistical instrument from the raw data without extra effort on developing software tools (awk scripts) or carefully extracting the required information. Though, one could extract the information needed for statistical analysis using the format used below with the sample of a trace file (raw data) as below.

1. s 5.000000000 _1_ AGT --- 0 cbr 1 [0 0 0 0] [energy 50.000000 ei 0.000 es 0.000 et 0.000 er 0.000]

we can interpret the above raw date (trace file data) gotten from the NS-2.34 simulation as; node 1 sent a cbr packet whose id is 0, and size 1Mb at time 5.0 second to application 0 on node o.

The initial energy is 50.0, energy consumption in idle state (ei) =0, energy consumption in sleep (es) =0, energy consumption in transmitting packet (et) =0, energy consumption in receiving packet (er) =0.

2. r 98.026409126 _6_ RTR --- 0 IEEABR 48 [0 ffffffff 3 800] [energy 49.927119 ei 0.070 es 0.000 et 0.001 er 0.002]

In the same way, the routing agent on node 6 received an IEEABR packet with mac address 0xff; routing packet whose id is 0 and size 48 at time 98.0264 seconds from node 3 and mac address 8. The total remaining energy on the path (residual energy) is 49.927J, energy consumed at the idle state =0.07, at the sleeping state =0, transmitting state =0.001, and receiving =0.002J.

This was used for the whole trace file gotten from the simulation and hence further processed (analyzed) for further used.

4.2 Simulations Environment

We use event driven network simulator-2 (NS-2) (NS2 installation, online) based on the network topology to be able to evaluate the implementation of the proposed ACO routing algorithm. This software provides a high simulation environment for wireless communication with detailed propagation, MAC and radio layers. AntSense (an NS-2 module for Ant Colony Optimization) (NS2 module for ANTSENSE) was used for the EEABR. The simulation environment was set with the same settings as that of the EEABR. Deployment of sensors node are randomly distributed over 200 x 200 square meters (10 nodes), 300 x 300 square meters (20 nodes), 400 x 400 square meters (30 nodes), 500 x 500 square meters (40 nodes) and 600 x 600 square meters when 50, 60, 70, 80, 90 and 100 nodes are used to monitor a static environment. Though, we also extend the number of deployed nodes to 120 to ascertain the behaviors of the network. The used traffic model is a CBR (constant bit rate) traffic model, size of data packet set to 1Mb. Each node had initial energy of 5 joules. We set the propagation model of wireless sensor network as a two-ray ground reflection model. The MAC (medium access control) protocol to IEEE 802.11 and the bandwidth of channel is set to 1Mbps. We assume that all nodes have no mobility since the nodes are fixed in application of most wireless sensor networks. Simulations were run for 5 minutes

(300 seconds) each time the simulation starts, and the remaining energy of all nodes were taken and recorded at the end of each simulation. The average energy calculated while also noting the minimum energy of the nodes. Figure 3(a) and (b) shows a screenshot of a NAM window of the simulation environment for 120 nodes randomly deployed and 10 nodes respectively.

4.3 Experimental Results

We present experimental results obtained for the three algorithms described in section 2 of this chapter; The Improved Ant based Routing Algorithm (IABR), Energy Efficient Ant Based Routing Algorithm (EEABR), and Improved Energy Efficient Ant Based Routing Algorithm (IEEABR). To better understand the performance and differences between the three algorithms, three performance parameters were used in the analysis. The parameters used are; 1. The Minimum Energy which gives the lowest energy amount of all nodes at the end of simulations, 2. The Average Energy which represents the average of energy of all nodes at the end of simulation, and 3. The Energy Efficiency which gives the ratio between the total consumed energy and the number of packets received by the sink node. The simulation was done on a static WSN, where sensor nodes were randomly deployed with objective to monitor a static environment. The location of the phenomenon and the sink node are not known. Nodes were responsible to monitor and send the relevant sensor data to the sink node in which nodes near the phenomenon will depreciate easily in energy as they will be forced to periodically transmit data. Simulations were run for 5 minutes (300 seconds) each time the simulation starts, and the remaining energy of all nodes were taken and recorded at the end of each simulation. The average energy calculated while also noting the minimum energy of the nodes. Figure 4, presents the results of the simulation for the studied parameters; The Average Energy, Minimum Energy, and Energy Efficiency of IABR, EEABR and IEEABR. As it can be seen from the results presented in the figures below, the IEEABR protocol had better results in both Average energy of the nodes and

Figure 3. (a) Graphical representation of the simulation environment in NS-2.34 with 120 Nodes; (b) Graphical representation of the simulation environment in NS-2.34 with 10 Nodes

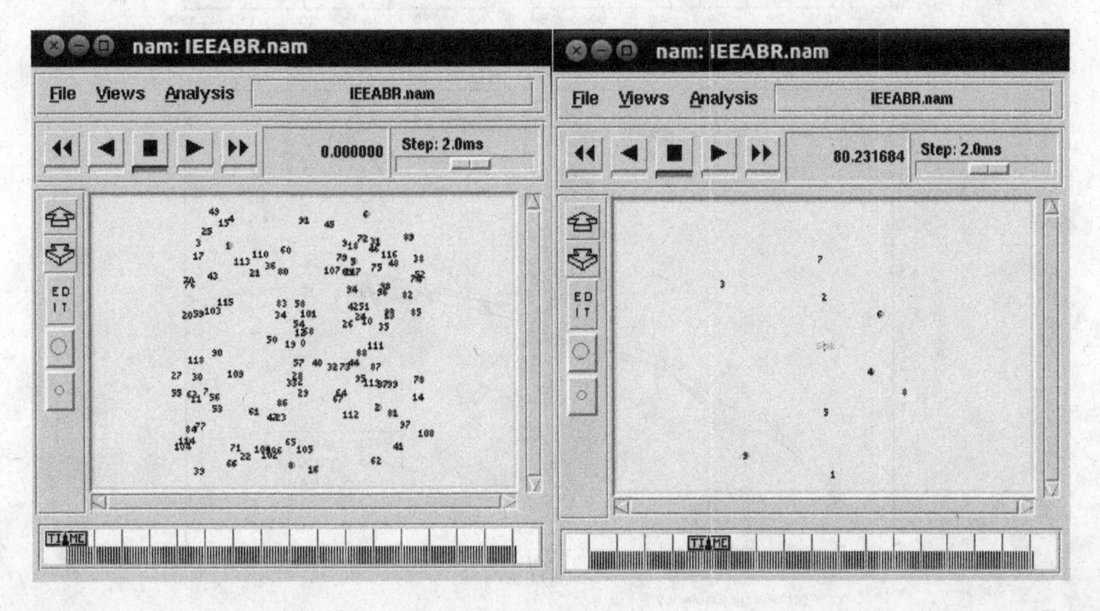

Figure 4. Performance analysis of IABR, EEABR and IEEABR energy efficient protocols

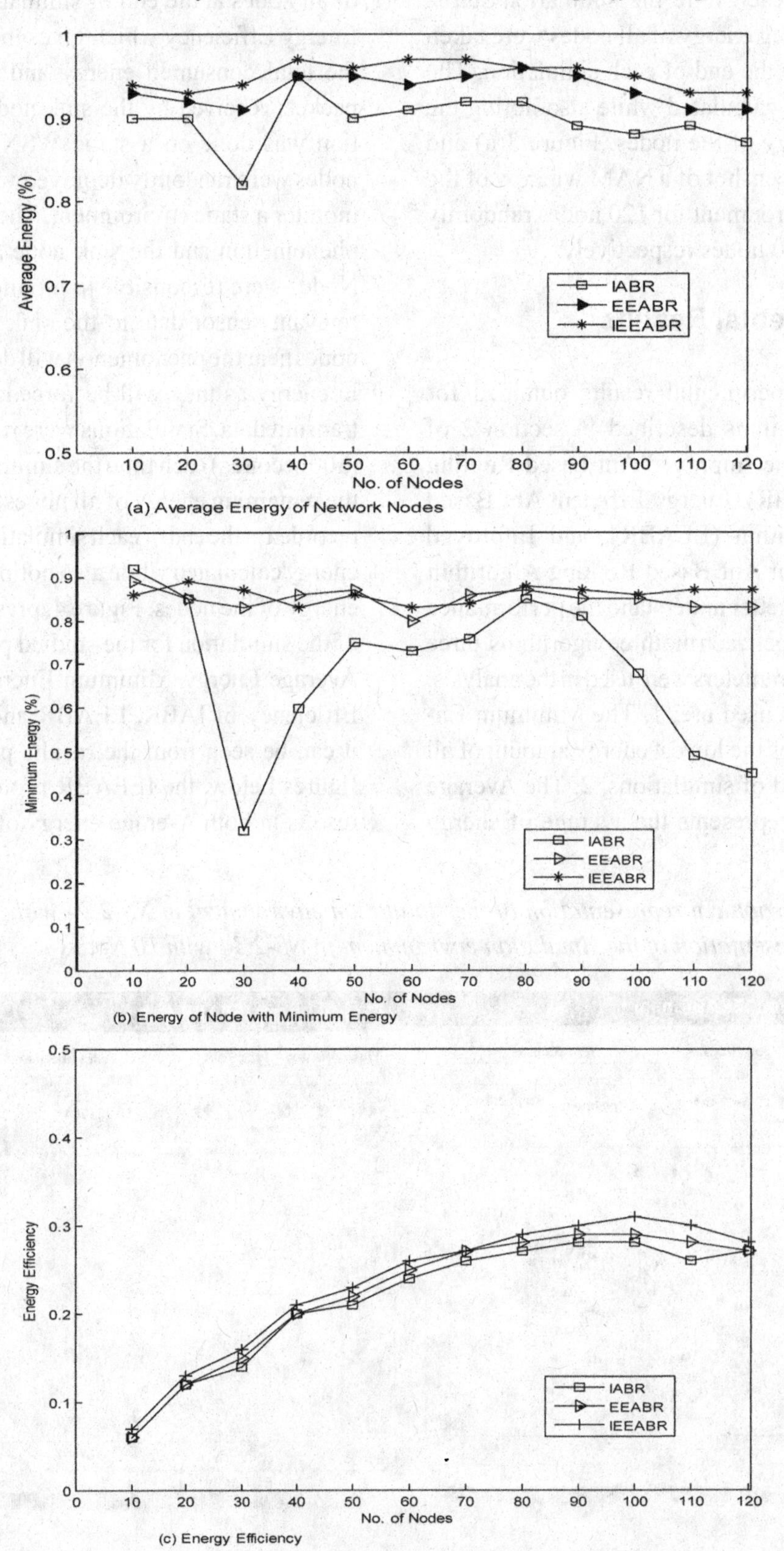

the Minimum energy of node experienced at the end of simulation. The IABR as compared to the EEABR perform worst in all cases. Though, both had a very good result in term of Energy efficiency since both Algorithms are Energy aware. Both Algorithms had a dead node at the end of simulation since it is practically not avoidable as the nodes were randomly distributed and only two nodes were responsible for connectivity between the source and the sink node. The results of the simulations are as shown in figure 4.

5 CONCLUSION AND FUTURE WORK

In this chapter, research based on the application of Ant Colony Optimization metaheuristic in solving routing problem in WSN was adopted. In this work we proposed an Improved Energy Efficient Ant Based routing Algorithm, which improves the lifetime of sensor networks. The improved version of the EEABR called the Improved Energy Efficient Ant-Based Routing (IEEABR), utilizes initialization of uniform probabilities distribution in the routing table while given special consideration to neighboring nodes which falls to be the destination (sinks) in other to save time in searching for the sink leading to reduced energy consumption by the nodes. The IEEABR approach has reduced energy usage and achieves a dynamic and adaptive routing which effectively balances the WSN node power consumption and increase the network lifetime. The experimental results showed that the algorithm leads to very good results in different WSNs. The protocol considers the residual energy of nodes in the network after each simulation period. Based upon the NS-2 simulation, the algorithm has been verified with very good performance in network life time and highly Energy efficient. Consequently, our proposed algorithm can efficiently extend the network lifetime without performance degradation. This Algorithm focused mainly on the lifetime of wireless sensor networks.

As future work, we intend to implement our routing algorithm on Waspmote, study a dual approach in the selection of sink, self destruction of the backward ants should there exist a link failure and an alternate means of retrieving the information carried by the backward ant to avoid lost of information. Some other protocols based on delay and quality of service may be needed for further studies, this will then increase the efficiency of the network performance.

REFERENCES

Akkaya, K., & Younis, M. (2005). A survey of routing protocols in wireless sensor networks. *Ad Hoc Networks*, *3*(3), 325–349. doi:10.1016/j.adhoc.2003.09.010

Akyildiz, I. F., Su, W., Sankarasubramaniam, Y., & Cayirci, E. (2002). Wireless sensor networks: A survey. *Computer Networks*, *38*(4), 393–422. doi:10.1016/S1389-1286(01)00302-4

Anandamoy, S. (2006). *Swarm intelligence based optimization of MANET cluster formation*. Unpublished Master thesis, The University of Arizona.

Baras, J., & Mehta, H. (2003). A probabilistic emergent routing algorithm for mobile ad hoc networks. In *WiOpt'03 Sophia-Antipolis*. France: PERA.

Bonabeau, E., Dorigo, M., & Theraulaz, G. (1999). *Swarm intelligence: From natural to artificial systems* (pp. 1–278). London, UK: Oxford University Press.

Buczak, A., & Jamalabad, V. (1998). Self-organization of a heterogeneous sensor network by genetic algorithms. In Dagli, C. H., Chen, C. L., & Akay, M. (Eds.), *Intelligent engineering systems through artificial neural networks* (*Vol. 8*, pp. 259–264). New York, NY: ASME Press.

Camilo, T., Carreto, C., Silva, J. S., & Boavida, F. (2006). An energy-efficient ant based routing algorithm for wireless sensor networks. In *Proceedings of 5th International Workshop on Ant Colony Optimization and Swarm Intelligence*, Brussels, Belgium, (pp. 49-59).

Chong, C.-Y., & Kumar, S. P. (2003). Sensor networks: Evolution, opportunities, and challenges. *Proceedings of the IEEE*, *91*(8), 1247–1256. doi:10.1109/JPROC.2003.814918

Di Caro, G., Ducatelle, F., & Luca, M. G. (2005). *AntHocNet: An adaptive nature-inspired algorithm for routing in mobile ad hoc networks*. Technical Report. IDSIA – Dalle Molle Institute for Artificial Intelligence Galleria, Switzerland.

Dorigo, M., & Di Caro, G. (1997). *AntNet: A mobile agents approach to adaptive routing*. Technical Report. IRIDIA- Free Brussels University, Belgium.

Dorigo, M., & Di Caro, G. (1998). AntNet: Distributed stigmergetic control for communications networks. *Journal of Artificial Intelligence Research*, *9*, 317–365.

Hussain, S., & Islam, O. (2008). Genetic algorithm for energy efficient trees in wireless sensor networks. *Advanced Intelligent Environments*, 1-14.

Iyengar, S., Wu, H.-C., Balakrishnan, N., & Chang, S. Y. (2007). Biologically inspired cooperative routing for wireless mobile sensor networks. *IEEE Systems*, *1*(2), 181–183.

Katz, R. H., Kahn, J. M., & Pister, K. S. J. (1999). Mobile networking for smart dust. In *Proceedings of the 5th Annual ACM/IEEE International Conference on Mobile Computing and Networking* (MobiCom99). Seattle, Washington, (pp. 271-278).

Kulik, J., Heinzelman, W., & Balakrishnan, H. (2002). Negotiation-based protocols for disseminating information in wireless sensor networks. *Wireless Networks*, *8*(2), 169–185. doi:10.1023/A:1013715909417

Liu, Z., Kwiatkowska, M. Z., & Constantinou, C. A. (2005). Biologically inspired QOS routing algorithm for mobile ad hoc networks. In *19th International Conference on Advanced Information Networking and Applications*, (pp. 426–431).

Min, R., Bhardwaj, M., Cho, S., Shih, E., Sinha, A., Wang, A., & Chandrakasan, A. (2001). Low power wireless sensor networks. In *Proceedings of International Conference on VLSI Design*, Bangalore, India, (pp. 221-226).

NS-2 installation on Linux. (n.d.). Retrieved from http://paulson.in/?p=29

NS-2 module for Ant Colony Optimization (AntSense). (n.d.). Retrieved from http://eden.dei.uc.pt/~tandre/antsense/

Rabaey, J. M., Ammer, M. J., Silver, J. L. D., Patel, D., & Roundy, S. (2000). PicoRadio supports ad hoc ultra low power wireless networking. *IEEE Computer*, *33*(7), 42–48. doi:10.1109/2.869369

Rodoplu, V., & Ming, T. H. (1999). Minimum energy mobile wireless networks. *IEEE Journal on Selected Areas in Communications*, *17*(8), 1333–1344. doi:10.1109/49.779917

Sohrabi, K., Gao, J., Ailawadhi, V., & Pottie, G. J. (2000). Protocols for self-organization of a wireless sensor network. *IEEE Personal Communications*, *7*(5), 16–27. doi:10.1109/98.878532

Uchhula, V., & Bhatt, B. (2010). Comparison of different ant colony based routing algorithms. *International Journal of Computer Applications* [Foundation of Computer Science.]. *Special Issue on MANETs*, *2*, 97–101.

White, T., & Pagurek, B. (1998). Toward multi-swarm problem solving in networks. *International Conference on Multi Agent Systems,* (pp. 333–340).

White, T., Pagurek, B., & Oppacher, F. (1998). Connection management using adaptive mobile agents. In *Proceeding of International Conference on Parallel Distributed Processing Techniques and Applications* (pp. 802-809). CSREA Press.

ADDITIONAL READING

Akkaya, K., & Younis, M. (2003). An Energy-Aware QoS Routing Protocol for Wireless Sensor Networks. In *Proceedings of the IEEE Workshop on Mobile and Wireless Networks*. Providence, Rhode Island.

Braginsky, D., & Estrin, D. (2002). Rumor routing algorithm for sensor networks. In *Proceedings of the 1st ACM International Workshop on Wireless Sensor Networks and Applications*. Atlanta, Georgia.

Camilo, T., Carreto, C., Silva, J. S., & Boavida, F. (2006). An Energy-Efficient Ant Based Routing Algorithm for Wireless Sensor Networks. In *Proceedings of 5th International Workshop on Ant Colony Optimization and Swarm Intelligence*, Brussels, Belgium, 49-59.

Chang, J.-H., & Tassiulas, L. (2000). Maximum Lifetime Routing in Wireless Sensor Networks. In *Proceedings of the Advanced Telecommunications and Information Distribution Research Program*. College Park, MD.

Chu, M., Haussecker, H., & Zhao, F. (2002). Scalable information-driven sensor querying and routing for ad hoc heterogeneous sensor networks. *International Journal of High Performance Computing Applications, 16*(3), 293–313. doi:10.1177/10943420020160030901

Dasgupta, K., Kalpakis, K., & Namjoshi, P. (2003). An Efficient Clustering-Based Heuristic for Data Gathering and Aggregation in Sensor Networks. In *Proceedings of the IEEE Wireless Communications and Networking Conference*. New Orleans, LA.

Estrin, D., Govindan, R., Heidemann, J., & Kumar, S. (1999). Next century challenges: scalable coordination in sensor networks. In *Proceedings of the 5th annual ACM/IEEE International Conference on Mobile Computing and Networking*. Seattle, Washington.

Finn, G. (1987). *Routing and Addressing Problems in Large Metropolitan-Scale Internetworks*. Technical Report, ISI Research Report ISI/RR, (pp. 87-180). University of Southern California.

Ganesan, D., Govindan, R., Shenker, C., & Estrin, D. (2001). Highly Resilient Energy Efficient Multipath Routing in Wireless Sensor Networks. *Mobile Computing and Communications Review, 5*(4), 11–25. doi:10.1145/509506.509514

Hedetniemi, S., & Liestman, A. (1988). A Survey Of Gossiping And Broadcasting in Communication Networks. *Networks, 18*(4), 319–349. doi:10.1002/net.3230180406

Heinzelman, W., Chandrakasan, A., & Balakrishnan, H. (2000). Energy-efficient communication protocol for wireless sensor networks, In *Proceeding of the Hawaii International Conference on System Sciences*, Hawaii, USA, 1-10.

Heinzelman, W., Kulik, J., & Balakrishnan, H. (1999). Adaptive protocols for information dissemination in wireless sensor networks. In *Proceedings of the 5th Annual ACM/IEEE International Conference on Mobile Computing and Networking*. Seattle, Washington.

Intanagonwiwat, C., Govindan, R., & Estrin, D. (2000). Directed diffusion: A Scalable and Robust Communication Paradigm for Sensor Networks. In *Proceedings of the 6th Annual ACM/IEEE International Conference on Mobile Computing and Networking*. Boston, MA, USA, 243-254.

Johnson, D. B., & Maltz, D. A. (2010). Dynamic Source Routing in Ad Hoc Wireless Networks. *International Journal on Applications of Graph Theory in Wireless Ad Hoc Networks and Sensor Networks*, 2(2), 1-15.

Kalpakis, K., Dasgupta, K., & Namjoshi, P. (2002). Maximum Lifetime Data Gathering and Aggregation in Wireless Sensor Networks. In *Proceedings of IEEE International Conference on Networking*. Atlanta, Georgia.

Karp, B., & Kung, H. T. (2000). GPSR: Greedy Perimeter Stateless Routing for Wireless Sensor Networks. In *Proceedings of the 6th Annual ACM/IEEE International Conference on Mobile Computing and Networking* (243-254). Boston, MA.

Krishnamachari, B., Estrin, D., & Wicker, S. (2002). Modeling data centric routing in wireless sensor networks. In *Proceedings of IEEE INFOCOM*, New York.

Li, L., & Halpern, J. Y. (2001). Minimum Energy Mobile Wireless Networks Revisited. In *Proceedings of IEEE International Conference on Communications*. Helsinki, Finland.

Liao, W.-H., Kao, Y., & Fan, C.-M. (2007). an Ant Colony Algorithm for Data Aggregation in Wireless Sensor Networks. In *International Conference on Sensor Technologies and Applications*, 101–106.

Lin, C. R., & Gerla, M. (1997). Adaptive Clustering for Mobile Wireless Networks. *IEEE Journal on Selected Areas in Communications*, *15*(7), 1265–1275. doi:10.1109/49.622910

Lindsey, S., & Raghavendra, C. S. (2002). PEGASIS: Power Efficient Gathering in Sensor Information Systems. In *Proceedings of the IEEE Aerospace Conference, Big Sky*. Montana, 4(7).

Lindsey, S., Raghavendra, C. S., & Sivalingam, K. (2001). Data gathering in sensor networks using the energy delay metric. In *Proceedings of the IPDPS Workshop on Issues in Wireless Networks and Mobile Computing*, San Francisco, USA.

Manjeshwar, A., & Agrawal, D. P. (2001). TEEN: A Protocol for Enhanced Efficiency in Wireless Sensor Networks. In *Proceedings of the 1st International Workshop on Parallel and Distributed Computing Issues in Wireless Networks and Mobile Computing*, San Francisco, USA.

Manjeshwar, A., & Agrawal, D. P. (2002). APTEEN: A Hybrid Protocol for Efficient Routing and Comprehensive Information Retrieval in Wireless Sensor Networks. In *Proceedings of the 2nd International Workshop on Parallel and Distributed Computing Issues in Wireless Networks and Mobile Computing*.

Nath, B., & Niculescu, D. (2003). Routing on a Curve. *ACM SIGCOMM Computer Communication Review*, *33*(1), 155–160. doi:10.1145/774763.774788

Niannian, D., Liu, P. X., & Chao, H. (2005). Data Gathering Communication in Wireless Sensor Networks Using Ant Colony Optimization. In *IEEE/RSJ International Conference on Intelligent Robots and Systems*, 697–702.

Noor, Z. (2010). *How to install NS2 with Ubuntu 10.04*; http://wsnsimulators.blogspot.com/2010/06/how-to-install-ns2-with-ubuntu-1004-by.html

Perkins, C., Elizabeth, M. B.-R., & Das, S. R. (2002in progress). (AODV) Routing. In *Internet Draft draft-ietf-manetaodv- 11.txt, work*. Ad Hoc on-Demand Distance Vector.

Sadagopan, N., Krishnamachari, B., & Helmy, A. (2003). The ACQUIRE Mechanism for Efficient Querying in Sensor Networks. In *Proceedings of the First International Workshop on Sensor Network Protocol and Applications*. Anchorage, Alaska, 149-155.

Schoonderwoerd, R., Holland, O., Bruten, J., & Rothkrantz, L. (1996). Ant-Based Load Balancing in Telecommunications Networks. *Hewlett-Packard Laboratories,* Bristol-England, *162*, 207.

Schurgers, C., & Srivastava, M. B. (2001). Energy efficient routing in wireless sensor networks. In *The MILCOM Proceedings on Communications for Network-Centric Operations*. McLean, VA, USA: Creating the Information Force. doi:10.1109/MILCOM.2001.985819

Shah, R., & Rabaey, J. (2002). Energy Aware Routing for Low Energy Ad Hoc Sensor Networks. In *Proceedings of the IEEE Wireless Communications and Networking Conference*. Orlando, FL, 350-355.

Subramanian, L., & Katz, R. H. (2000). Architecture for building self configurable systems. In *Proceedings of IEEE/ACM Workshop on Mobile Ad Hoc Networking and Computing*, Boston, Massachusetts.

Tian He, J. A. Stankovic, C. Lu, & Abdelzaher, T. F. (2003). SPEED: A Stateless Protocol for Real-Time Communication in Sensor Networks. In *Proceedings of International Conference on Distributed Computing Systems*. Providence, RI.

Xu, Y., Heidemann, J., & Estrin, D. (2001). Geography-Informed Energy Conservation for Ad Hoc Routing. In *Proceedings of the 7th Annual ACM/IEEE International Conference on Mobile Computing and Networking*. Rome, Italy.

Yao, Y., & Gehrke, J. (2002). The Cougar Approach to In-Network Query Processing in Sensor Networks, in ACM. *SIGMOD Record*, *31*(3), 9–18. doi:10.1145/601858.601861

Younis, M., Munshi, P., & Al-Shaer, E. (2003). Architecture for Efficient Monitoring and Management of Sensor Networks. In *Proceedings of the IFIP/IEEE Workshop on End-to-End Monitoring Techniques and Services*. Belfast, Northern Ireland.

Younis, M., Youssef, M., & Arisha, K. (2002). Energy-aware routing in cluster-based sensor networks. In *Proceedings of the 10th IEEE/ACM International Symposium on Modeling, Analysis and Simulation of Computer and Telecommunication Systems (MASCOTS2002)*, Fort Worth, Texas, 129-136.

Youssef, M., Younis, M., & Arisha, K. (2002). A Constrained Shortest path Energy-aware Routing Algorithm for Wireless Sensor Networks. In *Proceedings of the IEEE Wireless Communication and Networks Conference*. Orlando, Florida.

Yu, Y., Estrin, D., & Govindan, R. (2001). *Geographical and Energy Aware Routing. A recursive data dissemination protocol for wireless sensor networks*. UCLA Computer Science Department Technical Report (UCLA-CSD) TR-01-0023.

KEY TERMS AND DEFINITIONS

Ant Colony Optimization (ACO): ACO is a class of optimization algorithms modeled on the actions of an ant colony and a subset of Swarm Intelligence. The basic idea of the ant colony optimization is taken from the food searching behavior of real ants. ACO methods are useful in problems that need to find paths to goals. The simulating behavior of ant colony leads to optimization of network parameters for the WSN routing process to provide maximum network lifetime.

Ant-Based Routing: This is the act of moving information across an inter-network from a source to a destination in which along the way at least one intermediate node typically is encountered or the process of choosing a path over which to send the packets using the behavior of ants to find the shortest Path in a network using a probabilistic routing table.

Energy Efficiency: Energy Efficiency in WSNs is the ratio between the total consumed energy and the number of packets received by the sink node. In other words, Energy Efficient is the ability of the sensor networks performing the same services but using less power.

Network Animator (NAM): NAM is a visualization environment, is sometimes also useful when simulating Security mechanisms in NS-2 and also when to trace Congestion Window while working with TCP.

Network Simulator-2 (NS-2): NS-2 is a discrete event simulator targeted at networking research. NS-2 provides substantial support for simulation of TCP, routing, and multicast protocols over wired and wireless (local and satellite) networks.

Routing Table: A routing table is a set of rules, often viewed in table format that is used to determine where data packets traveling over a network will be directed. A routing table contains the information necessary to forward a packet along the best path toward its destination. Each packet contains information about its origin and destination and the next hop.

The Trace File: Trace is the most important file for network analysis, and useful when generating multiple trace files from a single TCL file. Normally use for Xgraph, and other analysis.

Wireless Sensor Networks: Wireless Sensor Networks (WSNs) are collections of compact-size, relatively inexpensive computational nodes that measure local environmental conditions or other parameters such as vibration, pressure, temperature, sound, motion or pollutant, and forward such information to a central point for appropriate processing.

Chapter 21
Event Based Data Gathering in Wireless Sensor Networks

Asfandyar Khan
Universiti Teknologi PETRONAS, Malaysia

Azween B Abdullah
Universiti Teknologi PETRONAS, Malaysia

Nurul Hasan
Universiti Teknologi PETRONAS, Malaysia

ABSTRACT

Wireless sensor networks (WSANs) are increasingly being used and deployed to monitor the surrounding physical environments and detect events of interest. In wireless sensor networks, energy is one of the primary issues and requires the conservation of energy of the sensor nodes, so that network lifetime can be maximized. It is not recommended as a way to transmit or store all data of the sensor nodes for analysis to the end user. The purpose of this "Event Based Detection" Model is to simulate the results in terms of energy savings during field activities like a fire detection system in a remote area or habitat monitoring, and it is also used in security concerned issues. The model is designed to detect events (when occurring) of significant changes and save the data for further processing and transmission. In this way, the amount of transmitted data is reduced, and the network lifetime is increased. The main goal of this model is to meet the needs of critical condition monitoring applications and increase the network lifetime by saving more energy. This is useful where the size of the network increases. Matlab software is used for simulation.

INTRODUCTION

A Wireless sensor network (WSN) is composed of a large number of tiny low powered sensor nodes and one or more multiple base stations (sinks). These tiny sensor nodes consist of sensing, data processing and communication components. The sensor nodes sense, measure and collect ambient environmental conditions, use their processing abilities to carry out simple computations and send partially processed sensed data to a base station either directly or through a gateway. The gateway can perform fusion of the sensed data in

DOI: 10.4018/978-1-4666-0101-7.ch021

order to filter out erroneous data and anomalies and to draw conclusions from the reported data over a period of time. A comprehensive overview of wireless sensor networks and their broad range of applications can be found in (Akyildiz et al. 2002; Younis et al. 2004). The sensor nodes in a wireless sensor network (WSN) are resource constrained, i.e., have limited energy and computation power (processor and memory), a short communication range and low bandwidth. A sensor node operates on limited battery power and it is very difficult or even impossible to recharge or replace it. When it is depleted of energy, it will die and disconnect from the network, which significantly affects network performance. The life of a sensor node determines the lifetime of the network. Maximizing the lifetime of the network involves energy conservation and harvesting. Energy is conserved through optimizing communication and minimizing energy usage (Akyildiz et al. 2002) (Younis et al. 2004). A sensor network is deployed with the objective of gathering information with the initial battery energy. It is desired that the network works continuously and transmits information for as long as possible. This is referred to as the maximum life time problem in sensor networks. In data gathering, nodes spend a part of their energy on transmitting, receiving and relaying packets. Hence, designing routing algorithms that maximize the lifetime until the first battery expires is an important consideration. One of the major tasks of a WSN is to detect events occurring in the sensing field as shown in Figure 1. In wireless sensor networks, whenever the event of interest occur the nodes should be aware and respond quickly without any delay. Delay is to be considered an important metric when detecting an event in real time applications such as surveillance (Biswas and Phoha 2006) (Tseng et al. 2005) or object tracking (Gui and Mohapatra 2004) (Wensheng and Guohong 2004).

Energy conservation in wireless sensor networks has been the primary objective; however, this constraint is not the only consideration for efficient working of wireless sensor networks. There are other objectives like scalable architecture, routing and latency. In most of the applications of wireless sensor networks, it is envisioned that they handle critical scenarios where data retrieval time is critical, i.e., delivering sensed information from each individual node to the base station as fast as possible becomes an im-

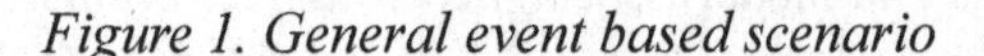

Figure 1. General event based scenario

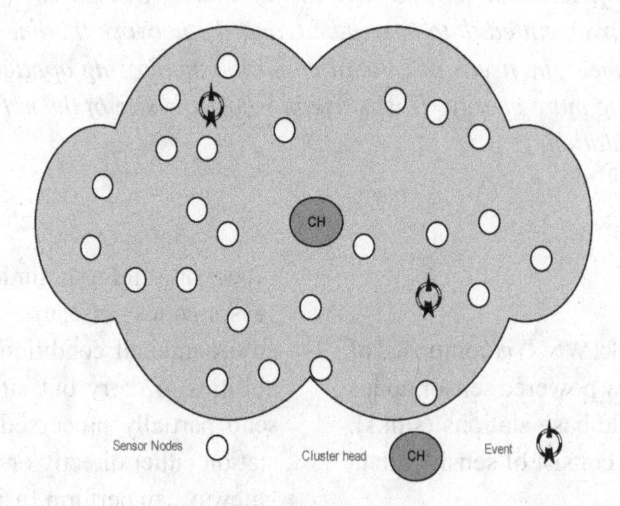

portant consideration. Since the data is usually time varying in sensor networks, it is essential to guarantee that data can be successfully received by the base station the first time instead of being retransmitted.

GROUND RESEARCH

Energy efficient communication is a matter of survival for wireless sensor networks. As a result, most of the research is focused towards energy efficient communication, energy conservation, maximizing network lifetime and energy harvesting. Various routing strategies have been proposed but cluster-based routing protocols are dominant and considered energy efficient and scalable. In (Haiming and Sikdar 2007), a framework to select the optimal probability with which a node should become a cluster head in order to minimize network energy consumption has been proposed. An appealing application area that can benefit from the WSANs is the monitoring of environments subject to emergency situations and conditions, such as fires, explosions, toxic gas leaks, etc. In a fire situation, for instance, it is very important to have reliable, accurate monitoring of the physical environment so that it is possible for the rescue teams to understand what is happening and take the best emergency actions in the environment, avoiding life and property loss. However, in order to understand what is going on in the environment and take the right actions, it is necessary to interpret the collected events accurately. In the same way that time is fundamental for our way of thinking (Lamport 1978) it is also very important for the correct interpretation of the state of an environment. Thus, event ordering is a fundamental requisite when the correctness of the interpreted information has to be guaranteed. Events that are captured by sensors may (or may not) be related to each other (correlation). Event correlation determines the need for interpretation. When two or more events are not correlated, each incoming event can be treated independently. However, when events are correlated, a mechanism for capturing and interpreting these events, jointly, is necessary. For instance, the correlation of the following three events: presence of oxygen, very high temperature and flammable material can lead to the interpretation of a fire condition (Boukerche et al. 2007). In Delaying Techniques (Mansouri-Samani and Sloman 1997; Shim and Ramamoorthy 1990), a certain time is waited before events are interpreted. So, delayed messages can arrive and be evaluated along with other messages that did not suffer any delay. By having a physical clock synchronism, it is assumed that the sending of a message from one node to another of the network takes a time D, i.e., D is the network delay. Offline event detection provides a model suitable for querying events from noisy and imprecise data. Both database systems (Deshpande and Madden 2006) and sensor networks (Deshpande et al. 2004) (Jain et al. 2004) (Chu et al. 2002) have explored model-based queries as a method for dealing with irregular or unreliable data. In the online case, sensor networks reduce the bandwidth requirements of data collection by suppressing results that conform to the model or compressing the data stream through a model representation. This has coincident benefits on resource and energy usage within the network. If sensors measure spatially correlated values, the values collected from a subset of nodes can be used to materialize the uncollected values from other nodes (Gupta et al. 2005; Kotidis 2005). One of the first techniques that come to mind for the chronological ordering between pairs of events is the use of synchronized physical clocks. For that, local timestamps (current time tags) are associated with the events collected by the sensors. With this, both the chronological order and the elapsed time between a pair of events can be determined by simply comparing the timestamps associated with the events.

SYSTEM MODEL

Our model of interest consists of a large number of tiny sensors densely and randomly scattered in a designated area, which generate and send sensor outputs at a constant rate. Note that the number of sensors needed to cover the same area is lower if a deterministic procedure is used for node placements in the network field. Since all sensors have similar capabilities and perform the same sensing task, and since sensing coverage of different sensors tend to overlap in a dense sensor network, not all of them are needed to carry out sensing, processing and communicating throughout the entire system lifespan. One effective approach to reduce the overall energy consumption is to turn off (off duty) some sensors and keep others active (on duty). An off-duty sensor has its CPU, sensing and radio modules all switched off, leaving only an active low power timer to resume duty at a later time. This approach is called node scheduling or sleeps scheduling, where sensors participate and take turns in monitoring the target area.

We present a new Energy-efficient event detection clustering approach that segments the sensor nodes in the group or clusters. Each cluster has a leader with a high remaining energy and a shorter distance from the base station (BS), and also has a node that covers the long range communication distance between the sensor nodes and base station (BS).

We modified the cluster based WSN architecture by introducing a candidate cluster head node and server node (SN). Second, we divided the sensor nodes into different clusters by the K-Mean and selecting the candidate cluster heads for each cluster by using the K-theorem. The details are as follows.

Cluster Based WSN Architecture

We have proposed a new node called the Server Node (SN) as shown in Figure 2. It has more computation resources (more processing power and memory) and energy. It also has a longer transmission range to reach base station. The SN is reachable by all the nodes in all cluster head. If it is not reachable, it is recommended to add another SN. It can be a special node, or even a device like a laptop or PDA which can be used for the purpose. The architecture tends to distribute the load of the CH to the SN. The SN is responsible for cluster head selection from the candidate nodes. The purpose of introducing the SN is to closely monitor the operation of sensor nodes in a cluster and command them for specific operations.

Cluster Formation and Node Deployment

The main purpose of clustering the sensor nodes is to form groups so as to reduce the overall energy spent in aggregation, and communicating the sensed data to the cluster head and base station. There are various algorithms proposed for clustering in literature, but the K-mean [26] has been found the most efficient and is used to solve the clustering problem. The procedure follows an easy way and defines the K centroid for each cluster. As the sensor nodes have different locations which cause different results, so the centroid should be placed in a careful way. So it is important that they be placed as far away from each other as possible and as often as possible. Once the centroids are determined, each sensor node is than associated to the nearest centroid. After assigning each node to the nearest centroid, the early grouping is done. From the first step, we need to recalculate a new K centroid. The same processes are repeated as in the first step and calculation of a new centroids is carried out for the sensor nodes until no more changes are possible. Figure 4 shows the formation of different clusters and the assignment of nodes to each cluster. The K-mean algorithm uses the Euclidean distance formula as shown in Equation 1 to calculate the distance between the centroid and other objects.

$$D_{(i,j)} = \sqrt{\left|x_{i1} - x_{j2}\right|^2 + \left|x_{i2} - x_{j2}\right|^2 + \cdots \left|x_{ip} - x_{jp}\right|^2} \quad (1)$$

The K-mean is computationally efficient and does not require the user to specify many parameters. The number of clusters and the nodes assigned to each cluster are shown in Table 1 and each cluster centroid is shown in Table 2.

Candidate Cluster Heads

The philosophy behind the K-theorem is to select candidate CHs based on a bunch of sensor nodes in a cluster. The working of the K-Theorem is quiet simple and was proposed to select the optimal server location. Table 3 shows the working details.

The sever node sets the value of ki for each cluster. The value of ki is relative to the node density in a cluster and the ratio, i.e., r of the cluster heads in a WSN. It is the product of the number of nodes in a cluster, i.e., ni and ratio r. The value of r can vary from 0.01 to 0.99 but it should not be more than 0.50. The less the value of ki is, the more the probability of getting the local optima is. The value of ki determines the ki number of best sensor nodes that can serve as cluster heads. The value of ki also provides the

Figure 2. A modified cluster-based architecture for wireless sensor networks where sensor nodes send the sensed information to the cluster-heads through Multihop routing. CHs aggregate the received information and transmit it to the Server Node (SN), which then forwards it to the base station.

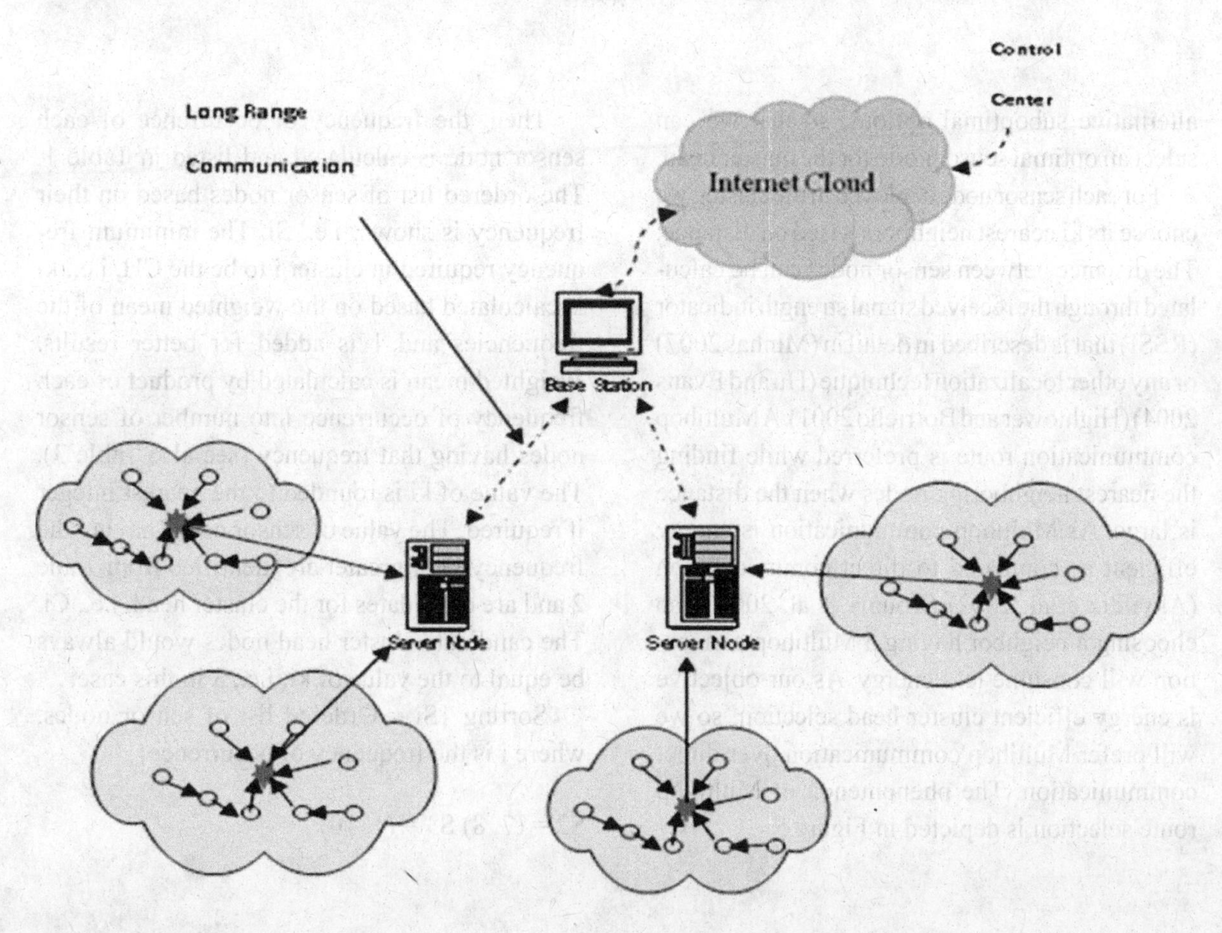

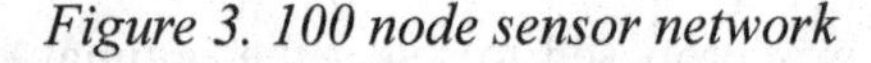
Figure 3. 100 node sensor network

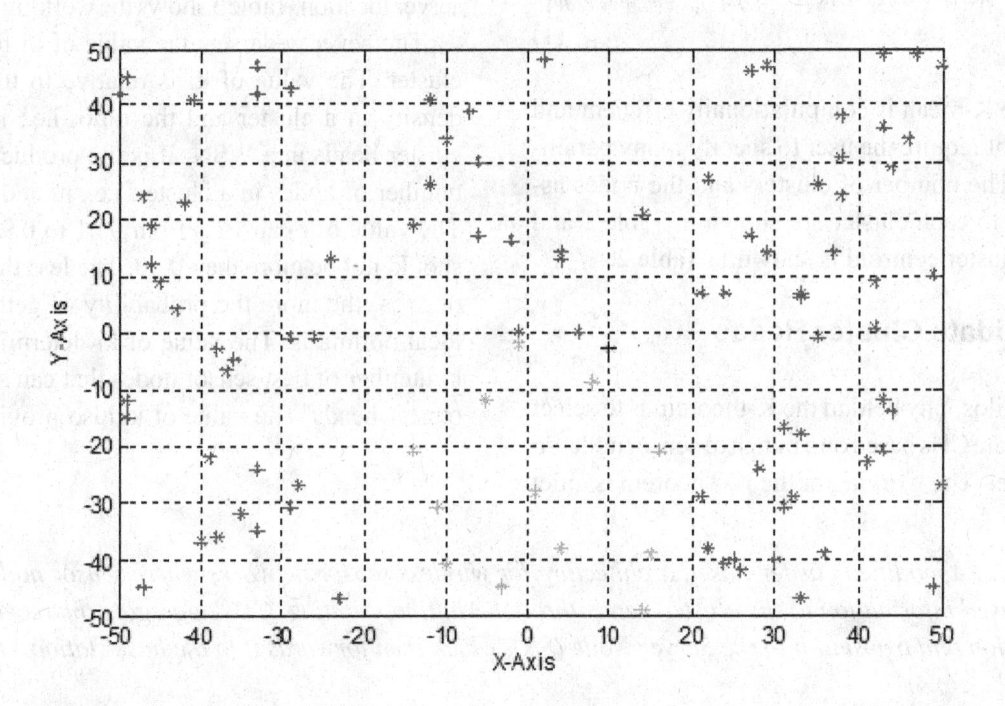

alternative suboptimal options, so that we can select an optimal sensor node for the cluster head.

For each sensor node deployed in the cluster, we choose its ki nearest neighbors based on distance. The distance between sensor nodes can be calculated through the received signal strength indicator (RSSI) that is described in detail in (Minhas 2007) or any other localization technique (Hu and Evans 2004) (Hightower and Borriello 2001). A Multihop communication route is preferred while finding the nearest neighboring nodes when the distance is large. As Multihop communication is energy efficient as compared to direct communication (Akyildiz et al. 2002) (Younis et al. 2004)], so choosing a neighbor having a Multihop connection will consume less energy. As our objective is energy efficient cluster head selection, so we will prefer Multihop communication over direct communication. The phenomenon of Multihop route selection is depicted in Figure 5.

Then, the frequency of occurrence of each sensor node is calculated and listed in Table 1. The ordered list of sensor nodes based on their frequency is shown, i.e., Si. The minimum frequency required in cluster i to be the CH, i.e., ki is calculated based on the weighted mean of the frequencies and 1 is added for better results. Weighted mean is calculated by product of each frequency of occurrence into number of sensor nodes having that frequency (see also Table 3). The value of ki is rounded to the nearest integer if required. The value of sensor nodes having the frequency ki or greater are identified from Table 2 and are candidates for the cluster head, i.e., Ci. The candidate cluster head nodes would always be equal to the value of ki, i.e., 3 in this case.

Sorting {Si = Ordered list of sensor nodes, where i is the frequency of occurrence}

S2 = (7, 8) S3 = (1, 10)

Figure 4. Cluster formation

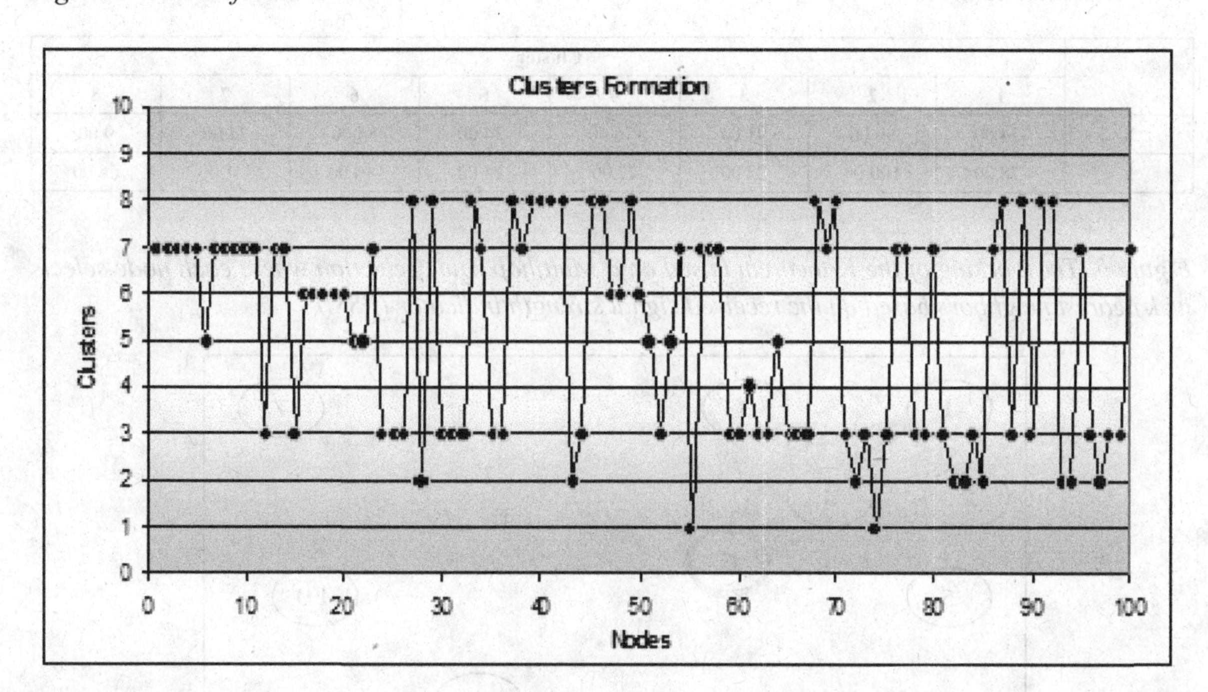

S4 = (2, 5, 9) S5 = (3)

S6 = (4) S7 = (6)

Ki = [Weighted Mean] + 1

= [(2*2)+(3*2)+(4*3)+(5*1)+(6*1)+(7*1) / 10] + 1

Table 1. Number of cases in each cluster

Cluster	Nodes
1	2
2	9
3	31
4	1
5	6
6	8
7	26
8	17
Valid	100
Missing	0

= [(4 + 6 + 12 + 5 + 6 + 7) / 10] + 1

= [(40) / 10] + 1 _ 4 + 1

Ki = 5

So, the best nodes for the candidate cluster head in cluster i are: Ci = {3, 4, 6} when Ki = 3.

Cluster Head Selection Method

Cluster heads are meant to perform some additional tasks such as organizing medium access within the cluster or performing some routing decisions. Due to some additional responsibility, cluster heads may drain their energy quickly. In cluster based architecture, we should take care of the selection process of cluster heads and share the responsibility among all the nodes. The role of cluster head sharing is in fact a good option to enable the network to work for a long period. We describe cluster-head selection algorithms which are energy efficient. Our proposed algorithm tends

Table 2. Cluster centroid

	Cluster							
	1	2	3	4	5	6	7	8
X	-34.00	99.00	91.00	-76.00	-34.00	54.00	34.00	9.00
Y	-78.00	100.00	22.00	22.00	-23.00	-90.00	-21.00	66.00

Figure 5. The working of the K-theorem based on a Multihop route selection where each node selects its k nearest neighbors based on the received signal strength indicator (RSSI)

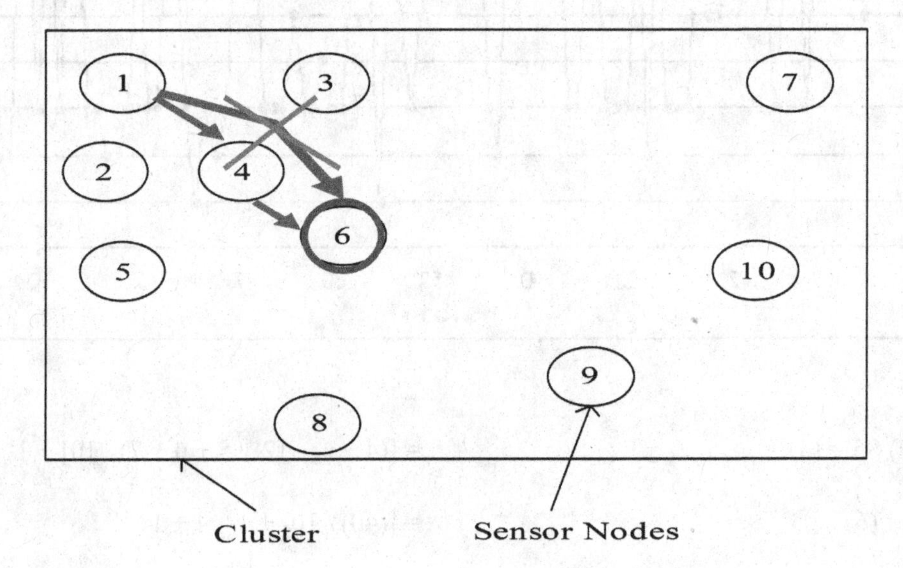

to balance energy in a cluster through energy efficient reliable cluster head selection. Suitable cluster-head selection makes the network efficient and increases the lifetime and data delivery of the networks. The flowchart below shows the process of cluster head selection within the candidate cluster heads by calculating the energy and distance from the base station (BS). The proposed algorithm tends to balance energy in a cluster through energy efficient reliable cluster head selection. The objective of the proposed algorithm is to improve the cluster head selection mechanism of LEACH. The operation of LEACH is divided into rounds. Each round consists of a setup phase and steady-state phase. In the setup phase, the server node (SN) selects the cluster head for each cluster. In the steady-state phase, sensor nodes sense the environment and transmit the sensed data to the corresponding CH for further onward transmission to the SN and BS. Our work assumes that clusters are already formed by the K-mean algorithm in the setup phase and the server node (SN) is aware of the cluster formation and information. The SN sets the value of k for the current round for each cluster based on the density of nodes in a cluster. It broadcasts the value of k to each corresponding cluster, i.e., ki. The value of k determines the k number of nearest neighbor nodes. The sensor nodes send their k number of nearest neighbors (based on distance) to the SN. The distance to the node can be calculated based on the received signal strength indicator (RSSI). The SN selects the candidate set of cluster heads, i.e., Ci for each cluster through the K-theorem. The value of ki is always equal to the number of candidate cluster heads in a cluster, i.e., Ci. The detailed working

Table 3. List of nodes with their K-nearest neighbors and their frequency of occurrence

Node ID	ki = 3 List of Terminals with its K-Nearest Neighbor	Frequency of Occurrence
1	1) 2, 3, 4	3
2	2) 1, 4, 5	4
3	3) 1, 4, 6	5
4	4) 2, 3, 6	6
5	5) 2, 4, 6	4
6	6) 3, 4, 5	7
7	7) 3, 9, 10	2
8	8) 5, 6, 9	2
9	9) 6, 8, 10	4
10	10) 6, 7, 9	3

of the K-theorem is described in the above. The SN requests that the candidate set of cluster heads in each cluster sends their combined rating (CR). Each candidate cluster head node calculates its own CR based on residual energy (RE) and the distance to the server node and sends it to the server node (SN). The server node selects a node as the cluster head from among the candidate set of cluster heads for each cluster based on the CR. The higher the CR a node has, the greater the chances of being a cluster head. The SN confirms each cluster about their CH. Each CH creates a TDMA (Time Division Multiple Access) schedule for intra-cluster communication. In the TDMA, a one-time slot is reserved for each cluster member to send or receive its data to/from the CH.

The residual energy of a node preferably is greater than the approximate energy dissipated in previous rounds by the cluster head. The equation for the residual energy of node i is described in (Tillapart et al. 2005). The nodes having less distance from the coordinator node should have a higher probability of becoming the cluster head. This is because energy consumption is directly proportional to the square of the distance. The distance to the server node for node i is expressed in (Tillapart et al. 2005). The energy distance ratio is calculated by the formula given below. As the role of the cluster head is very crucial for the successful operation of a wireless sensor network, if a cluster head stops working, the whole cluster becomes dysfunctional. A candidate cluster head sensor node may fail due to lack of energy, physical damage or environmental interference. Reliability deals with the continuity of service. The purpose of node reliability is to increase its trustworthiness. The reliability Ri (t) of a sensor node is modeled in (Hoblos et al. 2000) using the Poisson distribution to capture the probability of not having a failure within the time interval (0, t).

The lifetime of the network is greatly influenced by the mobility of a node. It can lead to higher topological changes and requires frequent cluster head reselection. Information about topological changes (due to dead nodes or node mobility) can be exchanged during the maintenance phase.

Event Detection Model

In many application areas of wireless sensor networks, the events that have occurred generate the same type of data repeatedly and the data will become redundant. In some scenarios all the sensor nodes are in an active state and drained of their energy quickly. It is suggested that the sensor nodes go to an active state whenever the event occurs and only those nodes will start sensing where the event takes place and during this time period the other nodes will be in the sleep mode. So, we can conserve sensor node energy and prolong the network lifetime. To eliminate the redundant data, in the model given below, we use the sum of the absolute differential function "SAD". The SAD equation (1) finds the similarity between two input data by performing the sum of absolute differences. Figure 7 shows the block diagram of the event detection model. When the first packets are detected, they are kept for some unit delay and compared with the second incoming packet, for example, N is compared with N+1, and N+1 is compare with N+2 until the last packet is received.

Figure 6. Flowchart of CH selection

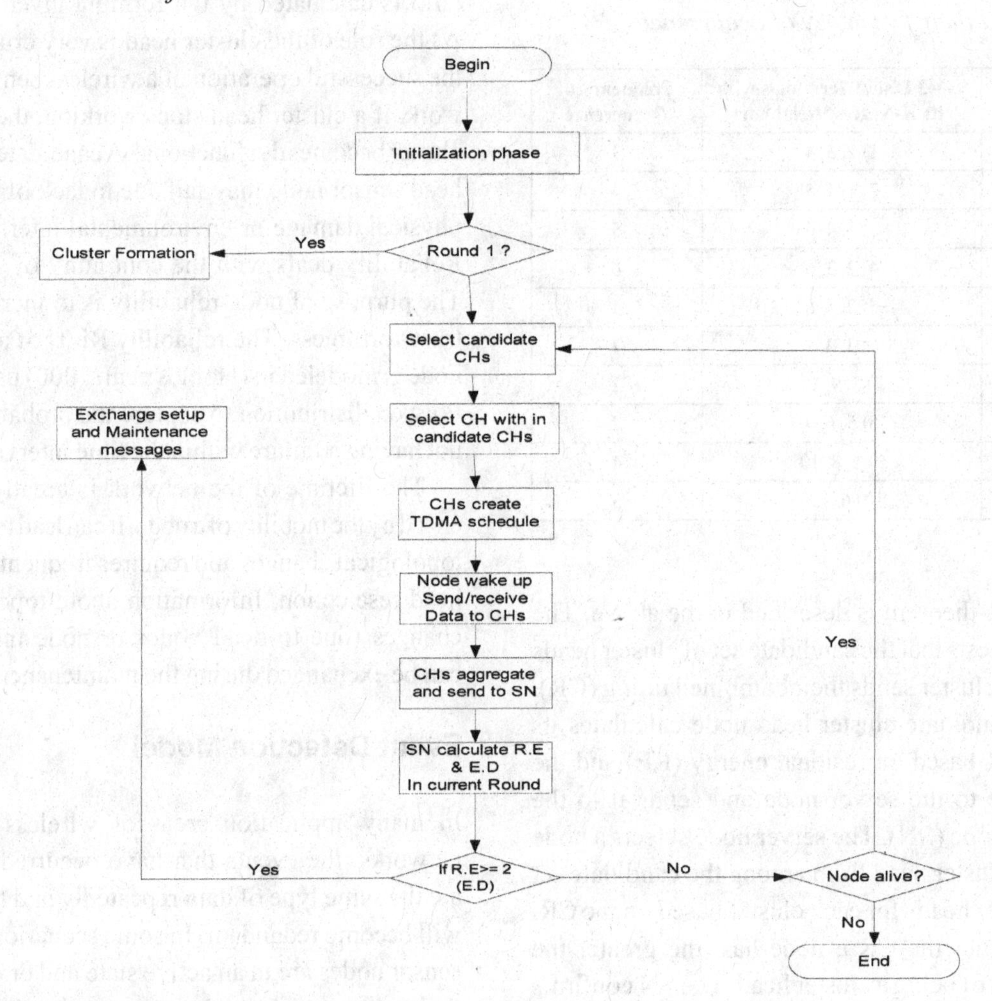

If the difference between N, N+1 is greater than the threshold values, then it will be recorded and the data will be saved; otherwise the packet will be discarded and the next incoming packets will be waited for.

$$c(j,k) = \sum_{m=0}^{(Mt-1)} \sum_{n=0}^{(Nt-1)} abs\left(I(m+j, n+k) - T(m,n)\right) \quad (2)$$

$$0 \le j < Mi - Mt + 1$$

$$0 \le k < Ni - Nt + 1$$

Proposed event based detection models are used for simulating when there are any significant changes occurring in the environment such as fires, explosions, toxic gas leaks, etc. In such an environment, we depute sensors for monitoring purposes, the sensors are continuously monitoring the activities and sending data to the Base station for further analysis and for the remedial action to be taken in case of finding discrepancies.

Figure 7. Event based recording model

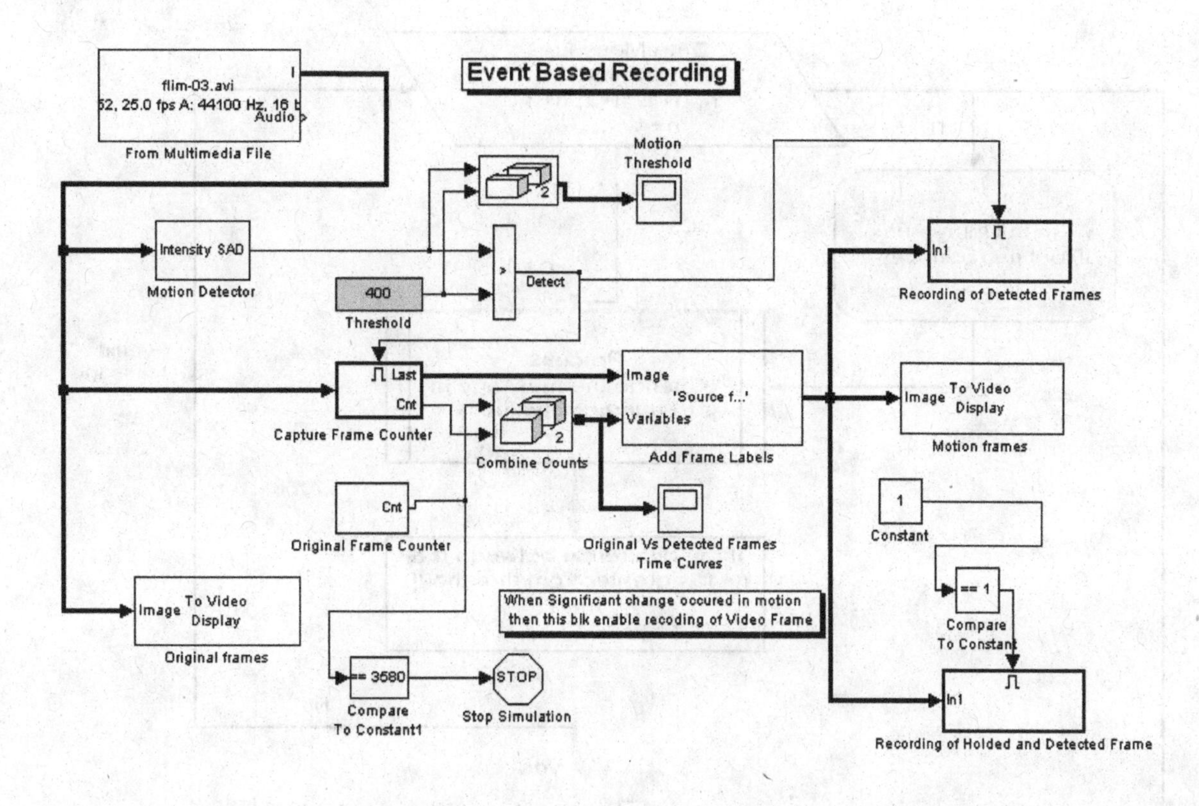

A sensor node consists of processor/controller unit; the main things in the unit of the sensor node are the transmitter, receiver and data processing units. This unit is generally powered up by small batteries having limited energy storage. The crucial unit of the sensor node is the power source, the best utilization of the power source in such a field area is very necessary for the long life of the system.

In some application areas of wireless sensor networks, some of the information sensed by the sensor nodes is closely related to one another and there is no need for the entire information to be transmitted to the base station. In most cases, all the sense data are transmitted to the base station, which greatly influences the network lifetime because 70% of the energy of the sensor nodes is exhausted during transmission. So, we should minimize the number of transmissions and send only that information to the base station which has some significant changes.

Our proposed system is the event based detecting system, which detects the events and takes action according to the program algorithm (Figure 8). This system "Event Based Detecting" is more reliable in terms of power savings and increasing the network lifetime. The proposed model is not saving or transmitting all the data to the base station; it only records and transmits data when there is some significant changes occurring in the environment.

Data Transmission Phase

Routing is a challenging task in a wireless sensor network due to its unique characteristics, dynamic nature, and architecture and design issues. There are various protocols proposed in the literature, but hierarchical or cluster based protocols are

Figure 8. Working flow of the event based detection model

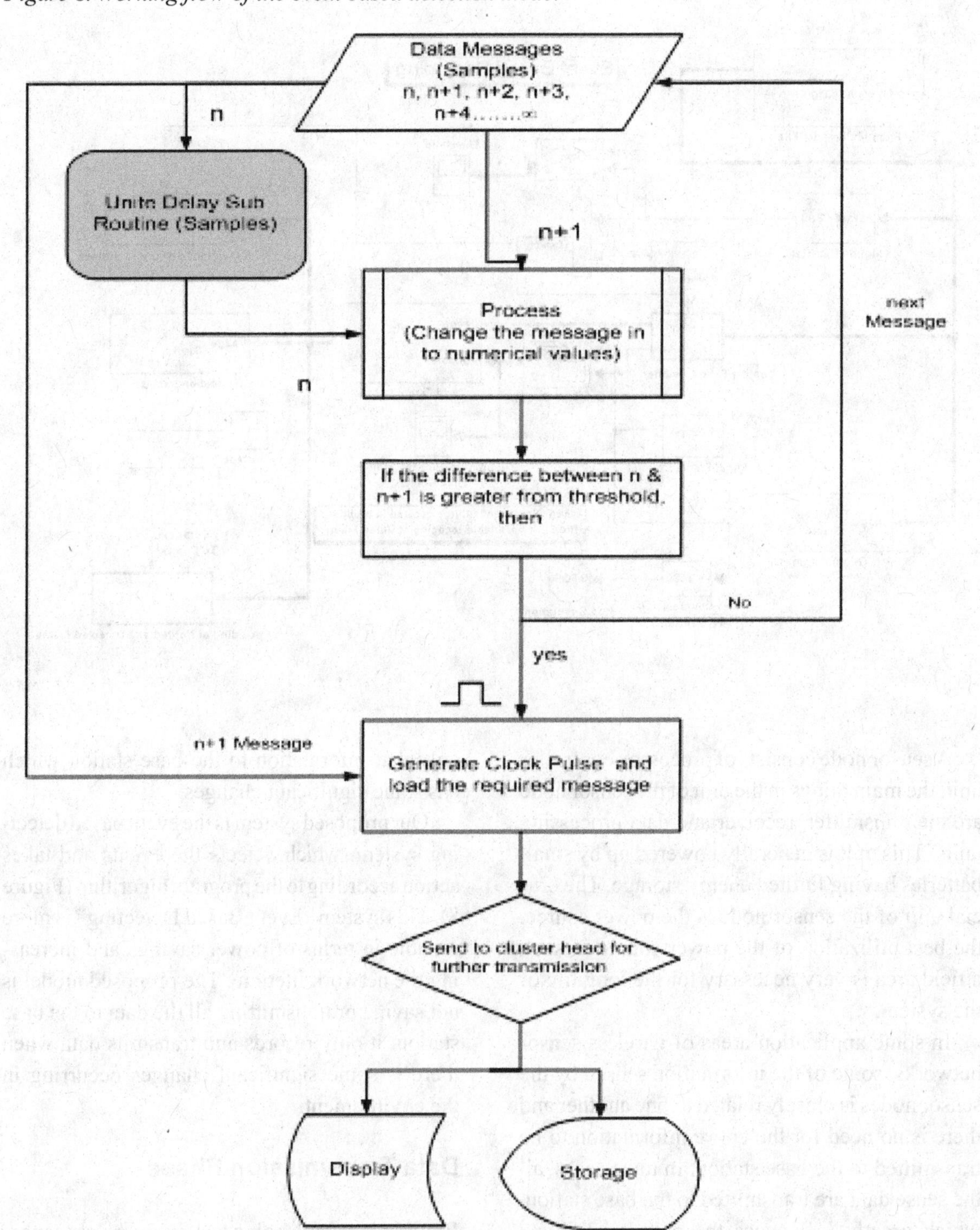

energy efficient, scalable and tend to prolong the network lifetime. They are summarized in (Akkaya and Younis 2005) (Abbasi and Younis 2007). The transmission of data is similar to the steady phase proposed in LEACH (Heinzelman et al. 2000). The data transfer path from the source node, where the events of interest occur is also a critical metric. During the operation in a hazardous environment, sensor nodes failure due to some external interference cannot be discounted especially for environmental monitoring and battle field sensor networks. So, sensor nodes must route strategic and time critical information via the most reliable path available. After the CH selection, routing paths are established; that is the part of this research. Each CH creates a TDMA (Time Division Multiple Access) schedule for intra-cluster communication. In the TDMA, one-time

Figure 9. Topology of shortest path selection

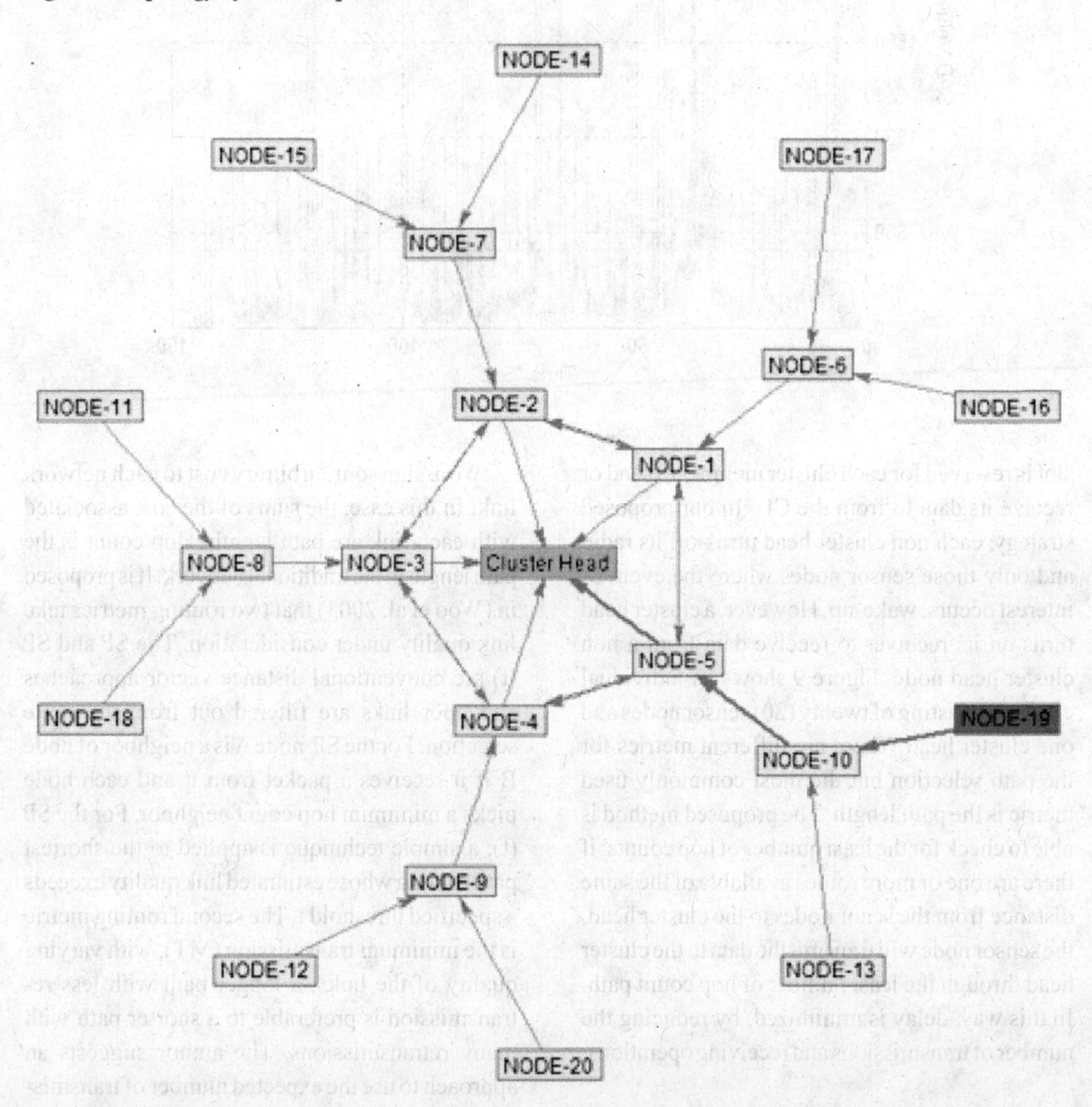

Figure 10.This graph shows the similarity between two input data by performing the sum of absolute differences "SAD". When the sum of absolute differences (SAD) value of a particular packet exceeds a threshold, it will record the data otherwise reject it.

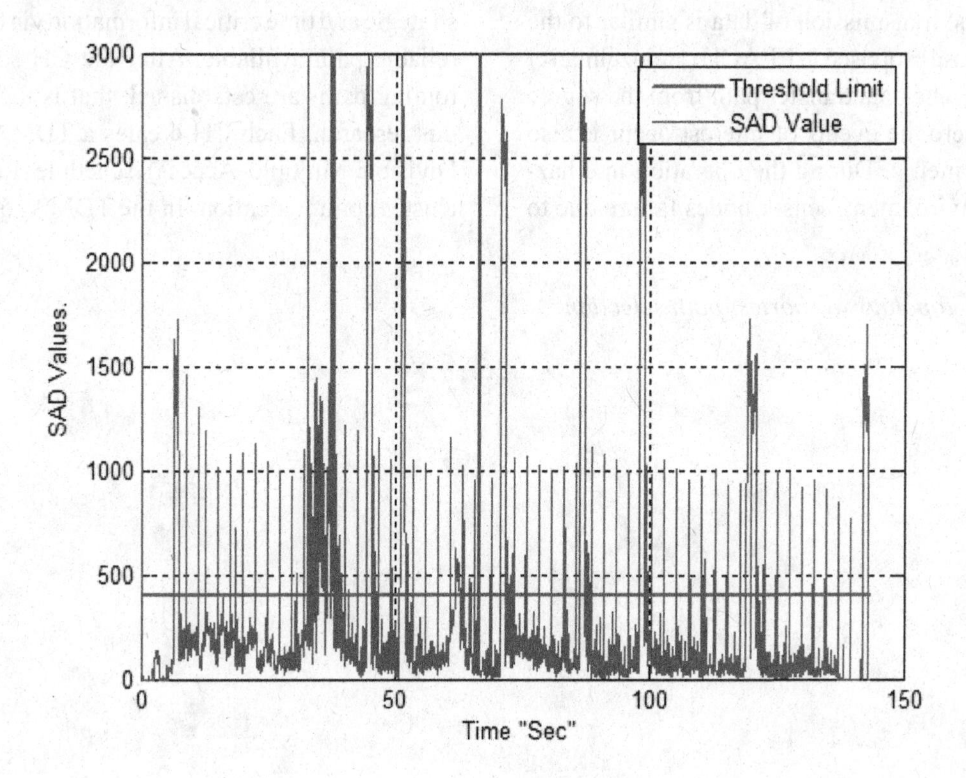

slot is reserved for each cluster member to send or receive its data to/from the CH. In our proposed strategy, each non cluster head turns-off its radio and only those sensor nodes where the event of interest occurs, wake-up. However, a cluster head turns on its receiver to receive data from a non cluster head node. Figure 9 shows an individual cluster consisting of twenty (20) sensor nodes and one cluster head. There are different metrics for the path selection but the most commonly used metric is the path length. The proposed method is able to check for the least number of hop counts; if there are one or more routes available of the same distance from the senor nodes to the cluster head, the sensor node will transmit the data to the cluster head through the least number of hop count path. In this way, delay is minimized, by reducing the number of transmissions and receiving operations.

We assign some arbitrary cost to each network link. In this case, the sums of the cost associated with each link are path length. Hop count is the path length in the traditional network. It is proposed in (Woo et al. 2003) that two routing metrics take link quality under consideration. The SP and SP (t) are conventional distance vector approaches and poor links are filtered out from the route selection. For the SP, node A is a neighbor of node B if it receives a packet from it and each node picks a minimum hop count neighbor. For the SP (t), a simple technique is applied to the shortest path routing whose estimated link quality exceeds a specified threshold t. The second routing metric is the minimum transmission (MT), with varying quality of the links; a longer path with less re-transmission is preferable to a shorter path with many retransmissions. The author suggests an approach to use the expected number of transmission as the cost metric for routing. So, the best path minimizes the total number of transmissions in delivering a packet over multiple potential hops to the destination. The constrained shortest path algorithm uses the distance between any two nodes as a metric for energy consumption and estimation for propagation delay among the nodes. The energy consumption is inverse with 'dn', where d is the distance between any two nodes and n is the value based on the system and application in use. Rumor uses the routing of queries to the node that has detected an event of interest; the algorithm uses a data retrieval technique based on events not using the addressing scheme. Each node in the network has its neighbors list and an event table with forwarding information to all the events it knows (Braginsky and Estrin 2002).

For communication within clusters, only those nodes wake-up where the event of interest occurs. The event detecting nodes broadcast a message to all the neighboring nodes; the set of nodes that are within the communication range of the original node (Event detecting node). For example, if node A hears from node B that it has a less number of hop counts and is a shorter distance from the cluster head, then node A itself sends the information to node B. This process is repeated until it reaches the cluster head. In this case, we focus on the minimum number of hop count path by minimizing the propagation delay and the smaller distance, minimizing the energy consumption during the transmission. On the other hand, a node will communicate with other nodes where the distance is less and the maximum number of hop counts occurs because most of the energy is consumed during transmission as compared to processing.

Figure 11. Total packets and the packets received by the base station. The base station only received those packets that have some significant changes.

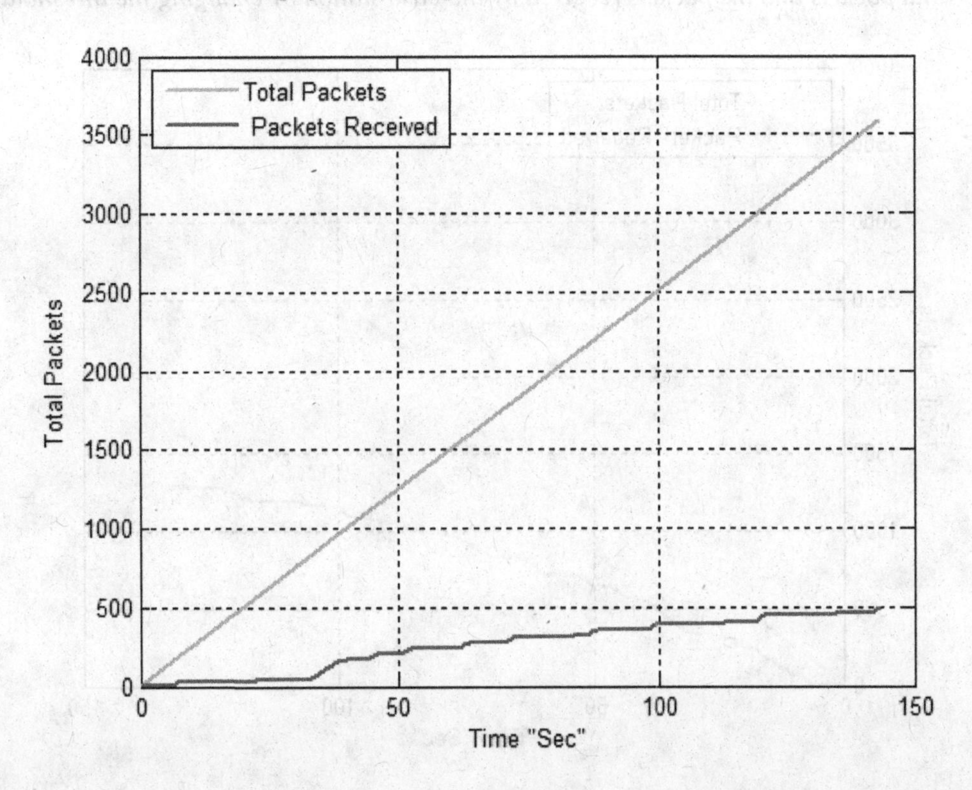

Experimental Results

In Figure 10, two values are plotted; one is the threshold limit and the second is the "SAD" (sum of absolute difference). These values are generated by the motion detector block. This graph is basically to show the detected and rejected packets. All data below the threshold values are rejected because the value of the "SAD" (sum of absolute differences) is less than the threshold, while the remaining data are above the threshold value and will therefore be recorded and transmitted to the base station for analysis.

Figure 11 shows the total packets that are received from the environment and the packets where significant change has occurred or the value was higher than the threshold value over time. This graph indicates that there are more than 3500 packets detected but only 500 packets are required to be sent to the base station because there is no need to send the same information repeatedly.

In Figure 12, the number of packets to be sent to the base station increases because of decreasing the value of the threshold limit. We can adjust the value of the threshold for some application areas because the event occurring ratio is known with respect to time.

CONCLUSION AND FUTURE WORK

The proposed model simulates the event based detection system; in this system, some parts of the nodes will be in an active state but some parts will be in a sleep state. Here we are showing that the data saving part of the system and receiver/transmitter of the system are in the sleep state position; it only works when the system detects some events, and then sends only the events of significant changes to the base station, not the en-

Figure 12. Total packets and the packets received by the base station by changing the threshold limit

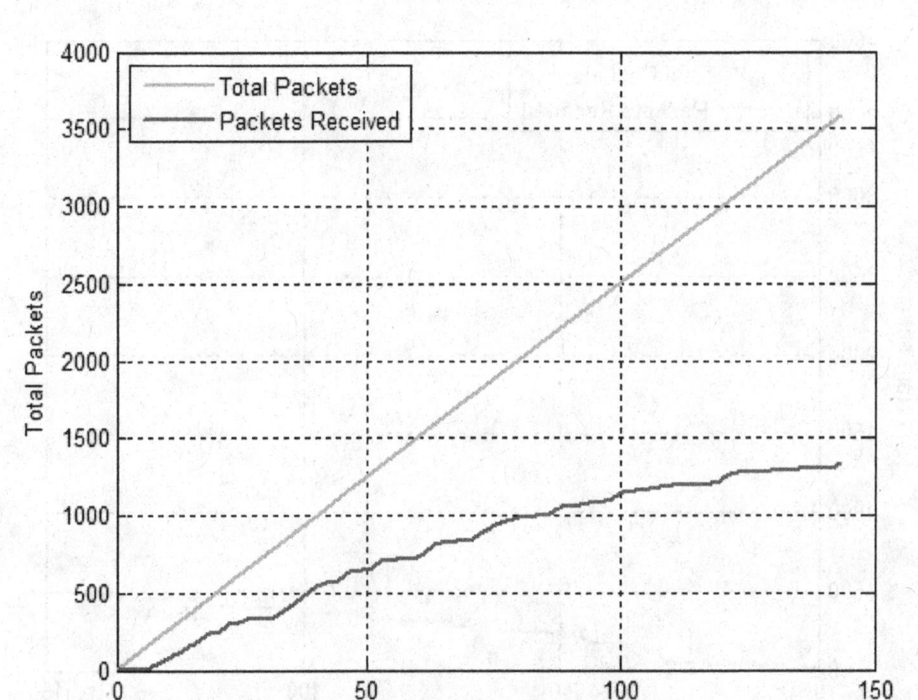

tire data. In this case, we are sending 500 packets instead of 3580 packets; by reducing the number of packets, we reduce the number of operations of transmitting data to the base station thereby increasing the network lifetime. Furthermore, the monitoring purpose is not affected due to such event based systems because we are only interested in events while they are occurring, like monitoring of forest fires. In this situation, we are interested in sending data of the event when the fire takes place. The event based detection model is also efficient in term of propagation delay. Sensor nodes select the best optimum path to reach the cluster heads. We plan to study some other latency techniques, which may well provide energy balance consumption. Our work can be extended to analyze the performance for other parameters, i.e., latency, throughput and efficient routing techniques.

REFERENCES

Abbasi, A. A., & Younis, M. (2007). A survey on clustering algorithms for wireless sensor networks. *Computer Communications*, *30*(14-15), 2826–2841. doi:10.1016/j.comcom.2007.05.024

Akkaya, K., & Younis, M. (2005). A survey on routing protocols for wireless sensor networks. *Ad Hoc Networks*, *3*(3), 325–349. doi:10.1016/j.adhoc.2003.09.010

Akyildiz, I. F., Su, W., Sankarasubramaniam, Y., & Cayirci, E. (2002). Wireless sensor networks: A survey. *Computer Networks*, *38*(4), 393–422. doi:10.1016/S1389-1286(01)00302-4

Biswas, P. K., & Phoha, S. (2006). Self-organizing sensor networks for integrated target surveillance. *IEEE Transactions on Computers*, *55*(8), 1033–1047. doi:10.1109/TC.2006.130

Boukerche, A., Araujo, R. B., & Silva, F. H. S. (2007). An efficient event ordering algorithm that extends the lifetime of wireless actor and sensor networks. *Performance Evaluation*, *64*(5), 480–494. doi:10.1016/j.peva.2006.08.009

Braginsky, D., & Estrin, D. (2002). Rumor routing algorthim for sensor networks. *Proceedings of the 1st ACM International Workshop on Wireless Sensor Networks and Applications*. Atlanta, GA: ACM.

Chu, M., Haussecker, H., & Zhao, F. (2002). Scalable information-driven sensor querying and routing for ad hoc heterogeneous sensor networks. *International Journal of High Performance Computing Applications*, *16*(3), 293–313. doi:10.1177/10943420020160030901

Deshpande, A., Guestrin, C., Madden, S. R., Hellerstein, J. M., & Hong, W. (2004). Model-driven data acquisition in sensor networks. *Proceedings of the Thirtieth International Conference on Very Large Data Bases*, vol. 30. Toronto, Canada: VLDB Endowment.

Deshpande, A., & Madden, S. (2006). MauveDB: Supporting model-based user views in database systems. *Proceedings of the 2006 ACM SIGMOD International Conference on Management of Data*. Chicago, IL: ACM.

Gui, C., & Mohapatra, P. (2004). Power conservation and quality of surveillance in target tracking sensor networks. *Proceedings of the 10th Annual International Conference on Mobile Computing and Networking*. Philadelphia, PA: ACM.

Gupta, H., Navda, V., Das, S. R., & Chowdhary, V. (2005). Efficient gathering of correlated data in sensor networks. *Proceedings of the 6th ACM International Symposium on Mobile Ad Hoc Networking and Computing*. Urbana-Champaign, IL: ACM.

Haiming, Y., & Sikdar, B. (2007). *Optimal cluster head selection in the LEACH architecture* (pp. 93–100).

Heinzelman, W. R., Chandrakasan, A., & Balakrishnan, H. (2000). Energy-efficient communication protocol for wireless microsensor networks. *Proceedings of the 33rd Hawaii International Conference on System Sciences*, vol. 8. IEEE Computer Society.

Hightower, J., & Borriello, G. (2001). Location systems for ubiquitous computing. *Computer, 34*(8), 57–66. doi:10.1109/2.940014

Hoblos, G., Staroswiecki, M., & Aitouche, A. (2000). *Optimal design of fault tolerant sensor networks* (pp. 467–472).

Hu, L., & Evans, D. (2004). Localization for mobile sensor networks. *Proceedings of the 10th Annual International Conference on Mobile Computing and Networking*. Philadelphia, PA: ACM.

Jain, A., Chang, E. Y., & Wang, Y.-F. (2004). Adaptive stream resource management using Kalman Filters. *Proceedings of the 2004 ACM SIGMOD International Conference on Management of Data*. Paris, France: ACM.

Kotidis, Y. (2005). *Snapshot queries: Towards data-centric sensor networks* (pp. 131-142).

Lamport, L. (1978). Time, clocks, and the ordering of events in a distributed system. *Communications of the ACM, 21*(7), 558–565. doi:10.1145/359545.359563

Mansouri-Samani, M., & Sloman, M. (1997). A generalised event monitoring language for distributed systems. *IEE/IOP/BCS. Distributed Systems Engineering, 4*(2). doi:10.1088/0967-1846/4/2/004

Minhas, A. A. (2007). *Power aware routing protocols for wireless ad hoc sensor networks*. Graz, Austria: University of Technology.

Shim, Y. C., & Ramamoorthy, C. V. (1990). *Monitoring and control of distributed systems* (pp. 672–681).

Tillapart, P., Thammarojsakul, S., Thumthawatworn, T., & Santiprabhob, P. (2005). *An approach to hybrid clustering and routing in wireless sensor networks* (pp. 1–8).

Tseng, Y.-C., Wang, Y.-C., & Cheng, K.-Y. (2005). An integrated mobile surveillance and wireless sensor (iMouse) system and its detection delay analysis. *Proceedings of the 8th ACM International Symposium on Modeling, Analysis and Simulation of Wireless and Mobile Systems.* Montreal, Canada: ACM.

Wensheng, Z., & Guohong, C. (2004). DCTC: Dynamic convoy tree-based collaboration for target tracking in sensor networks. *IEEE Transactions on Wireless Communications, 3*(5), 1689–1701. doi:10.1109/TWC.2004.833443

Woo, A., Tong, T., & Culler, D. (2003). Taming the underlying challenges of reliable multihop routing in sensor networks. *Proceedings of the 1st International Conference on Embedded Networked Sensor Systems.* Los Angeles, CA: ACM.

Younis, M., Akkaya, K., Eltoweissy, M., & Wadaa, A. (2004). *On handling QoS traffic in wireless sensor networks* (p. 10).

Section 3
Application, Theoretical and General

Chapter 22
A Game Theoretical Approach to Design: A MAC Protocol for Wireless Sensor Networks

S. Mehta
Wireless Communications Research Center, Inha University, Korea

B. H. Kim
Korea Railroad Research Institute, Korea

K.S. Kwak
Wireless Communications Research Center, Inha University, Korea

ABSTRACT

Game Theory provides a mathematical tool for the analysis of interactions between the agents with conflicting interests, hence it is a suitable tool to model some problems in communication systems, especially, to wireless sensor networks (WSNs) where the prime goal is to minimize energy consumption than high throughput and low delay. Another important aspect of WSNs are their ad-hoc topology. In such ad-hoc and distributed environment, selfish nodes can easily obtain the unfair share of the bandwidth by not following the medium access control (MAC) protocol. This selfish behavior, at the expense of well behaved nodes, can degrade the performance of overall network. In this chapter, the authors use the concepts of game theory to design an energy efficient MAC protocol for WSNs. This allows them to introduce persistent/non-persistent sift protocol for energy efficient MAC protocol and to counteract the selfish behavior of nodes in WSNs. Finally, the research results show that game theoretical approach with the persistent/non-persistent sift algorithm can improve the overall performance as well as achieve all the goals simultaneously for MAC protocol in WSNs.

DOI: 10.4018/978-1-4666-0101-7.ch022

INTRODUCTION

Communication in wireless sensor networks is divided into several layers. Medium Access Control (MAC) is one of those layers, which enables the successful operation of the network. MAC protocol tries to avoid collisions by not allowing two interfering nodes to transmit at the same time. The main design goal of a typical MAC protocol is to provide high throughput and QoS. On the other hand, wireless sensor MAC protocol gives higher priority to minimize energy consumption than QoS requirements. Energy gets wasted in traditional MAC layer protocols due to idle listening, collision, protocol overhead, and over-hearing (Heidemann et al., 2002; Dam et al., 2003). There are some MAC protocols that have been especially developed for wireless sensor networks. Typical examples include S-MAC, T-MAC, and H-MAC (Heidemann et al., 2002; Dam et al., 2003; Mehta et al., 2009). To maximize the battery lifetime, sensor networks MAC protocols implement the variation of active/sleep mechanism. S-MAC and T-MAC protocols trades networks QoS for energy savings, while H-MAC protocol reduces the comparable amount of energy consumption along with maintaining good network QoS. However, their backoff algorithm is based on the IEEE 802.11 Distributed Coordinated Function (DCF), which is based on Carrier Sense Multiple Access with Collision Avoidance (CSMA/CA) Mechanism. The energy consumption using CSMA/CA is high when nodes are in backoff procedure and in idle mode. Moreover, a node that successfully transmits resets it contention window (CW) to a small, fixed minimum value of CW. Therefore, the node has to rediscover the correct CW, wasting channel capacity and increase the access delay as well. So during the CSMA/CA mechanism, backoff window size and the number of active nodes are the major factors to have impact on the network performance and over all energy efficiency of MAC protocol. Hence, it is necessary to estimate the number of nodes in network to optimize the CSMA/CA operation. Furthermore, optimizing CSMA/CA operation is more challenging task for self-organizing and distributed networks as there are no central nodes to assign channel access in sensor nodes.

Furthermore, MAC protocol (or any other protocol) is designed under the assumption that all participating nodes are well behaved. While well-behaved nodes strictly obey the protocol operation, the misbehaving nodes may deviate from the standard or protocol rules to either cause unfairness problems or disrupt the network services. This misbehavior may be hard to differentiate from some normal cases. For example, when a node selects a smaller contention window, it is hard to distinguish whether this is due to an intentional choice or a random selection. In such distributed environment as sensor networks where coordination or punishment mechanisms could be expensive or in some cases even impossible to implement, so it is critical to evaluate the network performance under selfish behavior of nodes.

Recently lots of researchers have started using game theory as a tool to analyze the wireless networks. Their game theoretic approaches were proposed to the wide area of wireless communication right from the security issues to power control, etc., (Mehta et al., 2009; Agah et al., 2004; Kannan et al., 2004; Sengupta et al., 2005; Zhang et al., 2006). In sensor networks each node has a direct influence on its neighboring nodes while accessing the channel. So, these interactions between nodes and aforementioned observations lead us to use the concepts of game theory that could improve the energy efficiency as well as the delay performance of MAC protocol. As we mentioned earlier energy efficiency of MAC protocol in WSN is very sensitive to number of nodes competing for the access channel. It will be very difficult for a MAC protocol to accurately estimate the different parameters like collision probability, transmission probability, etc., by detecting channel. Because dynamics of WSN keep on changing due to various reasons like mobility of nodes, joining of some new nodes, and dying out of some exhausted nodes. Also, estimating about the other neighboring nodes information is too complex, as every node takes a distributed approach to estimate the current state of networks. Also, Detecting selfish behavior is not straightforward in a wireless network. The difficulty arises from the non-deterministic nature of the access protocol that does not lead to a straightforward way of distinguishing between a legitimate node and a selfish node. The open wireless medium and the different perceived channel conditions at different locations add difficulty to the problem. For all these reasons, game theoretical approach is the best way to design a MAC protocol for wireless sensor nodes, as game theory is the best powerful tool available in modeling interactions among the self interested users and predicting their choice of strategies. Moreover, a game theory could be a perfect candidate to optimize the performance of MAC protocol in sensor networks.

Game Theory

Game Theory is a collection of mathematical tools to study the interactive decision problems between the rational players1 (here it is sensor nodes). Furthermore it also helps to predict the possible outcome of the interactive decision problem. The most possible outcome for any decision process is "Nash Equilibrium." A Nash equilibrium is an out come of a game where no node (player) has any extra benefit for just changing its strategy one-sidedly (Choi et al., 2008; Mehta et al., 2009). From last few years, game theory has gained a notable amount of popularity in solving communication and networking issues. These issues include congestion control, routing, power control and other issues in wired and wireless communications systems, to name a few.

A game is set of three fundamental components: A set of players, a set of actions, and a set of preferences. Players or nodes are the decision takers in the game. The actions (strategies) are the different choices available to nodes. In a wireless system, action may include the available options like coding scheme, power control, transmitting, listening, etc., factors that are under the control of the node. When each player selects its own strategy, the resulting strategy profile decides the outcome of the game. Finally, a utility function (preferences) decides the all possible outcomes for each player. Table 1 shows typical components of a wireless networking game.

Games can be classified formally at many level of detail, here we in-general tried to classify the games for better understanding. As shown in the Figure 1, strategic games are broadly classified as co-operative and non-cooperative games. In non-cooperative games the player can not make commitments to coordinate their strategies. A non-cooperative game investigates answer for selecting an optimum strategy to player to face his/her opponent who also has a strategy of his/her own. Conversely, a co-operative game is a game where groups of player may enforce to work together to maximize their returns (payoffs). Hence, a co-operative game is a competition between coalitions of players, rather then between individual players. Furthermore, according to the

Table 1. A wireless networking game

Components of a game	Elements of a wireless network
Players	Nodes in the wireless network
A set of actions	A modulation scheme, transmit power level, etc.
A set of preferences	Performance metrics (e.g. Energy Efficiency, Delay, etc.)

players' moves, simultaneously or one by one, games can be further divided into two categories: static and dynamic games. In static game, players move their strategy simultaneously without any knowledge of what other players are going to play. In the dynamic game, players move their strategy in predetermined order and they also know what other players have played before them. So according to the knowledge of players on all aspect of game, the non-cooperative/co-operative game further classified into two categories: complete and incomplete information games. In the complete information game, each player has all the knowledge about others' characteristics, strategy spaces, payoff functions, etc., but all these information are not necessarily available in incomplete information game (Mehta et al., 2009).

MAC Protocol Games: Incomplete Corporative Game

To model WSNs problems into full information game theoretic problems is an extremely difficult task due to distributed nature of WSNs. In addition, full information sharing also results into additional energy and bandwidth consumption. So we use the concept of incomplete cooperative game theory to solve the aforementioned challenges. In this paper we present the basic idea of adjusting nodes' equilibrium strategy based on estimation of network conditions without full information. To the best of our knowledge, there is very little work on the incomplete cooperative game theory in wireless networks. In (Zaho et al., 2008) authors used the concept of incomplete cooperative game theory in wireless networks for first time and proposed the G-MAC protocol for

Figure 1. Classification of games

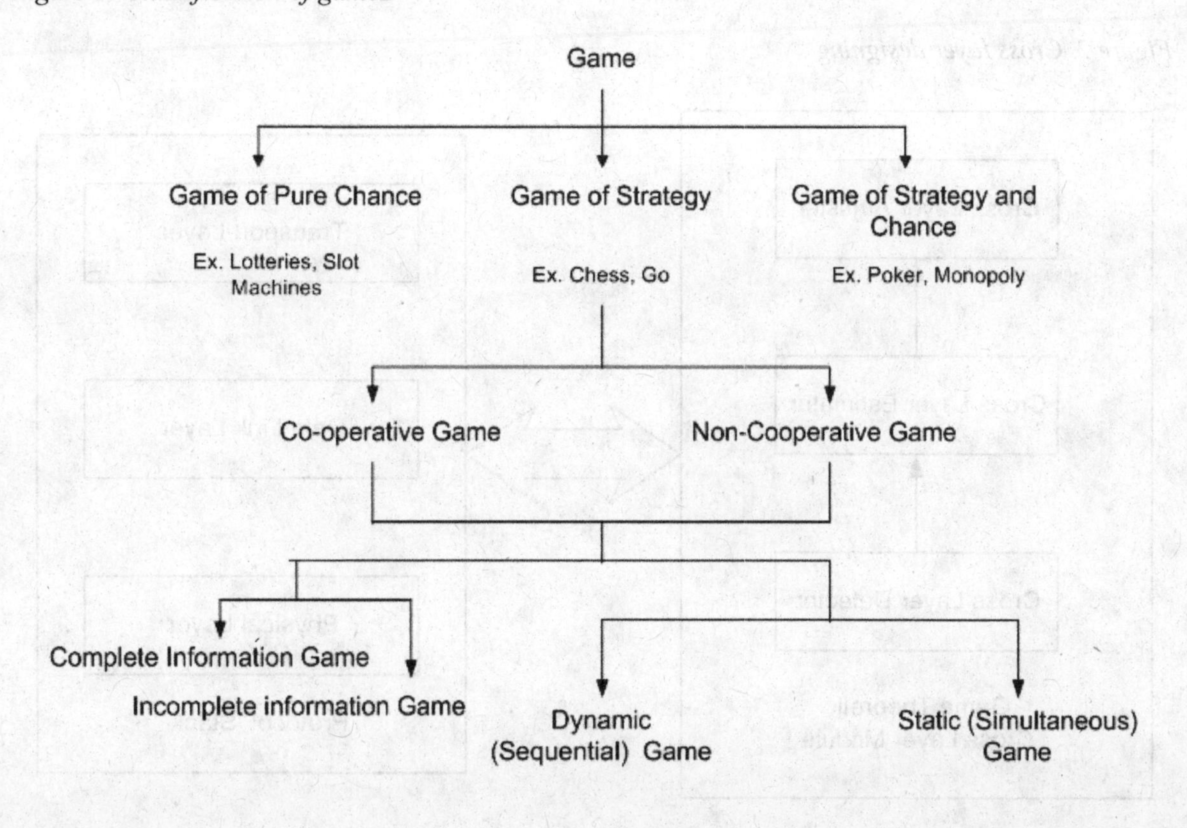

the same. However, their proposed scheme is not suitable for all traffic conditions, especially, non-saturation traffic condition which is most likely in sensor networks. In (Zaho et al., 2008) authors presented a virtual CSMA/CA mechanism to handle the non-saturation traffic condition which is too heavy and complex for the sensor networks. We also work on similar baseline and present our suboptimal solution for an energy efficient MAC protocol in wireless sensor networks. As shown in Figure 2, during the active part a node tries to contend the channel if there is any data in buffer and turn down its radio during the sleeping part to save energy.

In incomplete cooperative game, the considered MAC protocol can be modeled as stochastic game, which starts when there is a data packet in the node's transmission buffer and ends when the data packet is transmitted successfully or discarded. This game consists of many time slots and each time slot represents a game slot. As every node can try to transmit an unsuccessful data packet for some predetermined limit (Maximum retry limit), the game is finitely repeated rather than an infinitely repeated one. In each time slot, when the node is in active part, the node just not only tries to contend for the medium but also estimates the current game state based on history. After estimating the game state, the node adjust its own equilibrium condition by adjusting its available parameters under the given strategies (here it is contention parameters like transmitting probability, collision probability, etc.). Then all the nodes act simultaneously with their best evaluated strategies. In this game we considered mainly three strategies available to nodes: Transmitting, Listening, and Sleeping. And contention window size as the parameter to adjust its equilibrium strategy.

In this stochastic game our main goal is to find an optimal equilibrium to maximize the network performance with minimum energy consumption.

Figure 2. Cross layer designing

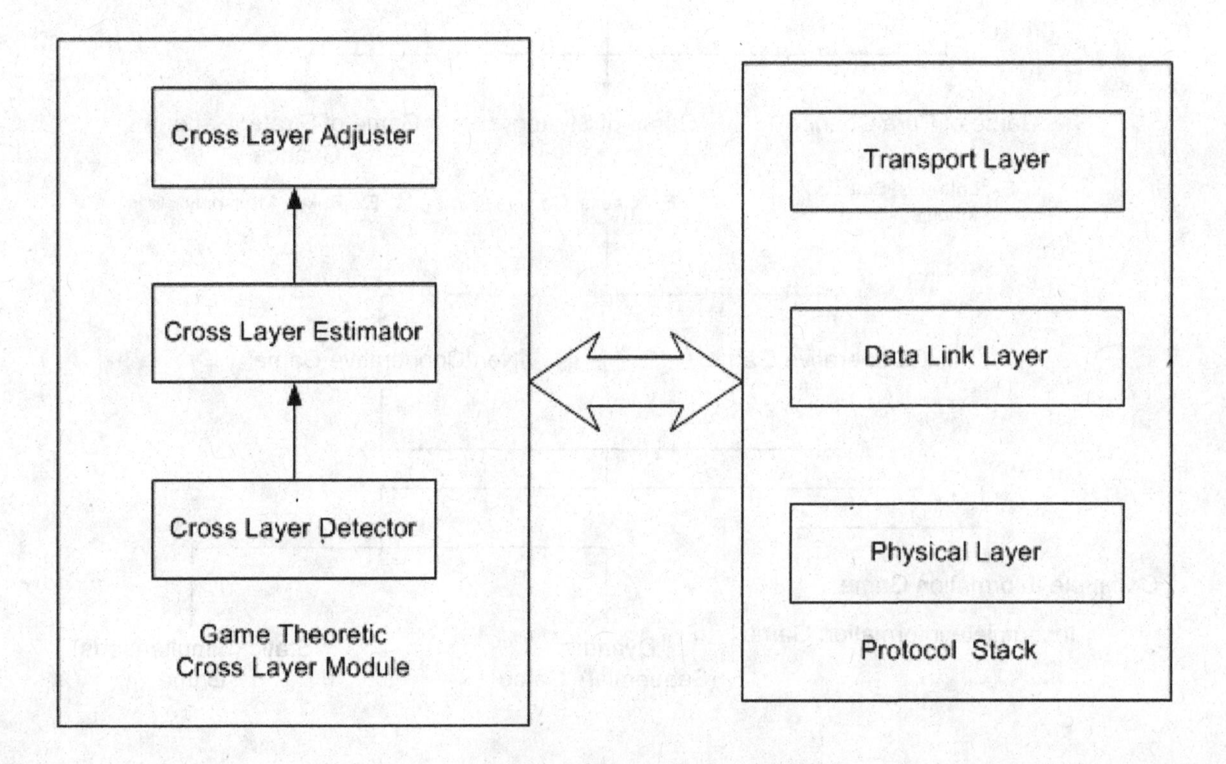

In general, with control theory we could achieve the best performance for an individual node rather than a whole network, and for this reason our game theoretic approach to the problem is justified. Based on the game model presented in (Zaho et al., 2008), the utility function of the node (node i) is represented by $\mu_i = \mu_i(s_i, \overline{s}_i)$ and the utility function of its opponents as $\overline{\mu}_i = \overline{\mu}_i(\overline{s}_i, s_i)$. Here, $s_i = (s_1, s_2, \ldots, s_{i-1}, \ldots, s_n)$ represents the strategy profile of a node and $\overline{s}_i$ of its opponent nodes, respectively. From the aforementioned discussion we can represent the above game as in Table 2.

As presented in [L.Zaho et al., 2008], we define P_i and $\overline{P}_i$ as the payoff for player 1 and 2 when they are listening, P_s and $\overline{P}_s$ when they are transmitting a data packet successfully, P_f and $\overline{P}_f$ when they are failed to transmit successfully, and P_w and $\overline{P}_w$ when they are in sleep mode, respectively. Whatever will be the payoff values, their self evident relationship is given by

$$P_f < P_i < P_w < P_S \tag{1}$$

and similar relationship goes for player 2. As per our goal we are looking for the strategy that can lead us to an optimum equilibrium of the network. As in [T.Rappot, 1996] we can define it formally as

$$\begin{cases} s_i^* = \arg\max_{s_i} \overline{\mu}_i(\overline{s}_i, s_i) \mid (e_i < e_i^*) \\ \overline{s}_i^* = \arg\max_{\overline{s}_i} \mu_i(s_i, \overline{s}_i) \mid (\overline{e}_i < \overline{e}_i^*) \end{cases} \tag{2}$$

where $e_i, e_i^*, \overline{e}_i$ and $\overline{e}_i^*$ are the real energy consumption and energy limit of the player 1 and 2, respectively. Now to realize these conditions in practical approach we redefine them as follows

$$\begin{cases} s_i^* = \arg\max_{(w_i, \tau_i)} \begin{matrix}[(1-\overline{\tau}_i)(1-\overline{p}_i)(1-\overline{w}_i)(1-w_i)\tau_i\overline{P}_s \\ +(1-\overline{\tau}_i)(1-\overline{w}_i)(1-w_i)\tau_i\overline{P}_i \\ +(1-\overline{p}_i)(1-\overline{w}_i)(1-w_i)\tau_i\overline{\tau}_i\overline{P}_f + \overline{\tau}_i\overline{p}_i(1-w_i)\overline{P}_f \\ +\overline{w}_i(1-w_i)\overline{P}_w] \mid (e_i < e_i^*)\end{matrix} \\ \overline{s}_i^* = \arg\max_{(w_i, \tau_i)} \begin{matrix}[(1-\overline{\tau}_i)(1-\overline{w}_i)(1-w_i)\tau_i P_s \\ +(1-\tau_i)(1-\overline{w}_i)(1-w_i)P_i \\ +\tau_i\overline{\tau}_i P_f + w_i(1-\overline{w}_i)P_w] \mid (\overline{e}_i < \overline{e}_i^*)\end{matrix} \end{cases} \tag{3}$$

Here, we define τ_i and $\overline{\tau}_i$ as the transmission probability of the player 1 and player 2, respectively. Similarly, w_i and $\overline{w}_i$ represents the sleeping probability of player 1 and player 2 while $\overline{p}_i$ is the conditional collision probability of player 2. As shown in Table 2 there are three strategies for both the players. First, player 1 transmits a packet with a probability $(1-\overline{\tau}_i)(1-\overline{w}_i)(1-w_i)\tau_i$, whose payoff is P_s. Second strategy of player 1 is listening with a probability $(1-\tau_i)(1-\overline{w}_i)(1-w_i)$, whose

Table 2. Strategy table

		Player 2 (all other *n* nodes)		
		Transmitting	Listening	Sleeping
Player 1 (Node *i*)	Transmitting	$(P_f, \overline{P}_f)$	$(P_s, \overline{P}_i)$	$(P_f, \overline{P}_w)$
	Listening	$(P_i, \overline{P}_s)$	$(P_i, \overline{P}_i)$	$(P_i, \overline{P}_w)$
	Sleeping	$(P_w, \overline{P}_f)$	$(P_w, \overline{P}_i)$	$(P_w, \overline{P}_w)$

payoff is P_i.Third strategy of player 1 is sleeping with a probability $w_i(1-\overline{w}_i)$, whose payoff is P_w. Finally, when both the players transmits simultaneously, their payoff are P_f, and $\overline{P}_f$, respectively. Similarly, we can also calculate the probabilities of different strategies for player 2.

From the strategy table and equation (3) we can see that every node has to play its strategies with some probabilities as here the optimum equilibrium is in mixed strategy form. In mixed strategy equilibrium, it is not possible to reach an optimum solution with one strategy so players have to mix two or more strategies probabilistically.

In addition, we can observe from the above equations that players can achieve their optimal response by helping each other to achieve their optimal utility. So the nodes have to play a cooperative game under the given constrained of energy. Here, the players can obtain the mixed strategy based optimum response by adjusting their transmission probabilities to the variable game states. The value of the transmitting probability can be adjusted by tuning contention parameters, such as the minimum contention window (CW_{min}), the maximum contention window (CW_{max}), retry limit (r), the maximum backoff stage (m), arbitrary interface spaces (AIFS), etc. For simplicity, we choose contention window (i.e. properly estimating the number of competing nodes) as tuning parameter for adjusting transmission probability of a node.

Estimation of Competing Nodes

In the proposed game, every node estimates the game state by anticipating the number of competing nodes from various parameters, especially, from transmitting probability p_{tr}. Many researchers have presented several performance and analysis models to calculate p_{tr}.However, majority of the work has neglected the contention counter freezing effect and considered only saturated traffic condition which is mostly suitable for WLAN and Ad-hoc networks than sensor networks. Arguably, non-saturation traffic condition is most likely traffic pattern in WSNs and need to be considered for a WSN MAC protocol designing as well. From (Mehta et al., 2010) and other previous analysis results we can show that the number (N) of competing nodes is the function of frame collision probability (p_c) of a competing node. By monitoring the channel all the nodes can independently measure the p_{tr} and p_c, hence, can estimate the value of n as well.

Cross Layering

Standardization of layered protocol stacks has enabled flexibility in system developments, but at the same time limited the performance of the overall system because of in-coordination among layers. This is very important issue in wireless networks, where the very physical nature of the transmission medium introduces several performance limitations (including time-varying behavior, limited bandwidth, and severe interference and propagation environments). As a result, the performance of higher layer protocols (e.g., TCP/IP) is severely limited. To overcome such limitations, a modification in layered protocol stacks been proposed, namely, cross-layer design, or "cross-layering." The core idea is to maintain the functionalities associated to the original layers but to allow coordination, interaction and joint optimization of protocols crossing different layers. In this chapter the term "cross-layer" means that the knowledge of the PHY, MAC and TCP/IP layers is used by the game theoretic cross layer module, to calculate the game state and implement the optimal strategy.

As seen in Figure 2, the framework of the game consists of three major components, a detector, an estimator and an adjustor. The first component, the detector, detects and records the experienced PHY-specific information, such as Signal-to-Noise Ratio (SNR), and MAC-specific information, such as frame transmission probability (p_{tr}), frame conditional collision probability (p_c), and TCP/IP-specific information, such as TCP Datagram Loss Ratio (DLR). The second component, the estimator, uses the above measurements to estimate the current game state, such as the number of competing nodes (n). The third component, the adjustor, makes a decision according to which strategy the player transmits its packets, and implements the optimal strategy by tuning contention parameters, such as TCP sliding window, MAC contention parameters (Mehta et al., 2009), and transmission power. Such cross-layer characteristics imply that to obtain the game state and to tune the equilibrium strategy in each timeslot, cross-layer detection, cross-layer estimation and cross-layer adjustment are needed. So we need to implement the detector, estimator, and adjuster to calculate the game state and implement the optimal strategy. However, this method presented is too complex and heavy (in terms of energy consumption, etc.) to implement in sensor networks.

This estimation mechanism gives good approximation but not the accurate results. There are some methods, especially (Bianchi, 2003; Garey et al., 1979), to name a few, to accurately predict the number of competing nodes in the networks. In (Vercauteren et al., 2007) authors, presented batch and sequential Bayesian estimators to predict the number of competing nodes. In (Garey et al., 1979) authors, presented two run time estimation methods named: 'auto regressive moving average (ARMA)' and 'Kalman Filters'. These two methods are very accurate in predicting the number of competing nodes in saturation as well as in non-saturation traffic conditions. However, all the methods presented in (Vercauteren et al., 2007; Xiao et al., 2006) are too complex and heavy (in terms of energy consumption, etc.) to implement in sensor networks.

MAC PROTOCOL GAMES: Selfish Behavior- 'MAC' Game

Node misbehaviors in Ad-hoc/sensor networks can be classified into main 2 categories; namely, selfish misbehavior (Guang et al., 2005; Kyasanur et al., 2004) and malicious misbehavior (Aad et al., 2004; Guang et al., 2005; Gupta et al., 2002). A selfish node can easily play around the MAC protocol to gain more network resources than well behaved nodes.

For example, in case of WLAN MAC Standard or BEB based MAC protocol, MAC protocol requires nodes competing for the channel to wait for backoff interval before any transmissions. A selfish node may choose to wait for a smaller backoff interval, thereby increasing its chance of accessing the channel and hence reducing the throughput share received by well behaved nodes. In some cases a selfish node may choose to wait for a longer backoff interval, there by increasing its chance to avoid the data forwarding and hence saving the energy. In this chapter our proposed solution is efficient regardless of the attack strategies, a smart attacker might apply, e.g., choosing smaller CW or larger CW. However, in this chapter we consider only the first case of selfish behavior for our analysis, i.e., a selfish node chooses smaller CW in order to improve its own throughput while deteriorating the performance of other well behaved nodes. In (Kyasanur et al., 2004), authors showed that such selfish misbehavior can seriously degrade the performance of the network and accordingly they proposed some modifications for the protocol to detect and penalize misbehaving nodes. Similarly, in (Raya et al., 2004, authors addressed the same problem and proposed a system, DOMINO, to detect greedy misbehavior such as backoff manipulations in IEEE 802.11.

Alternatively, malicious misbehavior aims primarily at disrupting the normal operation of the network. This includes malicious nodes that continuously send data (mostly fake or dummy packets) to each other in order to exhaust the channel capacity in their neighborhood and hence prevent other legitimate users from communicating (Zhou et al., 2004). In (Guang et al., 2005), authors presented a new class of vulnerabilities where a host could maliciously modify the protocol timeout mechanism (e.g. by changing SIFS parameter in IEEE 802.11) and force MAC frames to be dropped at well-behaved nodes. A node exploiting this vulnerability will completely cooperate in forwarding data packets but maliciously forces the forwarding operation to fail. Moreover, the attack also targets crossing flows (flows that traverse through a malicious node) by disrupting their communication and forcing the routing protocol to reroute packets around the misbehaved node.

From the above discussion we concluded that for any network protocol to design with complete stability possible only if there is a way to detect node's selfish behavior in a quick way. Therefore it is very critical for any network designer to understand the selfish behavior problem from root and design an efficient mechanism which can detect and protect the selfish behavior of nodes in MAC protocol as early as possible. Selfish behavior can classify in three main categories: naive, smart, and misguided behaviors, as shown in the Figure 3.

Naive Nodes: These nodes implement simple mechanism to achieve a selfish goal, e.g. improving throughput or conserving energy, by simply adjusting the MAC protocol parameters. Changing backoff contention window size is the most common and simple strategy adopted by naïve nodes.

Smart Nodes: These types of nodes know the complete operation or procedure of selfish-behavior detection system; however, it might not be capable of guessing the exact critical parameters, such as monitoring interval T and threshold Thresh as used in (Kyasanur et al., 2004) and (Raya et al., 2004). For more details on native and smart nodes attack readers are referred to (Kyasanur et al., 2004).

Misguided Nodes: Due to varying nature of channel condition or hardware malfunction or for any other reason sometimes nodes appears to behave like selfish nodes but in actual practice they are not. These kinds of situations some times create false alarm of being a selfish behavior node and unnecessarily get punished for the same. Along with detecting selfish nodes in the network it is also very important to reduce the number of false alarm by misguided nodes.

It is worth to note that Selfish behavior and fairness problem has a close association. Solving fairness problem can not guarantee protection from selfish behavior of a node but the other

Figure 3. Classification of misbehavior

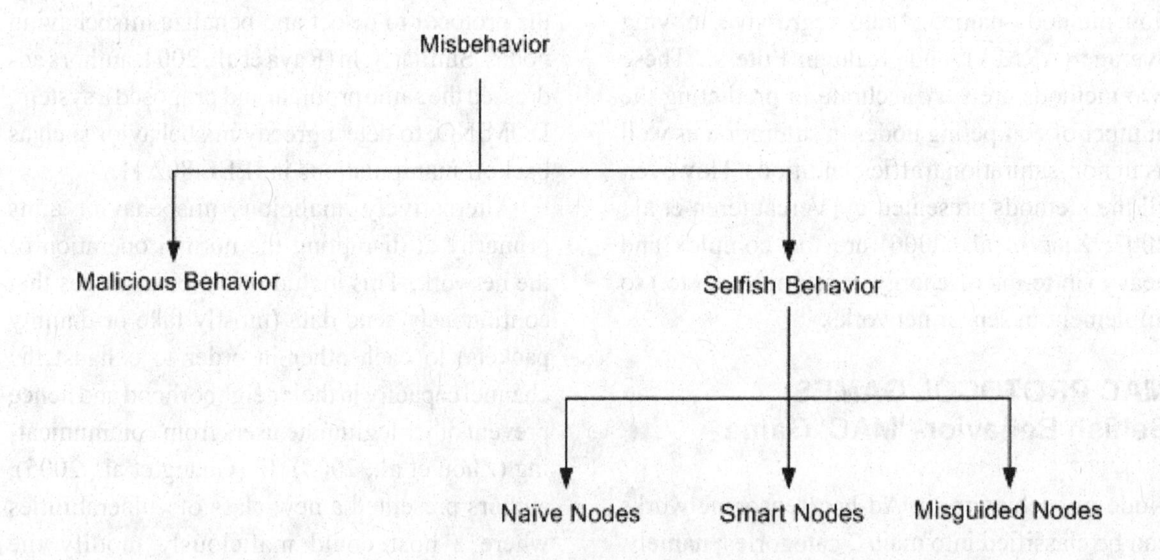

way around is possible, as shown in Figure 4. So solving selfish behavior of a node can also lead us to fairness problem solution.

Modeling BEB Based MAC Protocol with Selfish Nodes

In this chapter again we consider the BEB based MAC protocol for selfish behavior analysis. The BEB based MAC protocol can be easily associated with existing MAC protocols or MAC standards available in the research literature. We consider a wireless network consisting of a set $N = \{1,2,\ldots,n\}$ of selfish nodes within the same communication range, i.e., each node can hear any other node. By selfish we mean that each node can configure its own contention window (CW) value, network nodes do not double their CW value upon a collision. Since we assume that a selfish node's objective is to maximize his throughput, here we assume only the saturation condition, i.e., each node always has packets to send and the packets are of the same size. It is worth to note that selfish behavior is pointless under light or moderated load, where each station mostly gets all the bandwidth required. In this context, let W_i denote the CW value of node i, p_{tr} denotes the transmission probability of i in a random slot, p_c denotes the collision probability of i when it transmits a packet in a random slot, based on the Bianchi's model (Bianchi, 2003), we have:

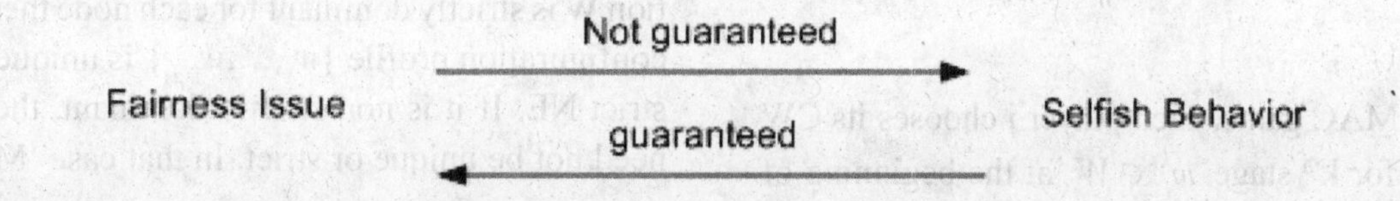

Figure 4. Fairness issue and selfish behavior

$$\begin{cases} p_{tr_i} = \dfrac{2}{W_i + 1} & \forall i \in N \\ p_{c_i} = 1 - (1 - p_{tr})^{n-1} \end{cases} \tag{4}$$

As an application, (4) can be used to calculate the normalized channel efficiency η_{BEB}, and is given by

$$\eta_{BEB} = \frac{T_S p_s}{T_i p_i + T_S p_s + T_c (1 - p_s)} \tag{5}$$

Here, T_i is the slot duration, constant and defined in the standard (of duration δ), T_s and T_c are the average time duration of successful transmission and collision, respectively. p_i and p_s represents probability of idle medium and successful transmission, respectively. T_s and T_c are given by

$$T_S = H_{P+M} + T_{Data} + SIFS + P + T_{ACK} + P + SIFS \tag{6}$$

$$T_C = H_{P+M} + T_{Data} + SIFS + P \tag{7}$$

Where H_{P+M} is the length of MAC and Physical layer frame headers and P is the propagation delay. T_{Data} is the duration of data packet.

Now, we establish a non-cooperative game theoretic model on the selfish MAC behavior in which all network nodes are selfish, rational and do not cooperative in managing their communication. Each node i chooses its CW value W_i to maximize its own benefit described by a utility function defined as (Chen et al., 2007)

$$u_i = \frac{p_{tr_i}[(1-p_{c_i})g_i - e_i]}{T_i p_i + T_S p_s + T_c(1-p_s)} \quad (8)$$

Where g_i is the gain of node i when successfully transmitting a packet, e_i is the cost of transmitting a packet. u_i, expressed as the expected gain during a slot time divided by the slot length, can be regarded as the expected payoff per unit time. To simplify the problem, we assume that g_i and e_i are the same for all $i \in N$, denoted respectively as g and e. Throughout this chapter, we assume that $e \ll g$.

As in (Chen et al., 2007), we can model the MAC protocol with selfish nodes as a repeated non cooperative game with unpredictable end time, meaning that the players cannot predict the end time of the game. This is often the case in strategic interactions, in particular networking operations. In game theory this can be modeled as an infinite multi-stage game with discount. The discount factor is usually very close to 1, indicating that the players are in general long-sighted. This is very important assumption for a 'MAC' game, as proven in (Chen et al., 2007) short sighted nodes can lead to network collapse. The game starts at time 0 and each stage lasts T. In our context, the players of the game are all the nodes in the network. The strategy set of the players is the CW value set $W = \{1, 2, \cdots, +\infty\}$. The strategy profile W^k played in k[th] stage is thus the n-tuple of individual player's stage game strategies, i.e.,

$$W^k = (w_1^k, \cdots, w_n^k) \quad w_i^k \in w. \quad (9)$$

We denote the correspondent transmission probability profile and collision probability profile in stage k as $p_{tr}^k = (p_{tr_1}^k, \cdots, p_{tr_n}^k)$ and $p_c^k = (p_{c_1}^k, \cdots, p_{c_n}^k)$.

In MAC game, each player i chooses its CW value for k[th] stage $w_i^k \in W$ at the beginning of the stage and operates on w_i^k for the whole stage. The decision of w_i^k is made based on previous actions of other players. Now, based on above discussion and as in (Chen et al., 2007) we can define the formal definition of MAC game as a MAC game G is a 3-tuple ($\{1,\ldots,n\}$, S, u), where $\{1,\ldots,n\}$ is the set of players (nodes), $S = \times_{i \in p} W$ is the strategy space, $u = \{U_1,\ldots,U_n\}$ is the utility function space where $U_i = \sum_{k=0}^{+\infty} \delta^k U_i^s(W^k)$ is the utility function expressed as the sum of the utility in each stage k, $U_i^s(W^k) = u_i(W^k)T$ is the stage utility function, δ is the discounting factor which is generally close to 1.

Nash Equilibrium of The 'MAC' Game

The most possible outcome for any game (in terms of game theory) is "Nash Equilibrium". Nash equilibrium (NE) is an outcome of a game where no node (player) has any extra benefit for just changing its strategy one-sidedly (Felegyhazi et al., 2006; Mehta et al., 2009). Generally, the Nash equilibrium concept offers a predictable, stable outcome of a game where multiple agents (nodes) with conflicting interests compete through self-optimization and reach a point where no player wishes to deviate; however, such a point does not necessarily exist. So it is important to first investigate the existence of NE in G. We can check the existence of NE by verifying strategy space. If strategy space is convex and compact then G is a concave n-person game defined as in (Rosen, 1965) and thus admits at least one NE. At a NE, each node selects a best replay to the other nodes' selected configurations, a likely outcome if all the players are rational and their rationality is common knowledge (Rosen, 1965). If a configuration W is strictly dominant for each node then the configuration profile $\{w_1,\ldots,w_{max}\}$ is unique and strict NE. If it is non-strictly dominant, the NE need not be unique or strict. In that case 'MAC'

Game will have many NEs, and usually all NEs are not good. It is important to remove those NEs that are not fair and Pareto efficient[2] from a global performance point of view. In nutshell a desired solution of the 'MAC' game should exhibit the following three properties.

1. **Uniqueness:** The solution should be unique. This is to avoid uncertainties with respect to what solution each player should choose.
2. **Fairness:** The solution should result in fair distribution of the system throughput.
3. **Pareto efficient:** The solution should result in a Pareto efficient for allocating the available bandwidth or channel capacity.

Now the most important question is how to reach to a desired solution of the 'MAC' game. A unique efficient NE $(W_c^*, \cdots, W_c^*)$, which maximizes both local and global payoff, is required to solve the 'MAC' game.However, to achieve such a unique efficient NE involve many practical challenges as follow

1. **Estimating Nodes:** Every player should know the number of participating nodes in the network to adjust its CW to optimize its output. If the number of nodes n in the network is known to players, the task becomes trivial in that the CW value of the efficient NE can be computed given n. In some cases, the network participants do not know the number of nodes in the network, so they cannot directly calculate W_C^*. Thus an algorithm or mechanism is needed to search W_C^*. This algorithm or mechanism put an extra burden on the nodes in terms of computational complexity, energy consumption, etc. Moreover, as shown in chapter 6 the node estimation mechanism gives good approximation but not the accurate results.
2. **Tit for Tat (TFT) mechanism3:** In 'MAC' game, players are self-interested and rational, thus they adopt the strategy that maximizes their own payoff. This could lead nodes into well known problem of "Tragedy of commons" (Rosen, 1965) where players will settle for highly undesirable payoff. For example in 'MAC' game players with greater CW values will decreases them according to their measurement so as not to be disfavored. Thus within the finite number of stages all player will operate on the same CW value which yields the same utility and throughput, i.e. zero payoff. Hence, punish or incentive mechanisms are needed to encourage nodes to adopt socially optimal behaviors. To protect game from this situation a most common and popular strategy is to implement TFT. TFT is well known strategy in non-cooperative environments and is the root of an ever growing amount of other successful strategies. The core idea of TFT is to cooperate for the first stage and then follow. Once again, to implement TFT players should also implement local measurement method, thus adding an extra burden of computational complexity, energy consumption, etc. on players.

Motivation for Sift (Persistent/ Non-persistent)

As we mentioned earlier, estimating the game state accurately and timely are the key obstacles in formulating the MAC protocol games. Every node changes its strategy by adjusting the contention window (i.e. properly estimating the number of competing nodes) and tries to achieve its optimal solution. However, according to (Garey et al., 1979) we cannot expect to find an algorithm that can give the theoretical optimum solution and runs in polynomial time, as the abovementioned problems have been proven to be NP-hard. Furthermore, we need a simple, light (in terms of energy and implementation) yet an effective suboptimal solution for the same. These challenges are the key motivation factors for us to introduce a persistent/non-persistent sift MAC protocol for Ad-hoc/Sensor networks, which can give suboptimal solutions to aforementioned the MAC protocol games.

Now based on our previous work and (Bianchi, 2000) and using the parameters listed in Table 5. In BEB based algorithm, we show the relation between throughput and contention window in Figure 5 with different number of nodes.

As shown in Figure 6, the value of throughput firstly increases and then decreases for given number of nodes (n) as the value of CW increases from 1 to 1000. For the small number of nodes first throughput is increasing and then decreasing while for large number of nodes throughput is increasing slowly before its maximum point. The reason behind this is very obvious, at the lower number of nodes, less waiting time in backoff procedure during low contention window size. At the higher number of nodes, at first CW is too small to adjust with the number of nodes hence high collision, but later it is adjusted with the number of nodes so less collision and less waiting time in backoff procedure. However, all the nodes in the network achieved nearly same maximum throughput as shown in Table 3.

Similarly, in Figure 7 we show the relation between average access time and contention window with different number of nodes. From Figure 7 we can observe that the average access delay time for different number of nodes is different. However, for given number of nodes, after certain length of contention window, the access delay time does not jitter and it is almost constant for rest of the contention window size. It is worth to note that the size of the superframe was kept fixed in order to obtain the results presented in Figure 5 and 6.

So, from the aforementioned explanation and results presented in Figure 5, 6 and Table 3, we can observe that if we can adjust the size of the

Figure 5. Relation between throughput and contention window

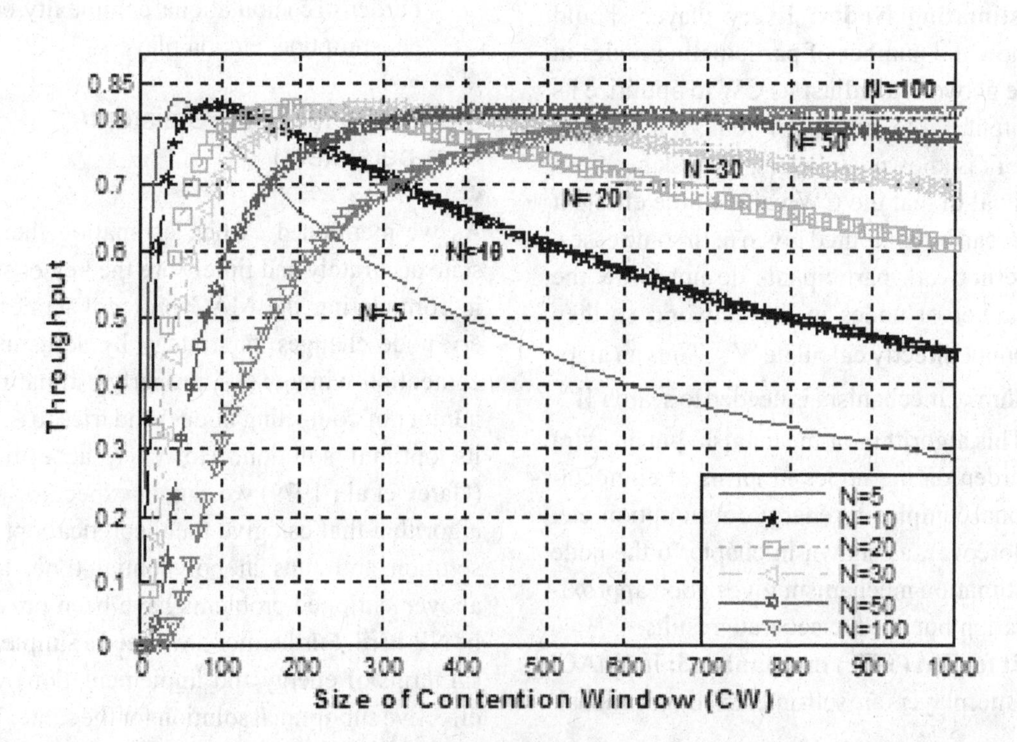

Figure 6. Relation between average access delay and contention window

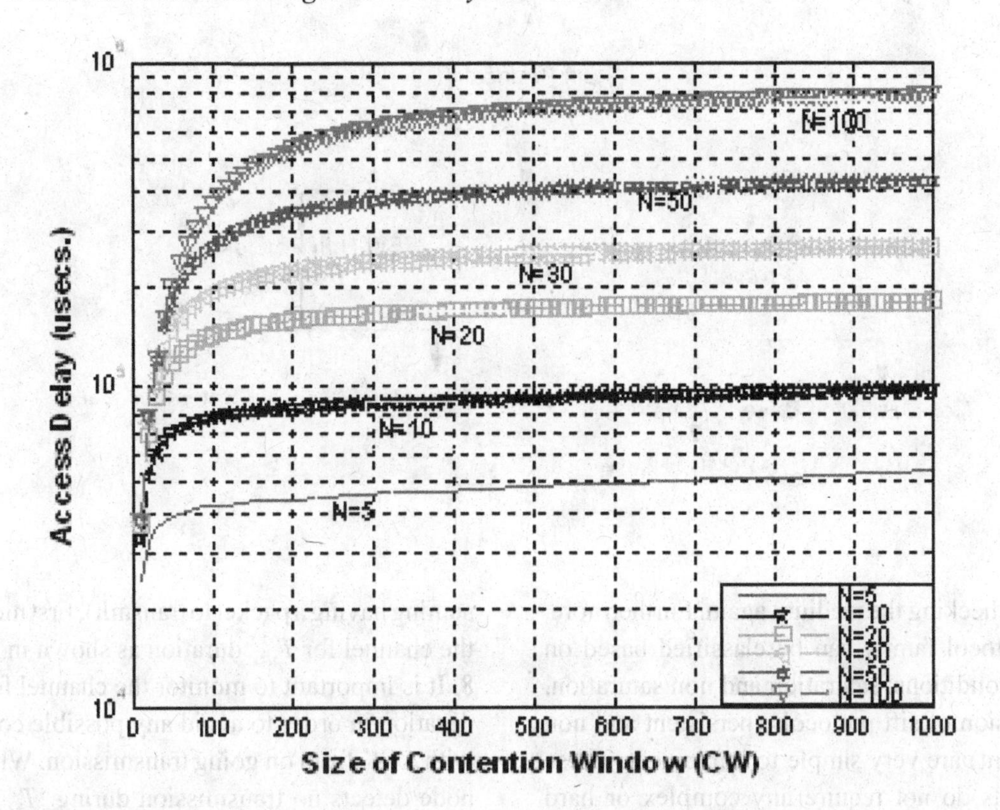

window or transmitting probability according to the number of competing nodes the maximum throughput can be achieved. This gives us an intuition to use persietnet/non-peristent scheme as sub-optimal and simple solution for MAC protocol games.

In this chapter, we use a fixed size contention window, but a non-uniform, geometrically-increasing probability distribution for picking a transmission slot (i.e transmitting probability) in the contention window interval instead of traditional4 backoff procedure.

Table 3. Number of nodes vs. maximum throughput

Number of Nodes (n)	Maximum Throughput	CW Size
5	0.825	45
10	0.820	97
20	0.819	195
30	0.818	300
50	0.817	450
100	0.816	800

Sift Protocol Mechanism

Before going further into persistent/non-persistent sift MAC, it is worth to briefly discuss about the sift scheme. In this section Figure 7 gives a glimpse of sift protocol family.

It is important to know the basic difference between persistent and non-persistent methods to understand the classification of sift protocol family. In persistent if the medium is idle, a node will transmit immediately but if the medium is busy a node will continue to listen until the medium becomes idle, and then transmit again. In non-persistent method if the medium is idle, a node transmit immediately but if medium is busy a node will wait for a random amount of time

Figure 7. Classification of sift protocol

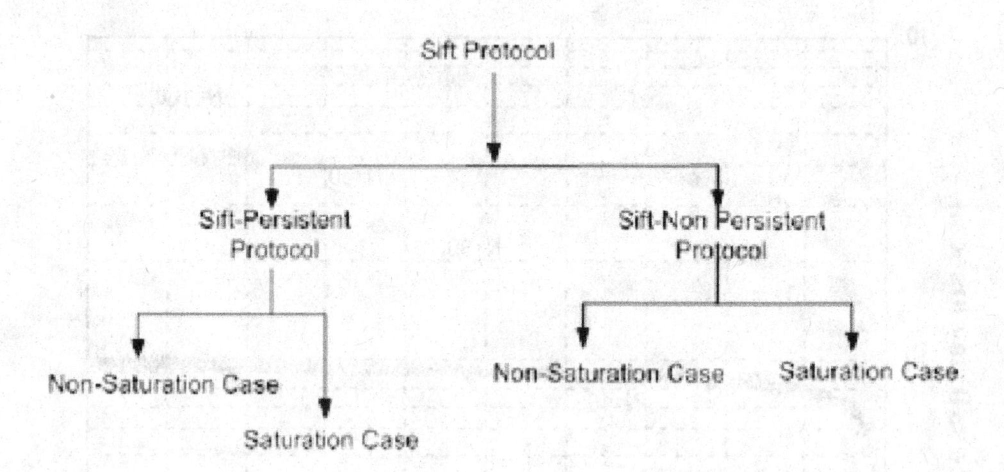

before checking the medium again. Furthermore, sift protocol family can be classified based on traffic conditions: saturation and non saturation. All version of sift protocols (persistent and non persistent) are very simple to implement. These protocols do not require any complex or hard method to estimate the number of nodes, nor do they need any extra overhead in the algorithm.

Sift-Persistent Mechanism

In persistent sift scheme, a node is randomly selecting a contention slot number i from the (1, CW) with probability p_i, as follow:

$$p_i = \frac{(1-\alpha)\alpha^{CW}}{1-\alpha^{CW}}\alpha^{-i} \ for \ i = 1,\ldots,CW \qquad (10)$$

The probability distribution p_i is increased slowly for initial slots and grows rapidly for the final slots. If the distribution parameter α is small, p_i is concentrated in the very last slot. Then, the number of contenders picking initial slots (i.e. early slots from CW) is drastically limited resulting in minimizing the probability of collision. The sift-persistent scheme operates in the following way. A node, running with sift-persistent scheme (assuming having a packet to transmit), first monitors the channel for T_{ACK} duration as shown in Figure 8. It is important to monitor the channel for T_{ACK} duration in order to avoid any possible collision with ACK from on going transmission. When the node detects no transmission during 'T_{ACK}' then node chooses a slot at random according to the non-uniform geometrically increasing probability distribution (p_i), as given in equation 10. Now in every slot node senses the channel, if the channel is still idle when the random delay expires, the node transmits. Otherwise, the node will waits till the end of ongoing transmission, and re-select a random slot according to p_i distribution. As shown in Figure 8 nodes pick a slot in the range of (1, CW) with the probability distribution p_i.

In the persistent sift algorithm a node has to waste unnecessary energy in channel listening or sensing during the backoff procedure, which might be expensive tradeoff between the latency performance and energy efficiency for some WSN applications. So, in this section we propose a non-persistent based sift algorithm for sensor networks. This proposed algorithm also uses the non uniform distribution (truncated increasing geometric distribution) to pick up the contention slot value from the range of (1, CW), where CW has the fixed value. The main difference between sift persistent and sift non persistent is whether node keeps to sense the channel or not.

Sift Non-Persistent Mechanism

A node running with sift-persistent algorithm always wakes up and power on all the time so that nodes collect more information on behavior of other's nodes but waste energy. However, in non-persistent sift, nodes wake up only when nodes want to transmit packets and power down in most of the time. So, nodes can save unnecessary energy waste. In non persistent-sift algorithm when a node has new packet or backlogged to send, the node immediately chooses a slot number i with probability p_i,as given in equation 10, and countdown until backoff counter reaches to slot number 1 regardless of channel activities. During this countdown procedure the node will turn down the power to save the energy. When backoff counter reaches to slot number 1, the node wakes up and performs CCA operation for once to check whether the channel is busy or not. Here, CCA operation is similar to CCA in IEEE 802.15.4. If the channel is busy, a node will restart from backoff procedure to select new contention slot. If the channel is idle, a node will transmit the packet. This transmission is either success or collision. A node will receive the ACK packet if transmission is successful and negative ACK packet in case of collision and repeats the aforementioned procedure. Figure 9(a) shows basic working flowchart of non persistent-sift mechanism, where Figure 9(b) shows the basic working of non persistent-sift mechanism in three casses: collisison, success, and busy, respectivly, during CCA operation. After ACK or negative ACK a node will stay in idle state if no packet is arrived. If any packet is arrived then node will begin from the backoff procedure. Here it is worth to note that ACK packet is sent immediately after turnaround time. Here turn around time duration is less than CCA time duration (T_{CCA}>$T_{trunaround}$), as shown in Figure 9(c). This assumption of turnaround time duration is very critical for the sucessful reception of a ACK packet. If a node performes CCA at the end of the packet then during the CCA node will listen ACK message and will not transmist a packet and so ACK packet is protected from any possible collision.

Figure 8. A time line of three nodes running sift –persistent Protocol (Jamieson et al., 2004)

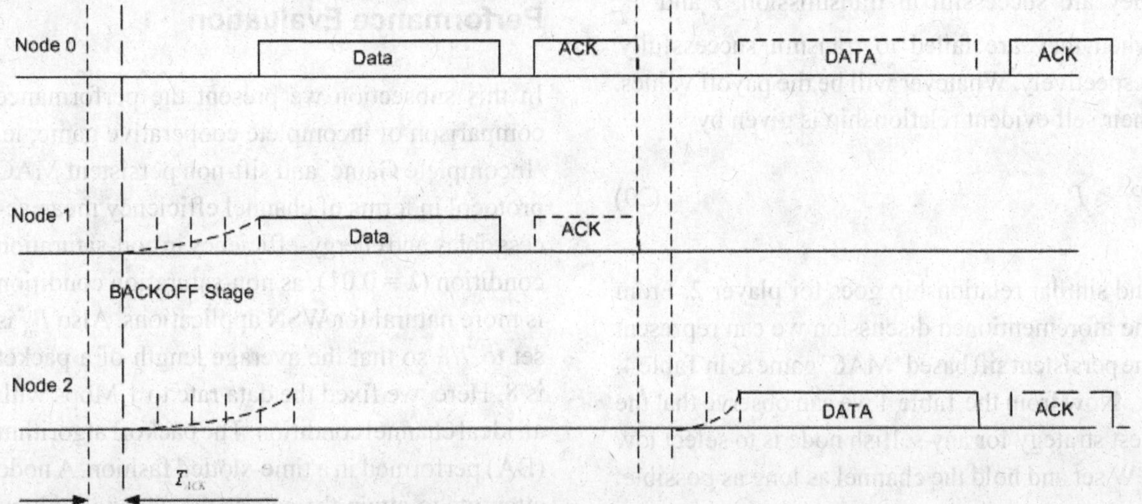

Linearly Increase and Linearly Decreased (LILD) Fairness Mechanism in lild, when any node experiences a collision or success, it increased contention slot by Δ. here Δ is a unit or a small step by which a node increase its present contention slot to select a new contention slot for next contention round (i.e. $cw_1' + \Delta \rightarrow cw_2'$). An overhearing node maintains its current contention slot for next contention round on collision, and decreases its contention slot by Δ(i.e $cw_2' - \Delta \rightarrow cw_1'$) for next contention round on success. thus, the operation of the lild method can be summarized as follows.

$$\begin{cases} CW \leftarrow \min(CW + \Delta, CW_{\max}) \text{ upon experiencing collisions or success} \\ CW \leftarrow \min(CW - \Delta, CW_{\max}) \text{ upon overhearing success} \\ CW \leftarrow \min(CW, CW_{\max}) \text{ upon overhearing collisions} \end{cases}$$

Now, considering persistent sift based 'MAC' game, every node has three set of CW values namely low, medium, and high. These sets also represent the strategy set of a node. Here, for simplicity and to show how selfish behavior can affect the MAC protocol, we assume that a collision will occur if any two or more nodes select the same set of CW values. We define P_i and $\bar{P}_i$ as the payoff for player 1 and 2, P_s and $\bar{P}_s$ when they are successful in transmission, P_f and $\bar{P}_f$ when they are failed to transmit successfully, respectively. Whatever will be the payoff values, their self evident relationship is given by

$$P_f < P_S \tag{20}$$

and similar relationship goes for player 2. From the aforementioned discussion we can represent the persistent sift based 'MAC' game as in Table 4.

Now from the Table 4 we can observe that the best strategy for any selfish node is to select low CW set and hold the channel as long as possible. Thus within finite number of stages all players will operate on the same CW set which results into undesired payoff of zero throughput, i.e. we will face a situation of "Tragedy of commons". So to counter-play with "Tragedy of commons", every node has to play its strategies with some probabilities as here the optimum equilibrium is in mixed strategy form. A mixed strategy is a probability distribution over pure strategies. Each player i chooses a probability distribution over his set of pure strategies S_i (independently of probability distributions of his opponents). In mixed strategy equilibrium, it is not possible to reach an optimum solution with one strategy so players have to mix two or more strategies probabilistically. Here, LILD mechanism gives a natural option for any player to play mixed strategies. It is worth to note that all players will strictly follow the LILD mechanism rules as they have long-sighted vision for the game. In next section we will show via numerical results that how IB with LILD scheme can obtain a sub-optimal solution for the 'MAC' game.

It is very important to note that we can not integrate LILD mechanism with sift-non persistent MAC, as a node running with sift-non persistent does not listen to the channel all the time. Because of this reason a node can not follow the LILD rules or any other mechanism to play mixed strategy.

Performance Evaluation

In this subsection we present the performance comparison of incomplete cooperative game, ie. 'Incomplete Game' and sift-non persistent MAC protocol in terms of channel efficiency mean access delay and energy-efficiency in non-saturation condition ($\lambda = 0.01$), as non-saturation condition is more natural for WSN applications. Also P_{tx} is set to 7/8 so that the average length of a packet is 8. Here, we fixed the data rate to 1 Mbps with an ideal channel condition. The backoff algorithm (BA) performed in a time-slotted fashion. A node attempts to attain the access the channel only at

Figure 9. Working of non persistent-sift (a) Flow diagram of non persistent-sift; (b) CCA operation in three cases: collision, success, and busy; (c) Turnaround time and CCA

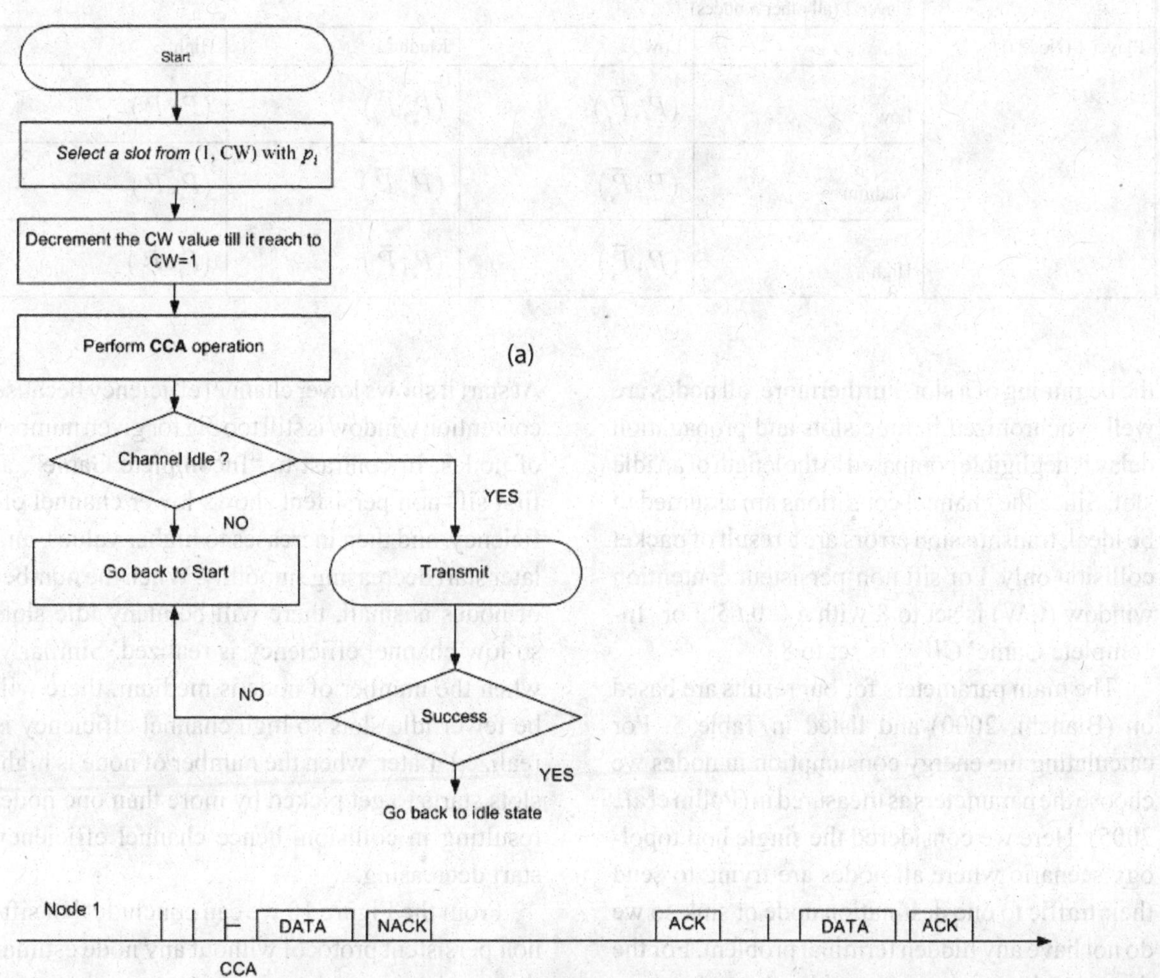

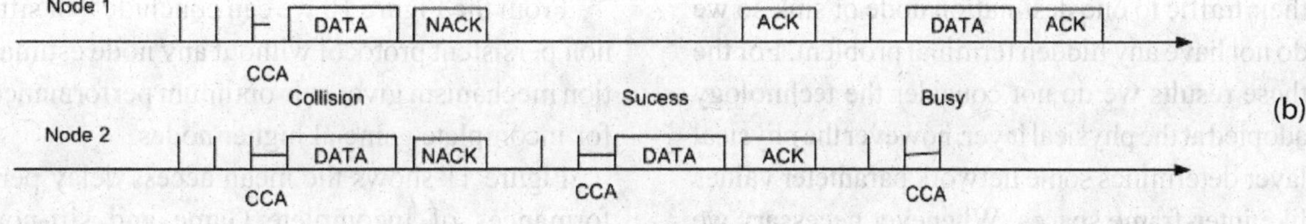

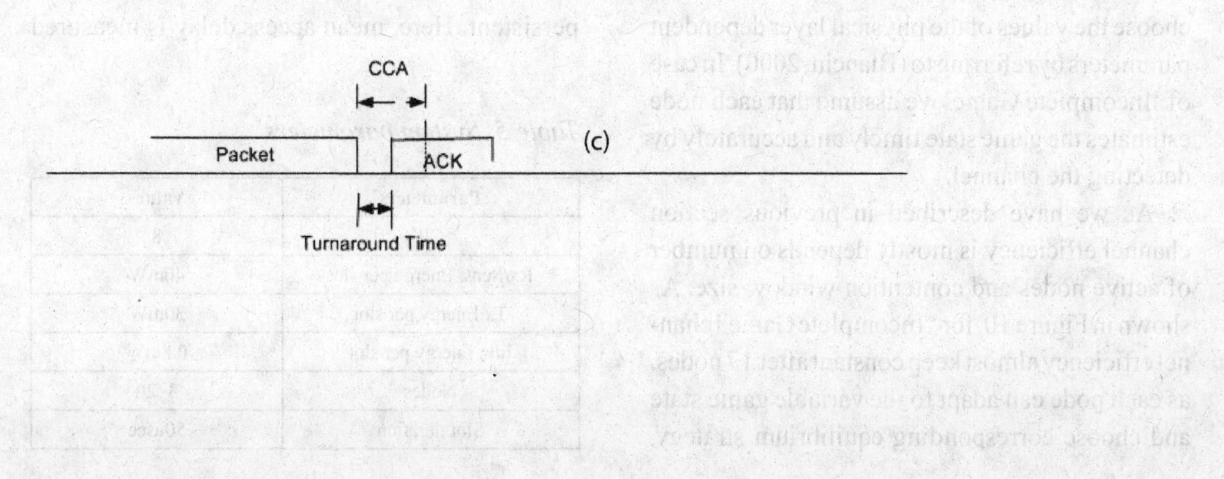

Table 4. Strategy table

	Player 2 (all other n nodes)			
Player 1 (Node i)		Low	Medium	High
	Low	$(P_f, \overline{P}_f)$	$(P_s, \overline{P}_s)$	$(P_s, \overline{P}_s)$
	Medium	$(P_s, \overline{P}_s)$	$(P_f, \overline{P}_f)$	$(P_s, \overline{P}_s)$
	High	$(P_s, \overline{P}_s)$	$(P_s, \overline{P}_s)$	$(P_f, \overline{P}_f)$

the beginning of a slot. Furthermore, all nodes are well synchronized in time slots and propagation delay is negligible compared to the length of an idle slot. Since the channel conditions are assumed to be ideal, transmission errors are a result of packet collision only. For sift non-persistent contention window (CW) is set to 8 with $\alpha = 0.65$. For 'Incomplete Game' CW_{min} is set to 8.

The main parameters for our results are based on (Bianchi, 2000) and listed in Table 5. For calculating the energy consumption in nodes we choose the parameters as measured in (Pollin et al., 2005). Here we considered the single hop topology scenario where all nodes are trying to send their traffic to one destination node or sink so we do not have any hidden terminal problem. For the these results we do not consider the technology adopted at the physical layer, however the physical layer determines some network parameter values like inter-frame spaces. Whenever necessary we choose the values of the physical layer dependent parameters by referring to (Bianchi, 2000). In case of 'Incomplete Game' we assume that each node estimates the game state timely and accurately by detecting the channel.

As we have described in previous section channel efficiency is mostly depends on number of active nodes and contention window size. As shown in Figure 10, for "Incomplete Game" channel efficiency almost keep constant after 17 nodes, as each node can adapt to the variable game state and choose corresponding equilibrium strategy. At start it shows lower channel efficiency because contention window is still too big for given number of nodes. In contrast to "Incomplete Game", at first sift-non persistent shows lower channel efficiency, and then increases to higher values, and later start decreasing smoothly. When the number of nodes is small, there will be many idle slots so low channel efficiency is realized. Similarly, when the number of node is medium, there will be fewer idle slots so high channel efficiency is realized. Later, when the number of node is high, slots starts to get picked by more than one node, resulting in collision, hence channel efficiency start decreasing.

From the Figure 10 we can conclude that sift-non persistent protocol without any node estimation mechanism gives sub-optimum performance for incomplete game at higher nodes.

Figure 11 shows the mean access delay performances of Incomplete Game and sift-non persistent. Here, mean access delay is measured

Table 5. System parameters

Parameters	Values
CW	8
Rx/Sens. Energy per slot	40mW
Tx Energy per slot	30mW
Idle Energy per slot	0.8mW
Nodes	3~20
Slot duration	50usec

in number of slots. In "Incomplete Game", mean access delay performance is far better than sift-non persistent, as it can easily adapt the variable game state and choose the corresponding equilibrium strategy by adjusting contention window according to number of nodes. In sift-non persistent, during the backoff, node does not listen to channel and have to wait for selected number of backoff slots to decrement till 1 before doing CCA operation. So sift-non persistent spends more average number of slots compared to 'Incomplete Game'.

Figure 12 illustrates the impact of CW on average energy consumption of 'incomplete game' and sift non persistent schemes.

From Figure 12 we can see that as number of nodes increases 'Incomplete Game' scheme consume more power as it spend good amount of energy for listing medium during the backoff procedure. In sift-non persistent scheme node does not listen to the channel before doing CCA so saves energy during the backoff procedure, hence better performance compared to 'Incomplete Game'. Accepting sift-persistent scheme can increase the overall performance of energy efficient MAC protocol to a large extends and we can also get the sub optimal solution for an incomplete cooperative game.

Now, we present the performance comparison of 'MAC' game MAC protocol, BEB based MAC protocol (BEB-MAC), and sift-persistent MAC in terms of channel efficiency. Here, BEB-MAC is implemented with nodes estimating mechanism and TFT strategy, while sift-persistent is able to play mixed strategy to avoid "Tragedy of commons". It is worth to note that selfishness problem is prominent only in saturation traffic condition so we consider saturation traffic condition for evaluation. As we mentioned before sift-non persistent MAC cannot listen to the channel during backoff procedure so it can not play mixed strategy, and will enter into "Tragedy of commons" problem. So there is no point to compare sift-non persistent performance with other protocols. Here, we fixed the data rate to 1 Mbps with an ideal channel condition. For the BEB-MAC protocol minimum contention window is set to 8 (also for sift-persistent MAC). The backoff algorithm (BA) performed in a time-slotted fashion. A node attempts to attain the access the channel only at the beginning of a slot. Furthermore, all nodes are well synchronized in time slots and propagation delay is negligible compared to the length of an idle slot. Incase of 'MAC' game (MAC-Game) MAC protocol, we assume that each node estimates the game state timely and accurately by detecting the channel.

Now, we study the comparison of all three MAC protocols for a single hop. Figure 13 shows the channel efficiency analysis of BEB-MAC, sift-persistent, and MAC-Game protocols. Here, MAC-game is based on the efficient NE when the CW value of all players is converged, i.e, ideal condition for 'MAC game'. We calculate the efficient NE from numerical methods. With help of node estimation and TFT mechanisms, BEB-MAC can find its own efficient NE and converge to best CW value possible for corresponding number of nodes. Thus, BEB-MAC achieves the global social optimality. However, there is big difference between BEB-MAC and MAC-Game protocol mainly due to some practical problems associated with node estimating and TFT mechanisms. In contrast to BEB-MAC, sift-persistent maintains high channel efficiency due to its unique quality in selecting contention value from non-uniform distribution. In sift-persistent, most of the nodes choose higher contention slots while very few nodes selects lower contention slots, hence, less or no collision, also low waiting time in backoff procedure. So high channel efficiency is realized for low number of nodes. Later, when the number of node is high, slots starts to get picked by more than one node, resulting in collision, hence channel efficiency start decreasing.

The above numerical results show that sift-persistent protocol can provide a sub-optimal solution to 'MAC' game without any complex or heavy (in terms energy consumption) mechanism such as node estimation scheme.

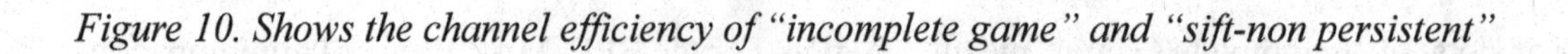

Figure 10. Shows the channel efficiency of "incomplete game" and "sift-non persistent"

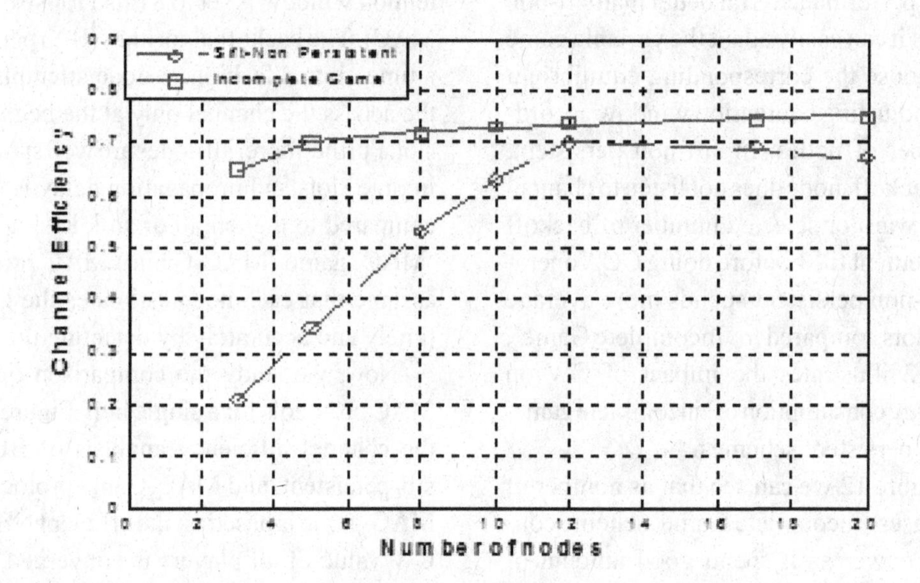

Figure 11. Mean access delay vs. number of nodes

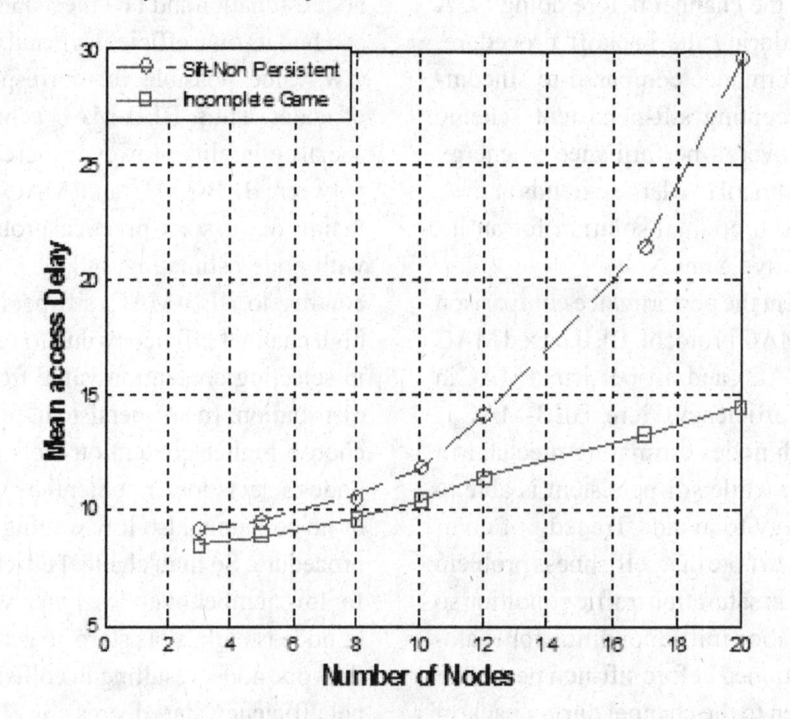

Figure 12. Average energy consumption vs. number of nodes

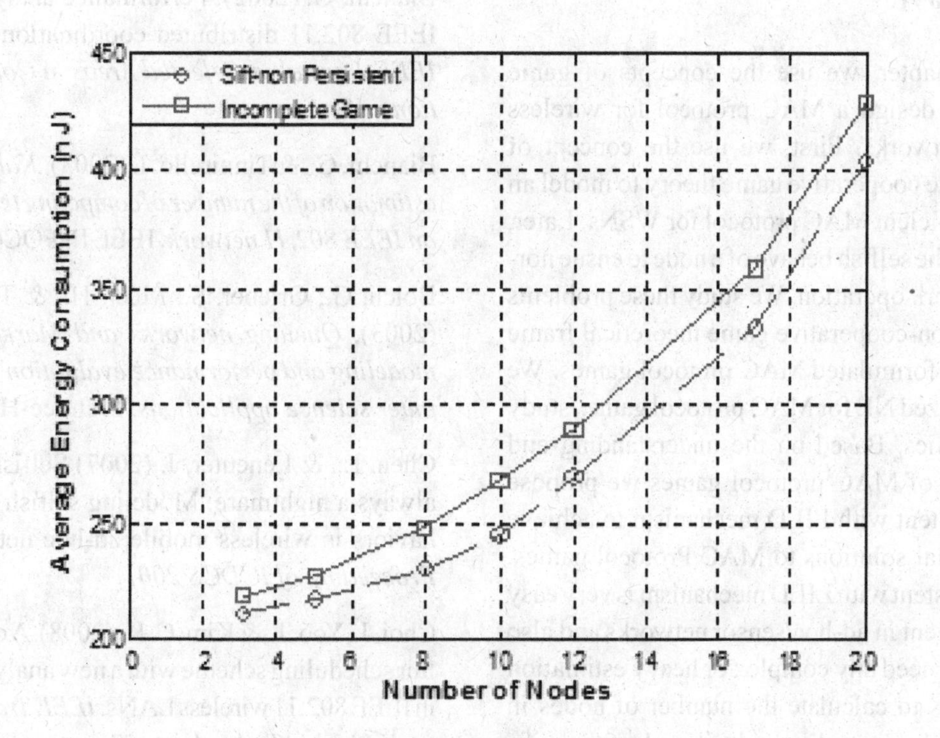

Figure13. Channel efficiency vs. number of nodes

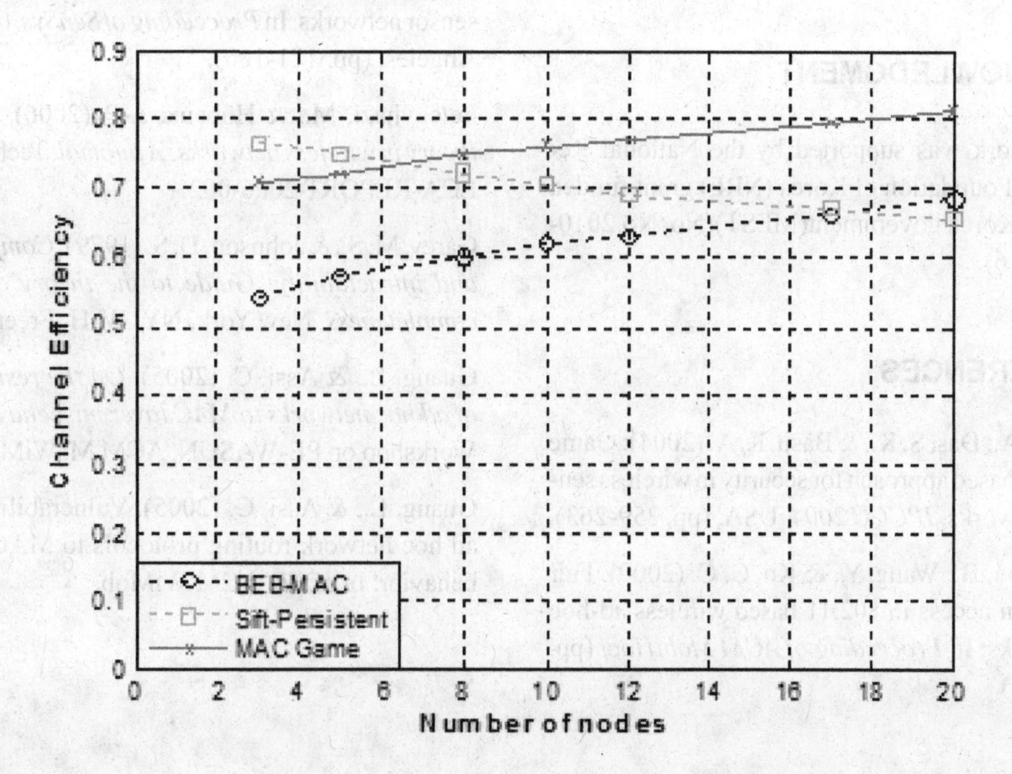

SUMMARY

In this chapter, we use the concepts of game theory to design a MAC protocol for wireless sensor networks. First, we use the concept of incomplete cooperative game theory to model an energy efficient MAC protocol for WSNs. Later, we study the selfish behave of a node to ensue normal network operation. We study these problems under a non-cooperative game theoretical frame work and formulated MAC protocol games. We characterized NE for MAC protocol games study its dynamics. Based on the understanding and dynamics of MAC protocol games we propose sift-persistent with LILD mechanism to achieve sub optimal solutions to MAC Protocol games. Sift-persistent with LILD mechanism is very easy to implement in ad-hoc/sensor networks and also we do not need any complex or heavy estimation algorithms to calculate the number of nodes in the network. From the results it is clear that sift-persistent with LILD mechanism can provide sub-optimal solutions to MAC Protocol games.

ACKNOWLEDGMENT

This work was supported by the National Research Foundation of Korea (NRF) grant funded by the Korea government(MEST)(No. No.2010-0018116).

REFERENCES

Agah, A., Das, S. K., & Basu, K. A. (2004). Game theory based approach for security in wireless sensor networks. *IPCCC 2004,* USA, (pp. 259-263).

Bensaou, B., Wang, Y., & Ko, C. C. (2000). Fair medium access in 802.11 based wireless ad-hoc networks. In *Proceeding of ACM MobiHoc,* (pp. 99-106).

Bianchi, G. (2002). Performance analysis of the IEEE 802.11 distributed coordination function. *IEEE Journal on Selected Areas in Communications*, 18.

Bianchi, G., & Tinnirello, I. (2003). *Kalman filter estimation of the number of competing terminals in an IEEE 802.11 network*. IEEE INFOCOM 2003.

Bolch, G., Griener, S., Meer, H., & Trivedi, K. (2003). *Queuing networks and Markov chains modeling and performance evaluation with computer science applications*. Prentice-Hall.

Chen, L., & Leneuter, J. (2007) Selfishness, not always a nightmare: Modeling selfish MAC behaviors in wireless mobile ad-hoc networks. In *Proceeding of ICDCS 2007*.

Choi, J., Yoo, J., & Kim, C. K. (2008). A distributed fair scheduling scheme with a new analysis model in IEEE 802.11 wireless LANs. *IEEE Transactions on Vehicular Technology*, *57*(5).

Dam, T. V., & Langendone, K. (2003). An adaptive energy-efficient MAC protocol for wireless sensor networks. In *Proceeding of SenSys '03*, Los Angeles, (pp. 171-180).

Felegyhazi, M., & Hubaux, J.-P. (2006). *Game theory in wireless networks: A tutorial*. Tech. Rep. LCA-REPORT-2006-002.

Garey, M. S., & Johnson, D. S. (1979). *Computers and intractability: Guide to the theory of NP-completeness*. New York, NY: W. H. Freeman.

Guang, L., & Assi, C. (2005). *On the resiliency of ad hoc networks to MAC layer misbehavior.* In Workshop on PE-WASUN, ACM MsWiM 2005.

Guang, L., & Assi, C. (2005). Vulnerabilities of ad hoc network routing protocols to MAC misbehavior. In *IEEE*. ACM WiMob.

Gupta, V., Krishnamurthy, S., & Faloutsous, M. (2002). Denial of service attacks at the MAC layer in wireless ad hoc networks. In *Proceedings of MILCOM*, 2002.

Heidemann, W., Ye, J., & Estrin, D. (2002). An energy- efficient MAC protocol for wireless sensor networks. In *Proceeding of INFOCOM 2002*, New York, (pp. 1567-1576).

Jamieson, K., Krishnan, B., & Tay, Y. C. (2006). Sift: A MAC protocol for event driven wireless sensor networks. In *Proceeding of EWSN 2006*, (pp. 260-275).

Kannan, R., Sarangi, S., & Lyengar, S. S. (2004). Sensor-centric energy constrained reliable query routing for wireless sensor networks. *Journal of Parallel and Distributed Computing*, *64*(7), 839–852. doi:10.1016/j.jpdc.2004.03.010

Kyasanur, P., & Vaidya, N. (2004). *Selfish MAC layer misbehavior in wireless networks*. IEEE Transactions on Mobile Computing.

Mehta, S., & Kwak, K. S. (2009). Game theory and Information Technology. In *Proceeding of ISFT 2009* Korea.

Mehta, S., & Kwak, K. S. (2009). H-MAC: A hybrid MAC protocol for wireless sensor networks. *International Journal of Computer Networks & Communications*, *2*(2).

Pollin, S., Ergen, M., Bougard, S. C., Perre, D. E., Cathoor, L. V., Morman, F., ... Variya, P. (2005). *Performance analysis of slotted IEEE 802.15.4 medium access layer*. Draft-jwl-tcp-fast-01.txt

Rapport, T. (1996). *Wireless communications: Principles & practice*. Prentice-Hall.

Raya, M., Hubaux, J. P., & Aad, I. (2004). DOMINO: A system to detect greedy behavior in IEEE 802.11 hotspots. In *Proceedings of ACM MobiSys*, June 2004.

Rosen, J. B. (1965). Existence and uniqueness of equilibrium points for concave n-person games. *Econometrica: Journal of the Econometric Society*, *33*, 520–534. doi:10.2307/1911749

Sengupta, S., & Chatterjee, M. (2004). *Distributed power control in sensor networks: A game theoretic approach* (pp. 508–519). India: IWDC.

Vercauteren, T., Toledo, A. L., & Wang, X. (2007). Batch and sequential Bayesian estimators of the number of active terminals in an IEEE 802.11 network. *IEEE Transactions on Signal Processing*, *55*(2), 437–450. doi:10.1109/TSP.2006.885723

Xiao, Y., Shen, X., & Jiang, H. (2006). Optimal ACK mechanism of the IEEE 802.15.3 MAC for ultra-wideband systems. *IEEE Journal on Selected Areas in Communications*, *24*(4).

Zhang, X., Cai, Y., & Zhang, H. (2006). *A game-theoretic dynamic power management policy on wireless sensor network*. China: ICCT.

Zhao, L., Zhang, J., Yang, K., & Zhang, H. (2008). An energy efficient MAC protocol for WSNs: Game theoretic constraint optimization. In *Proceeding of ICCS 2008* (pp. 114-118).

Zhao, L., Zhang, J., & Zhang, H. (2008). Using incomplete cooperative game theory in wireless sensor networks. In *Proceeding of WCNC 2008* (pp. 1483-1488).

Zhou, Y., Wu, D., & Nettles, S. (2004). *Analyzing and preventing MAC-layer denial of service attacks for stock 802.11 systems*. In Workshop on BWSA, BROADNETS '04.

ENDNOTES

1. In rest of the chapter, the authors keep using terms 'node' and 'player' interchangeably.
2. A situation is said to be pareto efficient if there is no way to rearrange strategies to

make at least one node better off without making any node worse off [88].

[3] In this chapter, the authors use words 'algorithm', 'scheme ', and 'method' 'interchangeably'.

[4] Here, traditional backoff procedure means CSMA/CA scheme with binary exponential backoff (BEB), unless and otherwise specified.

Chapter 23
Geometric Structures for Routing Decision in Wireless Sensor Networks

Alok Kumar
Indian Institute of Information Technology, India

Shirshu Varma
Indian Institute of Information Technology, India

ABSTRACT

This chapter surveys routing algorithms in Euclidean, virtual, and hyperbolic space for wireless sensor networks that use geometric structures for route decisions. Wireless sensor networks have a unique geographic nature as the sensor nodes are embedded and designed for employing in the geographic space. Thus, the various geometric abstractions of the network can be used for routing algorithm design, which can provide scalability and efficiency. This chapter starts with the importance and impulse of the geographical routing in wireless sensor networks that exploits location information of the nodes to determine the alternatives of the next hop node on the desired routing path. The scalability of geographical routing encourages more effort on the design of virtual coordinates system, with which geographical routing algorithms are built up and applied to route data packets in the network. The geometry of large sensor network motivates to calculate geometric abstractions in hyperbolic space. Thus the challenge is to embed the network virtually or hyperbolically, which affects the performance and efficiency in the geographical message delivery.

INTRODUCTION

Due to potential applications in the areas such as military, health, environmental, home and other, wireless sensor networks (WSNs) have came forth as a premier research area in current decade. Sensor networks consist of significantly large number of sensor nodes scatter over a geographical terrain (Akyildiz, 2002). These nodes are capable to perform sensing, processing and are additionally able of self-organize to interacting by means of a wireless network. The network will achieve a

DOI: 10.4018/978-1-4666-0101-7.ch023

larger sensing task in urban environments as well as inhospitable terrain, with coordination among these sensor nodes. The absolute numbers of these sensors and the varying dynamics in these environmental scenarios present unique challenges and limitations in the design of WSNs. Various research projects have been conduct to study for solve various problems in this new field. This book chapter assumes a WSN comprising of a set V of N nodes scattered over a routing space. Each node is content an omnidirectional antenna for physical connectivity with each other. This is significant as a single transmission of a node, taken to be a disk centered at the node, is hearable by many nodes within its vicinity. The radius of this omnidirectional disk is known as the transmission range of this sensor node. In other words, any node V can get the signal from sensor node u, if receiving node v lies in the transmission range of the sender node u, else, (u,v) nodes interact through multi-hop links while using nodes in between as relays of packet. Each node in the network also works as a router, which can forward data packets for other nodes. Through a proper scaling, all nodes are assumed to have the maximum transmission range, which is equal to unit. With information about the position, computational geometry techniques can be employed to solve some intriguing questions on sensor networks. Most geometric algorithms are formed for studying the structural properties, inclusion, searching or exclusion relations of a set of points or planes, or both. As an example, the structural properties include convex hull, intersections, triangulation, hyperplane arrangement, Voronoi diagram, and so on. In this chapter, the focus is on the use of some geometric structural properties for routing process in WSN.

This chapter surveys geographic routing algorithms which use various geometric structural properties for routing process in routing spaces for wireless sensor networks.

Geographic routing algorithms which are design for the Euclidian space are most common for researchers. Here, some consider the unit disk graph (UDG) as a communication model for the computation of as real network but UDG does not behave like real network. Many proposals also consider other models such as quasi disk graph and cross link detection protocol for model the network. Due to location unavailability and the presence of the localization error in the mechanism of node localization, geographic routing performs inefficiently. To overcome these problems these algorithm can be design on virtual coordinates, but these virtual coordinate assignment approaches affects the performance and efficiency in the geographical message delivery. Geographic routing using the physical coordinates of the sensor nodes has been considered to study due to its scalability and simplicity, and regained popularity in the research community in recent times with the increasing GPS-capable communication devices. Geographic routing based on node locations and distances in Euclidean space, has been proven to have a higher success rate, but fails when a packet reaches at node which is nearer to the location of the destination node than all of its straight neighbors even if a path exists to the destination, but it can provide higher success in hyperbolic space.

This chapter focuses on the design rational and the main theme in each algorithm. Please refer to the original literatures for implementation and experimental details for the validation of each. The typical situation is large homogeneous sensors network with the possibility of communication each other with nearby nodes and communicate with other using multi-hop routing. Since many of the routing protocols use network model and basic geometry for computation, we first survey background knowledge on communication network model and basic heuristics of geographic routing.

BACKGROUND

This section reviews definitions and concepts necessary for later discussion. It specifies routing space and the basic heuristics of the geographic routing algorithms.

Unit and Quasi Disk Graph

Consider a WSN consist of a set V of sensor nodes randomly distributed in a 2-dimensional plane. These nodes consider a unit disk graph (Clark et al., 1990) UDG(V) where an edge exists between two nodes if and only if the Euclidean distance between them is at maximum one. The quasi disk graph (Kuhn et al., 2004) describes as let $V \in R^2$ a set of points in the 2-dimensional plane and let $1 \in [0,1]$ be a parameter. The symmetric Euclidean graph G = (V,E), such that for any pair $(u,v) \in V$,

$$dist(u,v) \leq l \Rightarrow u, v \in E$$

$$dist(u,v) > l \Rightarrow u, v \in E$$

is called a l-quasi unit disk graph (d-QUDG). Quasi unit disk graph doesn't elaborate if there is an edge between two nodes u and v having distance $1 < dist(u,v) < 1$. A unit disk graph is a special case of a l-quasi unit disk graph for l=1 (see Figure 1)

Hyperbolic Space

The basic property of hyperbolic geometry (Anderson, 2005) is the exponential widening of space. Taking an example, in the hyperbolic plane, which is the two-dimensional hyperbolic space of negative curvature -1, the length of a circle and the area of a disc of radius R are $2\pi\sinh R$ and $2\pi(\cosh R-1)$, both growing as ~eR with R. The hyperbolic plane thus is metrically equivalent to an e-ary tree, i.e., a tree with the average branching factor equal to e. Indeed, in a b-ary tree, the analogies of the circle length or disc area are the number of nodes at distance exactly R or not more than R hops from the root. These numbers are $(b+1)b^{R-1}$ and $((b+1)b^R-2)/(b-1)$, both growing as ~bR. Informally, hyperbolic spaces can therefore be thought of as continuous versions of trees. The Hyperbolic model realizes hyperbolic space as a hyperboloid in $R^{n+1}=(x_0,\ldots,x_n)|x_i \in R$, i=0,1,…,n. The hyperboloid is the locus H^n of points whose coordinates satisfy $x_0^2 - x_1^2 - \ldots - x_n^2 = 1, x_0 > 0$. In this model a "line" (or geodesic) is the curve cut out by intersecting H^n with a plane through the origin in R^{n+1}. The hyperboloid model is closely related to the geometry of Minkowski space. The quadratic form $Q(x) = x_0^2 - x_1^2 - x_2^2 - \cdots - x_n^2$ which defines the hyperboloid polarizes to give the bilinear form B defined by $B(x,y) = (Q(x+y)-Q(x)-Q(y))/2 = x_0y_0 - x_1y_1 - \ldots - x_ny_n$. The space R^{n+1} equipped with the bilinear form B is an (n + 1)-dimensional Minkowski space Rn,1. From this perspective, one can link a notion of distance to the hyperboloid model, by setting the distance between two points x and y on H to be (x,y)= arccosh B(x,y). This function satisfies the axioms of a metric space. Moreover, it is saved by the action of the Lorentz group on $R^{n,1}$. Hence the Lorentz group acts as a transformation group of isometries on H_n.

Basic Geographic Routing Schemes

This subsection describes basic geographic routing protocols proposed in the networking and computational geometry literature.

Compass routing (Kranakis et al., 1999): Let d be the destination node. Current node u finds the next relay node v such that the angle $\angle vud$ is the smallest among all neighbors of u in a given topology.

Random compass routing (Kranakis et al., 1999): Let u be the current node and d be the destination node. Let v_1 be the node above line ud such that $\angle v_1ud$ is the smallest among all such neighbors of u. Similarly, define v_2 to be nodes

Figure 1. A example of unit disk graph

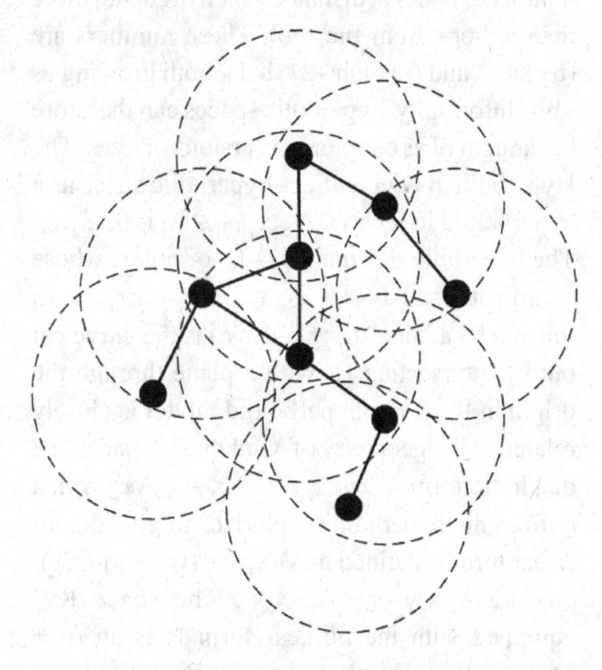

below line ut that minimizes the angle $\angle v_2ud$. Then node u randomly chooses v_1 or v_2 to forward the packet.

Greedy routing (Bose et al., 2001)**:** Let d be the destination node. Current node u finds the next relay node v such that the distance ||vd|| is least among all neighbors of u in a given topology. Greedy routing is unable to provide guaranteed delivery in every case, it fails when faces local minima condition (see figure).

Most forwarding routing (MFR) (Stojmenovic et al., 2001)**:** Current node u finds the next relay node v so that ||v'd|| is the least among all neighbors of u in a given topology, where v' is the projection of v on segment ud.

Nearest neighbor routing (NN): Let d be the destination node. Given a parameter angle a, node u finds the nearest node v as forwarding node among all neighbors of u in a known topology such that $\angle vud \leq a$.

Farthest neighbor routing (FN): Let d be the destination node. Given a parameter angle a, node u finds the farthest node v as forwarding node among all neighbors of u in a given topology such that $\angle vud \leq a$.

Greedy compass (Morin, 2001)**:** Let d be the destination node. Current node u first finds the neighbors v_1 and v_2 such that v_1 forms the smallest counterclockwise angle $\angle duv_1$ and v_2 forms the smallest clockwise angle $\angle duv_2$ among all neighbors of u with the segment ud. The packet is forwarded to the node of v_1,v_2 with minimum distance to d.

Face routing (Kranakis et al., 1999; Bose et al., 2001)**:** It follows the concept of faces, contiguous regions distinguished by the edges of a planar graph, which is a graph contains no two intersecting edges. It explores the face boundaries by employing the local right hand rule in analogy to following the right hand wall in a maze. On its way around a face, the algorithm maintains track of the points where it crosses the line ud connecting the source u and the destination d. Having totally surrounded a face, the algorithm returns to the one of these intersections lying closest to the destination (see Figure 2).

GEOGRAPHIC ROUTING IN WIRELESS SENSOR NETWORKS

A routing protocol is called geographic, if the routing decision is based on:

- The location information about source and destination nodes which contained in the header of the packet.
- The information gathered by the node from its one hop neighboring nodes.

Therefore, in the geographic routing, the source node has pre-existing knowledge or computes the current location of the destination node. Most applications of sensor networks usually have fixed destination or data collection node; hence, location service is not necessary. But in most of the ap-

Figure 2. Different variations of greedy routing

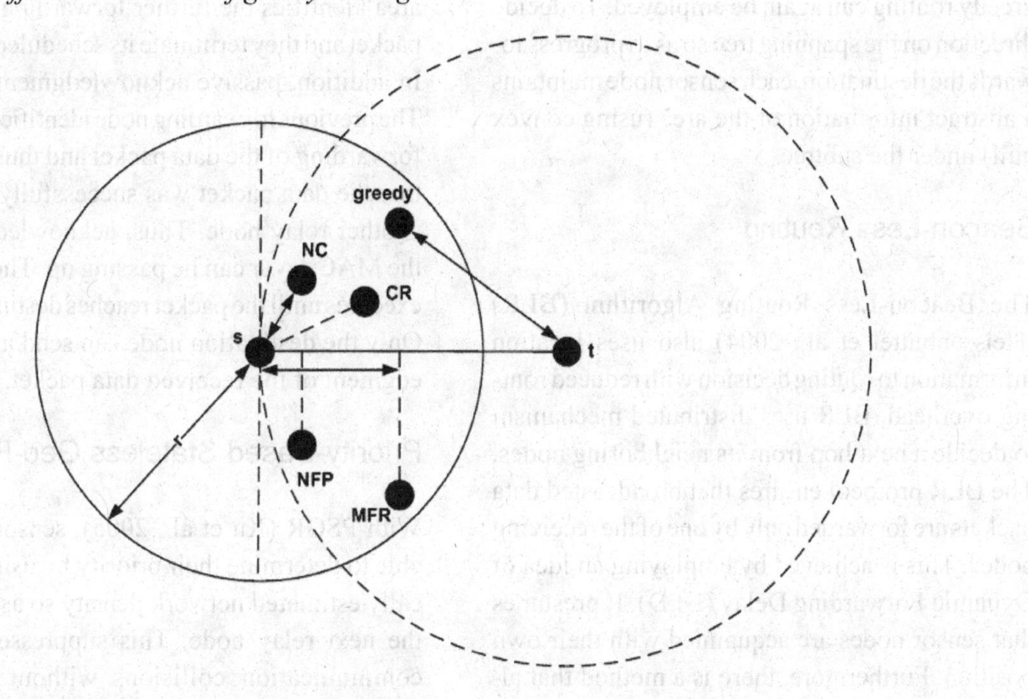

plications, the location service is needed. It would be hard to apply the centralized approach of location services in WSNs. The following subsection will provide the main idea of various geographic routing algorithms, which are categorized based on the routing space.

Geographic Routing in Euclidian Space

Greedy Stateless Perimeter Routing (GPSR)

Greedy Perimeter Stateless Routing (GPSR) (Karp et al., 2000)) routing algorithm comprises of two different forwarding schemes: greedy forwarding and routing around perimeter. In this approach, greedy forwarding method is applied whenever possible. If greedy forwarding fails, only then the forwarding approach is changed to the right hand rule (perimeter mode), in which the packet is moves along the faces of the planar graph. For considering the faces for routing, procedure require the 2-dimentional planer graph instead of 3-dimentional non-planer graph. The GPSR results in high communication and processing overhead due to huge number of beacon packets for maintaining routing table. GPSR performs well In case of dense networks and worsens with a low density of networks.

Greedy Distributed Spanning Tree Routing

Greedy Distributed Spanning Tree Routing (GDSTR) (Leong et al., 2006) uses a new data structure called hull trees to emulate planarization without really planarize the graph. Whenever feasible, the algorithm tries to route the packets in greedy way, but when void problem arises procedure route packets around the spanning tree. GDSTR handles the void situation by shifting to routing mode from greedy to spanning tree based routing until the packet has reached a node where greedy routing can again be employed. To decide direction on the spanning tree so as to progress towards the destination, each sensor node maintains a abstract information of the area (using convex hull) under the subtree.

Beacon-Less Routing

The Beacon-Less Routing Algorithm (BLR) (Heissenbuttel et al., 2004) also uses location information to routing decision with reduced routing overhead. BLR uses distributed mechanism to decide a next hop from its neighboring nodes. The BLR protocol ensures that broadcasted data packets are forwarded only by one of the receiving nodes. This is achieved by employing an idea of Dynamic Forwarding Delay (DFD). It presumes that sensor nodes are acquainted with their own position. Furthermore, there is a method that allows the source node to identify the approximate destination node's position.

To forward a packet, the source node firstly finds out the location of the destination and records its own present location along with the address (coordinates) of the destination in the header of packet. At every relay node, the current location in the header is changed by the location of the relay node before forwarding the packet. A sensor node does not have information about its neighboring nodes, so it broadcasts the data packet. When a sensor node obtains the data packet, the only available information a relay node has is its own location and the location of the previous and the destination node, which is in the header of the packet. Thus, a node can easily find out whether it is situated within a specific area relative to the previous relay node. Nodes located inside this forwarding area use DFD before forwarding the packet, whereas sensor nodes lying outside this region drop the received packet. The value of DFD is based on the relative location of current, previous, and destination node coordinates. Finally, the node which computes the minimum DFD sends the packet first. Every sensor node in the forwarding area identifies the further forwarding of the data packet and they terminate its scheduled broadcast. In addition, passive acknowledgments are used. The previous forwarding node identifies the further forwarding of the data packet and thus concludes that the data packet was successfully reached at another relay node. Thus, acknowledgments on the MAC-layer can be passing up. The algorithm executes until the packet reaches destination node. Only the destination node can send an acknowledgment of the received data packet.

Priority-Based Stateless Geo-Routing

With PSGR (Xu et al., 2005), sensor nodes are able to determine their priority by using dynamically estimated network density so as to serve as the next relay node. This suppresses potential communication collisions without increasing routing delays. In PSGR approach void problem is handled by using two alternative stateless methods, rebroadcast and bypass. It assumes that in a large field many sensor nodes are deployed in two-dimensional space. The sensor nodes can adjust their communication range R between $(0,R_{max}]$, here R_{max} is the maximum transmission range.

PSGR emphasizes on the route of packets from a node to a geographic location instead to a particular sensor node. The routing procedure is successful when either data packet reaches a node where destination exists within the range R_{max}, or the condition where destination is not reachable and therefore drops the packet. In order to effectively suppress competing acknowledgements and message collisions, it gives a priority-based autonomous acknowledgement mechanism based on online estimation of node degree, dynamic structure formation of forwarding zones, and minimized acknowledgement delay for each forwarding zone. The neighbors qualified for forwarding the data packet are called potential forwarders (PFs). The main idea in prioritized acknowledgement is to allocate acknowledgement precedence to all the PFs so that the PFs can react to a forwarding probe without contending with each other. An impulsive strategy is to assign unique acknowledgement precedence to every location point in the forwarding area in agreement with a total order relationship in between all the location points. Based on various heuristics, this total order relationship can be managed such as the distance to the destination. Its design chooses to assign an acknowledgement precedence value to a forwarding zone in place of a location point. It also proposes two alternative stateless strategies, namely rebroadcast and bypass as solutions for communication void problem.

On-Demand Geographic Forwarding

Self-Adaptive on Demand Geographic Routing (SOGR) (Xiang et al., 2007) Protocols adjusts the routing path as per topology dynamics and real data traffic requirements. Each node determines and adjusts the value of protocol parameter autonomously according to different network scenario, and the node's own requirement. The assumption it makes is that every mobile node is aware of its own location, a source can obtain the destination's position, and promiscuous mode is enabled on mobile nodes' network interfaces. SOGR works with two mechanisms.

Firstly SOGR with Hybrid Reactive Mechanism (SOGR-HR) that uses a geographic and topology-based mechanism to reactively search for the next-hop. In Geographic-based greedy forwarding current forwarding node attempts to forward a packet in greedy manner to a neighbor closest to destination and closer to destination than itself. With no next hop to destination cached, current node firstly stores the packet and broadcasts a request message with hop count to restrict the searching range to one-hop neighbors. The reply is send by a neighbor node which is closer to destination than current node. Current node will record the next hop to destination with transmission mode as greedy and packet is unicasted to recorded next hop. so as to avoid collisions, recorded next hop will wait for an amount of backoff period before the reply and the pending reply will be withdrawn if it overhears a reply from another neighbor closer to destination than itself. To make sure the neighbor closer to destination responds sooner and suppresses others' replies, the backoff time should be proportional to distance between current node and destination and bounded by the maximum value. In Topology-based recovery forwarding current node may not have neighbors closer to destination, resulting in a local "void". It uses a void handling strategy with expanded ring search, which is usually used in path searching in topology-based routing.

Second mechanism is SOGR with Geographic Reactive Mechanism (SOGR-GR) which depends only on one-hop neighbors' locations to comprise with greedy and face routing like other geographic routing algorithms. However, it considered a reactive beaconing mechanism. The periodic beaconing is activated only when a sensor node eavesdrop data traffic from its neighbors at first time, and the beaconing will be deactivated if no traffic is listened for a pre-defined period. When required to activate its neighbors' beaconing, a forwarding node may broadcast a request packet.

Geographic Node-Disjoint Path Routing (GNPR)

The geographic node-disjoint path-routing (GNPR) (Kumar et al., 2010) scheme exploits two hidden metric concepts for routing decision, i.e., direction and distance. These metric used in well-known routing approaches compass routing and greedy routing, respectively. Each sensor node knows; own coordinates, the coordinates of neighbors, and the coordinates of the destination d. Destinations coordinates are mention in the packet. Given these three location information, the node can find two sensor nodes (u,v) with smallest angle $\angle usd$ and $\angle vsd$ and route greedily by selecting its direct neighbor closest to the destination in (u,v) in the space. By this greedy-compass technique, node can discover all node-disjoint paths. These paths further can be optimized with end-to-end delay or path length, to route the multimedia data. When the greedy-compass technique fail to discover path while void condition occur, procedure revert back to previous hop node and start path discovery with another better option and summarized that node where void condition occur for further node-disjoint path discovery.

After discover sufficient node-disjoint paths, we optimize the minimum end-to-end delay path. Now, the path is available to route the packets from source node to destination node. All dead sensor nodes are not going to be a part of this routing path. So, again a procedure will be started to release all dead nodes which are not used in transmission of packet. This process will establish an optimum routing path between source node and destination node. Destination node will send an acknowledgment message to source node to start the process of transmission. When source node receives acknowledgment then it starts transmitting multimedia data packets.

Scalable Geographic Multicast Protocol

Scalable Geographic Multicast Protocol (RSGM) (Xiang et al., 2010) abides to two-level membership management and forwarding structure. A Zone structure is constructing based on location information at a lower level and on demand a leader node is chosen when a zone has group member nodes. A leader handles the group membership and gathers the locations of the member nodes in its representing zone. At the upper level, the leader nodes report the zone membership to the sources node directly along a virtual reverse-tree-based structure. A leader node which is not aware about source node address gets the information from the Source Home. A source node forwards packets to the zones which have group member nodes alongside the virtual tree rooted at the source node using the information of the member zones. When the packets reaches at a member zone, the leader node of the zone will further forward the data packets to the local member nodes in the zone alongside the virtual tree rooted at the leader node (see Figure 3).

The concerns related to zone management includes the mechanism for electing a zone leader node on demand and keeping the zone leader node during mobility, the managing of empty-zone problem, the method for Source-Home construction and maintenance, and the requirement to reduce packet loss during node progressing towards zones and various issues related to packet forwarding includes the method for virtual tree construction without the requirement of storing and tracking tree-state knowledge, and control of the reliable communication and multicast data packets without use alternative an external location server.

Geographic Routing with Virtual Coordinates

Beacon Vector Routing

Beacon vector routing (BVR) (Fonseca et al., 2005) which is landmark-based routing scheme, has received some significant attention. This routing scheme uses a potential function which depends on distances to the various k landmarks nearest to the destination whereas k is a parameter. The different k landmarks, the ones which are more near to the destination in comparison to the source simply impose a pulling force while in contrast the rest impose a pushing force. The potential function is just a combination of these two. For the destination node d, there is an assumption that $\tau(p, u_i)$ denotes the minimal hop count between d as well as a landmark u_i. Denote by $C_k(d)$ as d_s k nearest landmarks and denote by $C_k^+(p,d) \subseteq C_k(d)$ as a subset of some landmarks which are closer to d than p; while $C_k^-(p,d) \subseteq$

Figure 3. A reference zone structure used in RSGM

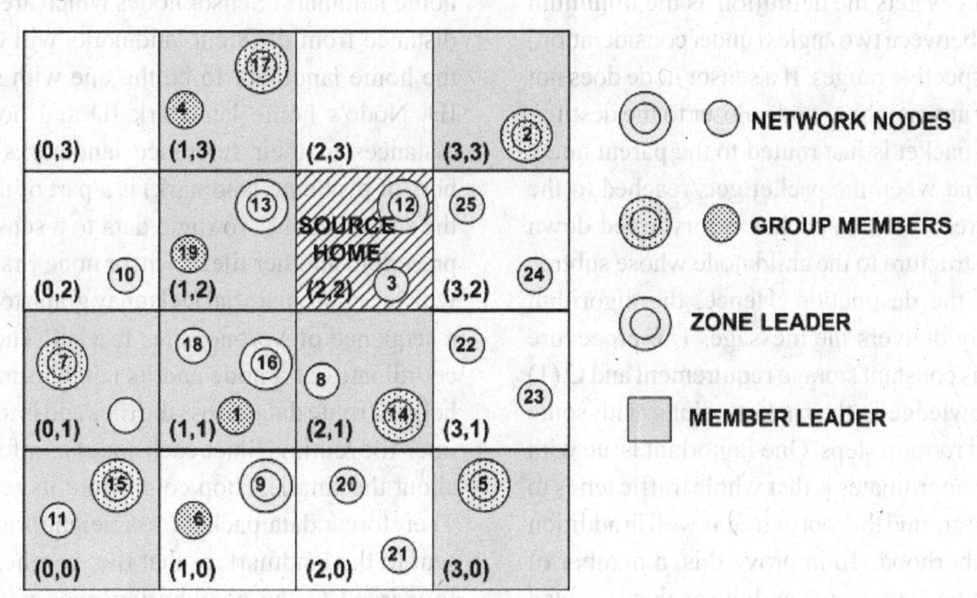

$C_k(d)$ as the subset of the landmarks which are additionally away from d in comparison to p. The potential function d(p,d) is simply defined as

$$\delta_k(p,d)=A\sum_{i\in C+k_k^+(p,d)}[\tau(p,u_i-\tau(d,u_i]+\sum_{i\in C+k_k^-(p,d)}[\tau(p,u_i-\tau(d,u_i]$$

where A is the parameter taken as 10. Additionally the greedy routing selects a neighbor which minimizes the considered potential function to the destination. Whenever the greedy routing gets stuck, the message which is being forward is delivered to a landmark which is closest to the destination, from where a limited flooding with the increasing scope is straightforwardly performed for delivering the packets. Although the design of potential function is a heuristic, the algorithm performs well as evaluated in the simulations and BVR has been opted in a number of various cases as a comparison benchmark.

Virtual Polar Coordinates Routing

Virtual polar-coordinate routing (GEM) (Chen et al., 2005) makes construction of a polar coordinate system as virtual coordinates for sensor nodes being considered. Exclusively, obtain a spanning tree on various sensor nodes as well as give each sensor node a polar coordinate in context to its position in the tree. A sensor node's virtual coordinate incorporates two components: first its level means lowest hop count to the root of tree in addition to its angle range. In addition to this the root of the tree is provided the largest angle range, e.g., from value 0 to 216-1. A node p's child is given with a subset of the overall range of p, directly proportional to the size of subtree. The ranges of angles for the two children do not overlap. With the consideration of this polar coordinate system the subsequent greedy routing scheme, named as virtual polar coordinate routing (VPCR), has been used for delivering all the messages from one sensor node to another. In this scheme a node simply checks its 1-hop neighbor nodes to look for a sensor node which is closer to the destination. So the closeness or the distance of those two angle ranges gets the definition as the minimum distance between two angles (under consideration) in the respective ranges. If a sensor node does not discover any neighbor node closer to the destination, the packet is just routed to the parent node. Notice that when the packet gets reached to the root of tree, the data packet is forwarded down the tree structure to the child node whose subtree includes the destination. Hence, the algorithm ultimately delivers the message. The procedure maintains constant storage requirement and O (1) state knowledge in the packets, along with some localized routing steps. One important issue with the polar coordinates is that whole traffic tends to mount up around the root of tree as well in addition its neighborhood. To improve this, a number of trees can be constructed and the packet is routed at random on one tree as well as the set of polar coordinates associated with it.

Gradient Landmark-Based Routing

Gradient landmark-based routing (glider) (Fang et al., 2005) is related with routing in a sensor network with some big holes or an asymmetrical shape. It differentiates global and local routing by coding the global topology of the sensor network with a compressed combinatorial graph. Greedy routing rules based on the landmark distances helps to realize the real route with global guidance. In this algorithm, landmarks are chosen and the sensor network is divided into Voronoi tiles so that all the sensor nodes in a particular tile are nearest to the same landmark. Two tiles are considered neighbors if there exists two neighboring sensor nodes in different Voronoi tiles. Combinatorial Delaunay graph denotes the tile adjacency graph, which is abstracted and distributed to all the sensor nodes, for global routing across the Voronoi tiles. For routing, every sensor node p is assigned a virtual coordinate. Once the Voronoi tiles are divided, every sensor node is a part of one of the Voronoi cell, this cell is identified as the occupant tile of node, and the landmark of this cell called the home landmark. Sensor nodes which are at same distance from different landmarks will consider the home landmark to be the one with smallest ID. Node's home landmark ID and hop count distances to their reference landmarks (neighbors of p's home landmark) is a part of names of the sensor nodes. To route data to a sensor node present in another tile, a sensor node first checks with the combinatorial Delaunay graph to identity a sequence of Voronoi tiles to visit. The virtual coordinates of a node and its neighboring nodes helps to route data across the tiles and is termed as inter-tile routing. Since each node has information about the smallest hop count from its reference. Therefore a data packet for a neighboring tile is sent to the landmark of that tile, in other words, forwarded to the neighboring node whose hop count distance is reduced from landmark of that tile. When a data packet enters the neighboring tile, it checks if the destination node is present in this tile or else, the data packet is routed to the next transit tile using inter-tile routing. To forward the data packet to a sensor node in the same tile, greedy descent on a probable function built with distances to a number of local landmarks is used to guide the data packet. If a destination q is known, then the greedy routing algorithm selects a neighbor r of p, such that d(r,q) is minimized i.e., packets are moved greedily by minimizing the Euclidean distance to the destination node, calculated in virtual coordinate system. Since this algorithm works only on the virtual coordinates of the neighboring nodes, hence it is local and efficient in nature.

Greedy Landmark Decent Routing

Greedy landmark descent routing (GLDR) algorithm (Nguyen et al., 2007), collectively with a landmark selection strategy, ensures data packet delivery in the continuous domain with bounded stretch. Selected landmarks are r-sampled i.e. any sensor node which is r hop count from a landmark. These landmarks are selected by parallely running a distributed algorithm on all the nodes. Every sensor node waits for some defined period of time and proclaims itself as a landmark, if and only if it has not been subdued by other landmarks. Once a node becomes a landmark it informs its r-hop neighbors and subdues all other non-landmark nodes. This creates an r-sampling and based on constrains of the waiting time selection of the number of landmarks is controlled. Similar to BVR, virtual coordinates of a node are defined with respect to addressing landmarks which are distance vector to a small set of nearest landmarks. Now, the source node selects a node from the destinations addressing landmarks which maximizes the fraction of the distances between source and destination to route a packet to destination, and make progress toward this landmark until it arrive at a node where the hop count to this landmark is same as the destination. At this position, the procedure is recurred and it chooses a different landmark until the destination is discovered. This scheme is always performs in the continuous domain and discover a path with constant bounded stretch. It may occur in a discrete network that the packet stuck in a loop, which can be find out by examine the last some landmarks toward where the packet make progress. When this occurs, the procedure use L_1 and L_∞ norms on the distance vectors of the addressing landmarks for forward the packet greedily to the destination. If path is still not discovered the destination, a scoped flooding is employed to deliver the packet.

Microscopic Geographic Greedy Routing

Macroscopic Geographic Greedy Routing (MGGR) (Funke et al., 2007) algorithm combines the previous two algorithms, namely glider and geographical routing, so as to solve the problem of small network holes, irregularities and inaccurate location information. In glider, we select a sparse group of landmarks to summarize the topological features of sensor networks for global routing. This algorithm selects a dense group of landmarks and geographical routing is applied on combinatorial Delaunay graph. It selects a subgraph from combinatorial Delaunay graph and embeds it in a plane. Every sensor node v is assigned a name including the landmark whose Voronoi tile has v, and its unique identifier. For intratile routing, the data packets are flooded to all the sensor nodes in the destination tile as there are large number of landmarks and every Voronoi tile has fixed number of sensor nodes. in case of intertile routing, the data packet is forwarded to the adjacent tile based on Euclidean distance between home landmark and the landmark of destination tile i.e. the subsequent landmark is selected based on greedy routing and data packet is routed to the next Voronoi tile using the concept of intertile routing as in case of glider. When the distance between neighboring landmark and destination is more than the Euclidian distance between the current home landmark does and destination, perimeter routing is applied to route the packets on planarized Delaunay graph.

Medial Axis-Based Routing

Bruck et al. (2005) proposed the Medial Axis-based routing. They consider that the wireless sensors are deployed uniformly in a geometrical region and it will contain all the possible obstacles and uneven shape. Also for global routing guidance, the overall topology of the sensor field is captured. The medial axis of sensor network is known as the set of sensor nodes as a minimum two closest nodes on the network border line. The medial vertex is defined as a point on the medial axis, if it has three or more neighboring nodes on the network border.

The medial axis is an outline of a section that holds all the topological positives, e.g. the total number of holes and their connections. For the routing purpose, the medial axis is denoted by a medial axis graph (MAG). This medial axis is known to all sensors. The nodes can be identified on the medial axis through local flooding. Each boundary node begins the flooding of a packet and holds its ID with a counter. The counter keeps account of the number of hops. It ends the forwarding this packet if a node receives a packet at a border node that is away from its current adjacent boundary node(s). The approach slashes down the number of messages delivered and remain the total communication rate low. This results, each node gains knowledge of its adjoining boundary node(s) and can decide that it itself is on the medial axis or not. An abstracted medial axis graph is constructed by pulling information regarding the medial axis. This graph is then transmitted to every sensor. In addition, every node on a medial edge retains information which medial edge it keeps on, its nearest nodes on that medial edge, and the number of hops from every endpoint of the medial edge. With the medial axis built, every node is given a unique name w.r.t. the medial axis graph.

Ring Based Routing

In virtual ring routing (VRR) (Caesar et al., 2006), random identifiers are given to the nodes and arrange these nodes in the cyclic increasing order of the IDs, into a virtual ring. Each node maintains r adjacent neighbors, called virtual neighbors of a node, on virtual the ring (r/2 neighbors clockwise and r/2 neighbors counter clockwise). These neighbors that are directly connected by wireless links are to be differentiated from its physical neighbors. The virtual neighbors are not physically close. In this, the routes to every virtual neighbors of a node from the node are set up and stored in the tables (routing tables) of the nodes on the paths. These paths or routes are called vset-paths and will be retains once the network topology varies. In particular, every node p keeps a routing table with the records (S_i,T_i, next hop to S_i, next hop to T_i), for each couple of virtual neighbors (S_i,T_i whose route goes through p).The packet will be transported to the destination according to the table entry, if the source node has a table entry for that destination. If not, the source s ensures its routing table and chooses the path endpoint q adjoining to the destination ID on the ring. The packet is forwarded to make progress towards the direction of q. This operation is repeated for every node. Due to the storage of virtual path, in the worst case too, the packet can follow ring path to reach destination. Naturally, on the virtual ring, a node is routed in the direction of the node adjoining to the destination in the ID space, even though that node might be in an entirely different route from the destination. Due to the routing table entries to the destination, the VRR sends a packet to other nodes with more information of the destination. It frequently happens that without having the path information towards the destination, a packet have not to visit the virtual neighbors. When the network is initialized, the virtual paths can be made incrementally without overflowing the network. The control messages are used to make a new virtual path from the current virtual path. Path failures repaired in the similar way.

Geography-Aided Multicast Zone Routing (GMZRP)

The protocol Geography-aided Multicast Zone Routing Protocol (GMZRP) (Cheng et al., 2008) uses a simple and efficient strategy for broadcasting the multicast route request (MRREQ) packets to remove as much as possible duplicate route queries. GMZRP is the primary hybrid multicast protocol which is suitable for both topological routing and geographical routing. It divides the network coverage area into small geographic zones. It assures that each geographic zone is queried just the once. GMZRP keeps a multicast forwarding tree at the zone granularity and the node granularity. It can help in recovering the link failure at the node level with the help of zone level information. It assumes all nodes in the network know its own location. GMZRP uses geographic separation to decrease the route discovery overhead. It accepts a network center based partition method. When the MANET is initialized, the center of the network coverage region is more or less estimated. The partition spreads in outward direction after starting at the network center. As shown in Figure, the area is divided into equal circular zones. The dotted lines show, the radius of the circular zones equals the side length of a hexagon that each circle contains. These hexagons are not been separated but can completely cover the entire network. By doing this, there exists a central sector at the network center and many other sectors spread around the central sector symmetrically. With the help of GMZPR, multicast sources and receivers setup and maintain forwarding state, which allows multicast communication. For the given multicast group G and source s, the multicast forwarding state is theoretically denoted as a loosely-structured multicast forwarding tree with root s. Each multicast packet is dynamically forwarded from s through the tree to the receiver members of the multicast group G. Source-based multicast forwarding trees requires at least one receiver and one source in the network. It can work with any geographic unicast protocol because it is designed to work separately with the geographic unicast protocol used in the network. GMZRP at this time operates only over bidirectional links.

Geographic Routing in Hyperbolic Space

Various communication networks especially wireless networks and Internet deployed and studies efficiently in the hyperbolic space and can be defined in terms of the location of nodes and area covered by it. This location information can help for the efficient decision to route the packets. The following subsection describes various schemes for node embedding and geographic routing in hyperbolic space.

Kleinberg

The algorithm proposed by the author in (Kleinberg, 2007) assigns a virtual coordinate to the each node in the hyperbolic plane, and performs greedy geographic routing with respect to these virtual coordinates. The embedding proposed guarantees that the greedy algorithm is always successful in finding a route to the destination. The hyperbolic embedding is a greedy embedding relied only on the fact that the hyperbolic greedy route is no longer than the spanning tree route between two nodes. Hence, a natural hypothesis is that the routes chosen by the greedy algorithm using the hyperbolic embedding may not be very different from the routes in the spanning tree, and they may inherit some of the undesirable properties such as high congestion on the popular nodes of the tree. A natural and simple distributed algorithm by nodes can compute their virtual coordinates for the greedy embedding. We assume that the network is capable of computing a spanning tree rooted at some node r. More specifically, we assume that the network has a distinguished node r and that each node $w \neq r$ has chosen one neighbor p(w) (the "parent" of w) such that the edges (p(w),w) form an arborescence rooted at r. Let d(w) be the degree of w, and let 0,1,…,d(w)-1 be a numbering of the neighbors of w such that 0 is assigned to p(w) and 0,1,…,d(w)-1 are assigned to the children of w in arbitrary order. Having computed a rooted spanning tree T(G) of G, we next compute the maximum degree of T(G), via a simple two-pass algorithm in which each node w reports to its parent the maximum degree of the sub tree rooted at w, and the root computes the maximum degree of T(G) and broadcasts this information to all other nodes. Finally, along every edge (p(w),w) of T(G), the parent p(w) transmits to its child w the coefficients of a Mobius transformation μ_w which maps f(p(w)) to u and f(w) to v, where u,v are the points of the Poincare disk. It also transmits to each child w' the coefficients of the Mobius transformation

$\mu_{w'} = b^i \circ a \circ \mu_w$, where i is the number which w assigned to w' when it labeled its children during the spanning tree computation. To establish the correctness of the algorithm, we must show that for each edge (p(w),w) of T, the function f maps p(w) and w to a pair of adjacent nodes in the greedy embedding of the infinite d- regular tree T described in Section II-C. Since μ_w is an automorphism of, it suffices to prove that μ_w(p(w)) and μ_w(f(w)) are adjacent nodes of T. But μ_w(f(w)) = $\mu_w((\mu_w^{-1}(v))$ = v by definition, so it suffices to show that μ_w(f(p(w))) = u. Note that $a \circ \mu_{p(w)}$ maps f(p(w)) to a(v) = u Moreover u is a fixed point of b, so $\mu_w = b^i \circ a \circ \mu_w$ maps f(p(w)) to u, as desired. In fact, the technique for proving that the hyperbolic embedding is a greedy embedding relied only on the fact that the hyperbolic greedy route is no longer than the spanning tree route between two nodes. Hence, a natural hypothesis is that the routes chosen by the greedy algorithm using the hyperbolic embedding may not be very different from the routes in the spanning tree, and they may inherit some of the same undesirable properties, such as high congestion on the popular nodes of the tree.

Papadopoulos

In this work (Papadopoulos et al., 2010), a hyperbolic network model is proposed which produces a scale free network topology. The paper demonstrates that a simple embedding of dynamically evolving networks in a hyperbolic metric space leads to the emergence of scale free topologies. The main benefit of these topologies is that greedy forwarding based simply on node coordinates in the metric space results in 100% reachability with near optimal path length even under dynamic condition, with link failures and node arrivals and departures. The paper uses two models

A simple model that builds scale free networks using hyperbolic geometries and the whole topology of network is constructed at once.

The second model of scale free networks grows in hyperbolic spaces.

To construct a network model, the two-dimensional hyperbolic plane of constant negative curvature is used. The N nodes are distributed uniformly over the disc of radius R this implies that the angular coordinates $\Theta \in [0, 2\pi]$ are assigned to nodes with uniform density, $f(\Theta) = 1/2\pi$, while as the density for the radial coordinate r∈[0,R] is exponential f® = sinhr/(coshR-1)≈e(r-R)~er. The connection used is the step function p(d) = q(R-d), implies that to connect a pair of nodes with polar coordinates (r, Θ) and (r', Θ') by a link only if the hyperbolic distance between them d ≤ R where d is given by the hyperbolic law of cosines: coshd = coshr coshr'-sinhr sinhr'cosΔq, with Δq = min(|q-q'|,2p-|q-q'|). The network generated has a strong clustering is due to result of the triangle inequality in the metric space. The model described produces graphs with power law node degree distribution:

$$P(k) \sim k^{-\gamma}, with\, \gamma = \begin{cases} 2\alpha + 1, & if\, \alpha \geq \frac{1}{2} \\ 2, & if\, \alpha \leq \frac{1}{2} \end{cases}$$

Two techniques of Greedy Forwarding are used for forward the packet, original (OGF) and modified (MGF). In OGF the packet is dropped at voids. While as MGF excludes the current hop from any distance comparisons and finds the neighbor that closer to the destination. The packet is dropped only if the neighbor is the same as the packet's previous hop. In the second model, the scale free network is formed that grow in hyperbolic space. The nodes arrive gradually over time. Initially there are 0 nodes. Each node is numbered by the order of its arrival. When a node i arrives to the system, it needs to know; the current number of nodes in the network, pre-specified parameter a for the node radial density, and a system prespecified constant c, which determines the average node degree.

Cvetkovski

The authors propose in this work (Cvetkovski et al., 2009), an embedding and routing scheme for arbitrary network connectivity graphs, based on greedy routing and utilizing virtual node coordinates. An algorithm is provided for online greedy graph embedding in the hyperbolic plane that enables incremental embedding of network nodes as they join the network without disturbing the global embedding. The paper proposes a simple but robust generalization of greedy distance routing called Gravity-pressure (GP) routing. The routing method always succeeds in finding a route to the destination provided that a path exists, even if a significant fraction of links or nodes is removed subsequent to the embedding. GP routing doesn't require pre-computation or maintenance of special spanning sub graphs and is particularly suitable for operation in tandem with our proposed algorithms for online graph embedding. GP routing normally forwards packets to the neighbor that provides most progress toward the destination. Gravity routing mode is analogous with a liquid flowing through a system of pipes in gravitational field of spherical symmetry toward the center located at the destination node. If a packet reaches a local minimum, GP forwards the packet to a next hop that provides the least negative progress with respect to the location of the destination. To ensure that the packet doesn't enter a loop and periodically returning to the same local lowermost point, concept of pressure as a second field that helps steer the packet out of the valley. This routing mode is called as gravity-pressure routing mode. In contrast to other proposed routing procedures that switch to a non greedy routing mode when a packet reaches dead-end, and sacrifice connectivity to achieve functionality by routing on a suitable graph. GP always retains the locally greedy disposition and works on the original network graph. Thus, GP routing can be viewed as a generalization of greedy distance routing as opposed to a hybrid, dual routing technique.

Zeng

Zeng et al. (2010) describe how to characterize the families of paths between two nodes s, t in a sensor network with holes. Two paths are said to be homotopy equivalent if the paths can be deformed to one another through local changes. Two paths are said to be homotopy types if they pass around holes in different ways. The paper proposes a distributed algorithm to compute an embedding of the network in hyperbolic space by using Ricci flow such that paths of different homotopy types are mapped naturally to paths connecting s with different images of t. The greedy routing to a particular image is guaranteed with success to find a path with a given homotopy type. This directs to simple greedy routing algorithms that are resilient to both local link dynamics and large scales jamming attacks and improve load balancing over previous greedy routing algorithms. The solution proposed by the authors is to embed the given network in hyperbolic space. Then compute the triangulation of the network from the connectivity graph by a distributed and local algorithm. The holes are modeled as non-triangular faces. The holes in the network are cut open to get a simply connected triangulation T. Using Ricci flow algorithm, embed T in a convex region S in hyperbolic space. Each node is given a hyperbolic space coordinate. Each edge uv has a length of d(u,v) as the geodesic between u,v in the hyperbolic space. Thus the greedy routing with hyperbolic metric has guaranteed delivery.

FUTURE RESEARCH DIRECTIONS

The work surveyed in this chapter revealed the excellent efficiency of routing using geometric concepts in sensor network. There are still some issues for future research on many fundamental challenges.

Models for communication: modeling of the wireless communication channels is a basic influencing factor in routing protocol design. Simplified communication models such as unit disk graphs have been adopted widely as it is simple yet have a rich set of geometric structures for routing decision. However, the difference from realistic situations makes some routing algorithms with unit disk graph model fail. More practical and realistic models have been proposed and adopted now days. But it still remains a challenge to develop various communication models that are more practical and realistic.

Routing space: designing of an efficient routing algorithm is highly depending on the routing space in which it will be deploying. It could be Euclidian, hyperbolic or spherical routing space.

Infrastructure: some supporting architecture components such as localization, synchronization, and topology understanding are often requires for applying the geometric ideas in routing algorithm design.

Mobility: most algorithms explained in this chapter presume static sensor networks. New thoughts are required on what geometric abstractions are needed in node mobility scenario and how that assists routing information for improving network efficiency.

Network structure: when a sensor network deploy in a physical environment, there is usual correlation between data from sensor nodes in the spatial and temporal aspects. Exploiting the geometry of sensor data can help with data aggregation, fusion and validation in routing protocol.

CONCLUSION

In the past decade, geometric ideas have been exploited extensively for the designing of wireless sensor networks. This chapter covers routing algorithms in the Euclidian, virtual and hyperbolic spaces that makes use of geometrical aspects of sensor network including the nodes' geographical locations as well as the global shape and topology of a sensor network scenario.

REFERENCES

Akyildiz, I. (2002). Wireless sensor networks: A survey. *Computer Networks*, *38*(4), 393–422. doi:10.1016/S1389-1286(01)00302-4

Anderson, J. W. (2005). *Hyperbolic geometry*. Springer.

Bose, P., Morin, P., Stojmenovic, I., & Urrutia, J. (2001). Routing with guaranteed delivery in ad hoc wireless networks. *Wireless Networks*, *7*(6), 609–616. doi:10.1023/A:1012319418150

Bruck, J., Gao, J., & Jiang, A. (2005). MAP: Medial axis based geometric routing in sensor networks. In *MobiCom '05: Proceedings of the 11th Annual International Conference on Mobile Computing and Networking*, (pp. 88-102). New York, NY: ACM Press.

Caesar, M., Castro, M., Nightingale, E. B., O'Shea, G., & Rowstron, A. (2006). Virtual ring routing: Network routing inspired by DHTs. *SIGCOMM Computer Communications Review*, *36*(4), 351–362. doi:10.1145/1151659.1159954

Chen, W., Kuo, S.-Y., & Chao, H.-C. (2006). Fuzzy preserving virtual polar coordinate space sensor networks for mobility performance consideration. *International Journal of Sensor Networks*, *1*(3/4), 179–189. doi:10.1504/IJSNET.2006.012033

Cheng, H., Cao, J., & Fan, X. (2008). GMZRP: Geography-aided multicast zone routing protocol in mobile ad hoc networks. *Mobile Networks and Applications*, *14*(2), 165–177. doi:10.1007/s11036-008-0135-4

Clark, B. N., Colbourn, C. J., & Johnson, D. S. (1990). Unit disk graphs. *Discrete Mathematics*, *86*(1-3), 165–177. doi:10.1016/0012-365X(90)90358-O

Cvetkovski, A., & Crovella, M. (2009). Hyperbolic embedding and routing for dynamic graphs. In *IEEE INFOCOM 2009 - The 28th Conference on Computer Communications*, (pp. 1647-1655). IEEE.

Fang, Q., Gao, J., Guibas, L. J., de Silva, V., & Zhang, L. (2005). GLIDER: Gradient landmark-based distributed routing for sensor networks. In *IEEE INFOCOM 2007 - IEEE International Conference on Computer Communications*, (pp. 339–350). IEEE.

Fonseca, R., Ratnasamy, S., Zhao, J., Ee, C. T., Culler, D., Shenker, S., & Stoica, I. (2005). Beacon vector routing: Scalable point-to-point routing in wireless sensornets. In *NSDI'05: Proceedings of the 2nd Conference on Symposium on Networked Systems Design & Implementation*, (pp. 329-342). Berkeley, CA: USENIX Association.

Funke, S., & Milosavljevic, N. (2007). Guaranteed-delivery geographic routing under uncertain node locations. In *IEEE INFOCOM 2007 - 26th IEEE International Conference on Computer Communications*, (pp. 1244-1252). IEEE.

Heissenbuttel, M. (2004). BLR: Beacon-less routing algorithm for mobile ad hoc networks. *Computer Communications*, *27*(11), 1076–1086. doi:10.1016/j.comcom.2004.01.012

Karp, B., & Kung, H. T. (2000). GPSR: Greedy perimeter stateless routing for wireless networks. In *MobiCom '00: Proceedings of the 6th Annual International Conference on Mobile Computing and Networking*, (pp. 243-254). New York, NY: ACM.

Kleinberg, R. (2007). Geographic routing using hyperbolic space. In *IEEE INFOCOM 2007 - 26th IEEE International Conference on Computer Communications*, (pp. 1902-1909). IEEE.

Kranakis, E., Singh, H., & Urrutia, J. (1999). *Compass routing on geometric networks*. In 11th Canadian Conference of Computational Geometry.

Kuhn, F., Moscibroda, T., & Wattenhofer, R. (2004). *Unit disk graph approximation*. In 2nd ACM Joint Workshop on Foundations of Mobile Computing (DIALM-POMC), Philadelphia, Pennsylvania, USA.

Kumar, A., & Varma, S. (2010). Geographic node-disjoint path routing for wireless sensor networks. *IEEE Sensors Journal*, *10*(6), 1135–1136. doi:10.1109/JSEN.2009.2038893

Leong, B., Liskov, B., & Morris, R. (2006). Geographic routing without planarization. In *NSDI'06: Proceedings of the 3rd Conference on 3rd Symposium on Networked Systems Design & Implementation*, (p. 25). Berkeley, CA: USENIX Association.

Morin, P. (2001). *Online routing in geometric graphs*. Ph.D. thesis, Carleton University School of Computer Science.

Nguyen, A., Milosavljevic, N., Fang, Q., Gao, J., & Guibas, L. J. (2007). Landmark selection and greedy landmark-descent routing for sensor networks. In *IEEE INFOCOM 2007 - 26th IEEE International Conference on Computer Communications*, (pp. 661 – 669). IEEE. Lin, X., & Stojmenovicm I. (1998). *Geographic distance routing in ad hoc wireless networks*. Technical report TR-98-10: SITE, University of Ottawa.

Papadopoulos, F., Krivoukov, D., Baguna, M., & Vahdat, A. (2010). Greedy forwarding in dynamic scale free networks embedded in hyperbolic plane. In *IEEE INFOCOM 2010: 29th Conference on Information Communication*, (pp. 2973-2981). San Diego, CA, USA. IEEE.

Stojmenovic, I., & Lin, X. (2001). Loop-free hybrid single-path/flooding routing algorithms with guaranteed delivery for wireless networks. *IEEE Transactions on Parallel and Distributed Systems, 12*(10). doi:10.1109/71.963415

Xiang, X., Wang, X., & Yang, Y. (2010). Stateless multicasting in mobile ad hoc networks. *IEEE Transactions on Computers, 59*(8), 1076–1089. doi:10.1109/TC.2010.102

Xiang, X., Zhou, Z., & Wang, X. (2007). Self-adaptive on demand geographic routing protocols for mobile ad-hoc networks. In *IEEE INFOCOM 2007 - 26th IEEE International Conference on Computer Communications*, (pp. 2296-2300). IEEE.

Xu, Y., Lee, W., Xu, J., & Mitchell, G. (2005). PSGR: Priority-based stateless geo-routing in wireless sensor networks. In *Proceedings of IEEE International Conference on Mobile Adhoc and Sensor Systems Conference,* (p. 680), Washington, DC: IEEE.

Zeng, W., Sarkar, R., Luo, F., Gu, X., & Gao, J. (2010). Resilient embedding of sensor network using hyperbolic embedding of universal covering space. In *IEEE INFOCOM 2010: 29th Conference on Information Communication,* (pp. 1-9). San Diego, CA: IEEE.

KEYTERMS AND DEFINITIONS

Ad Hoc Localization: A technique for determining the location of a node in a wireless ad hoc network.

Geographic Forwarding: The forwarding function of geographic unicast routing that forwards a packet from the sender to its next-hop node.

Geographic Multicast: A geographic routing technique that enables the routing of data packets from a single source to multiple destinations.

Geographic Routing: A routing approach that only exploits geographic information of nodes in the network to enable communications among nodes.

Geographic Unicast Routing: A geographic routing technique that enables the routing of data packets from a single source to a single destination.

Greedy Forwarding: A forwarding strategy that uses locally available information, usually the positions of the sender, its neighbors, and the destination, to make packet forwarding decisions in a greedy manner. Greedy forwarding is used wherever possible in geographic forwarding.

Planarization: A planar graph is a graph that can be embedded in the plane with no intersecting edges. Planarization is a technique used to extract a planar subgraph from the original graph. Distributed planarization algorithms are required in wireless ad hoc networks.

Void Handling: A communication void occurs when a sender cannot locate a nexthop node among its neighbors which is closer to the destination than itself. Void handling is a forwarding mode used to handle voids in geographic forwarding.

Chapter 24
An Outline of Security in Wireless Sensor Networks: Threats, Countermeasures and Implementations

M. Yasir Malik
Institute of New Media and Communications, Seoul National University, Korea

ABSTRACT

With the expansion of wireless sensor networks, the need for securing the data flow through these networks is increasing. These sensor networks allow for easy-to-apply and flexible installations, which have enabled them to be used for numerous applications. Due to these properties, they face distinct information security threats. Security of the data flowing through across networks provides the researchers with an interesting and intriguing potential for research. Design of these networks to ensure the protection of data faces the constraints of limited power and processing resources. The author provides the basics of wireless sensor network security in this chapter to help researchers and engineers in better understanding of this applications field. In this chapter, the author provides the basics of information security, with special emphasis on WSNs. The chapter also gives an overview of the information security requirements in these networks. Threats to the security of data in WSNs and some of their counter measures are also presented.

1. INTRODUCTION

Wireless sensor networks (WSNs) attract the attention of researchers and engineers thanks to their vast application scope. These allow for easy and flexible installation of wireless networks composed of large number of nodes. This gives WSN the capability to be used in unimaginable applications. They are finding their usages in habitat monitoring, manufacturing and logistics, environmental observation and forecast systems, military applications, health, home and office applications and a variety of intelligent and smart

DOI: 10.4018/978-1-4666-0101-7.ch024

systems. Multimedia wireless sensor networking is a relatively new branch in this domain, which can process multimedia content i.e. still images, audio and video to name a few.

Such a sensor network is typically composed of hundreds, and sometimes thousands of nodes. These nodes are capable of receiving, processing and transmitting information, as based on the assigned tasks. Information flowing through WSN may be susceptible to eavesdropping, retransmit previous packets, injection of redundant or causeless bits in packets and many other threats of diverse nature. To ensure that the data being received and transmitted across these networks is secure and protected, information security plays a vital role.

As contrary to the Moore's law, there has been not much development in the hardware capacity and computational capabilities of the sensors being deployed in wireless sensor networks. These networks are kept inexpensive, thus introducing many constraints in the performance parameters. Low cost sensors incorporate shortcomings in their storage capacity, power requirements and processing speed. This poses a unique dilemma for researchers as they have to design efficient and distinct information security schemes which work seamlessly with the resource constrained sensor networks.

Sensors in the network are mostly exposed to open environment as they have to interact with either other sensors or human beings. Physical security of these sensors is always vulnerable and thus poses an unprecedented threat to the overall security of the network. Advances in power analysis and time based attacks enable the malicious entities to perform various hazardous activities.

Wireless channels are still considered unreliable and the same is the case with wireless sensor networks, which may contain a very large number of nodes and sinks, thus giving rise to concerns about the validity of the communications in the network. Trust models for the nodes have to be developed to make sure that all the nodes taking part in the communications are trustworthy.

All these unique features of wireless sensor networks changes the way we look at their security. These networks face different kinds of threats from those of computer, wired, network or even the high-bandwidth wireless models. Thus, these intimidations are coped in distinctive manners.

This chapter will be beneficial in equipping the readers with the basic concepts of security and WSN security. Readers will be able to realize the strengths and weaknesses of WSN with respect to security. Some of the famous and latest attacks and their countermeasures will help in better understanding of the threats and our capabilities to cope with them. Readers with lesser or no prior knowledge of information security will be able to understand this chapter, because basic concepts needed for better apprehension of security issues will be defined.

We are hopeful that the basics provided in this chapter will help the readers to grasp the fundamental concepts of Wireless Sensor Network Security (WSNS), which will empower them to embark on their journey to further explore this ever-expanding field and to find new problems and their solutions in this interesting research and applications field.

General characteristics of WSN are presents in Section 2 of the chapter. These are the properties of these networks which make them the preferred solution in many applications, though they also present limitations on the viable solutions to the security issues in WSN. These attributes are studied with an emphasis on their importance in the security of WSN.

For reliable and secure communications in WSNs, there are some security qualifications that must be fulfilled. These security requirements are given in Section 3.

Threats in WSN are of diverse natures and kinds. Some of the important threats will be discussed in section 4 of this chapter. Countermeasures to some of the described attacks are presented in section 5.

With growing research work in this field, there are many new results that are benefiting us in making the WSN more resistant to attacks and more efficient in their secure implementations in terms of power and memory. Some of the latest research work and implementations of schemes and algorithms are provided in section 6.

Section 7, the last part of the chapter, concludes our discussion. It provides us with summary of the chapter and also outlines the research domains that can be pursued in the coming future related to WSN.

2. GENERAL CHARACTERISTICS OF WSN

Wireless sensor networks are unique in many of their features, which are discussed briefly here (Figure 1). These characteristics make them an attractive choice for many applications, and also present the researchers with distinct security challenges.

Figure 1. Sensor node components

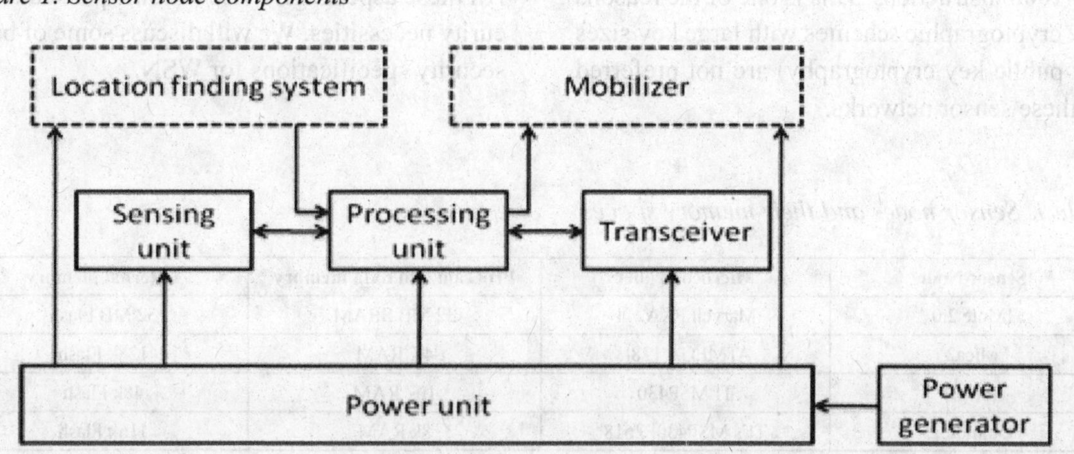

2.1 Compact Size

As discussed earlier, sensor network may contain hundreds or probably thousand of autonomous nodes. For such a huge network, size does matter. Sensors are kept small, which also limits the components on the main chip-board of the sensor and only the most crucial parts are installed on it.

Small sizes of sensors may be considered as a positive attribute, as sensors can be deployed so that they are not visible.

2.2 Physical Security

Sensors usually get information about the environment and perform their designated operations. They have to interact with exposed surroundings which pose hazards to the physical protection of the sensors.

2.3 Power

Sensors in WSN contain non-renewable power resources thus causing an energy starved wireless network. Sensors cannot be recharged because of the volume and distribution of the network, which makes recharging of the nodes a laborious and expensive task. Power limitations in WSN are considered the major constraint to the performance of the network. As all the nodes do local processing, they are always in need of power. Thus, the inclusion of security features like encryption, decryption, authentication etc comes at the price of decrease in the overall performance of the nodes because of the energy consumed during these cryptographic algorithms and schemes.

Security is vital for WSN, so there is always some compromise to make between the secure communication and allocation of energy resources for implementing cryptographic schemes.

2.4 Memory Space

Sensors have small memory space, which accounts for its low cost and power consumption. Memory is a precious asset for any sensor, thus keeping the size of the security algorithm source code small. Sizes of the keys that need to be stored are also kept at a minimum length because of scarcity of memory storage. Table 1 lists some of known sensor nodes and their memory spaces.

2.5 Bandwidth

WSN is a low bandwidth network and as compared to other wireless networks, the quantity of data transmitted and received by the nodes is very low. This helps the nodes in saving the crucial power for other functions. As an estimate, each bit transmitted consumes as much power as executing 800-1000 instructions. This is one of the reasons why cryptographic schemes with large key sizes (i.e. public key cryptography) are not preferred for these sensor networks.

2.6 Unreliable Communications

Like all other wireless communications, channels in the WSN are subject to unpredictable environmental conditions, state of channels, interference and many other factors that usually deteriorate the quality of service of the wireless links and induce errors in the information being transmitted.

Error correcting codes, MAC and cyclic redundancy check (CRC) are sometimes used to cope with these problems. They are widely being used in wireless links to ensure better service at the expense of extra bits added to the original messages.

3. SECURITY REQUIREMENTS IN WSN

WSN is a wireless network composed of sensors. Due to the attributes of being a network and utilizing wireless communications, the security demands for WSN are unique. Security requirements in WSN to ensure trustworthy and secure connections and communications are a combination of the specifications for computer network and wireless communication security. WSN has its own distinct features, as discussed in section 3, which make these networks unique. Their anomalous character is due to their large volume, pattern of distribution and resource restrictions. All these aspects give rise to some particular security necessities. We will discuss some of basic security specifications for WSN.

Table 1. Sensor nodes and their memory spaces

Sensor Node	Microcontroller	Program and data memory	External memory
IMote 2.0	Marvell PXA271	32 MB SRAM	32 MB Flash
Mica2	ATMEGA 128L	4K RAM	128k Flash
TelosB	TI MSP430	10k RAM	48k Flash
Ubimote2	TI's MSP430F2618	8k RAM	116k Flash

3.1 Data Confidentiality

Data is communicated between the sender and the recipient, sometimes being routed through many nodes. This data may also be kept in memory for further processing. This data can be sensitive enough to be known only by the sender and the recipient. Sometimes, the adversary can access this information by eavesdropping between wireless links, gaining admission to the storage or by other attacks. Data confidentiality means that the data can only be accessed, and thus utilized, by only those entities that are authorized for this purpose.

If any data is lost by negligence and weak security measures, it can lead to identity thefts, loss in business, privacy breaching and many other malicious activities. This makes data or message confidentiality the most important feature of any security protocol.

In WSN, data confidentiality can be observed by making sure that

1. Sensor network should not leak any data to other networks in vicinity, thus retaining the message completely within the network.
2. Data is sometimes routed through many nodes before reaching the destination node. This causes a rise in need for secure communication channels between different nodes and also between nodes and base stations.
3. Encryption is one of the most commonly used procedures to provide confidentiality of data. Critical information such as keys and user identities should be encrypted before transmission. Sensitive information can be characterized from the kind and type of protocol being used i.e. symmetric or asymmetric cryptography, mutual authentication, identity or nonce based encryption.
4. Steps can also be taken towards encrypting the sensitive data before storing them in memory. This is particularly important if the nodes are exposed to user interaction, or in military applications.

Mostly symmetric cryptography or stream ciphers are used for encryption and decryption in WSN, due to the high storage and computational costs associated with the public key cryptography. TinySec is a link layer security protocol that makes use of RC5 and Skipjack block ciphers in cipher block chaining (CBC) mode of operation. LLSP uses AES in CBC mode. LiSP utilizes stream cipher for providing encryption.

3.2 Data Integrity

Provision of data confidentiality stops the leakage of data, but it is not helpful against insertion of data in the original message by adversary. Integrity of data needs to be assured in sensor networks, which solidifies that the received data has not been altered or tampered with and that new data has not been added to the original contents of the packet. Environmental conditions and channel's quality of service can also change the primitive message.

Data integrity can be provided by Message Authentication Code (MAC). For this purpose, both sender and receiver share a secret key. Sender computes the MAC using this key and contents of message, and transmits the message along with the MAC to the receiver. The recipient re-calculates MAC by using the shared secret key and message. Absence of irregularity in composition of calculated MAC establishes integrity in the received message.

3.3 Data Authentication

Authentication is used in sensor networks to block or restrict the activities of the unauthorized nodes. Any disapproved agent can inject redundant information, or temper with the default packets carrying information. It is particularly important in case of decision making chunks of information. Nodes receiving the packets must make sure that the originator of packets is an accredited source. Nodes taking part in the communication must be capable of recognizing and rejecting the information from illegitimate nodes.

Although data or message authentication can be provided by incorporating calculation of MAC, this symmetric procedure is not recommended for multi-party communication.

Symmetric schemes normally use the calculation of MAC at the sender and receiver ends. It is usually done by the same technique as describes in 3.2 (previous part).

Multi-party communications or broadcasting makes use of asymmetric authentication schemes. Data authentication in broadcasting requires strong trust assumptions, thus giving rise to different categories of trust. For authentication purposes, both of the mutual authentication and one-way authentication method can be used based on trust requirements.

In SPINS (Perrig, Szewczyk, Tygar, Wen, & Culler, 2002), authors state that if a sender wants to send authentic data to mutually untrusted receivers, symmetric MAC is not secure since any one of the receivers already knows the MAC key and hence could impersonate itself as the original sender of the message. Then it can forge fake messages and send them to other receivers. SPINS constructs authenticated broadcast from symmetric primitives but it establishes asymmetry by the utilization of delayed key disclosure and one-way function key chains.

LEAP (Du & Chen, 2008), on the other hand uses a globally shared symmetric key for broadcast messages to the whole group. As the group key is shared among all the nodes in the network, steps are taken to update this key through rekeying mechanism if any node is compromised. LEAP exercises an efficient approach to get information about any compromised node.

3.4 Data Freshness

Some of the messages are critical enough that extra precautions need to be taken to ensure their correction. Confidentiality and Authentication may not be useful when any old message is replayed by any attacker. Data freshness implies that the received messages are recent, and previous messages are not being replayed. Importance of data freshness becomes evident in networks using shared key operations. During the time taken for transmission of shared key in WSN, replay attack can be carried out by adversary.

Data freshness is categorized into two types based on the message ordering; weak and strong freshness. Weak freshness provides only partial message ordering but gives no information related to the delay and latency of the message. Strong freshness, on the other hand gives complete request-response order and the delay estimation. Sensor measurements require weak freshness, while strong freshness is useful for time synchronization within the network.

To accommodate data freshness, nonce which is a randomly generated number or a time dependent counter can be appended to the data. Messages with previous nonce and old counter numbers are rejected. This guarantees acceptance of only recent data, and thus the freshness in data is achieved.

3.5 Availability

Introduction of security scheme in WSN comes at the expense of computational storage and energy costs. Security features in the network may be considered as extra feature by some because of the restrictions it can impose on the availability of the data. Insertion of security can cause earlier depletion of energy and storage resources, causing unavailability of data. Similarly, if security of any one node (especially in central point network management) is compromised or any Denial of Service (DoS) attack is launched, data becomes inaccessible.

Availability of data becomes an important security requirement because of the mentioned arguments. Security protocol should consume less energy and storage, which can be achieved by the reuse of code and making sure that there is minimum increase in communication due to the functioning of security protocols.

Processing within the networking and en-route filtering can be used to subsidize the effects of malicious attacks and other issues that may arise because of increase in communication due to utilization of security scheme. There is also a need to avoid central management scheme in sensor networks as they can affect the availability of data due to single point failures. These steps will also make the network robust against attacks.

3.6 Self-Organization in WSN

As mentioned in previous sections, one of the characteristics of WSN is their composition and distribution. A typical WSN may have hundreds of nodes performing different operations, installed at various locations. Ad-hoc networks are also sensor networks, having the same flexibility and extensibility. These otherwise attractive properties of WSN pose a serious threat to the overall security situation of the network, raising the importance of a self-organized and robust structure of network.

For using public key cryptography based scheme, an efficient design is needed that takes into account all the situations for sharing the key and is capable of trust management amongst different nodes. Keys can be redistributed between the nodes and base stations to provide key management. Schemes can use symmetric cryptography that applies key predistribution methods.

3.7 Secure Localization

WSN makes use of geographical based information for identification of nodes, or for accessing whether the sensors belong to the network or not. Some attacks work by analyzing the location of the nodes. Adversary may probe the headers of the packets and protocol layer data for this purpose. This makes the secure localization an important feature that must be catered during our implementation of security protocol.

4. ATTACKS ON WSN

Wireless sensor networks are power constraint networks, having limited computational and energy resources. This makes them vulnerable enough to be attacked by any adversary deploying more resources than any individual node or base station, which may not be a tedious task for the attacker. As described earlier, a typical sensor network may be composed of potentially hundreds of nodes which may use broadcast or multicast transmission. This mode of transmission results in a large volume wireless network with many potential receivers of the transmitted information. This makes a number of attacks such as packet alteration or new packet insertion, capturing of node, reply attacks, denial of service and traffic analysis possible to be performed on any sensor network.

WSN can be cooperatively attacked by colluding in which the adversary makes use of illegitimate nodes with the same capabilities as of network nodes. Deployed malicious nodes can work together to take control of any network node, which can be used further to make damages to the network or to amplify the scope of the attack.

The opponent may have highly capable communication links available to carry out any malicious activity, thus making the countermeasure an expensive task. This is a limitation to the security of WSN as we constantly need inexpensive and small devices as nodes in sensor networks.

Deployment of many nodes of WSN in open and harsh environment poses them another major threat. This compromises their physical security, and if the nodes are not temper-resistant, they can be mishandled and tempered with. Attacks on the physical security of the nodes can cause the node to give away the data stored on it, which may enable the attacker to gain access to critical information such as source code, key and other data which may be crucial for security protocol of the entire wireless network. Making these nodes temper resistant may be able to reduce the effects of side-channel attacks and to enhance the physical security of the network devices, but this may not be the feasible solution as the cost per node increases dramatically if we consider such defenses.

WSN are continuously being used in many critical and sensitive applications. WSN are popular thanks to their ability to incorporate in numerous applications in diverse fields. Health care, security, logistics and military applications are some of the areas of deployment of these wireless networks. It is evident that if the capabilities or functionalities of the sensor network are reduced or endangered, it may cause huge losses in terms of money, resources and may even result in human injuries or fatalities.

This section contains basics of some attacks on WSN, and effects of these attacks on the performance of the wireless networks.

Threat Models

An attacker may have access only to a few nodes which he or she has compromised. Such attacker is classified as mote class attacker. Alternatively an attacker may have access to more powerful devices such as laptops, hence the definition laptop class attacker. Such attackers have powerful CPUs, great battery power, high power radio transmitter and sensitive antennas at their disposal and pose a much larger threat to the network. For example a few nodes can jam a few radio links where as a laptop can jam the entire network.

Finally, attacks launched on a network may be insider or outsider attacks. In outsider attacks the attacker has no special access to the network. In insider attacks however, the attacker is considered to be an authorized participant of the network.

Such attacks are either launched from compromised sensor nodes running malicious code or laptops using stolen data (cryptographic keys & code) from legitimate nodes.

Now some of the major attacks on WSN are presented. Jamming and physical attacks affect the physical layer of the WSN structure. Collision, exhaustion and unfairness attack types belong to the attacks on data link layer of the WSN.

4.1 Denial of Service (DoS)

Jamming nodes of networks, sending continuous messaging without following the system communication protocol (link layer protocols) by any node, malicious attacks and environmental condition may cause resource exhaustion and failures of devices in the WSN. This causes degraded system performance and it is not able to function as expected. These are the forms of Denial of Service (DoS) attacks that intend to affect the functionality of WSN.

These attacks are carried out on the physical, link, routing and transport layers of the WSN architecture. Because of resource limitations of WSN, guarding against these attacks become very costly. Researchers put lot of effort to study these attacks and to devise the methods to minimize their impact on the network.

Now we briefly discuss some of the major types of DoS attacks according to the layers whom they affect. Jamming and physical attacks affect the physical layer of WSNs. Neglect and greed, homing, routing information alteration or spoofing, black holes and flooding belong to the type of attacks on network layer of the WSN architecture.

4.1.1 Jamming

Nodes in WSN utilize radio frequencies for the transmission of information, as these sensor networks use wireless channels for communications. Jamming is one of the basic yet detrimental attacks that intend to intervene in physical layer of the WSN structure. It is simply the transmission of the radio signals having the same frequencies as being used by the wireless network.

Jamming causes permanent or temporary suspension of message reception and transmission from the jammed node devices. WSN is widely distributed wireless network, which makes complete jamming an unfeasible attempt. Still jamming of a few nodes in WSN can lead to deterioration in effectiveness of many neighboring nodes.

4.1.2 Physical Attacks

As mentioned earlier, WSN devices may be deployed in vast geographical areas and in hostile and harsh environments. Moreover sensor nodes are kept cheap and light weight, which limits any effort to make them temper-proof, their ability to withstand harsh climate or conditions and to avoid or regulate any physical or more sophisticated side-channel attacks.

This makes the WSN nodes highly prone to any physical tempering or other attacks performed on its construction. Nodes can be modified to extract key and other important cryptographic parameters that are crucial for working of any security protocol. Similarly adversary can extract source code which eventually provides attacker the information about the network, which can modify the code to get access into the network. Attacker can replace the nodes with the illegitimate and malicious ones, thus compromising the operation of the whole sensor network.

Physical attacks gives the attacker the ability to alter the nodes and thus the network functioning. These attacks are hard to avoid due to the major characteristics of any WSN to be inexpensive and disperse.

4.2 Collisions

Collision is a type of link layer jamming, in which the efficiency of the network is reduced by using the fact that continuous transmission of messages can cause collisions in networks. Collisions cause retransmission of the collided messages and if it happens often then the energy resource of a node can be depleted. Another form of this attack can happen when some part packet is altered, which causes MAC mismatch at the receiver. The corrupted packets are transmitted again, increasing the energy and time cost for transmission. Such an attack when prolonged impels the decrease of network fruition.

4.3 Exhaustion

This attack drains the power resources of the nodes by causing them to retransmit the message even when there is no collision or late collision. A node can seek access to any channel deliberately and perpetually, forcing the neighboring nodes to respond continuously.

4.4 Unfairness

MAC protocols govern the communications in networks by forcing priority schemes for seamless correspondence. It is possible to exploit these protocols thus affecting the precedence schemes, which eventually results in decrease in service.

4.5 Neglect and Greed Attack

During communication between any two nodes in WSN, there may be need to route and re-route packets through many nodes. Transmission from source to destination depends on complete and successful routing of the destined packets. Malicious or compromised node in the way can influence multi-hopping in the network, either by dropping some of packets or by routing the packets towards a false node. This attack also disturbs the functioning of the neighboring nodes, which may not be able to receive or transmit messages.

4.6 Homing

Cluster head nodes or the base station neighboring nodes are the most important nodes in WSN. In homing attack, the adversary analysis the network traffic to judge the geographic location of cluster heads or base station neighboring nodes. It can then perform some other kind of attacks on these critical nodes, so as to physically disable them or to capture them which in turn can lead to major damages to the network.

4.7 Routing Information Alteration (Spoofing)

In this attack, routing information is altered and tempered with. This can create new routing paths, or lengthen or shorten existing routing paths thus increasing the end-to-end latency. It repels or attracts traffic decreasing the quality of service. It can also generate false error messages which disable or increase latency for nodes to access the channel.

4.8 Black Holes

In WSN, it is possible that nodes are not fully aware with the complete topology of the network because of the large volume of the network. If distance-vector-based protocols are used in these sensor networks, they are highly susceptible to the formation of black holes. Malicious nodes can advertise zero-cost routes to other nodes in the networks, which causes more traffic to flow toward these nodes. Malicious node's neighboring nodes compete for unlimited bandwidth, thus causing resource contention and message disruption. If this state continues, the neighboring nodes may as well exhaust causing a hole in the network. These attacks are also known as "sink hole" attacks.

4.9 Flooding

An attacker continuously sends connection establishment requests to a node in this type of resource exhaustion attack. Each of such requests makes the node allocate some resources to serve each request. Persist requests by a malicious node may drain the memory and energy resources of the node under attack.

4.10 De-Synchronization

In this attack, an adversary can fabricate messages containing any control flags or sequence numbers of previous frames, and transmit them to two connected nodes. These fake messages make the nodes realize as if they have lost their synchronization. Nodes retransmit the assumed missed frames, and if the adversary is capable of persistent transmission of forged messages then the resources of the nodes will be soon depleted. Moreover the connected nodes are not able to share any useful information during this attack, as they delve infinitely in synchronization-recovery protocols.

4.11 Interrogation

An interrogation attack exploits the two-way request-to-send/clear-to-send (RTS/CTS) handshake that many MAC protocols use to mitigate the hidden-node problem. An attacker can exhaust a node's resources by repeatedly sending RTS messages to elicit CTS responses from a targeted neighbor node.

4.12 Sybil Attack

In this interesting attack, a node can take multiple identities which lead to the failure of the redundancy mechanisms of distributed data storage systems in peer-to-peer networks. Sybil attack functions by its property of representing multiple nodes simultaneously. The Sybil attack is capable of damaging other fault tolerant schemes such as dispersity, multi path routing, routing algorithms, data aggregation, voting, fair resource allocation and topology maintenance. This attack also affects the geographical routing protocols, where the malicious node presents several identities to other nodes in the network and thus appears to be in more than one location at a time. Similarly, during the voting process the malicious node can create additional votes thanks to its ability to present several identities at a time. It can strike the routing algorithms by defining many routes through only one node. Resources of a node can be drained by requests from multiple entities which are in fact exhibited by a single malicious node.

4.13 Selective Forwarding

A node may drop partial or complete packets hopping through it, thus disturbing the quality of service in WSN. If all packets are dropped, the neighboring nodes become suspicious and may consider it to be malfunctioning thus finding new routes. Malicious node can selectively forward data to avoid suspicion. It can drop some of the data and passes all other to prevent issues that may arise concerning its performance. Malicious nodes may only allow the data transfer from some selective nodes, giving them the space to alter or suppress data from particular nodes.

This kind of attacks becomes very difficult to detect.

4.14 Worm Holes

Worm holes are formed by malicious nodes working in different parts of the network. In this attack, the attacker receives messages in one section of network over a low-latency link and sends them to another section of the network. These messages are then replayed in the other part of the network thus forming a worm hole in the present structure of the information flow in network. The impression can be detrimental if the adversary finds its presence near the base stations, giving the distant nodes the realization that they are in the vicinity of the base stations. Multi-hop nodes get the notion through wormholes that they are only one or two nodes away from the base station. Traffic flows to the low-latency route that the adversary provides to these distant nodes. This may cause congestion and further retransmissions of the packets by the legitimate nodes, dissipating their energy.

This attack when used in conjunction with Sybil and selective forwarding attacks becomes difficult to distinguish and evade.

4.15 Hello Flood Attacks

At the start of communication, node has to announce itself to the network by broadcasting hello message to their neighboring nodes. It also validates that the node sending hello message is in the vicinity. Adversary can exploit this feature by using a high-powered wireless link. It can assure every node in the network that he is their neighbor, thus starting communication with nodes. As obvious, by using this attack security of the information is compromised as the attacker gains access to the information flow in the network. If some puzzle scheme is used by the nodes to provide access to any node requesting for connection, then a variant of this attack can also be applied.

Adversary should possess enough resources to manage this attack, and should be able to provide high quality routing path to other nodes in network. Traffic will find this path attractive enough to send packets through it, creating data congestion and disturbing the hierarchy of the data flow in network.

4.16 Acknowledgement Spoofing

Acknowledgments play a vital role in determining the quality of service at any links and establishing further connections based on the this information. Adversary can alter acknowledgements to present to any transmitting node that any weak link is strong enough for reliable communication.

The packets that are sent on this link are partially or completely dropped, thus decreasing the overall attainment of the WSN.

4.17 Node Replication Attack

Sensor nodes have IDs as their identity (and indices of their location in geographical routing algorithms) in the WSN. An adversary can add new node to the sensor network by copying the ID of an already existing node and assigning it to the malicious node. This ensures presence of the adversary in the network allowing the malicious entity to induce destructive affects to the sensor network.

By using the replicated node, packets arriving through it can be dropped, misrouted or altered. This results in incorrect contents of information packet, loss of connection, data loss and high end-to-end latency. Adversary can gain access to the critical information (cryptographic key, source code or other security parameters) by practicing this attack, which brings about security implication of the whole sensor network.

Replicated nodes at specific location can be used to carry out coordinated attack to influence particular nodes or sections of the network.

5. COUNTERMEASURES TO ATTACKS ON WSN

5.1 Denial of Services (DoS)

5.1.1 Jamming

Jamming and its countermeasures depend on the resources of both the sensor nodes and that of the device used by attacker. One of the most obvious solutions to avoid jamming is spread spectrum, or code spreading as used in mobile communication. In these methods, several frequencies are utilized for transmission. Both of these spreading techniques are affective against jamming, as the simple jammer is usually not capable to jam wide band of frequencies or switch to the exact frequencies as being used in frequency hopping or spread spectrum. Implementing these procedures in hardware requires more space, and increases the overall complexity and cost of the device. Sensor devices are kept inexpensive and compact in size, which limits the prospect of deploying these methods in practice.

Jamming attacks can be characterized by high background noises which can be detected and reported by the neighboring nodes. If the jammed part of the WSN is identified, then a deviation in routing paths can help in avoiding this attack.

If jamming attack is found by the network, the sensor nodes under attack can be put to sleep for a long time. Low duty cycle can be applied to consume less power. This enables the nodes to conserve their already limited energy resources, which gives them opportunity to try to connect to WSN once the attack is over. Attacked sensor node can also send a high power message reporting the attack to neighboring nodes or base stations during the attack, if the attacker employs stuttering or interruptive jamming. Another efficient yet costly solution is the alternative use of optical or infra-red communications for sensor devices under jamming attack, but these modes are distance restricted and quite expensive.

5.1.2 Physical Attacks

Adversary can exploit physical weakness of motes to access crucial data stored on it, and is also capable of damaging or replicating the nodes. Steps that one must to ensure the physical safety of sensor nodes in WSN are based on the desired level of security. One cannot fully guarantee complete protection of hundreds or thousands of nodes, which are typically dispersed over large distance to form WSNs.

Nodes in hostile environments can be made temper-proof so that security of these motes is not compromised over cost. Camouflaging and hiding sensor nodes are other countermeasures against physical attacks.

Motes which handle critical data can use any erasure procedure which makes them remove any critical information i.e. cryptographic keys or codes, when they are tempered with.

5.2 Collisions

Altered packets of information can increase latency in networks, and results in dropping and discarding of packets once they are found corrupt thus degrading the service of the network. Collision detection and avoidance schemes can be employed to avert such situations. Cyclic redundancy check (CRC) of the messages can be computed on the transmitter and receiver ends to ascertain the integrity of the message. Similarly, error correcting codes can also be used for avoiding and corruption by outsider to the messages. Such codes, with high error correcting capabilities, come at the expense of extra bits that must be appended with the original message. This poses a limitation to the effectiveness of these codes as the malicious agents may be able to inject more errors in the message than the capabilities of the correcting codes. Cooperation between the communicating nodes can also avoid the corruption of the transmitted packets.

5.3 Exhaustion

Exhaustion of the power of the sensor due to retransmissions even though they are caused by late collisions, can be handled by use of time division multiplexing (TDM). TDM provides each sensor with a time slot to send its data which avoids collisions. This solves the infinite deference problem, which is caused by continuous retransmissions by nodes.

Allowing limited number of requests to access network at a time can also help in getting rid of collisions. Such a limitation is implemented by exercise of MAC admission control rate, which allows only specific number of requests to access the network.

5.4 Unfairness

Adversary exploits the cooperative MAC priority scheme by making sensors to miss their transmission deadlines. This attack affects the real-time users to a large extent. Use of small packets avoids this attack as each sensor node seizes the channel only for short time.

5.5 Neglect and Greed Attack

Due to partial drop of packets and unpredictable behavior of malicious node in this attack, it is not possible to detect this type of attack. The best step to avoid damage by neglect or greed of malicious sensor node is to define alternative routing paths. Another proposed solution is to use redundant messages that reduce the impairment by malicious node.

5.6 Homing

Adversary learns about the important nodes by analyzing the headers and contents of the messages flowing in the network. Encrypting the header and contents of message makes the task of adversary more difficult. Source and destination of the intercepted messages becomes discreet by using cryptography.

5.7 Routing Information Alteration (Spoofing)

Routing information included in the packets are altered or spoofed to divert the flow of traffic to the intended destinations. Node addresses can be changed and adversary can control the flow of traffic, which makes it possible for it to attack any particular node.

Packets construction can be made secure by using CRC or MAC schemes, which makes the detection of tempered packets easy. Similarly, link layer authentication also helps to avoid this attack. Only authorized nodes are allowed to take part in exchange of information.

Similarly, interrogation attacks can be handled by the use of authentication and antireplay protection schemes.

5.8 Black Holes

To counter the formation of black holes, similar steps to that of routing alteration (also termed as "misdirection" in some texts) are taken as this attack also functions by changes in the routing information of the traffic.

Requests for exchange of data should come only from authorized sensors, and an efficient authentication scheme must be deployed to ensure this. WSN can use public key cryptography to sign and verify the routing information and updates. Public key cryptography is quite costly and requires large overhead which makes its utilization for this purpose very difficult. Efficient certification and threshold based cryptography based schemes are advised to be used for authentication and trust management in WSN.

Neighboring nodes can monitor the activities of the node, and can analyze its behavior by sending dummy packets and checking whether it reaches its destination. Geography based probing do not require all nodes in the network to participate in monitoring activities. Physical topology of the network is analyzed by sending probe to detect any black holes and damaged regions.

5.9 Flooding

Flooding cause the allocation of resources to the requesting clients and limits the effectiveness of already resource starved sensor node. One method to void this attack is to limit the number of connections. This method has disadvantage of restraining the approval of connection to legitimate nodes at times.

Clients who wish to be connected can be presented puzzles to solve, to show their commitment. Adversary needs to allocate more resources to carry out this attack. Puzzle scheme takes more energy resources than usual of the sensors by it also makes the flooding attacks more studious for the attacker.

Legal nodes need to put more resources to establish connection, which comes as a drawback of this procedure.

5.10 De-Synchronization

Adversary forges the control fields and the transport layer header to cause retransmissions and eventually lose of synchronization between communicating nodes. Authenticating the critical parts for transportation of the packets provides counter to this kind of assault on motes.

The receiver end detects any fake messages and is able to ignore the instructions carried out by them.

5.11 Sybil Attack

Insider node cannot be prevented from launching this attack, but its activities can be restricted. In order to prevent an insider from communicating within the network and establishing shared keys with every node in the network, the base station limits the number of neighbors any sensor can establish connection with. If any node tries to exceed this limit, it results in occurrence of error. By using this scheme, a node when compromised, is limited to communication with only a limited number of nodes which tend to be in its vicinity.

Moreover identities of the nodes which request to establish connections are verified. Each node shares its unique key with the base station. Neighboring nodes exchange information between themselves using the shared key to verify the communication. Compromised node is able to communicate only with its neighbors, thus restraining the affect of this attack.

5.12 Selective Forwarding

Like route alteration attack, the step to eradicate or avoid this attack is the use of multipath routing. This measure ensures that the destination finally gets the message sent towards it, through some disjoint path of that of malicious node.

Regular monitoring of the network enables the WSN to track suspicious behavior by any node. Source routing that uses the geographical monitoring of the network can also be used as a prevention measure to this type of attack.

Similar preventions and counter-measures can be applied to other attacks on WSN, as they are also variants of the described attacks.

6. LATEST RESEARCH AND IMPLEMENTATIONS

6.1 User Authentication

Authentication is one of the foremost security features, and hence it finds its applications in WSN security at different levels. It may include authentication of client nodes to authorize the access to channel or exchange of information, Or it may be in form of signing and verifying messages so as to ensure that the contents of the received messages are intact, which saves network from many attacks.

Now we will briefly mention some of the important works that have been done for authentication process for WSN.

Jaing et al. (Jiang, Li, & Xu, 2007) presents a distributed user authentication scheme in wireless sensor networks. This scheme uses self-certified keys cryptosystem (SCK). They make use of Elliptic Curve Cryptography (ECC) to establish pair-wise keys in their user authentication scheme. This user authentication scheme provides less computational and communication overhead.

Taojun Wu et al. (Wu, Skirvin, Wener, & Kusy, 2006) propose a group-based peer authentication scheme for real-time sensor applications. Authenticity and integrity of messages received by base station are crucial in final tracking results. They designed a security component MultiMAC, which uses SkipJack implementation in TinySec as symmetric cipher. Each sensor node stores a different set of keys in its memory, pre-defined by a key mapping scheme. Multiple message authentication code (MAC)s of every message are calculated in SkipJack, using the key set assigned to the sensor node. The receiver authenticates the message by recalculating MACs using its shared keys, thus providing authenticity of received message.

Broadcast authentication limits the number of clients requesting to establish connection by giving access to authorized and trusted nodes, and thus proves to be an important security service. In WSN, digital signatures (sign and verify) and μTESLA-based methods provide broadcast authentication. Both of these techniques can be exploited by the adversary, which results in increase of cost for sensors. Signatures are too expensive to be applied for every connection request and packet forwarding can be used on μTESLA technique, thus making these two methods a weak choice. P. Ning et al. (Ning, Liu & Du, 2008) suggests the use of message-specific puzzle scheme to ensure broadcast authentication, which proves to be much better in terms of cost and effectiveness.

In (Nyang & Mohaisen, 2006), a scheme for cooperative distributed public key authentication scheme that does not require any cryptographic overhead is presented. Each node stores a few number of hashed keys for other nodes. When a public key authentication is needed, the nodes who store this key help in authenticating it in a distributed and cooperative manner.

K. Han et al. (Han, Kim, & Shon, 2010) in their paper in "Sensors" propose an untraceable node authentication and key exchange protocol. The protocol adds light overhead which intends to increase the lifetime of the sensors. The protocol insures untracebility of the nodes, and works well in dynamic environments.

6.2 Key Establishment

Key establishment among nodes of sensor network is an important security aspect. Key establishment is needed for authentication and encryption processes, which are crucial for securing the network against many attacks. Key management maintains stability between sensor nodes in spite of their low operational efficiency.

Key establishement is performed by using public key protocol like Diffie-Hellman (DH), Elliptic curve DH and by using El-Gamal public key scheme.

Q. Huang et al. (Huang, Cukier, Kobayashi, Liu, & Zhang, 2003) presents an authenticated key establishment protocol between a sensor and a security manager in a self-organizing sensor network. This hybrid technique uses symmetric key operations instead of public key protocols to reduce the burden on the resource constrained nodes.

In (Jeung & Lee, 2007), authors propose efficient hybrid key establishment protocol for sensor network self-organized with equal distribution between sensor nodes. This protocol is applicable to distributed environment without control of base station. The scheme combines elliptic curve Diffie-Hellman key establishment with implicated certificate and symmetric key encryption technology.

Efficient implementations of cryptographic key establishment for WSNs pose a challenge to the limited capability nodes. A light weight implementation of elliptic curve Diffie-Hellman (ECDH; see Figure 2) key exchange for ZigBee-compliant sensor nodes is given in (Lederer, Mader, Koschuch, Gro?ch?l, Szekely, & Tillich, 2009). This implementation uses ATmega128 processor running the TinyOS operating system and it perform 192 bit prime field elliptic curve cryptography.

6.3 Trust Management

Trust between the cooperating entities is an important issue in any networked environment. Trust can solve some problems beyond the capability of traditional cryptographic security. It can be used to judge the quality of service being provided by any sensor, which can further help in deciding about provision of access control to that node. In simple networked systems, where security was not deemed necessary, it was assumed that all the parties participating in the communication in the network are trusted ones. But this is not applicable to modern network systems and same is true for wireless sensor networks. We need good trust model within the network to be able to establish connections, exchange keys and information.

M. Momani et al. (Momani, Challa, & Aboura, 2007) presents a trust model based on the observed difference in monitoring events and reporting data. This model takes sensor reliability as a component of trust.

H. Chen (Chen, 2009) proposes a task-based trust management framework for WSNs, in which nodes maintain reputation for other nodes of several different tasks and use it to evaluate their trustworthiness. The sensor node maintains a trust rating for different tasks while cooperating with other nodes. The node considers this trust rating to decide its priority to cooperative with nodes with different operations and tasks. A watchdog technique observes the behavior in different task of these nodes and broadcast their trust ratings.

6.4 Implementations

Implementations of different cryptographic protocols are widely discussed and researched, due to the difference in their strength, key sizes and application abilities. Some of the latest research in this regard is described here.

R. Roman and C. Alcaraz (Roman & Alcaraz, 2007) discusses the possibility of using public key infrastructure in wireless sensor networks, as earlier public key systems were considered too expensive. The authors state that this notion has been partially changed due to development of new hardware and software prototypes based on Elliptic Curve Cryptography (ECC) and other PKC primitives. They point out the possibility

Figure 2. Elliptic Curve Diffie-Hellman (ECDH)

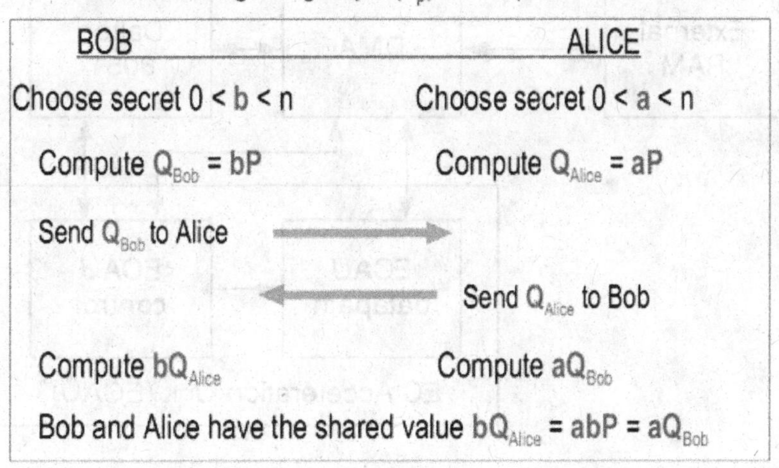

to incorporate public key infrastructure such as digital signatures, in the near future.

Hardware implementation of public key cryptosystems is given in (Wang & Li, 2006). The authors implement 1024-bit RSA and 160-bit ECC public key cryptosystems on Berkeley Motes. They achieve execution times of 0.79s for RSA public key operation and 21.5s for private operation, and 1.3s for ECC signature generation and 2.8s for verification. They also implement ECC on Telos B motes with signature time 1.60s and a verification time of 3.30s.In

In ECC, scalar multiplication takes most of the execution time. It has been estimated that nearly 80% of the time is taken by scalar multiplication step. Authors in (Huan, Shah & Sharma, 2010) suggest that there is a room to reduce the key calculation time to meet the potential applications, in particular for wireless sensor networks (WSN) by reducing the time needed for multiplications. They proposed that the positive integer in point multiplication may be recoded with one' complement subtraction to reduce the computational cost.

A. Liu and P. Ning (Liu & Ning, 2008) present the design, implementation scheme, and evaluation of TinyECC, which is a configurable library for implementation of ECC in wireless sensor networks. TinyECC provides a readymade, publicly available software package for ECC-based public key structures (Figure 3). Different optimization steps are included in TinyECC giving the developers the capability to utilize it on different platforms efficiently.

Author in (Mailk, 2010) states that most of the public-key cryptographic implemented on small devices are in conjunction with special purpose cryptographic hardware. Accelerators for many crypto functions are used along with small processors.

However in (Gurs, Patel, Wander, Sheueling, & Shantz, n.d.), authors implemented ECC without use of any special hardware. With the help of their new algorithm that reduces memory accesses, they achieved 160-bit ECC point multiplication on an Atmel ATmegal28 at 8MHz at 0.81s. This is the best known execution time for such an operation without using specialized cryptographic hardware.

Software and hardware co-design of ECC {GF(2^191)} is implemented in (Douceur, 2006) using Dalton 8051 and special hardware. The hardware consists of an elliptic curve acceleration unit (ECAU) and an interface with direct memory access (DMA) to enable fast data transfer between

Figure 3. System block diagram for software/hardware co-design of ECC

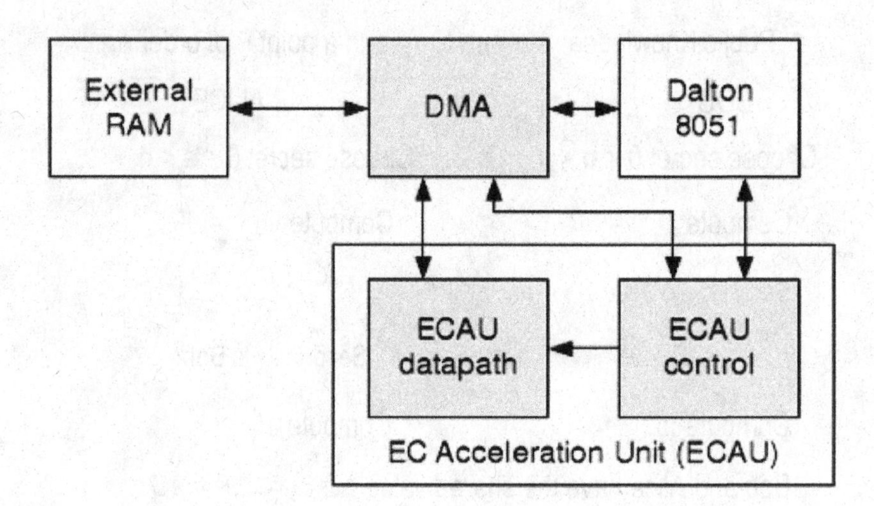

the ECAU and the external RAM (XRAM) attached to the 8051 microcontroller.

The special hardware and software combination enables the authors to perform the full scalar multiplication over the field GF (2^191) in about 118 msec, assuming that the Dalton 8051 is clocked with frequency of 12MHz.

Author in (Mailk, 2010) shows that ECC can be executed at 63.4 ms, by using TMS54xx type digital signal processors (DSP). With the decrease in the prices of DSP chips and their compactness, it is safe to think that these processors can be used in WSN sensors in near future.

7. CONCLUSION

This chapter serves as a text for researchers especially the beginners, and enables them to get an overview of this ever increasing area of research, wireless sensor networks. This chapter gives a brief yet extensive insight into intriguing world of sensors. Chapter contains many topics of interest, and many more can be found by investigating more deep into this research field.

Chapter has been divides into different sections, describing different aspects of WSN. Basic characteristics of WSN are discusses to give the readers an outline of WSN, which helps in understanding the attacks on WSN and their countermeasures. Some of the major attacks on WSN are given, along with their preventive and counter steps.

The challenges to the field of WSN are unique, and so are their security designs. In time to come, we must be ready to accept many more unique designs of WSN, more sophisticated attacks and their preventions.

REFERENCES

Chen, H. (2009). Task-based trust management for wireless sensor networks. *International Journal of Security and Its Applications*, *3*(2).

Douceur, J. (2002). The Sybil attack. In *Proceedings of the 1st International Workshop on Peer-to-Peer Systems* (IPTPS'02), February 2002. Retrieved from http://www.xbow.com/wireless home.aspx

Du, X., & Chen, H. (2008). Security in wireless sensor networks. *IEEE Wireless Communications,* 2008.

Gura, N., Patel, A., Wander, A., Sheueling, H. E., & Shantz, C. (n.d.). Comparing elliptic curve cryptography and RSA on 8-bit CPUs. Retrieved from http://www.research.sun.com/projects/crypto

Huang, X., Shah, P., & Sharma, D. (2010). Fast algorithm in ECC for wireless sensor network. *Proceedings of the International MultiConference of Engineers and Computer Scientists 2010,* vol.II, IMECS 2010, March 17 - 19, 2010, Hong Kong.

Jiang, C., Li, B., & Xu, H. (2007). An efficient scheme for user authentication in wireless sensor networks. *21st International Conference on Advanced Information Networking and Applications Workshops* (AINAW'07), (vol. 1, pp.438-442).

Lederer, C., Mader, R., & Koschuch, M. Groessschädl, J., Szekely, A., & Tillich, S. (2009). Energy-efficient implementation of ECDH key exchange for wireless sensor networks. *Information Security Theory and Practices, WISTP 2009,* (pp. 112–127). September 2009.

Liu, A., & Ning, P. (2008). TinyECC: A configurable library for elliptic curve cryptography in wireless sensor networks. In *Proceedings of the 7th International Conference on Information Processing in Sensor Networks* (IPSN '08), (pp. 245-256). Washington, DC: IEEE Computer Society.

Malik, M. Y. (2010). Efficient implementation of elliptic curve cryptography using low-power digital signal processor. In *Proceedings of the 12th International Conference on Advanced Communication Technology* (ICACT'10), (pp. 1464-1468). Piscataway, NJ: IEEE Press.

Momani, M., Challa, S., & Aboura, K. (2007). Modelling trust in wireless sensor networks from the sensor reliability prospective. *Innovative Algorithms and Techniques in Automation. Industrial Electronics and Telecommunications, 2007,* 317–321. doi:10.1007/978-1-4020-6266-7_57

Perrig, A., Szewczyk, R., Tygar, J. D., Wen, V., & Culler, D. E. (2002). Spins: Security protocols for sensor networks. *Wireless Networking, 8*(5), 521–534. doi:10.1023/A:1016598314198

Roman, R., & Alcaraz, C. (2007). Lecture Notes in Computer Science: *Vol. 4582. Applicability of public key infrastructures in wireless sensor networks. Public Key Infrastructure* (pp. 313–320).

Wang, H., & Li, Q. (2006). Lecture Notes in Computer Science: *Vol. 4307. Efficient implementation of public key cryptosystems on mote sensors. information and communications security* (pp. 519–528).

Wu, T., Skirvin, N., Werner, J., & Kusy, B. (2006). *Group-based peer authentication for wireless sensor networks.*

ADDITIONAL READING

Akyildiz, I. F., Su, W., Sankarasubramaniam, Y., & Cayirci, E. (2002, August). A survey on sensor networks. *IEEE Communications Magazine, 40*(8), 102–114. doi:10.1109/MCOM.2002.1024422

Anderson, R., & Kuhn, M. Tamper resistance - a cautionary note. In *The Second USENIX Workshop on Electronic Commerce Proceedings*, Oakland, California, 1996.

Anderson, R., & Kuhn, M. Low cost attacks on tamper resistant devices. In IWSP: International Workshop on Security Protocols, LNCS, 1997.

Aura, T., Nikander, P., & Leiwo, J. Dos-resistant authentication with client puzzles. In Revised Papers from the 8th International Workshop on Security Protocols, pages 170–177. Springer-Verlag, 2001.

Beresford, A. R., & Stajano, F. (2003). Location Privacy in Pervasive Computing. *IEEE Pervasive Computing / IEEE Computer Society [and] IEEE Communications Society*, *2*(1), 46–55. doi:10.1109/MPRV.2003.1186725

DaeHun Nyang and Abedelaziz Mohaisen. (2006). Lecture Notes in Computer Science: *Vol. 4159. Cooperative Public Key Authentication Protocol in Wireless Sensor Network* (pp. 864–873). Ubiquitous Intelligence and Computing.

D. Boyle and T. Newe,"Securing Wireless Sensor Networks: Security Architectures", *Journal of Networks*, 2008.

D. Braginsky and D. Estrin. Rumor routing algorthim for sensor networks. In *WSNA '02: Proceedings of the 1st ACM international workshop on Wireless sensor networks and applications*, pages 22–31, New York, NY, USA, 2002.

Han, K., Kim, K., & Shon, T. (2010). Untraceable Mobile Node Authentication in WSN. *Sensors (Basel, Switzerland)*, *10*(5), 4410–4429. doi:10.3390/s100504410

H. Chan, A. Perrig, and D. Song. Random key predistribution schemes for sensor networks. In *Proceedings of the 2003 IEEE Symposium on Security and Privacy*, page 197. IEEE Computer Society, 2003.

Hemanta Kumar Kalita and Avijit Kar, Wireless Sensor Network Security Analysis, *International Journal of Next-Generation Networks* (IJNGN),Vol.1, No.1, December 2009.

Hu, Y.-C., Perrig, A., & Johnson, D. B. *Wormhole detection in wireless ad hoc networks*. Department of Computer Science, Rice University, Tech. Rep. TR01-384, June 2002.

Huang, Q., Cukier, J. I., Kobayashi, H., Liu, B., & Zhang, J. "Fast Authenticated Key Establishment Protocols for Self-Organizing Sensor Networks", *International Conference on Wireless Sensor Networks and Applications (WSNA)*, ISBN: 1-58113-746-8, pp. 141-150, September 2003.

J. Granjal, R. Silva and J. Silva, Security in Wireless Sensor Networks. *CISUC UC*, 2008.

Koschuch, M., Lechner, J., Weitzer, A., Grobschadl, J., Szekely, A., Tillich, S., & Wolkerstorfer, J. Hardware/Software Co-Design of Elliptic Curve Cryptography o an 8051 Microcontroller, *CHES2006, LNCS4249*, pp. 430-444, 2006.

Liang, Z., & Shi, W. PET: A PErsonalized Trust model with reputation and risk evaluation for P2P resource sharing. In *Proceedings of the HICSS-38*, Hilton Waikoloa Village Big Island, Hawaii, January 2005.

Liang, Z., & Shi, W. *Analysis of recommendations on trust inference in the open environment*. Technical Report MIST-TR-2005-002, Department of Computer Science, Wayne State University, February 2005.

Liang, Z., & Shi, W. (2005). Enforcing cooperative resource sharing in untrusted peer-to-peer environment. *ACM Journal of Mobile Networks and Applications*, *10*(6), 771–783.

P. Albers and O. Camp. Security in ad hoc networks: A general intrusion detection architecture enhancing trust based approaches. In *First International Workshop on Wireless Information Systems, 4th International Conference on Enterprise Information Systems*, 2002.

Peng Ning, An Liu, and Wenliang Du. 2008. Mitigating DoS attacks against broadcast authentication in wireless sensor networks. *ACM Trans. Sen. Netw.* 4, 1, Article 1, February 2008.

Perrig, A., Stankovic, J., & Wagner, D. (2004). Security in wireless sensor networks. *Communications of the ACM*, *47*(6), 53–57. doi:10.1145/990680.990707

Mona Sharifnejad, Mohsen Shari, Mansoureh Ghiasabadi and Sareh Beheshti, A Survey on Wireless Sensor Networks Security, *SETIT 2007*.

Walters, J. P., Liang, Z., Shi, W., & Chaudhary, *V. Wireless Sensor Network Security: A Survey. Security in Distributed, Grid, and Pervasive Computing*. Yang Xiao, (Eds.) 2006.

Yoon-Su Jeong and Sang-Ho Lee. 2007. Hybrid Key Establishment Protocol Based on ECC for Wireless Sensor Network. In *Proceedings of the 4th international conference on Ubiquitous Intelligence and Computing (UIC '07)*, Jadwiga Indulska, Jianhua Ma, Laurence T. Yang, Theo Ungerer, and Jiannong Cao (Eds.). Springer-Verlag, Berlin, Heidelberg, 1233-1242.

Chapter 25
Wireless Sensor Network to Support Intelligent Transport Systems

H Ranganathan
Sakthi Mariamman Engineering College, India

ABSTRACT

The future generation of vehicles on the road is going to be driven by wire. To aid in this 'electronic' revolution in the vehicle, the role of wireless sensors and their interaction amongst themselves and with the environment is gaining importance. It is an area where the majority of research resources is allocated and being spent. For successful interaction of information from environment / vehicle, there is a need for wireless networking of the information from different sources. To keep pace with the development of wireless networks for intelligent transport systems, newer network architectures, protocols, and algorithms are being developed. This chapter sheds light on all these issues.

1. INTRODUCTION

The use of wireless sensor networks (WSN) in different fields is already discussed. One major area where WSN can be very useful is in transport. WSN can be effectively used to enhance the safety and mobility. We have been witnessing tremendous growth in Intelligent Transport Systems over the past three decades (Wang, 2010). Transport systems have become sophisticated with the introduction of electronics systems in vehicles. Growingly we find electronics finds increased use in vehicles. As per a study, the market for automobile electronics is growing @ 5.9% per year (http://www.oliverwyman.com/ow/pdf_files/1_en_PR_Automotive_Elektronics.pdf, n.d.). At this rate, it is expected that by the year 2015, electronics will contribute up to

DOI: 10.4018/978-1-4666-0101-7.ch025

30% of the value of the automobile. The major areas where the role of electronics is growing are safety, entertainment, information and comfort. We see that the growth potential is matched well with formidable technical challenges. In the current technology, we find increased transportation research and development are in the fields of Computer sciences, Control, Communication, Information Technology and many more emerging information science related fields (Wang, 2010). Modern Transportation Engineering encompasses all these fields along with traditional fields such as Civil, Mechanical and automobile engineering. All these developments are aimed towards making the journey pleasurable.

It can be found that majority of electronics components used in automobile are sensors and components used for intercommunication of information. There are attempts all over the world to standardize Intelligent Transportation Systems. The areas where sensors can aid the driver are Collision avoidance, Obstacle detection, Range detection, Reversing sensors, intelligent headlights and automatic breaking (Sawant, Tan, Yang, & Wang, 2004). Since there is a proliferation of attempts to increase the development efforts or to improve the existing design, we find there is a requirement for standardization of efforts so that the resources spent by various researchers are focused and not scattered. Network topologies, hardware and software designs are proposed for developing network protocols suitable for intelligent transport systems (Tao, Liu, & Ma, 2010).

To develop the wireless vehicle detection system, the California Department of Transportation (Caltrans) Division of Research and Innovation teamed up with the Partners for Advanced Transit and Highways (PATH) program, a research unit of the Institute of Transportation Studies at the University of California, Berkeley. Today the product is available on the market through Sensys Networks, a business founded by the researchers who developed the wireless vehicle detection system (http://www.techtransfer.berkeley.edu/newsletter/08-3/vehicle-detection-with-wireless-sensors.php, n.d.). The California Center for Innovative Transportation (CCIT) tested and reported on the wireless vehicle detector system on behalf of Caltrans DRI in October 2006.

The role of sensor networks in intelligent transportation systems can be between vehicles and between vehicle and stationary location such as vehicle to infrastructure or infrastructure to vehicle. The vehicle to vehicle (v2v) sensors are used for sharing information between vehicles. The concept of this direct communication is to send vehicle safety messages one-to-one or one-to- many vehicles via wireless connection (Iqbal, 2006). There are also systems for effective communication between the vehicles and the infrastructure. These include interactive information transfer between the vehicle and the traffic light controllers through cooperative technology. The technique is demonstrated by Audi which introduced Travolution vehicle to infrastructure communication system. The basic idea is to cut down on pollution and fuel consumption by reducing idling at stop signals and in some cases without the need to stop at all. The system also helps the drivers to avoid running the red signals by informing the driver about the status of the upcoming signal (will it be red or green) if the driver continues at current speed. The system helps the driver to keep track of traffic jams ahead and also payment at parking lots and gas stations (http://www.audi.in/sea/brand/in/company/news/company.detail.2010~06~audi_travolution_.html, n.d.).

The U.S. Department of Transportation, through the 1998 Intelligent Vehicle Initiative, identified eight areas where intelligent systems could "improve" or "impact" safety. The list includes four kinds of collision avoidances: rear end, lane change and merge, road departure, and intersection; two kinds of enhancements: vision and vehicle stability; and two kinds of monitoring: driver condition and driver distraction.

2. SYSTEMS IN USE FOR INTELLIGENT TRANSPORT

Currently the areas where wireless sensors networks can help the drivers are lane departure warning, road departure warning, curve over speed countermeasure, lane keeping assist, parallel parking assist, blind spot monitoring, lane change assist, roll over collision avoidance, night vision and park by wire. Let us discuss these in detail.

2.1 Lane Departure Warning

The lane departure warning systems (LDW) are intended for aiding the driver to keep the vehicle within the lane. These electronics systems monitor continuously the position of the vehicle within its lane. If the vehicle deviates from the lane and when the vehicle is about to deviate from its designated lane, LDW system alerts the driver. In the current technology, the systems are forward looking vision based systems. The systems interpret the images and use specified algorithms to estimate the vehicle state in terms of its lateral position and lateral velocity and roadway alignment in terms of lane width and road curvature. These systems are designed to minimize accidents by addressing the main causes of collisions: driving error, distraction and drowsiness.

US Department of Transportation has specified the Concept of Operations and Voluntary Operational Requirements for Lane Departure Warning systems on board commercial motor vehicles (http://www.fmcsa.dot.gov/facts-research/research-technology/report/lane-departure-warning-systems.htm, n.d.). LDW systems do not take any automatic corrective action in case the vehicle departs from its lane. So, it is the responsibility of the driver to maintain the vehicle's course and to make sure that safety of the other vehicles is maintained. Refer to Figure 1, we can understand the Lane Departure Warning system more closely:

Whenever the vehicle is travelling close to the center of the lane, it is said to be traveling in 'no warning zone' and so the system does not issue any warning to the driver. When the vehicle deviates from the no warning zone, the system calculates the time that may be taken for the vehicle to exit the lane. There are two warning lanes. They are 'earliest warning line' and the 'latest warning line' as shown in Figure 1. Whenever the vehicle departs from no warning zone in to the area be-

Figure 1. LDWS warning thresholds and warning threshold placement zones

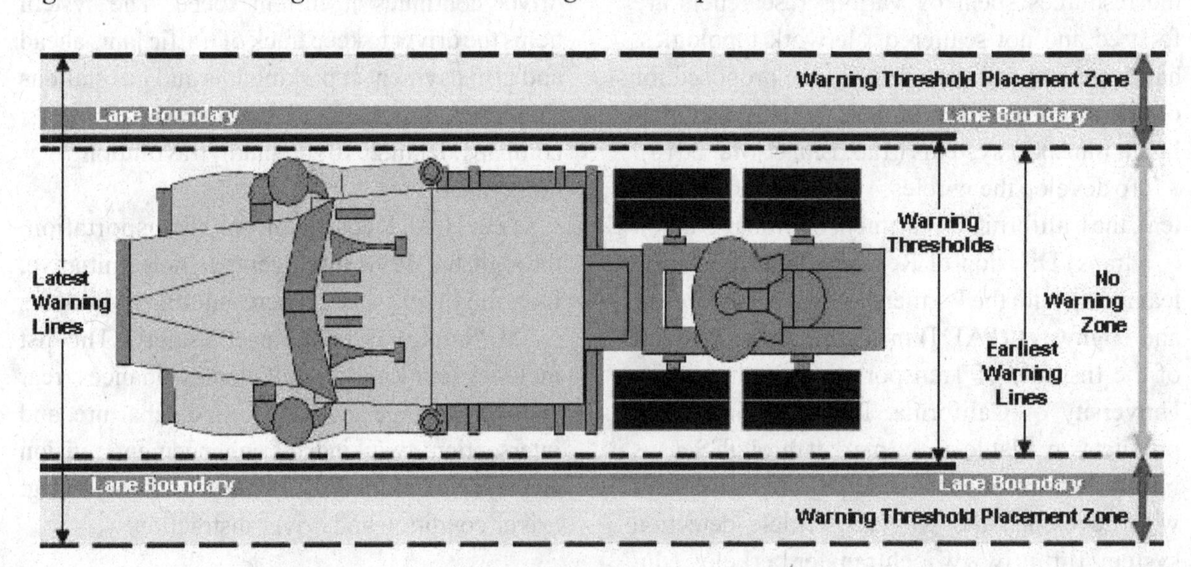

tween the earliest and the latest warning lines, the LDWS issues a lane departure warning. However, if the driver has indicated a turning by switching on the turn indicator, warning is not issued.

Figure 2 shows the hardware and software requirements for a LDWS. As shown, the Electronic Control Unit (ECU) takes data from the lane boundary sensors. The ECU checks the turn status of the vehicle and the engine power through Vehicle Network J1708 or J1709. If the vehicle is deviating from the lane, a warning is issued through driver vehicle interface.

2.2 Road Departure Warning

Simplest form of road departure warning can be in the form of rumble strips. Lane departure warnings mimic the sound of rumble strips. The sound comes from the side toward which the car veers. A waking driver can apply correction in the right direction instantly. Road Departure Warning system can be a combination of Lane Departure Warning system and Curve Departure Warning system (Barickman, 1995). Curve Departure Warning system is a technology to alert a driver that the vehicle is running too fast for negotiating an upcoming curve.

The Crewman's Associate for Path Control (CAPC) is a project funded and executed by US Army and Tank Automotive Command with assistance from Department of Mechanical Engineering and Applied Mechanics, University of Michigan (Ervin, et al., 1995). The system is an integrated system developed for prevention of road departure and implemented on a passenger car. The goal of CAPC is to limit the occurrence of inadvertent departure of motor vehicles from the travel lanes and intrusion in to adjacent road side risking rollover, collision with the fixed objects and possible uncontrolled reentry in to the traffic system. The system functions on the basis of the following operations:

1. Capturing of images of the road ahead generated through CCD camera
2. Prediction of vehicle's future position based on inertial sensing and model prediction
3. Computation of 'time to lane crossing (TLC) – beyond which the vehicle's mass center is expected to cross the lane
4. Decision making based on the above to decide if the warning is to be issued to the driver or to initiate an intervention control

The project is built on the interest of the Army in limited automation of driver function in military

Figure 2. Major functional (HW and SW) components of LDWS

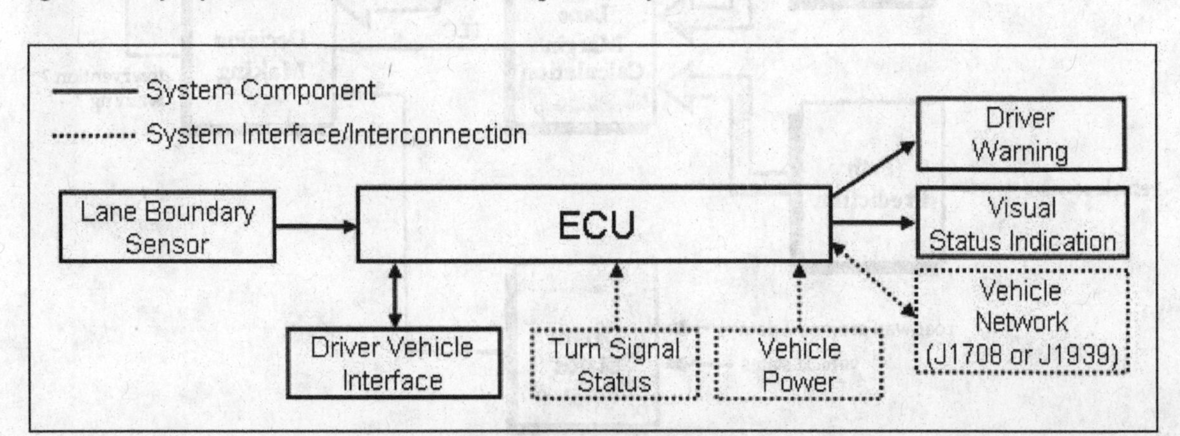

vehicles and the interest of Commercial Motor Vehicle Industry in safety technology in passenger vehicles. The system requirements are categorized in to three groups of requirements such as Driver activity, highway environment and warning and intervention process. The functional block diagram of CAPC is as shown in Figure 3.

The system observes the lane edge lines using the camera, converts the images in to geometric rendering of the roadway layout using the DSP hardware and software and compares this path with the predicted path of the vehicle in the near future time. Also considered are the time to lane crossing (TLC), the roadway crossing and the deduced status of the driver. The driver status is considered based on the driver activities in terms of effecting various controls in response to different roadway conditions. Audio, and visual warnings are issued and in case of necessity, differential braking is used as an intervention. For ensuring the smooth functioning of the system, different sensors are used. The various sub systems are as follows:

1. Lane marker sensor subsystem provides the lane marker data. This consists of a digital high resolution black and white CCD camera and associated image processing hardware and software to track lane marks.
2. Kalman filters are used for matching a model of roadway geometry with the perceived land mark data. Two Kalman filters are used, one for near - range data and one for far - range data to estimate near range and far range road geometry.
3. Driver status assessment system is used to assess the state of the driver. Driver state is used for deciding if a warning needs to be issued or action needs to be initiated or no action is required.

Figure 3. Functional block diagram of CPAC

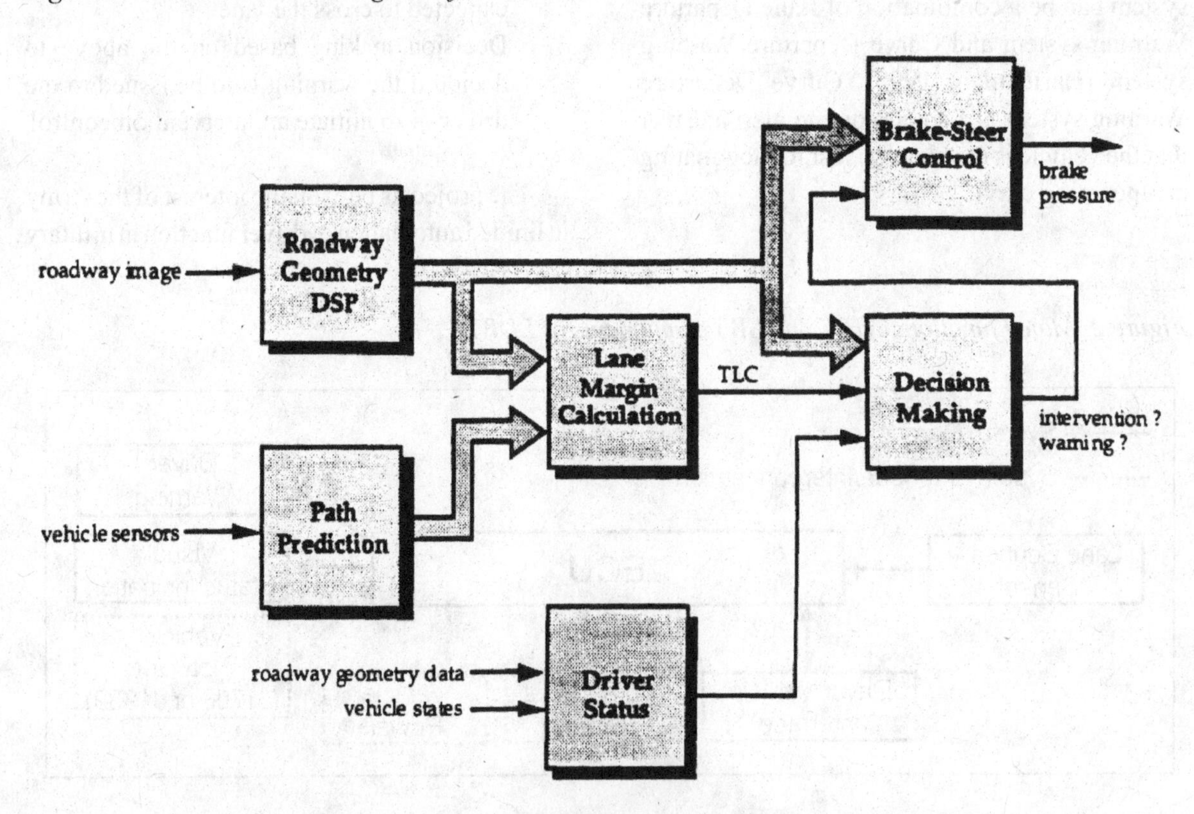

4. Brake – Steer controller that operates the rear brakes to correct the path of a vehicle if the CAPC system decides to intervene. This system includes ABS (antiskid Braking System), 4WS (4 Wheel braking System) and Yaw rate control using differential braking.
5. Design module to enable the system to make a decision based on the information available. The information is calculated TLC and driver state. The options available are:
   a. Warn only if the driver cannot perform lane keeping activity
   b. Intervene if the driver cannot keep the vehicle within the road
   c. Allow the driver the authority even at times of intervention
6. System simulation tool to combine all the sub systems in to a road departure avoidance system and to study the interaction among different sub systems.

2.3 Parallel Parking Assist

The Parallel Parking Assist takes the burden away from the driver when parallel parking. As most drivers can attest to, parallel parking is one of the hardest tasks that a driver has to learn early in his or her driving career. Otherwise, the chances of getting a parking space in busy cities will be near impossible. The Park Assist system uses ultrasonic based sensing system coupled with electric power steering to steer a vehicle into a parking slot. Electric power steering also helps to improve fuel efficiency by 5% and helps to save environment by reduced CO_2 emission. With the help of park assist system, the usually frustrating and stressful exercise of parking will become possible with the press of a single button (http://media.ford.com/article_display.cfm?article_id=29625, n.d.). Many commercial vehicles are now available with park assist feature.

Park assist system is available with variations. Some systems use camera and human interaction while the others use front and rear end ultrasonic sensors. The systems that use video camera require more driver interaction. Normally the sensor based system initiates the operation by the press of a button. After pressing the button, the driver has to only oversee the operation that is automatic. In this system, with the touch of a button, drivers can parallel park quickly, easily and safely without ever touching the steering wheel. The problems of parallel parking can be understood through Figure 4.

Active Park Assist works in tandem with other new technologies including Blind Spot Information System (BLIS™) and Cross Traffic Alert. BLIS employs a sensor on the outboard rear quarter panel that monitors the traditional blind spot area, and can notify the driver with a

Figure 4. Parallel parking of a car

warning indicator light in the corresponding side view mirror if the sensors in this optional system detect a vehicle in the blind spot. Cross Traffic Alert uses BLIS sensors to help detect cross traffic when backing out of a parking space.

2.4 Blind Spot Monitoring

On a motorway, a car which is far behind can be clearly seen in the rear view mirrors. However, as the car approaches, a point is reached where the car cannot be seen in either the interior or exterior mirrors. Typically this occurs when the car is just behind and to one side of the vehicle it is overtaking. It is a common mistake for drivers to change lanes when there is a vehicle in this so-called "blind spot", a maneuvers which causes many accidents on European motorways.

Several manufacturers have developed systems which monitor the blind-spot and help a driver to change lanes safely. Some systems are camera-based, others rely on radar. Either way, the area to one side and rearward to the vehicle is monitored and the driver is warned when there is a vehicle in a position where it may not be seen in the rear view mirrors.

A blind spot monitor is a vehicle-based sensor device that detects other vehicles located to the driver's side and rear. Warnings can be visual or audible. Increased warnings indicate potentially hazardous lane changes. Blind Spot Detection systems continuously monitor the rear blind spots on both sides of the vehicle. For example, before overtaking or changing lanes, the driver looks in the side mirror which confirms that the lane is free – but suddenly a car comes into the visual field from behind, just when the driver is about to change lanes. Such critical situations often arise in urban traffic and result in an accident if the vehicle in the blind spot is overlooked. When the turn signal is activated indicating that the driver is about to change lane, these systems warn the driver either visually or by discreet vibration of the steering wheel, if changing the lane is not safe at that moment. The concept of Blind Spot monitoring is explained by Figure 5.

Many commercial vehicle manufacturers offer the facility of blind spot monitoring.

Figure 5. Blind Spot monitoring

2.5 Night Vision

Night vision is the ability to see in a dark environment by biological or technological means. Night vision is made possible by a combination of two approaches, namely, sufficient spectral range, and sufficient intensity range. Technological night vision is accomplished by image intensification or active illumination or thermal imaging. With the proper night-vision equipment, one can see a person standing over 200 yards (183 m) away on a moonless, cloudy night.

3. ISSUES

We have seen the role of wireless sensors in building some intelligence in transport systems. However, the development in this area is fast and the role of sensors goes beyond providing some rudimentary support to the driver. We have other requirements for military and civil transport systems. Some of the issues are:

1. Road networks are under continuous surveillance in military operations. For the surveillance of these road networks, the localization of sensors is an important function. In such a case, unmanned aerial vehicles drop a large number of wireless sensors into road networks around a target area. For the localization, solutions have been proposed, using (i) precise range measurements (Priyantha, Chakraborty, & Balakrishana, 2000), (Wu, Liu, Pan, & Huang, 2007) or (ii) connectivity information (Bulusu, Heidemann, & Estrin, 2000), (He, Huang, Blum, Stankovic, & Abdelzaher, 2003) and (Lazos & Poovendran, 2004) between sensors for sensor localization. If large area in road networks needs to be covered, sensors have to be sparsely deployed to save costs. In this sparse deployment, since sensors cannot reach each other either through ranging devices or single-hop RF connectivity, the proposed solutions become ineffective.
2. WSN can be very effective in surveillance of infrastructure and sensitive areas. There are reports of surveillance in two dimensional spaces such as open battle fields (Tian & Geoganas, 2003), (Kumar, Lai, & Balogh, 2004). In the latest warfare, roadways are vantage areas for military surveillance and operations. Tracking of targets (say, vehicles) within these roadways is significantly different as the targets are moving continuously within the confines of road network that is clearly well known. The performance of systems as quoted above can be found to be less than optimal.
3. Vehicular Ad-hoc Networks (VANET), have emerged as one of the major research areas for providing safety and connectivity amongst vehicles, as we have seen in earlier section. They contribute to effective vehicle to vehicle or vehicle to infrastructure communication. These are accomplished mainly due to standardization of Dedicated Short Range Communication (DSRC) by IEEE (Carter, 2005). Simultaneously, GPS technology is becoming more popular. Though these technologies (DSRC and GPS) are popular individually on their own merits, there can be better benefits if these technologies can be integrated to make them to work together. GPS systems are used for navigation purposes and the sensors have the information such as road side reports (e.g., driving hazards – accidents, vehicle density and speed) and drivers video and audio data. Navigation becomes easier if the information also is available to the driver in addition to the GPS information. For smoother integration of technologies, there are requirements for communication between the vehicle and infrastructure. That is, vehicle data needs to be transferred to the nodes deployed along the roads (data forwarding) and information

needs to be transferred back to vehicles (reverse data forwarding). Some carry and forward algorithms for forward data transfer are proposed (Vahdat & Becker, 2000). These forwarding schemes could be inefficient especially in light traffic conditions as there could be delays in forwarding messages due to light traffic conditions. So, there is scope for improvement in such cases of data forwarding.

4. Similarly, the combination of GPS and DSRC technologies can be beneficial in reverse data forwarding. We have seen that many vehicles are using GPS based navigation techniques for selecting better driving paths both in terms of shorter route and less congestion routes. That is, in a roadway with sparsely deployed sensors for providing road-specific information to vehicles on physical road conditions and for warning vehicles about any accident or road block, infrastructure to vehicle data delivery makes use of Vehicular Ad-hoc Network (VANET). However, the vehicles are constantly on the move and so, there is a requirement for Disturbance Tolerant Network (DTN) for effective data delivery to individual vehicles on a continuous basis (Berry & Belmont, 1951). Earlier algorithms (Vahdat & Becker, 2000) are not designed for reverse data forwarding. The reverse data forwarding is more challenging as the data pocket destinations (vehicles) are on the move during the process of data delivery. For successful data delivery, an optimal rendezvous of the data pocket and vehicle destination point needs to be worked out.

We shall discuss different changes in wireless networks that are suggested mainly to accommodate the vagaries of transport systems. Wireless sensor networks are to be effectively used for the security and communications in the road networks as follows: (i) Localization for sensor location, (ii) Road Surveillance for vehicle monitoring, (iii) Data Forwarding for road sensing data delivery and (iv) Reverse Data Forwarding for road condition information sharing. The proposed network is expected to fit the requirements of the moving transport systems with the characteristics of Mobile Ad-hoc Network in such a way as to derive utmost benefits. The major characteristics that we expect from such network are predicting vehicle mobility within the roadways, representation of layouts of roadways into roadmaps, estimating vehicular traffic statistics and future vehicle mobility using GPS.

4. INTEGRATION OF INFORMATION FROM SENSORS – WIRELESS NETWORKS

Intelligent Transport Systems (ITS) include telematics and all types of communications in vehicles, between vehicles (e.g. car-to-car), and between vehicles and fixed locations (e.g. car-to-infrastructure). However, ITS are not restricted to Road Transport - they also include the use of information and communication technologies (ICT) for rail, water and air transport, including navigation systems. In general, the various types of ITS rely on radio services for communication and use specialized technologies. European Telecommunications Standards Institute (ETSI) (www.etsi.org, n.d.) produces standards for fixed telecom, mobile, radio, converged, broadcast and Internet technologies, supports the ITS domain with comprehensive standardization activities. Release 1 of a set of basic ITS standards has been published by them in early 2010 (http://www.etsi.org/WebSite/Technologies/Intelligent-TransportSystems.aspx, n.d.). This will lead to the harmonized and standardized development of ITS related products and their deployment on the market, responding to market demands. An overview is shown in Figure 6.

Figure 6. Overview of ITS

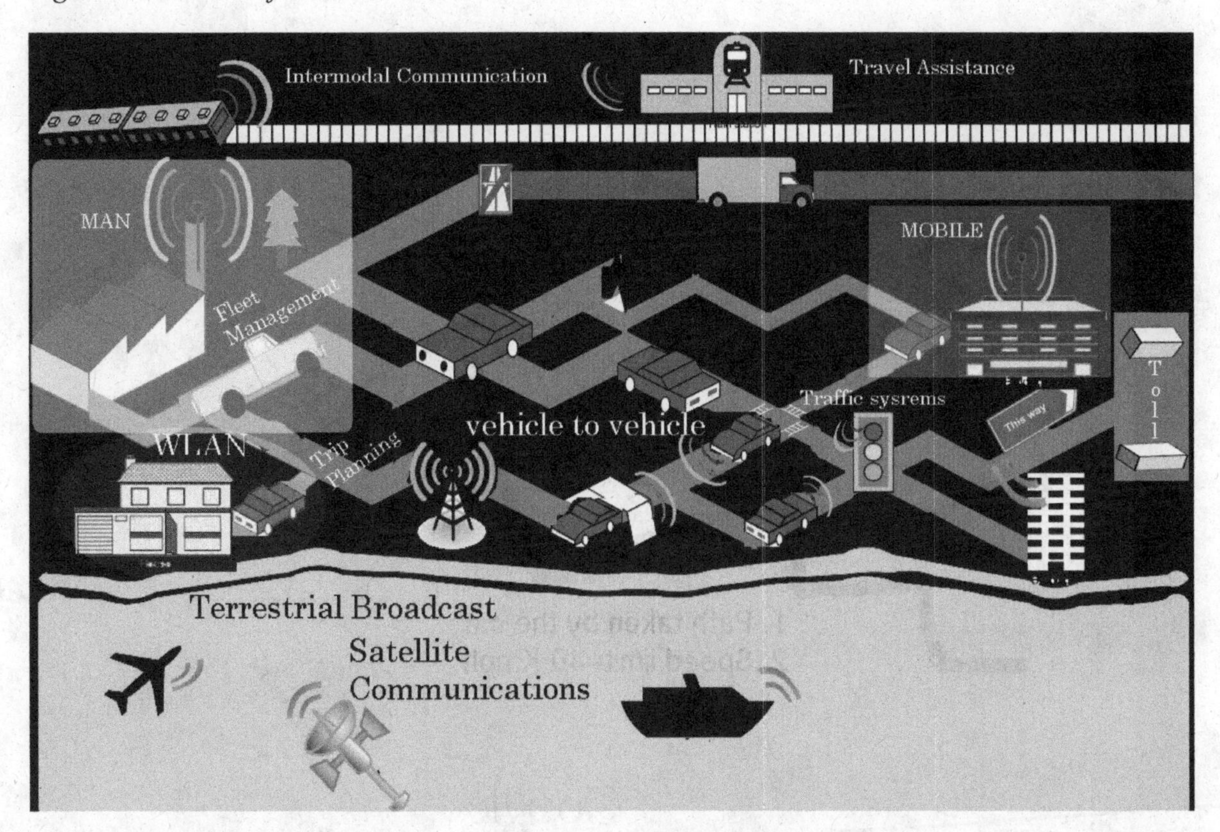

Based on the description of existing systems and their shortcomings, let us review the requirements again. Let us consider the characteristics of WSN for Intelligent Transport Systems. Let us reiterate the expected role of wireless sensor networks for Intelligent Transport Systems. The four characteristics of road networks namely, predictable vehicle mobility, road network layout, vehicular traffic statistics and vehicle trajectory are to be clearly understood before one can embark on improving the benefits of WSN in ITS (Jeong, 2009). Let us understand each of these as follows.

Vehicle Mobility refers to the possible speed with which the vehicles are expected to nove within a roadway. The limit may be fixed by the law or by the mechanics of traffic. Figure 7 shows the concept of vehicle mobility in detail.

Road network layout refers to the deployment sensors along a roadway of interest. As shown in Figure 8, the road network layout may be available as a road map. In this road network map, the vertices are intersections of road segments and the edges are road segments themselves.

Figure 9 explains the idea of vehicular traffic statistics. This is a measure of vehicle arrival time due to mobility of vehicles in the area of the road network and it can be collected from road segments or intersections.

In a specific road network area, each vehicle follows a specific path as it travels to its destination. Figure 10 shows the concept of vehicle trajectory for a single vehicle.

Using the four characteristics mentioned above and for addressing four main issues as stated in the earlier section, let us consider the four functions that need to be addressed for security and communications in transport sector for Military and civil applications. The four functions are: i)

Figure 7. Vehicle mobility

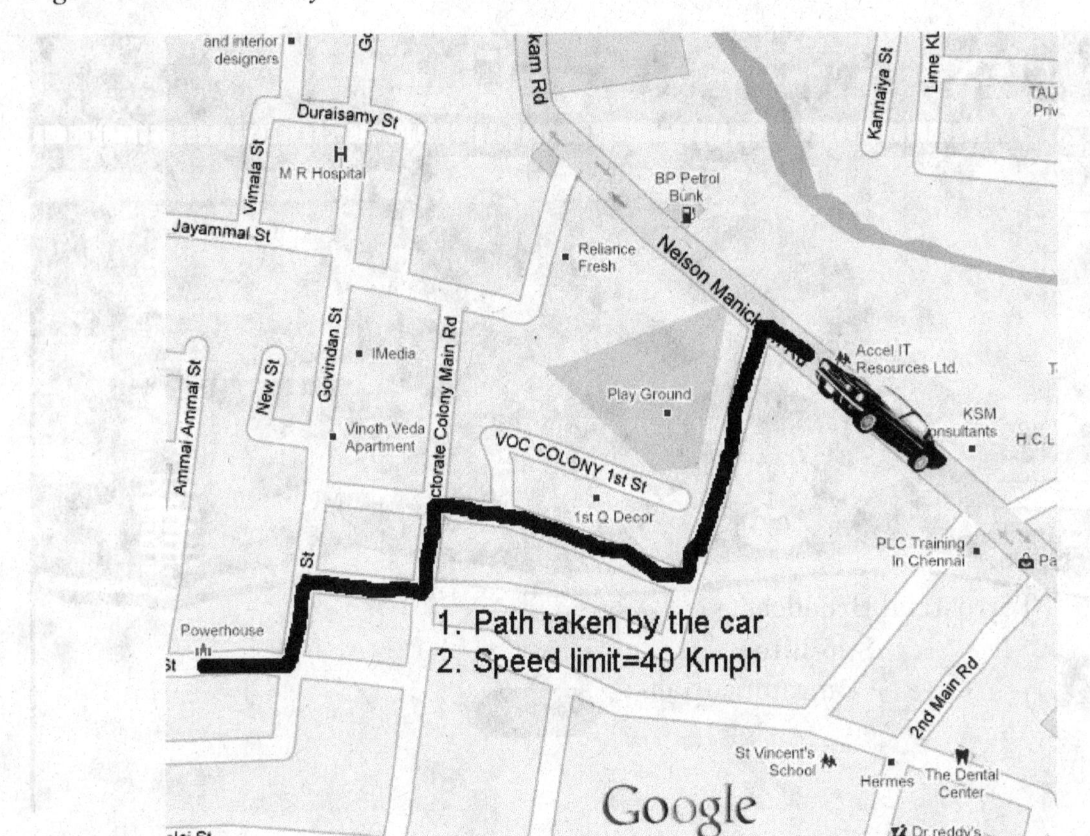

Localization for sensor location, (ii) Road Surveillance for vehicle monitoring, (iii) Data Forwarding for road sensing data delivery (iv) Reverse Data Forwarding for road condition information sharing. Let us discuss possible methods to effectively address the issues.

4.1 Localization of Sensor Location

We have seen that for large area localization of sensor location, the current schemes proposed do not meet the requirement. Some of the available research reports propose a number of algorithms to address this issue. Hu and Evans propose sequential Monte Carlo Localization method using mobile nodes and seeds and argue that it can exploit mobility to improve the accuracy and precision of localization (Hu & Evans, 2004). They reported that their approach does not require additional hardware on the nodes and works even when the movement of seeds and nodes is uncontrollable. Jonathan Bachrach and Christopher Taylor have presented the foundations of sensor network localization (Bachrach & Taylor, n.d.). They discussed localization hardware, issues in localization algorithm design, major localization techniques, and future directions. They have also summarized the tradeoffs and provided guidelines for choosing different algorithms based on context and available hardware. Mayuresh M. Patil, Umesh Shaha, U. B. Desai and S. N. Merchant propose a localization scheme plus a medium

Figure 8. Road network layout

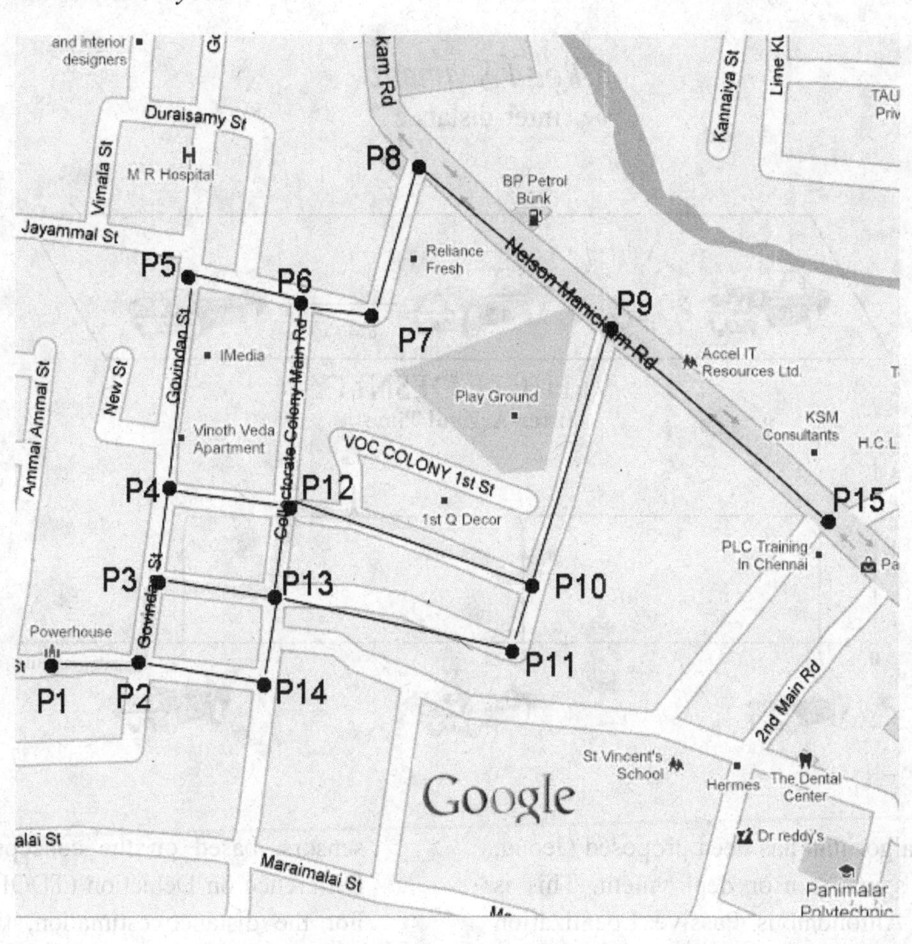

access protocol that (a) reduces power consumption, (b) does not require high power beacons, *(c)* provides better localization accuracy and (d) helps in reducing collisions (Patil, Shaha, Desai, & Merchant, 2005). The proposed localization scheme is based on received signal strength by the three masters. Moreover, it is a distributed algorithm. Vibha Yadav, Manas Kumar Mishra, A.K. Sngh and M. M. Gore discuss a range free localization mechanism for WSN that operate in a three dimensional space (Yadav, Mishra, Singh, & Gore, 2009). In this scheme, the sensor network is to be made up of mobile and static sensor nodes. Mobile sensor nodes are assumed to be equipped with GPS enabled devices and are expected to be aware of their position at any instance. These mobile nodes move in the network space and periodically broadcast beacon messages about their location. Static sensor nodes receive these messages as soon as they enter the communication range of any mobile node. On receiving such messages the static nodes calculate their individual position based on the equation of sphere. The proposed scheme gains in terms of computational and memory overhead as compared to existing approaches. This scheme assumes some nodes with expensive GPS hardware and they have shown only simulation results. It is found that each of these approaches has some limitations or other, major limitation being their inability to perform satisfactorily in sparsely deployed sensor network.

Figure 9. Vehicular traffic statistics

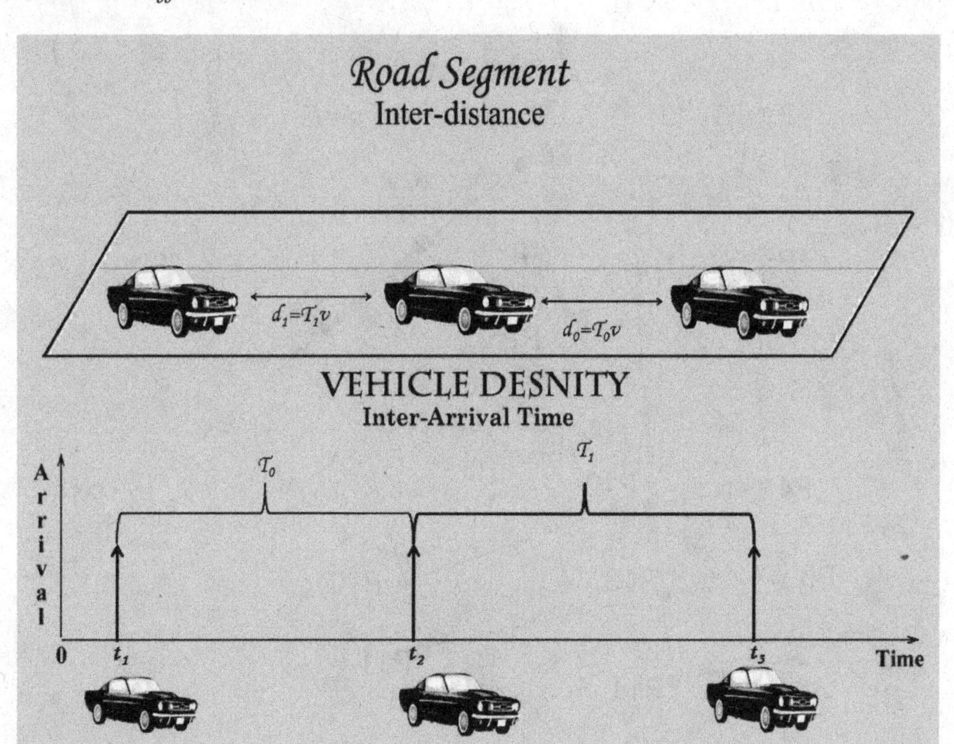

A new algorithm has been proposed (Jeong, 2009) for sparse sensor deployment. This is known as Autonomous Passive Localization (APL) algorithm. This scheme operates in three phases, namely, (i) the estimation of the distance between two arbitrary sensors in the same road segment; (ii) the construction of the connectivity of sensors on roadways; and (iii) the identification of sensor locations through matching the constructed connectivity of sensors with the graph model for the road map. The algorithm has the following characteristics:

- A new architecture for autonomous passive localization using only the binary detection of vehicles in the road networks is proposed. APL is designed especially for sparse sensor networks where long distance ranging is difficult.
- A statistical method to estimate road-segment distance between two arbitrary sensors, based on the concept of Time Difference on Detection (TDOD) is used. For the distance estimation, the TDOD operation uses the correlation between the timestamps of sensors geographically close to each other.
- A pre-filtering algorithm for selecting only robust edge distance estimates between two arbitrary sensors in the same road segment is proposed. Unreliable path distance estimates are filtered out for better accuracy.
- A graph-matching algorithm for matching the sensor's identification with a position on the road map of the target area is used. The graph matching uses the isomorphic structure between the road network and the sensor network.
- Considerations on practical issues, such as time synchronization error, vehicle detection missing, duplicate vehicle detection,

Figure 10. Vehicle trajectory for a single vehicle

and the sub-graph matching for the missing of sensor nodes or edge estimates make the algorithm robust.

Based on these strengths, APL system shows the encouraging results in the localization for the road network fully covered by vehicular traffic for the distance estimation along with one vehicle speed distribution. In order to enhance the APL for a more variety of road networks, other algorithms to deal with heterogeneous vehicular traffic density in a road network, considering heterogeneous vehicle speed distribution together can be developed.

4.2 Road Surveillance for Vehicle Monitoring

We have seen that the performance of existing algorithms for road surveillance has been sub-optimal. So, there is a requirement for a suitable algorithm that can detect target intrusion using the existing features of road network. Mihaela Cardei My T. Thai Yingshu Li Weili Wu suggest an energy efficient target coverage in wireless the sensor networks (Cardei, Thai, Li, & Wu, 2005). A critical aspect of applications with wireless sensor networks is network lifetime. Power-constrained wireless sensor networks are usable as long as they can communicate sensed data to a processing node. To cover a set of targets with known locations when ground access in the remote area is prohibited, one solution is to deploy the sensors remotely, from an aircraft. The lack of precise sensor placement is compensated by a large sensor population deployed in the drop zone that would improve the probability of target coverage. The data collected from the sensors is sent to a central node (e.g. cluster head) for processing. Here we propose an efficient method. Salma Begum, Nazma Tara and Sharmin Sultana propose Energy efficient target coverage in wireless sensor networks based on Modified Ant Colony algorithm (Begum, Tara, & Sultana, 2010) for increasing the life time of the wireless sensor network. They have suggested a method by selecting minimum working nodes that will cover all the targets. In their method, minimum working node is selected by modified Ant colony Algorithm. In their simulated results they have shown that their algorithm performs better compared to basic ant colony algorithm. Quan Zhao and Mohan Gurusamy propose Lifetime maximization for connected target coverage in wireless sensor networks based on Communication Weighted Greedy Cover (CWGC) algorithm (Zhao & Gurusamy, 2008). They have compared the performance of their simulated algorithm with basic algorithm and found that their algorithm performs better by 45%. They have also opined that the life time achieved by their algorithm is close to the upper bound. However, we have seen that the performance of these algorithms is sub-optimal. Moreover, the deployment of sensors in a road network needs be dense. So, we are in search of an algorithm that can offer optimum performance even in sparsely deployed sensor network.

Jeong proposes a sensing scheduling algorithm that uses the existing features of the road network (Jeong, 2009). This algorithm guarantees the detection of targets, entering from entrance points, before they reach one of protection points. One possible solution for road network surveillance is duty cycling, in which nodes wake up simultaneously for w seconds (the minimum working time before reliable detection can be reported) and then the whole network remains silent for T seconds. The detection is guaranteed if it takes more than T seconds for a target to travel along the shortest path between any pair of entrance points and protection points. Jeong proposes scan-based algorithm, which improves further energy efficiency of surveillance in road networks. In this, sensors wake up one by one for w seconds along road segments, creating waves of sensing activities, called virtual scanning. Waves propagate from one (or multiple) protection point P, split at the intersections, and merge along the route until they scan all of the road segments under surveillance. This scan-based method can achieve better performance than duty cycling algorithms. The concept of virtual scanning algorithm (VISA) is simple. However, in-depth design is very challenging due to a set of practical issues such as (i) how to optimize the network-wide silent duration T between scan waves, (ii) how to coordinate the working schedules of individual sensors during the scan, and (iii) how to deal with sensing holes due to unbalanced initial node deployment, node failure and the depletion of node energy over time. The design steps are as follows:

- A new architecture for surveillance in road networks. VISA is tailored for road networks, leading to orders-of-magnitude longer system life for target intrusion detection, using a novel scan-based algorithm.
- A sensing scheduling algorithm for an arbitrary road network. The working schedule of each sensor (i.e., when to wake up) is constructed in a decentralized way. The network- wide silent duration is computed by VISA scheduler and naturally disseminated along with sensing waves to the nodes in a network.
- An optimal sensing hole handling algorithm for uncovered road segments. The VISA scheduler deals with both the initial sensing holes at the deployment time as well as the sensing holes due to the hetero-

geneous energy budget among sensors by optimally labeling additional pseudo protection or entrance points.
- Considerations on two practical issues: (i) Detection failure probability and (ii) Time synchronization error. For each issue, we propose an optimal solution in terms of network lifetime

4.3 Data Forwarding for Road Sensing Data Delivery

We have seen that there are various efforts to use wireless sensors for safety and comfort in transportation. We have also seen that there is a requirement for combining DSRC and GPS technologies to make navigation smoother. Jing Zhao and Guohong Cao have suggested Vehicle-Assisted Data Delivery (VADD) in Vehicular Ad Hoc Networks to forward the packet with the lowest data delivery delay (Zhao & Cao, 2008). VADD is based on the idea of carry and forward. The most important issue is to select a forwarding path with the smallest packet delivery delay. Carter described the role of DSRC in crash prevention (Carter, 2005). However the reports do not envisage exploitation of combination of both the technologies of DSRC and GPS. It is found that the use of GPS for navigation is ever increasing. It is expected that 300 millions of portable GPS devices would have been being sold in 2009 alone (Yomogita, 2007). So, it is only natural to combine the two prominent technologies for better navigation.

Jeong proposed Trajectory-Based Data (TBD) Forwarding for Light-Traffic Vehicular Ad-Hoc Networks (Jeong, 2009). This work is motivated by the observed trend that a large number of vehicles have started to install GPS-receivers for navigation and the drivers are guided by these GPS-based navigation systems to select better driving paths in terms of the physically shortest path or the vehicular low-density traffic path. Therefore, the question is how to make the most of this trend to improve the performance of vehicular ad hoc networks. He proposes a data forwarding scheme utilizing the vehicles' trajectory information for light-traffic road networks. The initial challenge is how to use the trajectory information in a privacy-preserving manner, while improving the data forwarding performance. To resolve this challenge, he has designed an algorithm to compute expected data delivery delay (EDD) at individual vehicles to an access point, using private trajectory information and known traffic statistics. Only the computed delay is shared with neighboring vehicles. The vehicle with the shortest EDD is selected as the next packet carrier for its neighboring vehicles. The other challenge is how to model an accurate road link delay, which is the time taken for a packet to travel through a road segment using carry-and-forward. To resolve this challenge, he modeled road link delay, based on traffic density information obtained from the GPS-based navigation system. The steps are as follows:

- Modeling an analytical link delay model for packet delivery along a road segment that is much more accurate than that of the state-of-the-art solution. Besides serving as a critical building block of the TBD design, this link delay model is useful for other Vehicular Ad-hoc Network (VANET) designs, and the data dissemination through network-wide broadcast.
- An expected E2E delivery delay computation based on individual vehicle trajectory. The E2E delivery delay is estimated using both vehicular traffic statistics and individual vehicle trajectory. It turns out that this estimation provides a more accurate delivery delay, so vehicles can make better decision on the packet forwarding.

4.4 Reverse Data Forwarding

We have seen that the existing VANET algorithms are well suited for data delivery from vehicle to road sensor networks. There are times when the data from road sensor networks are to be transferred back to vehicle especially when the data regarding road conditions and information regarding any accident that may have occurred. Such reverse data transfer is more difficult when the vehicle that needs the data is on the move. For infrastructure-to-vehicle data delivery, the packet destination position needs to be accurately estimated considering the temporal-and-spatial meeting point of the packet and the destination vehicle. Jeong has investigated the reverse data forwarding based on the vehicle trajectory guided by GPS-based navigation systems. To ensure the meeting of a packet and a destination vehicle, an optimal target point is chosen as packet destination position in the road network in order to minimize the packet delivery delay while satisfying the user-required packet delivery probability. In order to search such an optimal target point, the idea is to use the two delay distributions: (i) the packet delivery delay distribution from the Access Pont (AP) to the target point and (ii) the vehicle travel delay distribution from the destination vehicle's current position to the target point. Once the target point is decided, TSF adopts the source routing technique, i.e., forwards the packet using the shortest-delay forwarding path as decided by multiple intersections in the target road network. The steps are as follows:

- A data forwarding architecture for the infrastructure-to-vehicle data delivery is proposed. The architecture adopts the stationary nodes (i.e., roadside units) for the reliable delivery.
- The distributions of the link delay and the E2E packet delay are modeled with the vehicular traffic statistics. The distribution of the vehicle travel delay is modeled with the destination vehicle's trajectory. These models are used for computing an optimal target point.
- With the packet delay distribution and the vehicle delay distribution, an optimal target point is selected to minimize the packet delivery delay while satisfying the user-required packet delivery probability

5. FUTURE WORK

With these four functions, the performance of the Intelligent Transportation Systems can be enhanced for the driving safety and efficient transportation on the road networks. These approaches open a new opportunity of the next-generation ITS using the characteristics observed on the road networks. A new research direction is the data dissemination of road condition information, such as accidents or obstacles related to the driving safety. With this data dissemination, the vehicles can change their movement paths for the better routes in terms of the travel time or gas consumption. An optimal data dissemination scheme can be devised considering the characteristics mentioned earlier. Another research direction is to utilize the trajectories of the vehicles fully or partially known to the infrastructures, such as Internet access points. With the known vehicle trajectories of the vehicles within a target road network, more efficient data forwarding or reverse data forwarding schemes can be devised.

Therefore, it can be envisioned that the future ITS will provide the road condition information to vehicles in real time and help vehicles avoid collisions at intersections or curved road spots, improving the driving safety. Also, it can be imagined that the roadside units are monitoring the accidents or collisions on the roadways or at intersections. With the help of wireless/ wired sensors and roadside units deployed on road networks, the ITS can disseminate this road information to the geographically adjacent vehicles before they enter the accident spots. This can reduce the traffic congestion due to the accidents by enabling the vehicles to select better routes beforehand for shorter travel time and less fuel energy consumption, leading to the green computing.

6. CONCLUSION

It is seen that the role of wireless sensor networks in Intelligent Transport Systems is on the increase and there are interesting research outcomes in the horizon. While there are some challenges in making the wireless sensor networks useful for Intelligent Transportation Systems, we also see that there are possible technologies to overcome the same. The result is that the future transportation systems will have all the popular technologies merged together to offer the best of the solutions towards safety, entertainment, information and comfort of the passenger. We also see that the progress is in the right direction for developing cutting edge architecture and formulating successful protocols for VANET to function well to serve the mankind.

REFERENCES

Audi India. (n.d.). *Audi travolution: Efficiently through the city.* Retrieved from http://www.audi.in/sea/brand/in/company/news/company.detail.2010~06~audi_travolution_.html

Bachrach, J., & Taylor, C. (2005). *Localization in sensor networks*. Retrieved from http://people.csail.mit.edu/jrb/Projects/poschap.pdf

Barickman, F. S. (2005). Lane departure warning systems. National Highway Traffic Safety Administration document, 2005.

Begum, S., Tara, N., & Sultana, S. (2010). Energy efficient target coverage in wireless sensor networks based on modified ant colony algorithm. *International Journal of Ad hoc, Sensor & Ubiquitous Computing, 1*(4).

Berkeley, I. T. S. (n.d.). *Vehicle detection with wireless sensors*. Retrieved from http://www.techtransfer.berkeley.edu/newsletter/08-3/vehicle-detection-with-wireless-sensors.php

Berry, D. S., & Belmont, D. M. (1951). Distribution of vehicle speeds and travel times. *Proceedings of the Second Berkeley Symposium on Mathematical Statistics and Probability.*

Bulusu, N., Heidemann, J., & Estrin, D. (2000). GPS-less low cost outdoor localization for very small devices. *OEEE Personal Communications Magazine*, 7(5), 28–34. doi:10.1109/98.878533

Cardei, M., Thai, M. T., Li, Y., & Wu, W. (2005). Energy efficient target coverage in wireless the sensor networks. *Proceedings - IEEE INFOCOM*, 2005.

Carter, A. (2005). *The status of vehicle-to-vehicle communication as a means of improving crash prevention performance.* Technical Report 05-0264, 2005. Retrieved from http://www.cs.umd.edu/class/fall2006/cmsc828s/PAPERS.dir/05-0264-W%5B1%5D.pdf

Ervin, R. D. (1995). *The Crewman's Associate for Path Control (CAPC): An automated driver function*. University of Michigan, Transportation Research Institute.

ETSI. (n.d.). *Intelligent transport systems*. Retrieved from http://www.etsi.org/WebSite/Technologies/IntelligentTransportSystems.aspx

ETSI. (n.d.). *Website*. Retrieved from www.etsi.org

Federal Motor Carrier Safety Administration. (n.d.). *Lane departure warning systems*. Retrieved from http://www.fmcsa.dot.gov/facts-research/research-technology/report/lane-departure-warning-systems.htm

Ford.com. (n.d.). *Ford makes parallel parking a breeze with new active parking assist*. Retrieved from http://media.ford.com/article_display.cfm?article_id=29625

He, T., Huang, C., Blum, B. M., Stankovic, J. A., & Abdelzaher, T. (2003). *Range-free localization schemes in large-scale sensor networks*. MOBICOM.

Hu, L., & Evans, D. (2004). Localization for mobile sensor networks. *Proceedings of MOBICOM*, Philadelphia, October.

Iqbal, Z. (2006). Self organizing wireless sensor networks for inter- vehicle communication. Thesis submitted at Halmstad University, School of Information Science, Computer and Electrical Engineering. Retrieved from http://urn.kb.se/resolve?urn=urn:nbn:se:hh:diva-230

Jeong, J. (2009). Wireless sensor networking for intelligent transport systems. Ph D thesis submitted to University of Minnesota, December 2009.

Kumar, S., Lai, T. H., & Balogh, J. (2004). On k-coverage in a mostly sleeping sensor network. *Proceedings of Mobicom*, Philadelphia.

Lazos, L., & Poovendran, R. (2004). SeRLoc: Secure range-independent localization for wireless sensor networks. *Proceedings of ACM 3rd Workshop on Wireless Security,* October.

OliverWyman.com. (2006). *Press release: Study on auto electronics*. Retrieved from http://www.oliverwyman.com/ow/pdf_files/1_en_PR_Automotive_Elektronics.pdf

Patil, M. M., Shaha, U., Desai, U. B., & Merchant, S. N. (2005). Localization in wireless sensor networks using three masters. *Proceedings of IEEE International Conference on Personal Wireless Communications*, New Delhi, January.

Priyantha, N. B., Chakraborty, A., & Balakrishnan, H. (2000). *The Cricket location-support system.* 6th ACM International Conference on Mobile Computing and Networking (ACM MOBICOM), Boston, MA.

Sawant, H., Tan, J., Yang, Q., & Wang, Q. (2004). Using Bluetooth and sensor networks for intelligent transportation systems. *Proceedings of IEEE Intelligent Transportation Systems Conference,* Washington DC, USA, 2004.

Tao, H., Liu, W., & Ma, S. (2010). Intelligent transportation systems for wireless sensor networks based on ZigBee. *Proceedings of IEEE International Conference Computer and Communication Technologies in Agriculture Engineering* (CCTAE), Chengudu, China.

Tian, D., & Geoganas, N. D. (2003). A node scheduling scheme for energy conservation in large wireless sensor networks. *Wireless Communications and Mobile Computing Journal, 3*(2).

Vahdat, A., & Becker, D. (2000). *Epidemic routing for partially-connected ad hoc networks.* Technical report, 2000. Retrieved from http://issg.cs.duke.edu/epidemic/epidemic.pdf

Wang, F.-Y. (2010). Parallel control and management for intelligent transportation systems: Concepts, architectures and applications. *IEEE Transactions on Intelligent Transportation Systems, 11*(3), 630–638. doi:10.1109/TITS.2010.2060218

Wu, K., Liu, C., Pan, J., & Huang, D. (2007). Robust range free localization in wireless sensor networks. *Mobile Networks and Applications, 12*(5). doi:10.1007/s11036-008-0041-9

Yadav, V., Mishra, M. K., Singh, A. K., & Gore, M. M. (2009). Localization scheme for three dimensional wireless sensor networks using GPS enabled mobile sensor nodes. *International Journal of Next-Generation Networks*, *1*(1).

Yomogita, H. (2007). *Mobile GPS accelerates chip development*. Retrieved from http://techon.nikkeibp.co.jp/article/honshi/20070424/131605/

Zhao, J., & Cao, C. (2008). VADD: Vehicle-assisted data delivery in vehicular ad hoc networks. *IEEE Transactions on Vehicular Technology*, *57*(3).

Zhao, Q., & Gurusamy, M. (2008). Lifetime maximization for connected target coverage in wireless sensor networks. *IEEE/ACM Transactions on Networking*, *16*(6).

Chapter 26
Middleware Systems for Sensor Network:
A Study on Different Middleware Systems

Nova Ahmed
North South University, Bangladesh

ABSTRACT

The middleware is a very important component in the wireless sensor network system. It has major challenges that are generic for any distributed systems as well as specific ones that are inherent to the physical nature of WSN systems due to the resource constraint nature. The author classifies WSN middleware systems as Event Based Middleware, System Abstraction Based Middleware, Application and Network Aware System, and Query Based Systems to gain a better understanding of such systems.

INTRODUCTION

Wireless sensor networks are gaining significant interest in recent years. The flexibility of sensor nodes to connect, communicate and compute opens up opportunities for many interesting applications ranging from monitoring of static, semi-static and dynamic environments. Researchers are able to gain insight from environments that are unreachable otherwise such as monitoring of animals in migration or situation in a battle field (Liu & Martonosi, 2003; Zebranet, 2004).

The wide varieties of applications require a proper management system of sensor nodes that are able to meet diverse set of characteristics. The application requirements comprise of real time query support, accuracy of sensor data, ability to support changes in sensor network system (eg, node failure or battery shortage), resource management etc.

DOI: 10.4018/978-1-4666-0101-7.ch026

Middleware systems that are designed specifically for sensor network systems are similar to middleware systems for other distributed, heterogeneous components – on top of the basic requirements, wireless sensor network middleware systems must take into account the resource limited sensor node characteristics. The middleware system must ensure efficient resource management ensuring maximum lifetime of the system.

We have looked at various wireless sensor network middleware systems and classified them according to the interactions among system components and operational strategies. We have taken a close look at each classes as well as the middleware systems supporting individual class We emphasize at the variations and requirements of applications. We consulted existing classification schemes as discussed in various literatures and surveys [Molla, M.M. and Ahamed, S.I., (2006), Hadim S. and Moamed N. (2006), Henricksen K. and Robinson R., (2006), Tubaishat, M. and Madria, S., (2003)] and created an extensive study that includes current and state of the art middleware systems.

We classify WSN middleware systems as Event Based Middleware, System Abstraction Based Middleware, Application and Network Aware System and Query Based Systems. Event based middleware are middleware systems allow events to trigger various system specific actions as it is implemented in systems like Impala (Liu & Martonosi, 2003; Zebranet, 2004) and DSWare (Li, Son, & Stankovic, 2004). System abstraction allows the middleware to hide specific details of the underlying sensor system for application development with lower degree of complexity, such as SENSORD/Stat (Sashima et al., 2008), Sensation (Hasiotis et al., 2005), Mate (Levis & Culler, 2002; Levis, Gay, & Culler, 2004), Magnet (Barr et al., 2002) and RF^{2}ID (Ahmed et al., 2007). Examples of query processing middleware are TinyDB (Madden et al., 2005), SINA (Shen, Srisathapornphat, & Jaikaeo, 2001) and Cougar (Johannes, Gehrke, & Seshadri, 2001) where query processing is inherent in the middleware system design. Application aware systems make desing and operational decisions based on sensor network applications. Examples of such systems are MiLan (Heinzelman et al., 2004), Middleware to Support Sensor, Agilla [Fok, Roman, & Lu, 2009), EnviroTrack (Abdelzaher et al., 2004) and AutoSec (Han & Venkatasubramanian, 2001). Classifying and discussing individual WSN middleware system provides insight to existing systems as well as future directions for WSN deployment and development.

We present the chapter in the following order, first we discuss examples of sensor network applications to present the importance of study on sensor network middleware systems, we follow the discussion by classifying the middleware systems for WSN in subsequent section, we discuss the contrast among various middleware classes after that and then we conclude our chapter.

WSN APPLICATIONS

We present various WSN applications as our motivation towards the study on various WSN middleware systems. The main category incorporates various monitoring schemes using sensor networks

Wildlife Behavior Monitoring

Wireless sensor nodes can be used to automate the study of mobile animals that would be difficult to study manually. For example, the ZebraNet (Liu & Martonosi, 2003; Zebranet, 2004) project attaches sensor nodes on wildlife like Zebras to study their migration behavior. The study places sensor nodes on Zebras at the Mpala Research Center at the Sweet Water Reserve in central Keya. The study enables interdisciplinary approach to take a close look at wildlife which are of interest to behavior researchers as well as researchers conducting sensor network studies.

Environmental Monitoring

Sensor nodes has been used to monitor the environment (Madden et al., 2005) such as Great Duck Island (Mainwaring et al., 2002), James Reserve, which is off the coast of Maine (Cerpa et al., 2001), and other locations (Brooke & Burrell, 2003; Madden, 2003) where sensor nodes are responsible to collect environmental parameters such as temperature, humidity etc. without human intervention.

Equipment Fault Detection and Monitoring

The redwood monitoring project (Madden, 2003) uses sensor nodes deployed on Intel fabrication plants to automatically detect and monitor equipment failures using signal vibration gathered from sensor nodes.

Surveillance System

Surveillance system (Sashima et al., 2008) is an emergency response system that is triggered by anomalies to check if there is any emergency situation. An example of emergency can be an even of fire. The system monitors the spatial distribution of temperature and detects a case of fire if the distribution is not adequate and takes proper actions.

Military Situation Awareness

In this application (sensor information networking arch) sensor nodes are dropped from airplane to find out about a disaster scenario for possible military applications.

Home and Office Security

It uses sensor nodes (Milan-app) to detect intrusion into the home and/ or working space along with the ability to detect any other alerts such as a fire alert.

Personal Health Monitor

Health monitor application (Milan –app) running on a mobile device such as a PDA are popular that are equipped with sensors like ECG, EMG, blood pressure etc. for people who need constant attention. Anomaly in sensor readings can trigger the proper personnel to take actions immediately.

Guidance and Monitoring

Sensors can be placed in the environment to enable guidance and monitoring capabilities for vulnerable population such as elderly people or people with visual impairment (GuarianAngel).

WSN MIDDLEWARE AND CHALLENCES

In general a middleware is a layer that acts as an intermediate abstraction to hide the hardware complexities making application development simple. A generic diagram of a middleware is presented in Figure 1.

WSN middleware systems have set of challenges due to the nature of the sensors being resource limited such as computational capabilities, limited battery power etc. The middleware system must be aware of the sensor limitations and complement such limitations as much as possible in the middleware system. We describe some of the key challenges faced by various WSN middleware systems. We classify the middleware challenges as general challenges and specialized challenges where the general challenges are applicable for any generic middleware system and the specialized ones are applicable to WSN middleware systems.

Figure 1. A generic middleware for WSN

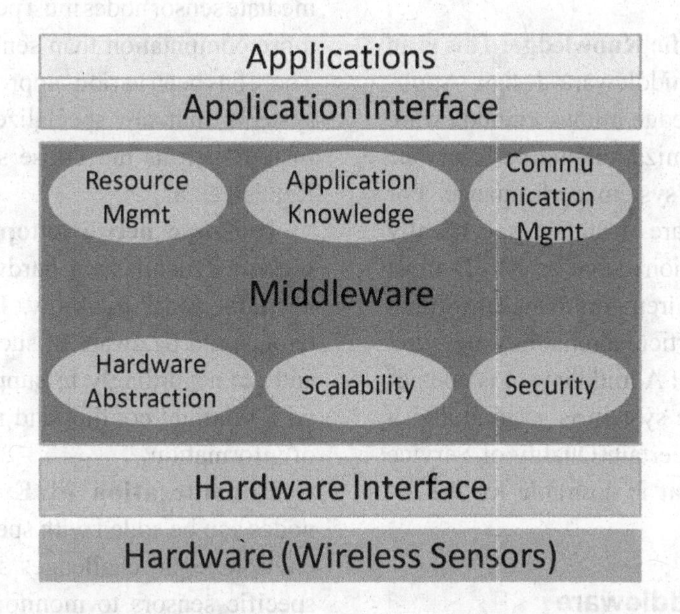

General Middleware Challenges

We discuss the important challenges that must be considered in most of the middleware systems that consider supporting various hardware systems for application development which are applicable for WSN systems as well.

Resource Management: Resource management is a feature desirable in different middleware support systems to make sure a middleware system is able to make proper resource management decisions to dynamically handle resource requirements. These challenges are stricter for WSN middleware systems as the resources are limited and in many cases may not be even physically accessible (e.g., sensors placed in a hazardous place dropped from an airplane). The resource constraint nature of WSN limits the compute and communication capabilities requiring careful design and deployment of middleware systems. The middleware system must support the application system so that the basic operations are supported without exhausting its resources.

The resource management is challenging due to its conflicting requirements where the sensors must stay dormant to conserve energy in its maximum possible time while making sure it is active enough to preserve connectivity. An optimal solution to meet these requirements is hard to achieve and a desirable component in middleware system considerations.

Hardware Abstraction: This challenge is considered in most of the middleware systems that deal with heterogeneous hardware. Different hardware present various platform, data acquisition and gathering interfaces which make the management task very challenging. A general abstraction to support all different hardware requires the knowledge of both the hardware and the application requirements. Abstracting the heterogeneity of network and hardware is considered as one of the major challenges in other literature as well (Molla & Ahamed, 2006).

Scalability: The middleware system must support addition or deletion of sensor nodes with minimal application involvement. The system must be able to meet increasing WSN size while ensuring system performance and proper management of sensor nodes.

Application Specific Knowledge: This is an important aspect of middleware design. Application specific knowledge allows a middleware to make specific optimization and performance decisions to enhance system performance. For example, a middleware designed specifically for real time applications such as RF^{2}ID must ensure the system requirements to meet real time performance with a particular reliability measure.

Security and QoS: A middleware is considered to make sure the system is secured and it is able to maintain a certain Quality of Service (QoS) requirement that is desirable for the set of applications.

WSN Specific Middleware Challenges

Special purpose challenges refer to the problems that are faced due to the physical and architectural properties of WSN itself. We discuss these challenges in the following subsections.

Dynamic network organization and location based information: Sensor networks require considerations based on limited resources such as compute capability, network bandwidth etc and the middleware system must take into account of such limitations when it organizes the network and provides routing mechanisms. The middleware system must be able to ensure the maximal lifetime of the network considering dynamic resource conditions and carefully designed algorithms Location based or location aware nodes can be very useful to make various system decisions and support applications based on proximity.

Data Fusion: Data fusion is the process of collecting sensor data and local processing of such data that contains more summary information of local sensor nodes. The data fusion across intermediate sensor nodes allows a way to eliminate unnecessary or extraneous information in the system and improve system performance. Data fusion requires intermediate processing and intermediate sensor nodes must perform comparatively more computation than sensor gathering nodes. There has been various approaches in middleware systems that are specialized in handling data fusion such as the Dfuse system presented by Rajnish et. al.

Dynamic network topology: WSN nodes present dynamism in hardware locations, connectivity and availability. The middleware system should be aware of such dynamic topology and act accordingly to support the applications with minimal conflict and maximal availability of information.

Consideration of Environment: WSN nodes can be added with special features to meet environmental challenges such as environment specific sensors to monitor, detect and trigger environmental parameters and the middleware must be able to perform accordingly.

CLASSIFICATION

Wireless sensor network has been studied in different literatures in a form of survey and studies (Hadim & Moamed, 2006; Henricksen & Robinson, 2006; Tubaishat & Madria, 2003). Our goal is to provide a comprehensive study of various systems and classify them for greater understanding which is complementary to existing studies. Several classifications of WSN middleware systems have been presented by literature mentioned above, we present a larger number of studied systems and classification that incorporates all the systems. We classify WSN based on how the system actually interacts among various system components such as

- Event Based Systems
- System Abstraction Based Systems
- Application Aware System
- Query Based Systems

Event Based Systems

Event based middleware systems allow events to trigger system specific actions. A general schematic of event based system is presented in Figure 2. The event based middleware systems are discussed in the following subsection.

Impala is an event based middleware system developed to support the Zebra project (Liu & Martonosi, 2003; Zebranet, 2004). Impala has been implemented in the ZebraNet efforts where sensors are attached to wildlife for long range migration studies. It acts as a light weight manager of event and device management for sensor nodes. It is designed specifically with the goal to support adaptive application software which takes into account various changes in application parameters. This particular property can be used to optimize system performances and it is one of the key properties of Impala. Impala supports adaptation of system components as sensor components can be prone to failure. The system uses wireless software update capability that arrive in modular pieces rather than a complete large complete software component. This property allows system flexibility that is required in WSN systems that are dynamic in nature.

The strength of the system lies in its ability to allow flexible adaptation and automated updating capabilities which are not possible in a monolithic system that is hardcoded along the line of application requirements. Impala allows a parallel system where system software running on sensor nodes can be updated in a very flexible way – it is able to handle partial updates which can be caused by many different environmental factors, it is able to progress the software upgrade in parallel to continuous execution of current system software which is very important where the update may take some time to complete. It uses effective versioning system for various updates of system software and the system provides methods for memory management schemes adequate for resource constraints sensor nodes.

The event based mechanism is implemented in Impala where every individual node acts as an event generator as well as even receiver. A sensor node may generate an event to send a message to a

Figure 2. Event based middleware

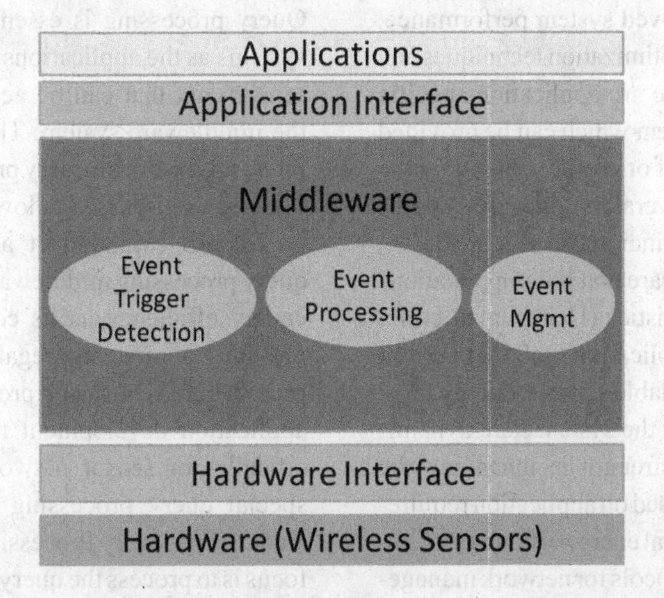

different sensor node which captures the event by its event handler and takes corresponding actions. The event based system is effective to handle the dynamic needs of the ZebraNet application.

The system provides many interesting features while it has no data fusion capability which limits certain optimization capabilities by reducing redundant data at intermediate points. It also does not consider heterogeneous components in the system.

DSWare (Li, Son, & Stankovic, 2004) is also a service oriented middleware which works on handling uncertain events in sensor network. It uses a notion of confidence to correlate sub-events in the system and figure out likelihood of a compound event. It manages a group of sensor abstracting failures of individual sensors. It uses an SQL like query language that is similar in approach as in Cougar. It has support for real time applications along with its support for data services. It can easily track failures. However the system does not support heterogeneity and mobility.

APPLICATION AWARE MIDDLEWARE

Application and network aware middleware systems take into account the application characteristics to for improved system performance. Application specific optimization techniques can take place only if there are application specific information in the system which can be provided by the application itself or can be profiled by the system. We discuss several middleware systems that incorporate this principle (see Figure 3).

MiLan is a middleware that links applications and network characteristics (Heinzelman et al., 2004). It considers application QoS and current system condition (available sensors, energy level of sensors etc.) to ajust the system state dynamically. MiLan has its strength in managing the network parameters based on application requirements to make an optimal energy efficient system. It uses the existing protocols for network management (e.g., service discovery) rather than proposing new ones which can become a shortcoming as it may not be suitable for resource constrained sensor networks.

Agilla (Fok, Roman, & Lu, 2009) is supports self adaptive applications using agent based technology. It structures the applications as mobile agents that can transfer from one sensor node to other sensor nodes. The system can allow a program to adapt to a dynamic change in the network. It also provides a scalable structure where the mobile agents can be replaced without affecting other agents.

EnviroTrack (Abdelzaher et al., 2004) considers physical events to be an addressable event in the system allowing greater flexibility in application development. This property allows monitoring and trigger applications to be developed without complexity.

AutoSec (Han & Venkatasubramanian, 2001) is a QoS aware middleware system driven by applications. It allows automatic service composition. It provides access control to maintain application level QoS.

QUERY PROCESSING MIDDLEWARE

Query processing is essential for sensor based systems as the applications often rely on specific sensor data that can be acquired, processed by the middleware system. There have been many interesting work on query processing middleware systems as discussed below (see Figure 4).

TinyDB (Madden et al., 2005) presents a query processing middleware system. It provides energy efficiency as its core functionality and provides a great aggregation model. TinyDB provides a great query processing interface for application development that is designed specifically for sensor network applications. The special query processing technique is named Acquisition Query Processing (ACQP) where the focus is to process the query based on sensor node

Figure 3. Application aware middleware

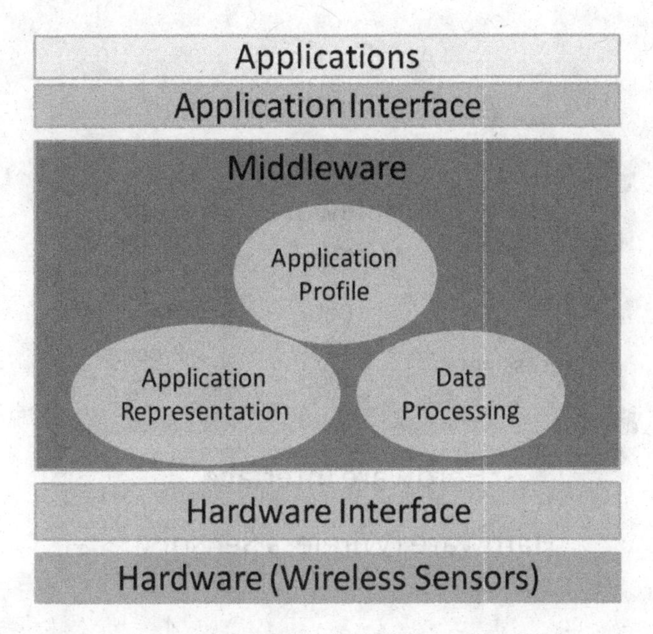

locations to minimize operational costs. TinyDB incorporates a distributed sensor processing infrastructure which runs a query processor on individual sensor nodes.

SINA (Shen, Srisathapornphat, & Jaikaeo, 2001) is a middleware system that runs on sensor nodes and allows applications to deliver queries, commands and collect results, monitor networks as a distributed database system. It automatically clusters sensor nodes to support energy efficient operations. It provides hierarchical clustering that takes into account the power level and proximity to form sensor clusters and attribute based naming for identification of sensor nodes. It uses a spreadsheet type of abstraction for data representation.

Cougar (Johannes, Gehrke, & Seshadri, 2001) provides a distributed database interface to support sensor related query applications. Energy efficiency is supported in Cougar by distributing the query among sensor nodes. It manages the sensor data as a form of a distributed database which is represented in a virtual relational database. It is suitable for large sensor collection of different network operations. It uses valuable resources for data transfer and uses a centralized optimization which may create a system bottleneck.

SYSTEM ABSTRACTION BASED MIDDLEWARE

System abstraction proposes hiding some of the complex system components and provides a simple set of API or components that are simpler to deal with from the application development perspective. It provides greater control over the underlying system and flexibility (see Figure 5).

The goal of this system is to hide the low level complexity and provide a common API named *SENSORD* (Event –Driven Service Coordination Middleware (SENSORD) (Sashima et al., 2008). The users are able to access sensor data using the spatio-temporal properties from SENSORD SENSORD/stat is an extension of SENSORD that cooperatively works on sensor data enabling statistical analysis techniques. It eases

Figure 4. Query processing middleware

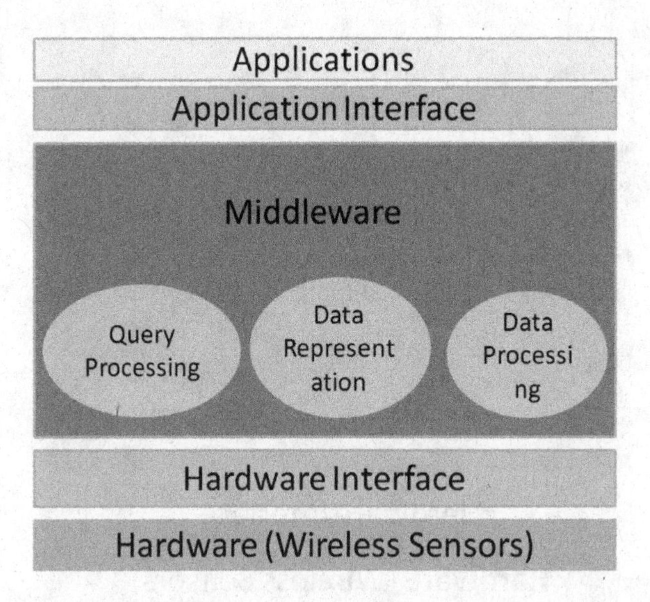

various complex application development such as surveillance in a way tat is desirable for application developers.

Sensation considers a middleware integration platform that coordinates among various heterogeneous middleware systems (Hasiotis et al., 2005). It aims to simplify the sensor related complexities for application developers. It enables low level handling, configuration of sensor network nodes that are prone to errors. It conceals the heterogeneity of different sensor networks and provides and appropriate programming model. It provides energy efficiency and scalability.

Mate (Levis & Culler, 2002; Levis, Gay, & Culler, 2004) provides a scalable infrastructure that uses active messages to update network protocols by injecting capsules where capsules are small code segments which are easy to inject in the network. It supports the application using the VM and provides heterogeneity, security and access control of network. It uses synchronous model that works when receives packets avoiding message buffering and large storage. However, the system requires management overhead when running high duty cycle programs as it has an overhead of instruction interpretation. It also provides a low level application development interface which is not easy to use and understand.

Magnet (Barr et al., 2002) uses virtual machine based system architecture. It is a power aware and adaptive middleware. It provides homogeneity for heterogeneous components of the system. It supports flexibility for programmers to adjust object placement and migration moving objects closer to the source. It is energy efficient but it has the overhead of instructions using java VM.

Global Sensor Networks (GSN) middleware aims to abstract the sensor specific details and provide a generic platform that hides the WSN heterogeneity. It supports the generic operations on sensor data such as query, filtering, aggregation of data in a distributed fashion The goal of this middleware is to support a large scale deployment of smaller sensor networks that are geographically distributed. Various standards and platform supports of sensors introduce complexity and cost in sensor network management. GSN aims to simplify by four design goals such as simplicity, adaptively, scalability and light weight implementation. It introduces virtual sensors that abstract the

Figure 5. System abstraction based middleware

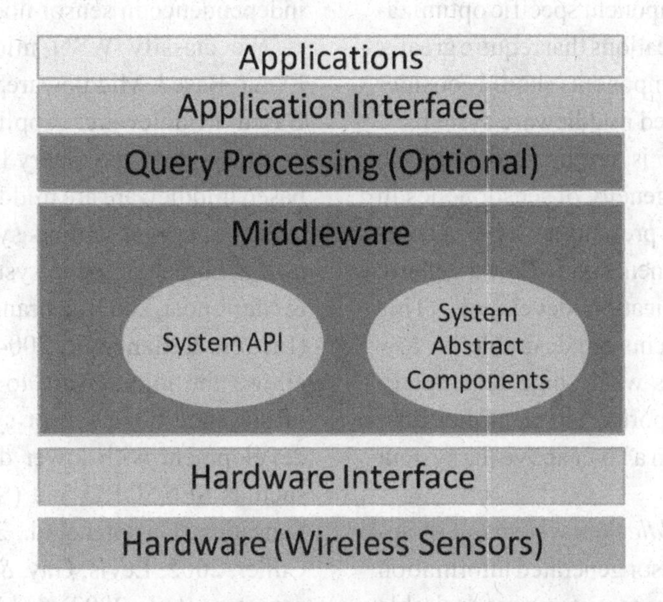

sensor specific details. It has been deployed using a combination of MICA 2 motes, RFID readers and wireless camera sensor network.

RF²ID (Ahmed et al., 2007) provides a system abstraction named virtual readers and virtual path that handles the computation and flow of sensor generated data respectively for RFID devices. It uses a distributed infrastructure to handle real time sensor generated data preserving system accuracy. The goal of the system is to provide system abstraction that understands the application requirement of data flow and incorporate system specific optimizations in the lower level to increase system performance in terms of response time and accuracy.

Dfuse (Kumar et al., 2003) is a middleware system that supports distributed data fusion techniques for sensor network systems. The system is able to dynamically make data fusion decisions based on current sensor status. It provides a set of APIs to support low level streaming data management generated by sensors. It presents a system abstract that enables the system to view it as a data flow graph.

DISCUSSION

It is a challenging task to compare among different middleware methodologies. Every technique has its own strength and weaknesses. The major challenge lies in the simple and limited resources of sensors that are to be handled very carefully. A combination of techniques cannot be used as it may increase the system complexity as well as drain the resources to operate such systems.

Event based middleware systems have the strength to capture the dynamic changes of the environment using resource constraint sensors and taking immediate actions by the event handlers. An event generator and event handler can be challenging to develop and deploy in sensor nodes requiring efficient design mechanisms and effective resource management in the middleware system.

Database oriented systems support various query mechanism in the sensor system. It is desirable for many application developers as it allows query specific operations and optimizations at low level. The query interface allows application developers to deal with query complexities while it limits the system component specific optimization capabilities. Applications that require greater control over system components should consider system abstraction based middleware systems.

System abstraction is required to hide the complexity and heterogeneity of sensor nodes in middleware systems. It presents a way to accommodate various components and allow a generic interface for the application developers. This set of middleware systems are desirable for low level tuning of systems where high level query processing can be supported either in the same middleware system or in a layer above the system abstraction layer.

Service oriented middleware systems allow applications to receive sensor generated information in a form of services. These systems are desirable in diverse sensor network where the sensor system has different services to offer and the subscription to services vary significantly.

Application oriented middleware systems are able to take into account various application specific parameters and it is able to incorporate optimizations based on application needs The system developed to support a single system is manageable However, a generic middleware system in this category can turn out to be extremely complex due to the various application requirements and possibilities available currently.

CONCLUSION

We have discussed various middleware systems designed to support WSN based applications. We have classified the middleware systems into .....

We have discussed various middleware that incorporate these classes. The classification and discussion allows us to take a comprehensive look at the state of the art middleware systems along with the applications supported by WSN and challenges faced by WSN. We assume the future research direction in this area would consider the current limitations and incorporate more independence in sensor nodes.

We classify WSN middleware systems as Event Based Middleware, System Abstraction Based Middleware, Application and Network Aware System and Query Based Systems. Event based middleware are middleware systems allow events to trigger various system specific actions as it is implemented in systems like Impala (Liu & Martonosi, 2003; Zebranet, 2004) and DSWare (Li, Son, & Stankovic, 2004). System abstraction allows the middleware to hide specific details of the underlying sensor system for application development with lower degree of complexity, such as SENSORD/Stat (Sashima et al., 2008), Sensation (Hasiotis et al., 2005), Mate (Levis & Culler, 2002; Levis, Gay, & Culler, 2004), Magnet (Barr et al., 2002) and RF^{2}ID (Ahmed et al., 2007).. Examples of query processing middleware are TinyDB (Madden et al., 2005), SINA (Shen, Srisathapornphat, & Jaikaeo, 2001) and Cougar (Johannes, Gehrke, & Seshadri, 2001) where query processing is inherent in the middleware system design. Application aware systems make desing and operational decisions based on sensor network applications. Examples of such systems are MiLan (Heinzelman et al., 2004), Agilla (Fok, Roman, & Lu, 2009), EnviroTrack (Abdelzaher et al., 2004) and AutoSec (Han & Venkatasubramanian, 2001). Classifying and discussing individual WSN middleware system provides insight to existing systems as well as future directions for WSN deployment and development.

REFERENCES

Abdelzaher, T., Blum, B., Cao, Q., Chen, Y., Evans, D., & George, J. ... Wood, A. (2004). EnviroTrack: Towards an environmental computing paradigm for distributed sensor networks. In *Proceedings of the 24th International Conference on Distributed Computing Systems (ICDCS'04)* (ICDCS '04), (pp. 582-589). Washington, DC: IEEE Computer Society.

Ahmed, N., Kumar, R., French, R. S., & Ramachandran, U. (2007). RF[2]ID: A reliable middleware framework for RFID deployment. *2007 IEEE International Parallel and Distributed Processing Symposium, IPDPS*, (p. 66).

Barr, R. (2002). On the need for system-level support for ad hoc and sensor networks. *Operating Systems Review*, *36*(2), 1–5. doi:10.1145/509526.509528

Brooke, T., & Burrell, J. 2003. From ethnography to design in a vineyard. In *Proceedings of the Design User Experiences (DUX) Conference*. Case study.

Cerpa, A., Elson, J., Estrin, D., Girod, L., Hamilton, M., & Zhao, J. (2001). Habitat monitoring: Application driver for wireless communications technology. In *Proceedings of ACM SIGCOMM Workshop on Data Communications in Latin America and the Caribbean*.

Deci, E. L., & Ryan, R. M. (1991). A motivational approach to self: Integration in personality. In R. Dienstbier (Ed.), *Nebraska Symposium on Motivation- Vol. 38: Perspectives on motivation* (pp. 237-288). Lincoln, NE: University of Nebraska Press.

Fok, C., Roman, G., & Lu, C. (2009). Agilla: A mobile agent middleware for self-adaptive wireless sensor networks. *ACM Transactions in Autonomic and Adaptive Systems,* 4(3).

Hadim, S., & Moamed, N. (2006). Middleware challenges and approaches for wireless sensor networks. *IEEE Distributed Systems Online, 7*(3).

Han, Q., & Venkatasubramanian, N. (2001). Autosec: An integrated middleware framework for dynamic service brokering. *IEEE Distributed Systems Online, 2*(7).

Hasiotis, T., Alyfantis, G., Tsetsos, V., Sekkas, O., & Hadjiefthymiades, S. (2005). Sensation: A middleware integration platform for pervasive applications in wireless sensor networks. *Proceedings of the Second European Workshop on Wireless Sensor Networks.*

Heinzelman, W., Murphy, A., Carvalho, H., & Perillo, M. (2004, January). Middleware to support sensor network applications. *IEEE Network Magazine Special Issue*.

Henricksen, K., & Robinson, R. (2006). A survey of middleware for sensor networks: State-of-the-art and future directions. In *Proceedings of the International Workshop on Middleware for Sensor Networks* (MidSens '06), (pp. 60-65). New York, NY: ACM.

Johannes, P. B., Gehrke, J., & Seshadri, P. (2001). Towards sensor database systems. Proceedings of the 2nd International Conference on Mobile Data Management, (pp. 3-14).

Kumar, R., Wolenetz, M., Agarwalla, B., Shin, J., Hutto, P., Paul, A., & Ramachandran, U. (2003). DFuse: A framework for distributed data fusion. In *Proceedings of the 1st International Conference on Embedded Networked Sensor Systems* (SenSys '03), (pp. 114-125). New York, NY: ACM.

Levis, P., & Culler, D. (2002). Mate: A tiny virtual machine for sensor networks. *Proceedings of the 10th International Conference on Architectural Support for Programming Languages and Operating Systems* (ASPLOSX), (pp. 85-95). ACM Press.

Levis, P., Gay, D., & Culler, D. (2004). Bridging the gap: Programming sensor networks with application specific virtual machines. *Proceedings of the 6th Symposium on Operating Systems Design and Implementation* (OSDI 04), 2004.

Li, S., Son, S., & Stankovic, J. (2004). Event detection services using data service middleware in distributed sensor networks. *Telecommunication Systems, 26*, 351–368. doi:10.1023/B:TELS.0000029046.79337.8f

Liu, T., & Martonosi, M. (2003). Impala: A middleware system for managing autonomic, parallel sensor systems. In *Proceedings of the Ninth ACM SIGPLAN Symposium on Principles and Practice of Parallel Programming* (pp. 107-118). New York, NY: ACM.

Madden, S. (2003). *The design and evaluation of a query processing architecture for sensor networks*. Ph.D. dissertation. University of California, Berkeley, Berkeley, CA.

Madden, S. R., Franklin, M. J., Hellerstein, J. M., & Hong, W. (2005). Tinydb: An acquisitional query processing system for sensor networks. *ACM Transactions on Database Systems, 30*(1). doi:10.1145/1061318.1061322

Mainwaring, A., Polastre, J., Szewczyk, R., & Culler, D. (2002). Wireless sensor networks for habitat monitoring. In *Proceedings of ACM Workshop on Sensor Networks and Applications.*

Molla, M. M., & Ahamed, S. I. (2006). A survey of middleware for sensor network and challenges. *International Conference on Parallel Processing Workshops*, ICPP, (p. 228).

Sashima, A., Ikeda, T., Inoue, Y., & Kurumatani, K. (2008). *SENSORD/Stat: Combining sensor middleware with a statistical computing environment.* International Conference of Networked Sensing Systems.

Shen, C.-C., Srisathapornphat, C., & Jaikaeo, C. (2001). Sensor information networking architecture and applications. *IEEE Personal Communications, 8*(4).

Tubaishat, M., & Madria, S. (2003). Sensor networks: An overview. *IEEE Potentials*, April-May 2003, 20-23.

Zebranet. (2004). *The ZebraNet wildlife tracker*. Retrieved December 21, 2010, from http://www.princeton.edu/~mrm/zebranet.html

ADDITIONAL READING

Ahmed, N., & Ramachandran, U. (2008). Reliable framework for RFID devices. In *Proceedings of the 5th Middleware doctoral symposium* (MDS '08). ACM, New York, NY, USA, 1-6.

Ekarna, S. (2010), Wildlife Photography, Retrieved on December 21, 2010 from http://www.sewildlife.com/Nature/African-wildlife/9982510_yn63s/2/731475023_DQpsS#731475023_DQpsS

Mpala (2009). Mpala. Retrieved December 21, 2010 from http://www.mpala.org/

Mpala Research Center. (2010). Mpala Research Center, Retrieved December 21, 2010 from http://www.princeton.edu/eeb/facilities/mpala-1/

Tiny, D. B. (2003), Retrieved on December 26, 2010 from http://telegraph.cs.berkeley.edu/tinydb/

Wikipedia. (2010) WSN. Retrieved December 21, 2010 from http://en.wikipedia.org/wiki/Wireless_sensor_network

Wikipedia. (2010) Middleware. Retrieved December 21, 2010 from http://en.wikipedia.org/wiki/Middleware

KEY TERMS AND DEFINITIONS

Event Based System: A system that has a mechanism to generate an event and take action when the event occurs using an event handler immediately.

Event Handler: A program that captures an event (e.g., new data captured by a sensor node) and takes corresponding actions (e.g., processing of sensor generated data).

Fusion Point: A point where sensor data are aggregated.

Middleware: A middle layer that hides hardware complexities and management issues for easy application development.

Mgmt: Management.

QoS: Quality of Service.

VM: Virtual Machine.

Wireless Sensors: Sensors enabled with wireless communicational capability and limited computational capabilities.

WSN: Wireless Sensor Network.

Chapter 27

A Grid-Based Localization Technique for Forest Fire Surveillance in Wireless Sensor Networks:
Design, Analysis, and Experiment

Thu Nga Le
Nanyang Technological University, Singapore

Xue Jun Li
Nanyang Technological University, Singapore

Peter Han Joo Chong
Nanyang Technological University, Singapore

ABSTRACT

This chapter presents a novel grid-based localization technique dedicated for forest fire surveillance systems. The proposed technique estimates the location of sensor node based on the past and current set of hop-count values, which are to be collected through the anchor nodes' broadcast. The authors' algorithm incorporates two salient features, grid-based output and event-triggering mechanism, in order to improve the accuracy while reducing the power consumption. The estimated computational complexity of the proposed algorithm is $O(N_a)$ where N_a is the number of anchor nodes. Through computer simulation, results showed that the proposed algorithm shows that the probability to localize a sensor node within a small region is more than 60%. Furthermore, the algorithm was implemented and tested with a set of Crossbow sensors. Experimental results demonstrated the high feasibility of good performance with low power consumption with the proposed technique.

DOI: 10.4018/978-1-4666-0101-7.ch027

INTRODUCTION

Wireless Sensor Networks (WSNs) provide unprecedented opportunities for monitoring areas of interests such as chemical factory, homes and offices, with low-cost, low-power and multi-functional sensors. As such, WSNs attract considerable amount of attention from researchers all over the world. Usually one should use a large number of sensor nodes to deploy a WSN because these sensors generally are small in size and can only communicate within short distances. Information can be collected from a WSN node through the base station. However, the collected information would be meaningless if we could not determine the location of a WSN node. Consequently, fast, efficient and low-cost localization techniques are highly desirable for WSNs applications.

The key idea of WSN localization is to allow some sensor nodes to know their own location at all time. Such nodes, usually called *anchors*, may be equipped with Global Positioning System (GPS) or be fixedly placed at pre-determined positions with known coordinates. For the sake of low cost, most sensor nodes do not know their locations. These nodes with unknown location information are called *non-anchor* nodes. Interestingly, their locations can be estimated by applying WSN localization techniques (Mao, Fidan, & Anderson, 2007).

Localization techniques in WSNs are classified into two groups: range-based and range-free techniques. Range-based techniques use sophisticated hardware to conduct complex measurements on distance or angle of signal arrival to obtain location estimates. Typical range-based localization schemes includes those using received signal strength (RSS) (Bahl & Padmanabhan, 2000), time of arrival (TOA) (Ward, Jones, & Hopper, 1997), angle of arrival (AOA) (Niculescu & Badri, 2003), and time difference of arrival (TDOA) (Priyantha, Chakraborty, & Balakrishnan, 2000). Noteworthily, range-based localization techniques are applicable only when the non-anchor node of interest is within communication range of the anchor nodes. Due to the expensive hardware requirement, range-based techniques are generally considered as high-cost solutions. Consequently, this shortcoming unfortunately hinders them from being applied for forest fire surveillance, which is normally formed by millions of sensor nodes.

Range-free algorithms estimate the location of a sensor only based on the connectivity between non-anchor nodes and anchors. Three typical existing range-free techniques are the DV-hop (Niculescu & Nath, 2001), Monte-Carlo Localization (MCL) (Hu & Evans, 2004) and Monte-Carlo Box (MCB) (Baggio & Langendoen, 2008) algorithms, which are revisited later. The general principle of these techniques is that localization can be estimated from the proximity constraints, which are defined by a sensor node of interest being in the transmission ranges of other sensor nodes. To the authors' best knowledge; none of aforementioned localization techniques is exclusively designed forest fire surveillance. Consequently, when the existing localization techniques were applied in forest fire surveillance systems, they would inevitably result in certain disadvantages such as high complexity, low efficiency and large power consumption. The reasons are that they normally return an exact point or coordination of the targeted sensor node's location. Thus, the algorithms are complicated in general. However, for forest fire detection, a precise location point may not be needed. In fact, we may need to find out an area of region inside a forest in where the fire is occurred.

This chapter aims to contribute a simple and novel localization technique for forest fire surveillance by monitoring and tracking groups of animals using WSN technology. The proposed technique combines the three existing range-free techniques and improves the accuracy. In brief, we propose to attach sensor nodes to selected animals. Whenever the temperature sensed at these animals' proximity rises beyond a predefined threshold, the localization module in the sensor would be activated and the subjects' motion paths are analyzed. The region of forest fire is estimated based on two indicators: (i) a group of animals are observed to run away from a certain area, and (ii) the temperature sensed around the animals' surrounding environment is higher than a predefined threshold.

BACKGROUND

Our proposed localization technique is range-free, which adopts similar assumptions as the existing three algorithms. In the following, we shall revisit the three existing range-free localization techniques in order to lay the foundation for the presentation of our proposed localization technique.

DV-hop Technique

Under DV-hop technique (Niculescu & Nath, 2001), each anchor node will broadcast an announcement message, which contains the anchor ID, position and a hop counter (initially set to 0). Each node also stores a hop-count table that records the least number of hops from itself to an anchor. Whenever a node receives an announcement message from a particular anchor node, it will update the corresponding entry of its hop-count table for that anchor node. Furthermore, it will modify the announcement message by increasing the carried counter value, and forward the resulted message to other nodes. At the end of the first network broadcast, all nodes will store the least number of hops from them to each anchor. The second broadcast is used to convert hop count to distance by multiplying the hop counts by an average hop distance. This technique can produce a relatively high accuracy in networks where sensor nodes are evenly distributed and the objects to be tracked are static.

MCL Technique

The Monte Carlo Localization (MCL) (Hu & Evans, 2004) is the first technique exclusively developed to track *mobile* sensor nodes. The algorithm calculates a set L_t of N location samples, each of which represents a possible location of the node to be tracked at time t. Initially, at t=0, MCL assumes that the node has no knowledge about its position; hence the first sample set L_0 consists of N random samples which are selected within the deployment area. At each time step, the set $\{l_t^i\}$ is updated based on possible movement of the node and new observations on the node's connectivity to the anchor nodes. This process can be divided into two phases:

1. **Prediction:** In this phase, the node uses its previous location and maximum velocity, v_{max} to predict its possible new location. For example, if the node was at location l_{t-1}^i at time t-1, its current location l_t^i should be within a circle with radius d_{max} from l_{t-1}^i, where d_{max} is the maximum distance that a mobile node can move within each time interval. Hence, from the old sample l_{t-1}^i the algorithm randomly selects a new sample l_t^i within the circle centered at l_{t-1}^i with radius d_{max}. By this way, from the previous sample set $L_{t-1} = \{l_{t-1}^i\}$, a new sample set $L_t = \{l_t^i\}$ can be predicted.
2. **Filtering:** In this phase, the node can eliminate some predicted samples obtained from the prediction phase based the connectivity between the node and the anchors which set up some space constrain to the node location. For example, if node M can hear an anchor A, its location must be within a distance r from A, where r is the radio range of the node/anchor. All location samples which fall out of this area ought to be eliminated. Consequently, the number of valid samples

Figure 1. Determine anchor box (Baggio & Langendoen, 2008)

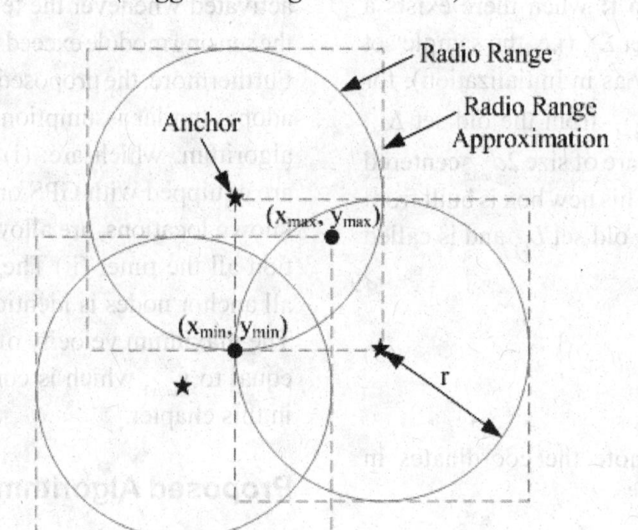

may drop below N due to elimination, hence re-sampling (repeating the prediction and filtering phase) is used to maintain N location samples at each time step.

Finally, the estimated location of the node at time t is the average of all N sample values in the sample set L_t.

MCB Algorithm

Monte-Carlo Localization Boxed (MCB) technique (Baggio & Langendoen, 2008) was developed based on the MCL technique. The major difference between MCB technique and MCL is on how to withdraw a new sample. In the prediction phase of MCB, new location samples are generated based on the following information: (i) information about the anchors heard by the mobile node, (ii) the maximum velocity v_{max} and (iii) the node's previous location. This would significantly reduce size of the area from which the new samples are withdrawn, thus improving the efficiency of prediction phase. Consequently, MCB reduces the number of re-sample iterations and speeds up the convergence. It is necessary to review the approach used to determine the area B from which location samples are withdrawn:

1. **Initialization:** At t=0, the node has no knowledge about its location. Let B_0 denote the initial 'anchor box' from which the first sample set L_0 is drawn.
   If the node is not connected to any anchor, $B_0 = \{(0,x_r); (0,y_r)\}$ where x_r and y_r is the maximum x and y coordinate of the deployment area. The first sample set L_0 consists of N samples selected randomly within the deployment area. Otherwise, B_0 is constructed from the location of all anchors that the node can communicate with:

$$B_0 = \left\{(x_{\min}, x_{\max}); \ (y_{\min}, y_{\max})\right\} \tag{1}$$

Let (x_j, y_j) denote the coordinates of the anchor j and N_a denote the total number of anchors heard:

$$\begin{cases} x_{\min} = \max(x_j - r), \ x_{\max} = \min(x_j - r) \\ y_{\min} = \max(y_j - r), \ y_{\max} = \min(y_j + r); \ j = 1 \ldots N_a. \end{cases} \tag{2}$$

2. **At each time step t:** when there exists a previous sample set L_{t-1} (i.e. the sample set is no longer empty as in Initialization), for each old sample l^i_{t-1} from the old set L_{t-1}. We construct a square of size $2d_{max}$ centered at the old sample. This new box is built from each sample in the old set L_{t-1} and is called a *sample box*:

$$B^i_t = \left\{(x^i_{\min}, x^i_{\max}), (y^i_{\min}, y^i_{\max})\right\} \tag{3}$$

Let $\left(x^i_{t-1}, y^i_{t-1}\right)$ denote the coordinates in l^i_{t-1}, we have

$$\begin{cases} x^i_{\min} = \max\left(x_{\min}, x^i_{t-1} - d_{\max}\right), \\ x^i_{\max} = \min\left(x_{\max}, x^i_{t-1} + d_{\max}\right), \\ y^i_{\min} = \max\left(y_{\min}, y^i_{t-1} - d_{\max}\right), \\ y^i_{\max} = \min\left(y_{\max}, y^i_{t-1} + d_{\max}\right), i = 1...N. \end{cases} \tag{4}$$

The area from which new samples are withdrawn would be the overlap of this square and the anchor box.

PROPOSED LOCALIZATION TECHNIQUE

The proposed localization technique reduces computational complexity and improves efficiency by the two features: (i) *Grid-based Output:* The algorithm divides the monitored area into an $n \times n$ grid structure and generates an output region as the estimated node location; which may consist of one or many neighboring grids; (ii) *Event-triggering Mechanism:* We proposed to adopt sleep-wake cycles for the localization module in the sensor nodes in order to save the scarce battery power of sensor nodes without sacrificing the algorithm's efficiency. In other words, the localization module does not operate continuously; instead, it is only activated whenever the temperature obtained by the sensing module exceeds a predefined threshold. Furthermore, the proposed localization technique adopts similar assumptions as the MCL and MCB algorithm, which are: (i) Anchor nodes, which are equipped with GPS or fixedly-placed at pre-known locations, are allowed to know their location all the time; (ii) The transmission range of all anchor nodes is identical and equal to *R*; (iii) The maximum velocity of animals is known and equal to v_{max}, which is considered to be 25km/h in this chapter.

Proposed Algorithm

Let the monitored area, e.g. the forest or nature reserve park, be bounded within x-coordinates $(0, X_s)$ and y-coordinates $(0, Y_s)$. The algorithm first divides the area into an $n \times n$ grid structure and places four anchor nodes at four corners. In particular, if the area is large or not of a square shape, we may repeat this arrangement over and place more anchors at corners of each $n \times n$ grid structure. For the sake of clarity, the smallest area unit is named as *grid* and each $n \times n$ grid structure as *square*. As shown in Figure 2, the dimension of each grid is $r \times r$, where $r = R \times cos45°$.

Similar to the DV-hop technique, our algorithm requires a set of distance information from a sensor node to each anchor node. Whenever the temperature sensed at a sensor node rises above a predefined threshold, the node would send a packet containing the set of hop-counts to the base station, which must store two sets of hop-counts: the current set of hop-counts and the last set of hop-counts. These two sets of hop-count values serve as the input for our algorithm. The current set of hop-counts is $H^c = \left\{h^c_1, h^c_2, ..., h^c_{N_a}\right\}$ where h^c_j is current number of hops from the sensor node to anchor node *j*; the past set of hop-counts is $H^p = \left\{h^p_1, h^p_2, ..., h^p_{N_a}\right\}$ where h^p_j is past number of hops from the sensor node to anchor node *j*;

Figure 2. A square consisting of 16 grids in a 4×4 structure

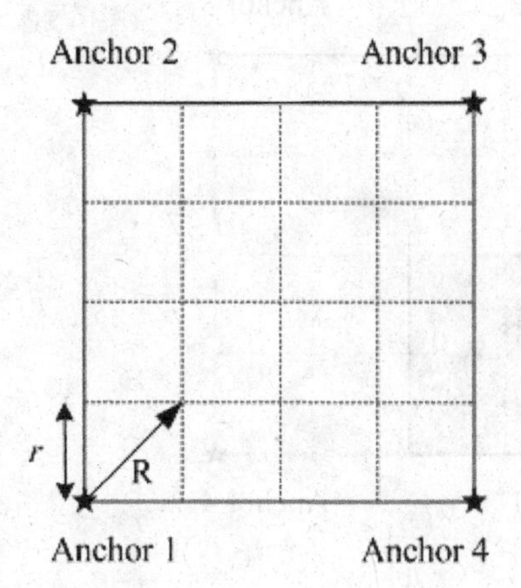

Figure 3. When h_1^c =2, M must be inside this 2×2 region

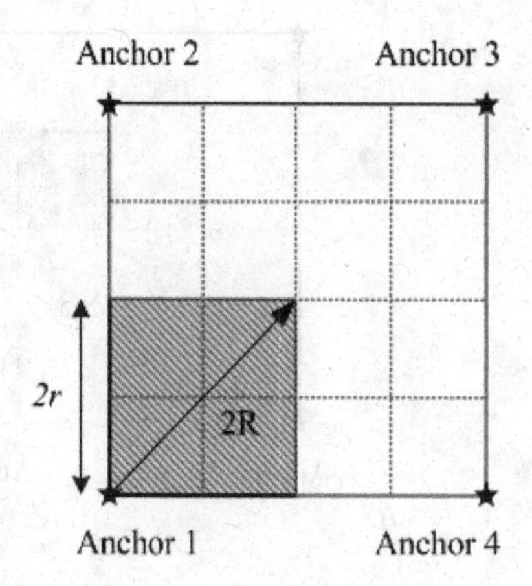

and N_a is the number of anchor nodes. The output of our algorithm is the estimated rectangular region-based location B_{est} of node M expressed in terms of left-bottom and right-top vertices' coordinates (x_{min}, y_{min}) and (x_{max}, y_{max}).

In brief, the localization process consists of two phases: phase I uses the *current* set of hop-counts to estimate node location. If the tentative output is smaller than or equal to a single grid, the algorithm terminates; otherwise, it proceeds to Phase II, which uses the *past* set of hop-counts and d_{max} to estimate the node location on a best-effort basis.

1. **Phase I:** In this phase, the tentative node location is determined based on H^c. For example, consider N_a=4 and $H^c = \{h_1^c, h_2^c, h_3^c, h_4^c\}$, then the algorithm estimates the location of sensor node M as follows:

   For each values h_j^c in H^c, the algorithm determines the region B_j^c which is the h_j^c-hop coverage area of anchor A_j, and B_j^c is a square with side length $h_j^c r$. The upper-bound and lower-bound for x and y coordinates of B_j^c are:

$$\begin{cases} x_{j,\min}^c = \max\{0,\ X_{A_j} - rh_j^c\} \\ x_{j,\max}^c = \min\{X_r,\ X_{A_j} + rh_j^c\} \\ y_{j,\min}^c = \max\{0, Y_{A_j} - rh_j^c\} \\ y_{j,\max}^c = \min\{Y_r,\ Y_{A_j} + rh_j^c\} \end{cases} \quad (5)$$

As shown in Figure 3, for instance, if h_1^c =2, node M must be located inside the shaded square consisting 4 grids, which represents the 2-hop coverage area of anchor A_1. Next, the algorithm finds the overlap of all regions, i.e., B_j^c, where j is 1, 2, ..., N_a. Then,

$$B^c = B_1^c \cap B_2^c \cap \cdots \cap B_{N_a}^c \quad (6)$$

The region B^c would be the tentative estimated grid-based location of sensor node M. The upper-bound and lower-bound for x and y coordinates of this rectangular region are:

$$\begin{cases} X_{\min}^c = \max\{x_{1,\min}^c, x_{2,\min}^c, \dots, x_{N_a,\min}^c\} \\ X_{\max}^c = \min\{x_{1,\max}^c, x_{2,\max}^c, \dots, x_{N_a,\max}^c\} \\ Y_{\min}^c = \max\{y_{1,\min}^c, y_{2,\min}^c, \dots, y_{N_a,\min}^c\} \\ Y_{\max}^c = \min\{y_{1,\max}^c, y_{2,\max}^c, \dots, y_{N_a,\max}^c\} \end{cases} \quad (7)$$

Figure 4. Estimated region for node M after phase I for (a) H^c=[3, 3, 2, 3]; (b) H^c=[2, 0, 0, 0]

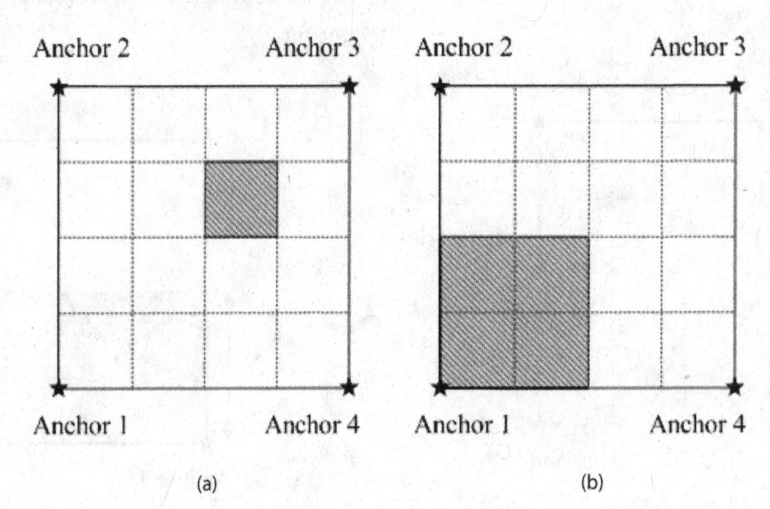

As show in Figure 4, B^c could be either a single grid (e.g. for input H^c ={3, 3, 2, 3} in Figure 4(a)) or larger (e.g. for H^c ={3, 3, 2, 3}in Figure 4(b)):

If the resulted B^c is null, the algorithm will return the entire monitored area as output of phase I:

$$\begin{cases} X^c_{\min} = 0, \; X^c_{\max} = X_r \\ Y^c_{\min} = 0, \; Y^c_{\max} = Y_r \end{cases} \tag{8}$$

2. **Phase II:** When the tentative estimation B^c returned after phase I is larger than a predefined threshold, e.g., a single grid, the algorithm proceeds to phase II. Similar to the current hop-count set H^c, the last hop-count set H^p also provides us with a tentative region for location of the node, but at the previous period. Let's denote this previous location as B^p, which is expressed in terms of upper-bound and lower-bound for x and y coordinates as $\left\{x^p_{\min}, x^p_{\max}, y^p_{\min}, y^p_{\max}\right\}$. With known B^p, the maximum node velocity v_{max} and the time interval between two consecutive time stamps, Δt, the current location of the node *M* can be inferred by expanding B^p in all dimensions with a change of $d_{max}=v_{max}\Delta t$. Specifically, the predicted current node location B^c_{new} could be defined as follows,

$$\begin{cases} X^c_{new,\min} = \max\{X^p_{\min} - d_{\max}, 0\} \\ X^c_{new,\max} = \min\{X^p_{\max} + d_{\max}, X_r\} \\ Y^c_{new,\min} = \max\{Y^p_{\min} - d_{\max}, 0\} \\ Y^c_{new,\max} = \min\{Y^p_{\max} + d_{\max}, Y_r\} \end{cases} \tag{9}$$

Consequently, the output should be a region that is the overlapped area of B^c and B_c^{new}. The output is limited by the following upper-bound and lower-bound for x and y coordinates:

$$\begin{cases} X_{\min} = \max\{X^c_{\min}, X^c_{new,\min}\} \\ X_{\max} = \min\{X^c_{\max}, X^c_{new,\max}\} \\ Y_{\min} = \max\{Y^c_{\min}, Y^c_{new,\min}\} \\ Y_{\max} = \min\{Y^c_{\max}, Y^c_{new,\max}\} \end{cases} \tag{10}$$

If there is not overlapping between B^c and B_c^{new}, the algorithm will output the region bounded by equation (8).

Figure 5. Percentage of inputs that result in a region whose area is smaller than 1 grid, 2 grids, 3 grids, 4 grids, 6 grids and 8 grids as a function of resolution

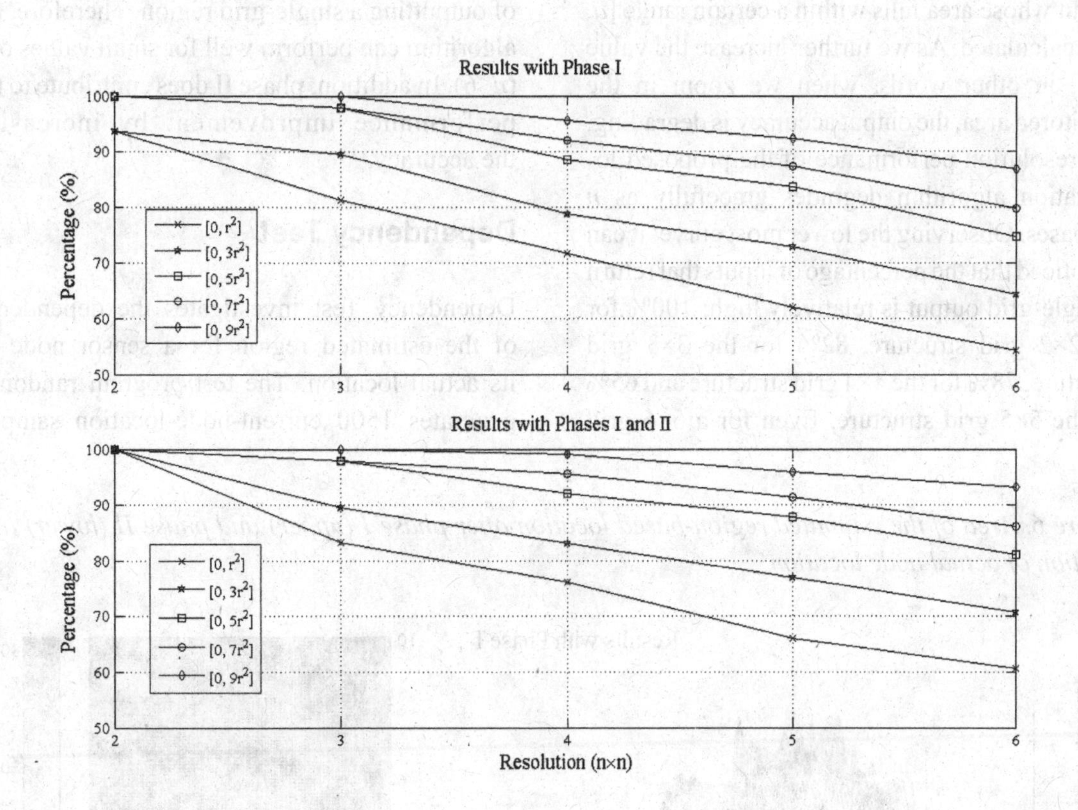

SIMULATION RESULTS

We have tested our proposed localization algorithm using computer simulations. In the following, we shall discuss resolution test, dependency test and computational complexity.

Resolution Test

The key performance metric of localization algorithms is the output accuracy. In general, the output is desirable to be a region with as small as possible area because it helps us to quickly locate the fire's actual position. Therefore, the smaller the estimated region's area is, the better the algorithm's accuracy is. In this resolution test, we aim to analyze relationship between the output and the distance of two adjacent anchor nodes. For simplicity, we denote the distance of two adjacent anchors as a multiple n of the grid edge length r, i.e., nr, where n is then referred as the resolution of a grid structure. For example, when n=2, we have the 2×2 grid structure. Computer simulation was done by varying the value of n from 2 to 6 and Figure 5 shows the result of resolution test.

In Figure 5, the x-axis indicates the value of n, whereas y-axis represents the percentage of inputs that result in an output of different area size. The lower-most curve represents the percentage of inputs that result in an outputted region which is smaller than or equal to a single grid (area=r^2) against n. Other curves represent the percentage of inputs that result in an outputted region which is smaller than or equal to 3 grids,

5 grids, 7 grids and 9 grids against *n*, respectively. The percentage of inputs that result in an output whose area falls within a certain range [*a*, *b*] is calculated. As we further increase the value of n, in other words, when we zoom in the monitored area, the output accuracy is degrading. The resolution performance of the proposed localization algorithm degrades gracefully as *n* increases. Observing the lower most curve, it can be noticed that the percentage of inputs that return a single-grid output is relatively high: 100% for the 2×2 grid structure, 82% for the 3×3 grid structure, 78% for the 4×4 grid structure and 65% for the 5×5 grid structure. Even for a 6×6 grid structure, which means there are in total 36 grids, the algorithm can still achieve a 60% probability of outputting a single-grid region. Therefore, the algorithm can perform well for small values of *n* ($n \leq 6$). In addition, phase II does contribute to the performance improvement by increasing the accuracy.

Dependency Test

Dependency Test investigates the dependency of the estimated region for a sensor node on its actual location. The test program randomly generates 1500 current-node-location samples

Figure 6. Area of the estimated region-based location after phase I (upper) and phase II (lower) as a function of actual node location

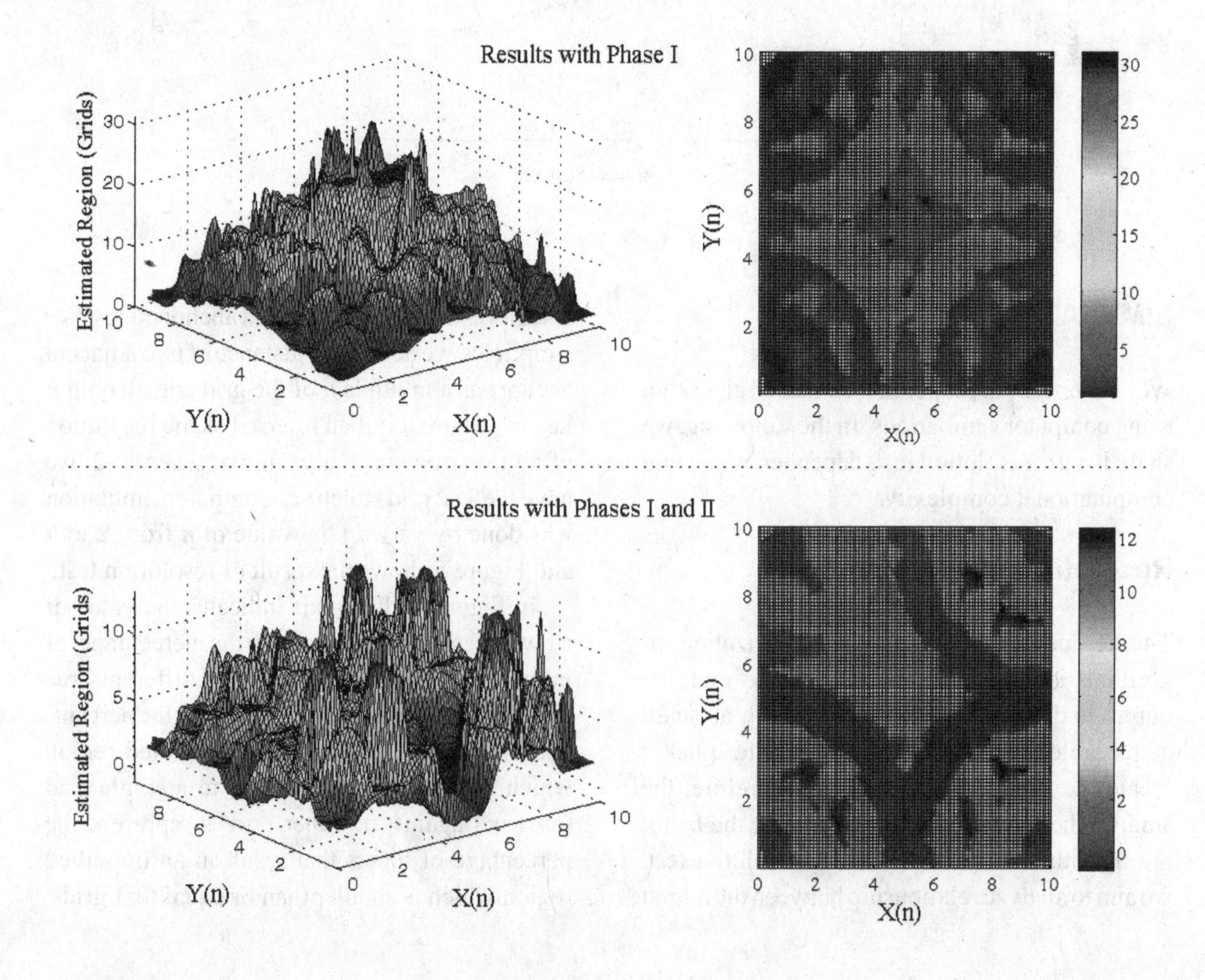

Figure 7. Layout of the monitoring area

with coordinates (*x*, *y*) together with a single last-node-location within the monitored area with $x \in [0,X_r], y \in [0,Y_r]$. The monitored area is divided into a 10×10 grid structure. Each actual node location (*x*, *y*) with respect the four anchor nodes is then transformed to the current and last set of hop-count values, which serve as the input for our algorithm. The test program subsequently estimates the grid-based location for each sample using our algorithm, calculates the area of the estimated region after phase I and II in grid unit, and plots the estimated region's area as a function of actual node location as shown in Figure 6.

Noteworthily, the color scale in Figure 6(a) and Figure 6(b) is different from that in Figure 6(c) and Figure 6(d). From Figure 6, we can observe the following two facts: (i) For the output obtained from phase I of our algorithm, the distribution of output is symmetrical about center of the monitored area. This is due to the fact that all anchor nodes are assumed to have exactly the same transmission range, and the radio signal is propagated omni-directionally. It can be observed that a large portion of the monitored area is covered in blue and green color, which corresponds to a small output's area. Specifically, on average, 59% of the 1500 sample points can return an output region smaller than 10 grids; 92% of them can return an output region smaller than 20 grids; and 100% returns an output region smaller than 30 grids. These observations indicate that the proposed algorithm is able to local a sensor node with satisfactory accuracy. (ii) For the output obtained from phase II of our algorithm, the maximum output region's area is considerably reduced from 30 grids to 12 grids. The region where output's area is reduced the most is the proximity of last node position (randomly-selected point), which is $[3.7961r, 9.8342r]$. Moreover, from the simulation results, after phase I, only 22% of the sample points can return an output region smaller than 5 grids; whereas after phase II, approximately 63% of the sample points results in an estimated region smaller than 5 grids. Thus phase II substantially improve the accuracy.

Table 1. Hardware list

Hardware	Model	Quantity	Image
Crossbow Sensor Nodes		N	
Gateway board	MB520	1	
Mote sensor data acquisition board	MTS300CA	3	
Mote	MPR2400CA	7	
Workstation	Any model	1	

Computational Complexity

Results have shown that under MCL and MCB algorithms, a set of 50 candidate locations (also referred as a *sample set*) must be estimated through prediction and filtering steps at every time period. A full sample set is required for MCL and MCB to function properly. However, under MCL the sample set may not be full even after 20,000 attempts; whereas under MCB on average 100 attempts are required to fill up one sample set (Baggio & Langendoen, 2008). This very large number of loops would significantly increase power consumption of sensor nodes and reduce efficiency. As compared the existing range-free localization techniques, our algorithm has a lower computational complexity of $O(N_a)$, where N_a is the number of anchors. Furthermore, with event-triggering mechanism, our algorithm helps reduce power consumption. Altogether, it is more suitable for application in forest fire surveillance systems.

Figure 8. Started at the field corner where Anchor 1 was placed

Figure 9. In the middle of Anchor 1 and Anchor 2

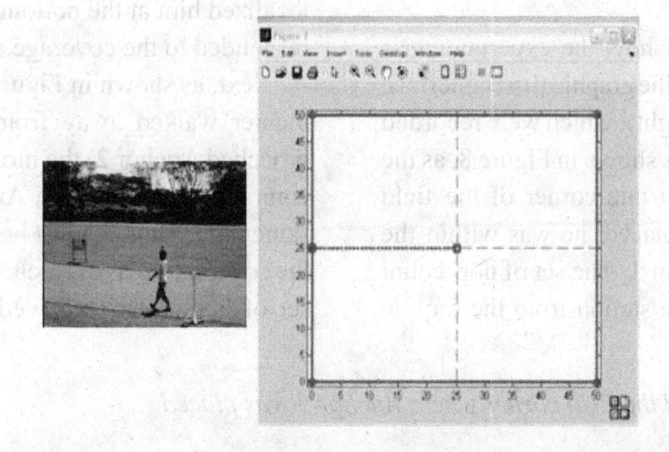

EXPERIMENTAL RESULTS

We have conducted an experiment to test the proposed localization technique. As shown in Figure 7, the monitored area is a 2×2 grid structure in 75×75m² and the algorithm was implemented in Crossbow's set of sensor motes. To simulate the animals' motion, an experimenter holding the sensor node was moving around the soccer field following the red arrow: started at Anchor 1, moved towards Anchor 2, turned to Anchor 3, then Anchor 4 and back to Anchor 1. The objective of this experiment was to determine the grid-based location of the mobile sensor node.

Hardware Requirements

Table 1 shows the hardware list during the implementation. For further information about each hardware component, Interested readers are referred to the Crossbow MICAz datasheet(Crossbow Tchnology, 2007a), MPR-MIB Users Manual (Crossbow Tchnology, 2007b) and MTS/MDA Sensor Board Users Manual (Crossbow Tchnology, 2007c).

Figure 10. At the field corner where Anchor 2 was placed

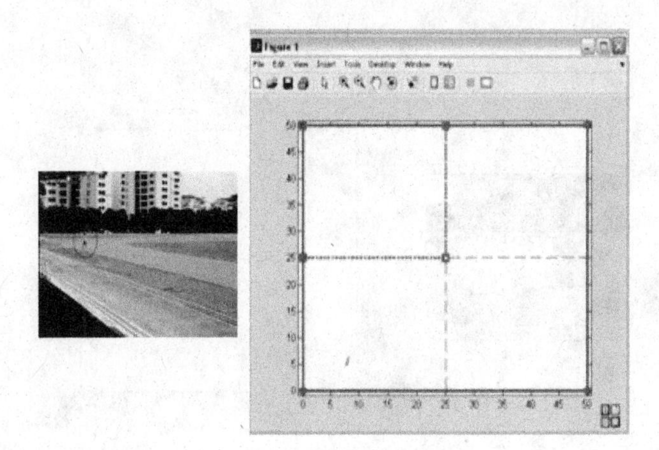

Results

Figure 8 to Figure 14 show the experimenter's motion (on the left) and the graphical user interface of our system (on the right), which were recorded synchronously. First, as shown in Figure 8, as the experimenter started at one corner of the field where Anchor 1 was placed, he was within the coverage area of Anchor 1. The set of hop-count values received at base station from the mobile node was $H=\{1, 0, 0, 0\}$. As a result, the system localized him at the bottom-left grid, which corresponded to the coverage area of Anchor 1.

Next, as shown in Figure 9, when the experimenter walked away from Anchor 1 and approached Anchor 2, the mobile node was able to communicate with both Anchor 1 and 2 at the same time. Thus, it must be inside the overlap of the coverage areas of Anchor 1 and Anchor 2. The set of hop-counts received by the base station

Figure 11. Approached the field corner where Anchor 3 was placed

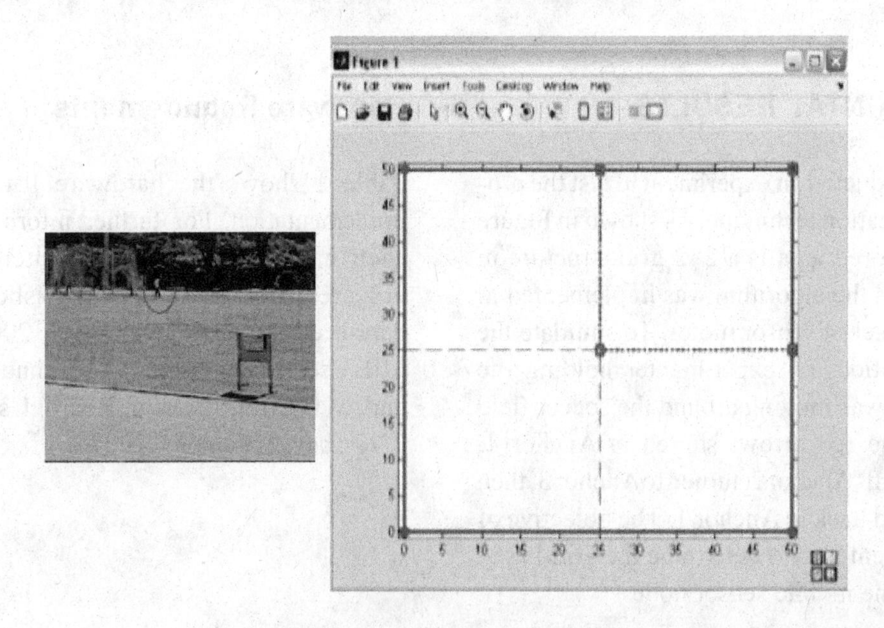

Figure 12. At the field corner where Anchor 4 was placed

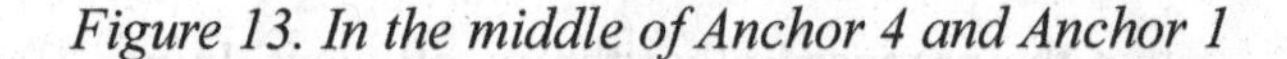

Figure 13. In the middle of Anchor 4 and Anchor 1

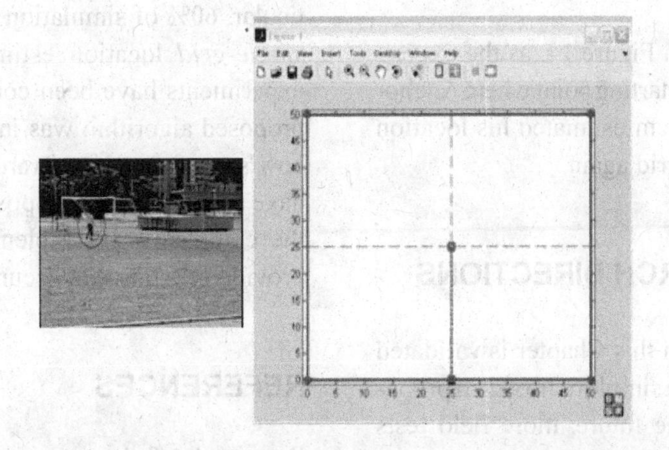

from the mobile node was H={1, 1, 0, 0}. Given this hop-count value set, the system output a line in between Anchor 1 and Anchor 2, which represented the overlapped region of their coverage areas.

After that, as shown in as Figure 10, the experimenter walked further away from Anchor 1 and completely left its coverage area, the mobile node could only communicate with Anchor 2. The set of hop-counts from the mobile node would become H={0, 1, 0, 0}. The system then estimated the node location as the top-left grid which corresponded to Anchor 2's coverage area.

Similarly, as shown Figure 11 and Figure 12, in as the mobile node was brought into the coverage area of Anchor 3 and 4 subsequently, the set of hop-counts changed to H={0, 0, 1, 0} and H={0, 0, 0, 1}, respectively. As a result, the system estimated its location as the top-right and bottom-right grid, respectively.

As shown in Figure 13, when the experimenter was in the middle of Anchor 4 and Anchor 1 (on the way back to Anchor 1), the sensor node he held could simultaneously communicate with both Anchor 4 and 1, thus it sent a hop-count set of H={1, 0, 0, 1} to base station. The system estimated the node location as a line in-between

Figure 14. Returned to the starting corner where Anchor 1 was placed

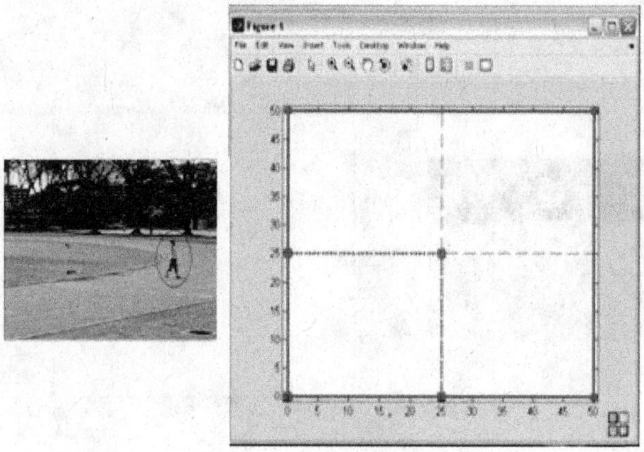

Anchor 4 and 1, which represents the shared area of their coverage area.

Lastly, as shown in Figure 14, as the experimenter returned to the starting point where Anchor 1 was placed, the system estimated his location to be the bottom-left grid again.

FUTURE RESEARCH DIRECTIONS

The work presented in this Chapter is validated through a field test with simple radio signal propagation channels. In the future, more field tests will be conducted and we would like to design an adaptive localization algorithm that is based on the radio channel conditions. Other than that, we will continue searching for better localization algorithms with lower computational complexity.

CONCLUSION

This Chapter presents a localization technique particularly tailored for forest fire surveillance systems. By adopting grid-based output and event-triggering mechanism, the proposed algorithm can reduce computational complexity, improve output accuracy and reduce power consumption. Computer simulations were conducted to verify that the proposed algorithm is efficient. In particular, 60% of simulation scenarios may return *single-grid* location estimation. Furthermore, experiments have been conducted in which the proposed algorithm was implemented in Crossbow's set of sensor hardware. Experimental results have shown that the proposed algorithm reduces the complexity of implementation, and it can provide considerable accuracy.

REFERENCES

Baggio, A., & Langendoen, K. (2008). Monte-Carlo localization for mobile wireless sensor networks. *Ad Hoc Networks*, *6*(1), 718–733. doi:10.1016/j.adhoc.2007.06.004

Bahl, P., & Padmanabhan, V. N. (2000, 26-30 March). *RADAR: An in-building RF-based user location and tracking system.* Paper presented at the IEEE/ACM INFOCOM'00, Tel Aviv, Israel.

Crossbow Technology. (2007a). *MICAz*.

Crossbow Technology. (2007b). *MPR-MIB user's manual*.

Crossbow Technology. (2007c). *MTS/MDA sensor board user's manual*.

Hu, L., & Evans, D. (2004, 26 September - 1 October). *Localization for mobile sensor networks.* Paper presented at the ACM MobiCom'04, Philadelphia, PA.

Mao, G., Fidan, B., & Anderson, B. D. O. (2007). Wireless sensor network localization techniques. *Computer Networks*, *51*(10), 2529–2553. doi:10.1016/j.comnet.2006.11.018

Niculescu, D., & Badri, N. (2003, 30 March - 3 April). *Ad hoc positioning system (APS) using AOA.* Paper presented at the IEEE/ACM INFOCOM'03, San Francisco, CA.

Niculescu, D., & Nath, B. (2001, 25-29 November). *Ad hoc positioning system (APS).* Paper presented at the IEEE GLOBECOM '01, San Antonio, TX.

Priyantha, N. B., Chakraborty, A., & Balakrishnan, H. (2000). *The cricket location-support system.* Paper presented at the ACM MobiCom'00, Boston, MA.

Ward, A., Jones, A., & Hopper, A. (1997). A new location technique for the active office. *IEEE Personal Communications*, *4*(5), 42–47. doi:10.1109/98.626982

KEY TERMS AND DEFINITIONS

Computational Complexity: A metric that is used to determine the inherent difficulty level of an algorithm and it is generally related to how much resource will be involved during its execution.

Dependence Test: A test of localization algorithm, which investigates the dependency of the algorithm's output as the estimated region for a sensor node on its actual location.

Forest Fire Surveillance: The mechanism to remotely monitor forest fire and take necessary action that depends on where, when and how a forest fire breaks out.

Global Positioning System: A space-based global navigation satellite systems, which was built to provide reliable location and time reference/information in all weather and at all times and anywhere on or near the Earth when and where there is an unobstructed line of sight to four or more satellites.

Localization Technique: A technique to determine the physical/relative location of a particular node in a system.

Range-Based Algorithm: A localization algorithm requires distance or angle information among sensor nodes.

Range-Free Algorithm: A localization algorithm that does not require distance or angle information among sensor nodes.

Resolution Test: A test of localization algorithm, which investigates the accuracy of the algorithm.

Wireless Sensor Network: A wireless sensor networks is a collection of sensor nodes, which are spatially distributed and configured to monitor certain physical or environmental conditions and then cooperatively to send their data through the network towards a sink node or remote data center.

Compilation of References

Abbasi, A. A., & Younis, M. (2007). A survey on clustering algorithms for wireless sensor networks. *Computer Communications, 30*, 2826–2841. doi:10.1016/j.comcom.2007.05.024

Abdelzaher, T., Blum, B., Cao, Q., Chen, Y., Evans, D., & George, J. … Wood, A. (2004). EnviroTrack: Towards an environmental computing paradigm for distributed sensor networks. In *Proceedings of the 24th International Conference on Distributed Computing Systems (ICDCS'04)* (ICDCS '04), (pp. 582-589). Washington, DC: IEEE Computer Society.

Acroname, I. (n.d.). *Garcia custom robot.* Retrieved October 25, 2009, from http://www.acroname.com/garcia/garcia.html

AEOLUS. (n.d.a). *SP6—AEOLUS.* Retrieved September 27, 2009, from http://aeolus.ceid.upatras.gr/sub-projects/sp6

AEOLUS. (n.d.b). *SP6/WSNtestbed - AEOLUS IST project.* Retrieved September 27, 2009, from http://ru1.cti.gr/aeolus/SP6/WSNtestbed

AEOLUS. (n.d.c). *Welcome to AEOLUS — AEOLUS.* Retrieved September 27, 2009, from http://aeolus.ceid.upatras.gr/

Agah, A., Das, S. K., & Basu, K. A. (2004). Game theory based approach for security in wireless sensor networks. *IPCCC 2004,* USA, (pp. 259-263).

Ahmed, A. A., Shi, H., & Shang, Y. (2005). Sharp: A new approach to relative localization in wireless sensor networks. In *Proceedings of IEEE ICDCS*, 2005.

Ahmed, N., Kumar, R., French, R. S., & Ramachandran, U. (2007). RF^{2}ID: A reliable middleware framework for RFID deployment. *2007 IEEE International Parallel and Distributed Processing Symposium, IPDPS*, (p. 66).

Ahn, G.-S., Miluzzo, E., Campbell, A. T., Hong, S. G., & Cuomo, F. (2006). *Funneling-MAC: A localized, sink-oriented MAC for boosting fidelity in sensor networks.* 4th ACM Conference on Embedded Networked Sensor Systems (SenSys'06). Boulder, Colorado, USA.

Akkaya, K., & Younis, M. (2003). An energy-aware QoS routing protocol for wireless sensor networks. In *Proceedings of the 23rd International Conference on Distributed Computing Systems,* 2003.

Akkaya, K., & Younis, M. (2005). A survey of routing protocols in wireless sensor networks. *Ad Hoc Networks*, *3*(3), 325–349. doi:10.1016/j.adhoc.2003.09.010

Akkaya, K., Younis, M. F., & Bangad, M. (2005). Sink repositioning for enhanced performance in wireless sensor networks. *Computer Networks*, *49*(4).

Akyildiz, I. F., Su, W., Sankarasubramaniam, Y., & Cayirci, E. (2002). A survey on sensor network. *IEEE Communications Magazine.*

Akyildiz, I. (2002). Wireless sensor networks: A survey. *Computer Networks*, *38*(4), 393–422. doi:10.1016/S1389-1286(01)00302-4

Akyildiz, I. F., Pompili, D., & Melodia, T. (2005). Underwater acoustic sensor networks: Research challenges. *Ad Hoc Networks*, *3*(3), 257–279. doi:10.1016/j.adhoc.2005.01.004

Akyildiz, I. F., Su, W., Sankarasubramaniam, Y., & Cayirci, E. (2002). A survey on sensor network. *IEEE Communications Magazine*, *40*(8), 102–116. doi:10.1109/MCOM.2002.1024422

Akyildiz, I. F., Su, W., Sankarasubramaniam, Y., & Cayirci, E. (2002). Wireless sensor networks: A survey. *Computer Networks*, *38*(4), 393–422. doi:10.1016/S1389-1286(01)00302-4

Akyildiz, I. F., & Varun, M. C. (2010). *Wireless sensor networks*. John Wiley and Sons Ltd. doi:10.1002/9780470515181

Akyildiz, I., Su, W., Sankarasubramaniam, Y., & Cayirci, E. (2002). A survey on wireless sensor networks. *IEEE Communications Magazine*, *40*(8), 102–114. doi:10.1109/MCOM.2002.1024422

Akyildiz, I., Su, W., Sankarasubramaniam, Y., & Cayirci, E. (2002). Wireless sensor networks: A survey. *Computer Networks*, *38*(4), 393–422. doi:10.1016/S1389-1286(01)00302-4

Akylidiz, F., Su, W., Sankarasubramaniam Y., & Cayirci, E. (2002 August). A survey on sensor networks. *IEEE Personal Communications Magazine*, 102-114

Al Rashidi, M. R., & El-Hawary, M. E. (2009). A survey of particle swarm optimization applications in electric power systems. *IEEE Transactions on Evolutionary Computation*, *13*(4), 913–918. doi:10.1109/TEVC.2006.880326

Alam, M. M., Rashid, M. O., & Hong, C. S. (2008). *WSNMP: A network management protocol for wireless sensor networks,* (pp. 17-20).

Alazzawi, L. K., Elkateeb, A. M., & Ramesh, A. (2008). Scalability analysis for wireless sensor networks routing protocols. *22nd International Conference on Advanced Information Networking and Applications - Workshops, AINAW 2008,* (pp. 139-144).

Ali, K., & Hassanein, H. (2008, 6-9 July 2008). *Underwater wireless hybrid sensor networks.* Paper presented at the IEEE Symposium on Computers and Communications, 2008.

Alippi, C., & Vanini, G. (2006). A RSSI-based and calibrated centralized localization technique for wireless sensor networks. In *Proceedings of Fourth IEEE International Conference on Pervasive Computing and Communications Workshops (PERCOMW'06),* Pisa, Italy, March 2006, (pp. 301-305).

Alippi, C., Anastasi, G., Di Francesco, M., & Roveri, M. (2009). *Energy management in wireless sensor networks with energy-hungry sensors* (pp. 16–23). Instrumentation & Measurement Magazine.

Al-Karaki, J. N., & Kamal, A. E. (2004). Routing techniques in wireless sensor network: A survey. *IEEE Wireless Communication*, December 2004.

Al-Khdour, T., & Baroudi, U. (2007). A generalized energy-aware data centric routing for wireless sensor network. *The 2007 IEEE International Conference on Signal Processing and Communications (ICSPC 2007),* (pp. 117–120). Dubai, United Arab of Emirates (UAE).

Al-Khdour, T., & Baroudi, U. (2009). A generalized energy-efficient time-based communication protocol for wireless sensor networks. [IJIPT]. *Special Issue of International Journal of Internet Protocols*, *4*(2), 134–146. doi:10.1504/IJIPT.2009.027338

Al-Khdour, T., & Baroudi, U. (2010). An energy-efficient distributed schedule-based communication protocol for periodic wireless sensor networks. *The Arabian Journal of Science and Engineering*, *35*(2B), 153–168.

Alsindi, N. A., Pahlavan, K., & Alavi, B. (2006). An error propagation aware algorithm for precise cooperative indoor localization. In *Proceedings of IEEE Military Communications Conference MILCOM 2006*, (pp. 1-7). Washington, DC, USA, October 2006.

Alvarez-Benitez, J. E., Everson, R. M., & Fieldsend, J. E. (2005). A MOPSO algorithm based exclusively on pareto dominance concepts. In C. Coello—Coello, et al. (Eds.), *Evolutionary Multi-Criterion Optimization, LNCS 3410*, (pp. 459-73).

Anandamoy, S. (2006). *Swarm intelligence based optimization of MANET cluster formation*. Unpublished Master thesis, The University of Arizona.

Anastasi, G., Conti, M., & Francesco, M. D., & Passarella, A. (2006). How to prolong the lifetime of wireless sensor networks. In L. T. Deng & M. K. Denko (Eds.), *Mobile ad hoc and pervasive communications.* American Scientific Publishers.

Anastasi, G., Conti, M., Francesco, M. D., & Passarella, A. (2009). Energy conservation in wireless sensor networks: A survey. *Ad Hoc Networks*, *7*(3), 537–568. doi:10.1016/j.adhoc.2008.06.003

Anderson, J. W. (2005). *Hyperbolic geometry.* Springer.

Anupama, K. R., Sasidharan, A., & Vadlamani, S. (2008, 27-28 August). *A location-based clustering algorithm for data gathering in 3D underwater wireless sensor networks.* Paper presented at the International Symposium on Telecommunications, 2008.

Arias, J., Santos, E., Marin, I., Jimenez, J., Lazaro, J., & Zuloaga, A. (2006). Node synchronization in wireless sensor networks. *International Conference on Wireless and Mobile Communications, ICWMC '06* (pp. 50-50).

Arora, A. (n.d.). *Exscal research group.* Retrieved October 25, 2009, from http://ceti.cse.ohio-state.edu/exscal/

Arora, A., Dutta, P., & Bapat, S. (2004). A line in the sand: A wireless sensor network for target detection, classification, and tracking. *Computer Networks*, *46*, 605–634. doi:10.1016/j.comnet.2004.06.007

Arora, A., Ertin, E., Ramnath, R., Leal, W., & Nesterenko, M. (2006). Kansei: A high-fidelity sensing test bed. *IEEE Internet Computing*, *10*(2), 35–47. doi:10.1109/MIC.2006.37

Audi India. (n.d.). *Audi travolution: Efficiently through the city.* Retrieved from http://www.audi.in/sea/brand/in/company/news/company.detail.2010~06~audi_travolution_.html

Awad, A., Sommer, C., German, R., & Dressler, F. (2008). Virtual cord protocol (VCP): A flexible DHT-like routing service for sensor networks. *2008 5th IEEE International Conference on Mobile Ad-Hoc and Sensor Systems, MASS 2008, September 29, 2008 - October 2,* (pp. 133-142).

Ayaz, M., & Abdullah, A. (2009, 16-18 Dec. 2009). *Hop-by-hop dynamic addressing based (H2-DAB) routing protocol for underwater wireless sensor networks.* Paper presented at the International Conference on Information and Multimedia Technology, 2009.

Ayaz, M., Abdullah, A., & Low Tang, J. (15-17 June 2010). *Temporary cluster based routing for underwater wireless sensor networks.* Paper presented at the 2010 International Symposium in Information Technology (ITSim).

Bachir, A., Barthel, D., Heusse, M., & Duda, A. (2007). O(1)-Reception routing for sensor networks. *Computer Communications*, *30*, 2603–2614. doi:10.1016/j.comcom.2007.05.054

Bachir, A., Dohler, M., Watteyne, T., & Leung, K. (2010). MAC essentials for wireless sensor networks. *IEEE Communications Surveys and Tutorials*, *12*(2), 222–248. doi:10.1109/SURV.2010.020510.00058

Bachrach, J., & Taylor, C. (2005). *Localization in sensor networks.* Retrieved from http://people.csail.mit.edu/jrb/Projects/poschap.pdf

Bachrach, J., Nagpal, R., Salib, M., & Shrobe, H. (2004). Experimental results for and theoretical analysis of a self-organizing a global coordinate system from ad hoc sensor networks. *Telecommunications System Journal*, *26*(2-4), 213–233. doi:10.1023/B:TELS.0000029040.85449.7b

Baggio, A., & Langendoen, K. (2008). Monte-Carlo localization for mobile wireless sensor networks. *Ad Hoc Networks*, *6*(1), 718–733. doi:10.1016/j.adhoc.2007.06.004

Bahl, P., & Padmanabhan, V. N. (2000, 26-30 March). *RADAR: An in-building RF-based user location and tracking system.* Paper presented at the IEEE/ACM INFOCOM'00, Tel Aviv, Israel.

Ballari, D., Wachowicz, M., & Callejo, M. (2009). Metadata behind the interoperability of wireless sensor networks. *Sensors (Basel, Switzerland)*, *5*, 3635–3651. doi:10.3390/s90503635

Bapat, S., Leal, W., Kwon, T., Wei, P., & Arora, A. (2009). Chowkidar: Reliable and scalable health monitoring for wireless sensor network testbeds. *ACM Transactions on Autonomous and Adaptive Systems*, *4*(1).

Baras, J., & Mehta, H. (2003). A probabilistic emergent routing algorithm for mobile ad hoc networks. In *WiOpt'03 Sophia-Antipolis*. France: PERA.

Barbancho, J., Leon, C., Molina, F. J., & Barbancho, A. (2007). Using artificial intelligence in routing scheme for wireless networks. *Computer Communications*, *30*, 2802–2811. doi:10.1016/j.comcom.2007.05.023

Bari, A., Teng, D., & Jaekel, A. (2009). Optimal relay node placement in hierarchical sensor networks with mobile data collector. *18th International Conference on Computer Communications and Networks*, (pp. 1–6). IEEE Computer Society.

Barickman, F. S. (2005). Lane departure warning systems. National Highway Traffic Safety Administration document, 2005.

Barndorff-Nielsen, O. (1977). Exponentially decreasing distributions for the logarithm of particle size. *Proceedings of the Royal Society of London*, *353*, 401–419. doi:10.1098/rspa.1977.0041

Barr, R. (2002). On the need for system-level support for ad hoc and sensor networks. *Operating Systems Review*, *36*(2), 1–5. doi:10.1145/509526.509528

Basagni, S., Petrioli, C., Petroccia, R., & Stojanovic, M. (2009, January). *Choosing the packet size in multi-hop underwater networks*. Northern University, Boston, MA.

Begum, S., Tara, N., & Sultana, S. (2010). Energy efficient target coverage in wireless sensor networks based on modified ant colony algorithm. *International Journal of Ad hoc, Sensor & Ubiquitous Computing, 1*(4).

Benkoczi, R., Hassanein, H., Akl, S., & Tai, S. (2005). QoS for data relaying in hierarchical wireless sensor networks. In *Proceedings of the 1st ACM International Workshop on Quality of Service & Security in Wireless and Mobile Networks*, 2005.

Bensaou, B., Wang, Y., & Ko, C. C. (2000). Fair medium access in 802.11 based wireless ad-hoc networks. In *Proceeding of ACM MobiHoc,* (pp. 99-106).

Berkeley, I. T. S. (n.d.). *Vehicle detection with wireless sensors*. Retrieved from http://www.techtransfer.berkeley.edu/newsletter/08-3/vehicle-detection-with-wireless-sensors.php

Berkely, U. C. (2007). *TinyOS*: *Tiki RSS feed for directory sites.* Retrieved from http://www.bwsn.net/tiki-directories_rss.php?ver=2

Berlekamp, E. R., Peile, R. E., & Pope, S. P. (1987). The application of error control to communications. *IEEE Communications Magazine*, *25*(4), 44–57. doi:10.1109/MCOM.1987.1093590

Berman, P., Calinescu, G., Shah, C., & Zelikovsly, A. (2005). Efficient energy management in sensor networks. In Xiao, Y., & Pan, Y. (Eds.), *Ad hoc and sensor networks*. Nova Science Publishers.

Berry, D. S., & Belmont, D. M. (1951). Distribution of vehicle speeds and travel times. *Proceedings of the Second Berkeley Symposium on Mathematical Statistics and Probability.*

Bertsekas, D. P., & Gallager, R. (1987). *Data networks*. Englewood Cliffs, NJ: Prentice-Hall.

Betz, V. (n.d.). *FPGA architecture for the challenge.* Retrieved October 25, 2009, from http://www.eecg.toronto.edu/~vaughn/challenge/fpga_arch.html

Beutel, J. (2010). *The sensor network museum* (SNM). Sensor Network Hardware Systems. Retrieved November 27, 2010, from http://www.snm.ethz.ch/Main/HomePage

Bhardwaj, M., & Chandrakasan, A. (2002). Bounding the lifetime of sensor networks via optimal role, assignments. *Proceedings of the INFOCOM*, *02*, 1587–1596.

Bhuse, V., & Gupta, A. (2005). *Anomaly intrusion detection in wireless sensor networks.* Western Michigan University, Kalamazoo, MI-49008 Report.

Bhuyan, B. (2009). Quality of service (QoS) provisions in wireless sensor networks and related challenges. *Wireless Sensor Network, 201*(2), 861–868.

Bianchi, G., & Tinnirello, I. (2003). *Kalman filter estimation of the number of competing terminals in an IEEE 802.11 network*. IEEE INFOCOM 2003.

Bianchi, G. (2002). Performance analysis of the IEEE 802.11 distributed coordination function. *IEEE Journal on Selected Areas in Communications*, 18.

Biaz, S., & Barowski, Y. (2004, April). GANGS: An energy efficient MAC protocol for sensor networks. *The 42nd Annual Southeast Regional Conference*. (pp. 82-87). Huntsville, AL, USA.

Bibyk, S. (n.d.). *ECE 582 class homepage: XScale mote.* Retrieved October 25, 2009, from http://www.ece.osu.edu/~bibyk/ee582/ee582.htm

Biswas, P. K., & Phoha, S. (2006). Self-organizing sensor networks for integrated target surveillance. *IEEE Transactions on Computers*, *55*(8), 1033–1047. doi:10.1109/TC.2006.130

Bi, Y., Li, N., & Sun, L. (2007). DAR: An energy-balanced data-gathering scheme for wireless sensor networks. *Computer Communications*, *30*(14-15), 2812–2825. doi:10.1016/j.comcom.2007.05.021

Blackwell, T., & Branke, J. (2006). Multiswarm, exclusion and anti-convergence in dynamic environments. *IEEE Transactions on Evolutionary Computation*, *10*(4), 459–472. doi:10.1109/TEVC.2005.857074

Blum, B. M., Stankovic, J. A., Abdelzaher, T., He, T., & Huang, C. (2005). Range-free localization and its impact on large scale sensor networks. In *ACM Transactions on Embedded Computing Systems*, (vol. 4, pp. 877-906). November 2005.

Blywis, B., Juraschek, F., Günes, M., & Schiller, J. (n.d.). *Design concepts of a persistent WSN testbed.* AG Computer Systems & Telematics - Freie Universität Berlin. Retrieved September 7, 2009, from http://cst.mi.fu-berlin.de/publications/

Bolch, G., Griener, S., Meer, H., & Trivedi, K. (2003). *Queuing networks and Markov chains modeling and performance evaluation with computer science applications*. Prentice-Hall.

Bonabeau, E., Dorigo, M., & Theraulaz, G. (1999). *Swarm intelligence: From natural to artificial systems* (pp. 1–278). London, UK: Oxford University Press.

Bondi, A. B. (2000). Characteristics of scalability and their impact on performance. *Proceedings Second International Workshop on Software and Performance WOSP2000, September 17, 2000 - September 20,* (pp. 195-203).

Bose, P., Morin, P., Stojmenovic, I., & Urrutia, J. (2001). Routing with guaranteed delivery in ad hoc wireless networks. [ACM/Kluwer.]. *Wireless Networks*, *7*(6), 609–616. doi:10.1023/A:1012319418150

Boukerche, A., Araujo, R. B., & Silva, F. H. S. (2007). An efficient event ordering algorithm that extends the lifetime of wireless actor and sensor networks. *Performance Evaluation*, *64*(5), 480–494. doi:10.1016/j.peva.2006.08.009

Boukerche, A., Cheng, X., & Linus, J. (2005). A performance evaluation of a novel energy-aware datacentric routing algorithm in wireless sensor networks. *Wireless Networks*, *11*(5). doi:10.1007/s11276-005-3517-6

Boulis, A., & Jha, S. (2005). Network management in new realms: wireless sensor networks. *International Journal of Network Management*, *15*(4). doi:10.1002/nem.569

Braginsky, D., & Estrin, D. (2002). Rumor routing algorthim for sensor networks. *Proceedings of the 1st ACM International Workshop on Wireless Sensor Networks and Applications.* Atlanta, GA: ACM.

Bray, T., Paoli, J., Sperberg-McQueen, C., Maler, E., & Yergeau, F. (2008). *Extensible markup language (XML) 1.0* (5th ed.). W3C recommendation.

Breslau, L., Estrin, D., Fall, K., Floyd, S., Heidemann, J., & Helmy, A. (2000). Advances in network simulation. *Computer*, *33*(5), 59–67. doi:10.1109/2.841785

Brooke, T., & Burrell, J. 2003. From ethnography to design in a vineyard. In *Proceedings of the Design User Experiences (DUX) Conference*. Case study.

Bruck, J., Gao, J., & Jiang, A. (2005). MAP: Medial axis based geometric routing in sensor networks. In *MobiCom '05: Proceedings of the 11th Annual International Conference on Mobile Computing and Networking*, (pp. 88-102). New York, NY: ACM Press.

Buczak, A., & Jamalabad, V. (1998). Self-organization of a heterogeneous sensor network by genetic algorithms. In Dagli, C. H., Chen, C. L., & Akay, M. (Eds.), *Intelligent engineering systems through artificial neural networks* (*Vol. 8*, pp. 259–264). New York, NY: ASME Press.

Bulusu, N., Bychkovskiy, V., Estrin, D., & Heidemann, J. (n.d.). *Scalable, ad hoc deployable RF-based localization.* Retrieved September 8, 2009, from http://lecs.cs.ucla.edu/~bulusu/papers/Bulusu02a.html

Bulusu, N., Heidemann, J., & Estrin, D. (2000). GPS-less low cost outdoor localization for very small devices. *OEEE Personal Communications Magazine*, *7*(5), 28–34. doi:10.1109/98.878533

Buratti, C., Conti, A., Dardari, D., & Verdone, R. (2009). *An overview on wireless sensor networks technology and evolution* (pp. 6869–6896). Open Access Sensors.

Buschmann, C., Pfisterer, D., Fischer, S., Fekete, S. P., & Kr\oller, A. (2005). SpyGlass: A wireless sensor network visualizer. *SIGBED Review, 2*(1), 1-6.

Buschmann, C., Pfisterer, D., Fischer, S., Fekete, S. P., & Kroller, A. (2004). SpyGlass: Taking a closer look at sensor networks. *Proceedings of the Second International Conference on Embedded Networked Sensor Systems, November 3, 2004 - November 5,* (pp. 301-302).

Caesar, M., Castro, M., Nightingale, E. B., O'Shea, G., & Rowstron, A. (2006). Virtual ring routing: Network routing inspired by DHTs. *SIGCOMM Computer Communications Review*, *36*(4), 351–362. doi:10.1145/1151659.1159954

Camilo, T., Carreto, C., Silva, J. S., & Boavida, F. (2006). An energy-efficient ant based routing algorithm for wireless sensor networks. In *Proceedings of 5th International Workshop on Ant Colony Optimization and Swarm Intelligence*, Brussels, Belgium, (pp. 49-59).

Cardei, M., Thai, M. T., Li, Y., & Wu, W. (2005). Energy efficient target coverage in wireless the sensor networks. *Proceedings - IEEE INFOCOM*, 2005.

Carlson, E. A., Beaujean, P. P., & An, E. (2006, 18-21 Sept. 2006). *Location-aware routing protocol for underwater acoustic networks.* Paper presented at the OCEANS 2006.

Carter, A. (2005). *The status of vehicle-to-vehicle communication as a means of improving crash prevention performance.* Technical Report 05-0264, 2005. Retrieved from http://www.cs.umd.edu/class/fall2006/cmsc828s/PAPERS.dir/05-0264-W%5B1%5D.pdf

Cerpa, A., Elson, J., Estrin, D., Girod, L., Hamilton, M., & Zhao, J. (2001). Habitat monitoring: Application driver for wireless communications technology. In *Proceedings of ACM SIGCOMM Workshop on Data Communications in Latin America and the Caribbean.*

Cerpa, D. G. A., Ye, W., Yu, Y., Zhao, J., & Estrin, M. D. (2004). Networking issues in wireless sensor networks. *Journal of Parallel and Distributed Computing*, *64*(7), 799–814. doi:10.1016/j.jpdc.2004.03.016

Chamam, A., & Pierre, S. (2007). Energy-efficient state scheduling for maximizing sensor network lifetime under coverage constraint. *The Third IEEE International Conference on Wireless and Mobile Computing, Networking and Communications*, (p. 63). Washington, DC, USA.

Chamam, A., & Pierre, S. (2009). On the planning of wireless sensor networks: Energy-efficient clustering under the joint routing and coverage constraint. *IEEE Transactions on Mobile Computing*, *8*(8), 1077–1086. doi:10.1109/TMC.2009.16

Chang, J., & Tassiulas, L. (2004, August). Maximum lifetime routing in wireless sensor networks. *IEEE/ACM Transactions on Networking*, 609–619. doi:10.1109/TNET.2004.833122

Changsu, S., Young-Bae, K., & Dong-Min, S. (2006). An energy efficient cross-layer MAC protocol for wireless sensor networks. *The Eighth Asia Pacific Web Conference,* (pp. 410–419), Harbin, China.

Chatterjee, M., Das, S. K., & Turgut, D. (2002). WCA: A weighted clustering algorithm for mobile ad hoc networks. *Cluster Computing*, *5*(2), 193–204. doi:10.1023/A:1013941929408

Chee-Yee, C., & Kumar, S. P. (2003). Sensor networks: Evolution, opportunities, and challenges. *Proceedings of the IEEE*, *91*(8), 1247–1256. doi:10.1109/JPROC.2003.814918

Chen, D., & Varshney, P. K. (2004). *QoS support in wireless sensor networks: A survey*. International Conference on Wireless Networks 2004.

Chen, L., & Leneuter, J. (2007) Selfishness, not always a nightmare: Modeling selfish MAC behaviors in wireless mobile ad-hoc networks. In *Proceeding of ICDCS 2007*.

Chen, S., Huang, Y., & Zhang, C. (2008). Toward a real and remote wireless sensor network testbed. *WASA '08: Proceedings of the Third International Conference on Wireless Algorithms, Systems, and Applications,* Dallas, Texas, (pp. 385-396).

Chen, Y., & Nasser, N. (2006, August). *Energy-balancing multipath routing protocol for wireless sensor networks*. The Third International Conference on Quality of Service in Heterogeneous Wired/Wireless Network, Waterloo, Ontario, Canada.

Chen, Y., Leong, H. V., Xu, M., Cao, J., Chan, K. C., & Chan, A. T. S. (2006). *In-network data processing for wireless sensor networks*. 7th International Conference on Mobile Data Management (MDM'06), IEEE Computer Society.

Chen, B., Jamieson, K., Balakrishnan, H., & Morris, R. (2002). SPAN: An energy-efficient coordination algorithm for topology maintenance in ad hoc wireless networks. *Wireless Networks*, *8*(5), 481–494. doi:10.1023/A:1016542229220

Cheng, K.-Y., Lui, K.-S., & Tam, V. (2007). Localization in sensor networks with limited number of anchors and clustered placement. In *Proceedings of Wireless Communications and Networking Conference, 2007 (IEEE WCNC 2007)*, March 2007, (pp. 4425–4429).

Chen, G., Li, C., Ye, M., & Wu, J. (2009). An unequal cluster-based routing protocol in wireless sensor networks. *Wireless Networks*, *15*(2), 193–207. doi:10.1007/s11276-007-0035-8

Cheng, H., Cao, J., & Fan, X. (2008). GMZRP: Geography-aided multicast zone routing protocol in mobile ad hoc networks. *Mobile Networks and Applications*, *14*(2), 165–177. doi:10.1007/s11036-008-0135-4

Cheng, Z., Perillo, M., & Heinzelman, W. B. (2008). General network lifetime and cost models for evaluating sensor network deployment strategies. *IEEE Transactions on Mobile Computing*, *7*(4), 484–497. doi:10.1109/TMC.2007.70784

Chen, H. (2009). Task-based trust management for wireless sensor networks. *International Journal of Security and Its Applications*, *3*(2).

Chen, P. Y., & Liestman, A. L. (2003). A zonal algorithm for clustering ad hoc networks. *International Journal of Foundations of Computer Science*, *14*(2), 305–322. doi:10.1142/S0129054103001741

Chen, S., & Nahrstedt, K. (1999). Distributed quality-of-service routing in ad-hoc networks. *IEEE Journal on Selected Areas in Communications*, *17*(8).

Chen, W., Hou, J., & Sha, L. (2004). Dynamic clustering for acoustic target tracking in wireless sensor networks. *IEEE Transactions on Mobile Computing*, 258–271. doi:10.1109/TMC.2004.22

Chen, W., Kuo, S.-Y., & Chao, H.-C. (2006). Fuzzy preserving virtual polar coordinate space sensor networks for mobility performance consideration. *International Journal of Sensor Networks*, *1*(3/4), 179–189. doi:10.1504/IJSNET.2006.012033

Chen, Y., & Zhao, Q. (2005). On the lifetime of wireless sensor networks. *IEEE Communications Letters*, *9*(11), 976–978. doi:10.1109/LCOMM.2005.11010

Chiang, C.-C., Wu, H.-K., Liu, W., & Gerla, M. (1997). Routing in clustered multihop, mobile wireless networks with fading channel. *IEEE Singapore International Conference on Networks (SICON)*, (pp. 197-211).

Chio, A. M. P., Sahagun, J. A., & Sabido, I. X. D. J. M. (2001). VLSI implementation of a (255,223) Reed-Solomon error-correction codec. *Proceedings of 2nd National ECE Conference*, 2001.

Chirdchoo, N., Wee-Seng, S., & Kee Chaing, C. (2009, 26-29 May 2009). *Sector-based routing with destination location prediction for underwater mobile networks*. Paper presented at the International Conference on Advanced Information Networking and Applications Workshops, WAINA '09.

Chitre, M., Shahabudeen, S., Freitag, L., & Stojanovic, M. (2008, 15-18 September). *Recent advances in underwater acoustic communications & networking*. Paper presented at the OCEANS 2008.

Choi, J., Yoo, J., & Kim, C. K. (2008). A distributed fair scheduling scheme with a new analysis model in IEEE 802.11 wireless LANs. *IEEE Transactions on Vehicular Technology*, *57*(5).

Chong, C.-Y., & Kumar, S. P. (2003). Sensor networks: Evolution, opportunities and challenges. [Invited Paper]. *Proceedings of the IEEE*, *91*(8), 1247–1256. doi:10.1109/JPROC.2003.814918

Chong, C.-Y., & Kumar, S. P. (2003). Sensor networks: Evolution, opportunities, and challenges. *Proceedings of the IEEE, 91*(8), 1247–1256. doi:10.1109/JPROC.2003.814918

Chraibi, Y. (2005). *Localization in wireless sensor network.* Master's thesis, KTH Stockholm, Sweden, 2005.

Chu, M., Haussecker, H., & Zhao, F. (2002). Scalable information-driven sensor querying and routing for ad hoc heterogeneous sensor networks. *International Journal of High Performance Computing Applications, 16*(3), 293–313. doi:10.1177/10943420020160030901

Chun-Hao, Y., & Kuo-Feng, S. (2008, November 30-December 3). *An energy-efficient routing protocol in underwater sensor networks.* Paper presented at the 3rd International Conference on Sensing Technology, ICST 2008.

Ci, S., Sharif, H., & Nuli, K. (2005, February). Study of an adaptive frame size predictor to enhance energy conservation in wireless sensor networks. *IEEE Journal on Selected Areas in Communications, 23*(2).

Clark, B. N., Colbourn, C. J., & Johnson, D. S. (1990). Unit disk graphs. *Discrete Mathematics, 86*(1-3), 165–177. doi:10.1016/0012-365X(90)90358-O

Coello, C. A. C., Pulido, G. T., & Lechuga, M. S. (2004). Handling multiple objectives with particle swarm optimization. *IEEE Transactions on Evolutionary Computation, 8*(3), 256–279. doi:10.1109/TEVC.2004.826067

COMMON-Sense Net. (n.d.). *COMMON-sense net home.* Retrieved August 29, from http://commonsense.epfl.ch

Costa, P., Cesana, M., Brambilla, S., Casartela, L., & Pizziniaco, L. (2008). *A cooperative approach for topology control in wireless sensor networks: Experimental and simulation analysis.* 9th IEEE International Symposium on Wireless, Mobile and Multimedia Networks, WoWMoM 2008, June 23, 2008 - June 26. IEEE Computer Society.

Cover, T., & Thomas, J. (2006). *Elements of information theory.* Wiley-Interscience.

Cranch, G. A., Nash, P. J., & Kirkendall, C. K. (2003). Large-scale remotely interrogated arrays of fiber-optic interferometric sensors for underwater acoustic applications. *Sensors Journal, IEEE, 3*(1), 19–30. doi:10.1109/JSEN.2003.810102

Crossbow (2008). *Product reference guide.* Retrieved from http://www.xbow.com/

Crossbow Technology. (2007a). *MICAz.*

Crossbow Technology. (2007b). *MPR-MIB user's manual.*

Crossbow Technology. (2007c). *MTS/MDA sensor board user's manual.*

Crossbow. (n.d.a). *Crossbow technology: Development kits.* Retrieved September 3, 2009, from http://www.xbow.com/Products/productdetails.aspx?sid=160

Crossbow. (n.d.b). *Crossbow wireless sensor networks interface - Crossbow technology.* Retrieved October 10, 2009, from http://www.xbow.com/Technology/UserInterface.aspx

Crossbow. (n.d.c). *Technical support - Software downloads - Crossbow technology.* Retrieved October 10, 20069, from http://www.xbow.com/support/wSoftwareDownloads.aspx

Crowley, E., et al. (1998). *A framework for QoS based routing in the Internet.* Internet-draft, draft-ietf-qosrframework-06.txt, Aug. 1998.

Cui, S., Madan, R., Goldsmith, A., & Lall, S. (2005, May). Joint routing, MAC, and link layer optimization in sensor networks with energy constraints. *IEEE International Conference on Communications, ICC 2005, Vol. 2,* (pp. 725–729). Seoul Korea.

Cui, S., Goldsmith, A. J., & Bahai, A. (2004). Energy-efficiency of MIMO and cooperative MIMO techniques in sensor networks. *IEEE Journal on Selected Areas in Communications, 22*(6), 1089–1098. doi:10.1109/JSAC.2004.830916

Cvetkovski, A., & Crovella, M. (2009). Hyperbolic embedding and routing for dynamic graphs. In *IEEE INFOCOM 2009 - The 28th Conference on Computer Communications,* (pp. 1647-1655). IEEE.

Daeyoup, H., & Dongkyun, K. (2008, 15-18 September). *DFR: Directional flooding-based routing protocol for underwater sensor networks.* Paper presented at the OCEANS 2008.

Dai, S., Li, L., & Xu, D. (2006, October). A novel cluster formation approach based on the ILP for wireless sensor networks. *The First International Conference on Communications and Networking in China*, (pp. 1-5, 25-27). China.

Dalton, A. R., & Hallstrom, J. O. (2009a). Carolina WSN testbed: An interactive, source-centric, open testbed for developing and profiling wireless sensor systems. *International Journal of Distributed Sensor Networks*, *5*(2), 105–138. doi:10.1080/15501320701863403

Dalton, A. R., & Hallstrom, J. O. (2009b). An interactive, source-centric, open testbed for developing and profiling wireless sensor systems. *International Journal of Distributed Sensor Networks*, *5*(2), 105–138. doi:10.1080/15501320701863403

Dam, T. v., & Langendoen, K. (2003). An adaptive energy-efficient MAC protocol for wireless sensor networks. *1st ACM Conference on Embedded Networked Sensor Systems (SenSys)*, (pp. 171 - 180).

Dargie, W., & Poellabauer, C. (2010). *Fundamentals of wireless sensor networks: Theory and practice*. West Sussex, UK: John Wiley & Sons Ltd. doi:10.1002/9780470666388

Dario Pompili, T. M., & Akyildiz, I. F. (2006). *A resilient routing algorithm for long-term applications in underwater sensor networks.* Paper presented at the MedHocNet.

Dataman Papers. (n.d.). *TopDisc NW management index of dataman papers.* Retrieved September 7, 2009, from http://www.cs.rutgers.edu/dataman/papers/

De, P., Raniwala, A., Krishnan, R., Tatavarthi, K., Modi, J., Syed, N. A., et al. (2006). MiNT-m: An autonomous mobile wireless experimentation platform. *MobiSys '06: Proceedings of the 4th International Conference on Mobile Systems, Applications and Services,* Uppsala, Sweden (pp. 124-137).

De, P., Raniwala, A., Sharma, S., & Chiueh, T. (2005). MiNT: A miniaturized network testbed for mobile wireless research. *IEEE INFOCOM 2005, March 13, 2005 - March 17,* (pp. 2731-2742).

Deb, B., Bhatnagar, S., & Nath, B. (2001). *A topology discovery algorithm for sensor networks with applications to network management*. (Technical Report dcs-tr-441). Rutgers University, May 2001

Deb, B., Bhatnagar, S., & Nath, B. (2003). Multi-resolution state retrieval in sensor networks. *IEEE International Workshop on Sensor Network Protocols and Applications* (pp. 19-29).

Deb, B., Bhatnagar, S., & Nath, B. (2003). ReInForM: Reliable information forwarding using multiple paths in sensor networks. In *Proceedings of the 28th Annual IEEE International Conference on Local Computer Networks,* 2003.

Deb, K., Pratap, A., Agarwal, S., & Meyarivan, T. (2000). A fast and elitist multi-objective genetic algorithm: NSGA-II. *IEEE Transactions on Evolutionary Computation*, *6*(2), 182–197. doi:10.1109/4235.996017

Deci, E. L., & Ryan, R. M. (1991). A motivational approach to self: Integration in personality. In R. Dienstbier (Ed.), *Nebraska Symposium on Motivation- Vol. 38: Perspectives on motivation* (pp. 237-288). Lincoln, NE: University of Nebraska Press.

Demirkol, I., Ersoy, C., & Alagöz, F. (2006). MAC protocols for wireless sensor networks: A survey. *IEEE Communications Magazine.*

Deshpande, A., & Madden, S. (2006). MauveDB: Supporting model-based user views in database systems. *Proceedings of the 2006 ACM SIGMOD International Conference on Management of Data.* Chicago, IL: ACM.

Deshpande, A., Guestrin, C., Madden, S. R., Hellerstein, J. M., & Hong, W. (2004). Model-driven data acquisition in sensor networks. *Proceedings of the Thirtieth International Conference on Very Large Data Bases*, vol. 30. Toronto, Canada: VLDB Endowment.

Developer Zone Tutorials, N. I. (2010). *Wireless sensor network topologies and mesh networking*. Retrieved December 28, 2010, from http://zone.ni.com/devzone/cda/tut/p/id/11211

Dezhgosha, K., & Angara, S. (2005). Web services for designing small-scale Web applications. *IEEE International Conference on Electro Information Technology*, (pp. 1-4).

Di Caro, G., Ducatelle, F., & Luca, M. G. (2005). *AntHocNet: An adaptive nature-inspired algorithm for routing in mobile ad hoc networks*. Technical Report. IDSIA – Dalle Molle Institute for Artificial Intelligence Galleria, Switzerland.

Dimitriou, T., Kolokouris, J., & Zarokostas, N. (2007). SenseNeT: A wireless sensor network testbed. *MSWiM'07: 10th ACM Symposium on Modeling, Analysis, and Simulation of Wireless and Mobile Systems, October 22, 2007 - October 26,* (pp. 143-150).

Di, T., & Georganas, N. D. (2003). *A node scheduling scheme for energy conservation in large wireless sensor networks. Wireless Networks and Mobile Computing* (pp. 271–290). Wiley.

Djuric, P., Kotecha, J. Z. J., Huang, Y., Ghirmai, T., Bugallo, M., & Miguez, J. (2003, September). Particle filtering. *IEEE Signal Processing Magazine*, *20*, 19–38. doi:10.1109/MSP.2003.1236770

Domingo, M. C., & Prior, R. (2007, 3-7 Sept. 2007). *A distributed clustering scheme for underwater wireless sensor networks.* Paper presented at the IEEE 18th International Symposium on Personal, Indoor and Mobile Radio Communications, PIMRC 2007.

Dong, W., Liu, X., Chen, C., He, Y., Chen, G., Liu, Y., & Bu, J. (2010, March). Dynamic packet length control in wireless sensor networks. In *IEEE INFOCOM 2010*. San Diego: DPLC. doi:10.1109/INFCOM.2010.5462063

Donoso, Y., Fabregat, R., & Marzo, J. (2004). Multi-objective optimization algorithm for multicast routing with traffic engineering. *Proceedings of 3rd International Conference on Networking IEEE ICN'04* Pointe-à-Pitre, Guadeloupe, French Caribbean.

Dorigo, M., & Di Caro, G. (1997). *AntNet: A mobile agents approach to adaptive routing*. Technical Report. IRIDIA- Free Brussels University, Belgium.

Dorigo, M., & Di Caro, G. (1998). AntNet: Distributed stigmergetic control for communications networks. *Journal of Artificial Intelligence Research*, *9*, 317–365.

Douceur, J. (2002). The Sybil attack. In *Proceedings of the 1st International Workshop on Peer-to-Peer Systems* (IPTPS'02), February 2002. Retrieved from http://www.xbow.com/wireless home.aspx

Dovis, F., Fantini, R., Mondin, M., & Savi, P. (2001). 4G communications based on high altitude stratospheric platforms: Channel modeling and performance evaluation. *Global Telecommunications Conference, GLOBECOM '01 IEEE, 1,* (pp. 557-561).

Drane, C. J. (1968). Useful approximations for the directivity and beamwidth of large scanning Dolph-Chebyshev arrays. *Proceedings of the IEEE.*

Drupal. (2007). *PlanetLab - An open platform for developing, deploying, and accessing planetary-scale services.* Retrieved October 07, 2009, from http://www.planet-lab.org

DS099. (1999). *Xilinx DS099 Spartan-3 FPGA family data sheet.* Retrieved from www.xilinx.com/support/documentation/data_sheets/ds099.pdf

Du, X., & Chen, H. (2008). Security in wireless sensor networks. *IEEE Wireless Communications,* 2008.

Du, X., & Lin, F. (2005). *Designing efficient routing protocol for heterogeneous sensor networks*. Performance, Computing, and Communications Conference, 2005. IPCCC 2005. 24th IEEE International, 51-58.

Dust Networks. (n.d.). *Products: Dust networks.* Retrieved September 29, 2009, from http://www.dustnetworks.com/products

Dutta, P. (n.d.). *Epic: An open mote platform.* Retrieved September 12, 2009, from http://www.cs.berkeley.edu/~prabal/projects/epic/

Dutta, P., & Culler, D. (2008b). Epic: An open mote platform for application-driven design. *International Conference on Information Processing in Sensor Networks, IPSN '08,* (pp. 547-548).

Dutta, P., Hui, J., Jeong, J., Kim, S., Sharp, C., Taneja, J., et al. (2006). Trio: Enabling sustainable and scalable outdoor wireless sensor network deployments. *Fifth International Conference on Information Processing in Sensor Networks, IPSN '06, April 19, 2006 - April 21,* (pp. 407-415).

EAAST. (2009). *EAAST workshop volume 17: Kommunikation in verteilten sytemen 2009. Retrieved* September 6, from http://eceasst.cs.tu-berlin.de/index.php/eceasst/issue/view/24

El-Hoiydi, A., & Decotignie, J. (2004). WiseMAC: An ultra low power MAC protocol for the downlink of infrastructure wireless sensor networks. *Ninth IEEE Symposium on Computers and Communication, ISCC04*, (pp. 244 - 251).

Eliasson, J. (n.d.). *The Mulle - Bluetooth & Zigbee wireless networked sensor node.* Retrieved October 24, 2009, from http://www.csee.ltu.se/~jench/mulle.html

Emulab. (n.d.). *Emulab.net - Emulab - Network emulation testbed home.* Retrieved September 27, 2009, from http://www.emulab.net/index.php3?stayhome=1

Epson. (2008, September 28). *Epson develops non-contact power transmission module capable of transmitting 2.5 W with a 0.8mm coil.* Newsroom/Seiko Epson Corp. Retrieved April 26, 2010, from http://global.epson.com/newsroom/2008/news_20080922.htm

Er, I. I., & Seah, W. K. G. (2004). *Mobility-based d-hop clustering algorithm for mobile ad hoc networks*. IEEE Wireless Communications and Networking Conference (WCNC), Atlanta, USA.

Ergen, S., & Varaiya, P. (2006). Optimal placement of relay nodes for energy efficiency in sensor networks. In *Proceedings of the IEEE International Conference on Communications,* (vol. 8, pp. 3473–3479).

Erol, M., & Oktug, S. (2008, 13-18 April). *A localization and routing framework for mobile underwater sensor networks.* Paper presented at the INFOCOM Workshops 2008, IEEE.

Erol, M., Vieira, L. F. M., & Gerla, M. (2007). *Localization with Dive'N'Rise (DNR) beacons for underwater acoustic sensor networks*. Paper presented at the Second Workshop on Underwater Networks.

Ertin, E., Arora, A., Ramnath, R., Nesterenko, M., Naikt, V., Bapat, S., et al. (2006). Kansei: A testbed for sensing at scale. *Fifth International Conference on Information Processing in Sensor Networks, IPSN '06, April 19, 2006 - April 21,* (pp. 399-406).

Ervin, R. D. (1995). *The Crewman's Associate for Path Control (CAPC): An automated driver function.* University of Michigan, Transportation Research Institute.

Esensors, Inc. (n.d.). *Esensors, Inc. - Networked sensors.* Retrieved September, 30, 2009, from http://www.eesensors.com/

Estrin, D., Govindan, R., Heidemann, J., & Kumar, S. (1999). Next century challenges: Scalable coordination in sensor networks. *Proceedings of 5th Annual IEEE/ACM International Conference on Mobile Computing and Networking*, (pp. 263-270).

ETSI. (n.d.). *Intelligent transport systems*. Retrieved from http://www.etsi.org/WebSite/Technologies/IntelligentTransportSystems.aspx

ETSI. (n.d.). *Website*. Retrieved from www.etsi.org

Ettus, M. (1998). System capacity, latency, and power consumption in multihop-routed SS-CDMA wireless networks. *Proceedings of RAWCON, 98*, 55–58.

Evans, J. J. (2007). Undergraduate research experiences with wireless sensor networks. *Frontiers in Education Conference - Global Engineering: Knowledge without Borders, Opportunities without Passports, 2007. FIE '07. 37th Annual,* (pp. S4B-7-S4B-12).

Fall, K. (1999). Network emulation in the VINT/NS simulator. *IEEE Symposium on Computers and Communications Proceedings,* (pp. 244-250).

Fang, Q., Gao, J., Guibas, L. J., de Silva, V., & Zhang, L. (2005). GLIDER: Gradient landmark-based distributed routing for sensor networks. In *IEEE INFOCOM 2007 - IEEE International Conference on Computer Communications,* (pp. 339–350). IEEE.

Fasolo, E., Rossi, M., Widmer, J., & Zorzi, M. (2007). In-network aggregation techniques for wireless sensor networks: A survey. *Wireless Communications*, *14*(2), 70–87. doi:10.1109/MWC.2007.358967

Federal Motor Carrier Safety Administration. (n.d.). *Lane departure warning systems*. Retrieved from http://www.fmcsa.dot.gov/facts-research/research-technology/report/lane-departure-warning-systems.htm

Felegyhazi, M., & Hubaux, J.-P. (2006). *Game theory in wireless networks: A tutorial*. Tech. Rep. LCA-REPORT-2006-002.

Felemban, E., Lee, C. G., & Ekici, E. (2006). MMSPEED: Multipath multi-speed protocol for QoS guarantee of reliability and timeliness in wireless sensor networks. *IEEE Transactions on Mobile Computing, 5*(6). doi:10.1109/TMC.2006.79

Fielding, R., Gettys, J., Mogul, J., Frystyk, H., Masinter, L., Leach, P., & Berners-Lee, T. (1999). *Hypertext transfer protocol -- HTTP/1.1. RFC editor, USA,* (pp. 1-176).

Fielding, T., & Taylor, N. (2002). Principled design of the modern Web architecture. *ACM Transactions on Internet Technology*, 115–150. doi:10.1145/514183.514185

Fok, C., Roman, G., & Lu, C. (2009). Agilla: A mobile agent middleware for self-adaptive wireless sensor networks. *ACM Transactions in Autonomic and Adaptive Systems,* 4(3).

Fonseca, R., Ratnasamy, S., Zhao, J., Ee, C. T., Culler, D., Shenker, S., & Stoica, I. (2005). Beacon vector routing: Scalable point-to-point routing in wireless sensornets. In *NSDI'05: Proceedings of the 2nd Conference on Symposium on Networked Systems Design & Implementation,* (pp. 329-342). Berkeley, CA: USENIX Association.

Ford.com. (n.d.). *Ford makes parallel parking a breeze with new active parking assist.* Retrieved from http://media.ford.com/article_display.cfm?article_id=29625

Forouzan, B. A. (2007). *Data communications and networking* (pp. 267–301). New York, NY: McGraw-Hill.

Freshchi, F., Coello, C. A. C., & Repetto, M. (2008). Multiobjective optimization and artificial immune system: A review. In Mo, H. (Ed.), *Handbook of research on artificial immune systems and natural computing: Applying complex adaptive technologies* (pp. 1–21). Hershey, PA: IGI Global.

Frye, L., & Cheng, L. (2007). *Network management of a wireless sensor network.* Technical Report LU-CSE-07-003. Lehigh University.

Funke, S., & Milosavljevic, N. (2007). Guaranteed-delivery geographic routing under uncertain node locations. In *IEEE INFOCOM 2007 - 26th IEEE International Conference on Computer Communications,* (pp. 1244-1252). IEEE.

Furrer, S., Schott, W., Truong, H. L., & Weiss, B. (2006a). The IBM wireless sensor networking testbed. *2nd International Conference on Testbeds and Research Infrastructures for the Development of Networks and Communities, TRIDENTCOM 2006, March 1, 2006 - March 3,* (pp. 42-46).

Gama, J., & Gaber, M. (2007). *Learning from data streams: Processing techniques in sensor networks.* New York, NY: Springer-Verlag Inc.doi:10.1007/3-540-73679-4

Ganek, A., & Corbi, T. (2003). The dawning of the autonomic computing era. *IBM Systems Journal, 42*(1), 5–18. doi:10.1147/sj.421.0005

Ganesan, D., Cerpa, A., Ye, W., Yu, Y., Zhao, J., & Estrin, D. (2003). Networking issues in wireless sensor networks. *Journal of Parallel and Distributed Computing, 64*(7).

Garey, M. S., & Johnson, D. S. (1979). *Computers and intractability: Guide to the theory of NP-completeness.* New York, NY: W. H. Freeman.

Garg, V. (2007). *Wireless communications and networking.* Elsevier - Morgan Kaufmann Publishers.

Gatzianas, M., & Georgiadis, L. (2008). A distributed algorithm for maximum lifetime routing in sensor networks with mobile sink. *IEEE Transactions on Wireless Communications, 7*(3), 984–994. doi:10.1109/TWC.2008.060727

Gay, D., Levis, P., Behren, R., Welsh, M., Brewer, E., & Culler, D. (2003). The nesC language: A holistic approach to networked embedded systems. In *Proceedings of the ACM SIGPLAN Conference on Programming Language Design and Implementation (PLDI '03),* (pp. 1-11). New York, NY: ACM.

GENI. (n.d.). *GENI: Geni - Trac.* Retrieved September 26, 2009, from http://groups.geni.net/geni

GenSym. (n.d.). *Gensym: Business rule management.* Retrieved September 30, 2009, from http://www.gensym.com/

Gerla, M., & Tsai, J. T. C. (1995). Multicluster, mobile, multimedia radio network. *Wireless Networks, 1*(3), 255–265. doi:10.1007/BF01200845

Gilreath, W. F., & Laplante, P. A. (2003). *Computer architecture: A minimalist perspective.*

Gilreath, W. F., & Laplante, P. A. (2003). *Subtract and branch if negative (SBN). In Computer architecture: A minimalist perspective* (pp. 41-42). Springer.

Girod, L., Ramanathan, N., Elson, J., Stathopoulos, T., Lukac, M., & Estrin, D. (2007). Emstar: A software environment for developing and deploying heterogeneous sensor-actuator networks. *ACM Transactions in Sensor Networks, 3*(3), 13. doi:10.1145/1267060.1267061

Gnawali, O., Fonseca, R., Jamieson, K., Moss, D., & Levis, P. (2009). *Collection tree protocol*. In ACM SenSys '09, Berkeley, CA, USA.

Goense, D., Thelen, J., & Langendoen, K. (2005). *Wireless sensor networks for precise phytophthora decision support*. 5th European Conference on Precision Agriculture (5ECPA). Uppsala, Sweden.

Goodman, T., & Benaissa, M. (2006). Very small FPGA application-specific instruction processor for AES. *IEEE Transaction on Circuits Systems, 53*(7).

Google. (n.d.). *Google Maps*. Retrieved October 25, 2009, from http://maps.google.com/

Gracanin, D., Eltoweissy, M., Wadaa, A., & DaSilva, L. A. (2005). A service-centric model for wireless sensor networks. *IEEE Journal on Selected Areas in Communications, 23*(6), 1159–1166. doi:10.1109/JSAC.2005.845625

GSN. (n.d.). *GSN*. Retrieved September 9, 2009, from http://sourceforge.net/apps/trac/gsn/

Guang, L., & Assi, C. (2005). *On the resiliency of ad hoc networks to MAC layer misbehavior*. In Workshop on PE-WASUN, ACM MsWiM 2005.

Guang, L., & Assi, C. (2005). Vulnerabilities of ad hoc network routing protocols to MAC misbehavior. In *IEEE*. ACM WiMob.

Gui, C., & Mohapatra, P. (2004). Power conservation and quality of surveillance in target tracking sensor networks. *Proceedings of the 10th Annual International Conference on Mobile Computing and Networking*. Philadelphia, PA: ACM.

Gupta, H., Navda, V., Das, S. R., & Chowdhary, V. (2005). Efficient gathering of correlated data in sensor networks. *Proceedings of the 6th ACM International Symposium on Mobile Ad Hoc Networking and Computing*. Urbana-Champaign, IL: ACM.

Gupta, V., Krishnamurthy, S., & Faloutsous, M. (2002). Denial of service attacks at the MAC layer in wireless ad hoc networks. In *Proceedings of MILCOM*, 2002.

Gura, N., Patel, A., Wander, A., Sheueling, H. E., & Shantz, C. (n.d.). Comparing elliptic curve cryptography and RSA on 8-bit CPUs. Retrieved from http://www.research.sun.com/projects/crypto

Gurkan, D., Yuan, X., Benhaddou, D., Figueroa, F., & Morris, J. (2007). *Sensor networking testbed with IEEE 1451 compatibility and network performance monitoring*. Paper presented at the 2007 IEEE Sensors Applications Symposium, SAS, Febrary 6, 2007 - Febrary 8, IEEE Instrumentation and Measurement Society.

Gutierrez, J., Callaway, E., & Barrett, R. (2003). *Low-rate wireless personal area networks: Enabling wireless sensor networks with IEEE 802.15.4*. New York, NY: IEEE Press.

Haapola, J., Shelby, Z., Pomalaza-Raez, C., & Mahonen, P. (2005). Cross-layer energy analysis of multi-hop wireless sensor networks. *European Conference on Wireless Sensor Networks, EWSN'05*, (pp. 33-44). Istanbul, Turkey.

Hac, A. (2003). *Wireless sensor network design*. Wiley. doi:10.1002/0470867388

Hackmann, G., Chipara, O., & Lu, C. (2008, November). Robust topology control for indoor wireless sensor networks. In *6th ACM Conference on Embedded Networked Sensor Systems, ACM SenSys '08*, Raleigh, NC, USA.

Hadim, S., & Moamed, N. (2006). Middleware challenges and approaches for wireless sensor networks. *IEEE Distributed Systems Online, 7*(3).

Haensel. (n.d.). *An FDL'ed textbook on sensor networks*. Retrieved September 2, 2009, from http://www.informatik.uni-mannheim.de/~haensel/sn_book/

Haibo, Z., & Doshi, P. (2009). Towards automated RESTful Web service composition. *IEEE International Conference on Web Services ICWS '09*, (pp. 189-196).

Haiming, Y., & Sikdar, B. (2007). *Optimal cluster head selection in the LEACH architecture* (pp. 93–100).

Han, Q., & Venkatasubramanian, N. (2001). Autosec: An integrated middleware framework for dynamic service brokering. *IEEE Distributed Systems Online, 2*(7).

Handziski, V., Kopke, A., Willig, A., & Wolisz, A. (2006). TWIST: A scalable and reconfigurable testbed for wireless indoor experiments with sensor networks. *REALMAN 2006 - 2nd International Workshop on Multi-Hop Ad Hoc Networks: From Theory to Reality, May 26, 2006 - May 26,* (pp. 63-70).

Han, G., Shu, L., Ma, J., Park, J., & Ni, J. (2010). Power-aware and reliable sensor selection based on trust for wireless sensor networks. *The Journal of Communication, 5*(1), 23.

Hansen, W. W., & Woodyard, J. R. (1938). A new principle in directional antenna design. *Proceedings IRE, 26*(3).

HART Communication Foundation. (n.d.). *HART communication protocol - Wireless HART technology.* Retrieved October 24, 2009, from http://www.hartcomm.org/protocol/wihart/wireless_technology.html

Hasiotis, T., Alyfantis, G., Tsetsos, V., Sekkas, O., & Hadjiefthymiades, S. (2005). Sensation: A middleware integration platform for pervasive applications in wireless sensor networks. *Proceedings of the Second European Workshop on Wireless Sensor Networks.*

Haykin, S. (2002). *Adaptive filter theory* (4th ed.). New York, NY: Prentice-Hall.

He, T., Huang, C., Blum, B., Stankovic, J., & Abdelzaher, T. (2003). Range-free localization schemes in large scale sensor networks. In *Proceedings of the Ninth Annual International Conference on Mobile Computing and Networking (MobiCom '03),* September 2003, San Diego, CA, USA, (pp. 81-95).

He, T., Stankovic, J. A., Lu, C., & Abdelzaher, T. (2003). Speed: A stateless protocol for real-time communication in sensor networks. In *Proceedings of the 23rd International Conference on Distributed Computing Systems.*

He, T., Stoleru, R., & Stankovic, J. A. (2006). *Range free localization.* Technical report, University of Virginia, 2006.

Hedetniemi, S., & Liestman, A. (1988). A survey of gossiping and broadcasting in communication networks. *NETWORKS: Networks: An International Journal, 18*(4), 319–349. doi:10.1002/net.3230180406

Heidemann, J., Bulusu, N., Elson, J., Intanagonwiwat, C., Lan, K., & Xu, Y. … Govindan, R. (n.d.) *Effects of detail in wireless network simulation.* Retrieved September 14, 2009, from http://www.isi.edu/~johnh/PAPERS/Heidemann01a.html

Heidemann, J., Silva, F., & Estrin, D. (2003). Matching data dissemination algorithms to application requirements. *SenSys '03: Proceedings of the 1st International Conference on Embedded Networked Sensor Systems,* Los Angeles, California, USA, (pp. 218-229).

Heidemann, J., Wei, Y., Wills, J., Syed, A., & Yuan, L. (2006, 3-6 April). *Research challenges and applications for underwater sensor networking.* Paper presented at the Wireless Communications and Networking Conference, WCNC 2006. IEEE.

Heidemann, W., Ye, J., & Estrin, D. (2002). An energy-efficient MAC protocol for wireless sensor networks. In *Proceeding of INFOCOM 2002,* New York, (pp. 1567-1576).

Heinzelman, W. R., Chandrakasan, A., & Balakrishnan, H. (2000). Energy efficient communication protocol for wireless microsensor networks. *Proceedings of the 33rd Annual Hawaii International Conference on System Sciences,* 4-7 January 2000.

Heinzelman, W., Kulik, J., & Balakrishnan, H. (1999, August). Adaptive protocols for information dissemination in wireless sensor networks. *The 5th Annual ACM/IEEE International Conference on Mobile Computing and Networking (MobiCom'99)* (pp. 174-185). Seattle, WA.

Heinzelman, W., Murphy, A., Carvalho, H., & Perillo, M. (2004, January). Middleware to support sensor network applications. *IEEE Network Magazine Special Issue.*

Heinzelman, W., Chandrakasan, A., & Balakrishnan, H. (2002). An application-specific protocol architecture for wireless microsensor networks. *IEEE Transactions on Wireless Communications, 1*(4). doi:10.1109/TWC.2002.804190

Heissenbuttel, M. (2004). BLR: Beacon-less routing algorithm for mobile ad hoc networks. *Computer Communications*, *27*(11), 1076–1086. doi:10.1016/j.comcom.2004.01.012

Henricksen, K., & Robinson, R. (2006). A survey of middleware for sensor networks: State-of-the-art and future directions. In *Proceedings of the International Workshop on Middleware for Sensor Networks* (MidSens '06), (pp. 60-65). New York, NY: ACM.

He, T., Huang, C., Blum, B. M., Stankovic, J. A., & Abdelzaher, T. (2003). *Range-free localization schemes in large-scale sensor networks*. MOBICOM.

He, T., Krishnamurthy, S., Luo, L., Yan, T., Gu, L., & Stoleru, R. (2006). VigilNet: An integrated sensor network system for energy-efficient surveillance. *ACM Transactions in Sensor Networks*, *2*(1), 1–38. doi:10.1145/1138127.1138128

He, Y., Lee, I., & Guan, L. (2009). Distributed algorithms for network lifetime maximization in wireless visual sensor networks. *IEEE Transactions on Circuits and Systems for Video Technologies*, *19*(5), 704–718. doi:10.1109/TCSVT.2009.2017411

Hightower, J., & Borriello, G. (2001). Location systems for ubiquitous computing. *Computer*, *34*(8), 57–66. doi:10.1109/2.940014

Higuera, J., & Polo, J. (2010). Understanding the IEEE 1451 standard in 6loWPAN sensor networks. *IEEE Sensors Applications Symposium (SAS)*, (pp. 189-193).

Higuera, J., Polo, J., & Gasulla, M. (2009). A Zigbee wireless sensor network compliant with the IEEE 1451 standard. *IEEE Sensors Applications Symposium (SAS)*, (pp. 309-313).

Hill, J., Horton, M., Kling, R., & Krishnamurthy, L. (2004). The platforms enabling wireless sensor networks. *Communications of the ACM*, *47*(6), 41–46. doi:10.1145/990680.990705

Hoblos, G., Staroswiecki, M., & Aitouche, A. (2000). *Optimal design of fault tolerant sensor networks* (pp. 467–472).

Hoesel, L. V., Nieberg, T., Wu, J., & Havinga, P. J. (2004). Prolonging the lifetime of wireless sensor networks by cross-layer interaction. *IEEE Wireless Communication*, *11*(6), 78–86. doi:10.1109/MWC.2004.1368900

Hong, X., Gerla, M., Bagrodia, R., Kwon, T. J., Estabrook, P., & Pei, G. (2001). *The Mars sensor network: Efficient, energy aware communications*

Hong, X., Gerla, M., Pei, G., & Chiang, C. (1999). A group mobility model for ad hoc wireless networks. In *Proceedings of ACM/IEEE MSWiM*, Seattle, WA.

Hong, X., Gerla, M., Yi, Y., Xu, K., & Kwon, T. J. (2002). Scalable ad hoc routing in large, dense wireless networks using clustering and landmarks. In *Proceedings of the IEEE International Conference on Communications. (ICC'02), 25*(1), 3179 –3185.

Hong, X., & Liang, Q. An access-based energy efficient clustering protocol for ad hoc wireless sensor network. *15th IEEE International Symposium on Personal, Indoor and Mobile Radio Communications*, vol. 2, (pp. 1022-1026).

Hongwei, H., Hongke, Z., Yanchao, N., Shuai, G., Zhaohua, L., & Sidong, Z. (2007). *MSRLab6: An IPv6 wireless sensor networks testbed.* Paper presented at the 8th International Conference on Signal Processing, ICSP 2006, November 16, 2006 - November 20.

Hou, T.-C., & Tsai, T.-J. (2001). An access-based clustering protocol for multihop wireless ad hoc networks. *IEEE Journal on Selected Areas in Communications*, *19*(7), 1201–1210. doi:10.1109/49.932689

Hou, Y., Shi, Y., Sherali, H., & Midkiff, S. (2005). On energy provisioning and relay node placement for wireless sensor networks. *IEEE Transactions on Wireless Communications*, *4*(5), 2579. doi:10.1109/TWC.2005.853969

Hu, L., & Evans, D. (2004). Localization for mobile sensor networks. *Proceedings of the 10th Annual International Conference on Mobile Computing and Networking*. Philadelphia, PA: ACM.

Hu, X., & Eberhart, R. C. (2002). Multiobjective optimization using dynamic neighborhood particle swarm optimization. *Proceedings of Conference on Evolutionary Computation,* Honolulu, HI, (pp. 1677-1681).

Huang, R., Zaruba, G. V., & Huber, M. (2007). Complexity and error propagation of localization using interferometric ranging. In *Proceedings of IEEE International Conference on Communications ICC 2007*, (pp. 3063-3069). Glasgow, Scotland, June 2007.

Huang, X., Shah, P., & Sharma, D. (2010). Fast algorithm in ECC for wireless sensor network. *Proceedings of the International MultiConference of Engineers and Computer Scientists 2010,* vol. II, IMECS 2010, March 17 - 19, 2010, Hong Kong.

Huang, Y. L., Tygar, J. D., Lin, H. Y., Yeh, L. Y., Tsai, H. Y., Sklower, K., et al. (2008). SWOON: A testbed for secure wireless overlay networks. *CSET'08: Proceedings of the Conference on Cyber Security Experimentation and Test,* San Jose, CA, (pp. 1-6).

Hussain, S., & Islam, O. (2008). Genetic algorithm for energy efficient trees in wireless sensor networks. *Advanced Intelligent Environments*, 1-14.

IBM. (n.d.). *IBM research: Wireless sensor networking.* Retrieved August 29, 2009, from http://domino.watson.ibm.com/comm/research.nsf/pages/r.communications.innovation2.html

IDEALS. (n.d.). *Debugging wireless sensor networks using mobile actors: IDEALS @ illinois.* Retrieved September 10, 2009, from http://www.ideals.uiuc.edu/handle/2142/4607

IEEE Std. 802.15.4. (2006). *IEEE standard for Information Technology-Telecommunications and information exchange between systems-Local and metropolitan area networks-Specific wireless medium access control (MAC) and physical layer (PHY) specifications for low rate wireless personal area networks -Amendment of IEEE Std 802.15.4-2003.* IEEE Std 802.15.4-2006.

IEEE. (1999). *IEEE 802.11 standards, part 11: Wireless medium access control (MAC) and physical layer (PHY) specifications. Technical report.* The Working Group for WLAN Standards.

IEEE. (2007). *IEEE standard for a smart transducer interface for sensors and actuators - Wireless communication protocols and transducer electronic data sheet (TEDS) formats*

IEEE. (n.d.). *Welcome to IEEExplore 2.0: IEEE standard for a smart transducer interface for sensors and actuators - Common functions, communication protocols, and transducer electronic data sheet (TEDS) formats.*

IEEE. 802.15.4. (2006). *MAC and PHY specifications for LR-WPANs.* Retrieved from http://ieee802.org/15/pub/TG4.html

IEEE. 802.15.4. (2006). *Wireless medium access control (MAC) and physical layer (PHY) specifications for low rate wireless personal area networks* (LR-WPANS). Standard, IEEE.

Ikki, S., & Ahmed, M. (2007). Performance analysis of cooperative diversity wireless networks over Nakagami-m fading channel. *IEEE Communications Letters*, *11*(4), 334. doi:10.1109/LCOM.2007.348292

Ilyas, M., & Mahgoub, I. (2006). *Sensor network protocols.* Taylor and Francis Group.

Iniko. (2010). *PureEnergy solutions takes battery charging to a whole new level.* Shop Pure Energy. Retrieved September 15, 2009, from http://www.shoppureenergy.com

Intanagonwiwat, C., Govindan, R., & Estrin, D. (2000). Directed diffusion: A scalable and robust communication paradigm for sensor networks. *Proceedings of the Sixth Annual International Conference on Mobile Computing and Networking* (pp. 56-67)

Intanagonwiwat, C., Govindan, R., & Estrin, D. (2000, August). Directed diffusion: A scalable and robust communication paradigm for sensor networks. *The 6th Annual ACM/IEEE International Conference on Mobile Computing and Networking (MobiCom'00)*, (pp. 56-67). Boston, MA.

Intanagonwiwat, C., Govindan, R., Estrin, D., Heidemann, J., & Silva, F. (2003). Directed diffusion for wireless sensor networking. *IEEE/ACM Transactions on Networking*, *11*(1), 2–16. doi:10.1109/TNET.2002.808417

Internet Engineering Task Force. (n.d.). *IPv6 over low power WPAN (6LoWPAN).* Retrieved October 24, 2009, from http://www.ietf.org/dyn/wg/charter/6lowpan-charter.html

Iqbal, Z. (2006). Self organizing wireless sensor networks for inter- vehicle communication. Thesis submitted at Halmstad University, School of Information Science, Computer and Electrical Engineering. Retrieved from http://urn.kb.se/resolve?urn=urn:nbn:se:hh:diva-230

Iranli, A., Maleki, M., & Pedram, M. (2005). Energy efficient strategies for deployment of a two-level wireless sensor network. In *Proceedings of the 2005 International Symposium on Low Power Electronics and Design,* (p. 238).

ISA. (n.d.). *ISA 100, wireless systems for automation.* Retrieved October 24, 2009, from http://www.isa.org/MS-Template.cfm?MicrositeID=1134&CommitteeID=6891

Iyengar, S., Wu, H.-C., Balakrishnan, N., & Chang, S. Y. (2007). Biologically inspired cooperative routing for wireless mobile sensor networks. *IEEE Systems, 1*(2), 181–183.

Jaballah, W. B., & Tabbane, N. (2009). Multi path multi speed contention window adapter. *International Journal of Computer Science and Network Security, 9*(2).

Jain, A., Chang, E. Y., & Wang, Y.-F. (2004). Adaptive stream resource management using Kalman Filters. *Proceedings of the 2004 ACM SIGMOD International Conference on Management of Data*. Paris, France: ACM.

Jamieson, K., Krishnan, B., & Tay, Y. C. (2006). Sift: A MAC protocol for event driven wireless sensor networks. In *Proceeding of EWSN 2006*, (pp. 260-275).

Jenkins, J. (1973). Some properties and examples of random listening arrays. *IEEE Oceans, 5.*

Jeong, J. (2009). Wireless sensor networking for intelligent transport systems. Ph D thesis submitted to University of Minnesota, December 2009.

Jeong, J., Sharafkandi, S., & Du, D. H. C. (2006). Energy-aware scheduling with quality of surveillance guarantee in wireless sensor networks. In *Proceedings of the 2006 Workshop on Dependability Issues in Wireless Ad Hoc Networks and Sensor Networks*.

Jiang, C., Li, B., & Xu, H. (2007). An efficient scheme for user authentication in wireless sensor networks. *21st International Conference on Advanced Information Networking and Applications Workshops* (AINAW'07), (vol. 1, pp.438-442).

Jiang, S.-F., Yang, M.-H., Song, H.-T., & Wang, J.-M. (2007). An enhanced perimeter coverage based density control algorithm for wireless sensor network. *Third International Conference on Wireless and Mobile Communications,* ICWMC '07, (p. 79).

Jiang, X., Taneja, J., Ortiz, J., Tavakoli, A., Dutta, P., & Jeong, J. … Shenker, S. (2007). *An architecture for energy management in wireless sensor networks*. International Workshop on Wireless Sensor Network Architecture.

Jiang, Z. (2008). Underwater acoustic networks- Issues and solutions. *International Journal of Intelligent Control and Systems, 13*, 152–161.

Jin, J., Wang, Y., Tian, C., Liu, W., & Mo, Y. (2007). Localization and synchronization for 3D underwater acoustic sensor networks. In *Ubiquitous Intelligence and Computing 2007*, (pp. 622-631). Hong Kong, China, July 2007.

Jin, W., de Dieu, I. J., Jose, A. D. L. D., Lee, S., & Lee, Y.-K. (2010). *Prolonging the lifetime of wireless sensor networks via hotspot analysis*. In International Workshop on Computing Technologies and Business Strategies for u-Healthcare (CTBuH2010). Seoul, Korea.

Jinming, C., Xiaobing, W., & Guihai, C. (2008, 24-26 Oct. 2008). *REBAR: A reliable and energy balanced routing algorithm for UWSNs.* Paper presented at Seventh International Conference on the Grid and Cooperative Computing, GCC '08.

Joe, I. (2005, May). Optimal packet length with energy efficiency for wireless sensor networks. In *IEEE International Symposium on Circuits and Systems: Vol. 3* (pp. 2955-2957).

Johannes, P. B., Gehrke, J., & Seshadri, P. (2001). Towards sensor database systems. Proceedings of the 2nd International Conference on Mobile Data Management, (pp. 3-14).

Johnson, D., Stack, T., Fish, R., Flickinger, D. M., Stoller, L., Ricci, R., et al. (2006). *Mobile Emulab: A robotic wireless and sensor network testbed.* Paper presented at the INFOCOM 2006: 25th IEEE International Conference on Computer Communications, April 23, 2006 - April 29, Keithley Instruments Inc. (n.d.). *2701 - Integrated DMM/ Switch.* Keithley Instruments Inc. Retrieved October 25, 2009, from http://www.keithley.com/products/switch/dmmswitch/?mn=2701

Johnson, D. B., Maltz, D. A., & Broch, J. (2001). *DSR: The dynamic source routing protocol for multihop wireless ad hoc networks. Ad Hoc Networking* (pp. 139–172). Addison-Wesley Longman Publishing Co., Inc.

Johnson, D. H., & Dudgeon, D. E. (1993). *Array signal processing: Concepts and techniques*. P T R Prentice-Hall Inc.

Jornet, J. M., Stojanovic, M., & Zorzi, M. (2008). *Focused beam routing protocol for underwater acoustic networks*. Paper presented at the Third ACM International Workshop on Underwater Networks.

Jothi, R., & Raghavachari, B. (2005). Approximation algorithms for the capacitated minimum spanning tree problem and its variants in network design. [TALG]. *ACM Transactions on Algorithms*, *1*(2), 265–282. doi:10.1145/1103963.1103967

Juang, B., & Rabiner, L. (1985). A probabilistic distance measure for hidden Markov models. *AT & T Bell Laboratories Technical Journal*, *64*(2), 391–408.

Julier, S. J., & Uhlmann, J. K. (1997, July). New extension of Kalman filter to nonlinear systems. In I. Kadar (Ed.), *AeroSense '97, Vol. 3068., Signal Processing, Sensor Fusion, and Target Recognition VI, Multisensor Fusion, Tracking, And Resource Management II,* (pp. 182-193). Orlando, FL, USA.

Julier, S., & Uhlmann, J. (2004). Unscented filtering and non-linear estimation. *Proceedings of the IEEE*, *92*, 401–422. doi:10.1109/JPROC.2003.823141

Jun-Hong, C., Jiejun, K., Gerla, M., & Shengli, Z. (2006). The challenges of building mobile underwater wireless networks for aquatic applications. *Network*, *20*(3), 12–18.

Kadayif, I., & Kandemir, M. (2004). Tuning in-sensor data filtering to reduce energy consumption in wireless sensor networks. Design, Automation and Test in Europe Conference and Exhibition. *Proceedings*, *2*, 852–857.

Kai Chen, Y. Z., & He, J. (2009). A localization scheme for underwater wireless sensor networks. *International Journal of Advanced Science and Technology, 4*.

Kalpakis, K., Dasgupta, K., & Namjoshi, P. (2002). *Maximum lifetime data gathering and aggregation in wireless sensor networks*. Presented at The IEEE International Conference on Networking (NETWORKS '02), Atlanta, GA.

Kan, M. A., Afzal, S., & Manzoor, R. (2003). Hardware implementation of shortened (48,38) Reed Solomon forward error correcting code. *7th International Multi Topic Conference 2003*, (pp. 90-95).

Kannan, A. A., Mao, G., & Vucetic, B. (2006). Simulated annealing based wireless sensor network localization. *Journal of Computers*, *1*(2), 15–22. doi:10.4304/jcp.1.2.15-22

Kannan, R., Sarangi, S., & Lyengar, S. S. (2004). Sensor-centric energy constrained reliable query routing for wireless sensor networks. *Journal of Parallel and Distributed Computing*, *64*(7), 839–852. doi:10.1016/j.jpdc.2004.03.010

Kaplan, L. (2006). Global node selection for localization in a distributed sensor network. *IEEE Transactions on Aerospace and Electronic Systems*, *42*(1), 113–135. doi:10.1109/TAES.2006.1603409

Karl, H., & Willig, A. (2006). *Protocols and architectures for wireless sensor networks*. John Wiley and Sons Ltd.

Karp, B., & Kung, H. T. (2000). GPSR: Greedy perimeter stateless routing for wireless networks. In *MobiCom '00: Proceedings of the 6th Annual International Conference on Mobile Computing and Networking,* (pp. 243-254). New York, NY: ACM.

Katz, R. H., Kahn, J. M., & Pister, K. S. J. (1999). Mobile networking for smart dust. In *Proceedings of the 5th Annual ACM/IEEE International Conference on Mobile Computing and Networking* (MobiCom99). Seattle, Washington, (pp. 271-278).

Kawadia, V., & Kumar, P. R. (2005). *A cautionary perspective on cross-layer design*. IEEE Wireless Communication Magazine.

Kennedy, J. (1997). Minds and cultures: Particle swarm implications. *Socially Intelligent Agents. 1997 AAAI Fall Symposium,* (pp. 67-72). Technical Report FS-97-02. Menlo Park, CA: AAAI Press.

Kennedy, J. (1997). The particle swarm: Social adaptation of knowledge. In *Proceedings of IEEE International Conference on Evolutionary Computation*, Indianapolis, Indiana, (pp. 303-308).

Kennedy, J. (1998). The behavior of particles. In *Proceedings of 7th Annual Conference on Evolutionary Programming*, San Diego, USA.

Kennedy, J., & Eberhart, R. C. (1995). Particle swarm optimization. In *Proceedings of IEEE International Conference on Neural Networks*, Perth, Australia, (vol. 4, pp. 1942-1948).

Kim, H. S., Abdelzaher, T. S., & Kwon, W. H. (2005). Dynamic delay-constrained minimum-energy dissemination in wireless sensor networks. *Transactions on Embedded Computing Systems, 4*(3).

Kim, S.-J., Wang, X., & Madihian, M. (2006). Joint routing and medium access control for lifetime maximization of distributed wireless sensor networks. *IEEE International Conference on Communications* (pp. 3467-3472). Istanbul, Turkey.

Kleinberg, R. (2007). Geographic routing using hyperbolic space. In *IEEE INFOCOM 2007 - 26th IEEE International Conference on Computer Communications*, (pp. 1902-1909). IEEE.

Klopfer, M. (2005). *Interoperability & open architectures: An analysis of existing standardization processes & procedures*. Open Geospatial Consortium. OGC document number 05-049r1.

Klues, K., Hackmann, G., Chipara, O., & Lu, C. (2007). *A component-based architecture for power-efficient media access control in wireless sensor networks*. 5th International Conference on Embedded Networked Sensor Systems (SenSys'07). Sydney, Australia.

KonTest. (n.d.). *KonTest: WSN testbed.* Retrieved September 19, 2009, from http://www.few.vu.nl/~agaba/publications/

Kosucu, B., Irgan, K., Kucuk, G., & Baydere, S. (2009). FireSenseTB: A wireless sensor networks testbed for forest fire detection. *IWCMC '09: Proceedings of the 2009 International Conference on Wireless Communications and Mobile Computing,* Leipzig, Germany, (pp. 1173-1177).

Kotidis, Y. (2005). *Snapshot queries: Towards data-centric sensor networks* (pp. 131-142).

Kranakis, E., Singh, H., & Urrutia, J. (1999). *Compass routing on geometric networks*. In 11th Canadian Conference of Computational Geometry.

Kuhn, F., Moscibroda, T., & Wattenhofer, R. (2004). *Unit disk graph approximation*. In 2nd ACM Joint Workshop on Foundations of Mobile Computing (DIALM-POMC), Philadelphia, Pennsylvania, USA.

Kuhn, F., Wattenhofer, R., & Zollinger, A. (2003). Worst-case optimal and average-case efficient geometric ad-hoc routing. *The 4th ACM International Conference on Mobile Computing and Networking*, (pp. 267-278). Dallas, TX, USA.

Kulik, J., Heinzelman, W., & Balakrishnan, H. (2002). Negotiation-based protocols for disseminating information in wireless sensor networks. *Wireless Networks*, *8*(2), 169–185. doi:10.1023/A:1013715909417

Kumar, P., Gunes, M., Almamou, A. A., & Hussain, I. (2008). *Enhancing IEEE 802.15.4 for low-latency, bandwidth, and energy critical WSN applications*. 4th International Conference on Emerging Technologies (IEEE ICET 2008). Rawalpindi, Pakistan.

Kumar, P., Gunes, M., Mushtaq, Q., & Schiller, J. (2010). *Performance evaluation of AREA-MAC: A cross-layer perspective*. Fifth International Conference on Mobile Computing and Ubiquitous Networking (ICMU 2010). Seattle, USA.

Kumar, R., Wolenetz, M., Agarwalla, B., Shin, J., Hutto, P., Paul, A., & Ramachandran, U. (2003). DFuse: A framework for distributed data fusion. In *Proceedings of the 1st International Conference on Embedded Networked Sensor Systems* (SenSys '03), (pp. 114-125). New York, NY: ACM.

Kumar, S., Lai, T. H., & Balogh, J. (2004). On k-coverage in a mostly sleeping sensor network. *Proceedings of Mobicom*, Philadelphia.

Kumar, A., & Varma, S. (2010). Geographic node-disjoint path routing for wireless sensor networks. *IEEE Sensors Journal*, *10*(6), 1135–1136. doi:10.1109/JSEN.2009.2038893

Kumar, P., Gunes, M., Mushtaq, Q., & Blywis, B. (2009). A real-time and energy-efficient MAC protocol for wireless sensor network. [IJUWBCS]. *International Journal of Ultra Wideband Communications and Systems, 1*(2), 128–142. doi:10.1504/IJUWBCS.2009.029002

Kuorilehto, M., Kohvakka, M., Suhonen, J., Hämäläinen, P., Hännikäinen, M., & Hamalainen, T. (2007). *Ultra-low energy wireless sensor networks in practice: Theory, realization and deployment*. Hoboken, NJ: Wiley. doi:10.1002/9780470516805

Kyasanur, P., & Vaidya, N. (2004). *Selfish MAC layer misbehavior in wireless networks*. IEEE Transactions on Mobile Computing.

Labrador, M. A., & Wightman, P. M. (2009). *Topology control in wireless sensor networks*. Springer Science and Business Media B.V.

LabVIEW drivers for wireless sensor networks - developer zone - national instruments. (n.d.). Retrieved December 17, 2009, from http://zone.ni.com/devzone/cda/tut/p/id/5435

Lamport, L. (1978). Time, clocks, and the ordering of events in a distributed system. *Communications of the ACM, 21*(7), 558–565. doi:10.1145/359545.359563

Langendoen, K. G. (2008). Medium access control in wireless sensor networks. In Wu, H., & Pan, Y. (Eds.), *Medium access control in wireless networks* (pp. 535–560). Nova Science Publishers, Inc.

Laplante, P., & Gilreath, W. (2004). One instruction set computers for image processing. *The Journal of VLSI Signal Processing, 38*, 45–61. doi:10.1023/B:VLSI.0000028533.41559.17

Lazarevic, A., Ertoz, L., Kumar, V., Ozgur, A., & Srivastava, J. (2003). A comparative study of anomaly detection schemes in network intrusion detection. In *Proceedings of the Third SIAM International Conference on Data Mining*, (pp. 25–36).

Lazos, L., & Poovendran, R. (2004). SeRLoc: Secure range-independent localization for wireless sensor networks. *Proceedings of ACM 3rd Workshop on Wireless Security,* October.

Lazos, L., & Poovendran, R. (2005). SeRLoc: Robust localization for wireless sensor networks. In *ACM Transactions on Sensor Networks,* (pp. 73-100). August 2005.

Lederer, C., Mader, R., & Koschuch, M. Gro?ch?l, J., Szekely, A., & Tillich, S. (2009). Energy-efficient implementation of ECDH key exchange for wireless sensor networks. *Information Security Theory and Practices, WISTP 2009*, (pp. 112–127). September 2009.

Lee, S., Choe, H., Park, B., Song, Y., & Kim, C.-K. (2009). Luca: An energy-efficient unequal clustering algorithm using location information for wireless sensor networks. *Wireless Personal Communications*, (pp. 1–17). DOI:10.1007/s11277-009- 9842-9

Lee, S., Kim, C., & Kim, S. (2006). Constructing energy efficient wireless sensor networks by variable transmission energy level control. *The Sixth IEEE International Conference on Computer and Information Technology,* CIT '06, (pp. 225-225).

Lee, W. L. (n.d.). *WSN network management survey.* Winnie Louis Lee's Homepage. Retrieved September 7, 2009, from http://www.csse.uwa.edu.au/~winnie/

Lee, K., & Song, E. (2008). Understanding IEEE 1451-Networked smart transducer interface standard - What is a smart transducer? *Instrumentation & Measurement Magazine, 11*(2), 11–17. doi:10.1109/MIM.2008.4483728

Lee, W. C. (1995). *Routing subject to quality of service constraints integrated communication networks. IEEE Network*. July/August.

Lee, W. L., Datta, A., & Oliver, R. C. (2006). Network management in wireless sensor networks. In Deng, L. T., & Denko, M. K. (Eds.), *Mobile ad hoc and pervasive communications.*

Leong, B., Liskov, B., & Morris, R. (2006). Geographic routing without planarization. In *NSDI'06: Proceedings of the 3rd Conference on 3rd Symposium on Networked Systems Design & Implementation*, (p. 25). Berkeley, CA: USENIX Association.

Leong, W., & Yen, G. G. (2006). Dynamic population size in PSO-based multiobjective optimization. *IEEE Congress on Evolutionary Computing*, Vancouver, BC, Canada, (pp. 1718-1725).

Le, T., Hu, W., Corke, P., & Jha, S. (2009). ERTP: Energy-efficient and reliable transport protocol for data streaming in wireless sensor networks. *Computer Communications, 32*(7-10), 1154–1171. doi:10.1016/j.comcom.2008.12.045

Levis, P., & Culler, D. (2002). Mate: A tiny virtual machine for sensor networks. *Proceedings of the 10th International Conference on Architectural Support for Programming Languages and Operating Systems* (ASPLOSX), (pp. 85-95). ACM Press.

Levis, P., Gay, D., & Culler, D. (2004). Bridging the gap: Programming sensor networks with application specific virtual machines. *Proceedings of the 6th Symposium on Operating Systems Design and Implementation* (OSDI 04), 2004.

Levis, P., & Gay, D. (2009). *TinyOS programming*. Cambridge, UK: Cambridge University Press.

Levis, P., Lee, N., Welsh, M., & Culler, D. (2003). Accurate and scalable simulation of entire TinyOS applications. In *ACM SenSys, '03*. Los Angeles, CA, USA: TOSSIM.

Lewis, F. L., & Donnell, M.-O. (2006). *Wireless sensor networks: Issues, advances, and tools.* Automation & Robotics Research Institute (ARRI). The University of Texas at Arlington, Sponsored by IEEE Singapore Control Chapter.

Lewis, F. L. (2004). Wireless sensor networks. In Cook, D. J., & Das, S. K. (Eds.), *Smart environments: Technologies, protocols, and applications*. New York, NY: John Wiley.

Li, C., Ye, M., Chen, G., & Wu, J. (2005). An energy-efficient unequal clustering mechanism for wireless sensor networks. *IEEE International Conference on Mobile Ad Hoc and Sensor Systems*, (p. 604).

Li, H., Shenoy, P., & Ramamritham, K. (2004). Scheduling communication in real-time sensor applications. In *Proceedings of the 10th IEEE Real-Time and Embedded Technology and Applications Symposium*, 2004.

Li, J., & Lazarou, G. (2000, January). A bit-map-assisted energy-efficient MAX scheme for wireless sensor networks. *Hawaii International Conference on Systems Sciences*, (pp. 3005–3014). Maui, Hawaii.

Li, J., Kao, H., & Ke, J. (2009). *Voronoi-based relay placement scheme for wireless sensor networks.*

Li, X., Mao, Y., & Liang, Y. (2008). A survey on topology control in wireless sensor networks. *2008 10th International Conference on Control, Automation, Robotics and Vision, ICARCV 2008, December 17, 2008 - December 20,* (pp. 251-255).

Li, Y., Chen, C. S., Song, Y.-Q., & Wang, Z. (2007). *Real-time QoS support in wireless sensor networks: A survey*. 7th IFAC International Conference on Fieldbuses and Networks in Industrial and Embedded Systems. Toulouse, France.

Li, Z., Li, S., & Xingshe, Z. (2009). PFMA: Policy-based feedback management architecture for wireless sensor Networks. *5th International Conference on Wireless Communications, Networking and Mobile Computing,* (pp. 1–4).

Liang, J. J., Qin, A. K., Suganthan, P. N., & Baskar, S. (2006). Comprehensive learning particle swarm optimizer for global optimization of multimodal functions. *IEEE Transactions on Evolutionary Computation, 10*(3), 281–295. doi:10.1109/TEVC.2005.857610

Liang, Q. (2007). Radar sensor networks: Algorithms for waveform design and diversity with application to ATR with delay-doppler uncertainty. *EURASIP Journal on Wireless Communications and Networking*, (1): 18–18.

LibeLium Waspmote. (n.d.). *Wireless sensor networks mote.* Retrieved September 17, 2009, from http://www.libelium.com/products/waspmote

LibeLium. (n.d.). *Main page - SquidBee.* Retrieved September 12, 2009, from http://www.libelium.com/squidbee/index.php?title=Main_Page

Lin, C. R. (2000). On demand QoS routing in multihop mobile networks. *IEICE Transactions on Communications*, July.

Lin, P., Qiao, C., & Wang, X. (2004, March). Medium access control with a dynamic duty cycle for sensor networks. *The IEEE Wireless Communications and Networking Conference (WCNC)*, (pp. 1534-1539). Atlanta, Georgia, USA.

Lindsey, S., & Raghavendra, C. S. (2002). PEGASIS: Power-efficient gathering in sensor information systems. *Aerospace Conference Proceedings, IEEE,* (vol. 3, pp. 1125-1130).

Lindsey, S., Raghavendra, C., & Sivalingam, K. (2001, April). Data gathering in sensor networks using the energy*delay metric. *The IPDPS Workshop on Issues in Wireless Networks and Mobile Computing*, (pp. 2001-2008). San Francisco, CA.

Lin, E.-Y., Rabaey, J., & Wolisz, A. (2004). *Power-efficient rendezvous schemes for dense wireless sensor networks* (pp. 3769–3776). IEEE ICC.

Lin, P., Qiao, C., & Wang, X. (2004). *Medium access control with a dynamic duty cycle for sensor networks* (pp. 1534–1539). IEEE WCNC.

Lin, S., & Costello, D. J. (2004). *Error control coding* (2nd ed.). Upper Saddle River, NJ: Pearson Education, Inc.

Li, S., Son, S., & Stankovic, J. (2004). Event detection services using data service middleware in distributed sensor networks. *Telecommunication Systems*, *26*, 351–368. doi:10.1023/B:TELS.0000029046.79337.8f

Liu, A., & Ning, P. (2008). TinyECC: A configurable library for elliptic curve cryptography in wireless sensor networks. In *Proceedings of the 7th International Conference on Information Processing in Sensor Networks* (IPSN '08), (pp. 245-256). Washington, DC: IEEE Computer Society.

Liu, A.-F., Ma, M., Chen, Z.-G., & hua Gui, W. (2008). Energy-hole avoidance routing algorithm for WSN. *International Conference on Natural Computation, 1*, 76–80.

Liu, T., & Martonosi, M. (2003). Impala: A middleware system for managing autonomic, parallel sensor systems. In *Proceedings of the Ninth ACM SIGPLAN Symposium on Principles and Practice of Parallel Programming* (pp. 107-118). New York, NY: ACM.

Liu, Y., Elhanany, I., & Qi, H. (2005). *An energy-efficient QoS-aware media access control protocol for wireless sensor networks*. In IEEE International Conference on Mobile AdHoc and Sensor Systems Conference, 2005.

Liu, Z., Kwiatkowska, M. Z., & Constantinou, C. A. (2005). Biologically inspired QOS routing algorithm for mobile ad hoc networks. In *19th International Conference on Advanced Information Networking and Applications*, (pp. 426–431).

Liu, J., Reich, J., & Zhao, F. (2003). Collaborative in-network processing for target tracking. *EURASIP Journal on Applied Signal Processing*, *2003*, 378–391. doi:10.1155/S111086570321204X

Li, X. Y. (2008). *Wireless ad hoc and sensor networks, theory and applications*. Cambridge University Press. doi:10.1017/CBO9780511754722

Li, Y., Ye, W., & Heidemann, J. (2005). *Energy and latency control in low duty cycle MAC protocols*. IEEE WCNC.

Li, Z., Zhou, X., Li, S., Liu, G., & Du, K. (2005). Issues of wireless sensor network management. *ICESS, LNCS*, *3605*, 355–361.

Lloret, J., Tomas, J., Garcia, M., & Canovas, A. (2009). A hybrid stochastic approach for self-location of wireless sensors in indoor environments. *Sensors (Basel, Switzerland)*, *9*(5), 3695–3712. doi:10.3390/s90503695

Lodder, M., Halkes, G. H., & Langendoen, K. G. (n.d.). *A global-state perspective on sensor network debugging*. Retrieved September 9, 2009, from http://www.st.ewi.tudelft.nl/~koen/papers/globalstate.pdf

Lotfifar, F., & Shahhoseini, H. (2006). A mesh-based routing protocol for wireless ad-hoc sensor networks. *The International Wireless Communication and Mobile Computing Conference (IWCMC'06)*, (pp. 115-120). Vancouver, British Columbia, Canada.

Louridas, P. (2006). SOAP and Web services. *IEEE Software*, *23*(6), 62–67. doi:10.1109/MS.2006.172

Luke, H. (1999). The origins of the sampling theorem. *IEEE Communications Magazine*, 106–108. doi:10.1109/35.755459

Luo, L., Hei, T., Zhou, G., Gu, L., Abdelzaher, T. F., & Stankovic, J. A. (2006). *Achieving repeatability of asynchronous events in wireless sensor networks with EnviroLog*. Paper presented at the INFOCOM 2006: 25th IEEE International Conference on Computer Communications, April 23, 2006 - April 29, M2M Magazine. (n.d.). *The business value of sensors*. Retrieved October 6, 2009, from http://www.m2mmag.com/issue_archives/story.aspx?ID=7890

Luo, X., & Giannakis, G. (2008). Energy-constrained optimal quantization for wireless sensor networks. *EURASIP Journal on Advances in Signal Processing, 2008*, 1–12. doi:10.1155/2008/462930

Ma, Q., & Steenkiste, P. (1997). Quality-of-service routing with performance guarantees. *Proceedings of the 4th IFIP Workshop on Quality of Service*, May 1997.

Madan, R., Member, S., Cui, S. L., & Goldsmith, A. J. (2007). Modeling and optimization of transmission schemes in energy-constrained wireless sensor networks. *IEEE/ACM Transactions on Networking, 15*(6), 1359–1362. doi:10.1109/TNET.2007.897945

Madden, S. (2003). *The design and evaluation of a query processing architecture for sensor networks*. Ph.D. dissertation. University of California, Berkeley, Berkeley, CA.

Madden, S., Franklin, M., Hellerstein, J., & Hong, W. (2005, March). TinyDB: An acquisitional query processing system for sensor networks. *ACM Transactions on Database Systems, 30*(1), 122–173. doi:10.1145/1061318.1061322

Magistretti, E., Jiejun, K., Uichin, L., Gerla, M., Bellavista, P., & Corradi, A. (2007, 11-15 March). *A mobile delay-tolerant approach to long-term energy-efficient underwater sensor networking.* Paper presented at the Wireless Communications and Networking Conference, WCNC 2007. IEEE.

Mainwaring, A., Polastre, J., Szewczyk, R., Culler, D., & Anderson, J. (2002). *Wireless sensor networks for habitat monitoring*. First ACM International Workshop on Wireless Sensor Networks and Application (WSNA). Atlanta, GA, USA.

Malik, M. Y. (2010). Efficient implementation of elliptic curve cryptography using low-power digital signal processor. In *Proceedings of the 12th International Conference on Advanced Communication Technology* (ICACT'10), (pp. 1464-1468). Piscataway, NJ: IEEE Press.

Manjeshwar, A., & Agrawal, D. (2001, April). TEEN: A protocol for enhanced efficiency in wireless sensor networks. *The 1st International Workshop on Parallel and Distributed Computing Issues in Wireless Networks and Mobile Computing,* (pp. 2009-2015). San Francisco, CA.

Manjeshwar, A., & Agrawal, D. (2002, April). APTEEN: A hybrid protocol for efficient routing and comprehensive information retrieval in wireless sensor networks. *The 2nd International Workshop on Parallel and Distributed Computing Issues in Wireless Networks and Mobile computing*, (pp. 195 – 202). Ft. Lauderdale, FL.

Mansouri, M., Ouchani, I., Snoussi, H., & Richard, C. (2009). *Cramer-Rao Bound-based adaptive quantization for target tracking in wireless sensor networks*. In IEEE/SP Workshop on Statistical Signal Processing, SSP'09.

Mansouri-Samani, M., & Sloman, M. (1997). A generalised event monitoring language for distributed systems. *IEE/IOP/BCS. Distributed Systems Engineering, 4*(2). doi:10.1088/0967-1846/4/2/004

Manyika, J., & Durrant-Whyte, H. (1995). *Data fusion and sensor management: A decentralized information-theoretic approach*. Upper Saddle River, NJ: Prentice Hall PTR.

Mao, G., Fidan, B., & Anderson, B. D. O. (2007). Wireless sensor network localization techniques. *Computer Networks, 51*(10), 2529–2553. doi:10.1016/j.comnet.2006.11.018

Maroti, M., Kusy, B., Balogh, G., Volgyesi, P., Nadas, A., & Molnar, K..... Ledeczi, A. (2005). Radio interferometric geolocation. In *Proceedings of 3rd International Conference on Embedded Networked Sensor Systems (SenSys)*, (pp. 1-12). San Diego, California, USA, November 2005.

Martins, M., Cheung So, H., Chen, H., Huang, P., & Sezaki, K. (2008). Novel centroid localization algorithm for three-dimensional wireless sensor networks. In *IEEE Transactions on Wireless Communication, Networking and Mobile Computing*, (pp. 1-4). October 2008.

Mavaddat, F., & Parhami, B. (1988). URISC: The ultimate reduced instruction set computer. *International Journal of Electrical Engineering Education, 25*, 327–334.

Ma, Y. W., Chen, J. L., Huang, Y. M., & Lee, M. Y. (2010). An efficient management system for wireless sensor networks. *Sensors (Basel, Switzerland), 10*, 11400–11413. doi:10.3390/s101211400

McKay, M. R., & Collings, I. B. (2006). Improved general lower bound for spatially-correlated Rician MIMO capacity. *IEEE Communications Letters*, *10*(3). doi:10.1109/LCOMM.2006.1603371

Meertens, L., & Fitzpatrick, S. (2004). *The distributed construction of a global coordinate system in a network of static computational nodes from inter-node distances.* Kestrel Institute Technical Report KES.U.04.04, Kestrel Institute, Palo Alto, 2004. Retrieved from ftp://ftp.kestrel.edu/pub/papers/fitzpatrick/LocalizationReport.pdf

Mehta, S., & Kwak, K. S. (2009). Game theory and Information Technology. In *Proceeding of ISFT 2009* Korea.

Mehta, S., & Kwak, K. S. (2009). H-MAC: A hybrid MAC protocol for wireless sensor networks. *International Journal of Computer Networks & Communications*, *2*(2).

Melodia, T., Vuran, M. C., & Pompili, D. (2005). Lecture Notes in Computer Science: *Vol. 3883*. *The state of the art in cross-layer design for wireless sensor networks. Proceedings of EuroNGI Workshops on Wireless and Mobility*. Springer.

Merico, D., & Bisiani, R. (2006). *Positioning, localization and tracking in wireless sensor network.* Technical report, DISCo, NOMADIS, March, 2006.

MeshNetics. (n.d.). *Development tools.* Retrieved December 17, 2009, from http://www.meshnetics.com/dev-tools/

Mgillis. (2009). *PlanetLab team brings collaboration and policy expertise to GENI.* GENI. Retrieved September 17, 2009 from http://www.geni.net

Mhatre, V., & Rosenberg, C. (2004). Homogeneous vs heterogeneous clustered sensor networks: a comparative study. 2004 IEEE International Conference on Communications, vol. 6, pp. (646 - 3651).

MicroStrain. (n.d.). *2.4 GHz G-LINK wireless accelerometer node.* Retrieved December 17, 2009, from http://www.microstrain.com/g-link.aspx

Milowski, A. (2010). *Poster Firefox extension.* Retrieved from https://addons.mozilla.org/en-US/firefox/addon/2691/

Min, R., Bhardwaj, M., Cho, S., Shih, E., Sinha, A., Wang, A., & Chandrakasan, A. (2001). Low power wireless sensor networks. In *Proceedings of International Conference on VLSI Design,* Bangalore, India, (pp. 221-226).

Minhas, A. A. (2007). *Power aware routing protocols for wireless ad hoc sensor networks*. Graz, Austria: University of Technology.

Mishar, S., & Nasipuri, A. (2004). An adaptive low power reservation based MAC protocol for wireless sensor networks. *The IEEE International Conference on Performance, Computing, and Communications,* (pp. 731–736). Phoenix, Arizona.

Misra, I., Dolui, S., & Das, A. (2005). *Enhanced-efficient adaptive clustering protocol for distributed sensor networks*. ICON, 2005.

Misra, S., Woungang, I., & Misra, S. C. (2009). *Guide to wireless sensor networks*. Springer-Verlag London Limited.

Missoula Fire Sciences Laboratory. (n.d.). *FireModels.org - FARSITE.* Retrieved October 25, 2009, from http://www.firemodels.org/content/view/112/143

Miyashita, M., Nesterenko, M., Shah, R. D., & Vora, A. (2005). TOSGUI visualizing wireless sensor networks: An experience report. *2005 International Conference on Wireless Networks, ICWN'05, June 27, 2005 - June 30,* (pp. 412-419).

Modiano, E. (1994, October). *Data link protocols for LDR MILSTAR communications*. Communications Division Internal Memorandum. MIT Lincoln Laboratory, Lexington, MA.

Modiano, E. (1999, July). An adaptive algorithm for optimizing the packet size used in wireless ARQ protocols. *Wireless Networks*, *5*(4), 279–286. doi:10.1023/A:1019111430288

Modules, S. (n.d.). *Manual data sheet for MICAz*. Retrieved from http://www.xbow.com

Mohammed, A., Arnon, S., Grace, D., Mondin, M., & Miura, R. (2008). Advanced communications techniques and applications for high-altitude platforms. *Editorial for a Special Issue, EURASIP Journal on Wireless Communications and Networking.*

Mohammed, A., & Yang, Z. (2010). Next generation broadband services from high altitude platforms. In Adibi, S., Mobasher, A., & Tofighbakhsh, M. (Eds.), *Fourth-generation wireless networks: Applications and innovations* (pp. 249–267). Hershey, PA: Information Science Reference. doi:10.4018/978-1-61520-674-2.ch012

Molla, M. M., & Ahamed, S. I. (2006). A survey of middleware for sensor network and challenges. *International Conference on Parallel Processing Workshops*, ICPP, (p. 228).

Moller, S., Newe, T., & Lochmann, S. (2009). Review of platforms and security protocols suitable for wireless sensor networks. *IEEE Sensors Conference* (pp. 1000-1003). Christchurch, New Zealand: IEEE.

Momani, M., Challa, S., & Aboura, K. (2007). Modelling trust in wireless sensor networks from the sensor reliability prospective. *Innovative Algorithms and Techniques in Automation. Industrial Electronics and Telecommunications, 2007*, 317–321. doi:10.1007/978-1-4020-6266-7_57

MonSense. (n.d.). *MonSense - FEUP WSN group.* Retrieved September 12, 2009, from http://whale.fe.up.pt/wsnwiki/index.php/MonSense

Moore, D., Leonard, J., Rus, D., & Teller, S. (2004). Robust distributed network localization with noisy range measurements. In *Proceedings of the Second ACM Conference on Embedded Networked Sensor Systems (SenSys'04)*, November 2004, Baltimore, MD, (pp. 50-61).

Morcos, H., Matta, I., & Bestavros, A. (2004). *BIPAR: Bimodal power-aware routing protocol for wireless sensor networks*. Presented at the 1st International Computer Engineering Conference New Technologies for the Information Society (ICENCO). Cairo, EGYPT.

Morin, P. (2001). *Online routing in geometric graphs.* Ph.D. thesis, Carleton University School of Computer Science.

Mostaghim, S., & Teich, J. (2004). Covering pareto-optimal fronts by subswarms in multi-objective particle swarm optimization. In *Congress on Evolutionary Computation*, Portland, USA, (pp. 1404-1411).

MSN. (n.d.). *MSN virtual earth.* Retrieved October 25, 2009, from http://local.live.com.qe2a-proxy.mun.ca

Mui, E. N. C. (2007). *Practical implementation of Rijndael SBox using combinational logic.*

Nakamura, E. F., Loureiro, A. A. F., & Frery, A. C. (2007). Information fusion for wireless sensor networks: Methods, models, and classifications. *ACM Computing Surveys, 39*(3).

Nakamura, E. F., Ramos, H. S., Villas, L. A., de Oliveira, H. A. B. F., de Aquino, A. L. L., & Loureiro, A. A. F. (2009). A reactive role assignment for data routing in event-based wireless sensor networks. *Computer Networks, 53*(12), 1980–1996. doi:10.1016/j.comnet.2009.03.009

Nath, S., Liu, J., Miller, J., Zhao, F., & Santanche, A. (2006). SensorMap: A web site for sensors world-wide. *SenSys'06: 4th International Conference on Embedded Networked Sensor Systems,* (pp. 373-374).

National Institute of Standards and Technology. (2001). *Advance encryption standard AES, Federal information processing standards publication.* FIPS 197, 2001. Retrieved from http://csrc,nist.gov/publications/fips

National Science Foundation. (n.d.). *NSF network research testbed workshop report: Executive summary.* Retrieved September 2, 2009, from http://www-net.cs.umass.edu/testbed_workshop/exec_summary_html.htm

Neiswender, C. (2009). What is a controlled vocabulary? In *The MMI guides: Navigating the world of marine metadata.* Retrieved November 27, 2010, from http://marinemetadata.org/guides/vocabs/vocdef

NESTbed. (n.d.). *NESTbed- Project hosting on Google Code.* Retrieved October 25, 2009, from http://code.google.com/p/nestbed/

Nguyen, A., Milosavljevic, N., Fang, Q., Gao, J., & Guibas, L. J. (2007). Landmark selection and greedy landmark-descent routing for sensor networks. In *IEEE INFOCOM 2007 - 26th IEEE International Conference on Computer Communications*, (pp. 661 – 669). IEEE. Lin, X., & Stojmenovicm I. (1998). *Geographic distance routing in ad hoc wireless networks.* Technical report TR-98-10: SITE, University of Ottawa.

NI. (n.d.). *Wireless sensor network (WSN) starter kit - National Instruments.* Retrieved December 17, 2009, from http://sine.ni.com/nips/cds/view/p/lang/en/nid/206916

Nicolaou, N., See, A., Peng, X., Jun-Hong, C., & Maggiorini, D. (2007, 18-21 June). *Improving the robustness of location-based routing for underwater sensor networks.* Paper presented at the OCEANS 2007 - Europe.

Niculescu, D., & Badri, N. (2003, 30 March - 3 April). *Ad hoc positioning system (APS) using AOA.* Paper presented at the IEEE/ACM INFOCOM'03, San Francisco, CA.

NIST. (2010). *IEEE 1451 standard.* Retrieved November 27, 2010, from http://ieee1451.nist.gov/

NS-2 installation on Linux. (n.d.). Retrieved from http://paulson.in/?p=29

NS-2 module for Ant Colony Optimization (AntSense). (n.d.). Retrieved from http://eden.dei.uc.pt/~tandre/antsense/

NSLU2. (n.d.). *NSLU2-Linux - Main home page browse.* Retrieved October 25, 2009, from http://www.nslu2-linux.org/wiki/Main/HomePage

Obaisat, Y. A., & Braun, R. (2006). *On wireless sensor networks: Architectures, Protocols, applications, and management*. AusWireless Conference.

OGC. (n.d.). *Welcome to the OGC website OGC®.* Retrieved September 17, 2009, from http://www.opengeospatial.org

OliverWyman.com. (2006). *Press release: Study on auto electronics*. Retrieved from http://www.oliverwyman.com/ow/pdf_files/1_en_PR_Automotive_Elektronics.pdf

Ong, J. J., Ang, L.-M., & Seng, K. P. (2010). Implementation of (255,223) Reed Solomon minimal instruction set computing using Handel-C. *3rd IEEE International Conference on Computer Science and Information Technology 2010,* (vol. 5, pp. 49-54).

Oracle. (2011). *Metro Web services overview.* Retrieved January 17, 2011, from http://www.oracle.com/technetwork/java/index-jsp-137004.html

Osais, Y., St-Hilaire, M., & Yu, F.-R. (2009). On sensor placement for directional wireless sensor networks. *IEEE International Conference on Communications*, (pp. 1-5). Dresden, Germany.

Ou, C.-H., & Ssu, K.-F. (2008). Sensor position determination with flying anchor in three dimensional wireless sensor networks. *IEEE Transactions on Mobile Computing*, (September): 1084–1097.

Ouferhat, N., & Mellouck, A. (2006). QoS dynamic routing for wireless sensor networks. In *Proceedings of the 2nd ACM International Workshop on Quality of Service & Security for Wireless and Mobile Networks*, 2006.

Paek, J., Chintalapudi, K., Govindan, R., Caffrey, J., & Masri, S. (2005). A wireless sensor network for structural health monitoring: Performance and experience. *The Second IEEE Workshop on Embedded Networked Sensors, EmNetS-II*, (pp. 1-10).

Pai, S., Bermudez, S., Wicker, S. B., Meingast, M., Roosta, T., & Sastry, S. (2008). Transactional confidentiality in sensor networks. *IEEE Security and Privacy*, *6*(4), 28–35. doi:10.1109/MSP.2008.107

Papadaki, K., & Vasilis, F. (2008, March). Joint routing and gateway selection in wireless mesh networks. *The IEEE Conference of Wireless Communications and Networking, (WCNC)* (pp. 2325-2330). Las Vegas, USA.

Papadopoulos, F., Krivoukov, D., Baguna, M., & Vahdat, A. (2010). Greedy forwarding in dynamic scale free networks embedded in hyperbolic plane. In *IEEE INFOCOM 2010: 29th Conference on Information Communication*, (pp. 2973-2981). San Diego, CA, USA. IEEE.

Papoulis, A., & Pillai, S. (2002). *Probability, random variables and stochastic processes. McGraw-Hill Education*. India: Pvt Ltd.

Parameswaran, A., Husain, M. I., & Upadhyaya, S. (2009). *Is RSSI a reliable parameter in sensor localization algorithms? An experimental study.* In Field Failure Data Analysis Workshop (F2DA'09).

Park, S., Savvides, A., & Srivastava, M. B. (2000). SensorSim: A simulation framework for sensor networks. *MSWIM '00: Proceedings of the 3rd ACM International Workshop on Modeling, Analysis and Simulation of Wireless and Mobile Systems,* Boston, Massachusetts, United States, (pp. 104-111).

Parrott, D., & Xiaodong, L. (2004). Locating and tracking multiple dynamic optima by a particle swarm model using speciation. *IEEE Transactions on Evolutionary Computation, 10*(4), 440–458. doi:10.1109/TEVC.2005.859468

Parsopoulos, K. E., Tasoulis, D. K., & Vrahatis, M. N. (2004). Multiobjective optimization using parallel vector evaluated particle swarm optimization. *IASTED International Conference on Artificial Intelligence and Applications,* Innsbruck, Austria, (pp. 823-828).

Patil, M. M., Shaha, U., Desai, U. B., & Merchant, S. N. (2005). Localization in wireless sensor networks using three masters. *Proceedings of IEEE International Conference on Personal Wireless Communications*, New Delhi, January.

Patra, C. (2010). Using Kohonen's self-organizing map for clustering in sensor networks. *International Journal of Computers and Applications, 1*(24), 80–81.

Patra, C., Roy, A. G., Chattophaday, S., & Bhaumik, B. (2010). Designing energy-efficient topologies for wireless sensor network: Neural approach. *International Journal of Distributed Sensor Networks, 2010*, 216716. doi:10.1155/2010/216716

Patwari, N., & Hero, A. O. (2006). Indirect radio interferometric localization via pairwise distances. In *Proceedings of 3rd IEEE Workshop on Embedded Networked Sensors (EmNets 2006)*, (pp. 26-30). Boston, MA, May 30-31, 2006.

Pei, J., Ivey, R. A., Lin, H., Landrum, A. R., Sandburg, C. J., & Ferzli, N. A. (2009). An experimental investigation of applying Mica2 motes in pavement condition monitoring. *Journal of Intelligent Material Systems and Structures, 20*(1), 63–85. doi:10.1177/1045389X08088785

Peng, S., Seah, W. K. G., & Lee, P. W. Q. (2007, 17-20 April). *Efficient data delivery with packet cloning for underwater sensor networks.* Paper presented at the Underwater Technology and Workshop on Scientific Use of Submarine Cables and Related Technologies, 2007.

Peng, X., Zhong, Z., Nicolas, N., Andrew, S., Jun-Hong, C., & Zhijie, S. (2010). *Efficient vector-based forwarding for underwater sensor networks*. Hindawi Publishing Corporation.

Perillo, M., & Heinzelman, W. (2004). *DAPR: A protocol for wireless sensor networks utilizing an application-based routing cost.*

Perillo, M., Cheng, Z., & Heinzelman, W. (2005). An analysis of strategies for mitigating the sensor network hot spot problem. In *Proceedings of the Second International Conference on Mobile and Ubiquitous Systems*, (pp. 474–478).

Perrig, A., Szewczyk, R., Tygar, J. D., Wen, V., & Culler, D. E. (2002). Spins: Security protocols for sensor networks. *Wireless Networking, 8*(5), 521–534. doi:10.1023/A:1016598314198

PHP REST SQL. (2010). *A HTTP REST interface to MySQL written in PHP*. Retrieved November 28, 2010, from http://phprestsql.sourceforge.net/download.html

Pillutla, L. S., & Krishnamurhty, V. (2005). *Joint rate and cluster optimization in cooperative MIMO sensor networks*. IEEE 6th Workshop on Signal Processing Advances in Wireless Communications.

Pokraev, S., Quartel, D., Steen, M. W. A., & Reichert, M. (2006). A method for formal verification of service interoperability. *ICWS '06 International Conference on Web Services*, (pp. 895-900).

Pokraev, S., Reichert, M., Steen, M., & Wieringa, R. (2005). Semantic and pragmatic interoperability: A model for understanding. *Open Interoperability Workshop on Enterprise Modelling and Ontologies for Interoperability*, (pp. 1-5).

Polastre, J., Hill, J., & Culler, D. (2004). Versatile low power media access for wireless sensor networks. *2nd ACM Conference on Embedded Networked Sensor Systems (SenSys 2004)*, (pp. 95 - 107). Baltimore, MD.

Pollin, S., Ergen, M., Bougard, S. C., Perre, D. E., Cathoor, L. V., Morman, F., ... Variya, P. (2005). *Performance analysis of slotted IEEE 802.15.4 medium access layer*. Draft-jwl-tcp-fast-01.txt

Pollin, S., Ergen, M., Ergen, S., Bougard, C., Van der Perre, L., & Catthoor, F. … Varaiya, P. (2006, November). Performance analysis of slotted carrier sense IEEE 802.15.4 medium access layer. In *49th IEEE Global Telecommunications Conference, GLOBECOM 2006*. San Francisco, CA, USA.

Polze, A., Richling, J., Schwarz, J., & Malek, M. (1999). Towards predictable CORBA-based Web-services. *Proceedings 2nd IEEE International Symposium on Object-Oriented Real-Time Distributed Computing (ISORC '99)*, (pp. 182-191).

Pompili, D. (2007). *Efficient communication protocols for underwater acoustic sensor networks*. Georgia Institute of Technology.

Potdar, V., Sharif, A., & Chang, E. (2009). Wireless sensor networks: A survey. In *WAINA '09: Proceedings of the 2009 International Conference on Advanced Information Networking and Applications Workshops*, (pp. 636–641). Washington, DC: IEEE Computer Society.

Powercast corporation. (n.d.). Retrieved April 26, 2010, from http://www.powercastco.com/

Priyantha, N. B., Chakraborty, A., & Balakrishnan, H. (2000). *The Cricket location-support system.* 6th ACM International Conference on Mobile Computing and Networking (ACM MOBICOM), Boston, MA.

Priyantha, N., Balakrishnan, H., Demaine, E., & Teller, S. (2003). *Anchor-free distributed localization in sensor networks.* MIT Laboratory for Computer Science, Technical Report TR-892, April 2003. Retrieved from http://citeseer.ist.psu.edu/681068.html

Project, E. U. (n.d.). *WISEBED - Wireless sensor network testbeds - EU project.* Retrieved August 29, 2009, from http://www.wisebed.eu/

Qiao, D., Choi, S., & Shin, K. G. (2002, October). Goodput analysis and link adaptation for IEEE 802.11a wireless LANs. *IEEE Transactions on Mobile Computing, 1*(4), 278–292. doi:10.1109/TMC.2002.1175541

Qing, Z., & Tong, L. (2005). Energy efficiency of large-scale wireless networks: Proactive versus reactive networking. *IEEE Journal on Selected Areas in Communications, 23*(5), 1100–1112. doi:10.1109/JSAC.2005.845411

Querin, R., & Orda, A. (1997). QoS-based routing in networks with inaccurate information: Theory and algorithms. In *Proceedings of IEEE INFOCOM'97*, Japan, (pp. 75-83).

Rabaey, J. M., Ammer, M. J., Silver, J. L. D., Patel, D., & Roundy, S. (2000). PicoRadio supports ad hoc ultra low power wireless networking. *IEEE Computer, 33*(7), 42–48. doi:10.1109/2.869369

Rabiner, L. (1989). A tutorial on hidden Markov models and selected applications inspeech recognition. *Proceedings of the IEEE, 77*(2), 257–286. doi:10.1109/5.18626

Radmand, P., Talevski, A., Petersen, S., & Carlsen, S. (2010). Comparison of industrial WSN standards. *4th IEEE International Conference on Digital Ecosystems and Technologies (DEST)*, (pp. 632-637).

Rajendran, V., Obraczka, K., & Garcia-Luna-Aceves, J. J. (2003). *Energy-efficient, collision-free medium access control for wireless sensor networks*. Los Angeles, CA: ACM SenSys.

Rapport, T. (1996). *Wireless communications: Principles & practice*. Prentice-Hall.

Rashid-Farrokhi, F., Tassiulas, L., & Liu, K. J. R. (1998). Joint power control and beamforming in wireless networks using antenna arrays. *IEEE Transactions on Communications, 46*(10). doi:10.1109/26.725309

Ratnasamy, S., Karp, B., Shenker, S., Estrin, D., Govindan, R., & Yin, L. (2003). Data-centric storage in sensornets with GHT, a geographic hash table. *Mobile Networks and Applications, 8*(4), 427–442. doi:10.1023/A:1024591915518

Raya, M., Hubaux, J. P., & Aad, I. (2004). DOMINO: A system to detect greedy behavior in IEEE 802.11 hotspots. In *Proceedings of ACM MobiSys*, June 2004.

Reddy, A. M. V., Kumar, A. V. U. P., Janakiram, D., & Kumar, G. A. (2009). Wireless sensor network operating systems: A survey. *International Journal of Sensor Networks, 5*(4), 236–255. doi:10.1504/IJSNET.2009.027631

Reed, I. S., & Solomon, G. (1960). Polynomial codes over certain finite fields. *Journal of the Society for Industrial and Applied Mathematics, 8*, 300–304. doi:10.1137/0108018

Rentel, C. H., & Kunz, T. (2005). Mac coding for QoS guarantees in multi-hop mobile wireless networks. In *Proceedings of the 1st ACM International Workshop on Quality of Service & Security in Wireless and Mobile Networks*, 2005.

Reverter, F., & Pallas-Areny, R. (2005). *Direct sensor-to-microcontroller interface circuits: Design and characterization.* Barcelona, Spain: Editorial Marcombo S.A.

Rhee, I., Warrier, A., Aia, M., & Min, J. (2005). *ZMAC: A hybrid MAC for wireless sensor networks.* 3rd ACM Conference on Embedded Networked Sensor Systems (SenSys'05).

Rick, H., Pin-Han, H., & Shen, X. (2006). Cross-layer application-specific wireless sensor network design with single-channel CSMA MAC over sense-sleep trees. *Computer Communications, 29*(17), 3425–3444. doi:10.1016/j.comcom.2006.01.019

Rodoplu, V., & Meng, T. H. (1999). Minimum energy mobile wireless networks. *IEEE Journal on Selected Areas in Communications, 17*(8), 1333–1344. doi:10.1109/49.779917

Roman, R., & Alcaraz, C. (2007). Lecture Notes in Computer Science: *Vol. 4582. Applicability of public key infrastructures in wireless sensor networks. Public Key Infrastructure* (pp. 313–320).

Romer, K., Frank, C., Marron, P. J., & Becker, J. (2004). Generic role assignment for wireless sensor networks. *Proceedings of the 11th Workshop on ACM SIGOPS.*

Romer, K., & Mattern, F. (2004). The design space of wireless sensor networks. *IEEE Wireless Communications, 11*(6), 54–61. doi:10.1109/MWC.2004.1368897

Ronen, O., Rohlicek, J., & Ostendorf, M. (1995). Parameter estimation of dependence tree models using the EM algorithm. *IEEE Signal Processing Letters, 2*(8), 157–159. doi:10.1109/97.404132

Rooney, S., Bauer, D., & Scotton, P. (2005). *Edge server software architecture for sensor applications.*

Rosen, J. B. (1965). Existence and uniqueness of equilibrium points for concave n-person games. *Econometrica: Journal of the Econometric Society, 33*, 520–534. doi:10.2307/1911749

Rothery, S., Hu, W., & Corke, P. (2008). An empirical study of data collection protocols for wireless sensor networks. *3rd Workshop on Real-World Wireless Sensor Networks, REALWSN 2008, April 1, 2008 - April 1,* (pp. 16-20). Retrieved from http://dx.doi.org/10.1145/1435473.1435479

Rouvroy, G., Standaert, F.-X., Quisquater, J.-J., & Legat, J.-D. (2004). Compact and efficient encryption / decryption module for FPGA implementation of the AES Rijndael very well suited for small embedded applications. *International Conference on Information Technology: Coding and Computing* (ITCC'04) (volume 2, p. 583).

Ruiz, L., Nogueira, J., & Loureiro, A. (2003). MANNA: A management architecture for wireless sensor networks. *Communications Magazine, IEEE, 41*(2), 116–125. doi:10.1109/MCOM.2003.1179560

Saad, L., & Tourancheau, B. (2009). Multiple mobile sinks positioning in wireless sensor networks for buildings. *The Third IEEE International Conference on Sensor Technologies and Applications (SENSORCOMM)*, (pp. 264-270). Athens, Greece.

Sadagopan, N., Krishnamachari, B., & Helmy, A. (2003, May). The ACQUIRE mechanism for efficient querying in sensor networks. *The First International Workshop on Sensor Network Protocol and Applications*, (pp. 149-155). Anchorage, Alaska.

Sadek, A. K., Su, W., & Liu, K. J. R. (2007). Multinode cooperative communications in wireless networks. *IEEE Transactions on Signal Processing, 55*(1), 341–355. doi:10.1109/TSP.2006.885773

Safwati, A., Hassanein, H., & Mouftah, H. (2003, April). Optimal cross-layer designs for energy-efficient wireless ad hoc and sensor networks. *The IEEE International Conference of Performance, Computing, and Communications* (pp. 123 – 128).

Samaras, I., Gialelis, J., & Hassapis, G. (2009). Integrating wireless sensor networks into enterprise information systems by using Web services. *Third International Conference on Sensor Technologies and Applications SENSORCOMM*, (pp. 580-587).

Sanchez, J., Ruiz, P., & Stojmenovic, I. (2007). Energy-efficient geographic multicast routing for sensor and actuator networks. *Computer Communications, 30*, 2519–2531. doi:10.1016/j.comcom.2007.05.032

SANE. (2009). *Global WSN, SANE, Fraunhofer FOKUS.* Retrieved from http://www.fokus.fraunhofer.de/en/sane/projekte/laufende_projekte/global_wsn/index.html

Sankarasubramaniam, Y., Akyildiz, I. F., & McLaughlin, S. W. (2003, May). Energy efficiency based packet size optimization in wireless sensor networks. In *1st IEEE Int. Workshop on Sensor Network Protocols and Applications (SNPA'03)*, Anchorage, USA.

Santi, P. (2005). Topology control in wireless ad hoc and sensor networks. *ACM Computing Surveys*, *37*(2), 164–194. doi:10.1145/1089733.1089736

Sashima, A., Ikeda, T., Inoue, Y., & Kurumatani, K. (2008). *SENSORD/Stat: Combining sensor middleware with a statistical computing environment.* International Conference of Networked Sensing Systems.

Satoh, A., Morioka, S., Takano, K., & Munetoh, S. (2001). A compact Rijndael hardware architecture with s-box optimization. In *Proceedings of ASIACRYPT'01, LNCS vol. 2248*, (pp. 239-254).

Savarese, C., Rabaey, J., & Beutel, J. (2001). Locationing in distributed ad-hoc wireless sensor networks. In *Proceedings of IEEE International Conference on Acoustics, Speech, and Signal Processing (ICASSP '01)*, May 2001, Salt Lake City, Utah, USA, (vol. 4, pp. 2037-2040).

Savvides, A., Han, C.-C., & Strivastava, M. B. (2001). Dynamic fine-grained localization in ad-hoc networks of sensors. *Proceedings of the 7th Annual International Conference on Mobile Computing and Networking,* Rome (pp. 166-179).

Savvides, A., Park, H., & Srivastava, M. (2002). The bits and flops of the n-hop multilateration primitive for node localization problems. In *Proceedings of the 1st ACM international Workshop on Wireless Sensor Networks and Applications (WSNA'02),* September 2002, Atlanta, Georgia, USA, (pp. 112-121).

Sawant, H., Tan, J., Yang, Q., & Wang, Q. (2004). Using Bluetooth and sensor networks for intelligent transportation systems. *Proceedings of IEEE Intelligent Transportation Systems Conference,* Washington DC, USA, 2004.

Schurgers, C., & Srivastava, M. (2001). Energy efficient routing in wireless sensor networks. *The MILCOM Proceedings on Communications for Network-Centric Operations: Creating the Information Force*, (pp. 357 - 361). McLean, VA.

Schurgers, C., Tsiatsis, V., & Srivastava, M. B. (2002). STEM: Topology management for energy efficient sensor networks. *Aerospace Conference Proceedings,* vol. 3, (pp. 1099-1108).

Schurgers, C., Tsiatsis, V., Ganeriwal, S., & Srivastava, M. (2002b). *Topology management for sensor networks: Exploiting latency and density.* MOBIHOC'02, Lausanne, Switzerland, ACM, June 9-11 2002.

Schurgers, C., Tsiatsis, V., Ganeriwal, S., & Srivastava, M. (2002). Optimizing sensor networks in the energy-latency-density design space. *IEEE Transactions on Mobile Computing*, *1*(1), 70–80. doi:10.1109/TMC.2002.1011060

Schwartz, M. (1988). *Telecommunication networks*. Addison Wesley.

Seah, W. K. G., & Tan, H. P. (2006). *Multipath virtual sink architecture for wireless sensor networks in harsh environments*. Paper presented at the First International Conference on Integrated Internet Ad Hoc and Sensor Networks.

Selavo, L. (n.d.). *Open source wireless sensor networks.* Retrieved September 12, 2009, from http://www.openwsn.com/

Sengupta, S., & Chatterjee, M. (2004). *Distributed power control in sensor networks: A game theoretic approach* (pp. 508–519). India: IWDC.

SensorMap. (n.d.). *SensorMap home.* Retrieved from http://atom.research.microsoft.com/sensewebv3/sensormap/

SensWiz. (n.d.). *SensWiz: Dreamajax products.* Retrieved from http://www.senswiz.com/index.php?page=shop.browse&category_id=9&option=com_virtuemart&Itemid=54&vmcchk=1&Itemid=54

Sesay, S., Xiang, J., He, J., Yang, Z., & Cheng, W. (2006). Hotspot mitigation with measured node throughput in mobile ad hoc networks. In L. Li (Ed.), *The 6th* IEEE *International Conference on ITS Telecommunications (ITST 2006)*, (pp. 749–752).

Shah, R. C., & Rabaey, J. M. (2002). Energy aware routing for low energy ad hoc sensor networks. *Wireless Communications and Networking Conference*, WCNC2002, (vol. 1, pp. 350-355).

Shahbazi, H., Araghizadeh, M. A., & Dalvi, M. (2008). Minimum power intelligent routing in wireless sensors networks using self organizing neural networks. *IEEE International Symposium on Telecommunications*, (pp. 354-358).

Shahzad, W., Khan, F. A., & Siddiqui, A. B. (2009). Clustering in mobile ad hoc networks using comprehensive learning particle swarm optimization (CLPSO). *Communication and Networking, CCIS, 56*, 342–349. doi:10.1007/978-3-642-10844-0_41

Shang, Y., Ruml, W., Zhang, Y., & Fromherz, M. (2003). Localization from mere connectivity. In *Proceedings of ACM Symposium on Mobile Ad Hoc Networking and Computing (MobiHoc '03)*, June 2003, Annapolis, Maryland, USA, (pp. 201-212).

Sha, S. S., Yaqub, S., & Suleman, F. (2001). *Self-correcting codes conquer noise part 2: Reed-Solomon codecs* (pp. 107–120). EDN.

Shen, C.-C., Srisathapornphat, C., & Jaikaeo, C. (2001). Sensor information networking architecture and applications. *IEEE Personal Communications, 8*(4).

Shih, E., Cho, S. H., Ickes, N., Min, R., Sinha, A., Wang, A., & Chandrakasan, A. (2001, July). Physical layer driven protocol and algorithm design for energy-efficient wireless sensor networks. In *ACM MobiCom'01* (pp. 272-286). Rome, Italy.

Shim, Y. C., & Ramamoorthy, C. V. (1990). *Monitoring and control of distributed systems* (pp. 672–681).

Shi, Z. J. (2006). *Architectural challenges in underwater wireless sensor networks*. Department of Computer Science and Engineering University of Connecticut.

Sichitiu, M. L. (2004). Cross-layer scheduling for power efficiency in wireless sensor networks. *INFOCOM 2004, Twenty-third Annual Joint Conference of the IEEE Computer and Communications Societies,* vol. 3, (pp. 1740-1750). Hong Kong, China.

Simic, S., & Sastry, S. (2002). *Distributed localization in wireless ad hoc networks.* Technical report UCB/ERL M02/26, UC Berkeley, 2002. Retrieved from http://citeseer.ist.psu.edu/simic01distributed.html

Simon, G., Maroti, M., Ledeczi, A., & et-al. (2004). *Sensor network-based countersniper system.* 2nd ACM Conference on Embedded Networked Sensor Systems (SenSys). Baltimore, MD, USA.

Simon, M., & Alouini, M. (2005). *Digital communication over fading channels.* IEEE, 2005.

Singh, M., & Prasanna, V. K. (2003). *A hierarchical model for distributed collaborative computation in wireless sensor networks*. IEEE International Parallel and Distributed Processing Symposium.

Singh, S., & Raghavendra, C. (1998, July). PAMAS: Power aware multi-access protocol with signalling for ad hoc networks. *ACM Computer Communications Review, 28*(3), 5–26. doi:10.1145/293927.293928

Sivakumar, R., et al. (1998). *Core extraction distributed ad hoc routing (CEDAR) specification.* IETF Internet draft draft-ietf-manet-cedar-spec-00.txt, 1998.

Sivavakeesar, S., & Pavlou, G. (2004). Stable clustering through mobility prediction for large-scale multihop intelligent ad hoc networks. In *Proceedings of the IEEE Wireless Communications and Networking Conference (WCNC'04),* Georgia, USA, (vol. 3, pp. 1488-1493).

Skonnard, J. (2003). *Understanding SOAP*. Retrieved January 17, 2011, from http://msdn.microsoft.com/en-us/library/ms995800.aspx

Sleman, A., & Moeller, R. (2008). Integration of wireless sensor network services into other home and industrial networks.*3rd International Conference on Information and Communication Technologies ICTTA*, (pp. 1–5).

Slipp, J., Changning Ma, Polu, N., Nicholson, J., Murillo, M., & Hussain, S. (2008a). *WINTeR: Architecture and applications of a wireless industrial sensor network testbed for radio-harsh environments.*

Sliva, I. J. (2008). Technologies used in wireless sensor networks. *15th International Conference on Systems Signals and Image Processing IWSSIP*, (pp. 77-80).

Snoussi, H., & Richard, C. (2006). Ensemble learning online filtering in wireless sensor networks. In *10th IEEE Singapore International Conference on Communication Systems, ICCS 2006,* (pp. 1–5).

Sohrabi, K., Gao, J., Ailawadhi, V., & Pottie, G. J. (2000). Protocols for self-organization of a wireless sensor network. *IEEE Personal Communications*, *7*(5), 16–27. doi:10.1109/98.878532

Sohraby, K., Minoli, D., Znati, T., & John. (2007). *Wireless sensor networks: Technology, protocols, and applications.* Retrieved from http://www.wiley.com/WileyCDA/WileyTitle/productCd-0471743003.html

Sorniotti, A., Gomez, L., Wrona, K., & Odorico, L. (2007). Secure and trusted in-network data processing in wireless sensor networks: A survey. *Journal of Information Assurance and Security*, *2*(3), 189–199.

Sozer, E. M., Stojanovic, M., & Proakis, J. G. (2000). Underwater acoustic networks. *IEEE Journal of Oceanic Engineering*, *25*(1), 72–83. doi:10.1109/48.820738

Stankovic, J. A., Abdelzaher, T. F., Lu, C., Sha, L., & Hou, J. C. (2003). Realtime communication and coordination in embedded sensor networks. *Proceedings of the IEEE*, *91*(7), 1002–1022. doi:10.1109/JPROC.2003.814620

Std, I. E. E. E. 1451.0. (2007). Standard for a smart transducer interface for sensors and actuators - Common functions, communication protocols, and transducer electronic data sheet (TEDS) formats. *IEEE Std 1451.0-2007*, (pp. 1-335).

Std, I. E. E. E. 1451.5. (2007). Standard for a smart transducer interface for sensors and actuators wireless communication protocols and transducer electronic data sheet (TEDS) formats. *IEEE Std 1451.5-2007*, (pp. C1-236).

Std, I. E. E. E. 1451.7. (2010). Standard for smart transducer interface for sensors and actuators--Transducers to radio frequency identification (RFID) systems communication protocols and transducer electronic data sheet formats. *IEEE Std 1451.7-2010*, (pp. 1-99).

Steffan, J., Cilia, M., & Buchmann, A. (2004). *Scoping in wireless sensor networks*. Workshop on Middleware for Pervasive and Ad-Hoc Computing. Toronto, Canada: ACM Press.

Stojanovic, M. (2005, June). Optimization of a data link protocol for an underwater acoustic channel. In *IEEE Oceans'05*. Brest, France: Conference.

Stojmenovic, I., & Lin, X. (1999, November). *GEDIR: Loop-free location based routing in wireless networks.* Paper presented at In International Conference on Parallel and Distributed Computing and Systems, Boston, MA, USA.

Stojmenovic, I., & Lin, X. (2001). Loop-free hybrid single-path/flooding routing algorithms with guaranteed delivery for wireless networks. *IEEE Transactions on Parallel and Distributed Systems*, *12*(10). doi:10.1109/71.963415

Strat, X. X. (2008). *StratXX near space technology*. Retrieved from http://www.stratxx.com/products/

Su, W., & Lim, T. L. (2006, June). Cross-layer design and optimization for wireless sensor networks. *The Seventh ACIS International Conference on Software Engineering, Artificial Intelligence, Networking, and Parallel/Distributed Computing* (pp. 278 – 284). Las Vegas, Nevada, USA.

Subramanian, L., & Katz, R. (2000, August). An architecture for building self configurable systems. The *IEEE/ACM Workshop on Mobile Ad Hoc Networking and Computing*, (pp. 63 – 73). Boston, MA.

Subramanian, S., Shakkottai, S., & Gupta, P. (2007, May). On optimal geographic routing in wireless networks with holes and non-uniform traffic. *The 26th IEEE International Conference on Computer Communications. INFOCOM 2007*, (pp. 1019-1027). Anchorage, Alaska, USA.

Suh, C., & Ko, Y. (2005, May). A traffic aware, energy efficient mac protocol for wireless sensor networks. *IEEE International Symposium on Circuits and Systems ISCAS 2005,* vol. 3, (pp. 2975 – 2978). Kobe, Japan.

Sun, Y., Gurewitz, O., & Johnson, D. B. (2008). *RI-MAC: A receiver-initiated asynchronous duty cycle MAC protocol for dynamic traffic loads in wireless sensor networks*. 6th ACM Conference on Embedded Networked Sensor Systems, SenSys'08. Raleigh, NC, USA.

Swami, A., Zhao, Q., & Win Hong, Y. (2007). *Wireless sensor networks signal processing and communications perspectives*. West Sussex, UK: John Wiley & Sons Ltd.

Sylvester, J. (2001, January). Reed Solomon codes. *Elektrobit*.

Takai, M., Martin, J., & Bagrodia, R. (2001). Effects of wireless physical layer modeling in mobile ad hoc networks. *Proceedings of the 2001 ACM International Symposium on Mobile Ad Hoc Networking and Computing: MobiHoc 2001,* October 4, 2001 - October 5, (pp. 87-94). Retrieved from http://dx.doi.org/10.1145/501416.501429

Tang, L., & Guy, C. (2009). Radio frequency energy harvesting in wireless sensor networks. *IWCMC '09: Proceedings of the 2009 International Conference on Wireless Communications and Mobile Computing,* Leipzig, Germany (pp. 644-648). Retrieved from http://doi.acm.org/10.1145/1582379.1582519

Tao, H., Liu, W., & Ma, S. (2010). Intelligent transportation systems for wireless sensor networks based on ZigBee. *Proceedings of IEEE International Conference Computer and Communication Technologies in Agriculture Engineering* (CCTAE), Chengudu, China.

TechnologyReview. (2003, February). 10 emerging technologies that will change the world. *Technology Review, 106,* 33-49.

Teng, J., Snoussi, H., & Richard, C. (2007). Prediction-based proactive cluster target tracking protocol for binary sensor networks. In *2007 IEEE International Symposium on Signal Processing and Information Technology,* (pp. 234–239).

Teng, J., Snoussi, H., & Richard, C. (2007b). Binary variational filtering for target tracking in sensor networks. In IEEE/SP 14th Workshop on Statistical Signal Processing, SSP'07, (pp. 685–689).

Thanigaivelu, K., & Murugan, K. (2009). Impact of sink mobility on network performance in wireless sensor networks. *International Conference on Networks and Communications*, (pp. 7–11).

The Data Fusion Server. (n.d.). *Workshop on the theory on belief functions.* Retrieved from http://www.data-fusion.org/article.php?sid=70

Thornton, J., Grace, D., Capstick, M. H., & Tozer, T. C. (2003). Optimizing an array of antennas for cellular coverage from a high altitude platform. *IEEE Transactions on Wireless Communications, 2*(3), 484–492. doi:10.1109/TWC.2003.811052

Tian, D., & Geoganas, N. D. (2003). A node scheduling scheme for energy conservation in large wireless sensor networks. *Wireless Communications and Mobile Computing Journal, 3*(2).

Tian, H., Stankovic, J. A., Chenyang, L., & Abdelzaher, T. (2003, May). SPEED: A stateless protocol for real-time communication in sensor networks. *The International Conference on Distributed Computing Systems*, (pp. 46-55). Providence, RI.

Tillapart, P., Thammarojsakul, S., Thumthawatworn, T., & Santiprabhob, P. (2005). *An approach to hybrid clustering and routing in wireless sensor networks* (pp. 1–8).

Tolle, G., Polastre, J., Szewczyk, R., et al. (2005). *A macroscope in the redwoods*. 3rd ACM Conference on Embedded Networked Sensor Systems (SenSys). San Diego, CA, USA.

Tonghong, L. (2008, 25-27 May). *Multi-sink opportunistic routing protocol for underwater mesh network.* Paper presented at the International Conference on Communications, Circuits and Systems, ICCCAS 2008.

Townsend, C., & Arms, S. (2004). *Wireless sensor networks: Principles and applications* (pp. 441–442). MicroStrain, Inc.

Tse, D., & Viswanath, P. (2005). *Fundamentals of wireless communication*. Cambridge University Press.

Tseng, Y.-C., Wang, Y.-C., & Cheng, K.-Y. (2005). An integrated mobile surveillance and wireless sensor (iMouse) system and its detection delay analysis. *Proceedings of the 8th ACM International Symposium on Modeling, Analysis and Simulation of Wireless and Mobile Systems.* Montreal, Canada: ACM.

Tseng, Y.-C., & Wang, Y.-C. (2008). Efficient placement and dispatch of sensors in wireless sensor network. *IEEE Transactions on Mobile Computing*, (February): 262–274.

Tseng, Y.-C., & Wang, Y.-C. (2008b). Distributed deployment scheme in mobile wireless sensor networks to ensure multi level coverage. *IEEE Transactions on Parallel and Distributed Systems*, (September): 1280–1294.

Tubaishat, M., & Madria, S. (2003). Sensor networks: An overview. *IEEE Potentials*, 20-23.

Turgut, D., Das, S. K., Elmasri, R., & Turgut, B. (2002). Optimizing clustering algorithm in mobile ad hoc networks using genetic algorithmic approach. In *Proceedings of GLOBECOM'02*, Taipei, Taiwan, (pp. 62– 66).

Tutornet. (n.d.). *Tutornet.* Retrieved from http://enl.usc.edu/projects/tutornet/

Uchhula, V., & Bhatt, B. (2010). Comparison of different ant colony based routing algorithms. *International Journal of Computer Applications* [Foundation of Computer Science.]. *Special Issue on MANETs*, *2*, 97–101.

Ucinski, D., & Patan, M. (2007). D-optimal design of a monitoring network for parameter estimation of distributed systems. *Journal of Global Optimization*, *39*(2), 291–322. doi:10.1007/s10898-007-9139-z

Uichin, L., Wang, P., Youngtae, N., Vieira, L., Gerla, M., & Jun-Hong, C. (2010, 14-19 March). *Pressure routing for underwater sensor networks.* Proceedings IEEE INFOCOM.

Vahdat, A., & Becker, D. (2000). *Epidemic routing for partially-connected ad hoc networks.* Technical report, 2000. Retrieved from http://issg.cs.duke.edu/epidemic/epidemic.pdf

Valenzuela, R. (1987). A statistical model for indoor multipath propagation. *IEEE Journal on Selected Areas in Communications*, *5*(2), 128–137. doi:10.1109/JSAC.1987.1146527

Valle, Y., Venayagamoorthy, G. K., Mohagheghi, S., Hernandez, J.-C., & Harley, R. G. (2008). Particle swarm optimization: Basic concepts, variants and applications in power systems. *IEEE Transactions on Evolutionary Computation*, *12*(2).

Venkitasubramaniam, P., Adireddy, S., & Lang, T. (2003, Oct.). Opportunistic ALOHA and cross layer design for sensor networks. *Military Communications Conference IEEE,* Vol. 1, (pp. 705 – 710). Boston, MA, USA.

Vercauteren, T., Toledo, A. L., & Wang, X. (2007). Batch and sequential Bayesian estimators of the number of active terminals in an IEEE 802.11 network. *IEEE Transactions on Signal Processing*, *55*(2), 437–450. doi:10.1109/TSP.2006.885723

Vermaak, J., Lawrence, N., & Perez, P. (2003). Variational inference for visual tracking. In *Proceedings of the 2003 IEEE Computer Society Conference on Computer Vision and Pattern Recognition*, vol. 1.

Vesanto, J., & Alhoniemi, E. (2000). Clustering of self organizing map. *IEEE Transactions on Neural Networks*, *11*(3), 586–358. doi:10.1109/72.846731

Vieira, M. S. (2005). *A reconfigurable group management middleware service for wireless sensor networks.* 3rd International Workshop on Middleware for Pervasive and Ad- Hoc Computing, Grenoble, France.

VigilNet. (n.d.). *UVA VigilNet: Home.* Retrieved from http://www.cs.virginia.edu/wsn/vigilnet/

Vincent, P. J., Tummala, M., & McEachen, J. (2006, April 2006). *An energy-efficient approach for information transfer from distributed wireless sensor systems.* Paper presented at the IEEE/SMC International Conference on System of System Engineering, Los Angeles, CA, USA.

Vlajic, N., & Stevanovic, D. (2009). *Sink mobility in wireless sensor networks: a (mis)match between theory and practice* (pp. 386–393). IWCMC.

Vuran, M. C., & Akyildiz, I. F. (2008, April). *Cross-layer packet size optimization for wireless terrestrial, underwater, and underground sensor networks*. In IEEE INFOCOM '08 Phoenix, Arizona.

Wade, G. (2000). *Coding techniques, an introduction to compression and error control*. New York, NY: Palgrave.

Wagenknecht, G., Anwander, M., Braun, T., Staub, T., Matheka, J., & Morgenthaler, S. (2008). MARWIS: A management architecture for heterogeneous wireless sensor networks. *6th International Conference on Wired/Wireless Internet Communications, WWIC 2008, May 28, 2008 - May 30, 5031 LNCS* (pp. 177-188). Retrieved from http://dx.doi.org/10.1007/978-3-540-68807-5_15

Wang, H., Dong, B., Chen, P., Chen, Q., & Kong, J. (2007). LATEX DSL: A coverage control protocol for heterogeneous wireless sensor networks. *Second International Conference on Systems and Networks Communications*, ICSNC 2007, (p. 2).

Wang, L., & Xiao, Y. (2005). Energy saving mechanisms in sensor networks. *2nd International Conference on Broadband Networks* (pp. 724–732).

Wang, Q., Xu, K., Takahara, G., & Hassanein, H. (2005). *Locally optimal relay node placement in heterogeneous wireless sensor networks.*

Wang, Z., & Crowcraft, J. (1996). QoS-based routing for supporting resource reservation. *IEEE Journal on Selected Area of Communications*, September.

Wang, F.-Y. (2010). Parallel control and management for intelligent transportation systems: Concepts, architectures and applications. *IEEE Transactions on Intelligent Transportation Systems*, *11*(3), 630–638. doi:10.1109/TITS.2010.2060218

Wang, H., & Li, Q. (2006). Lecture Notes in Computer Science: *Vol. 4307*. *Efficient implementation of public key cryptosystems on mote sensors. information and communications security* (pp. 519–528).

Wang, Y. (2008). *Topology control for wireless sensor networks* (pp. 113–147). Wireless Sensor Networks and Applications.

Ward, A., Jones, A., & Hopper, A. (1997). A new location technique for the active office. *IEEE Personal Communications*, *4*(5), 42–47. doi:10.1109/98.626982

Wattenhofer, R., & Zollinger, A. (2004). XTC: A practical topology control algorithm for ad-hoc networks. *Proceedings of the 18th International Parallel and Distributed Processing Symposium*, 2004, 26-30 April (p. 16).

Wei, L., Haibin, Y., Lin, L., Bangxiang, L., & Chang, C. (2007, 5-8 August). *Information-carrying based routing protocol for underwater acoustic sensor network.* Paper presented at the International Conference on Mechatronics and Automation, ICMA 2007.

Welcome to Accsense.com. (n.d.). Retrieved from http://www.accsense.com/

Wensheng, Z., & Guohong, C. (2004). DCTC: Dynamic convoy tree-based collaboration for target tracking in sensor networks. *IEEE Transactions on Wireless Communications*, *3*(5), 1689–1701. doi:10.1109/TWC.2004.833443

Werner-Allen, G., Swieskowski, P., & Welsh, M. (2005). MoteLab: A wireless sensor network testbed. Paper presented at the *4th International Symposium on Information Processing in Sensor Networks, IPSN 2005, April 25, 2005 - April 27,* (pp. 483-488). Retrieved from http://dx.doi.org/10.1109/IPSN.2005.1440979

Werner-Allen, G., Lorincz, K., Welsh, M., Marcillo, O., Johnson, J., Ruiz, M., & Lees, J. (2006). Deploying a wireless sensor network on an active volcano. *IEEE Internet Computing*, ▪▪▪, 18–25. doi:10.1109/MIC.2006.26

White, B., Lepreau, J., & Guruprasad, S. (n.d.). *Lowering the barrier to wireless and mobile experimentation.* Retrieved May 9, 2009, from http://www.cs.utah.edu/flux/papers/barrier-hotnets1-base.html

White, T., & Pagurek, B. (1998). Toward multi-swarm problem solving in networks. *International Conference on Multi Agent Systems,* (pp. 333–340).

White, T., Pagurek, B., & Oppacher, F. (1998). Connection management using adaptive mobile agents. In *Proceeding of International Conference on Parallel Distributed Processing Techniques and Applications* (pp. 802-809). CSREA Press.

Wicker, S. B., & Bhargava, B. K. (1994). *Reed-Solomon codes and their applications*. IEEE Press.

Wireless Accelerometers - Digital accelerometers - Vibration monitoring - Wireless CBM - Wireless sensor. (n.d.). Retrieved from http://www.techkor.com/industrial/wireless.htm

Wireless industrial sensor network testbed for radio-harsh environments (WINTeR). (n.d.). Retrieved from http://winter.cbu.ca/

WiSense. (n.d.). *WiSense.* Retrieved from http://wisense.ca/

Witricity. (n.d.). *Wireless power Transfer| Electricity Transmission |Nikola Tesla.* Retrieved from http://www.witricitynet.com/

Woo, A., Tong, T., & Culler, D. (2003). Taming the underlying challenges of reliable multihop routing in sensor networks. *SenSys '03: Proceedings of the 1st International Conference on Embedded Networked Sensor Systems,* Los Angeles, California, USA (pp. 14-27). Retrieved from http://doi.acm.org.qe2a-proxy.mun.ca/10.1145/958491.958494

Wood, A. D., & Stankovic, J. A. (2010). Security of distributed, ubiquitous, and embedded computing platforms. In Voeller, J. G. (Ed.), *Wiley handbook of science and technology for homeland security*. Hoboken, NJ: John Wiley & Sons. doi:10.1002/9780470087923.hhs449

Wood, A. D., Stankovic, J. A., Virone, G., Selavo, L., He, Z., & Cao, Q. (2008). AlarmNet context-aware wireless sensor networks for assisted living and residential monitoring. *IEEE Network*, *22*(4), 26–33. Retrieved from http://dx.doi.org.qe2a-proxy.mun.ca/10.1109/MNET.2008.4579768doi:10.1109/MNET.2008.4579768

Woungang, I., Subhas, C., & Cheng, L. (2009). Topology management for wireless sensor networks. In Misra, S., Woungang, I., & Misra, S. C. (Eds.), *Guide to wireless sensor networks*. London, UK: Springer.

WU WSN Wiki. (n.d.). *TinyOS 1.x installation on Windows XP - WSN*. Retrieved from http://www.cs.wustl.edu/wsn/index.php?title=TinyOS_1.x_Installation_on_Windows_XP

Wu, D., Ci, S., Sharif, H., & Yang, Y. (2007). Packet size optimization for goodput enhancement of multi-rate wireless networks. In *IEEE Wireless Communications and Networking Conference, WCNC 2007* (pp. 3575-3580). Hong Kong.

Wu, T., Skirvin, N., Werner, J., & Kusy, B. (2006). *Group-based peer authentication for wireless sensor networks.*

Wu, X., & Chen, G. (2006). On the energy hole problem of nonuniform node distribution in wireless sensor networks. In *Proceedings of The IEEE International Conference on Mobile Adhoc and Sensor Systems (MASS)*, (pp. 180–187).

Wu, K., Liu, C., Pan, J., & Huang, D. (2007). Robust range free localization in wireless sensor networks. *Mobile Networks and Applications*, *12*(5). doi:10.1007/s11036-008-0041-9

Xhang, X., & Parhi, K. K. (2004). High-speed VLSI architectures for the AES algorithm. *IEEE Transactions on Very Large Scale Integration (VLSI). Systems*, *12*(9), 957–967.

Xia, L., & Chen, X. (2005). *Embedded software and systems*. Berlin, Germany: Springer.

Xiang, X., Zhou, Z., & Wang, X. (2007). Self-adaptive on demand geographic routing protocols for mobile ad-hoc networks. In *IEEE INFOCOM 2007 - 26th IEEE International Conference on Computer Communications*, (pp. 2296-2300). IEEE.

Xiang, X., Wang, X., & Yang, Y. (2010). Stateless multicasting in mobile ad hoc networks. [IEEE.]. *IEEE Transactions on Computers*, *59*(8), 1076–1089. doi:10.1109/TC.2010.102

Xiao, Y., Shen, X., & Jiang, H. (2006). Optimal ACK mechanism of the IEEE 802.15.3 MAC for ultra-wideband systems. *IEEE Journal on Selected Areas in Communications*, *24*(4).

Xie, P., Cui, J.-H., & Lao, L. (2006). *VBF: Vector-based forwarding protocol for underwater sensor networks. Networking 2006. Networking Technologies, Services, and Protocols; Performance of Computer and Communication Networks; Mobile and Wireless Communications Systems* (*Vol. 3976*, pp. 1216–1221). Berlin, Germany: Springer.

Xing, G., Sha, M., Hackmann, G., Klues, K., Chipara, O., & Lu, C. (2009). Towards unified radio power management for wireless sensor networks. *Wireless Communications and Mobile Computing, 9*(3), 313-323. Retrieved from http://dx.doi.org/10.1002/wcm.622

Xing, L., & Shrestha, A. (2006). *QoS reliability of hierarchical clustered wireless sensor networks*. In 25th IEEE International Performance, Computing, and Communications Conference, 2006.

Xi-Rong, B., Zhi-Tao, Q., Xue-Feng, Z., & Shi, Z. (2009). An efficient energy cluster-based routing protocol for wireless sensor networks. In *CCDC '09: Proceedings of the 21st Annual International Conference on Chinese Control and Decision*, (pp. 4752–4757). Piscataway, NJ, USA: IEEE Press.

Xu, Y., Bien, S., Mori, Y., Heidemann, J., & Estrin, D. (2003). *Topology control protocols to conserve energy in wireless ad hoc networks*. Technical Report 6, University of California, Los Angeles, Center for Embedded Networked Computing, January 2003.

Xu, Y., Heidemann, J., & Estrin, D. (2001). Geography-informed energy conservation for ad hoc routing. *MobiCom '01: Proceedings of the 7th Annual International Conference on Mobile Computing and Networking,* Rome, Italy (pp. 70-84). Retrieved from http://doi.acm.org/10.1145/381677.381685

Xu, Y., Lee, W., Xu, J., & Mitchell, G. (2005). PSGR: Priority-based stateless geo-routing in wireless sensor networks. In *Proceedings of IEEE International Conference on Mobile Adhoc and Sensor Systems Conference*, (p. 680), Washington, DC: IEEE.

Yadav, V., Mishra, M. K., Singh, A. K., & Gore, M. M. (2009). Localization scheme for three dimensional wireless sensor networks using GPS enabled mobile sensor nodes. *International Journal of Next-Generation Networks*, *1*(1).

Yan, H., Shi, Z. J., & Cui, J.-H. (2008). *DBR: Depth-based routing for underwater sensor networks*. Paper presented at the 7th International IFIP-TC6 Networking Conference on Ad Hoc and Sensor Networks, Wireless Networks, Next Generation Internet.

Yang, J., & Zhang, D. (2009). *An energy-balancing unequal clustering protocol for wireless sensor networks*. Retrieved from http://www.scialert.net/pdfs/itj/2009/57-63.pdf

Yang, Q., & Miao, C. (2007). Semantic enhancement and ontology for interoperability of design information systems. *IEEE Conference on Emerging Technologies and Factory Automation*, (pp. 169-176).

Yang, Z., & Mohammed, A. (2008). *On the cost-effective wireless broadband service delivery from high altitude platforms with an economical business model design*. IEEE 68th Vehicular Technology Conference, 2008. VTC 2008-Fall.

Yang, Z., Mohammed, A., Hult, T., & Grace, D. (2007). *Assessment of coexistence performance for WiMAX broadband in high altitude platform cellular system and multiple-operator terrestrial deployments*. 4th IEEE International Symposium on Wireless Communication Systems (ISWCS'07).

Yao, K., Hudson, R., Reed, C., Chen, D., & Lorenzelli, F. (1998). Blind beamforming on a randomly distributed sensor array system. *IEEE Journal on Selected Areas in Communications*, *16*(8), 1555–1567. doi:10.1109/49.730461

Yao, Y., & Gehrke, J. (2002). The cougar approach to in-network query processing in sensor networks. *SIGMOD Record*, *31*(3), 9–18. doi:10.1145/601858.601861

Ye, M., Li, C., Chen, G., & Wu, J. (2005). EECS: An energy efficient clustering scheme in wireless sensor networks. 24th IEEE International Performance, Computing, and Communications Conference, (pp. 535- 540).

Ye, W., Heidemann, J., & Estrin, D. (2002). An energy-efficient MAC protocol for wireless sensor networks. *Twenty-First Annual Joint Conference of the IEEE Computer and Communications Societies Proceedings*, (vol. 3, pp. 1567-1576).

Ye, W., Silva, F., & Heidemann, J. (2006). *Ultra-low duty cycle MAC with scheduled channel polling*. 4th ACM Conference on Embedded Networked Sensor Systems (SenSys'06).

Yeditepe University. (n.d.). *Scalable querying sensor protocol (SQS)*. Retrieved from http://cse.yeditepe.edu.tr/tnl/sqs.php?lang=en

Ye, W., Heidemann, J., & Estrin, D. (2004). Medium access control with coordinated adaptive sleeping for wireless sensor networks. *IEEE/ACM Transactions on Networking*, *12*(3), 493–506. doi:10.1109/TNET.2004.828953

Yick, J., Mukherjee, B., & Ghosal, D. (2008). Wireless sensor network survey. *Computer Networks*, *52*(12), 2292–2330. Retrieved from http://dx.doi.org/10.1016/j.comnet.2008.04.002doi:10.1016/j.comnet.2008.04.002

Yomogita, H. (2007). *Mobile GPS accelerates chip development*. Retrieved from http://techon.nikkeibp.co.jp/article/honshi/20070424/131605/

Younis, M., Youssef, M., & Arisha, K. (2002, October). Energy-aware routing in cluster-based sensor networks. *The 10th IEEE/ACM International Symposium on Modeling, Analysis and Simulation of Computer and Telecommunication Systems (MASCOTS2002)*, (pp. 129–136). Fort Worth, TX.

Younis, O., Fahmy, S., & Santi, P. (2004). *Robust communications for sensor networks in hostile environments*. Twelfth IEEE International Workshop on Quality of Service, (pp. 10-19).

Younis, M., Akkaya, K., Eltoweissy, M., & Wadaa, A. (2004). *On handling QoS traffic in wireless sensor networks* (p. 10).

Younis, O., & Fahmy, S. (2004). HEED: A hybrid, energy-efficient, distributed clustering approach for ad hoc sensor networks. *IEEE Transactions on Mobile Computing*, *3*(4), 366–379. Retrieved from http://doi.ieeecomputersociety.org/10.1109/TMC.2004.41doi:10.1109/TMC.2004.41

Yu, L., Wang, N., & Meng, X. (2005). Real-time forest fire detection with wireless sensor networks. *Proceedings of the International Conference on Wireless Communications, Networking and Mobile Computing*, (pp. 1214-1217).

Yu, M., Mokhtar, H., & Merabti, M. (2006). A survey of network management architecture in wireless sensor network. *Annual Postgraduate Symposium on the Convergence of Telecommunications, Networking & Broadcasting* (pp. 221-225).

Yu, Y., Estrin, D., & Govindan, R. (2001, May). *Geographical and energy-aware routing: A recursive data dissemination protocol for wireless sensor networks*. UCLA Computer Science Department Technical Report, UCLA-CSD TR-01-0023.

Yu, R., Sun, Z., & Mei, S. (2006). A power aware and range free localization algorithm for sensor network. *IEEE Transactions on Communications*, (August): 1–5.

Yurish, S. (2010). Sensors: Smart vs. intelligent. *Sensors & Transducers Journal*, *114*(3), I–VI.

Zaman, N., & Abdullah, A. B. (2011). *Position responsive routing protocol PRRP*. The 13th International Conference on Advanced Communication Technology, Phoenix. *Korea & World Affairs*, (February): 2011.

Zebranet. (2004). *The ZebraNet wildlife tracker*. Retrieved December 21, 2010, from http://www.princeton.edu/~mrm/zebranet.html

Zeng, W., Sarkar, R., Luo, F., Gu, X., & Gao, J. (2010). Resilient embedding of sensor network using hyperbolic embedding of universal covering space. In *IEEE INFOCOM 2010: 29th Conference on Information Communication,* (pp. 1-9). San Diego, CA: IEEE.

Zhang, B., & Li, G. (2008). Analysis of network management protocols in wireless sensor network. *International Conference on Multimedia and Information Technology* (pp. 546–549).

Zhang, B., & Li, G. (2009). Survey of network management protocols in wireless sensor network. *International Conference on E-Business and Information System Security*, (pp. 1-5).

Zhang, L., et al. (1993). RSVP: A new resource reservation protocol. *IEEE Network*, September.

Zhang, Y., & Shu, F. (2009, April). Packet size optimization for goodput and energy efficiency enhancement in slotted IEEE 802.15.4 networks. In *IEEE Wireless Communications and Networking Conference, WCNC '09* (pp. 1-6). Budapest.

Zhang, H., & Shen, H. (2009). Balancing energy consumption to maximize network lifetime in data-gathering sensor networks. *IEEE Transactions on Parallel and Distributed Systems*, *20*(10), 1526–1539. doi:10.1109/TPDS.2008.252

Zhang, W.-B., Xu, H.-F., & Sun, P. G. (2010). *A network management architecture in wireless sensor network. Communications and Mobile Computing* (pp. 401–404). CMC.

Zhang, X., Cai, Y., & Zhang, H. (2006). *A game-theoretic dynamic power management policy on wireless sensor network*. China: ICCT.

Zhang, Z. (2009). Resource prioritization of code optimization techniques for program synthesis of wireless sensor network applications. *Journal of Systems and Software*, *82*(9). doi:10.1016/j.jss.2009.05.018

Zhao, G., Liu, X., & Sun, M. (2007, Feb). *Anchor based geographic routing for sensor networks using projection distance.* Paper presented at the IEEE International Symposium on Wireless Pervasive Computing (ISWPC), San Juan, Puerto Rico.

Zhao, L., Zhang, J., & Zhang, H. (2008). Using incomplete cooperative game theory in wireless sensor networks. In *Proceeding of WCNC 2008* (pp. 1483-1488).

Zhao, L., Zhang, J., Yang, K., & Zhang, H. (2008). An energy efficient MAC protocol for WSNs: Game theoretic constraint optimization. In *Proceeding of ICCS 2008* (pp. 114-118).

Zhao, Q., & Tong, L. (2003). *QoS specific medium access control for wireless sensor networks with fading.*

Zhao, J., & Cao, C. (2008). VADD: Vehicle-assisted data delivery in vehicular ad hoc networks. *IEEE Transactions on Vehicular Technology*, *57*(3).

Zhao, Q., & Gurusamy, M. (2008). Lifetime maximization for connected target coverage in wireless sensor networks. *IEEE/ACM Transactions on Networking*, *16*(6).

Zheng, G., Colombo, G., Bing, W., Jun-Hong, C., Maggiorini, D., & Rossi, G. P. (2008, 23-25 January). *Adaptive routing in underwater delay/disruption tolerant sensor networks.* Paper presented at the Fifth Annual Conference on Wireless on Demand Network Systems and Services, WONS 2008.

Zheng, J., & Jamalipour, A. (2009). *Wireless sensor networks: A networking perspective*. Hoboken, NJ: John Wiley and Sons Inc.doi:10.1002/9780470443521

Zhou, G., He, T., Krishnamurthy, S., & Stankovic, J. A. (2004). Impact of radio irregularity on wireless sensor networks. *MobiSys 2004 - Second International Conference on Mobile Systems, Applications and Services,* June 6, 2004 - June 9, (pp. 125-138). Retrieved from http://dx.doi.org/10.1145/990064.990081

Zhou, Y., Wu, D., & Nettles, S. (2004). *Analyzing and preventing MAC-layer denial of service attacks for stock 802.11 systems.* In Workshop on BWSA, BROADNETS '04.

Zhou, X., Zhang, L., & Cheng, Q. (2006). Landscape-3D: A robust localization scheme for sensor networks over complex 3D terrains. In *Proceedings of the 31st IEEE International Conference Local. Computer Networks*, (November): 239–246.

Zhu, C., & Corson, M. S. (2002). QoS routing for mobile ad hoc networks. *Proceedings - IEEE INFOCOM*, 2002.

Zhu, J., Papavassiliou, S., & Yang, J. (2006). Adaptive localized QOS-constrained data aggregation and processing in distributed sensor networks. *IEEE Transactions on Parallel and Distributed Systems*, *17*(9).

ZigBee Alliance. (n.d.). *Our mission.* Retrieved from http://www.zigbee.org/About/OurMission/tabid/217/Default.aspx

Zuo, L., Niu, R., & Varshney, P. (2007). A sensor selection approach for target tracking in sensor networks with quantized measurements. In *Proceedings of the 2007 IEEE International Conference on Acoustics, Speech, and Signal Processing*, II, (pp. 2521 – 2524).

Zurich, E. T. H. (n.d.). *BTnodes - A distributed environment for prototyping ad hoc networks: Main - overview browse.* Retrieved October 10, 2009, from http://www.btnode.ethz.ch/

About the Contributors

Noor Zaman acquired his Degree in Engineering in 1998, and Master's in Computer Science at the University of Agriculture in Faisalabad in 2000. His academic achievements further extended with a PhD in Information Technology at UTP, University Technology PETRONAS Malaysia. He is currently working as a Faculty member in the College of Computer Science and IT, King Faisal University, in Saudi Arabia. He takes care of versatile operations including teaching, research activities, Information Technology management, and leading ERP projects. He headed the Department of IT, and administered the prometric center in the Institute of Business and Technology (BIZTEK), in Karachi Pakistan. He has worked as a consultant for Network and Server Management remotely in Apex Canada, a USA based Software house, and call center. Noor Zaman has authored more than 35 research papers, and edited two books, has many publications to his credit. He is an Associate Editor and reviewer for reputed international research journals around the world. He has completed several research grants and currently involved with funded projects in different courtiers. His areas of interest include wireless sensor network (WSN), networks, artificial intelligence, telecommunication, mobile computing, software engineering, Unix, and Linux.

Azween Abdullah obtained his Bachelor's degree in Computer Science in 1985, Master in Software Engineering in 1999, and his PhD in Computer Science in 2003. His work experience includes twenty years in institutions of higher learning and as a director of research and academic affairs at two institutions of higher learning, Dean of his faculty, and fifteen years in commercial companies as Software Developer and Engineer, Systems Analyst, and IT/MIS and educational consultancy and training. He is currently an Associate Professor at Universiti Teknologi PETRONAS, Malaysia. He spent more than a decade with leading technology firms and universities as a process analyst, senior systems analyst, and project manager. He has participated in and managed several software development projects, including management Information Systems, software process improvement initiatives design and implementation, and several business application projects. He has many years of experience in the application of IT in business, engineering, and research and has personally designed and developed a variety of computer software systems, including business accounting systems, software for investment analysis, website traffic analysis, determination of hydrodynamic interaction of ships, and computation of environmental loads on offshore structures. His area of research specialization includes computational biology, system survivability and security, autonomic computing, self-healing and regenerating systems, formal specifications, and network modeling. His contributions include publishing several academic books and in the development of programs to enhance minority involvement in bridging the ICT digital gap.

K. Ragab is an Assistant Professor at Department of Computer Sciences, College of Computer Sciences and Information Technology, King Faisal University, Saudi Arabia. Moreover, he is on leave Assistant Professor of Computer Science at Department of Mathematic, Computer Science division, Ain Shams University, Cairo, Egypt. He joined Department of Computer Science, Tokyo University in 2005 as postdoctoral position. He was born in 1968 and received his B.Sc. and M.Sc. degrees in Computer Science from Ain Shams University Cairo, Egypt in 1990 and 1999, respectively, and Ph.D. degree in Computer Science from Tokyo Institute of Technology in 2004. He has worked in Ain Shams University, Cairo Egypt in 1990-1999 as Assistant Lecturer. He has worked as Research Scientist in Computer Science Dept., Technical University of Chemnitz, Germany in 1999-2001. His research interests include autonomous decentralized systems, peer-to-peer systems, overlay networks, wireless sensor networks, Web-services, and application-level multicast.

* * *

Tayseer Al-Khdour received his B.Sc. and M.S. degree in Computer Engineering from Jordan University of Science and Technology (JUST), Irbid, Jordan, in 1994 and 1998, respectively. He worked as Lecturer in King Faisal University, Al-Ahsa, Saudi Arabia from 1999 to 2004. In 2009 he received his Ph.D. degree in Computer Science and Engineering from King Fahd University of Petroleum and Minerals (KFUPM), Dhahran, Saudi Arabia. Since 2009, he has been an Assistant Professor in the department of Communication and Network, King Faisal University. He has published several papers in refereed international journals and conferences. His research interest is wireless sensor networks and cross layer optimizations in wireless network.

Muneer Ahmad obtained his Bachelor's degree in 1997, Master in Mathematics in 1999, Master in Computer Science in 2001, and MPhil Computer Science in 2006. His academic achievements further extended with a PhD in Information Technology at UTP, University Technology Malaysia. He is currently working as a Faculty member in the College of Computer Science and IT, King Faisal University, in Saudi Arabia. He takes care of versatile operations including teaching, research activities, Information Technology management, and leading ERP projects.

Muneer Ahmad has several publications in bio-inspired computing / medical informatics. He is a reviewer for reputed international journals around the world. He has completed several research grants and currently involved with funded projects at different places.

Nova Ahmed is an Assistant Professor in North South University, Dhaka, Bangladesh. She has completed her PhD from Georgia Institute of Technology in 2010 May, worked in Georgia Tech Research Institute until February 2011 and came back to Bangladesh to serve her own country. Her research interests include distributed systems, sensor systems, and various applications considering distributed sensing systems, such as healthcare systems for elderly care. She has worked particularly on RFID enabled distributed sensing systems and is currently enjoying interesting applications in the context of her own country.

Hamid Ali is currently pursuing his Ph.D. in Computer Science the department of Computer Science, National University of Computer and Emerging Sciences (NUCES), Islamabad, Pakistan. He did his BS in Computer Science from Virtual University of Pakistan and MS in Computer Science from National University of Computer and Emerging Sciences (NUCES), Islamabad, Pakistan in 2007 and 2009, respectively. His research is mainly focused on wireless ad hoc and sensor networks, evolutionary algorithms, and multi-objective optimization. He is currently working on various applications of ad hoc and sensor networks using evolutionary algorithms.

Li-Minn Ang is an Associate Professor in the School of Electrical & Electronic Engineering and the head of Electronics, Communications and Intelligent Mathematical techniques research division at The University of Nottingham Malaysia Campus. He completed his Bachelor of Engineering and Ph.D at Edith Cowan University in Perth, Australia in 1996 and 2001 respectively. He then taught at Monash University before joining The University of Nottingham Malaysia Campus in 2004. His research interests are in the fields of visual processing, embedded system and wireless sensor networks.

Adeel Ansari is currently an Associate Consultant at A. F. Ferguson & Co., a member firm of PricewaterhouseCoopers, UK. He has also worked at Habib Bank Limited as an Assistant Policy Manager for a period of three years and was a Trainee Engineer and an Application developer at Siemens Pakistan. He has done his MPhil. in Software Engineering and MBA MIS from PAF-KIET. He did his BSc(Hons) in Computing from Staffordshire University, UK with a first class honors degree. He has written several research papers as part of his research and thesis. His areas of interests are multimedia, software engineering, data mining, and data warehousing.

Seema Ansari is an Associate Professor / Head of Curriculum and New Programs, at College of Engineering, Pakistan Air Force – Karachi Institute of Economics and Technology (PAF-KIET), Karachi, Pakistan. She has more than 50 research papers publications in both national and international conferences and journals. She is also a thesis / research Supervisor of MS/MPhil students. During her 25-years career, she has held various positions in the field of education and management. She also served as Dean of Student Affairs and Director, Asia Pacific Institute of Information Technology, Karachi-PK, (a joint venture of APIIT Malaysia and affiliated with Staffordshire University U.K). She is a PhD (Telecommunication Engg.) candidate at Hamdard University, Karachi, MS/CS-Telecomm. from University of Missouri, KC, USA and an Electronic Engineer from NED University of Engineering & Technology, Karachi, Pakistan.

Muhammad Ayaz received his MS degree in Computer Science from SZABIST Islamabad Pakistan in 2006. During MS, his specialization area was in Networks and Communication, and he passed with distinction as highest grades in batch (88%). He worked as a full time Lecturer at Computer Science Department of Federal Urdu University of Arts Science and Technology, Islamabad, Pakistan from 2006-08. In the same year, he got PETRONAS fellowship and joined to Universiti Teknologi PETRONAS, Malaysia as a full time PhD candidate. His research interests include mobile and sensor networks, routing protocols, and underwater acoustic sensor networks.

Sheikh Tahir Bakhsh is currently student of Ph.D. in Department of Computer and Information Sciences, Universiti Teknologi PETRONAS, Malaysia. He received the Master's degree in Computer Science from COMSATS Institute of Information Technology Abbottabad, Pakistan. In the recent past, he has directed 4 undergrad projects. His current research interests are Bluetooth networks, wireless sensor networks (WSN), and mobile ad hoc networks (MANET).

Uthman A. Baroudi received his B.Sc. and M.S. degrees from King Fahd University of Petroleum and Minerals (KFUPM), Dhahran, Saudi Arabia in 1988 and 1990, respectively, and in 2000, he received his Ph.D. from Concordia University, Montreal, Canada, all in Electrical Engineering. In 2000, he joined Nortel Networks, Ottawa, Canada, to work in R&D for next generation wireless networks. Since January 2002, he has been an Assistant Professor in the Department of Computer Engineering at KFUPM. He has published several papers in refereed international journals and conferences. His research interests lies in the areas of radio resource management (RRM) and QoS provisioning for the next generation wireless networks, wireless ad hoc, and sensor and actuator networks. Dr. Baroudi has over 30 publications in journal and conference proceedings, and 1 US patent.

Parama Bhaumik is a PhD in Engineering from Jadavpur University (2009), and M.Tech & B.Tech in Computer Science & Engineering from Calcutta University, India (2002). She is currently working as Assistant Professor in Jadavpur University. She has more than 12 research publications in international conferences and journals of repute. Her present research activities include: wireless sensor network, mesh networking, opportunistic networks, and topology management using intelligent methods.

Matangini Chattopadhyay is currently an Associate Professor at the School of Education Technology, Jadavpur University, India. She obtained her Ph.D from the department of Computer Science & Engineering and an author of over 25 papers in reputed journals and conferences. She has a decade of industrial experience of working in senior positions in leading multi-national organizations such as Lucent Technologies (Bell Labs Development Center), Bangalore, Computer Associates TCG Software Pvt. Ltd. etc. Her current research interest include Middleware, Mobile Technology, Optimization Algorithms.

Peter H. J. Chong received the B.Eng. (with distinction) in Electrical Engineering from the Technical University of Nova Scotia, Halifax, NS, Canada, in 1993, and the M.A.Sc. and Ph.D. degrees in Electrical Engineering from the University of British Columbia, Vancouver, BC, Canada, in 1996 and 2000, respectively. Between July 2000 and January 2001, he worked in the Advanced Networks Division at Agilent Technologies Canada Inc., Vancouver, BC, Canada. From February 2001 to May 2002, he was with the Radio Communications Laboratory at Nokia Research Center, Helsinki, Finland, and was involved in research on WCDMA and standardization for HSDPA. During his stay in Finland, he has taught part of a graduate course in WCDMA at the Helsinki University of Technology, Helsinki, Finland. Since May 2002, he has been with the School of Electrical and Electronic Engineering, Nanyang Technological University (NTU), Singapore, where he is now an Associate Professor. He was a Technical Program Committee Chair for Mobility Conference 2005 and 2006, a General Chair of Mobility Conference 2007 and 2008, and a General Co-chair of Mobility Conference 2010. He served as a Guest Editor of *Journal of Internet Technology* in 2006, *International Journal of Ad Hoc and Ubiquitous Computing* in 2007, and *IEEE Communications Magazine* for the Special Issue on 'Technologies in Multihop Cellular Network' in 2007. He is currently a lead Guest Editor of *IEEE Wireless Communications* for the Special Issue on 'Technologies for Green Radio Communication Networks' in 2011. He is an Editorial Board Member of *Security and Communication Networks, International Journal of Wireless Communications and Networking, Wireless Sensor Network,* and an Editor of *Far East Journal of Electronics and Communications.* His research interests are in the areas of mobile communications systems including, channel assignment schemes, radio resource management and multiple access, MANETs, and multihop cellular networks.

Alan Downe is a Senior Lecturer in the Department of Management & Humanities at Malaysia's Universiti Teknologi PETRONAS. Originally from Canada, he completed his Bachelor's degree (Psychology/Biology) in 1980, his M.Sc. (Educational Psychology) in 1984, and his PhD in Management in 2009. Earlier in his career, he worked as a senior manager and advisor in a range of government positions, and has now held faculty appointments for 11 years. His current research interests include bio-mimetics related to problem-solving in hybrid intelligence and digital infrastructure, user acceptance of technology, and service industry operations and strategy. He is the co-leader of UTP's Hybrid Intelligence & Digital Infrastructure research group.

Khalid El-Darymli is a PhD candidate in the Faculty of Engineering and Applied Science at Memorial University of Newfoundland, Canada. He received his MSc with distinction in Computer and Information Engineering from the International Islamic University of Malaysia; and his BSc in Electrical Engineering with Honours from Garyounis University of Libya. He is a recent recipient of the Ocean Industries Student Research Award of the Research & Development Corporation (RDC) in Newfoundland, Canada. His research interests include wireless sensor networks, target detection, and automatic target recognition (ATR) in Synthetic Aperture Radar (SAR) imagery.

Ibrahima Faye completed his MSc and PhD in Mathematics at University of Toulouse and his Specialized MSc in Engineering of medical and biotechnological data at Ecole Centrale Paris. He is currently a Senior Lecturer at the Department of Fundamental and Applied Sciences of Universiti Teknologi PETRONAS. His research interests include engineering mathematics, image processing, and communication systems.

Kong Jia Hao received his Bachelor of Engineering degree (with honours) in the field of Electrical and Electronics from the University of Nottingham Malaysia Campus in 2009. He is currently a PhD Research Student at The University of Nottingham Malaysia Campus. His research interest is in the field of reconfigurable crypto-coprocessors, computer architecture, cryptography and wireless sensor network.

Fernando Henrique Gielow is M.Sc. student in Informatics Department at Federal University of Parana, Brazil. Fernando is a member of the research group NR2, and obtained his B.Sc. in Computer Science by the Federal University of Parana. His research interests include routing and transport on wireless sensor networks. He is currently working with wireless multimedia sensor networks.

Mesut Günes is head of the working group Distributed, Embedded Systems (DES) at the Institute of Computer Science, Freie Universität Berlin, Germany. Dr. Günes studied Computer Science at RWTH Aachen University, where he also got his PhD in 2004. After being a research fellow at the International

Computer Science Institute (ICSI), Berkeley, USA and research and teaching assistant at the Department of Computer Science, RWTH Aachen University; he is an Assistant Professor at Freie Universität Berlin since 2007.His research focus is on wireless and mobile systems, communication protocols, embedded devices, the Internet of Things (IoT), and quality of service aspects in communication systems. Furthermore, he is interested in nature and biological based solutions for distributed communication systems like swarm algorithms and their application.

Nurul Hasan has received PhD from UNSW in 2003 (Chemical Engineering) on Multiphase Flow. He worked for CSIRO minerals, Centre for Multiphase. Nanoparticle, shock, wave energy, EOR, CO2 removal, nucleation, Brownian diffusion, light metal production, artificial diamond, oil and gas simulation, process modelling, and biomedical engineering are his major interests. He has 13 years experience in CFD and process flow modelling and molecular dynamics. He does consulting for oil and gas, aerospace & maritime, automotive & transportation, environmental flows-water, environmental flows - atmospheric, air-conditioning & industrial ventilation, turbo machinery, pumps & appliances, combustion & power generation, biomedical flows, chemical & process, electronics & semiconductors, food & beverage, mineral processing, and refining & smelting.

Syeda Fariha Hasnain is currently associated with PAF-KIET, where she is an Assistant Professor in the College of Engineering, at Pakistan Air Force – Karachi Institute of Economics and Technology (PAF-KIET). She completed her MS in Telecommunication Engineering from PAF-KIET in 2010, and a Bachelor's degree in the field of Electronic Engineering from Sir Syed University of Engineering and Technology. She is a member of IEEE since 2010. She has numerous publications on wireless communication topics. She is also supervising final year projects of B.E students. Her topics of interest are WiMAX, UMTS, co-operative diversity, and cognitive network. At present she is a PhD candidate at PAF-KIET.

Jorge Higuera received the Ph.D. degree in electronics from Universitat Politècnica de Catalunya (UPC), Barcelona, Spain, in 2011. Also he received the B.Sc and M.Sc in electronics from the Industrial University of Santander, Bucaramanga, Colombia in 2002 and 2006, respectively. He is currently working in the Castelldefels School of Telecommunications and Aerospace (EETAC), Universitat Politècnica de Catalunya UPC, Castelldefels, Barcelona, Spain. His research interests are in the field of electronic instrumentation, in particular, the environmental monitoring with wireless sensor networks, using interoperable IEEE1451 smart-sensors systems and wireless communication onboard high speed trains.

Tommy Hult received his M.Sc. degree in Electrical Engineering with emphasize on signal processing from Blekinge Institute of Technology, Sweden in 2002 and his Ph.D. in 2008. Since 2009 he is working as a research fellow in radio systems at the communications group of the Department of Electrical and Information Technology, Lund University, Sweden. He is also the representative and management committee member of Sweden in five EU COST actions: COST280, COST296, COST297, IC0802, and IC0902. He is the author of more than 60 conference and journal papers in the area of telecommunication. His research interests are mainly in the area of smart antenna systems, wave propagation, channel modelling, channel measurements, wireless sensor networks, and cognitive radio.

Ong Jia Jan received his Master of Engineering from the University of Nottingham Malaysia Campus in 2009. He is currently pursuing his PhD at the University of Nottingham Malaysia Campus. His research interests are in the fields of error correction, image, vision processing, and wireless sensor network.

Asfandyar Khan received his Msc degree in Computer Science from University of Peshawar Pakistan in 2004. During Msc, his specialization area was in data communication. He worked as a Lecturer at Computer Science Department of University of Science & Technology Bannu Pakistan. In 2008, he got PETRONAS fellowship and joined Universiti Teknologi PETRONAS, Malaysia as a full time PhD candidate. His research interests include mobile and sensor networks.

Sana Khan has received her BS degree in Information Technology from University of Agriculture Peshawar Pakistan and MS degree in Computer Science from COMSATS Institute of IT Abbottabad Pakistan in 2006 and 2008, respectively. She has the experience of software development, software maintenance, software testing, and requirement engineering in Sungi development foundation and COMSATS Information Technology Centre Abbottabad Pakistan.

Lyes Khoukhi is an Associate Professor at the University of Technology of Troyes (France), since 2009. In 2008, he was researcher at the Computer Sciences department of the University of Montreal (Canada). He received PhD degree in Electrical and Computer Engineering from the University of Sherbrooke (Canada) in 2007, and M.Sc degree in Computer Engineering from University of Versailles (France) in 2002. During 2003-2007, he stayed in INTERLAB Communications Research Laboratory, Sherbrooke University. His research interests include wireless communications, multimedia networking, quality of service, and intelligent systems.

Pardeep Kumar received Bachelor of Engineering (B.E.) in Computer Systems and Master of Engineering (M.E.) in Communication Systems and Networks from Mehran University of Engineering and Technology, Jamshoro, Pakistan. He is currently working toward a Ph.D. Degree at the Institute of Computer Science, Freie Universität Berlin, Germany. His research interests include MAC and routing protocols for wireless networks with special focus on energy and delay aspects for wireless sensor and ad hoc networks.

Thu Nga Le received her Bachelor Degree (with First Class Honors) from School of Electrical and Electronics Engineering, Nanyang Technological University, Singapore in 2009. She is currently pursuing her Master's Degree in Communication Software and Networks in Nanyang Technological University. Her research interests include localization techniques and applications in wireless sensor networks, WLANs, and channel allocation in 4G cellular communication networks.

Xue Jun Li received the B.Eng. (with First Class Honors) and Ph.D from Nanyang Technological University, Singapore, in 2004 and 2008, respectively. From November 2007 to August 2008, he worked as a Research Engineer, and later a Research Fellow, at Network Technology Research Centre. From August 2008 to September 2008, he worked in Temasek Laboratories @ NTU as a Research Scientist. Currently, he is an academic staff with the School of Electrical and Electronic Engineering, Nanyang Technological University, Singapore. His research interests include design / analysis of wireless networking protocols, modeling / design of radio frequency integrated circuits, computer network design and implementation, system optimization, and genetic algorithm analysis /development.

Nauman Israr received his BCS from University of Peshawar, Pakistan. He joined the Performance Modelling and Engineering Research Group, University of Bradford in 2005. He received his PhD from the Department of Computing, Bradford University, UK (2008). During his PhD studies, he developed energy efficient communication protocols for Wireless Sensor Networks. After completing his PhD, he joined the Electronics, Communications, and Information Technology research centre of Queens University Belfast as a Research Fellow. At Queens University he worked on the Active Aircraft project. The aim of the project was to develop smart skin of sensors and actuators for the future Airbus Aircrafts to reduce the skin friction of the aircrafts. In 2009 he joined the School of Computing, Teesside University as a Senior Lecturer and is the module coordinator for Networks and System Administrator and Computer Networks. His research interests include wireless sensor networks, wireless networked control systems, and wireless coordinated UAVs.

Low Tang Jung obtained his Bachelor's degree in Computer Technology from Teesside University, UK in 1989 and MSc IT from National University of Malaysia in 2001. Low has been in the academic line for the past 20 years as Lecturer in various public and private institutes of higher learning. He teaches various Engineering and ICT courses. He is currently a Senior Lecturer in Computer and Information Sciences Department, Universiti Teknologi PETRONAS, Malaysia. His research interest includes wireless technology, embedded systems, wireless sensor networking, and robotics. Some of his current ongoing R&D projects and other completed projects have been recognized at national as well as international level by winning medals and awards at various national and international exhibitions/competitions. Low has his research works published in various conference proceedings and journals.

Farrukh Aslam Khan is currently working as an Associate Professor at the Department of Computer Science, National University of Computer and Emerging Sciences (NUCES), Islamabad, Pakistan. He received his B.Sc. and M.Sc. degrees in Computer Science from University of Peshawar, Pakistan. He did his MS in Computer System Engineering from GIK Institute of Engineering Sciences and Technology, Topi, Pakistan and Ph.D. in Computer Engineering from Cheju National University, Jeju, South Korea, in 2003 and 2007, respectively. His research interests include routing, security, management, optimization, and performance analysis of Wireless Ad hoc and Sensor Networks. Dr. Khan has served as TPC member of several international conferences and workshops including ICSI 2011, WIAS 2011, IUPT 2011, FGCN 2010, ACN 2010, ICET 2009/2010, SUComS 2010/2011, and FGCN/ACN 2009. Currently, he is the TPC Co-chair of ICET 2011 and Publicity Co-chair of WIAS 2011. He has also served as reviewer of international journals including *Swarm Intelligence* (Springer), *Data and Knowledge Engineering* (Elsevier), and *Sensors* (MDPI). He is currently serving as Guest Editor of a special issue of *Journal of Supercomputing* (Springer).

B.H. Kim received the B.S., M.S., and Ph.D. degrees from the Inha University, Inchon, Korea in 1994, 1996, and 2003, respectively, under the Korean Air Line Scholarship Grants. Since 2003 he has been with Train Control and Communication Research Department, Korea Railroad Research Institute, as a Senior Researcher. His research interests include multiple access communication systems, railway communication, train control, and PRT(Personal Rapid Transit) network operation.

Alok Kumar received the Bachelor of Engineering degree in Information Technology from Jai Narain Vyas University, Jodhpur, India and completed Master of Technology degree in Computer Science & Engineering from Center for Development of Advanced Computing, Noida, India. He is currently pursuing the PhD degree in Information Technology, from Indian Institute of Information Technology, Allahabad, India. He has published about 11 papers in leading international journals and conferences. His PhD work mainly focuses on the geometrical aspects of networks and complexity of quality of service (QoS) routing under both static and dynamic stateless networks. He is a student member of the IEEE and ACM.

K. S. Kwak received the B.S. degree from Inha University, Korea in 1977, and the M.S. degree from the University of Southern California in 1981 and the Ph.D. degree from the University of California at San Diego in 1988, respectively. From 1988 to 1989 he was a Member of Technical Staff at Hughes Network Systems, San Diego, California. From 1989 to 1990 he was with the IBM Network Analysis Center at Research Triangle Park, North Carolina. Since then he has been with the School of Information and Communication, Inha University, Korea as a Professor. He had been the Chairman of the School of Electrical and Computer Engineering from 1999 to 2000 and the Dean of the Graduate School of Information Technology and Telecommunications from 2001 to 2002 at Inha University, Inchon, Korea. He is the current Director of Advanced IT Research Center of Inha University, and UWB Wireless Communications Research Center, a key government IT research center, Korea. He has been the Korean Institute of Communication Sciences (KICS)'s President of 2006 year term. In 1993, he received Engineering College Young Investigator Achievement Award from Inha University, and a distinguished service medal from the Institute of Electronics Engineers of Korea (IEEK). In 1996 and 1999, he received distinguished service medals from the KICS. He received the Inha University Engineering Paper Award and the LG Paper Award in 1998, and Motorola Paper Award in 2000. His research interests include multiple access communication systems, mobile communication systems, UWB radio systems and ad-hoc networks, and high-performance wireless Internet.

Aldri Luiz dos Santos is Professor at Federal University of Parana, Brazil. Aldri received his Ph.D. in Computer Science from Department of Computer Science of Federal University of Minas Gerais, Belo Horizonte, Brazil. He is one of leaders of the research group in wireless and advanced networks (NR2). His research interests include network management, dependability, security and wireless networks. He is member of Brazilian Computing Society and IEEE Society.

M. Yasir Malik has been involved in information security for four years, and is currently working as a researcher at Institute of New Media and Communications, Seoul National University. He is currently team member of a research group working on Next Generations Vehicle (NGV) program of Hyundai-Kia Corporation. He is also associated with the Coding and Cryptography Lab of Department of Electrical Engineering and Computer Sciences. Prior to that he worked for two years at reputed R & D centers in Pakistan, in the areas of embedded and communications security. There he worked on teams catering various multi-million dollar projects. The author has also published his works at IEEE conferences, including IEOM, WISA, and IACR e-prints. He is serving as reviewer and committee member of ACM-FIT, IEEE ICACT, and the journal *IAJIT*. The author completed his B.Sc in EE from University of Engineering and Technology Taxila, Pakistan and is currently a postgraduate candidate at Seoul National University.

Mansouri Majdi was born November 11, 1982 in Kasserine, Tunisia. He received the Engineer Diploma in "Telecommunications" from High School of Communications of Tunis (SUP'COM) in 2006 and the Master Diploma in "Signal and Image Processing" from High school of Electronic, Informatique and Radiocommunications in Bordeaux (ENSEIRB) in 2008. He is PhD student at Troyes University of Technology, France, since October 2008. His current research interests include statistical signal processing and Wireless Sensors Networks. He is the author of over 28 papers.

S.Mehta received the B.E., M.S., and Ph.D. degrees in Electronics Engineering from Mumbai University, Mumbai, India, Ajou University, Korea, and Inha University, Korea, in 2002, 2005, and 2011, respectively. His research interests are in performance analysis of wireless networks and RFID systems.

Abbas Mohammed is a Professor of Telecommunications Theory and the Head of Telecommunication and Radio Navigation Research at the School of Engineering, Blekinge Institute of Technology, Sweden. He was awarded the PhD degree from Liverpool University, UK, in 1992 and the Swedish Docent degree in Radio Communications and Navigation from Blekinge Institute of Technology in 2001. He was the recipient of the *Blekinge Research Foundation Award "Researcher of the Year Award and Prize"* for 2006. From 1993 to 1996, he was a Research Fellow with the Radio Navigation Group, University of Wales (Bangor), UK. From 1996 to 1998, he was with the University of Newcastle, UK, working on a European collaborative project within the ACTS (Advance Communications Technologies and Services) FP4 Research Programme that investigated 3G Satellite-UMTS systems. He was also employed by Ericsson AB, where he consulted on power control standardization issues for 3G mobile communication systems. He has been a visiting Lecturer to several Swedish universities. He is a *Fellow* of The Institution of Electrical Engineering.

Arindam Mondal is a student of Master of Engineering (M.E) at the Jadavpur University in Software Engineering under the Department of Information Technology. He has completed his Bachelor of Technology (B.Tech) from B.P. Poddar Institute of Management and Technology under the West Bengal University of Technology in Information Technology.

Michele Nogueira is Ph.D. in Computer Science by the University of Pierre et Marie Curie, LIP6, France, and Professor at the Department of Informatics of Federal University of Parana, Brazil. Michele received her M.Sc in Computer Science at Federal University of Minas Gerais, Brazil, 2004. She has worked at security area for many years; her research interests are security, wireless networks and dependability. She is member of the IEEE Communication Society (ComSoc) and the Association for Computing Machinery (ACM).

Chiranjib Patra is a candidate of PhD at the Department of Information Technology, Jadavpur University. He is presently the Senior Lecturer at the department of Information Technology, Calcutta Institute of engineering and Management, Kolkata. His interests include the application of soft computing in wireless sensor networks. He has about 5 publications in international journal/ conferences.

Jose Polo received the M.E. degree in Telecommunication Engineering and the Ph.D. degree from the Universitat Politècnica de Catalunya (UPC), Barcelona, Spain, in 1990 and 2000, respectively. From

1993 to 1999, he was Professor with the UPC. From 1999 to 2006, he was a Professor with the Universitat Pompeu Fabra, Barcelona. Since 2006, he has been again a Professor with the UPC. He has authored more than 20 papers in international conferences and journals, and a regular section in a Spanish magazine about sensors. He has formed part of the organization committee of four international conferences. Initially, his research interests were electronic commerce, author right management, and services. At present, his research interests include sensor networks, mainly wireless networks, and communication standards for wireless sensor network and direct interfaces for sensors.

Vasaki Ponnusamy is a PhD student at Universiti Teknologi PETRONAS working in the area of wireless sensor networks. Her research mainly focuses on energy efficient self-healing sensor network looking using nature-inspired concepts for the deployment of mobile entities in sensor network. Previously she was on the faculty of Sunway University College, Malaysia as a Senior Lecturer in the School of Information Technology teaching networking, mobile technology, artificial intelligence, and image processing. Her primary research is focused on networking and, more specifically, on mobile and wireless networks.

S. R. S. Prabaharan is currently a full Professor of Electronics in the University of Nottingham in its Malaysia Campus. Prabaharan graduated from The American College, an autonomous affiliate of Madurai Kamaraj University, India and all of his degrees are from Madurai Kamaraj University. He earned his PhD in Solid State Devices in 1992. After university, he was employed at one of the CSIR research labs (National Aerospace Laboratory) in Bangalore, India where he started working on Ionic/electronic devices using Raman spectroscopy. He then moved to CECRI, another CSIR research lab in Karaikudi, India where he was hired under Quick Hire Scheme for Young Scientists, CSIR, Govt. of India. Later, he became a Research Scientist under DST (Department Science and Technology) Young Scientist Research Award programme. He then moved to Universiti Malaya, Malaysia and later joined the Applied Sciences Faculty at Universiti Teknologi PETRONAS. In 1999, he joined the Faculty of Engineering, Multimedia University as a Senior Lecturer before moving to The University of Nottingham Malaysia Campus as an Associate Professor. Dr. Prabaharan has been invited as visiting Senior Fellow/Scientist in different academic institution to foster research and initiate joint research activities with academic/research institutions which include: UPMC, Paris (1998), Tokyo Institute of Technology; Japan (2003), Southern University, USA (2005), and University of Sheffield, UK (February 2006). He is also a visiting research consultant to Southern University, Baton Rouge, USA where he was invited under US Army Research Project. He is a peer reviewer for numerous international journals published by Elsevier, Springer, and Hindawi, and has been the guest editor for *Solid State Journal* published by Springer (2007). His research interest are in the field of supercapacitors; hybrid power sources; lithium-ion batteries; nanotechnology of clean energy; solid state devices; semiconductor gas sensors; modeling and simulation of power electronic circuits (UPS, Mini power grids and Solar PV MPP devices); and supercapacitors for power assist applications.

Kah Phooi Seng received her Ph.D and Bachelor degree (first class honours) from University of Tasmania, Australia in 2001 and 1997 respectively. She is currently a full Professor in the School of Computer Technology at Sunway University, Malaysia. Her research interests are in the fields of intelligent visual processing, biometrics and multi-biometrics, artificial intelligence and signal processing.

H Ranganathan completed his B E and M Sc (Engg) from College of Engineering, Guindy, Madras, India in 1975 and 1978, respectively, specializing in Electronics and Communication Engineering. He has had experience in the Computer Industry in Hardware and software fields supporting major clients all over India. He started teaching in the year 2000. He taught in the Gulf for a brief period of little over a year. He completed his PhD from Anna University, Chennai (Madras), India in the year 2007. He has authored many papers in referred international journals and international conferences. He has been the reviewer for journal papers and Ph D theses. His interests include wireless communication networks, antennas, medical electronics, and artificial neural networks. Currently he is the Professor and Head, Electronics and Communication Engineering Department in Sakthi Mariamman Engineering College, Chennai, India.

Cédric Richard received the Dipl.-Ing. and the M.S. degrees in 1994 and the Ph.D. degree in 1998 from the University of Technology of Compiegne, France, all in Electrical and Computer Engineering. From 1999 to 2003, he was an Associate Professor at the University of Technology of Troyes, France. From 2003 to 2009, he was a Full Professor at the Institut Charles Delaunay (CNRS FRE 2848) at the UTT, and the supervisor of a group consisting of 60 researchers and Ph.D. In winter 2009 and autumn 2010, he was a Visiting Researcher with the Department of Electrical Engineering, Federal University of Santa Catarina (UFSC), Florianopolis, Brazil. Cédric Richard is a junior member of the Institut Universitaire de France since October 2010. Since September 2009, Cédric Richard is a Full Professor at Fizeau Laboratory (CNRS UMR 6525, Observatoire de la Cote d'Azur), University of Nice Sophia-Antipolis, France. His current research interests include statistical signal processing and machine learning. Prof. Cédric Richard is the author of over 100 papers. He was the General Chair of the XXIth Francophone Conference GRETSI on Signal and Image Processing that was held in Troyes, France, in 2007, and of the IEEE International Workshop on Statistical Signal Processing that will be held in Nice in 2011. Since 2005, he is a member of the board of the federative CNRS research group ISIS on Information, Signal, Images and Vision. He is a member of GRETSI Association Board and of the EURASIP Society, and Senior Member of the IEEE. Cédric Richard serves as an Associate Editor of the *IEEE Transactions on Signal Processing* since 2006, and of the *EURASIP Signal Processing Magazine* since 2009. In 2009, he was nominated liaison local officer for EURASIP, and member of the Signal Processing Theory and Methods (SPTM) Technical Committee of the IEEE Signal Processing Society.

Waseem Shahzad did his MS and PhD in Computer Science from the Department of Computer Science, National University of Computer and Emerging Sciences, Islamabad, Pakistan in 2007 and 2010, respectively. Currently, he is working as an Assistant Professor at National University of Computer and Emerging Sciences, Islamabad, Pakistan. His research interests include evolutionary computation, machine learning, data mining and data warehousing. He has several research publications in reputed international journals and conferences.

Hichem Snoussi was born in Bizerta, Tunisia, in 1976. He received the Diploma degree in Electrical Engineering from the Ecole Superieure d'Electricite (Supelec), Gif-sur-Yvette, France, in 2000. He also received the DEA degree and the Ph.D. in Signal Processing from the University of Paris-Sud, Orsay, France, in 2000 and 2003, respectively. He has obtained the HdR from the University of Technology of Compiègne in 2009. Between 2003 and 2004, he was postdoctoral researcher at IRCCyN, Institut de Recherches en Communications et Cybernétiques de Nantes. He has spent short periods as visiting scientist at the Brain Science Institute, RIKEN, Japan and Olin Neuropsychiatry Research Center at the Institute of Living in USA. Between 2005 and 2010, he was Associate Professor at the University of Technology of Troyes. Since September 2010, he has been appointed a Full Professor position at the same university. He is in charge of the regional research program S3 (System Security and Safety) of the CPER 2007-2013 and the CapSec platform (wireless embedded sensors for security). He is the principal investigator of an ANR-Blanc project (mv-EMD), a CRCA project (new partnership and new technologies) and a GDR-ISIS young researcher project. He is partner of many ANR projects, GIS, strategic UTT programs. He obtained the national doctoral and research supervising award PEDR 2008-2012.

Shirshu Varma graduated in Electronics and Communication Engineering from Allahabad University, Allahabad, India, post-graduated in Communication Engineering from BIT Mesra Ranchi, India, and received the Ph.D. from the University of Lucknow, India. He worked as Lecturer, Senior Lecturer, IT Consultant at BIT Mesra Ranchi, IET Lucknow, and C-DAC Noida. Presently he is working as an Associate Professor at IIIT, Allahabad. He has published about 35 papers in international and national journals and conferences of repute. He also serves as a member of editorial board in various journals of repute. His areas of interest are wireless sensor networks, mobile computing, digital signal processing, and optical communication systems. Dr. Varma is a Member of IEEE and ISTE. He has been a recipient of many national awards.

Zhe Yang received the M. Sc. degree with distinction in communications from the Department of Electronics, University of York, UK in 2006. He is working toward the Ph.D. degree in the Department of Electrical Engineering, Blekinge Institute of Technology, Sweden. In 2007, 2008, and 2009, he was an invited guest researcher in Ben-Gurion University, Israel, University of York, UK, and Peking University, China. He is a member of IEEE and the author of publications including book chapters, journal, and conference papers. He currently participates in European project developing broadband communications from high-altitude platforms. His research interests include radio communication techniques, communications techniques from high-altitude platforms, signal processing techniques, and economical models of infrastructure deployment and analysis.

Adamu Murtala Zungeru received his B.Eng. degree in Electrical and Computer Engineering from the Federal University of Technology (FUT) Minna, Nigeria in 2004, and M.Sc. degree in Electronics and Telecommunication Engineering from the Ahmadu Bello University (ABU) Zaria, Nigeria in 2009. He is a Lecturer Two (LII) at the Federal University of Technology Minna, Nigeria from 2005-present. He is a registered Engineer with the Council for the Regulation of Engineering in Nigeria (COREN), and Member of the Institute of Electrical and Electronics Engineers (IEEE). He is currently a PhD candidate in Electrical and Electronics Engineering at the University of Nottingham Malaysia Campus. His research interests include energy efficient routing, energy harvesting, storage, and management in wireless sensor networks.

Index

S

T

U

V

W